The Digital Flood

THE DIGITAL FLOOD

The Diffusion of Information Technology Across the U.S., Europe, and Asia

James W. Cortada

OXFORD
UNIVERSITY PRESS

OXFORD
UNIVERSITY PRESS

Oxford University Press is a department of the University of Oxford.
It furthers the University's objective of excellence in research,
scholarship, and education by publishing worldwide.

Oxford   New York
Auckland   Cape Town   Dar es Salaam   Hong Kong   Karachi
Kuala Lumpur   Madrid   Melbourne   Mexico City   Nairobi
New Delhi   Shanghai   Taipei   Toronto

With offices in
Argentina   Austria   Brazil   Chile   Czech Republic   France   Greece
Guatemala   Hungary   Italy   Japan   Poland   Portugal   Singapore
South Korea   Switzerland   Thailand   Turkey   Ukraine   Vietnam

Oxford is a registered trade mark of Oxford University Press in the
UK and certain other countries.

Published in the United States of America by
Oxford University Press
198 Madison Avenue, New York, NY 10016

Library of Congress Cataloging-in-Publication Data
Cortada, James W.
The Digital Flood: The Diffusion of Information Technology
across the U.S., Europe, and Asia / James W. Cortada.
     p.   cm.
Includes bibliographical references and index.
ISBN 978-0-19-992155-3 (cloth : alk. paper)
1. Electronic digital computers—History.
2. Electronic digital computers—Social aspects.
3. Technology transfer. I. Title.
QA76.17.C67   2012
004—dc23   2011051399

9 8 7 6 5 4 3 2 1

Printed in the United States of America
on acid-free paper

To Grady and Levi, who will live in a connected world

CONTENTS

PREFACE

Before today's wave recedes we must catch the wave of the future and let the winds of change breathe new life into the business. Nor should we imagine that tomorrow's business will be a simple linear extension of past trends. We must draw lessons from the past, then elaborate on them to prepare for the future.

—Koji Kobayashi, *1989*
Chairman, NEC Corporation[1]

Well over a third of the world's population use cell phones—a tiny digital computer—yet these devices did not exist a quarter of a century ago, while sales of personal computers and iPads are counted in the millions. How did such items spread so fast around the world, even to regions where the poorest people on earth live? Commercially available computers came into their own in the early 1950s, and within a quarter of a century, virtually every large corporation around the world relied on them, as did most governments and, by the late 1990s, small and medium-sized businesses in the "advanced" and "developing" economies of the world. Today large computers—called mainframes in pre-Internet times and now more often known as servers—power the movement of massive quantities of data through the Internet, a "human nervous system," accessed by over a third of the world's population and by as many as ten billion devices ranging from security cameras to water purification systems.

What was it about the technology that caused so many managers and public officials to embrace it? We are normally told that declining costs for the technology and increased capabilities and improved reliabilities made diffusion possible. But do these features tell the whole story? I think not. The great speed and extent of adoption are features of the diffusion of computing that is so undeniable and possibly unique in the annals of the history of technologies, that economic and technological answers cannot be enough to address what is an ongoing very complex process. Yet, to be sure, using the language and models of the economist and technologist are useful tools for engaging in a conversation about this complex process that, on the one hand, is about diffusion and, on the other hand, about adoption. In one's lifetime an individual has witnessed an enormous surge in the variety of digital devices that were developed just in the past six decades and that became so integrated into the fabric of our lives and the work we do. For those of us who entered the workforce in the early 1970s, as an example, we saw the arrival of the networked computer, online processing, personal computers, cell phones, and the Internet. Mainframes went from room sized to handheld, from software that seemed like an ancient language to simple commands in all widely used human languages. For those who became teenagers in the late 1990s, the Internet and PCs had always been around, just as for their parents electricity and airplanes. For the Generation X and their children downloading was the routine way to acquire music and later games. All three generations experienced the explosive growth of social networking tools and Internet sites; their parents innovated, too, having computer chips implanted in household pets to keep track of them. Grandparents are also extensive users of computers, which they call hearing aids, telephones, and laptops. All share one feature of computing in common, each has integrated the use of information technologies into almost all aspects of their lives from the great events of war and politics to the mundane tasks of work and play.

Scientists and futurists are almost unanimous in their affirmation that the evolution of information technologies into new forms will not slow down, despite the fact that some business observers might argue computing is so ubiquitous that it no longer confers competitive advantages. Nonsense! History teaches us that when anything changes, for whatever reason, opportunities and risks appear. Information technology's circumstance is no exception.

With people increasing their expenditures and reliance on digital technologies, both in absolute terms as the number of humans populating the world expands and in relative terms as one's reliance on this technology increases, it is no wonder that we would need to understand the process more fully. For technologists such information enlightens their decisions regarding the development of new forms or in their research; for the business manager how and when to deploy the technology for some form of competitive advantage or in support of cost reduction and growth in profits; for public officials in how they serve their citizens and protect their nation. In short, "customers" for this sort of insight are broad and extensive. Their need goes far to explain the continued publication of studies on trends, about patterns of diffusion of all manner of technologies and scientific knowledge, regarding the power and nature of innovative activities of all kinds, others concerning construction of new organizational forms (what managers call business models), all coming from myriad sources: economists, sociologists, political scientists, physicists, chemists, and, of course, historians. Hardly any institution or sector of society ignores the issue. The discussion about emerging technologies extended back to before World War I, driven by the enormous number of scientific, techno-logical, engineering, and organizational changes that came with the arrival of the First Industrial Revolution by the nineteenth century in Western Europe and in North America by mid-century, from the development of electricity to the creation of the modern corporation, from synthetic rubber to nuclear bombs, from railroads to space ships and the World Trade Organization (WTO). It has been a long and wide-ranging interest.

The purpose of this book is to contribute insight toward a global view of the spread of computing around the world: its adoption, reasons for this diffu-sion, and extent of deployment. It identifies some of the shared and unique patterns of behavior—the dynamics involved—as they manifested them-selves over the course of its first five to seven decades. Yes, computers have been in use for nearly three-quarters of a century! Enough time has passed to begin integrating a number of aspects of the story into a more cohesive ac-count, one that goes beyond simply discussing technology and how it deter-mined life in the late twentieth century. It is hard to believe that computing has been with us so long as we are bombarded constantly with stories about the newness of the technology. We will deal with more than just how com-puters spread, addressing successes and constrained adoptions, what they were used for, by whom, and how behavior changed over time. This book argues that there have really been two waves of adoption, and I will focus on the first one, which extended from the early 1940s to roughly the early 1990s in the pan-Atlantic world and in East Asia into the early years of the next cen-tury. Much material is presented that takes the story to a second wave, which I believe to have commenced in the 1990s in some societies and still not yet in others. To understand the first wave, however, what I call Wave One, requires that I provide some information about the characteristics of the second wave for context and consequences, what I call Wave Two. The two are contrasted the most in the first and last chapters of this book, with additional discussion about Wave Two in the second appendix. But the core of this book is a discus-sion of how computing came to be used on both sides of the northern ends of the Atlantic Ocean—United States and Europe—and by a number of Asian societies. The two combined regions account for the most teledense societies in the World and half the population of our planet.

The majority of existing studies about the arrival and diffusion of inven-tions and technologies focused on things other than computing. It has only been since the late 1970s that students of the process began to pay attention to computers and to the patterns of their deployment. As illustrated in the chapters ahead, often looking at the issue parodied in the old joke about the seven blind men asked to describe an elephant in the park. Each touches the elephant and responds with an analysis of what they experienced. Touching its belly on the side, one says "It's a rock." Another petting its nose, says "It's a tree limb," and so it goes. Additionally, the vast majority of the literature on the spread of technology is devoted to things other than computers. Historians, for instance, have been fascinated with pottery, bronze then iron tools, sailing, gun powder, rifles, airplanes, medicine, atomic bombs, and so forth. Their accumulated literature runs into the tens of thousands of citations. To be sure, the number about computers runs into the thousands too, yet like the seven blind men, they often deal with one facet of the story and indeed, often, one incident or case study. So a librarian publishes an article on the introduction of the Internet in one state government agency, while an historian describes the development of a specific technology (or computer), and a think tank produces an account of regulatory practices with respect to telecommunications in one country during one decade.

But, somehow, we need to pull together the whole story to the extent possible, because there were—and are—forces at work that make each of those examples related to one another. One of the key findings of this study is that diffusion did not occur in isolation from events in other countries. The story told here is that the world of scientists, engineers, mathematical institutes, universities, and "high-tech" vendors of information technologies was a great deal more global and interconnected many decades earlier than today's pundits acknowledge. The flow of information about the technology knew few or no borders over the past century, regardless of censorship, world wars, Iron Curtains, and limited budgets. Engineers and computer scientists wrote and read a great deal about computing and talked to each other regardless of what their political leaders were advocating. Equipment crossed borders legally and illegally and the inability of one country to deploy information technologies contributed mightily to its collapse.

If we take some blindfolds off, we can discover that it is an elephant in the park, not a rock or a tree trunk. We find that one can learn a great deal about the nature of the technology itself, its effects on society in general, and how best to go forward with its continuing transformation. Combining all these elements into a more comprehensive description of its evolution is the primary objective of this book. This can be accomplished because so much good work already exists by not-so-blind observers in the information technology park. This book should be seen as a first attempt to provide a history of information technology that spans large swaths of the earth, providing a more complete view of the three issues of how much information technology spread, why as comprehensively as it did (and not yet concluded), and why this happened so quickly.

The reader will be surprised, perhaps shocked, at how little space is devoted to the evolution of the technology itself. Relying on the excuse that so much historical research has already been done on that aspect that we can bypass is not the reason. Rather, one of the findings of this study is that the forces at work affecting the spread of computing were profoundly political, economic, social, cultural, strategic, military, and philosophical. The Soviet experience, for instance, teaches that the age or newness of the technology was largely irrelevant to diffusion, while Marxist–Leninist world views and style of management profoundly gated the speed of diffusion for nearly four decades. How a company was organized in the Netherlands played a greater role in determining when computers came into an enterprise than some technical improvement. Organizations run in a highly centralized fashion liked big computers in data centers, while decentralized ones wanted small distributed systems and were not willing to install "big iron" even if economic and managerial wisdom suggested they should.

A surprise finding that bears more investigation is how central computers came to be in many nations, surprising because historians had only documented that fact for a few countries and only in recent years. Military leaders relied upon them for rockets, airplane avionics, and, most recently, smart bombs. The best example known the longest by historians was the nuclear bomb, because the threat of their use helped to shape the features of the Cold War. However, we learn that senior leaders in a number of countries bet their political success on the technology working. When their nations failed to leverage it well, their power diminished and, in one case, may have led to the demise of a regime, as happened in the German Communist state long before the Egyptian government fell in 2011 in part owing to the use of social networking tools. Every major political leader in the world has had to deal with computers within the context of national security and economic development since at least the mid-1950s in the West and from the 1960s in the East, with the exceptions proving to be just that, exceptions concentrated largely in Africa, parts of Latin America, and in some areas of Southeast Asia.

What made this finding a surprise was the extent to which computers were integral to the main events of a nation's history during the second half of the twentieth century, so much so that the conclusion I reach is that we no longer can write the history of information technology isolated from the broad considerations of mainstream historiography: politics, economics, social and cultural issues, role of children, women, and minorities. Conversely, political, social, and economic historians will have to take into consideration the role of computers prior to the arrival of the personal computer or the Internet. The story is bigger, more complex, and more woven into the fabric of human activity going back further in time than what we are told today about the role of the Internet. So the role of the technology will now have to be more fully integrated into the mainstream of a nation's modern historical narrative. To present the evidence, I concentrate overwhelmingly on patterns of diffusion during Wave One.

Diffusion occurred in phases (some scholars prefer the word waves) over time, beginning with early projects in the 1940s, commercialization of mainframe computing in the 1950s, and expanded adoptions in advanced economies in the 1960s through the 1980s in most parts of the world. Then came another surge—wave—driven worldwide by miniaturization of technologies and communications in the 1980s, and Internet adoption beginning in wide form in the 1990s. By the end of the century, this last wave had become a substantial global phenomenon well underway. These waves of adoption have common features, but different timings. They all are in response to technological innovations, changing costs of computing, and, even more important than the machines, new uses, effects of local economic conditions, social attitudes, and expanding globalized activities with respect to trade, communications, and information sharing. The story told in this book is analogous to what historians have learned about the diffusion of other technologies that they have had more time to study, such as that of the relatively similar case of the clock.

To a large extent, the insights offered here fit nicely into the larger widely held beliefs regarding the general evolution and receptivity by people of technologies over hundreds of thousands of years. That is a comforting message because when I began to look at information technology as the evolving deployment of new tools and toys, I was not sure what would emerge. I was not sure if information technology represented an exception, something so new that it departed from well-understood patterns of adoption of other old and modern technologies. My concern is the same shared by historians in general when exploring the specific circumstances of, say, a computer device in a particular year and place. It is, of course, still a unique story. While we will not have space to compare and contrast precomputer technologies with this newer class of artifacts, it was my reading of literature on the earlier history of such things as ancient pots, weapons, steam ships, use of nineteenth and twentieth century forms of transportation, and so forth that led to the list of issues I discuss.

Nearly 40 years of business experience in working with information technologies at IBM and with many clients, and writing about their use and later history, reinforced my conclusion that experience with information technologies (IT) extends the arc of humankind's involvement with all manner of technologies, even though, just as with any specific class of these, IT also have their unique characteristics. One example will have to suffice to illustrate the point. While bronze, iron, and steel are malleable when heated to make swords, furnaces, and fences, their shared feature of malleability is different than that of IT. Computer chips and how they are put together to form such objects as large mainframes or tiny cell phones share the characteristic that their technology is malleable, that is to say, can be made into all kinds of forms. But, computer chips and the programmable software languages used to fill them with instructions represent many classes of activities different than those of a blacksmith pounding a piece of metal into only a sword, knife, or cooking pot.

While much of what is discussed in this book is about information technologies in varied forms, one does not need to have knowledge of the technology to understand its diffusion. In other words, the reader does not need to know how computers work, about the technical infrastructure that makes the Internet seem so magical to most of us, or anything concerning the intricacies of software. If there is a feature of a particular form of information technology that must be understood, I explain it in terms that should make sense to the "person on the street," and only as much as needed to appreciate the discussion at hand. This is an important design point for this book because I hope many readers are non-IT management in corporations and governments who could use some of the findings in this book with which to perform their duties. I also want scholars in many fields to understand the breadth of the computer's acceptance and influence so in my narrative I assume many too are not technical at their core and have no desire to be. For both audiences this technology has already become part of an era in which hardly any significant activity can be accomplished without the aid of some form of digital tools. It is a remarkable statement to make, but very true in nearly all industries in the majority of nations, and completely so for some industries in all countries (such as banking and telephony).

My intention is not to be encyclopedic. Rather, I describe patterns of adoption and use at a relatively high level, so that future historians, economists, and political scientists at a minimum who study individual countries either discussed in this or other books will have a framework to assist them in defining issues to look at and gaps that need to be filled to enhance our understanding, and, for the nonhistorian, appreciation about how such a complex technology comes into society and is used. To do this, I rely largely on the growing body of monographic literature on the role and history of computing and the evergrowing collections of data on diffusion statistics produced by such groups as the United Nations, Organization for Economic Cooperation and Development (OECD), and the World Bank. To a large extent, in the language of the academic, this will be a synthetic history, concentrating on a few specific themes about diffusion and deployment. Emphasis changes in the story when we do this. The U.S. experience, which has always dominated the narrative about the rise and use of computers, receives less attention than it has in the past in order to shed light on the role of other parts of the world, most notably Europe, which I define as from Ireland to the far eastern edges of Russia, taking into account Western, Central, Eastern Europe, and all of the Russian state, all of which have been active adopters of this technology in spite of World War II, the Cold War, and myriad economic ups and downs. Asia receives, of course, considerable attention. There are areas of the world left out of the discussion about diffusion that will not be discussed in this book, specifically Latin America and dozens of countries in Africa and the Middle East. I begin to tell the global story by discussing the most extensive users of information technologies.

To write one book on such a broad topic I had to focus the text on the key elements of the story of diffusion. Readers will have their favorite criticisms of what I left out, largely based on their world views of the subject. But let me be transparent on the matter. I devote less attention to the history of telecommunications except when it was central to the diffusion of computing. It is a far more important technological component of Wave Two diffusion because of the Internet and cell phones, less so for Wave One computing. I will be criticized by some who will argue that data went over telephone lines back in the late 1950s and massively so in corporations in the United States in the 1960s. But, that was not the case in so many other places, such as in the huge Soviet Union's civilian agencies and businesses until, effectively, the late 1980s. China today tries to censor politically sensitive information over its telecommunications network yet has more individual users of the Internet to conduct commerce and personal communications than all of Europe combined; it has more users than even all of North and South America. Throughout the second half of the century the number of nations doubled in number; I cannot write one volume covering 200 countries, let alone even all countries in the two regions of the world I do cover. I picked nations to describe that were emblematic of diffusionary practices. My apologies go to the Canadians, Australians, Norwegians, Danes, Austrians, Spaniards, Argentinians, and Chileans—all extensive users of information technologies. To historians in these countries, I challenge them to write further about their national histories drawing from this book what lessons they find of use to them, building on work they have already done. However, I can assure the reader that prior to deselecting a country I studied its IT history and collected materials on each that are part of the records spun off of this book and that I will deposit in an archive so that future historians can continue the process of filing in gaps. I alone take responsibility for selecting which nations to include in or exclude from this book. In the end I did not want to burden the reader with a multivolume history of the subject as I did with my prior study of the computing experience of the United States in *The Digital Hand*. Let that happen once additional research has been conducted by many historians in the years to come and after computing has been more fully integrated into the overall history of the late twentieth century.

To help future scholars, I commented on neighboring countries that were not specific targets of my concern for two reasons: neighboring countries affected the nature of the diffusion that occurred in a selected nation, such as all the Nordics on each other, including on Sweden about which I do write; and second I discuss the amount of information (or lack thereof) about a region or country, such as the dearth of quality data on diffusion across all of the Communist countries of eastern Europe and parts of Asia. I pick up the problem again in the bibliographic essay at the back of the book.

I recognize that this book is only a first step on the path to some more comprehensive view of the history of computing if for no other reason that because of all the topics I did not discuss, and of those that I did, or not well or thoroughly enough to satisfy everyone. I discuss the path taken and other options cast aside in some detail in Appendix A because this is not going to be the last multinational history of IT to appear. Because over time even some of the language will become dated; let me explain what I mean because even phrases carry messages of their times. Throughout this book I use the phrase IT to mean precisely information technology(ies) and specifically I mean computers, their peripheral equipment (such as printers, terminals, and tape drives), and all manner of software. Some individuals are more liberal in their definitions, and include processes, documentation, attitudes, and so forth; all of which is fine but without the artifact installed and in use, none of the other components of IT are of any importance. I concentrate on installation and use of the object and quantities count in my assessments. I also use the phrase ICT on occasion, which means information *and* communications technologies, a term embraced earliest by West Europeans and largely after they started using telecommunications with their computers. When I use the term ICT, it is because I am including in that sentence or paragraph discussion about both computers and telecommunications. In some instances, for example, a body of data includes both and I have no choice, so I inform the reader, while in other instances, I want to include the two because they were co-joined in some relevant discussion. When I use a phrase like data processing instead of information management or knowledge management, I mean what I say. Data processing—a term widely used in the 1950s to mid 1970s—speaks to the issue of collecting, collating, and presenting large bodies of data, normally for business applications (uses). Knowledge management—a term that came into vogue in the 1980s and sadly possibly now almost out of fashion—is less about computers (although the technology is a part of it) and more about the use of data, information, knowledge, and even wisdom. In short, I have attempted to intend what terms I used consistently throughout this book.

HOW THIS BOOK IS ORGANIZED AND CAN BE READ

This book is organized so that it can be read in its entirety from front to back, or in pieces. I structured it so that a reader can get the full story I am telling but also use it as a reference in support of other uses. Thus, if you are interested in how diffusion of IT swept like a wave across the industrialized economies of the world, reading it cover to cover makes sense; on the other hand, if you are concerned about events in one country only, you can read that section (or chapter) and not be dependent on having to examine others to make sense of that part of the book of greatest interest to you. The same applies to the bibliographic essay, which is organized by country and region. Within each chapter, the first section describes the scope of the chapter and the few key points. These observations are amplified and summarized in a short concluding section in each chapter.

To obtain a full discussion of the major findings of the book without having to read all the country examples, I suggest five chapters. The first is crucial because it describes the key themes and findings of the book. Chapter 2 summarizes the U.S. experiences, while chapter 5 provides an overview of all of Western Europe's encounters with computers. The same approach to reviewing all of Asia's experience with IT is contained in chapter 11, while the last chapter summarizes key messages and, more important, implications of the world's experience in appropriating IT by audience: scholars, business managers, and public officials.

So, this is either a very long book or one in which the key messages from across all chapters amount to less than 50 printed pages. Tables are designed to present a great deal of information to make a point but not to be definitive. I have assembled the kind of data and narratives that would be of use to someone looking at economics, politics, and business. The chapters are longer than one normally would expect in a history book to cover a topic in adequate detail, but short enough that if one is only interested in a particular country or region, it is de facto a short book. But every chapter is synchronized with all others so that messages are aligned, common issues are discussed from one to another, and comparisons and contrasts made. To facilitate the reference quality of this book, it includes a detailed index. Appendices are for those interested in how this project was implemented. Finally, if one is only interested in country-specific discussions, they can skip the first chapter. At one point I even considered making it an appendix but decided that too many readers would want an overarching statement about the scope and significance of the subject before wading into the volume.

ACKNOWLEDGMENTS

Writing and publishing a book is a team sport, requiring the collaboration of many individuals and organizations. A number of people were of particular importance in this project. Jeffrey Yost and Tom Misa at the Charles Babbage Institute, University of Minnesota, helped me conceptualize the scope and structure of this project over many conversations. Stephanie Crowe, their archivist, was enormously helpful in finding me hundreds of documents to examine, uncovering important gems along the way that I would have overlooked on my own. Paul Lasewicz, IBM's archivist, not only made IBM's corporate archives available to me, but also sought out materials he thought essential to this project. This company's records are often the most significant available in English on the acquisition of computers in many countries. Dawn Stanford, the Reference Archivist at IBM's Corporate Archives, was extraordinarily prompt and thorough in her support of my research efforts. Honor Sherlock at IBM helped me determine how many computers there were in the world, and identified sources of materials for me to consult. Peggy Kidwell, a distinguished historian at the Smithsonian Institution, obtained access for me to the computing archives and library of the National Museum, while finding materials crucial to this project. When I ran into language problems and frustrations in finding materials relevant to Sweden's experiences, Lars Arosenius stepped in and helped. I deeply appreciate the assistance and materials given to me by economists and officials in the South Korean government, World Bank, Organization for Economic Cooperation and Development, and in the government of Singapore. Without their assistance, several chapters would simply have been impossible to write with any sense of authority. Richard W. Judy, an expert on Soviet computing, and now chairman of IT Workforce Associates, gave me a good collection of materials on the topic and the benefit of his insights. Honghong Tinn, while working on a Ph.D. degree looking at Taiwanese technologies, found the time to share useful information as well. Ray Kurzweil graciously allowed me to use several of his iconic charts.

Conceiving of computing as a worldwide integrated phenomenon of the second half of the twentieth century was made possible by my working with clients and IBM colleagues all over the world. In particular, I want to thank the group of consultants and experts on all manner of business and economic issues related to information technologies and modern management working at the IBM Institute for Business Value, my home for over a half dozen years. Their sense of the global nature of what happened with computing shaped many of the issues addressed in this book, while their personal contacts in dozens of countries made it possible for me to validate my ideas and findings in over two dozen countries with local nations who were colleagues, clients, or nationally recognized experts on the subject. There is no better park bench in the world to sit on to observe, indeed influence, the unfolding of Wave Two than this institute, especially now as IT makes its way into the farthest corners of our planet.

This is my fifth project with Oxford University Press, where I am always treated as if I were the only author they are focused on. My editor, Terry Vaughn, supported the project from the minute I proposed it to him, while associate editor, Joe Jackson, worked quietly behind the scenes to facilitate publication. The production team, led by Leslie Johnson, took a very large manuscript and made it into a very nice book, earning in the process my deep gratitude. I want to acknowledge as well three anonymous reviewers of this very long manuscript and for their wonderful suggestions for improving it. Additionally, I want to point out that the views expressed in this book, and any errors of weaknesses, are of my own making and do not necessarily reflect the views or work of IBM.

Finally, I once again must thank my wife, Dora, who made it possible for me to find enough time to write this book, and my two grandsons, Grady and Levi, for not eating the manuscript or doing other things little boys can do to boxes of files and books.

James W. Cortada

The Digital Flood

CHAPTER 1

How Much Computing Is in The World?

We are not necessarily evolving toward a single world culture, nor must we become subservient to . . . intelligent machines.

—David E. Nye[1]

We no longer really know the answer to what should seem an obvious question to ask, but at one time economists, public officials, some computer scientists, and historians did. That was a long time ago, largely in the 1940s and 1950s, but not after the 1960s, when the number expanded around the world and digital technologies went from just being computers to becoming components in other devices, thereby complicating the answer to our question. Specifically by the end of the 1990s, integrated circuits, small computers, cell phones, portable sound systems, and, of course, software and microprocessors embedded in all manner of machinery blurred any definition of computer systems by continuing to extend diffusion from machines that only did data processing computing to myriad devices that have become "intelligent," with computational capabilities part of the functions of those devices. In the late 1940s and early 1950s in North America and in Western Europe, inventories of such systems were routinely kept and published, made possible because there were so few such devices to track.[2] But with the introduction of minicomputers in the 1960s, personal computers in the 1970s, and the rapid diffusion of digital telephone switches and portable telephones in the 1980s, for example, tracking populations of digital technologies in their ever-expanding variety of forms (what technologists call configurations, platforms, and systems) had become very difficult, if not impossible to do. Personal computers, for instance, sold in the thousands per year in the mid to late 1970s, in the tens of millions within a decade, while various industries acquired computer chips in the billions.

In the early 2000s, most regions of the world were spending between 5 and 7 percent of the Gross Domestic Product (GDP) on various information technologies (IT). In 2011, the world's GDP was $65.6 trillion of which $3.9 trillion went to IT—a massive quantity by any measure. Even that number is low because IT industry associations place 2011's total IT expenditures at $4.4 trillion. A third of the world's population had access to the Internet with over half the populations in North America, Europe, and East Asia online. Out of 7 billion people on earth 4.6 billion had access to mobile cellular telephones. All the statistics on various types of adoption indicate that each year the rate of adoption of various technologies is actually speeding up worldwide.[3]

Thus, it would seem superfluous to want to know how many computers there are in the world because so many people might already have concluded that computing is ubiquitous, with these devices in use everywhere. Computers have been around for seven decades. Anyone under the age of 25 years in most countries cannot remember a time when the Internet did not exist, and with respect to personal computers (PCs), probably anybody under the age of 45. But, as this book demonstrates, diffusion of this technology occurred unevenly around the world and, at the time this chapter was being written (2012), there still existed wide disparities in the deployment and use of the technology all over the planet. North America, East Asia, and all of Europe became extensive users of computing, but Sub-Saharan African used less of this technology than did the most advanced economies even 50 years ago. So, the answer to the question is not obvious. It is sufficiently variegated that to appreciate how this technology spread around the world we should really begin with a brief understanding of what the answer might look like. To be sure, as well, finding the answer poses many difficulties as there are no convenient, universal sources that completely catalog how many exist, let alone how they are used, and by whom.

Yet we know several things. Over the past seven decades digital technologies evolved into a large array of devices, from mainframes to iPods, many thousands of types of instruments and equipment, in fact. Second, the number of these increased continuously, particularly as they became smaller and more affordable for individuals. We went from several thousand expensive machines affordable only by corporations, universities, and government agencies in the 1950s to the point where in the first decade of the twenty-first century over a third of the world's population had access to cell phones and the Internet, while in the most advanced economies Apple Computer reported selling over 150 million iPods between October 2001 (when it introduced these) and September 2007.[4] Third, digital technologies became embedded in so many devices that were not computers that over time these too could be viewed as computerlike. There probably is no longer any large manufacturing machine and any mechanized transportation vehicle (e.g., airplanes, automobiles, trains, and trucks) that are not controlled and operated with the use of computers, if manufactured after 1975. Additionally, that statement also applies to machines in all industries for how services are provided in many other industries (e.g., consulting, banking, tourism).[5] Fourth, we are increasingly learning more about what percentages of a nation's Gross Domestic Product (GDP) and Gross National Product (GNP) are going toward funding the acquisition and use of information technologies (IT) and its companion, telecommunications.[6] The link between GDP and extent of adoption of IT has now been irrefutably established; the higher the GDP, the more one spends on IT, hence the larger the amount of IT installed and used in a nation.[7]

Ultimately the whole purpose of using IT anywhere was to access information, be it data, text, video, and conversation, or a transaction. There is a debate going on among academics about how much information there is in the world and while we will brush up against that conversation from time-to-time in this book, we will focus on diffusion. However, the debate is relevant in that it is mounting evidence that humankind is moving a vast amount of information into computers at a very rapid rate, indicating the rapidly growing dependence on digitized information with which to go about their daily lives. Just as important, however, is the increased amount of digitized data people need with which to do their work and to play. For most people, the calculations are beyond reasonable comprehension, such as "in 2008, the world's servers [computers] processed 9.57 zettabytes of information, almost 10 to the 22nd power, or ten million million gigabytes," which meant "12 gigabytes of information daily for the average worker, or about 3 terabytes of information per worker per year."[8] One terabyte is the equivalent of 4.5 million books each with 200 pages, so an average worker then used 13.5 books' worth of information, or 233 DVDs per terabyte, which amounts to over 700 DVDs per worker. A zettabyte is more in the range of 5 billion books, each 200 pages in length, or 251 thousand DVDs. The volumes have been increasing for decades, indicating a profound shift has already occurred in how humankind handles information.[9]

This chapter describes the scope and history of computing's diffusion around the world in an introductory manner to establish a sense of the magnitude of the deployment and use of these various forms of digital technologies. In short, the numbers are in support of the rationale for why one needs a book on the subject. The statistics presented in this chapter serve as testimony that computing became relatively ubiquitous in many parts of the world, an important feature of contemporary life. This statement is based on the assumption that people would not go out and buy computers, or, commit their companies and government agencies to acquire these technologies unless they were going to use them, because they were not free and most frequently still expensive. In fact, computers cost a great deal, contrary to an enormous amount of marketing hype presented to customers over the past half century; therefore, they had better be used. Even today, with what we are told are inexpensive mobile phones, hundreds of millions of people still gather as communities of users in villages to determine if they can afford just one, much the way Americans and Europeans did in the richest economies in the 1970s to fret over the cost of $700 handheld Hewlett-Packard calculators. To be sure, the equipment has been most productive but it is not free and so looking at deployment numbers begins to suggest, first, the value of such technology to whole societies willing institutionally and personally to invest in this equipment and software, and second about the kinds of uses and effects. That is why statistics on use and penetration are useful as a way to begin examining the spread of this technology.

Additionally, this chapter introduces aspects of patterns of deployment evident at this time, which the evidence leads us to conclude seem to be the most important contributing forces at work in the rapid diffusion of these classes of technologies. To be sure, there may well be others not taken into consideration, largely because they are not evident today. Also, as historians understand, when one examines the specific experience of a single nation, or of one class of digital devices, new factors become critically important, such as the role of individual business executives (e.g., the late Steve Jobs at Apple Computer or Bill Gates at Microsoft), specific actions of government leaders (e.g., Plan Calcul in France in the 1960s and 1970s), and local economic circumstances (e.g., capitalism and entrepreneurship in the United States, central planning in Communist states, and poverty in Africa). This chapter ends with an introduction to the major themes explored in this book.

WHAT ARE DIGITAL, OR, INFORMATION TECHNOLOGIES?

Definitions are important for a number of obvious reasons, not the least of which is that they help clarify the scope of the discussion regarding diffusion of information technologies all over the world during a period of just over a half century. But we also need clarity because one of the reasons digital technologies diffused so fast into so many places was due to inherent features of the base technology itself—computer chips—about which more will be said throughout this book. For IT has undergone profound and rapid transformations since the original concepts of modern computing were developed by such individuals as Alan M. Turing, John von Neumann, J. Prespert Eckert, Jr., John W. Mauchly, and Konrad Zuse between the 1930s and the early 1950s.[10]

Its single most enduring feature and the one, perhaps, more than most explains why definitions are important, is its malleability. IT is a general purpose technology, meaning it could be shaped into all manner of forms and uses; in other words, it was very flexible. In the 1950s there existed big Univac and IBM computers, with a great deal of peripheral equipment the size of modern refrigerators and, of course, early uses of software; but, by the early 1980s, Apple, IBM, and Compaq, and scores of other computers were small enough to sit on a desk, while many people walked around with pocket sized calculators from Hewlett-Packard and Texas Instruments (TI) that could per-form engineering calculations and calculus more easily than large computers in the 1940s and 1950s. By the late 1980s people were using diminutive cell phones, 50 percent smaller than earlier models, and by the mid-1990s, yet another 25 percent tinier, and the latter with many functions, not the least of which, by the late 1990s, included the ability to take and transmit photo-graphs, access the Internet, and host electronic video games. By the early 2000s, these phones had become, in the parlance of computing, the "plat-forms" that carried uses (applications) that once were only available on PCs and laptops. Likewise, laptops and handheld calculators in the late 1990s had also acquired more "horse power" to hold data and to perform calculations and instructions than even the large boxy computers of the 1960s and early 1970s. Very few earlier technologies had these kinds of considerable flexi-bility of form, other than size. One could make a large or small hammer, a big sail boat or a small one, but a user could not make a hammer that sailed the ocean, or that also performed the functions of a saw or a drill. But, in essence, one could metaphorically do such varied information handling tasks with digital technologies, using a combination of computer chips, programming languages, and software, which is why computers were such an important general purpose technology.

The evolution of the malleability of IT shows no end in sight. For example, IBM's engineers and other scientists are experimenting with microscopic-sized nanotechnology-based computers, while others at many institutions around the world are beginning to wonder if living cells can be created to serve as com-putational and data storage devices.[11] To put a finer point on this feature of IT—malleability—one can think of its breadth of reconfigurability as exten-sive as the ability of a cook to take meat, potatoes, and beans and convert them into myriad dishes. Our analogous cook in the 1950s might only have been able to make big and small steaks, boiled or fried potatoes, and possibly baked beans. Push the analogy still further and today those items can be served inde-pendently of each other, be parts of various stews, sandwiches, appetizers, and, of course, be boiled, fried, grilled, or microwaved—all simultaneously with various types of sauces, condiments, and cooking technologies from camp fires to programmable bread making machines. If one thinks of the evolution of IT in that analogous way, it becomes easier to understand the effect of this class of technology on users. The concept of flexibility—malleability—is essential to any understanding of at least the speed with which digital technol-ogies spread around the world. Ease of use and convertibility contributed to speed of adoption, not just to considerations of cost. The price of a technology for an end user was only an issue once someone had converted a specific piece of technology into some useable form, such as into a cell phone or a GPS system using satellites and computers embedded in an automobile that gave drivers directions to a nearby restaurant. So, one has to think of the flexibility of IT's forms as central to any appreciation for what was diffused and how.

One other characteristic of IT to keep in mind as we define the technology is that during the period covered by this book (essentially 1940–early 2000s) all the various technologies that composed IT were developed. While the first systems were built out of existing electronic components and radio parts, very quickly new ones, base technologies, and subassemblies emerged tailored to the needs of computing, such as early memory systems for computers and, of course, integrated circuits (what many refer to simply as computer chips). Each of the base components evolved rapidly over the entire period, some eventually displaced by radically newer core technologies (silicon chips displacing transistors, C++ programming language replacing COBOL, which earlier had superseded the Assembler programming language).[12] The creation of technologies occurred at the same time as computing diffused, and was affected in what forms it took partly in response to the needs and desires of its users, not simply because of what engineers concocted in some isolated laboratory. The vast majority of IT came into existence and into the market through the work of profit-seeking corporations, which paid particular attention to the wishes of their potential customers and investors.

Officials and academic computer scientists also played enormously important roles in the process, particularly during the infancies of all major—we should say successful—technological innovations, but it was in response largely to market realities that led to the general form technologies took and to their acceptance by users. Therefore, throughout this book, the views and actions of users and customers largely take primacy over the supply side of the story, but not to the exclusion of the role of IT vendors, scientists, and governments, because all played enormously important roles.

All new technologies undergo a period of rapid evolution before their creators begin stabilizing them in forms that substantially remain in use for a long time. With respect to computing, that pattern proved essentially no different than what happened with the telephone, internal combustion engine, or aircraft, for example. This pattern of change and stabilization by computing and communications is still ongoing. One would be hard pressed to find a computer scientist today who believes these various technologies have reached some technological equilibrium. In fact, they believe the opposite to be the case; computer scientists and other experts do not see transformations slowing down.[13] To be sure there are contrarian views but they come from individuals who are not technologists.[14] Even a crude measure of innovation, such as applications for patents, bears this out. IBM, for example, increased its number of patents filed annually all through the period and leads all companies in all industries around the world in the quantity obtained.[15] More than kudos for IBM, this is evidence of the continuing rapid evolution of the technology, because other major suppliers of IT are also still transforming their products and technologies, as evidenced in the rising number of patents at such places as Hewlett-Packard (H-P), Intel, Stanford University, and the Massachusetts Institute for Technology (MIT). In a dramatic example, we have at MIT the now highly publicized case of Professor Nicholas Negroponte's $100 laptop (One Laptop Per Child) which, by 2010 had been deployed in various countries around the world for use by students, initially in countries with low gross domestic products (GDP) per capita, and which did not exist a decade before.[16]

To simplify definitions, table 1.1 lists types of computing devices and software available in various time periods. It is not a definitive categorization, of course, but it points out the major technological components that composed IT, or digital technologies, in each historical period. All those that came after 1950 relied on the base technologies of transistors (1950s–early 1960s) and subsequently digital integrated circuits (computer chips), which continue to house instructions and data in whatever device one wants an IT function, be it a computer, an automobile motor, airplane avionics, or a wrist watch. As these technologies evolved over time and, along with the language used to describe them, definitions and lists of such technologies did too; we know this by simply looking at the contents of introductory text books on the subject of computer science and more advanced texts on computer architectures and software (usually programming, database management, and operating systems). Note that the variety of ways IT could be put together increased over time. These machines—systems of devices and software—all shared common building parts: computer chips; parts of devices that calculate, move information about, and store data; software called operating systems that synchronized the activities of programs and all manner of hardware, hence the term *system* to describe an IT device; programming languages with which to give a computer device directions on what to do; application software which are programs that do specific things, such as allow one to write books, or spreadsheets to track and calculate numbers; some form of display or output unit (e.g., a printer) to present information (think in terms of answers) that are intelligible to humans, such as a printed report or a TV-like screen on a PC or iPad to display data.

All of these basic building blocks of digital devices changed over time, often at different speeds and in multiple forms, a subject studied extensively by historians.[17] We will not retell that story here; however, in subsequent chapters specific changes are called out when they caused a spurt in deployment. For example, when declines in the cost of computer chips occurred simultaneously with improvements in their speed of operation, or expanded their capacity to store and move data, adoptions increased. Greater reliability in performance

Table 1.1 INFORMATION TECHNOLOGIES AND ARTIFACTS, 1940–2012

Technologies and Artifacts	Era
Mainframe computers	1940s–1950s
Transistors	1940s–1950s
Programming languages	1950s–1960s
Integrated circuits and microchips	1960s
Online systems	1960s
Minicomputers	1960s–1970s
ATMs appear in banking	1970s
Intel's microprocessors	1970s
Personal computers	1970s–1980s
Compact discs	1980s
Video games	1980s
Microsoft Windows	1980s
Internet with World Wide Web	1990s
Cell phones	1990s
DVDs	1990s
iPods, iPhones, iPads	2000s

Because technologies, hardware, and software often were in simultaneous use from one period of time to another, Era refers largely to when an item was initially widely deployed, not to when it was first developed.

often made it possible to lower the cost of a device to such a point that yet another cluster of users could afford to use the technology. The ability to re-package the technology in ever smaller increments proved just as important, such as from mainframes, to minis, then to desktop microcomputers, next to handheld devices. The move from mainframes to PCs in the 1980s, for instance, meant going from systems that cost hundreds of thousands of dollars to lease or buy and maintain, to systems one could purchase for a few thousands of dollars. The move from wired to wireless communications and simultaneously to newer, smaller components and batteries in the late 1990s facilitated the evolution of increasingly reliable, less expensive, more practical cell phones, and further stimulated use of the Internet. Cell phone usage, in particular, spread in poor countries that did not have a prior nationally wired telephone infrastructure, while the Internet penetrated rapidly and widely across Central and Eastern Europe.

If one had to pick a base technology that became cheaper, faster, and better it is the computer chip. Figure 1.1 presents data on how dramatically that tech-nology decreased in cost over time, often the one component of IT economists cite routinely as the most important reason why computing was so rapidly adopted. One team of economists conveniently documented the decline in the cost of computing and the increased capacity and speed of semiconductor components (hence computers):

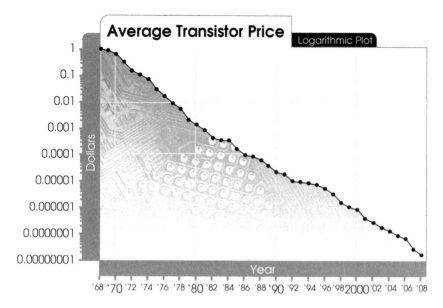

Figure 1.1
Decline in the cost of computing over time: The example of transistors and chips
Source: http://singularity.com/images/charts/AverageTrans.jpg

- First logic chip (1971) comprised 2,300 transistors; the Pentium 4 (2000) had 42 million
- Number of transistors increased by 34 percent per year between 1971 and 2000
- Memory chips declined in cost by a factor of 27,270 times (40.9 percent per year) between 1974 and 1996
- Price of logic chips decreased by 54.1 percent per year between 1985 and 1996.[18]

Daniel E. Sichel, a highly respected economist and expert on the economics of computing, mirrored the views of many of his colleagues when he argued that "the single factor that most drives the rapid expansion in computer spending is the rapid and relentless decline in price."[19] As I argue throughout this book, while dramatic changes in price performance of computer chips proved quite important, along with sharp improvements in what one could do with this technology, there were many other causes for how computing spread.[20] Thus, the economic orthodoxy that the influence of chip economics is the paramount influencer of adoption rates—widely accepted for over three decades as the primary explanation—is proving inadequate for explaining more fully the diffusion of computing as historians and others learn more about the role of computers—a point I reiterate repeatedly. Yet, cost economics of hardware cannot be dismissed as irrelevant, because the economic productivity of the technology proved very substantial and compelling; it just is not the whole explanation.

Nor should we embrace fully the logic for our acceptance of computing put forth before the economic argument, namely, that the technology improved so magnificently thanks to engineers who built it such that IT did work while other machines and methods could not do as well. As with cost economics, technological imperatives indeed were quite important; without computer chips, for example, we would not have had the extensive diffusion of computing that occurred. Additionally, costs of integrated circuits (computer chips) have always been inextricably linked to economic considerations. One should recognize the profound importance of the transitions documented by figure 1.1. Figure 1.2 shows how dramatically the costs dropped per activity done (transaction) by the device. Figure 1.3 illustrates Moore's Law, which essentially holds that every new generation of computer chips had roughly twice as many transistors (or functions) as the previous one and that each new generation appeared roughly every 18 to 24 months, a pattern that has essentially continued since the early 1960s. From an economic (or cost) perspective this meant the price of computer chips kept dropping over the years at relatively predictable rates.[21] One could see the results of the declining cost of this basic building block of computing in a variety of parts of a computer. In figure 1.4 we see the effects of declining costs of computer chips when converted into memory in computers, where data are actually stored. Over time, these digital building blocks became physically smaller, which allowed more of them to be packed into a given space (such as a computer, car, or camera). Yet, as one can sense without a dissertation on chip design, more work could be done, first within a chip, second at faster speeds, and third, in the collaboration among chips linked together to handle ever-larger more complex work. The result was absolutely remarkable. It was largely why, for example, in those countries where labor costs rose sharply over time, it became more cost effective to move (automate) work into computers, providing the technology could do the work well enough, thereby requiring less human labor in such countries as the United States and in many parts of Western Europe.[22]

However, to put the role of computing chips into historical perspective, it should also be noted that their declining costs over many decades alone did not account for the wide acceptance of computers. As the remainder of this book demonstrates, other factors were at play as well, such as the actual uses of computers to do the practical work of governments, companies, and individuals.

Before moving to an introduction of the key concepts explored in this book, defining information technologies as notions for how they were used is as important as any definition of the specific machines and other components. As the technologies evolved over time, so did their definitions and how one cataloged these artifacts, uses, and roles. In the 1940s, computers were large

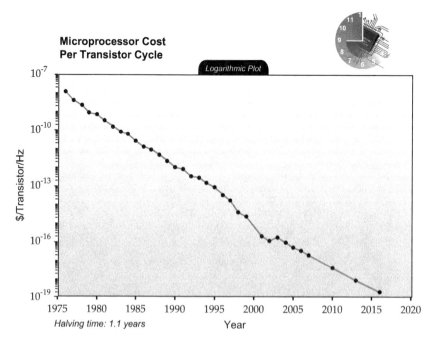

Figure 1.2
Microprocessor cost per transistor cycle http://singularity.com/images/charts/MicroProcessCostPerTrans.jpg

mainframe systems, consisting of various input, processing, and output devices; they were highly experimental and used largely for scientific and military applications.[23] As the technologies became more reliable and affordable in the 1950s and continued to improve in the 1960s, the collection of numerous hardware and software configurations became known as electronic data processing (EDP).[24] They largely did mathematical calculations and stored information in various ways, using electronics unlike earlier systems that had moving parts and sometimes were electrified, what were known then as electromechanical devices, such as punch card readers and tabulators. In the 1960s and 1970s, the "e" was dropped and the clusters of technologies were called data processing (DP).[25]

Beginning at the end of the 1970s and extending into the early 1990s, the term management information systems (MIS) became the label used widely to describe what now were myriad managerial and operational activities involving the use of IT, as they became increasingly physically pervasive in many parts of an organization, and were not simply locked up in a data processing center, the latter often referred to as a "glass house."[26] Over time, managers of these various functions were called DP Managers or MIS Directors (yes, they continued to rise in the corporate hierarchy as the amount of IT in an organization increased). By the early 1990s, a new term had come into use—CIO—to

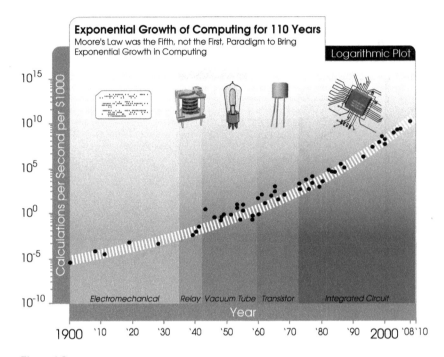

Figure 1.3
The effect of Moore's Law over time Moore's Law The Fifth Paradigm http://singularity.
com/images/charts/MooresLaw.jpg

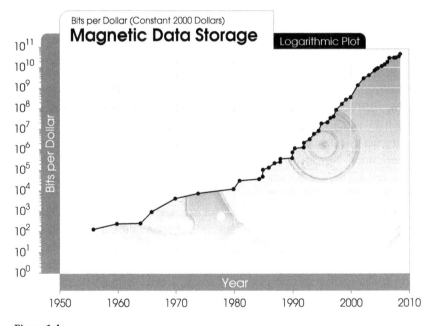

Figure 1.4
Magnetic data storage (bits per dollar, constant 2000 dollars) http://singularity.com/
images/charts/MagneticDataStorage.jpg

describe chief information officers. Computing was now about information and not merely gathering, sorting, storing, and reporting data.[27] A subsidiary cluster of functions and perspectives centered on the Internet, which became widely deployed around the world in the 1990s (although actually established at the start of the 1970s for military and academic use in the United States), such as webmaster, to describe the individual responsible for establishing and maintaining websites for an organization. In short, labels reflected the changing views of the role of IT over time.

Finally, a few comments are in order about terms already used in this chapter. Just as data processing and computer science (or computer technologies) came into wide use, beginning in the 1950s, so too did others later. Information technology(ies) as a phrase first gained currency in Europe, reflecting the shift underway in thinking of data as information, particularly as data stored in computers became less numeric and now included text and simple graphics, and by the late 1980s a rising tide of richly textured visual images, such as photographs and videos. All the base technologies embedded in the chips, and even earlier in the architecture of most computer systems dating from the 1950s, involved use of digital methods to convert information and instructions into electronic pulses that a computer could understand; hence, the frequent use of the word *digital*, rather than some other term, such as *information*. Beginning in the 1960s and massively so by the end of the 1980s, computers were connected to telephones and other forms of communication technologies that made it possible for computers to send information from one machine to another, such as over the Internet, or earlier in the 1960s–1980s over dial-up telephone lines or private networks. Once again Europeans were the first to provide a term reflecting that change: information and communications technology(ies), or ICT. Today, people use the terms IT and ICT interchangeably all over the world, although the North Americans rely on the phrase IT, while most of the rest of the world prefers ICT.

Telecommunications as a term was used in the 1950s and to a certain extent still is; however, the shorter word, communications, also came into vogue in the 1980s and is often used today as well. Both mean the same, namely, the communications among computers made possible by wired telephony, wireless telephony, satellite transmissions, and even underwater telephone and data cables. Copper wiring, or twisted pair wires, simply refers to wired telephone lines, which have been around for over a century; cable is a variant of that technology that can transmit data as well as voice; glass or optical fiber is the technology for telephony that came into use starting in the 1980s in the most advanced economies to perform the same work as voice and data communications, but with the distinct advantage of taking up less room than wire and being able to transmit orders of magnitude greater amounts of information through a line.

HOW MANY DIGITAL DEVICES ARE THERE IN THE WORLD?

It would not make sense to discuss the diffusion of IT if the number of computers used around the world did not reach an appreciable level, thereby justifying an examination of the issue. So, we still cannot avoid the question, but
it is one that can be dispensed with relatively quickly, even if complete inventories by decades across the world do not yet exist, nor all the data sets
required for scholars to create such findings, particularly for Central and Eastern Europe and Latin America. However, enough information exists on the
relatively current diffusion of digital technologies to suggest the magnitude
of digital diffusion. This can be examined by looking briefly at six basic sets of
information covering mainframes, minicomputers, personal computers,
Internet access, cell phones, and employment of IT for the early years of the
twenty-first century.

The largest systems—often called mainframes or "Big Iron"—but which
also include super computers that have massively more horsepower and
memory (think of these as either sports cars or the large earth moving equipment used by mining companies), are traditionally used by the Fortune 1000
firms and the largest universities, and city, provincial, state, and national government agencies. There are absolutely no accurate statistics on how many
were produced and used; however, extant evidence in the early 2000s would
put the number worldwide in the tens of thousands, each with an average
original purchase cost of $1 million or more. Population figures approaching
40,000 would not be unreasonable to expect with roughly half the systems
installed in North America, and nearly a similar amount in Western Europe
and East Asia, with a few hundred systems scattered among such places as
South Africa, Mexico, Argentina, and Brazil.

For minicomputers, that is to say, systems that are not as big as mainframes, but larger than personal computers, comprehensive data are also not
available. However, in the case of the IBM AS/400, IBM, in 1998 publicity on
the occasion of its tenth anniversary, let out that over 250,000 had been sold.
If one extrapolated that this number represented roughly half the market, and
because such systems tended to remain in use for a minimum of three years,
one could conclude that there were at least 500,000 in operation around
the world. By 2008, the global population of minicomputers had probably
doubled.[28]

With the diffusion of personal computers (PCs) beginning largely in the
United States in the late 1970s, and that spread across the most advanced
economies in the 1980s then around the world in the 1990s and beyond, we
see powerful computing reaching the masses. Because hundreds of millions of
these devices have been purchased, used, and replaced over the past three
decades, it makes more sense to measure their diffusion by the number of PCs
available per 100 inhabitants. Table 1.2 shows how many PCs there were per

Table 1.2 NUMBER OF PERSONAL COMPUTERS PER 100 INHABITANTS
IN MOST ADVANCED ECONOMIES, 2005

Switzerland	86.2	United Kingdom	60.0
United States	76.2	Hong Kong	59.3
Sweden	76.1	France	57.9
Australia	68.9	Norway	57.2
Netherlands	68.5	Germany	54.5
Denmark	65.5	Japan	54.2
Luxembourg	62.4	South Korea	53.2
Austria	61.1		

Source: International Telecommunications Union (ITU).

Table 1.3 NUMBER OF PERSONAL COMPUTERS PER 100 INHABITANTS
IN LATE DEVELOPING AND DEVELOPING ECONOMIES, 2005

Ireland	49.7	Costa Rica	23.1
Estonia	48.9	Kuwait	22.3
Iceland	48.3	Croatia	19.1
Finland	48.2	Dominica	18.2
New Zealand	48.2	Qatar	17.9
Slovenia	41.1	Malta	16.6
Slovakia	35.7	Grenada	15.7
Saudi Arabia	35.4	Lithuania	15.5
Belgium	34.7	Mexico	13.1
Cyprus	30.9	Russia	12.1

Source: International Telecommunications Union (ITU).

100 people in 2005 in those countries that were historically the most extensive users of computing of all kinds. Table 1.3 shows the same kind of data for
a sampling of an economically emerging group of nations as measured by GDP,
largely from Europe and Latin America, to suggest how many PCs another
cluster of not-so-poor but not the most extensive users had acquired. The accuracy of the data can be accepted as quite reliable. While the density of deployment varied widely for Asian and Latin American countries, and less so
for Europe, Africa overwhelmingly ranged between less than 4–5 to 0.02 PCs
per 100 people. In fairness, we should point out that in the over 50 countries
in Africa several ranked well above their peers, for example, Mauritius (16.2),
Seychelles (19.8), and South Africa (8.4).[29] The obvious conclusion to draw
from this information is that inhabitants in the countries with the highest
standards of living, as conveniently measured by GDP, had become the most
extensive owners of PCs.

Next, we consider the Internet, one of the most phenomenal technologies to burst onto the world stage in modern times, and which now surpasses personal computers in usage, but not yet the even more widely diffused mobile phones. During the 1970s and 1980s, the Internet remained the private preserve of American academics, a few interlopers living on the "digital frontier," and most largely by the military, national government officials, and defense contractors, concentrated at the time in North America but increasingly also in Western Europe. With the development and deployment of browsers in the early 1990s making the Web user friendly, diffusion of the Internet entered a period of rapid adoption first in North America, parts of Asia (notably South Korea), and Western Europe.[30] By the early 2000s, the Internet had spread around the world, doubling in subscribers within all OECD countries in the period 2000–2006; in that latter year the number reached 309 million, up from 158 million in 2000. In 2000, 35 percent of subscribers lived in Europe, 47 percent in North America (overwhelmingly USA), and the remaining 28 percent in Asia and the Pacific. In the next half dozen years, Europe's share expanded to 44 percent, largely due to rapid adoptions in Central and Eastern Europe, while Asia etched up a fraction, with North America's share now at 38 percent. When we look at the number of computers (called hosts) attached to the Internet—the fundamental way that one went online until mobile telephones and other more specialized devices made access possible too[31]—there were 542 million systems connected in OECD countries by 2008, up 13 fold from 1999. By 2009, OECD analysts had estimated that nearly 60 percent of all households in their member countries had access to the Internet, with South Korea already at 94 percent in 2007—the highest in the world—tiny Iceland sat at 84 percent and the Netherlands with a similar rate of penetration. As a result of the massive adoption that took place worldwide, the United States no longer served as one of the leading users, with only some 60 percent of households equipped with access. OECD now considered the United States as only "average." In 2007 OECD analysts determined that on average 95 percent of medium and large businesses used the Internet, making 95 percent "average," while the lowest rate of penetration, in Hungary, still with an impressive 85 percent. As for non-OECD countries, what were their rates of adoption? The results proved essentially the same, even in small businesses with less than 10 employees each also experienced 85 percent rates of penetration.[32]

One other way to look at the massive deployment of the Internet is to examine the number of users per thousand inhabitants in a country, which gives us further appreciation of the ubiquitous feature of this technology around the world. Using data for 2005, a year in which the surge continued and, therefore, understates the extent of subsequent deployment (if the reader wonders about events in later years), gives us a sense of the rate of deployment underway. Table 1.4 shows the top 10 most extensive users of the

Table 1.4 LARGEST NUMBER OF INTERNET USERS PER 1000 INHABITANTS

Netherlands	739	Japan	668
Norway	735	United States	630
Australia	698	Slovenia	545
South Korea	684	Finland	534
New Zealand	672	Estonia	513

Source: Table 5.11, *2007 World Development Indicators*, March 25, 2007, http://sitesources.worldban...STICS/ Resources/table5_11.pdf (last accessed 4/20/2009).

Table 1.5 LEAST NUMBER OF INTERNET USERS PER 1,000 INHABITANTS

Afghanistan	1	Niger	2
Iraq	1	Sierra Leone	2
Tajikistan	1	Bangladesh	3
Democratic Rep.Congo	2	Cambodia	3
Ethiopia	2	Chad	4
Myanmar	2	Lao PDR	4

Source: Table 5.11, *2007 World Development Indicators*, March 25, 2007, http://sitesources.worldban...STICS/ Resources/table5_11.pdf (last accessed 4/20/2009).

Internet. While dozens of countries had access in the range of between 300 and 500 per 1,000 inhabitants, the least adoption occurred in Africa where deployment of any measurable amount barely occurred, home to the digital divide so many commentators discuss writ large (see table 1.5). As with all other information technologies, nations with higher per capita incomes embraced this and other digital technologies earlier and more extensively than poorer nations. Just to make the point before discussing cell phones, for 2005 the European Union (a wealthy group of nations) had 439 users per 1,000 and when the wealthiest nations around the world are added to that group, the number climbs to 527. Low-income nations reflected more limited adoptions, such as those in Sub-Saharan Africa (29 per 1,000), Middle East, and North Africa (89 per 1,000).[33]

As remarkable as are the large numbers for Internet access, deployment of PCs, and even for large mainframe computer systems, these statistics pale in significance when compared to cell phones, which, by the nature of their technology, are handheld computers that use wireless technologies to communicate, increasingly to access web sites, are used as digital cameras, calculators, e-mail servers, and to host games, store photographs, and display reading materials. They also cost far less to buy, lease, and use than any of the IT devices reviewed so far. By the end of 2007 there were probably in excess of 2.6 billion cell phone subscribers in the world and given the fact that many people often share the same subscription and device, the number of users was undoubtedly substantially larger, certainly half of all humans. This is such a remarkably diffused device that it has exceeded the extent to which any technology in recent times became available to individuals, surpassing television, radio, wireless communications, digital watches, bicycles, and just about any other complex consumer product one can cite. People buy roughly one billion cell phones each year and there are probably two billion non-functioning devices in desk drawers, landfills, lost in kitchen cabinets, and hidden in old brief cases.[34] No digital device has been so variously adopted than cell phones, with subscription services more diverse, for example, than for land-line telephones, television, or PCs. This technology has been studied so extensively since it became widely used in the 1980s (although first introduced in 1979 in Japan) that it is quite tempting to devote many pages of this chapter to the subject.[35]

However, it is more important to understand its deployment briefly as further testimony that as digital technologies became smaller, more versatile, and less expensive, the more people embraced these in most corners of the world, evidence that digital technologies are still working their way into ever smaller crevices of our lives around the globe. Much like PC adoption patterns of the 1980s in which acceptance increased substantially once widely available, so too cell phones in the 1990s, with worldwide subscriptions increasing from less than a million as late as 1992 to roughly 200 million in 1997, doubling next in number within 2 years, exceeding a billion in 2001 and 1.5 billion sometime in 2003.[36]

This is the only digital device in some countries that is available in greater numbers than the human population itself, because some people have more than one. Table 1.6 offers some examples, along with numerical information on other high rates of deployment. All drawn from highly developed cell phone markets, these account cumulatively for the activities of 909 million people, or roughly 15 percent of the world's population in 2006. In 25 countries in the European Union that year, on average 80 percent of their citizens owned a cell phone.[37] Table 1.7 shows the most remarkable trends, that of some of the poorest countries in the world, those in Africa. To put Africa in context, the same data show that East Asia in 2004 had 257 digital cell phone subscriptions per 1,000 inhabitants, Europe and Central Asia 512, and Latin America and the Caribbean 337 among the developing nations. Yet, there still existed room for much growth because in 2006, for example, 923 million people lived in the African continent (as compared to 6.6 billion on Earth) and used 7.2 percent of the world's digital cell phones, supporting 198 million subscriptions.[38]

New devices of potentially enormous significance for how people receive useful information and entertain themselves keep appearing that in time will be adopted, first, by people in the wealthiest, most literate societies, but later in other less prosperous nations. As this book was being written, one such innovation was just starting to unfold, the e-book. For two decades various

Table 1.6 NATIONS IN HIGHLY DEVELOPED MARKETS WITH MORE
THAN 100 PERCENT CELL PHONE PENETRATION, 2006

Italy	138	Finland	114	New Zealand	98
Greece	127	Austria	113	Hungary	96
Portugal	122	Norway	109	South Korea	83
Czech Republic	118	Ireland	108	France	79
Israel	118	Spain	106	Japan	78
U.K.	117	Germany	104	USA	77
Sweden	116	Denmark	104	Canada	58

Source: Various statistical sources; IBM Institute for Business Value.

Table 1.7 DIFFUSION OF CELLULAR PHONE SUBSCRIPTIONS IN
AFRICA, PER 1000 INHABITANTS, 1995–2004

Region	2004	Percentage Change (1995–2004)
Analog Subscriptions		
Middle East and North Africa	13	58
Sub-Saharan Africa	8	61
Digital Subscriptions		
Middle East and North Africa	142	70
Sub-Saharan Africa	83	47

Source: World Bank; World Development Indicators.

companies and inventors tried to develop an electronic device that could displace books, magazines, and newspapers. They all had various technological shortcomings that we do not need to discuss; suffice it to say they were not more useful, cost effective, or convenient to use than the older paper-based medium used for centuries to deliver text and other information. However, in November 2007 Amazon.com introduced the Kindle in the United States, a device that finally began overcoming technical and operational problems of prior attempts to make an e-book. It also received its content wirelessly. While the company did not reveal how many it had sold when this book went to press, speculation within the IT community over a number of years had the figure at well over a million by 2010. In time, such a device might cost only as much as a cell phone and also function much better than initial models, suggesting yet another class of digital devices being widely used around the world, provided by various producers of consumer electronics.

Next we look briefly at the employment of IT workers around the world because these constitute a class of users, on the one hand, hence demonstration points for various digital technologies in their homes, work, and communities— carriers if you will within a society—and, on the other hand, indicative of the existence of the necessary economic circumstances required to sustain this type of a work force. It was not uncommon in the 1990s, and as recently as 2007, for European nations to have between 2 and 3.5 percent of their workforce in information technology industry jobs, for the original European Union 15 countries 2.61 percent in 1995 and 3.06 percent in 2007. The United States hovered at 3.29 percent in 1995 but expanded to 4.24 percent in 2007, which should be of no surprise because, historically, it was a major provider and user of IT.[39] East Asian percentages remained lower, and far more so in Latin America and Africa, in the latter instance measurably almost nonexistent, with the small exception of South Africa.

We might ask a related question: How big is the world market for IT products? Such information suggests yet another way to see how diffusion continued to progress. There is no clear consensus, but a few numbers suggest that a great many digital devices permeate the world. If one includes communications equipment, computer and peripheral hardware, electronic components (such as computer chips), audio and video products, and other IT-related equipment, and just leave out of the equation digital cameras and other consumer goods for the moment, we see that the global market has been expanding for some time. Picking up with 1996—a year widely recognized along with 1995 as the great takeoff in global adoption of the Internet—and extending through the boom period of the 1990s for IT, past the IT recession of the early 2000s, and the tragedies of 9/11 to our currently available data, one can see that demand for digital products grew substantially throughout the period. Figure 1.5 summarizes the data. Note that worldwide trade in the IT sector essentially doubled between 1996 and 2003 and by the start of 2008—the first full year of the next recession—was well on its way to a nearly a third more growth before slowing. Finally, one might note that if we added software, IT services, and consumer electronics into the mix, the total dollar volumes double in size.

MAJOR THEMES IN THE DIFFUSION AND DEPLOYMENT OF INFORMATION TECHNOLOGIES

The central purpose of this history of computing is to describe the spread of information technologies across North America, all of Europe, and large portions of Asia and to begin identifying the reasons why people and whole nations embraced the technology so quickly. Cumulatively, these regions account for the bulk of IT diffusion, so far, around the world and defined the patterns of adoption currently being embraced by the rest of the world, notably Latin America, Africa, Central Asia, and the Middle East. Commentators have noted for the past two decades the speed with which this class of technology has been spreading globally, resulting in the massive quantities of its artifacts

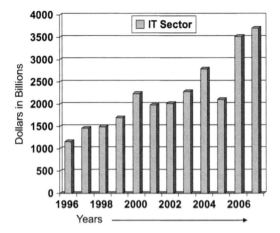

Figure 1.5
Global IT sales in billions US dollars
Source: OECD Information Technology Outlook, 2008.

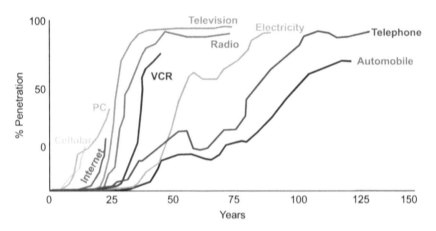

Figure 1.6
Rate of adoption of various technologies to over 75 percent

suggested earlier in this chapter. Figure 1.6 provides a simple snapshot of the rate of deployment of some commonly available devices in modern times to demonstrate the relative speed of adoption of IT; we do not need to linger on the issue as these types of data have been circulated widely in historical, economic, and scientific literature in various forms for several decades. It is presented to remind us that digital technologies have spread quicker than others.

The story of how computing (IT) spread can be described as two mega-waves of diffusion, the first from the 1940s to the end of the 1990s for many nations, then a second one for the most extensive users of ICT. Each wave was caused by a combination of factors that interplayed at different times and places. The elements were straightforward. First, there was the technology itself: emerging new functions, increased reliability of components, improved ease of use, continuous miniaturization of components and of whole families and generations of devices (from large to small computers), and their declining unit costs. As Everett M. Rogers, the great expert on diffusion of all types of technologies reminded us, speeds of adoption varied enormously as well, "cellular telephones and VCRs required only a few years to reach widespread adoption in the United States, other new ideas, such as using the metric system or seat belts in cars, require decades to reach complete use."[40] He explained that these various rates of adoption were influenced by the characteristics of specific innovations. Relative advantage of one product or idea over another held by consumers and users were very important, for example. As long as an older technology, for instance, an earlier version of Microsoft Windows, was seen as working just fine, users were reluctant to move to a new edition, the problem this firm experienced with its Vista operating system. Compatibility was always profoundly too important an issue, particularly for large computers as it was—and is—more expensive to convert one's software to fit on a different technical standard than to buy a compatible upgrade to a machine. Rogers reminded us as well that complexity was also influential: the more complex something people perceived it to be, the less likely one was to adopt it. An influence Rogers called out that had varying, but limited influence on consumer IT, yet a great deal on commercial and industrial computing, involved the notion he called *trialability*, which is the ability to experiment or to try a technology before committing fully to it. Finally he highlighted the characteristic of *observability* of a technology. This means the extent to which others can observe the successful functioning of a technology, such as when, for instance, someone shows all his or her friends how easy it is to use an iPhone, or an Apple computer versus one running Microsoft Windows.[41]

Second, fundamental economic circumstances in various countries either made possible more rapid deployment, or slowed it down; either way macroeconomic realities affected profoundly diffusion. The industrialized economies sustained the highest levels of GDP relative to other nations, and had large corporations and government agencies that became some of the earliest adopters, regardless of whether in capitalist economies or behind the Iron Curtain run by Communist governments. Invariably, nations needed such economic attributes. That economic reality proved most important to large enterprises that could afford computers, particularly in the 1950s through the 1980s, when their costs were so high in comparison to those IT systems of later years, which were then often both more tailored to one's needs and less expensive. Economists have known for many years that higher GDP meant earlier and more extensive adoption of computing, largely because countries that had high GDPs tended to encourage and sustain competition, which in turn fostered innovation and the hunt for productivity, all crucial to the advancement of any expensive technology. Nations have created essential economic and social infrastructures crucial to the process, such as enforcement of contracts, protection of patents and copyrights, regulatory practices supportive of enterprises, and flexible labor laws. They educated their citizens, at least to the point where they could either operate IT or maintain a high-enough standard of living that made it possible for them to buy and use the technology and that caused other parts of the economy to do the same, such as manufacturing and media.[42]

Third, at different times and for varying purposes and ways, public policy encouraging development of local IT industries and uses of computers always influenced the speed and direction of adoptions within a nation. This aspect of economic development has been the subject of considerable attention as well, especially on the part of economists and historians.[43] In this volume we encounter their work in all countries examined. What that literature tells us is that economic development since the 1950s has increasingly had an IT stimulus component, either to create IT industries (e.g., Great Britain in the 1950s–1960s, France in the 1960s and 1970s, Japan and Korea in the 1970s and 1980s, India in the 1990s and 2000s), and to raise the standard of living in developing countries, such as much of the work of the World Bank in Latin American, Asian, and African nations, particularly since the 1970s.[44] Today, no nation is considered modern or competitive in the global economy if it does not have a healthy diffusion of all manner of IT within its businesses and public institutions, and in private use.[45] Over time, public policy "best practices" increasingly converged, such that by the end of the 1980s, one could see common patterns of execution in various stages of development. These included government infusion of capital into emerging industries, and regulations stimulating local demand.[46] While initially protectionist, the successful ones evolved quickly into market-driven capitalist forms that valued competition, sharing of information, protection of patents, and development of internal IT capabilities.[47] Successes such as those enjoyed by Japan (1960s–1980s), South Korea (1980s–2000s), United States (1940s–2000s), Ireland (1990s), and India (1990s–2000s) also demonstrated that there were many forms: India majored in software and IT services, Ireland in manufacturing, South Korea in manufacturing and broadband implementation. The list is long and varied, suggesting that each nation has its own story, with some common players and circumstances: regulators, bankers, educators, businesses, and government agencies that used IT, increasing standards of living, and expanded access to global markets.[48]

Fourth, international vendors of software and hardware played a crucial role in introducing one generation after another of new products to emerging and mature markets. IBM was the most important of these vendors for the majority of the period. It introduced ever larger mainframe systems to more governments, universities, and private corporations than any other supplier, motivating a few political leaders to launch national policies to counter its local influence, as occurred, for example, with Charles de Gaulle in France in the 1960s and 1970s, and even earlier by several prime ministers in Great Britain.[49] The power of the marketplace prevailed, however, and IBM's influence on large computer markets remained profound for over a half century as did Intel's with computer chips. The same occurred with other vendors as well, although for shorter periods of time, such as Apple and Dell with PCs, and Nokia with cell phones. No history of the influx or adoption of digital tools can ignore the influence of major vendors.

Fifth, international corporations often exported to their offices and factories in other countries business practices and the underlying technologies in support of these that originated in the home country of the enterprise. E-mail systems linking a global corporation together around the world is one obvious example, so too the requirement often imposed on local suppliers to an international corporation to use specific types of equipment and software in order to conduct business with it. In addition to this practice of introducing specific digital tools into many countries, the habits of international corporations also led to the diffusion of knowledge about IT to these secondary enterprises, which, in turn, then began using these with their other customers and suppliers for the same reasons as large corporations. This is such a well recognized phenomenon that by the late 1990s, governments were beginning to insist that vendors and citizens interact with public agencies via computing not only to simplify operations and lower costs, but also as a way of stimulating increased use of IT as an economic development mechanism.[50]

Sixth, specific political and social circumstances also influenced deployment of the technology. More than any other factor for the longest period of time, the Cold War served as an important engine of technological innovation, exposure to IT, and deployment in such countries as the USSR, Great Britain, France, Australia, and to the greatest extent, the United States from the 1940s to at least the start of the 1990s. Consumer electronics, including cell phones, spread rapidly across Western Europe, into Japan and South Korea for largely non-military reasons, driven either by public policy, as in the case of the South Koreans when the government facilitated deployment of broadband connections for use with PCs, telephones, and the Internet, and with photography and other digital consumer products in Japan where, for instance, limited living space placed a premium on the use of small electronic devices.

Seventh, infrastructures of various kinds were essential to the adoption of computers. To begin with, the combination of reliable and cost-effective supplies of electricity is the minimum, or *sine qua non* requirement. Telecommunications available to a nation proved almost as important, which is why in every country that used IT as an economic stimulus focused on creating telephone and broadband availability after they had established reliable sources of electricity. In every decade since the 1970s all countries that enjoyed extensive use of IT had telecommunications infrastructures and today all underdeveloped countries have as part of their economic strategies the requirement to invest earlier in telecommunications (wired and wireless) than even in such other crucial areas as education, road construction, and more than occasionally in the provision of adequate health care to their citizens.[51] Economists have argued that other types of infrastructures were also required and present: rule of law, standard enforcement practices for contracts, shared technical standards, common regulatory practices (e.g., about e-pollution in Europe), and languages (English, for example). One of the leading economists of technological economic development, Richard R. Nelson, has argued "that institutions matter, and the ones that are appropriate for economic growth tend to change as the key driving technologies change." He also observed that economists did not have a good theory about how this happens; historical examples can contribute to the deployment of one.[52]

Linked to issues concerning economic development and use of IT is the fundamental role of information in the work and play of individuals, often called knowledge management (KM) in business and government. Economists have long understood the importance of "knowledge work." Recently, historians have started to turn their attention to the issue as well, recognizing that it is a subject that transcends computers and in general opens new ground for research.[53] In a recent study of knowledge's role by economic historian Joel Mokyr, however, we see the challenge faced by the authors of such studies, one that we face in this book on diffusion:

> The growth of human knowledge is one of the deepest and most elusive elements in history. Social scientists, cognitive psychologists, and philosophers have struggled with every aspect of it, and not much of a consensus has emerged. The study of what we know about our natural environment and how it affects our economy should be of enormous interest to economic historians. The growth of knowledge is one of the central themes of economic change, and for that reason alone it is far too important to be left to the historians of science.[54]

Economists and technologists tied the role and diffusion of IT to the expanded awareness of the use of data, information, and knowledge in the most advanced societies of the world.[55] This is an important realization because a major reason why computing spread around the world derived from the ability it made possible for workers to use larger quantities of data (information) in their work. As the percentage of the workforce moved from farms to factories, and then into offices, the amount of information they needed in their work increased. This is a process underway for well over a century, but, which, nonetheless, increased demand for data processing technologies to facilitate work and improve productivity.[56]

More narrowly tied to our book, global institutions also spread information about IT throughout the world. Note in the endnotes the numerous citations from the World Bank, which for decades codified best practices concerning economic development, often commenting on the role of IT and telecommunications. Already mentioned are international corporations that exported their practices—IBM, Phillips, Exxon, Toyota, and ABB—the list is endless and examples come from many parts of the world. So the combination of tracking activities and the actions of multinational enterprises, even governments (such as the use of IT by NATO and the US military) contributed to the diffusion of digital tools around the world.

Conversely, the absence of these various factors resulted in slowed rates and extent of adoption in any country in any period. Low GDP levels, poor education, and inadequate infrastructures, for instance, retarded the flow of technologies into Sub-Saharan Africa and parts of Asia and Latin America. Yet, one should be cautious in over-generalizing about these negative factors because, as technologies became less expensive, they were adopted in nations that had lower levels of education, GDP, and economic development than might otherwise make sense. A poor agricultural village in India might adopt one cell phone so that farmers could check on the price of crops in the 2000s, yet in one of that nation's cities, the average citizen might be able to afford his or her own cell phone or personal computer. A citizen working for a corporation in Central Europe might be able to afford a laptop. Therefore, circumstances varied by degree over the years and even within a country.[57]

The story of diffusion, while frequently defined as occurring in macro waves in this book, also has as a subtext the constant interplay and dynamic shifting of the relative influences of each of the eight variables introduced above. Generalizing in this book is intended to provide a macro view of the "big picture," as a first step in helping others describe in more detail the specific flow of events within a given country or community. As one moves forward in time from one decade to another, an increasing number of these variables are in play and more extensively than in earlier years. Each also evolved over time and could be seen as trends, or waves, particularly as the interconnections of variables across geographies became more the case as time passed, into what we routinely call the global economy. Therefore, the story is complicated. That complexity is fundamentally retained because the vast majority of studies are country-centric, fewer regional, and even more scarcely global. A careful examination of the bulk of the literature cited in our endnotes, for example, would reveal that major studies of global or regional diffusion are anthologies of chapters devoted to individual countries written by separate authors with only a minimal integrative patina of thinking. That has to change if we are to gain an understanding of global trends. Country case studies remain crucial, and should continue to be written because the story of diffusion is unique at the national level, but it is time to begin consolidating the findings of hundreds of these case studies, and to form a comparative view of the diffusion process, because no society did this in isolation from activities occurring elsewhere in the world.

Before discussing further the role of such concepts as diffusion and deployment—ideas that will permeate every chapter—it is relevant to ask about the role of wave theories of history in this project. To someone not an historian, it is tempting to think that history repeats itself. Both World War I and World War II were world wars; economic recessions and depressions come and go all the time; people make the same mistakes every day. In about every decade since the late nineteenth century there emerged resurgences of academic discussions about waves, usually stimulated by scholars who are not historians (often economists and sociologists), but in the process they provided useful insights, perspectives that often influenced the behavior of business executives and government officials.

An example illustrates the issues. The distinguished professor of geography who long studied urban and regional planning, Peter Hall, wrote a fascinating book called *The Carrier Wave: New Information Technology and the Geography of Innovation, 1846–2003*, in which he described how various forms of IT emerged during the first three Kondratieff economic waves, interspersing the ideas of another great European economist, Joseph Schumpeter and his notions of business cycles and economic "creative destruction" brought about by the emergence of new technologies.[58] In the process he did two things. First, he described a pattern of innovation and adoption from the telegraph to the PC in an elegant construct that would almost lead one to believe there was some unseen economic or technological determinism at work. Second, in the process, he made the story of the evolution of various technologies link together in some causal fashion. It is fruitful, if difficult, reading. But, historians would argue that no two events are ever quite the same. World War I and World War II were global, yet the specific events of each differed. The fighting at Verdun in the first war proved different than that at Normandy in the second war, yet both were fought in France with the same nations in combat across enemy lines. Nor do technologies do things, people do.[59] In fairness to Peter Hall and his colleagues, they were more interested in defining some features of long waves of historical evolution than in characterizing the fate of IT as some predetermined process. To quote him and his co-author, Paschal Preston, "We believe that they [clusters of key interrelated technologies] tend to come forward at points when the returns from existing investments are declining."[60] Their focus was on identifying how patterns of innovations surfaced, implying without saying so that there are some "rules of the road." It is a wonderful book, despite the problems for the historians, as it attempts to define some common patterns of behavior in a thoughtful manner.

Another influential scholar, also not an historian, is Carlota Perez, who looked at the dynamics of bubbles and Golden Ages, defining five of them with specific technological revolutions associated with each one. The first begins with the Industrial Revolution of the 1700s in Great Britain with steam, water power, and machinery, while her fifth in 1971, which she called the "Age of Information and Telecommunications" and was (is) dominated by information technologies, such as microelectronics, telecommunications, and software. She characterizes this one by the late 1900s as entering a phase in which the basic technologies have been discovered and are now being exploited in myriad ways, and will continue to do so for the first several decades of the twenty-first century.[61] Perez also is more forthcoming in using her fifth age to forecast what will occur:

- Information-intensity (microelectronics-based ICT)
- Decentralized integration/network structures
- Knowledge as capital/intangible value added
- Heterogeneity, diversity, adaptability
- Segmentation of markets/proliferation of niches
- Economies of scope and specialization combined with scale
- Globalization/interaction between global and local
- Inward/outward cooperation/clusters
- Instant contact and action/instant global communications.[62]

We will not debate the veracity of her forecasts, but her tables and charts on the five waves or revolutions (to use her word) have captivated the thinking of many executives in the IT world, and thus influenced their behavior.

Like Hall and others, Perez's intellectual scaffolding is useful for presenting what otherwise would be an eclectic mass of facts about IT; but these perspectives are not universal "rules of the road," hence the historical perspective that the experiences of various countries in embracing IT, while sharing many common practices, remain unique stories of adoption. Like most historians, I find it very difficult to accept one class of causation dominating a period of history—in this case the notion that we are living in an Information age—because a great deal of other things are going on as well: effects of birth control on the expanding role of girls and women, and the political and military clashes of cultures manifesting themselves in what the Americans called the "war on terror," to mention just two other tectonic waves of influence on contemporary society.[63]

In the chapters ahead, I accept inherently the notion that there are patterns of similar, but multifaceted, behaviors in any particular era, because we know that people who bought and used computers did not do so in isolation; they duplicated or partially mimicked what they saw others do. We know that, because in countless memoirs and reports, they have either sought out "best practices" or admitted to adopting the ideas of others.[64] The best concept of this, and which is well understood by economists, is the idea that a later adopter of a technology simply leapfrogs over various prior steps to adopt the latest devices, leveraging what is known as the "late-comer advantage."[65] The poster child example of this continues to be African countries that, for sound economic reasons, bypass the expense of installing telephone lines and instead go straight to wireless telephony, while the more "advanced" economies remain straddled with the burden of maintaining the now-older, more technologically primitive wired infrastructure with their ugly wooden telephone polls, while also deploying wireless telephony.[66] We will accept notions of waves and eras as facilitative scaffolding upon which to hang our tale, but caution that the story is the most important focus, not the elegant scaffolding itself.

Because in this book we intersperse notions of innovation with statements that these are byproducts of some activity in a nation (such as creation of the Internet in the United States) or as sources of new activities (as in the use of cell phones earliest in Japan and Europe), some definitions are in order. Specifically, we must understand concepts of diffusion and deployment. Both are related to each other and to many of the themes of this book.

Diffusion, as defined by its most eminent authority, Everett M. Rogers, is "the process in which an innovation is communicated through certain channels over time among the members of a social system."[67] This is a very formal, even abstract, definition, which, however, folds reasonably well into discussions about something that has been developed, manufactured, or turned into an activity introduced into a society. That can be IBM selling a new generation of computers in France in the 1960s or the American government funding the installation of PCs in classrooms in the 1990s connected to the Internet. When examining diffusion, economists and historians explore economic incentives for introducing and using a technology, either by individuals or organizations.[68] To a large extent, scholars view as a set of actions initiated on the supply side of the story, that is to say, steps taken by governments to encourage creation of innovative products and services, and others taken by companies to encourage people to buy their offerings. The story of how digital computing spread around must take into account these kinds of steps because governments, in particular, have played very proactive roles in encouraging their institutions, companies, and citizens to embrace computers and all its derivative byproducts, perhaps more so than with any other recent technology, with the possible exception of automobiles. While Rogers' definition may (or may not) apply to consumer digital electronics, his characterization of diffusion is used to explore even this aspect of digital's spreading footprint, because it proved so important, particularly in very poor countries that could more afford a cell phone than a mainframe.

Before continuing further with discussions of terms, it is imperative to remember that these words are laden with implications, values, and untold details. At best they are quick shorthand ways of calling up massive quantities of tasks that cumulatively result in such things as diffusion. Seasoned historian of technologies of all kinds, Rudi Volti, reminded us of that reality:

> The movement of technologies from one place to another is a process fraught with difficulties and uncertainties. Terms with physical referents like the "transfer," "diffusion," and "absorption" of technology merely hint at the complex processes that accompany the successful movement of a technology from one place to another.[69]

This observation applies as much to information technologies as to such others as the wheel, sailing, gun powder, and electricity. In a number of the country-case studies in this book, the breadth of activities behind such terms are displayed, such as in what it took for users of IBM's S/360 to embrace the technology, or organizations in the Netherlands to be part of their nation's process of IT diffusion. Volti's admonition helps us to remain conscious of the work that we historians glibly acknowledge through our language and nowhere is this more so than in discussions about deployment.

Deployment normally refers to tasks done in order for a technology to be adopted. Use of third- or fourth-generation programming languages (tools of the 1950s–1970s and 1980s–2000s, respectively) to create and apply a technology fits into deployment; so too tasks one might perform to train people how to use a social networking software tool. Deployment consists more about the activities done by an end user than just by a company.[70] To be sure, IT companies have to do things in order to make it possible for a user to receive their product, such as to invent, manufacture, and ship it to a retail store; or, to provide training on how to run a company or use a cell phone. Historians have largely studied the tasks and strategies of IT companies and of governments, but I think of those actions as diffusionary in nature. Deployment in this book is presented more as the activities of the customer, the user of the technology, because if they did not embrace it, all the diffusionary activities in the world would not have resulted in this class of technology becoming so embedded in a society. Children and teenagers had to buy and use video games, all age groups the mobile phone, companies computers to track inventory and send out invoices, military personnel digital electronics for smart bombs and avionics, warehouse managers robots to move goods, barcodes by retail store managers to control stocks of goods. Diffusion from the perspective of the end user is crucial to the deployment of a technology because where it was absent, and yet a technology existed, that device or innovation failed and, therefore, of no importance to the history of mankind. Examples of the lack of acceptance by end users (customers) exist in all ages with many technologies: the elegant mousetrap that costs too much or is too complicated to use when compared to existing technologies,[71] electric typewriters in the late nineteenth century (they finally appealed to users in the mid-1930s), AT&T's

Picturephone in the 1960s (beginning to be used in the early 2000s with laptops in combination with Skype).[72]

For our purposes, think of the two notions as almost synonymous enough because both exist simultaneously where digital technologies are embraced and used, the yin/yang of our story. So, while we will use these terms almost interchangeably, because both are about moving a technology from their sources into and through a society in all its forms, they are meant to indicate that there are two aspects to every acceptance of a technology: vendor/customer, supplier/user, cost/benefit, relevant/outdated, fashionable/out-of-style, safe/dangerous, and big/small. Often the technologies and their acceptance display many of these contrary features, but always others that are unique to a nation, such as rich/poor, high GDP/low GDP, environmentally clean/polluting, and educated workforce/uneducated workforce. What makes the story of any nation's experience with IT unique is that each variable is both specific to a society and also changes fairly rapidly from one decade to another, yet are also evident from one nation to another. I acknowledge unique features but I also qualify these by generalizing as much as possible about patterns, because in doing so we can create an inventory, even a mental model, of specific issues future historians, economists, business executives, and public officials can leverage as a checklist of considerations they can use as they diffuse, deploy, and understand future technologies. One way to track the extent of activity and acceptance will be by counting digital devices, as we have done in this chapter, for example, with the assumption that people are acquiring them for some purpose, so the concept of diffusion and deployment—yin/yang—can serve as a surrogate method for seeing what people are doing with their digital devices. Other words come into our vocabulary, and are defined as needed, because, as the last chapter in this book argues, there are characteristics of digital technology that facilitated its speedy adoption. Finally, we should note that the words *adoption* and *application* are used as the most proactive terms that we have to describe what some user acquired and used, such as a consumer buying an iPhone and then calling friends or finding a good Italian restaurant.

IMPLICATIONS OF THE SPREAD OF DIGITAL TECHNOLOGIES

Why tell a story of how IT spread around the world, and, indeed, so quickly? For historians of information technology the tale itself is of interest, because it is about the growing influence of a class of technology based on its features and uses, so that would be like preaching to the converted the case for religious belief. For social, economic, military, and political historians the story should be of interest because of the profound influence this collection of technologies had on the affairs of nations and cultures in the second half of the twentieth century and that are extending into the new century. By looking at the historical record we may find insights that will help product developers, economists, business managers, political leaders, military organizations, media, teachers, and professors understand an underlying development still very much underway and far from reaching some stasis. If one again examines figure 1.6 (p. 23), note that today every major class of technology that moves rapidly into the hands of consumers has IT components: VCRs, TVs, automobiles, radios, telephones, and so forth. As the IT content of these and so many other devices increases, as they surely will and clearly have, can we expect those and other future products and services to mimic increasingly the patterns of diffusion identified in this book for purely IT devices, software, and services? One can prudently conclude that the answer is probably yes, so the story of how IT spread around the world has many urgent lessons to teach all of us.

Another reason why this story is important comes from an unanticipated, and as yet not studied, consequence of so much IT infusing into the economies and social fabrics of people living all over the world. For lack of a better phrase to describe it, think of it as a Consumer's Moore's Law. Moore's Law states that IT (specifically computer chips) will double their computing capacity and speed roughly every 18–24 months, which means that the cost of doing something with a computer chip essentially continues to drop at a predictable rate. Intel, the world's largest developer and manufacturer of computer chips, has often set its annual business goals based on that premise, which leads to conversations about whether the law is real, or is a self-fulfilling prophesy. However, independent of that interesting notion is the effect on consumers (retail, corporate, and government). Over the past three decades customers, for example, bought microwave ovens, TVs, portable music systems, personal computers, laptops, games, cell telephones, digital cameras, and, most recently, flat screen TVs, iPhones, and e-books. In each instance, the initial introduction of a product that had a high content of IT cost more than in subsequent years and often performed worst in its first year, continuously improving over the next 3–5 years. Those twin patterns of prices dropping and performance improving have not been lost on consumers. Roughly 10 percent of consumers buy a digital product when initially available, because of its novelty and attraction of the technology, or sometimes they urgently need some new function, as often happens with military weapons systems. But the 10 percent rule driven by novelty has been observable for many decades in both consumer and commercial markets. The other roughly 90 percent acquire the new device later in waves over time, with "waves of time" defined as between 1 and 3 years, 3 and 5 years, and up to 10 years. Consumers experienced in buying digital products understand either consciously or subconsciously that the cost of a desired product will decline in price by 20–30 percent each year, for example, and as that happens, more individuals buy it.

One can track this practice through sales and pricing of product after product, even in the sale of new models of automobiles. This is the industry in which I first observed the phenomenon in 1975–1980, when the large infusion of new technologies appeared in engines of vehicles to optimize gas usage and nearly simultaneously the introduction of whole new generations of electronics to lock doors, automatically raise and lower windows, operate radios, and soon after, balance vehicle brake systems. IT professionals who acquired systems in government and industry also exhibited this pattern of behavior, although over a longer period of time than consumers, since they had more decades in which to acquire computing (at least back to the late 1950s), with roughly 10 percent early adopters in their industries; with others following over time, again in the 3–5 year range from early, to intermediate, then late adopters. Therefore, as the IT content of industrial and consumer products and services increases over time, can we expect a Consumer Moore's Law to apply? I think yes, once again justifying the relevance of our history of the diffusion of IT.

Finally, I want to draw the reader to a higher level of generalization to understand the spread of computing, for there is a general truth about the pattern diffusion and deployment that occurs when a class of technologies matures, that is to say, has acquired its general form and its uses have settled in, much as occurred with swords, hammers, forks, and knives. We can expect that in time, as with all earlier mature technologies, that how IT continues to diffuse and deploy around the world will increasingly converge into some common set of behaviors and practices. Just as people essentially use a hammer and nails the same way around the world in all countries, so too can we someday expect to see that happen with various types of IT, remaining stable until some subsequent transition, possibly revolution in science or technology.[73] To be sure, just as there are specialized big and small hammers, designed for specific nailing activities, so too is that the case with IT. As a technology increases in complexity, so too does the possibility that uses will vary more than for simpler devices. That said, however, convergence of diffusion and deployment tend to occur and we can expect that to be the case for IT.

There is already a hint of that process underway with IT. In an earlier research project intended to inform the structure of this book, I studied various patterns of deployment of IT with an eye cast toward the actions of governments trying to encourage their citizens to use the technology. I identified eight clear patterns of deployment used by governments, companies, and users over the past seven decades. But most startling was that the closer one looked at the evolution of these approaches, the more it became quite evident that by the arrival of the twenty-first century they were rapidly converging into one or two global mega-processes clearly in evidence in the most advanced economies. Furthermore, only one of the eight was driven by a technological innovation so compelling that "everyone had to have it"; rather, nontechnical influences proved more important, such as economic justification and governments' encouragement.[74] That occurred as part of, and as a consequence of, the simultaneous linking of those facets of national economies into a global, integrated set of economic activities that proved most dependent on the use of computing, such as financial systems and logistics, hence trade. It was no accident that the economies of the world globalized and integrated at the same time as computing diffused around the world into those nations that most closely linked to the global economy. In other words, there was a process of homogenization, of standardization, or of "best practices," that emerged for how IT spread around the world. This unfolding set of activities occurred at the same time that many other forms of harmonization were occurring, such as the spread of English, universal business practices, and even expanded deployment of democratic forms of government.[75] So, the story of the spread of computing is explicitly bonded to that of the evolution of many societies, almost, if you will, as part of the social, economic, and political plumbing of late twentieth and early twenty-first century nations.

We live in an age when computing has become so iconic an image of our time that it is quite easy to fall into the possible trap of anointing information technologies with the mantle of exceptionalism, of being different than previous technologies. To be sure it has some features that have facilitated its diffusion, discussed above, but is it really all that different from previous technological innovations? A new generation of scholars is addressing that question, such as Diego Comin and Bart Hobijn, for example, who use the methods of the economist to look at 100 technologies and their patterns of diffusion in 157 countries in the period 1750–2003. All the obvious technologies were included in the study, such as those in transportation (e.g., steam shipping, rail, vehicles, aviation), communications (telegraph, telephone, radio, television, cable, Internet, mobile phones), manufacturing (including spindle, steel, electrification, and synthetic textiles), and medical (for instance, X-ray, dialysis, various major types of surgery, CAT scans). They reached two fundamental conclusions that historians of IT can no longer ignore. First, diffusion of technologies actually increased over time, and that this observation applied to these technologies around the world. Second, the amount of time it took for a technology to reach 25 percent deployment in a society also shrank over time, providing the kind of data we so frequently see about how long it took for a particular technology to be used by a certain percentage of a nation's population. Interestingly, those nations that tended to be successful in quickly adopting one technology also adopted others successfully and quicker than others too.[76] This last point is one probed throughout this book, and ties into notions introduced suggesting consumers and institutional buyers of IT have learned subconsciously how to acquire IT quickly, as I suspected, the existence of an implicit Moore's Law behavior at work in the marketplace. Additionally—demonstrated in subsequent chapters—the richer a country is (read, has a higher GDP) the more it is going to embrace computing technology. In fact, for earlier technologies, that also proved to be the case. In short, while much has gone on very quickly with IT diffusion, it is not a candidate for exceptionalism, rather it fits into a broader trend that spans at least two and half centuries, and probably more.

ARC OF IT DIFFUSION: WAVE ONE, 1940–1990S/EARLY 2000S

When a variety of factors that influenced the diffusion of computers are collectively taken into account in reconstructing the story of how IT spread around the world so fast, the chronology of events begins shifting. Most notably, economic factors weigh in, such as the ebbs and flows of recessions and the cost of capital. Major political events do too, such as regime changes, for example, in Germany and the USSR with the fall of Communism, or earlier political policies motivated by the exogenous influences of Cold War priorities and events. Add in business factors, such as market conditions, quality of a firm's management, and then finally the one factor that always seems to get attention in isolation from the others just mentioned—technology—and we begin to see a more complicated, richer story emerge. Now further complicate the narrative by the variety of each factor, from tribal politics in Africa to political elections in Europe and North America, to Asian economic developments, and the timings of events and patterns of behavior become less arbitrary.

The merger of computing and communications represents a story onto its own, yet to be fully documented, which may be the biggest human project underway. ICT diffusion has surfaced as one of the largest, if not biggest, investments made collectively by humankind in recorded history. Time will tell if I exaggerated, but if so, not by much. But if one wants to take a less bullish stand on the matter, the historical evidence presented in this book unequivocally demonstrates that IT played a major role in a wide variety of ways in many nations' political, economic, and social history during the second half of the twentieth century. Therefore, the history of IT can no longer be isolated to the realms of narrow subfields, such as the history of computing, technology, or science. Because it is such a large story, its diffusion must be better understood within the context of a more general history of recent human activities, and most specifically that since at least the 1970s.

If we step back and look at the long history of technology and that of science and nations, we can begin to see that perhaps the development and early history of computing—what I want to refer to as Wave One of the process—extended from the late 1930s or early 1940s to the end of the 1990s or beyond, not to the end of the 1950s when second then third generation computers appeared, not to some period between 1975 and 1981 when PCs became available, or to some other time when yet another technology appeared. Smashed together, all these technologies evolved along a fairly evident line, what economists would argue reflect a case of path dependency, although never fully predictable or inevitable. Failures and dead ends abounded to be sure. Uses of these actually help define the wave more precisely as being the application of the technology to do the fundamental and often preexisting work of large institutions during the period. That perspective leads us more closely to the notion that an initial wave ran for roughly a half century. To be sure, many aspects of what one might consider to be a subsequent wave were clearly in evidence in Wave One, such as the use of massive telecommunications and the early and profound shift of computing from just being in organizations to others added that were personal and individual to people. The world in the 1950s through the 1970s slowly became linked through networks, a process that took off in the 1980s, but clearly nowhere close to the profound extent evident in the 1990s and early 2000s. Globalization of how firms and government, even, operated began changing in the 1980s, far beyond simply globalized trade. The list of how the world functioned before and after roughly the late 1990s suggests that a shift occurred that can be measured by looking at what and how much got deployed in the world of computing. For those reasons, I believe we can reorient the chronology of events in a more informed way, beginning with the acceptance that a first wave took a half century globally, although the start dates varied by region and country. Regardless of when the process began, that was the same amount of time the world took to maximize its use of punch card tabulating equipment, roughly 1900 to 1950 with an overlap with early uses of computing in the 1940s and earlier in the 1890s with embryonic uses of the technology, also in use with other forms of data processing, such as calculators, adding machines, and typewriters.

If Wave One existed—and that is my contention—with all the elements of an era that an historian would want in evidence, then do various nations live in Wave Two today? The short answer is yes in some societies, most notably in the pan-Atlantic community and in parts of Asia, but clearly not in many others. The Cold War took place during Wave One, other wars in Wave Two, with unique features, such as the use of laptops by enemies of the United States to communicate, cell phones and garage door openers for what Westerners characterize as "primitive" terrorists and tribal folk that they used to blow up advanced American military vehicles. Some of the technological features just now evident of Wave Two are the Internet and cell phones, clearly dominant features of post Wave One IT, even serving as the emblematic iconography of its time. Massive interconnection among individuals and groups real-time is a second feature already evident as well. Viewed through the prism of diffusionary practices, we can even argue that Wave Two is occurring so rapidly, perhaps at twice the speed of Wave One. That typology of events also begs the question, what are the future frontiers between Wave Two and Wave Three? Futurists have started to comment about the human–machine, robotic, and approach of the singularity of computing and human biological forms.[77]

To situate, and describe, Wave One we need to understand not only what happened during that first half century but also demonstrate that the period that came afterward was different. Providing that perspective means while the bulk of this book tells the story to roughly the late 1990s–early 2000s, it must also briefly outline the contours of Wave Two, something done in two ways. Often in presenting evidence of Wave One diffusion, I offer some data about Wave Two at the same time so that one can see the arc of the diffusionary activity and differences in patterns of deployment. The reader experienced that technique applied straightaway earlier in this chapter. The second approach is to provide an appendix that provides a brief overview of Wave Two in order to draw more sharply the contrast with Wave One and to provide future students of the subject an intellectual framework for facilitating their work. What is not done is to do the same with the period prior to Wave One largely because I think that task has already been accomplished through the work of several other historians.[78]

The reader will immediately note that my names for these two waves are quite uninformative, and certainly lacking literary panache. That is on purpose. Naming eras is great sport, but if Wave One is such an era, or contributes mightily to one, we do not know enough to dare give it a name. The same applies to Wave Two, simply calling it the Internet Age or something like that denies the significance of over a billion people rising out of poverty into the middle class now underway, or the political developments following the Cold War. We are simply not ready to do that, but I needed to differentiate one from another and because we seem to be living in a period in which numbers play such an important role as descriptors of things and events at the expense of adjectives, I chose one and two. I mean no disrespect to those who have named Wave One as the Age of Computers, the Networked Age, the Digital Age, or other things; the historical evidence simply suggests such an action is more useful for selling books than in defining our most recent past.[79]

The evidence presented in this book highlight a number of features of Wave One. During this wave, a nation first learned about the existence of computers, normally in the 1950s or early 1960s. This unfolding awareness occurred typically at universities and other technical institutions, the result of exchange of information by engineers and early computer scientists talking to each other all over the world, starting in the 1940s and 1950s. In the West and in Eastern Europe governments sponsored initial work on computing, largely by the military as part of a broad arms race and intelligence gathering efforts due to the Cold War, which then led to civilian versions of the technology that spread across manufacturing, government, financial industries, and finally into the services industries. A similar pattern emerged, but with less military involvement, in Asia. By the 1960s in the pan-Atlantic community, diffusion had increasingly become a private sector initiative, such as by electronics and office appliance companies. In Communist Europe it remained a government-driven effort. In East Asia, governments sought to launch massive economic development initiatives and by the end of the 1970s, recognized that they had to create and promote local information technology industries and by the end of the 1980s, deployment of this technology in various parts of their societies if they were to develop economically. The pan-Atlantic community operated in a market-driven style almost from the beginning, while the East more in a developmental-driven form, with government agencies leading the charge until the 1990s by which time it was rapidly transforming into a Western model, even in Communist China. Meanwhile, Wave One computing went from standalone mainframes and minicomputers to the adoption of personal computers as well, and to early efforts to connect all of these devices via telecommunications. I demonstrate that part of the reason why telecommunications—phone systems—improved all over the world after 1970 was caused by both the availability of digital technologies to move conversations and data around more effectively, but also by the growing worldwide demand for connectivity among computers. Users more than governments and vendors forced the convergence, beginning in the West but perhaps most passionately embraced by the East, particularly by individuals using handheld devices.

During Wave One, first public then private organizations embraced computers, beginning with large institutions, but quickly spreading to ever smaller ones, and finally to consumers. Late Wave One nations had already also embraced the use of the Internet, albeit largely for looking up information, as a place to position data, and to start the process of business and communications transactions. Wave One nations all were ranked in the top quartile of GDP per capita among all the countries of the world and over half its citizens had access to the Internet. By late Wave One, intelligent handheld digital devices were beginning to appear, and commentators from many walks of life were discussing the emergence of some Information Society or Information Age.

Less clear is when a society moves into Wave Two. However, several features suggest the makings of an answer to the question of when. First, they have gone through Wave One. Second, the use of digital, intelligent devices in a networked fashion can be described as ubiquitous, as we demonstrate in the case of South Korea and likely already in Singapore and across the Nordic countries in Europe, and certainly in many urban centers in both the pan-Atlantic and Pacific Rim. Third, and the most difficult but perhaps the most critical juncture, the work and private pursuits of workers and citizens has been altered fundamentally by the extensive use of digital technologies, affecting how tasks are done, how people spend their time, and their attitudes toward myriad topics, including expectations of how businesses and governments will operate. The process of migration to Wave Two is ragged, and thus does not lend itself to the clean, neat charts and descriptions we would want. For example, in poor villages, Indians use digitized fingerprints for identification because so many are still illiterate; while Africans check prices for crops using cell phones—potential Wave Two uses of IT—but neither would be said to be living in Wave Two. On the other hand, has Spain's society changed fundamentally due to the extensive use of IT in Iberia, particularly in Barcelona and Madrid by "everyone"? One would hardly conclude that the answer is yes. The same happened within industries. In the United States, stock brokers are extensive users of computing, school teachers are still virtual luddites. City dwellers are very wired but less so rural communities—this is true in all countries. So we have more to learn; but, what can be said is that some nations have fundamentally changed enough that can be attributed to the use of sufficient amounts of information technologies to conclude that something has changed. We will come back to this issue in the final chapter when discussing implications of IT's diffusion in many parts of the world.

Coming back to the question that titles this chapter, despite the difficulties of answering it, experts keeping trying, even though every set of answers differ

Table 1.8 NUMBER OF INFORMATION TECHNOLOGY AND KNOWLEDGE WORKERS WORLDWIDE IN 1980S AND 2005

	1980s	2005
People with college degrees	82 million	212 million
Ph.D.s in science and engineering	57,217	154,710
R&D researchers	1.9 million	5.1 million
Published scientific articles	466,419	698,726
Licensing revenue	$10.8 billion	$109.8 billion
Personal Computers	131 million	898 million
PCs per 1000 people	19	140
Landline telephones	333 million	1.2 billion
Mobile telephones	11.2 million	2.7 billion
Countries connected to Internet	20	209
Internet web sites	9,300	110 million
Host computers connected to Internet	313,000	395 million
Portable memory storage (megabytes)	1.44	16,384
Broadband subscribers	0	217 million
Internet users	2.6 million	1.02 billion
Internet users per 1,000 people	.5	157

Dates for data cited for 1980s varies, but relies whenever possible on information from 1980–1984. Data on knowledge workers covers all fields, not just IT, but most use computing. Earliest data on Internet, capacity, and communications is circa 1990. Data derived from a larger pool of such information in U.S. Federal Reserve Bank of Dallas, *The Best of All Words: 2006 Annual Report* (Dallas, Tex.: U.S. Federal Reserve Board, 2006): 4–9.

from those of others. But, having at least a reasonable idea in mind as we launch through a global tour of IT's arrival is helpful. Work done by economists at the U.S. Federal Reserve Bank of Dallas helps us understand how big the topic is. Table 1.8 can serve that purpose.

The World Bank has tried to answer the question by looking at ICT spend. The global GDP in 2008 hovered at $60.6 trillion dollars, and of that they estimated $3.6 trillion went to ICT (computers and communications), or about 7 percent of GDP in the United States, 6 percent for other high income countries, and closer to 5 percent by the lowest income nations. In short, the world spends about 6 percent of its income on information technologies each year.[80] IBM's internal tracking that I have seen suggests similar amounts, with slow growths in expenditures—not percentage of expenditures—in the subsequent two years, which is how one arrives at the current widely accepted figure that humans spend annually about $4 trillion.

To begin telling the story of how computing spread around the world in roughly a half century, we looked at the early experiences of those regions of the world where computing emerged from the handiwork of engineers and tinkerers, academic laboratories and government officials in North America and Europe, then wrap around the world in the rough order in which the first wave of diffusion spread. As we review the introduction and diffusion of computing around the world, we do this by examining developments in countries that played seminal roles (such as the United States) and others that represented a type of diffusion or economic circumstance shared by others (such as a group of small European states and the Asian Tigers), examining in each case the various facets discussed earlier in this chapter, from development of the technology to the role of public policy, economic realities, and the roles of computer vendors and their customers. The comparative analysis of the Atlantic and Asian world is a technique that highlights the diversity and commonalities of one's technology diffusion. We end with a discussion of the current experiences of China and India, where the most extensive diffusion of IT is currently unfolding, making their experience the largest examples of Wave One diffusion underway. In the process we will have built up a composite picture of global events. To start, we turn to the earliest major epicenter of digital computers, the United States.

CHAPTER 2

Diffusion of Computing Starts in the United States

There was, first of all, the long-standing strength in mass production industries that grew out of unique conditions of resource abundance and large market size. There was, second, a lead in "high-technology" industries that was new and stemmed from investments in higher education and in research and development, far surpassing the levels of other countries at that time.

—Richard R. Nelson[1]

The origins of the computer, the community of early users, and the dawn of a commercial computer market began largely in the United States. As this chapter and the next several describe, computational projects and development of the earliest computers and initial installations of this technology were also underway early in a number of countries, most notably in Great Britain, France, and Germany. But, the Americans surged forward in development, diffusion, and deployment beyond that of any other nation in the years following World War II, maintaining that wide lead into the twenty-first century. As time passed European and East Asian installations and uses of computing advanced rapidly too, but did not reach levels of deployment similar to the American experience until late in the century.[2] Additionally, we are learning more about important deployment activities behind the "Iron Curtain," which proved to be far greater than understood previously by historians, a realization which will undoubtedly lead future scholars to refine further their understanding of the global situation in the second half of the twentieth century. Of all the national histories one could write about computing, the most documented and frequently studied is that of the United States, and for the good reason that deployment occurred the fastest and most extensively in this country in the early stages of the technology's evolution. It is also why this chapter is devoted entirely to one country, unlike many subsequent ones, which discuss whole regions or multiple countries. If we approach the subject by using our loose notion of waves of development and, more importantly, adoption, and if we also want to argue that the first wave—Wave One—took place between the 1940s and the mid-1990s, then we can conclude that this is a story significantly influenced by American activities.

Because the boundaries between Wave One and Wave Two are perhaps earliest and most clearly evident in the experience of the United States, this chapter describes events in this country only through Wave One, in other words, to the early 1990s. The next wave, with the Internet possibly its most iconic technology (although developed by the start of the 1970s in the United States), appears so complicated that it would deserve its own chapter, which is beyond the scope of this book. However, historians have already started to examine that period in American technological history and it is proving to be a massively complex effort.[3] A similar, but not as stringent, limitation is imposed on Western Europe's history with computing. Because the boundaries between waves are less obvious in East Asia, where computing came later, quickly, and included elements of Wave Two diffusion during its Wave One period of adoption, that region's story was brought into the twenty-first century.

The American narrative has various facets that are essential to our understanding of the global diffusion of information technologies (IT). These include development of the technologies themselves; their implementation in the form of specific machines and software; uses and users; role of government regulators and funders; the rise of a commercial computer industry, with all the institutional and behavioral accouterments of an industry; and data about the numbers of systems installed, users, and, applications. This chapter focuses less on the various inventions of the period, a subject that has been extensively studied by historians and others, and more on the diffusion and reception of IT in American society. That emphasis lays the foundation for much of what occurred in other countries, such as the export of American computing by its manufacturers and users. In parallel, observers, economists, and historians of the creation and use of the computer also reflect the emergence and flowering of this technology and so a simultaneous tip of the hat to their views also is warranted for their work gives us insights into the history of computing, beginning with the epigram at the start of this chapter by a distinguished American economist who has long studied technology-driven economic development, Richard R. Nelson. Experts in the field of American IT history will see much that is familiar in the pages ahead, but also far less than they might want. By compressing the story of the origins of American computing to this one chapter, we encourage them to begin recalibrating their perspectives about the global experience with computing to recognize that it proved far greater and more diverse than any of us acknowledged in earlier research. This chapter suggests a modified definition of proportionality of the diffusion, propelling us closer to the reality of the bigger, richer story of world-wide information processing.

ORIGINS OF COMPUTERS IN AMERICAN SOCIETY

Two fundamental sources existed for telling the story of the early American flowering of computing. The first was a series of circumstances and activities that made possible this growth of IT. The second, and a logical consequence of those events, was the publication of many histories, economic analyses, and memoirs of those activities that have essentially swamped the historical record with a tsunami of accounts, stories that appeared over the years to explain what happened. As with any nation, local circumstances profoundly influenced events. Many of these realities were not directly of computing itself, an important point to make, because so many of the early descriptions of the origins of American computing focused largely on the technology *per se* with only a minimum acknowledgment of other factors, and when doing so concentrated largely on the profoundly exogenous influences of World War II and the Cold War.

While historians agree that both analog and digital computers were first built—invented if you will—in the 1940s, in particular digital machines during World War II, much preliminary work took place in various countries in earlier decades that made it possible to construct these early systems during that war. Most of these preliminary activities took place in the United States, Great Britain, and if we count cryptographic technologies, also in Poland.[4] But it was a set of circumstances and events in the United States that existed in the 1930s and, more importantly in the 1940s, that positioned the nation most fruitfully over the long run to leverage this new technology. These involved the existence of concentrated clusters of technical and academic experts in electronics and radio; the ability and willingness of the national government to invest heavily in the development of computers over a very long period of time (early 1940s right through the 1970s, indeed one could argue to the present), the stimulative effects of the Cold War in general and specifically the Korean War (1950–1953) and to a lesser extent the Vietnam War (1959–1975); a large and growing economy populated with office appliance and electronics firms capable of commercializing computers; and a set of potential customers eager to use the technology to address issues of productivity, expanding scope of work, and the high cost of labor, many of whom had deep experiences and knowledge with prior forms of IT, such as tabulating equipment and calculators, and several hundred with relay calculators. Still others understood both network and differential analyzers, many built by Vannevar Bush and his students at the Massachusetts Institute of Technology (MIT) in the 1920s and 1930s. If one imposed on all those environmental realities an ever improving technology that, in hindsight, proved stunning in its capabilities, one begins to see that the wind was to the back of the United States pushing deployment further and faster than might otherwise have been the case.

The most important technical community involved in the development of computers and subsequently myriad IT products in the 1930s through the 1960s were not chemists, materials scientists, chess players, mathematicians, or computer scientists—most all those people only began affecting the technologies in the mid- to late 1950s. Rather, beginning in the 1930s and 1940s, engineers (largely electrical), physicists, and even ham radio operators played the predominant role in IT development, because they were able to lash together various existing parts to create new devices. They used vacuum tubes (a radio component), electricity, and other electronic instrumentation, and subsequently developed more specialized parts and sub-assemblies that later became essential elements of computers.[5] To be sure there were exceptions; the most important were scientists at Bell Labs who studied silicon in the 1930s and who were later able to develop transistors in the 1940s, which became essential building blocks of computers in the 1950s.[6]

Electrical engineers and experts in radio electronics were extensively clustered together in major urban centers and often worked at large universities and at important research facilities. Early concentrations of engineers and academics existed on both coasts and in the middle of the nation: California at Los Angeles and near San Francisco, where universities and electronics manufacturers existed; and at a few mid-Western universities, such as in those in Chicago, Champaign–Urbana, Minneapolis, Madison, and Ann Arbor. In the case of Chicago electrical engineers worked for office appliance firms (Felt & Tarrant and its successor firm, Victor), while others lived in Dayton, Ohio (at NCR) and in Detroit, Michigan (Burroughs).[7] However, the largest concentrations of engineers and radio experts worked on the East Coast, specifically in Philadelphia (radio capital of the United States), New York state (home of IBM), and, most important, in the Boston area.[8] The latter city proved quite important because the Massachusetts Institute of Technology (MIT) was located there, where in the 1930s and 1940s Vannevar Bush built differential calculators and in the process exposed many electrical engineers to early computing. At nearby Harvard University, Howard Aiken examined computational issues at the start of the 1940s, leading to both the construction of a computer during the war years in collaboration with IBM engineers, and as a by-product, to the training of many engineers and others who went on to create new technologies themselves in the 1940s through the 1960s, such as An Wang.[9] Philadelphia, site of the University of Pennsylvania and its Moore School of Electrical Engineering, became the locale of important developments during the second half of World War II. There John Mauchly and J. Presper Eckert built what is widely recognized as the first American digital computer, the ENIAC, introduced to the world in 1946.[10] These centers of electrical engineering were not trivial. Both in Philadelphia and Boston such institutions employed hundreds of individuals who, by the early 1950s, were organized into teams of engineers building computational devices. Finally, we should mention Princeton, New Jersey, where John von Neumann, who is often credited with crystallizing the concept of the modern computer in the 1940s, went on to construct and advise others on computational devices in the 1940s and 1950s.[11]

In short, the United States had several dozen universities and nearly as many cities that were homes to about 1,000 technical individuals who eventually built computers and developed new information technologies.[12] While total numbers are not easy to come by, we know there were several thousand capable individuals who, by the end of the 1940s, were gainfully employed at universities or companies with sufficient budgets and technical management to focus on computers. That far exceeds the concentrations of electrical engineers and academics in European and Japanese places at that time that had focused on computers. In fact, not until the Soviets began their initiatives in computing essentially during the 1950s could one find similar clusters of experts necessary to build computers and who could evolve its underlying technologies.[13]

During and after World War II the largest electronics and office appliance manufacturers each hired hundreds of engineers who had learned about the most advanced electronics while in military service during the war, concentrating them in specific locations devoted to the application of their newly learned knowledge. For example, IBM hired electrical engineers to work at its relatively new combined plant and laboratory in Poughkeepsie, north of New York City, near other IBM facilities. In the 1950s and 1960s, they developed over a dozen major computers, ranging from military machines to the all-important System 360 family that so transformed computing around the world, beginning in the 1960s.[14]

By themselves engineers and scientists were never enough, however; many countries had technically skilled individuals who could have built computers in the 1930s and 1940s, as did Conrad Zuse in his parent's living room in Germany.[15] Funding for scale and scope was a second requirement after possessing requisite expertise, and that support came from the U.S. government. All economists and historians who have looked at computing are in agreement that no government—with the exception of the USSR later—invested as much on research and development (R&D) of computer technology as did the American government. In fact, one cannot speak of that government's initiatives in monolithic terms because it was not one agency, rather nearly a dozen that individually took the initiative to fund major multimillion dollar projects for the development and use computers. Table 2.1 catalogs some of the earliest government-funded projects, clearly the largest collection of such initiatives in the "Free" World at that time. In the case of SAGE, guided by employees of MIT, hundreds of engineers and others worked on the project, which ultimately cost the American government billions of dollars. To give one a sense of the relative size of that investment, the British, who also were investing

Table 2.1 US GOVERNMENT SUPPORTED COMPUTER
PROJECTS, 1945–1950

Project	Source of Funding	Initial Operation
ENIAC	Army	1945
Harvard Mark II	Navy	1947
BINAC	Air Force	1949
Harvard Mark III	Navy	1949
SEAC	Air Force	1950
ERA 1101	Navy	1950

Source: Kenneth Flamm, *Creating the Computer: Government, Industry, and High Technology* (Washington, D.C.: Brookings Institution, 1988): 76. Flamm identified 27 other computer projects supported by the U.S. Government in the 1950s, ibid., 76–77.

heavily in IT development, only expended annually a few tens of millions of dollars.[16]

Federal commitments to computing began during World War II, when both the Army and Navy sought help in the development of accurate firing tables and to support calculations required to help develop the atomic bomb. The uniformed services had been users of punched card equipment since before World War I, thus many officers and soldiers had personal experience with data processing. During World War II thousands of engineers and scientists were assigned to many scientific projects, so computing became part of a larger investment. Officials recruited firms and universities to do their scientific and engineering work, providing cash for these initiatives. Areas of R&D included radar, proximity fuse, battlefield medicine, nuclear energy, weather, and fire-control systems. By the late 1940s, the U.S. government was funding about 50 percent of all scientific research being done in the country and, by the early 1960s, two-thirds of all American R&D.[17] With respect to computers, Kenneth Flamm observed that, "between 1945 and 1955 the U.S. Government dominated computer development. All major computer technology projects in the United States were supported by government and military users," while in subsequent years the percentage of federal funding declined as the private sector increasingly invested in R&D to convert these earlier projects into commercial products.[18] However, the U.S. government continued to invest in R&D in the more than six decades following World War II. Key participants in that funding included the Department Defense (DoD), established in 1947, followed in importance by the Atomic Energy Commission (and subsequent iterations of this organization), Department of Energy (DoE), National Aeronautics and Space Administration (NASA), and the National Institutes of Health (NIH). Details of their investments need not detain us as these have been recited many times by others.[19] Table 2.2, drawn from Flamm's research, illustrates the near complete coverage of federal support for all the early systems,

Table 2.2 US GOVERNMENT FUNDED COMPUTER
PROJECTS, 1940S–1950S BY AGENCY

Agency	Number
Army	7
Navy	11
Air Force	11
National Security Administration	4
Atomic Energy Commission	5

Source: Kenneth Flamm, *Creating the Computer: Government, Industry, and High Technology* (Washington, D.C.: Brookings Institution, 1988): 76–77. This list represents every major project in the US, but not all, because some universities, for example, were experimenting with small systems, such as the University of Wisconsin and there were minor side-bar projects underway at private companies, particularly in the second half of the 1950s as many of the earlier systems were converted into commercial versions.

which in their totality, made it possible for the United States to create the earliest and most extensive base of knowledge about computing and a commercial industry to support sales of computer systems.

In addition to the military and the other agencies just cited, the National Science Foundation (NSF) became the most important source of support for all manner of scientific research in the United States during the second half of the twentieth century. Established in 1950 for that purpose, the NSF supported academic work on computing and all manner of science, while the DoD and other federal agencies tended to do that too; but, they contracted more directly with companies and universities to work on specific projects of immediate interest to them. In the case of NSF, by the late 1950s, it had quickly become an important source of support to myriad computer projects, increasing funding to substantial levels over the years. Kenneth Flamm's research shows that during the second half of the 1950s annual support often exceeded two billion dollars, then in the first half of the 1960s that amount doubled. In the second half of the 1960s, funding ranged annually from nearly 9 to 12 billion dollars.[20] Overall, while private funding expanded too in the 1950s, and far in excess of anything historians are aware of in the rest of the world by other firms, government investments still dominated, with funding in the years 1949 through 1959 accounting on average for 59 percent of all American investments in computer R&D. To put a fine point on the numbers, in total during that crucial decade in the emergence of the computer industry and its base technologies, the entire nation (private and public sectors) invested nearly $7 billion.[21]

With the exception of the Soviet Union, no government between 1945 and at least the mid-1960s had the same sense of urgency and motivation to invest in this technology as the American government, which led the "Free World" in its Cold War against the "Iron Curtain" nations led by the Soviet regime. That commitment to urgency on the part of both was shaped by military concerns. While it is difficult now to recall the intensity of the rivalry of the democracies and communist regimes, it lasted for nearly a half century. That contest always carried with it the real threat of the destruction of humankind, because both the US and the USSR had armed themselves with massive weapons of destruction, including several generations of atomic bombs. Their intense arms race involved the use of ever-more "high tech" weapons systems of all kinds, many tested in dozens of proxy wars, often using computing technology for guidance systems, development of new weapons, massive battlefield operations, management modeling, command-and-control, and data processing to track purchases of arms and goods and to deploy military personnel and materiel. The leading authority on the role of computers in America's side of the Cold War, Paul N. Edwards, has persuasively demonstrated that "computers controlled vast systems of military technology central to the globalist aims and apocalyptic terms of Cold War foreign policy," arguing that no major military system operated without relying on this technology, and that these even served as metaphors by which "such systems constituted a dome of global technological oversight, a *closed world* [his italics], within which every event was interpreted as part of a titanic struggle between the superpowers."[22] Indeed, that pervasive presence, influenced by the technology's characteristics, and uses, remained important to the end of the Cold War in the late 1980s. In Edwards' words, "computers made the closed world work simultaneously as technology, as political system, and as ideological mirage."[23]

Side-by-side with the exigencies of the Cold War, economic realities made it affordable for the American government to invest massively in the development of this technology, the creation and highly successful expansion of what came to be known as the American computer industry on a global scale, and its diffusion within U.S. society so extensively not only during the period covered by this chapter, but across the entire second half of the twentieth century. Just as there were technological pre-conditions for diffusion, namely creation of the technologies themselves to such a point that they actually worked reliably enough and proved sufficiently cost effective in comparison to older technologies or human labor, so too there were economic preconditions facilitating this diffusion. The American economy was strong and expansive enough in the early decades of the development of computers, important because the technology proved expensive to invent. The US economy grew dramatically by any measure one chooses to use. For example, if one uses the US Gross National Product (GNP) as a test, between 1950 and 1960 it expanded by 38 percent, clearly a by-product of the United States not suffering physical destruction of its cities, factories, and economy the way most other participating countries in the war had, notably Europe, the European portion of the USSR, and Japan. GNP grew another 48 percent during the decade of the 1960s, and by another 32 percent in the 1970s. Similar statistics could be cited for the rest of the century.[24] Second, the number of firms and workers grew proportionally in the second half of the century as well, which meant that there were more potential users of the technology. The expense of this labor force in the expanding economy provided economic incentives for corporations to use technologies that improved labor productivity while facilitating their scaling up of operations to address growing demands for products and services. The latter issue—scaling—came at a time when other new technologies and processes for doing work became more complex, requiring the operating features of computers to help get things done.[25]

An additional prerequisite evident by the end of World War II, and more essential and obvious over time, was the transformation of the American economy into a form that lent itself to the use of computing more than might otherwise have been the case. Initially, its transformation from an agricultural-based economy of the late nineteenth century into a heavily manufacturing-based one was spurred on by the needs of World War II and global greater demand for American goods following the conflict. Roughly 21 percent of the GNP in each decade of the second half of the century came from manufacturing, which could most benefit from the technology in its first several generations. Financial industries, such as banking, insurance, and brokerage to a lesser extent, also expanded in this period, representing 10 percent of GNP in 1950, 13 percent in 1970, and toward the end of the century over 18 percent. If one looks at government and the service sector combined, we see that the two grew as well—areas that could use computers in every form in which they emerged between the middle of World War II and the present—comprising 25 percent of GNP in 1950 and over 32 percent by the end of the century.[26]

A third economic prerequisite was the increased size of enterprises, because the larger the firm (or agency), the more it could benefit from computer-based automation. These could afford what was, quite candidly, always an expensive technology, despite the spectacular improvements in its declining unit costs over time. In the 1950s, roughly one out of every four employees worked for one of the 200 largest enterprises in the nation. In turn these enterprises affected over half the economy through employment, networks of subcontractors, or sales.[27] By 1963, the top 200 firms owned 41 percent of all the economic "value add" and 56.3 percent of the economy's assets.[28] That trend of economic concentration spread across most industries throughout the second half of the century.[29] Many of the largest had thrived during World War II as primary contractors to the US government and thus had all manner of technological, financial, and managerial momentums in their favor going into the post war period, the wind to their back referred to at the start of this chapter.[30]

Another feature of the American economy that facilitated rapid diffusion was the existence of a vibrant collection of electronics and office appliance companies that had developed good working relations with institutional customers over the previous half century, many as far back as the 1880s and

1890s, conditioning both communities to use technology in data processing (typewriters, calculators, tabulating equipment, cash registers, billing equipment, telegraph, and telephones). Many customers were also the same individuals and firms that acquired significant new state-of-the-art knowledge about electronics during World War II. Most office appliance vendors had global operations, although these were largely disrupted and damaged (or destroyed) during the war, but re-established fairly quickly. In particular, office appliance firms proved able to develop computers first for the government (mainly military) and then commercially, placing these into American firms, such as the 200 just cited, subsequently into government agencies in the United States and many countries, and, even later, quickly within European enterprises, overcoming competition from local rivals.[31] Leading lights of the post-World War II period included Remington Rand (later Sperry Rand), NCR, Burroughs, and IBM among office appliance providers; among the old electronics firms, most notably General Electric (GE).[32] They had not been destroyed by the war, already called on ideal customers for computers, and bent the technology to meet needs which they understood. Finally, they had access to investment capital, which increased as time passed giving American firms a unique competitive advantage.[33]

In the early 1950s, and extending into the next century, thousands of new firms emerged, reflecting the introduction of new technologies and products, a circumstance that has been examined by many historians and economists.[34] Rather than plow through the specifics of the success of all these firms, several statistics suggest their results. In 1954, the entire computer industry generated revenues for computer hardware of $10 million; in 1958, this climbed to $250 million; in 1963, to $1.5 billion; and a decade later, to $7 billion. Note, however, that some of those revenues came from non-US markets.[35] Looked at as a percentage of GNP spent on computing, one can see that by the late 1950s—slightly less than a decade after the computer went commercial—less than 2 percent of American GNP went toward this and other office technologies, a percentage that grew slowly during the 1960s and 1970s reaching about 3 percent of the GNP in 1980s; yet remember that these percentages reflected massively expanded absolute cash expenditures, because the nation's GNP also expanded.[36]

Finally, we should ask, what about software, as much of the original commercial products made out of software originated in the United States, how was it received? Accurate numbers are difficult to come by, as its leading historian, Martin Campbell-Kelly, has frequently noted.[37] As a collection of products, these first came onto the market in the second half of the 1950s and by roughly 1964, were generating upward of $275 million in revenue each year, a quantity that reached a half billion dollars in 1970. Within a half dozen years, that volume had doubled and doubled again by the end of 1980, just as the massive surge in deployment of personal computers began, triggering another round of truly massive increases in the sale of software, this time for use with PCs.[38]

Before turning to customers and the public's reception of this new technology, the American experience with IT diffusion needs to be understood within the context of a more international perspective. Picking 1972, a year when computing in many other countries had taken off enough to be measurable, we see some interesting patterns. That year, the US spent about 2.59 percent of its GNP for all manner of data processing, while on average other industrialized nations hovered at 1.29 percent. Individual leading adopters of computers outperformed this average: United Kingdom at 2.13 percent; Japan and Canada at 1.4 percent; France at 1.36 percent; and West Germany at 1.29 percent. In decades to come Switzerland and Denmark would be seen as "high tech," and already one could see that by 1972 they too had become extensive investors in this technology, with their GNPs that year measured at 1.63 and 1.71 percent, respectively.[39]

How did customers and the American public meet and greet the computer? It is an important question because we know today that images of the value of a technology influence their rate of adoption; pure functionality and attractive prices are not enough. For example, the American public has long been wary of the value of nuclear energy to provide electricity, regardless of its proven benefits, while the French are most comfortable with it. I mentioned in the first chapter how a sense of Moore's Law pervaded the market for IT, which influenced the rate at which new devices were accepted. Misunderstandings and concerns about possible dangers of color television sets and microwave ovens had their effects too, TVs in the 1960s, microwave ovens in the 1970s, and cell phones in the 1990s. As advertising experts in IT firms believed in any decade, and certainly in the years under review in this chapter, perceptions were enormously important.[40]

Engineers, military officers, and funding agencies alone did not make computers a visible part of American society. Nor were the salesmen of IBM, RCA, or Burroughs alone the heroes of capitalism, because they still needed customers to agree to embrace the technology—the American public. In the literature on the adoption and history of computing, customers' voices were often too muted. Technical staffs usually received more attention from historians, women workers as well by historians.[41] While the voices of these influential communities were important to document, theirs' represented only some of the constituents of computing. In the period 1940s to the early 1980s at least, most consumers of computing were professional data processing staffs, workers in enterprises and public agencies, and their management; but, the next generation of users of IT were factory workers, logistics agents, truck drivers, clerks in supermarket check out lines, bank clerks, and many of their children.[42] But before discussing the public and customers in more detail, we explore the role of the creators of the technology because they were some of the earliest users too.

CONTRIBUTION OF ENGINEERS AND SCIENTISTS TO EARLY DIFFUSION OF INFORMATION TECHNOLOGIES

Individuals who designed and created components for computing built machines, then developed software in the years after World War II, worked through the late 1960s to the present, contributing directly to the diffusion of IT in the United States. They did this largely along the east and west coasts of the nation, and in the Midwest (particularly in southern Minnesota). To a large extent they did not know that they were serving as diffusionary agents for these technologies. In fact, most did not know that what they were doing would ever turn out to be of such a great importance to the nation. John Backus, the father of FORTRAN, the most widely used scientific programming language of the period, recalled life in the 1950s as a world in which "recognition in the small programming fraternity was more likely to be accorded for a colorful personality, an extraordinary feat of coding, or the ability to hold a lot of liquor well," than for any brilliance or sense of historic destiny.[43] To be sure, the group was small in the beginning. In the 1940s and early 1950s, there were only several dozen projects underway at various universities, companies, and government agencies. As part of their work, engineers and academics found it possible to share information with each other through conferences, such as the Moore School lectures of 1946, visits to each other's projects, and other lecture series.[44] Transmission of information and advice flowed within and outside of government agencies, such as the help contributed to other agencies and the private sector about computing by the Bureau of the Census and the US Navy Department. These various early activities involved more diffusion of information about computers than computers themselves.

In the 1950s and 1960s, their diffusion of knowledge about computing continued transforming and expanding dramatically the awareness of computers. Four fundamental activities stimulated the diffusion of information technologies. First, beginning in the 1950s, companies that sold computers hired engineers to design, build, sell, install, and train users on the machines. As table 2.3 illustrates, numerous firms appeared in the early 1950s and, indeed throughout the period, staffed and managed largely by individuals steeped in knowledge of this new technology.[45] Second, in the 1950s, the subject of computing itself evolved into a formal academic field, which academics and practitioners called Computer Science, manifested in the creation of professorships and whole departments in universities devoted to the subject. They trained students who in turn went on to develop new computer technologies and to deploy these.[46] Table 2.4 lists some of the early institutional reflections of this trend, while others that taught electrical engineering also began training experts in computing.

Two examples illustrated the process underway. At the University of Wisconsin engineers built early experimental machines, such as the WISC, and trained engineers, most famously Gene Amdahl, to whom the University of

Table 2.3 US COMPUTER VENDORS AND
MANUFACTURERS, EARLY 1950S

Major Suppliers	Other Manufacturers
Burroughs	AT&T
IBM	Bendix Aviation
NCR	Electrodata
RCA	Marchant Calculators
Sperry Rand	Minneapolis-Honeywell
	Monroe Calculating Machine
	Raytheon Manufacturing
	Underwood

Source: Kenneth Flamm, *Creating the Computer: Government, Industry, and High Technology* (Washington, D.C.: Brookings Institution, 1988): 82. This list does not include the academic and government projects also underway at the time.

Table 2.4 MAJOR CENTERS OF EARLY US COMPUTER SCIENCE, 1950S

Harvard University
Massachusetts Institute of Technology
University of California at Berkeley
University of California at Los Angeles
University of Chicago
University of Illinois at Urbana-Champaign
University of Michigan
University of Minnesota
University of Pennsylvania
University of Wisconsin

Wisconsin awarded a Ph.D. in 1952 based on the work he did designing this system. He went on to be a lead designer of IBM's S/360 in the early 1960s, and later established Amdahl Corporation (founded in California in 1970), which sold mainframe computer systems in various countries in the 1970s and 1980s.[47]

The second example, involving MIT, is important because of a rich collection of information about the experiences of people involved there in computing. Individuals who worked on the Whirlwind project spread across the country in subsequent years. Thanks to their forming an informal alumni community that surveyed its members in 1987, we know by name what 114 individuals did. All 114 reported that the project was important for the development of information technology; 110 pursued careers in IT, with about half working for a combination of MITRE, MIT, and other firms in the Boston metropolitan area. More specifically, however, is the fact that members of this cohort went on to work for almost every important computer company that came into existence between the early 1950s and the late 1980s. About 25 percent went into teaching at the university level or worked at such government agencies as the National Aeronautics and Space Administration (NASA) and the US Department of Defense. Their history also showed that they worked in every region of the United States. The Whirlwind group was statistically large enough that one can conclude that the other hundreds of individuals who worked at MIT in the early days of computing probably diffused into the larger American IT world in similar fashion. All had started out largely right out of graduate school or were early in their careers as engineers and mathematicians, many with advanced technical degrees. Clearly, they played an important role in disseminating knowledge about IT all over the United States.[48]

Third, these technologists also established societies to serve their professional needs in this new discipline. These functioned like other professional associations: they held annual meetings, hosted seminars, and published magazines and journals. All of these activities were designed to introduce fellow computer engineers to each other and to share information about the subject, just as in any other field. Leading associations are listed in table 2.5, but not all from the 1940s–1960s. Most of the important publications that disseminated information about computing also came along quickly (cited in table 2.6), becoming primary sources of information read all over the country, indeed in many nations. One way we know this is by looking at subscriptions, now made easier by examining online catalogs of universities and colleges, which often state when their subscriptions began, or how far back their holdings went, indicative of the period when interest in the topic began. While such an exercise is increasingly becoming difficult to do, because so many university libraries cancelled their subscriptions in the early 2000s and, instead, moved to online versions, such as those of the Institute of Electrical and Electronic Engineers (better known as the IEEE) and the Association for Computing Machinery (ACM); nonetheless, we know that, by the end of the 1950s, scores of libraries were beginning to accumulate documentation about computers.[49]

Fourth, experts in computing and its technologies moved from one organization to another, taking with them knowledge that they shared. This occurred, for example, with Gene Amdahl who, after graduating from the University of Wisconsin, moved to the east coast to work for IBM. But there were thousands of such cases, particularly beginning in the second half of the 1950s, extending all through the 1960s and beyond, to the point where today tens of thousands of such people migrate from one job to another each year. A particular feature of their behavior most evident in the United States was the practice of engineers moving from an academic setting to the private sector to commercialize computing, followed by members of an upstart firm moving to yet another company. Three examples illustrated the process at work, a trend continuing to the present, most recently epitomized by the thousands of firms caught up in

Table 2.5 LEADING US ASSOCIATIONS OF COMPUTER SCIENCE
AND USERS GROUPS, 1940S–1960S

Organization	Year Established
Association for Computing Machinery (ACM)	1947
Data Processing Management Association (DPMA)	1949
IEEE Computer Society (IEEE-CS)	1951
Society for Computer Simulation (SCS)	1952
SHARE	1955
UNIVAC Scientific Exchange (USE)	1955
GUIDE	1956
Joint Users Group (JUG)	Late 1950s
International Federation for Information Processing (IFIP)	1960
American Federation of Information Processing Societies (AFIPS)	1961
DECUS	1961
CUBE	1962

Other associations and user groups sprang up in these and in every subsequent decade.

Table 2.6 MAJOR US PUBLICATIONS ON DATA PROCESSING AND
COMPUTER SCIENCE, 1943–1958

Publication	Year Established
Mathematical Tables and Other Aids to Computation	1943 (demised 1960)
Digital Computer Newsletter	1948
Computers and Automation	1951
Association for Computing Machinery Journal	1954
Datamation	1957
IBM Journal of Research and Development	1957
Computer Journal	1958

Hundreds of other publications appeared on a continuous basis into the next century. The most widely circulated of all publications, *Computerworld*, first began publication On June 21, 1967, with a story announcing its targeted subscription market numbered 300,000, and presented itself as the industry's weekly newspaper.

the Dot.com bubble at the start of the new century.[50] The examples involve the engineers who built the ENIAC in the early 1940s in Pennsylvania, a second group working in Minnesota who established the Engineering Research Associates computer (better known as ERA), and the most spectacular example of diffusion, the experience of Fairchild in California, which led to the creation of dozens of spinoff technology firms. The experiences of these three firms reflected those of thousands of small and large enterprises that, by the end of the 1970s and early 1980s, existed all over the nation, involving the manufacture of components (especially computer chips), specialized software products, digital instruments, or parts of other machines (for example, automotive engines in Detroit, Michigan), and services (such as ADP's time sharing).[51]

This first generation—those that came into existence in the quarter century following World War II—set the pattern followed by so many to the present. Many historians have looked at the experiences of these firms—part of the supply-side story of how the computer industry came into existence so early in the United States—and so we understand many features of this early wave. Nobody has more thoroughly studied some of these early firms than Arthur L. Norberg, one of the pioneering professional historians who looked at the history of American computing. His work confirmed that the stories of the Eckert–Mauchly Computer Company (EMCC) and that of Engineering Research Associates (ERA) were emblematic of the events that occurred in this first generation of the computer industry.[52] While Norberg and most historians concluded that these and many other early computer firms failed to be successful business enterprises on their own, when looked at from the perspective of the diffusion of computing, they clearly played important, indeed, transformative, roles on behalf of the proto-computer industry.

During the war, the US Navy created a team of scientists, engineers, and others to perform intelligence operations that became involved extensively with computing projects. As the war's end came in sight, so too did the prospect of many of the personnel involved leaving the military to return to civilian life. Navy officials became concerned because they needed further work by this group. They began exploring the possibility of encouraging some of these people to form a private corporation to continue their activities on behalf of the Navy, particularly that of electronic engineers. Some of these Navy personnel were also eager to exploit their recent experiences in a commercial setting, most notably Howard T. Engstrom, William Norris, and John Parker, among others. With colleagues from the war years they established Engineering Research Associates, Inc. (ERA), set up operations in St. Paul, Minnesota, hired ex-Navy personnel, and in time the Navy awarded them an initial contract necessary to keep them going. By November, 1946, some 42 individuals were active employees of the company and they went about the task of designing and building their first computer. The story of the firm has been told by others, but the point to make here, as Norberg did in his study of the firm, is that it moved quickly into computing. It did important work in the St. Paul–Minneapolis area such that by mid-1947 ERA had accumulated the necessary "engineering skills to pursue research and development on an organized and sustained basis, and to carve out for an area of computing machine development that would have an important influence on the field."[53] By 1948, ERA had become a computer manufacturer. The year before the Navy awarded ERA a contract to construct a digital computer, called the Atlas, and soon after the firm received permission to build a commercial version, named the ERA 1101, designed for scientific computing. When office appliance firm Remington Rand wanted to enter the commercial computing field quickly in the early 1950s, it did so by buying fledgling firms, and ERA became one of these.[54]

We can end the story here because the point to be made is that ERA illustrated how government projects could—and did—transform into civilian commercial ones and equally important, how quickly in two respects. First, there were individuals who saw and pursued commercial possibilities. They proved able to pull together financing, identify customers, and proceed with the work of developing, building, and selling machines. Second, they did this quickly, with greater speed than European firms and with multiple products coming out one after another in rapid succession. ERA contributed mightily to the dissemination of computing knowledge and use of the technology in the Minneapolis–St. Paul area, if for no other reason than it hired people to work in the field, while other feeder enterprises interacting with ERA sprung up as well. Over time employees from ERA moved on to other companies or set up their own.[55]

Simultaneous with the emergence of ERA was that of another firm that also proved instrumental in diffusing knowledge and people about computing directly into the American economy and, like ERA, did it well-enough and quickly to help establish the commercial computer industry in the United States. As with ERA, the story of Eckert–Mauchly Computer Company (EMCC) has been told many times in various ways by historians and others,[56] although the most thorough study of the firm was done by Arthur L. Norberg.[57] To get to our main point, recall that John Mauchly and J. Prespert Eckert, as part of a team of engineers and physicists at the Moore School of Electrical Engineering, proposed to the US Army the construction of a computer to conduct military calculations, such as firing tables. The Army let a contract for this work and in 1945 these engineers completed construction of the most important early digital computer in the United States, known as the ENIAC, which historian Jeffery Yost characterized as "the first useful large-scale digital computer project in the United States."[58] Eckert and Mauchly realized quickly the commercial possibilities of their work, moved in 1946 to establish EMCC, filed for patents, and started looking for financial support. Like ERA, they began small—they only had a dozen engineers on staff—and started building a series of machines, initially the BINAC, which they constructed for Northrop Aircraft Corporation (a military contractor) to generate cash flow and subsequently the UNIVAC, which became the first major commercially successful computer product sold in the United States.

In the early 1950s, EMCC proved more successfully than ERA, and began to receive orders for its system. Remington Rand came in with the capital needed to scale up, and bought the firm. With ERA and EMCC it now had quickly become a major supplier of computing both to the US government in the

Washington DC–Baltimore area but also able to serve commercial customers in other parts of the United States.[59] These two firms—ERA and EMCC—were not the only ones to get into the business in the early 1950s; nearly two dozen enterprises entered the market, including, of course IBM, with the latter's concentration of engineers in the Mid-Hudson Valley, in New York. Each enterprise in its own way trained through the work of building systems, becoming computer experts, and learning how to manufacture and run these machines. As they sold these to agencies and companies, the population of people familiar with the technology expanded and spread geographically across many organizations and industries all through the 1950s and 1960s.[60]

So far, the story told has been about diffusion of computing in the Washington DC area, the Boston–New York corridor, Philadelphia, and the St. Paul–Minneapolis region, with spillover into nearby regions (e.g., Rochester, Minnesota by IBM in the late 1950s).[61] Each in time recruited people who went on to work either in other firms or as users of the technology. However, another very important locale for computing in the 1950s and 1960s that extended into the next century developed from nearly one end of California to the other and in the process leading, of course, to the Silicon Valley concentration of thousands of firms in that region during the second half of the twentieth century. But one company in particular emerged as the exemplar for how computing diffused quietly into the American economy and thus is worth introducing briefly: Fairchild Semiconductor. Recall that AT&T had invented transistors in the 1940s, then signed production licenses with some 250 other firms, giving them rights to manufacture these all over the United States in the early 1950s, even to a few enterprises outside the United States. This licensing made it possible for the new technology to displace vacuum tubes in computers in the 1950s and 1960s. Demand for these components grew quickly and a group of manufacturers supplied these in volume. Some of the early most important ones included AT&T's Western Electric, Raytheon, General Electric, RCA, and Texas Instruments (TI). Note that they were scattered around the United States on both coasts and in Texas. Diffusion of advanced electronics spread. In 1958, Jack Kilby at TI and Robert Noyce, founder of Fairchild Semiconductor, independently developed new transistors that combined multiple circuits on a single chip. Both filed for patents for what we today call integrated circuits (ICs). Leaving aside the story of the development of computer chips, which many scholars have already examined,[62] we turn our attention to Fairchild the enterprise.

William Shockley, co-inventor of the transistor at AT&T in New Jersey in 1946 (introduced in 1947), which eventually led to his shared Nobel Prize, established in Palo Alto, California, the Shockley Semiconductor Laboratory. It proved to be a financial and managerial failure. Eight of his engineers left in 1957 and established Fairchild Semiconductor with funding from Fairchild Camera and Instrument Company. In time, employees of this firm left to

establish their own enterprises in the region, leading in large part to its designation as Silicon Valley. The firm's alumni spun off yet other semiconductor
companies that in turn did the same throughout the 1960s and beyond, including Intel, established in 1968, and that eventually became the largest
American manufacturer of computer chips. Many observers have noted that
hardly any important engineer working in semiconductors in the late 1950s
and 1960s had *not* worked at Fairchild, or for that matter any organization
established by an ex-employee of the firm.[63] Innovation in semiconductor
technology largely created the business opportunities for new startups, for
venture capital to pour into its industry, and for the dissemination of all
manner of IT. The economic growth that ensued led to the employment of
tens of thousands of computer-literate people all over California and beyond
by the end of the 1960s. In the process, that economic transformation made
California the epicenter for so many of the technological innovations in computing that surfaced in the United States during the rest of the century. The
diffusion of these innovations extended as chip manufacturing expanded to
over three dozen states in one fashion or another in subsequent decades and
spurred the growth of the software industry with a footprint in every state,
and in many countries, most notably Japan and South Korea.[64]

It would be difficult to underestimate the importance of these kinds of
companies in jump starting the computer industry in the United States
between the 1940s and the 1970s. They were largely small groups of engineers,
computer scientists, software experts, and IT service providers who saw the
commercial potentials of computing, found capital funding from government
and later private sources, and moved quickly, driven as much by the innovation of the technology as by the business opportunities and rivalries they
spawned. This role proved particularly the case after major electronics and
office appliance firms entered the market during the second half of the 1950s
and raced through the 1960s with a new generation of computing that finally
made the technology ubiquitous in the largest agencies and enterprises in the
United States. Thus, one can acknowledge the contributions innovations in
the technology made to the American dominance in computing during the
middle decades of the century.

EVOLUTION AND SPREAD OF AMERICAN COMPUTER TECHNOLOGIES

Computing emerged from an almost nonexistent base before World War II to
a point three decades later where it was becoming difficult to imagine a large
American corporation or government agency not reliant on this technology.
We now have such a deep understanding of the degree of dependence on this
technology in the United States that it should be of no surprise that US industries were the earliest users of computing and certainly the most extensive

during the period examined in this chapter.[65] Economists and historians of technology and science point generally to several macro forces at work that contributed substantially to this development. First, in the United States, but also across Europe since the eighteenth century, science and technology merged together closely enough in the emergence of practical uses of both classes of knowledge such that their union resulted in the generation of much new-found insights and practices derived from each. In the case of computing we saw AT&T studying the properties of silicon in the 1930s (science) and in the 1940s construction of a transistor based on that knowledge (technological), following in this long pattern of behavior.[66]

Technological innovations alone, however, do not result in wide use. In modern times diffusion normally follows when economic factors are in play, such as the potential of a large market; availability of financial support, as occurred with the US government and later through venture capital; vendors to promote their use and support users; and, finally, cost justification that emerges from such factors as high labor costs and complexity of operations and goods that can be mitigated, changed, or controlled with a new tool. All of these factors were in play in the United States, providing the combination of motives and applications of existing knowledge to develop computing technologies that made diffusion in this country so rapid. Through the next half century, the majority of innovations in the technology came out of the United States, with a few notable exceptions, such as development of the browser for the Internet (Swiss) and CDs (Netherlands).[67]

We have already introduced three major innovations that served as the building blocks of the new industry: transistors, computer chips, and early commercial computers, replete with the economic underpinnings of firms and financial support necessary to move laboratory experimentations and innovations to commercial products. These innovations combined due to the physical proximity of developers, manufacturers, and vendors, clustering in descending order in California (Silicon Valley near San Francisco and in Los Angeles), Connecticut, Massachusetts, Texas, and Minnesota. Proximity within cities allowed people to share information, move from one company to another, and to attract capital.[68]

Table 2.7 lists major technological milestones achieved in the United States in the period of the 1940s to the early 1960s. These ranged widely from components to hardware and software. There were many, all were important, each introduced to the commercial market, about which we will say more throughout this book. Most of these systems were built almost as a "one-of-a-kind" in a neoexperimental environment in the early 1950s with findings and best practices seeping out incrementally as machines were put together one after another. The biggest of these projects was the SAGE Defense System, directed by employees of MIT, but with work distributed to a number of electronics and office appliance firms.

Table 2.7 MAJOR US TECHNOLOGICAL MILESTONES IN IT
HARDWARE, 1940S–EARLY 1960S

Transistor tubes	1946
Electromechanical and electronic components	1948
Mercury delay lines	1951
Mercury delay lines and drum	1952
Magnetic core memories	1950s
Flipflop switching circuits and CRTs	1952
Williams tubes and plastic tapes deployed	1953
Drum memories	1953
Ferrite cores	1954
Disk drives	1957
Transistors in computers	1958
S/360 compatible systems	1964

Paul Ceruzzi, a leading authority on the history of American computer technology, summarized the scope and complexity of this project:

> The Air Force's SAGE (Semi-Automatic Ground Environment), a system that combined computers, radar, aircraft, telephone lines, radio links, and ships, was intended to detect, identify, and assist the interception of enemy aircraft attempting to penetrate the skies over the United States. At its center was a computer that would coordinate the information gathered from far-flung sources, process it, and present it in a combination of textual and graphical form . . . to have multiple copies of this computer in operation around the clock. . . .[69]

Building on work already done at MIT on the Whirlwind computer system, IBM became the prime manufacturer of what eventually were some 30 systems built in the 1950s and early 1960s.[70] By any measure, SAGE evolved into a massive undertaking, involving thousands of engineers and other specialists, and proved relatively successful in meeting its objectives. The only other nation in the world that potentially had the capability of launching a project of such magnitude was the Soviet Union, which did have important projects underway.[71] In the 1950s, SAGE generated some half billion dollars in revenue for IBM, while exposing its engineers and others in many companies to the latest electronic and digital technologies under development in the country. This project proved so beneficial to IBM, in particular, that the earlier, faster moving Remington Rand with its UNIVAC I and soon after UNIVAC II, was overrun in the market by IBM.[72] In the process, SAGE spun off new memory systems, improvements in cathode ray tubes (better known as CRTs), use of transistors, enormous improvements in computer architecture, operating systems, telecommunications, software, and software utility programs, to mention a few.

Many individuals involved in this work learned how to manage very large R&D and manufacturing of computers.[73] One way to describe in nontechnical terms what happened is to say that the "look-and-feel" of what constituted a digital, online system, with mainframes and peripheral equipment that became common for decades to come, had been solidified during this project, and built the organizational infrastructures required to design, construct, diffuse, and service these. This wider emergence of new knowledge and creation of necessary sociomanagerial infrastructures to facilitate this work followed a pattern of managerial practice evident in many other industries that historian Alfred D. Chandler, Jr. recognized as institutional "capabilities," "organizational learning," and "paths of learning."[74]

By the late 1950s, IBM dominated the computer industry in the United States and certainly by the end of the 1960s around the world. Two new developments in the 1950s proved crucial to IBM's success. The first involved learning how to manufacture machines in quantity, and second, its development of the disk drive. Both of which, in combination with experiences gained working on SAGE and other projects, made possible development of the S/360 family of computers in the early 1960s that quickly and fundamentally changed the face of computing around the world.

ERA, EEMC, Remington Rand, and other early developers of computers actually did not manufacture these systems; they built them one at a time, fabricating them out of existing or newly developed components. Engineers constructed families of machines in tiny quantities, even the highly visible UNIVAC I, considered in the early 1950s to be the progenitor of the commercial computer industry, resulted in less than 70 orders for such systems and only 46 were built.[75] Even IBM's early military systems were nearly one-offs and recall that it only put together less than three dozen computer systems for SAGE.

For many years IBM had also located on the same campus research, product development, and manufacturing. The largest center for the development and construction of computers in the 1950s through the 1980s at IBM in the United States was located in Poughkeepsie, New York, north of New York City. That combination of development lab and manufacturing facilities, which had grown out of how IBM developed and manufactured tabulating equipment at its plant in Endicott, New York, for nearly a half century made it possible to design machines that could be manufactured (meaning, in quantity in a production line manner), compressing manufacturing costs, while placing a premium on quality and what came to be a new word used at IBM for decades, *manufacturability*.

The exemplar of this combined process at work surfaced with the 650, introduced in the mid-1950s as a commercial product with a standard configuration. It was relatively affordable, and reliable. It became the most widely used system of the 1950s; IBM built more than 1800 of these, using a hybrid mass-production form of manufacturing.[76] Researchers and engineers at IBM had to demonstrate to their managers that specific combinations of technology would lead to profitable products for which markets existed, or were emerging soon that would be ready by the time a new machine appeared. To be sure, the 650 emerged from a long heritage of engineering and product manufacturing, drawing on successes dating back to the 1930s; nonetheless, innovation was always driven in large part by market considerations, a point that cannot be stressed too much. In the case of the 650, for instance, IBM management thought the market for it would be as a scientific calculator; instead, customers started using it for all manner of applications, largely in accounting and manufacturing. The next IBM system, called the 1401, continued building on the momentum started with the 650. So, the firm shifted quickly its marketing and subsequent innovations in response to that reality.

Often IBM's customers concluded that there existed opportunities for improvements with these systems, and brought their observations back to IBM, which either included these suggestions in its next products or ignored them, but always with purpose in mind. This attention to customer opinions dated back to at least the 1890s. No system in the post-war period came out of IBM without formal feedback and significant influence by customers. IBM funded many early R&D projects through government contracts. These included the 701 (1951 Defense Calculator), 702 (1954 outgrowth of the 701), 704 (1955 using SAGE technology), 705 (1956 also using SAGE technology), 709 (1958 outgrowth of the 704), and the 7070/7090 (1959 from a US Air Force project). The other 79 American vendors of computers and other IT components in the 1950s more or less followed a similar strategy of relying extensively on government contracts.[77] The one major exception was IBM's Project STRETCH, which the company funded in the 1950s to develop a variety of new technologies, "to take a giant step" forward, and which in the 1950s and 1960s was viewed internally within the firm as a relatively expensive failure. However, many of the "nuts-and bolts" developments regarding components, architectures, and software were ported over to what became the company's most spectacular product in its century-long history, the S/360, one that also changed over time as a result of customer complaints (particularly about bugs in the operating system), compliments, and suggestions for changes.

But before examining the S/360, there is the matter of IBM's development of the disk drive, more widely known as DASD (Direct Access Storage Device). Ceruzzi is not an historian who waxes effusively in his descriptions of technology nor does he mince words, but he labeled the development of DASD as "computing triumphant," and for good reason.[78] From the days of the first computer, data that went into a machine, or was stored in a digital format, normally resided on 80-column punched cards using IBM equipment, while data centers increasingly kept data on magnetic tape by the late 1950s. To get to a piece of information on tape, one had to read all the information before the data they wanted. Disk storage was a true revolution because one could go directly to the data they wanted, unlike with tape, saving time. One could also update a file at the same or nearby spot with the computer remembering how to get to it again. Data were stored on what looked like huge CDs called disks, stacked into groups inside disk drives. IBM developed the first DASD product and began shipping it to customers in 1957, providing for the first "random access," which meant that now users could create applications and write software that allowed one to interact with data real time, just as one does today on their laptop when they reach out into Wikipedia or Google for a piece of information.[79] Until then one had to submit a request for information to a data center, usually in person, then an employee mounted a tape retrieved from a tape library onto a tape drive, read the data, printed it on cards or paper, and turn it over to the user. The first customer for IBM's DASD was an American firm, United Airlines in Denver, Colorado, for use with its passenger flight reservation system. It opened up possibilities of doing computing the way it has been done since the 1960s: online and in real time. While having undergone many changes in over a half century of existence, it remains a signature technology that helped profoundly in the diffusion of computing of all types around the world.

One would be hard pressed to find a more successful "high tech" product in American history than the IBM System 360 family of computers. Introduced in 1964, after a half decade of frightfully complicated and expensive development at a cost of some $7 billion dollars (in 1960s purchasing power), it remains a profoundly important innovation in modern history. IBM's System 360 is in the same class as the Ford Model T automobile or the birth control pill in its transformative effects. The S/360 consisted of five mainframes and a full compliment of peripheral equipment and software, totaling some 150 products when introduced on April 7. In one fell swoop, customers who had been using seven different lines of computers, all popular, many running out of capacity, and in need of replacement with bigger machines capable of processing more data faster, were now offered a set of products that were compatible with each other. Compatibility meant that a data center could begin using the product line starting with a small system (typically the Model 20) and as a company's or government agency's inventory of uses and volume of data processing increased, it could upgrade to a larger model without having to rewrite old software or replace with newer and faster tape drives, DASD, control units, and printers.

It is difficult today to imagine how profoundly important this upgradeable compatibility was for all concerned or how much that function fostered a massive expansion in the use of mainframes in all the developed economies. Prior to S/360, data processing centers moving from one machine to another needed to do a near total rewrite of all application software, often in different programming languages, and install a new operating system that also had to be learned, a combination of expenses that usually exceeded that of whatever new proposed system was going to cost either in money or disruption to business operations. So, migrating to a bigger computer usually did not make sound operational or economic sense unless absolutely necessary. The only other option was to install another copy of the currently installed computer system, which often led to increasing costs of additional operations staff, while blocking the opportunity of linking together applications and files resident on multiple systems, and often in the process acquiring more computing power than needed. The move to a new noncompatible system disrupted all the users of existing data-processing services, putting a company at risk of losing business, customers, or even entering bankruptcy. Data-processing managers wanted upgradable systems that were compatible. Computer scientists and engineers had been discussing how best to provide that for years.[80] IBM and other vendors had the high cost of maintaining and improving multiple technologies and product lines in a period when competition from other suppliers was increasing; so they all needed a breakthrough.[81]

The history of mainframes in the 1960s is largely the story of how vendors reacted to the S/360 as IBM went from just one of many vendors to the one holding a commanding position that, depending on whose numbers one uses, amounted to between 60 and 80 percent market share with customers installing machines in far greater quantity than before.[82] They were price-effective and included new functions, reliability in addition to compatibility, and financial results tell the story. IBM's revenues in 1964 (the last year before it collected rental fees for its new class of systems) amounted to $3.24 billion. In 1965, revenues climbed to $3.58 billion, and the following year to $4.25 billion. In 1967, with additional models and peripherals continuing to appear, revenues made a huge jump to $5.35 billion and in 1968, to $6.9 billion. It closed out the next year with $7.2 billion in revenues and the following year these increased to $7.5 billion. In 1964, users had some 11,000 IBM computers in their data centers. In 1970, that population of systems had grown to 35,000, nearly two-thirds of which where S/360s. One other statistic is most telling: 7,400 S/360s Model 20 were installed in the United States.[83] What makes this last piece of information insightful is that the Model 20 was an entry level S/360 (the smallest one in the collection), indicative that (a) migration to a system that would undoubtedly be upgraded were being installed and (b) that many first time users of computers (corporatewide or within a division of a firm) finally were embracing in-house computing. To put even those numbers into further context, in 1960 there were between 5,500 and 6,000 systems installed from all vendors; by the end of 1965 there were 22,000, again from all vendors; in 1970, that population had climbed to 70,000.[84] In short, an explosion in deployment had taken place, lifting all vendors' boats, not just IBM's.

A replacement product line called the S/370 had a similar effect on revenues. IBM ended the next decade extending to year-end 1980 with revenues of over $26 billion.[85] IBM's personnel grew in number as well. In the United States, in 1965 IBM employed 96,532 people, another third in other countries; in 1970, the American base had grown to 156,859, with 110,000 in other nations; it ended 1980 with a U.S. employee population of 194,423, with another third in other countries.[86]

That one core set of products—S/360 and its sequel the S/370—fundamentally took IBM and its customers to a new level of number, sophistication, size of IBM, and diffusion of computing. The distinguished business historian, Alfred D. Chandler, Jr. after a half century of studying various corporations and industries concluded that the S/360 was truly "extraordinary" and "immediately defined the computer industry worldwide."[87] Kenneth Flamm, in his study of American computing drew a similar conclusion, it (and its compatibility) "created a unified market that greatly stimulated the commercial use of computers," calculating that IBM won 70 percent of the market in the United States.[88]

Historians have not yet examined in sufficient detail the effect on diffusion beyond just the number of machines installed or of IBM's new products on competitors (they too introduced IBM-compatible systems but most left the market by the end of the 1960s), or on their customers. As these IBM systems began appearing around the world, whole data-processing staffs learned how to operate them, to use their software, and to forget rapidly how to run alternative systems. By the end of the 1960s, there were hundreds of thousands of people around the world who knew how to work with these systems, over half located in the United States. They wanted to do so as well, because it made them more marketable as they moved from one firm to another and, as they did, also displaced older systems not compatible with the S/360s, thereby demonstrating once again the power of technical standards in stimulating diffusion of a technology.[89] The issue of compatibility and inventory of "legacy systems" proved so important that even in the early years of the twenty-first century, despite many product changes and innovations, much of the core IBM operating systems remained largely compatible with what had been introduced in the 1960s, even with some of its component's names (operating systems for instance, MVS, OS) and architecture.[90]

Prior to S/360, to be sure it was a messy affair for IBM to merge systems, use pieces of Stretch, fixed software problems, and trained users, which is why almost every historian of the product and every memoirist of the firm spoke uniformly about how this was a "bet your company" play. IBMers who worked in the company several decades later still spoke of the experience almost the way veterans of warfare did: the tensions, long days that stretched into years, fear of failure, the challenge of solving problems never faced by humans before, motivation to help companies and agencies do positive things never before possible, over 200 weekends spent working in crisis mode, and so forth. If IBM had failed to produce a successful product after its massive financial and emotional investment over the course of nearly a decade to build and make it work, this company would not have emerged as the giant it became. Quite possibly, it might have disappeared as an independent firm by the end of the 1960s or 1970s, due to the enormous expense in developing this system, much the way ERA and EEMC had when they folded into Remington Rand. But that did not happen, with the result that its technological success led to the substantial increased use of computing in all industrialized economies. Not until a similar wave of innovation came in the form of IBM's personal computer standard in the 1980s did another surge in diffusion occur of a technology of comparable significance; one surprisingly repeated yet again by a plethora of vendors and hundreds of millions of users with the cumulative adoption of Internet, beginning largely in the 1990s.[91]

CONTRIBUTION OF COMPUTER VENDORS TO DIFFUSION

Inventing important hardware and software in the United States alone was not enough to promote the fast diffusion of computing that took place. Without effective vendors the technology would never have been developed or deployed. Many historians and defenders of IBM and most of its critics, while acknowledging that it proved to be the most successful American firm in the industry during the years of Wave One, have mostly argued that it was not always the technological innovations that made computers attractive, but good (or aggressive) marketing and salesmanship. Thomas J. Watson, Jr., president of the company when it transitioned into the computer business in the 1950s and throughout the years of the 1960s, reflected the widely held belief in IBM, recognized as far back as when the firm had to fight UNIVAC competition, that "we constantly outsold people who had better technology because we knew how to put the story before the customers, how to install the machines successfully, and how to hang on to the customers once we had them."[92] Other sales forces believed the same as well, such as at Burroughs who shared a common style of selling with IBM in addressing its competition in the 1950s and 1960s. Historian Arthur Norberg criticized management at Remington Rand and EEMC for failing to grow their businesses with the industry, precisely because it did not have the sales and managerial skills to promote effectively its products and the general use of computers.[93] It is still a lesson not lost on executives who run companies like IBM, Apple, Microsoft, and Intel in the twenty-first century. No technology sold itself, despite its appeal, so any discussion of diffusion of IT in any country must account for the role of vendors.[94]

There were always large numbers of them in the computer industry in the United States selling systems, peripheral equipment, software, or components for other products. In the 1950s and 1960s these came from three sources: old-line office appliance firms (such as IBM, Remington Rand/Sperry,

NCR, Burroughs), existing electronics and consumer electronics enterprises (such as General Electric and RCA), and new companies, particularly in semi-conductors, of course, since ICs comprised a completely new line of products in the 1950s and 1960s, from Fairchild to Intel. Not until the 1970s did new firms operate in the American economy that sold large mainframes (Amdahl) and, of course, again a new set of products—midrange computers—of which there were dozens of vendors (DEC or Wang, for example), not to mention hundreds of other suppliers selling peripheral equipment and a great deal of software. The American industry has been extensively studied for the entire second half of the century, making it possible to summarize with confidence some of its key diffusionary activities. By the early 1960s, a handful of vendors controlled less than 5 percent of the large computer market in the United States, while IBM, which had surged ahead during the second half of the 1950s, enjoyed a market share in the early 1960s of about 60 percent, a margin that by the end of the decade probably exceeded three-fourths.[95] Because of its domi-nance, its competitors had been nicknamed the "seven dwarfs" by newspaper reporters, a pejorative term also used by customers and industry experts. These comprised Burroughs, Sperry Rand (into which ERA and EEMC had folded), NCR, RCA, General Electric, and newcomers Control Data Corporation (CDC—based out of Minneapolis) and Honeywell (also based in Minneapolis).[96]

Because IBM dominated the market, a few additional comments are in order about its role in diffusing computing. In the 1950s and 1960s, IBM had over 100 sales offices located in major American cities, ranging in size from a score of salesmen—and in those days they were all men—along with com-puter maintenance field engineers in larger offices that could each have over one hundred employees. Sales personnel were assigned to specific accounts, were well trained in selling techniques, "managing" accounts, and, of course, informed properly about the products. They established ongoing relations with data-processing managers, became embedded in the communities in which they lived as members of the local elite social structures.[97] Other vendors had a similar branch office coverage strategy as well, but not as extensively as IBM. Having that infrastructure across the country, populated with local sales people, who had established relations with all the key firms and government agencies, and who talked with them every week about business issues and the potential uses of IT, played an enormous role in teaching potential users on how best to leverage computing in their daily work. Indeed, they followed suc-cessfully the same practice in over a dozen countries. Despite all the grand strategies of the various vendors—including IBM—ultimately as Watson noted, this network of employees persuaded a great number of managers to use computers, teaching them which ones to install and use, and why within a context that made sense to the firm and the managers and technicians within it. In the case of RCA, product strategy proved important too, because once the S/360 standards had won the day, it introduced less expensive, look-alike computers as well that were "IBM compatible," which is to say, IBM's software operating system and application software (most of which were sold by other vendors or were written by programmers working for IBM's customers) designed to run on S/360s and S/370s ran on these machines. But RCA failed for essentially two reasons: IBM's highly effective sales strategy and performance, and second, the rapid introduction of new models all through the 1960s and 1970s, the rapidity of the latter which also devastated European rivals slower to market with new offerings (a story discussed later in the book).

Users of IBM's computers have long received short shrift by students of the American computer industry, but ultimately they were the people who made decisions to acquire computers and spread their use within their own enterprises. In recent years, we have begun to learn more about their important role. Scholars such as Tom Haigh and Nathan Ensmenger have started examining the role of this IT community.[98] Already suggested in this chapter, once skilled in the use of IBM's versions of IT, a technician became a marketable savvy resource that could move from one firm to another and *de facto* promoted the only thing they knew, IBM technologies and IBM's ways of doing things. That community grew in size as IBM's market share expanded, creating tens of thousands of programmers, systems analysts, computer operators, data-input clerks, and data-processing managers knowledgeable about IBM's products, indeed often only about IBM's systems. That population expanded into the millions by the end of the 1970s. User groups had been formed in the 1950s, most notable SHARE and GUIDE, which held annual conferences where users of IBM's products discussed how best to deploy these and debated the costs and benefits involved. They also discussed best practices, and heard IBMers talk about the company's products. These communities of users reinforced amongst themselves the wisdom of using IBM's equipment and software, and helped each other with new implementations. Not infrequently, they recruited IBM personnel away from the firm (already known as "Big Blue") to work for them. In fact, by the early 1970s, it had become customary for IBM sales personnel to track who these people were in their accounts, because IBM salesmen viewed them essentially as potential allies, individuals sympathetic to IBM's offerings, and who could serve as "inside salesmen."[99] It is quite probable that someday an historian will convincingly argue the case that all users of the company's products—IBM called them customers—did more to promote the diffusion of computers than even IBM's own salesmen.

Observers of the industry noted a number of activities on the part of customers that facilitated rapid diffusion of IT in the American economy. There were always a few enterprises willing to be early adopters of a new product, because they either had a data-processing manager who wanted to be early or due to an immediate need satisfied by the availability of a new application or product. Most adopters waited to see if the earlier ones proved successful, learning often about these successes at user conferences or through the industry press (most notably *Datamation* and *Computerworld*), before embracing a new computer or use. But when they did, it happened quickly, often within 24–36 months of a new product being delivered. Positive reports by trade publications within one's own industry were highly influential and appeared frequently.[100] Modifications of early versions of a use of computing facilitated greater deployment by other firms, meeting their tailored needs. Uses came in waves, almost as a fashion. For example, users started to convert batch systems to online versions by the end of the 1960s, using DASD, online terminals, and new software. Because of tax incentives and depreciation schedules, among others, economic and accounting reasons led to the practice of swapping out old hardware and software on a regular basis, usually in long spurts of 4–7 years, a pattern in evidence by the end of the 1960s as well. Already in the 1960s, but more so in later years, work transformed through the use of computers made it difficult, if not impossible, to return to an earlier way of doing business. That reality, in turn, led to additional uses of computing that built on preexisting applications of IT. What they were used for also came in waves: data collection and accounting in the 1950s, inventory control and manufacturing applications in the 1960s.[101]

Using sales revenues of machines as a rough surrogate measure of diffusion, we can see how active American customers were in these years. Already in the late 1950s total expenditures on the technology amounted to millions of dollars. Yet by the end of 1968, these exceeded annually $5 billion. Earlier I characterized the 1950s as the startup of the commercial and governmental uses of computing with a significant uptick in the second half of the 1950s. We can say that the 1960s represented momentum building upward at a pace that exceeded that of the 1950s and the 1970s (when an economic recession in the early part of the decade slowed acquisitions), but nonetheless continued. In fact, to carry the evidence forward in support of the notion of momentum, expenditures more than doubled again by the end of 1978, and during the 1980s actually surged still higher, with all manner of hardware generating nearly $25 billion a year in revenue during the second half of that decade.[102] Right through the middle of the 1970s, the lion's share of the revenue came from mainframes, their peripheral equipment, and software. Not until the second half of that decade did revenues from minicomputers reach appreciably measurable levels; the same held for microcomputers (PCs), which, in the 1980s, individuals and enterprises adopted at a faster rate than firms had mainframes in the 1950s through the 1970s.[103]

Users came from all sectors of the American economy. Economist Daniel E. Sichel documented this breadth of participation by so many sectors. During the first three decades of deployment (1950s–1970s), the leading adopters were goods-producing industries—manufacturing firms to a large extent—often consuming half the available inventory. That percentage began declining slowly as other parts of the economy also embraced computing, with transportation and public utilities, for example, consuming about 17 percent of all manner of IT in the early 1950s, 13 percent in the early 1960s, and by the start of the 1970s, closer to 7 percent. Wholesale and retail trade increased their absorption decade over decade, while as a relative percent of consumption governments remained constant.[104] One should keep in mind that these percentages were dollar volumes that increased rapidly all through these years.[105]

One other way of examining American diffusion is by looking at the percentage of expenditures for all manner of data processing hardware by who paid for this class of technology. Based on extant evidence we can conclude that by the late 1940s, it was nearly 100 percent by government, yet by the end of 1958, just over 90 percent by businesses. Government expenditures as a percent of total grew from 1 percent in 1958 to 6 percent in 1967, then leveled out to 4 percent over the next few years; in contrast, business as a whole contributed 90 percent in 1963; then over the next decade hovered at about 84 percent of the total. The rest of the expenditures in all decades came from miscellaneous groups, such as consumers of office appliances (for example, typewriters) and non-profit organizations.[106] Many hands pulled the oars; no single sector, industry, vendor, or government investment program made that happen. Diffusion was broad as measured by business sector, region, number of users, and dollars expended on IT, and so it is to the broader story to which we next turn our attention.

HOW AMERICANS MET AND EMBRACED COMPUTERS

Part of this issue has already been addressed: vendors introduced specific machines and uses to people in a position to acquire them; over time data-processing staffs learned about IT, then in turn advocated use of computing where they worked; universities and governments developed the field of computer science and developed continuously new technologies, largely funded by federal agencies; and briefly, we suggested the enormously important role of user groups, such as SHARE and GUIDE for IBM users, but others existed using equipment from various vendors. There were other ways as well in which Americans learned about computing that seeded demand and, thus, facilitated diffusion worth recognizing.

Industry and trade associations provided much important information. Almost every industry in the United States had at least one trade association. These were largely an American business phenomenon, dating back to the nineteenth century, but which mostly came into existence in many industries in the years following World War I. They provided a forum for discussing industry-specific issues, held conventions for social networking, exhibited new products and methods, and trained members of their industry. They published weekly or monthly newspapers or magazines about their issues. Almost without exception, these associations supported and promoted use of all new technologies and methods of operation that appeared in the twentieth century, and so it was with computers. For many members of an industry, their initial introduction to a particular computer, software package, use, or managerial best practices related to IT came from such an organization, which explained their relevance in their specific industry's terms, role of computers for banking, others about shop floor data collection in manufacturing, and so forth. Furthermore, their endorsement of a particular computer, software, or application often proved to be the impetus for wide acceptance of a particular IT.

From the 1950s to the present the evidence abounds in the pages of their flagship publications and in the proceedings and agendas of its annual meetings. For example, the American Banking Association (ABA), APICS (now known as the Association for Operations) in manufacturing, and LOMA in insurance routinely, sometimes in most issues of their magazines, carried positive accounts about the use of computing. At their annual conferences these associations held training sessions and hosted presentations about what was new and working with respect to computers. In one study of how computers were used in roughly three dozen American industries between the 1950s and early 2000s, the history of each industry's use of IT rested largely on the publications of their appropriate association.[107] Americans and Canadians are very tribal, and a given that if one had a company or agency, it belonged to its industry's association. These groups could thus reach out to the majority of their industry's members in a routine, relatively low cost, methodical way to promote data processing's benefits, a remarkable feature of the North American scene, one yet to be studied in depth by business historians.

The cross-industry business literature of the day also routinely discussed computing in positive terms. This practice applied to every major publication one could select: *Business Week, Dun's Review, Fortune, U.S. News & World Report, Business Horizons, Harvard Business Review*, and even more general publications, such as *New York Times Magazine, Newsweek, Time*, and *Reader's Digest*. In some instances, subscriptions exceeded a million, most in this period hundreds of thousands and then, as today, more people read a journal than subscribed to it so one can assume an untold additional number of readers were introduced to computing in this period, particularly within a department of an enterprise where it was common to have one shared subscription which circulated around the office in a prescribed routine.[108]

Additionally, there were book publishers. We do not yet know fully how many books they published about computers in the United States between the late 1940s and the start of the twenty-first century, let alone around the world; however, two facts stand out: first, more books were originally published in the United States on computing in all decades than in all other countries put together between the 1940s and the present; second, in the years between the late 1940s and the end of the 1970s, the number of books that appeared on all aspects of computing technology, user guides, applications, management, history, economics, and social effects exceeded 10,000.[109] All the major trade publishers of business and scientific books participated. It was not uncommon for a publisher to offer series of books on computing, as did Addison-Wesley, Harper & Brothers, John Wiley & Sons, McGraw-Hill, Prentice-Hall, and Reinhold Publishing Company. All of these (or their present corporate incarnations) continued to publish books on computing. Along with thousands of publications from computer vendors—prior to the 1990s IBM was considered the world's largest publisher, outpacing even the US Government Printing Office—equipment suppliers were the main source of technical literature on how computers worked, what they should be used for, and their management, including an endless supply of "how to" books.[110] More than even university presses, commercial publishers and some vendors also published the vast majority of the computer science books that appeared in these years, most notably Prentice-Hall and John Wiley & Sons.[111] These books, like journals, often presented case studies and glowing reports of successful new uses of computing.

On occasion, but not frequently, a "best seller" volume appeared that drew larger public attention that stimulated interest. Two important examples appeared at the dawn of the commercial computer. Edmund C. Berkeley authored the first one, *Giant Brains or Machines That Think* (1949), published by John Wiley & Sons.[112] The title delivers an obvious message popular at the time, that computers were a new tool becoming as smart as people. The second major publication came from a business management consultant, John Diebold, *Automation: The Advent of the Automatic Factory* (1952), published by D. Van Nostrand Company in Princeton, New Jersey, at the time a hotbed of early computing projects at nearby Princeton University and other local organizations.[113] Diebold's book had the additional distinction of being the first volume to publish the word *automation*, although the phrase had been used in conversations (at least) at Ford Motor Company since the late 1940s. Diebold's book was aimed squarely at the business manager, the first of seven he wrote over the next quarter century introducing computing to the business and public sector communities.[114] The majority of the best seller volumes one thinks of in computing appeared initially in the United States and, most notably, beginning in the 1970s, although their authors began thinking about, commenting on, and writing concerning computing in the 1960s, or earlier. The most obvious example for Americans was Alvin Toffler's *Future Shock* (1970) published by one of the largest, most distinguished trade publishers, Random House. Until at least the 1980s, it remained the single best selling book on computing and the future of society, exposing hundreds of thousands of potential users to the technology while suggesting how it would change society, largely in positive ways. Toffler continued to write such books on a regular basis, publishing one roughly every decade into the early years of the new century. Not discussed in our book is sci-fi literature, which proved very popular all through the period in the United States. While this later genre of thousands of publications focused largely on nuclear energy, atomic wars, and encounters with aliens from other planets, by the 1960s these also included fantastic stories about the role of computers, a theme they frequently came back to for decades.[115]

Finally, we should acknowledge the power of electronic media all through this period. No one event more emphatically introduced Americans to computers and led to serious consideration about the benefits of their use in many circles than what happened during the national presidential elections of 1952. CBS News used a UNIVAC I on election night, November 4, 1952, on live television to predict the outcome of the election. On the basis of early returns from 27 states (3.4 million voters of an expected 60 million) at 8:30 P.M. eastern standard time, the computer "predicted" the outcome correctly to within four electoral votes. The impression left with people proved to be extraordinary; smart brains had arrived in American society.[116] It is difficult today to imagine the impact of that event because it is routine to model election outcomes, but at the time IBM's sales offices, for example, saw a surge of queries by potential customers for information about computing.[117] In hindsight we know that it was one of those events that helped Americans to anticipate the impending role of this technology; reinforced in subsequent years by movies, novels, and the first arrival of humans on the Moon in 1969 in a space ship the public was told navigated under the control of computers.[118]

FROM DIVERSIFYING TECHNOLOGY TO DIFFUSED DEPLOYMENT, COMPUTING IN THE 1980S

The decade of the 1980s proved as remarkable for the diffusion of computing in the United States as any before or since. New forms of computing emerged, along with novel uses and additional users, further setting in place a pattern of use and adoption that characterized how Americans relied on computing right into the next century. Four patterns of use epitomized IT in this remarkable decade. Each in time was mirrored elsewhere around the world in different forms and at varying rates of adoption as well.

First, with the initial availability of minicomputers in the 1970s now well in place, their diffusion across many industries continued the process of distributing data processing to engineering departments in manufacturing companies, to shop floors in all kinds of manufacturing and services, and into stores and warehouses as well. Linked to mainframes by way of dial-up or dedicated telephone lines, these machines extended the reliance of small and large enterprises on computing and to each other.[119] New telecommunications equipment and software continued to appear on the market, making distributed computing more sophisticated as well. Cisco was founded in 1984 while IBM in 1986 introduced its token-ring technology for controlling LAN (local area network) traffic to manage the activities among mainframes and remote systems of different sizes.[120] Distributed computing took two forms. The first encompassed distribution of data processing within an enterprise or public institution to departments or various physical sites, localities away from the location of the centralized IT departments, such as to factories or bank branches. A second mode, largely using PCs, involved the use of IT in many large and small communities across the nation by individuals either as part of their work or for personal reasons, such as for games and home financial management.

I have described elsewhere the most remarkable innovation in the use of commercial computing that occurred in the 1980s, which can be briefly summarized here. It is a story that ties directly to the physically dispersed use of computing. In the 1980s, companies linked together their various suppliers and distributors into large integrated networks to speed up the manufacture, movement, and sale of goods, using private networks and computers to such an extent that by the end of the 1980s one could begin to speak about supply chains as transnational networks of business activities. Thus, for example, a large automotive manufacturer could link together all its suppliers of components and subassemblies around the world to such an extent that manufacturing plants required their suppliers of components, such as, for example, car wheels, to have a certain number and style delivered on the shop floor each shift, leaving the responsibility to determine how to get that done to the supplier who would have access to the production schedule of the factory in question. Or, a large retailer of women's clothing could link to its suppliers in Asia and Israel so that as inventory of a particular product, say a blue dress declined, suppliers in other countries would know to ship more to the United States right away. Known in business circles as "just-in-time" inventory control, the concept spread to all manner of commercial activities, engaging both large and small enterprises.[121] When observers of the world's economy began speaking about globalization—and that most certainly became quite pronounced in this decade—it was the linkage of computing and networking on a global basis that made this process possible, and it occurred in barely a decade. That use of computing by American firms drew in other companies (suppliers and customers) in many countries, thereby extending the process first evident in the early 1960s of American companies introducing digital applications into other nations.[122]

A second development was diffusion of the personal computer, which occurred first and most intently in the United States during this decade. While Apple and a few other firms had introduced the earliest PC-like products in the late 1970s, it was not until the 1980s that diffusion took off on a near large-scale basis. IBM's introduction of a microcomputer in 1981, which it named the Personal Computer (or PC), had the effect of giving Corporate America "permission" to use this class of technology, the story often told by historians and contemporary observers.[123] However, in the previous several years it had become increasingly evident to business users that spreadsheet software first, then word processing products next, could extend computing to the individual, and equally important, to ever-smaller enterprises that previously could not afford expensive mainframes and minicomputers.[124] They did not wait for IBM to legitimize use of this class of computers in business. During the 1980s, Americans acquired tens of millions of machines; dozens of vendors of these devices came and went and, in the process, a modern icon of American business came into full flower, Microsoft. This new firm provided the operating system for more than 90 percent of all PCs by the end of the decade; Apple dominated much of the remaining PC software operating systems' market.[125] Initially, Americans acquired stand-alone PCs, that is to say, they were not connected to networks (roughly 1977–1984), followed by a second era that extended to 1994, when increasingly they bought machines that could communicate over telephone lines to mainframes to access data, and to each other, and in the process extending use of e-mail outside the work walls of academia, government, and large corporations.

After 1994, they moved rapidly to the Internet, a story more a part of Wave Two's diffusion of computing that occurred in the 1990s.[126] In the first period hobbyists and gamers bought mostly Apples and now moribund systems, while business users IBM and other versions. In the 1980s and early 1990s, uses extended substantially to schools, self-study, further extensive deployment of spreadsheets, and myriad specialized applications, such as recipes. By 1984, 8 percent of all US households had PCs (roughly 7 million) and by the early 1990s, that number had tripled.[127] One could personally use (hence control) computing for entertainment and self-improvement, and to get things done (such as to prepare tax returns). Online services also came into their own to feed this market that now sought information from commercial databases. It was in this period, for example, that Compuserve, Prodigy, and America Online began operations. By the late 1980s, everything was in place for the transition to the Internet that would occur in the next decade when it became a user-friendlier technology to use than dial-up telephone networks and private databases. People who had PCs linked together and to sources of data over telephone lines. Individuals knew about the technology and appreciated their value as "tools for Modern Times," to quote an IBM advertising tagline of the time.[128] They appreciated this value despite their often frustrating experiences in using these early systems that mimicked the kinds of difficulties faced by early users of the automobile in the early 1900s.[129]

A third trend, and one that tends to be overlooked by observers of how the PC spread, was the emergence of powerful desktop computers that were physically as small as PCs, but had the power of minicomputers at far less cost. Their arrival essentially destroyed demand for many classes of minicomputers by the early 1990s. For example, in 1981 Apollo Computer brought to market the DN100, one of the first such workstations, which helped engineers and scientists who needed inexpensive, fast, muscular computing power. In 1982, one of the suppliers that in time dominated large swaths of this new class of computing came into existence, Sun Microsystems, which that year introduced its first product. In that same year H-P did the same as did DEC in 1984. Over the course of the 1980s, newer more powerful versions appeared from various suppliers, all with the ability to operate independently or in some networked fashion. Between these devices and PCs, Americans now had access to computing in a very rich variety of forms, prices, and functions. It was not uncommon in many American corporations in the late 1980s to have as much computing power sitting on the desks of their employees as they did ensconced in mainframes controlled by data processing staffs, all interconnected through various forms of telecommunications networks. The diffusion of IT outside the "glass house" reached such a level in the 1980s that MIS executives began complaining that they no longer had mastery over the use of computing in their corporations that they had been held responsible for during the previous three decades.

Fourth, and to a point made earlier in this chapter, computer chips—basic building blocks of computers—became more embedded in various noncomputational devices than ever before. In fact, that was so much the case that after two decades, historians do not yet have a complete picture of where all these chips went, despite many industry surveys and contemporary economic studies. But here is what we know. Using 1978 as a base year, when sales rose sharply around the world thanks to yet another new generation of chips making their way into the world market, 56 percent of all integrated circuits sold in the United States went into computers manufactured in the United States, another 9 percent into telecommunications equipment, and 11 percent into various industrial machines. Just five years later (1983) the proportion going into computers had dropped to 48 percent, now 13 percent into telecommunications equipment, and 9 percent into testing and other industrial machines. Consumer products absorbed 9 percent in 1978 and 11 percent in 1983, suggesting that devices other than PCs had already started to use computer chips, such as microwave ovens and automobile engines. Regarding automotive users, in 1978 vehicle manufacturers consumed 2 percent of all chips but by 1983, 6 percent. In this 5-year period, government uses accounted for about 13 percent.[130] During the same period the quantities of chips consumed tripled as well, so the numbers consumed behind each percentage had grown too. To put all these statistics into some global context, by 1983, Americans were consuming about 38 percent of the world's supply of chips, Japan 27 percent, Western Europe 26 percent, and everyone else just over 8 percent.[131] All through the 1980s, the percentages varied somewhat while the actual number of computer chips consumed rose sharply around the world, with the Americans using the largest amount. The purpose of presenting all these data is to suggest that diffusion seeped deeply into many corners of the American economy and around the world into "high-tech" industries, such as transportation and industrial equipment. Thus, chips serve us as useful digital footprints to track where computing capabilities went in this or any other society.

One could argue that there actually was a fifth development in the 1980s, that of the diffusion of digital games and by the end of the decade a sixth, portable telephones. The spread of mobile telephones in the United States remained quite minimal in comparison to diffusion in Western Europe; it proved to be an important American story in the 1990s and beyond. However, games became more digitized in the 1980s, with memorable examples in the form of PONG and Pacman, and Nintendo and Gameboy devices.[132] This development introduced some children to computing beyond what they might have experienced with their parent's PCs at home or their Apples in a few schools. However, the volumes of games shipped, while the highest in the world in the 1980s occurring in the United States, were low, and really a phenomenon of enormous importance in the 1990s. In fact, when observers of the American scene wrote in the 1990s and later about children growing up comfortable with digital devices, they referred largely to their experiences with games and PCs.[133]

ECONOMIC EXPLANATIONS FOR DEPLOYMENT OF COMPUTERS

Economists, public officials, and members of the computer industry in many countries have displayed a growing interest in the economic effects of computing in technologically advanced societies since at least the 1970s in the United States, and beginning largely in the early 1980s in Asia and Europe. By the late 1990s, those economists and others interested in national economic development and on the role of technologies were writing extensively about computing and its diffusion, attempting to measure its impact on individual national economies from nations with high GDPs to the poorest. The intellectual epicenter of the discussion was the "productivity paradox" in which economists in the United States were asking: If computers have been so useful (read, productive) to companies using them, why don't we see the productivity gains reflected in traditional national economic productivity data (read, labor productivity improvements)? They began noticing a disconnection in the national data from the rhetoric of computer vendors and their customers as early as the 1970s, with traditional economic statistics showing no effects, yet users proclaiming and documenting productivity gains, hence the paradox. By the early 2000s, they had enough measures and sources of data to conclude that a sufficient amount of IT had been installed by the 1990s to start appearing in both new and traditional economic measures.[134]

But much earlier, certainly by the 1980s, officials and economists were accepting the notion that part of how one modernizes an economy was by injecting both greater use of computing in all manner of business and governmental operations and also by fostering development of local specialized IT and telecommunications industries, such as officials did with semiconductor manufacturing in South Korea, advanced electronics in Japan, and software in India. Only large, advanced economies could expand the scope of their IT industries, such as the United States, while smaller ones that tried, often failed miserably, as occurred in the Communist German Democratic Republic (GDR), but not all (for example, Sweden and Taiwan). By the early to mid-1990s, a central feature of almost every nation's economic development strategy included a heavy dose of IT.[135] Inspirations for such activities were the activities unfolding in the United States, the source of the earliest widespread experience with computing. The orthodoxy about the value of IT in economic activities dates to the 1980s; certainly the productivity paradox conversation did. But, what have the economists learned about the American experience covering the years from the 1940s through to at least the early 1970s, before the conversation about productivity paradoxes began?

Part of the answer can be dispensed with quickly, because it has already been discussed in chapter one, the notion that the cost of computing was so low that it proved to be a competitive class of investments for productivity improvement superior to those which could be made in labor or with other technologies. The conventional economic storyline holds further that the cost of computing dropped dramatically during the second half of the twentieth century. The data in chapter one on costs of computer chips, discussion about Moore's Law, and so forth spoke to that issue. In that logic, as computing became less expensive, more people adopted it, particularly as the technology became more reliable as well (for example, IBM's 650 and S/360s and S370s). That story, and the specific data on unit costs of semiconductors, and so forth, has been extensively documented, indeed it dominates the narrative presented by most economists and historians. The growth of the new IT industries have also been the subject of much discussion and research, although more for the period late in this first wave (post early 1970s), when there were millions of individuals working for IT companies or with the technology in other firms.[136] But, what do we know about the economic impact of computing in the early years—the demand side of the narrative—from the 1940s to the early 1970s, when economists and public officials around the world began to wake up to the notion that maybe computing did affect national economies and specific industries?

Very basic to our understanding of diffusion is an appreciation of what IT people acquired, specifically in the first three to four decades of the existence of computers when only enterprises and large public institutions could afford their cost and had the need for such technologies. Extant evidence demonstrates that they voted with their pocketbooks, spending an enormous amount on computing. Table 2.8 documents the total invested in hardware and software

Table 2.8 PRIVATE FIXED INVESTMENTS IN HARDWARE, SOFTWARE
AND COMMUNICATIONS EQUIPMENT, 1960–2010, SELECTED YEARS

(Billions of dollars)

Year	Combined Computing	Communications	Combined Total
1960	0.3	2.3	2.6
1970	5.0	6.5	11.6
1980	22.3	24.3	46.6
1990	86.3	45.2	131.5
2000	287.7	123.3	411.0
2010	351.8	83.8	435.6

Source: U.S. Bureau of Economic Analysis, "National Economic Accounts," Data from Table 5.5.5 "Private Fixed Investment in Equipment and Software by Type," http://www.bea.gov/national/nipaweb/TableView. asp?SelectedTable=157&ViewSeries=NO (last accessed 11/1/2011).

over many decades. For a more comprehensive view of investments, the table also includes expenditures on telecommunications hardware because computing became increasingly online, networked, and dispersed, beginning in the 1960s and extending substantially all through the 1970s and beyond. The table provides a sense of the magnitude of what was involved. To have a complete picture of the costs of computing in this period one should add between 100 and 150 percent additional costs to cover the expense of staffs and data centers in which all this technology resided. The data tell us several things. First, it is quite clear that in the beginning enterprises and agencies spent more on telephone and telegraph systems than on computers, which makes sense because the former were around for nearly a century and the latter barely 20 years. Second, as the two classes of technologies either improved (computers) then merged with ICT (Communications), there occurred a massive and rapid takeoff in adoption of computing, which each decade outdid the previous 10 years by multiples.[137] Hidden in the data is yet more evidence of adoption because the dollar amounts are actual expenditures, not recalibrated to one year's standard to account for inflation (which proved quite extensive in the 1970s and early 1980s). All through these decades, the cost of computing and communications equipment dropped, in many years on average by roughly 15 percent or more, washing away any statistical masquerading that inflationary effects might otherwise have; indeed, the increased price performance accelerated much faster than inflation. We can thus accept at face value that Americans invested massively in this technology.[138] Or, to put it in the language of one economist who studied these, expenditures "increased much more rapidly than real gross domestic product."[139]

Use of economist Joseph Schumpeter's perspective on technological innovations and his famed "creative destruction" notions that he had articulated in the years before World War II did not influence economic thinking about computing until late in the century. So, to apply them retrospectively in some manner to the second half of the twentieth century feels like an anachronistic exercise, but a very useful one for understanding so much growth in the adoption of this technology. His explanation has in hindsight proven useful to historians looking at the broader theme of how technologies emerged, displaced others, and brought new sources of value to an economy. Using that framework one sees in the American experience new technologies emerging (such as transistors then ICs in the 1940s–1960s), many new IT firms (such as ERA and EEMC), and the transformation of older ones (such as IBM and GE) in the 1950s–1960s, all generating new economic value in the economy as measured by revenues, employment of vendors and users, creation of new patents and products, additional ones in the 1970s and 1980s (Microsoft, Compaq), and yet others in the 1990s (Google.com). These were always measurable even in those early years.[140]

Less measurable, yet part of the story, was the development of new skills and innovations in the way work was done. Research on the use of computing in the United States demonstrated that in some industries in the 1950s and beyond reliance on computing proved already so extensive that the Schumpeterian features of innovation and transformed behavior of firms were observable, most notably in various manufacturing, transportation, banking, insurance, and brokerage industries, and in large, complex public sector work, such as naval and air defense, census, and other statistical activities, and in tax collection and accounting.[141] But, at the same time, the presence of computing proved less in other industries of this period, most notably in all the media and information handling industries, higher education, primary and secondary education, and many services industries.

In one of the earliest economic studies of the American service economy, led by economist Victor R. Fuchs for the National Bureau of Economic Research in the 1960s, there is no mention of computers, which suggests that the technology was not yet visible enough, that economists could not yet measure its effects, or that IT was simply clustered extensively in manufacturing, financial, and some public sector corners of the economy at that time.[142] In fairness to Fuchs, even in the early 2000s, when economists had access to an enormous amount of information about the role of IT on dozens of industries, they were still complaining that "concepts and data are still inadequate," that "anomalies need to be resolved," and ultimately that "there are still significant measurement problems."[143]

However, as Kenneth Flamm and others have long argued, governmental intervention into the economy can—and did—stimulate development and deployment of various technologies that were perceived at the time as beneficial to the nation both economically and militarily. Economist Vernon W. Ruttan, using the language of economics, acknowledged the existence of this awareness, citing the example of the development of atomic energy: "intervention by a central authority to ensure that sufficient resources are devoted to the exploration of alternative technologies could reduce the probability of lock-in of an inferior technology."[144] So, in addition to creating an IT world in support of Cold War exigencies, and the spill-out of commercial products that built nicely on the use of earlier data processing, is Ruttan's point about keeping the flow of new technologies constantly coming into the Schumpeterian mix of innovations, something clearly evident in the American experience from the 1940s–1960s forward.

The economic issue more central to our study of diffusion is that economists only began to examine the dynamics of technological diffusions of all kinds in a serious way in the 1950s as the American experience with computers also started. The earliest area of interest demonstrated by economists concentrated on agriculture, which had undergone profound technological changes over the prior century, followed by their interest in some industrial and transportation transformations, such as the diffusion of new steel production methods and use of diesel locomotives. Economists like Ruttan and Sichel viewed the emergence of American computing in these years essentially through the lens of the supply side of this story. They largely hung their comments on the base technological artifacts themselves. Ruttan, for instance, spoke of "at least four generations, identified by reference to the technology of the basic components," ticking off the role of vacuum tubes, transistor-based computers (such as IBM's 7000 series introduced in 1959), integrated circuits in computers in the early 1970s, and finally very large-scale integration (VLSI) technologies in computers, first introduced in the 1970s.[145]

Overwhelmingly, economists who have looked at the early American experience focused on this supply side aspect of the story, and one economist reflecting on the work of his colleagues observed their "consistent underestimation of latent demand by even the most optimistic scientists, engineers, entrepreneurs, and managers who developed the computer, semiconductor, and software industries."[146] The point not to lose in this discussion is that underestimating latent demand was not just some failing of the economists, or even of the suppliers and inventors of the technologies, but also of its users in the period 1940s–1990s, and realistically one could say a pattern of thinking which extends to the present with respect to new byproducts of innovations. In other words, the act of underestimating rates of adoption occurs routinely in each decade during the early stages of the arrival of a new technology into the American economy.

Yet much activity occurs in such an early phase in the American economy that is documented as it unfolds. Economists chronicled the number of member firms in the various IT industries all through the period. Thus, we learned that by the mid-1960s, there were over 50 firms in the United States manufacturing semiconductors; that Western Electric and Bell Laboratories applied for 40 percent of all patents in the period awarded on this tech-nology.[147] Similar "nose counting" of suppliers of computers and peripheral equipment also hinted at the growing importance of this technology.[148] Econ-omists awarded primacy to semiconductors in the pantheon of early comput-ing's diffusion, linking its evolution in both product and industry to the overall attractiveness of American computing technology in the early years, and later to the national economic vibrancy of those nations that specialized in the manufacture and sale of ICs.[149] That linkage made sense. Thirty percent of all sales of semiconductors in the United States in 1960 went to computer manufacturers; another 50 percent to military and aerospace devices, which one could argue were largely also computational. That combined 80 percent shrank to 60 percent by the end of 1965, as ICs began appearing in such indus-trial products as process control machinery, data-collection and text-processing equipment, and myriad telecommunications and office devices, accounting for 26 percent of sales of semiconductors. By the end of 1972, percents had contin-ued some shifting: 27 percent to computers, 18 percent to consumer goods, 30 percent to industrial devices, and 25 percent to military and aerospace.[150]

To round out our description of their supply-side paradigm in the way economists looked at IT diffusion of the 1940s–early 1970s, one should call attention to the fact that they paid less attention to software as both a supply-side product and industry. This is actually quite explainable. While software was "invented" in the 1950s, its commercial manifestation in the form of revenue-generating products, firms, and service providers took off in the 1960s–1970s as a result of several events. First, the introduction of the S/360 facilitated—indeed encouraged by the new technological, economic, and marketing possibilities—independent software firms in developing prod-ucts that could be sold to much larger numbers of customers than might oth-erwise have been the case with earlier systems. Second, in 1969, IBM unbundled the delivery of software from its systems; in prior years most software that ran on its computers came "free" with the system, that is to say, its expense buried in the rental fees for the hardware. After 1969, other software vendors could offer products that competed with IBM's utility and application programs, for example. Third, in the late 1960s and more importantly in the 1970s, the ar-rival of a robust minicomputer business in the United States spawned many more software products from new suppliers that, in combination with the first two causes, led to the rapid emergence of a vibrant software industry in the United States that grew at about the same rate as the hardware business. The same occurred in a similar fashion in the 1980s with PCs. In short, and to use economic phrasing, the software industry in the United States had "disinte-grated" in the 1960s and diffused.[151]

The bulk of the dialogue among economists with respect to American eco-nomic growth and the role of computers concentrated on the post 1980 period, when both the volume of uses of computing by organizations and the explosive growth of an institutional and consumer market for PCs, then for other digital electronics, became so obvious, pervasive, and measurable. Yet, they do acknowledge that in the 1960s sales and uses of computing in the United States grew at a compounded rate of 20 percent per year, beginning in 1960; furthermore, this trend continued roughly unabated to the start of the 1980s.[152] Economists have just started to quantify patterns of economic performance for the 1940s or 1950s. One economist stated the obvious: "Until at least the early 1970s investment in computer and related information technology was so small, as a share of producers' durable equipment investment, that it could not be expected to have a significant impact on economic growth," a circumstance that, by the early 1990s, had completely reversed.[153] No economist has yet been able to quantify adequately the economic value of computing to the nation's economy of the 1950s–1980s. The closest to do so was a team of economists that defined total factor productivity from 1948 to 2002. For the years 1948 to 1973, which they lumped together, the best they could do was to suggest that IT contributed about 0.15 percent to capital productivity. To put that in some perspective, for the years 1973–1989, they pegged it at nearly 0.40 percent and in the years 1989–1995, at a half of a percent. Then came the large bump in productivity (1995–2002), which they calculated exceeded 0.90 percent.[154] Flamm was able to conclude that the national government's return on its investment in IT R&D in these early decades proved to be quite impressive, in the 50–70 percent range.[155]

Computing has often been set within the context of much larger economic developments, such as the general role of technological innovation in increasing a nation's productivity. Already by the late 1970s, a rough consensus had emerged among economists that technology in general improved productivity, and, therefore, one should expect the same to occur at some point with computers when they were sufficiently diffused. In other words, productivity was already increasing more due to the use of new technologies (not limited just to computers) more dramatically than as a result of implementing such earlier strategies for increasing productivity, as in investing additionally in capital or in improving the quality of labor. Table 2.9 provides a snapshot of how influential experts on American productivity viewed the relative value of technologies in general, not the least of which was IT. Already one could see the presaging of arguments that came by the end of the century that technologies had a far larger impact on modern economies than might have been thought perhaps when Schumpeter studied the issue in the 1930s and 1940s.

The general observations of the economic community, then, for the period 1940s–1980s, is that (a) technological developments occurred first on the demand side, induced by proactive substantial initiatives and investments by government; (b) subsequent rates of improvements in the function and price/performance of American products emerged out of the inherent characteristics of the technologies themselves; and (c) the ability of American suppliers to distribute these products to willing users so rapidly proved crucial to the

Table 2.9 HOW ECONOMISTS RANKED THE EFFECTS OF
TECHNOLOGY ON ECONOMIC PRODUCTIVITY AS PERCENT OF
EFFECT, CIRCA LATE 1970S

Economist	Technology	Capital	Labor Quality
Denison	62	20	18
Kendrick	72	18	10
Christensen & Jorgenson	44	42	14
Brookings	44	16	12*

*Does not total 100% because of other factors introduced into the productivity discussion (Better resource allocation 12%, Economies of scale 16%, Education 12%
Source: "Productivity and the Economy," *Bulletin of the Bureau of Labor Statistics* no. 1926 (1977): 63; see also *Spectrum* (October 1978): 46.

story. Spillover effects of this technology on other scientific and technological activities in the economy during these early years have not been studied; but, we can reasonably assume that while not absent, they did not play as significant a role in the 1940s–1970s as in the 1980s and beyond, particularly in such other fields as astronomy, biology, pharmaceutical research (now usually called biotechnology), and medicine.

Even before, and while, the economists were passing judgment on the effects of computing, participants in the American economy continued their investment in IT. In 2000, American business and consumer expenditures in IT equipment combined amounted to $135.9 billion, and if we step over the decline in such investments during the recessionary years of 2001–2003, the investments picked up again, with business expenditures going from a low of $80.3 billion in 2004 to $97.3 billion in 2007. Consumers went from spending $39.3 billion in 2004 to $5.0 billion in 2007—substantial volumes by any measure.[156]

IMPLICATIONS FOR THE UNITED STATES AND THE REST OF THE WORLD

The biggest initial wave in the development and deployment of computing occurred in the United States. As demonstrated in this chapter, this occurred directly because of the massive investments made by the American government in the development of the technology in response to the exigencies of World War II and the long Cold War that followed. By relying on academics and existing office appliance and electronics firms, knowledge about computing diffused into the larger economy and permitted business opportunities to create a vibrant and rapidly developing commercial computer market by the late 1950s, building on the emergence of a proto-computer industry as early as two to three years after the end of World War II. These commercial developments, coupled to a preexisting infrastructure that made the sale of IT an important export business dating back to the decade of the 1890s, facilitated diffusion of American computing products to other countries. This diffusion occurred at such a speed that American firms were able to overtake local embryonic computer rivals, many of which also had the added burdens of emerging out of the wrecked economies from World War II and functioning in national markets that were tiny in comparison to that of the United States. Back in the United States, the expanding community of American experts on IT, coupled with the massive scale of some industries and their key enterprises, made adoption of computing compelling. This adoption occurred so fast that we can speak of a definable Wave One as occurring in the United States that kept rolling into the early 1990s.

The rest of the world began to experience its first wave later, beginning largely in the mid- to late 1950s and for many realistically in the early 1960s, often with some features different from those of the American experience. While Western Europeans and some Asian nations were entering their first wave of adoptions, the Americans were, in effect, filling in their gaps of adoption across various industries, much along the lines described by such observers as Flamm and Perez. High drama occurred in the 1960s with important technological innovations; the mundane work of installing and using the new technologies consumed all the energies of the IT community in the United States in the 1970s and 1980s, similarly in Western Europe, Japan, and South Korea in the 1970s, 1980s and 1990s.

Fast forwarding to the early 2000s, when the effects of these investments in IT, along with the use of the Internet, were becoming clear, one can identify these as early hints of the possible developments in many other countries. The central issue for users, managers, vendors, and public officials was the impact of IT on labor. Jobs polarized, that is to say, some were automated or exported in most industries with the result that low-end jobs began disappearing while those requiring technical skills, strong cognitive abilities, and post secondary education rose. Routine work was increasingly automated, first in the office and the shop floor of a manufacturing industry, later across service industries too. The highly skilled workers in this new environment saw their wages rise while those for low-skilled labor declined in most countries that had become extensive adopters of computing. The influence of labor unions in improving salaries of their workers waned, and where that proved a slow process, national productivity gains as a result of automation and use of IT proved less extensive than in the United States. The supply of the kinds of skills needed in a highly digitized economy has always lagged demand, a condition that remains so in the second decade of the twenty-first century. These kinds of issues affect unemployment recoveries during and after recessions, and simultaneously push up salaries for the kinds of workers now needed, regardless of macroeconomic conditions.[157]

As a testament to the global diffusion of IT originating in large part from American experiences, came historians in other nations examining the notion that the Americans had their own style of computing. It is a topic worthy of exploration because as we learn more about what happened in specific countries, it becomes increasingly apparent that diffusion of information technologies—really of ICT—acquired their own local personalities. The first historians to explore this issue were Europeans. Gerard Alberts, a Dutch historian, initiated a pan-European discussion of the issue in 2010, calling for an understanding that transcends specific European experiences, what he called the "clash of narratives," or even the Americanization of European computing for a more composite view of global computing, an objective of my book.[158]

The story told in this chapter suggests that as time passed, diffusion became more a narrative of users stepping onto the stage to adopt the technologies that so many historians, economists, and engineers have written about. Users became the main players by the end of the 1960s, acquiring knowledge, applying it to the work of their industries, and transforming the way things were done in many agencies and enterprises. It is the interaction of supply- and demand-side activities set within the context of extensive government support and the background of a strong, large, and prosperous economy that accounts for the enormous surge in American adoptions, and that goes far to explain the rate and types of diffusions that occurred subsequently in so many other nations.

This chapter ends with hardly a word about the history of the Internet, a largely American made development of the past four decades of the twentieth century. Some readers will be critical that an overview of American use of IT would exclude such a discussion. It is as if all discussions are about the Internet with barely a tip of the hat to all that came before it. This chapter demonstrates that so much computing had been adopted by organizations and individuals prior to the Internet and before the wide deployment of that new network that essentially took off in the mid-1990s such that we can reasonably come to two conclusions. First, the United States led the world in the extent of adoption of IT before the effects of the Internet are factored into the assessment that made it clearly a dense thickly IT populated Wave One adopter. Second, that adoption of all manner of IT proved essential to the nation's ability to embrace the Internet in an effective, affordable manner, if for no other reason than its citizens needed to have already implemented ways to connect to it (telephone and PCs), reasons for doing this (to access information to augment preexisting activities), and with a base of essential knowledge (such as how to use a PC and software). Discussing the Internet would have run the risk of obfuscating these essential points. The American experience is so influential on other nations that it was essential to tell the history of United States diffusion in its broadest terms.

In subsequent chapters we will see others paths to adoption, but even these were always affected by the prior experiences of the United States, as in the

USSR, Japan, and India. This certainly proved so in Western Europe, chronologically the area that after the United States became the next area of the world to embrace computers. It is to its experience that we next turn our attention. By the end of the first wave of adoptions in Western Europe, there existed a transnational IT industry of sufficient size that even the rate of continued diffusions of computing in the United States was now in part influenced by events in other countries.

CHAPTER 3

West European Deployment Begins

Great Britain, France, and West Germany

The "computer revolution" is not the only technological innovation of recent years, but it does constitute the common factor that speeds the development of all others. Above all, insofar as it is responsible for an upheaval in the processing and storage of data, it will alter the entire nervous system of social organization.

—Simon Nora and Alain Minc, *1978*[1]

Our understanding of the diffusion of computing in Europe is nearly as fragmented as that of the world at large. Almost every study conducted on the role and history of digital technologies has been on a country-basis, rather than of Europe as a whole, and that includes telecommunications, with minimal attempts to compare and contrast national experiences. As a result, it has been very difficult to determine if there were common, interconnected patterns of behavior with respect to the introduction, use, and diffusion of computing on the Continent. The reasons are not hard to find. Like Latin America and Africa, Europe comprises many countries. Its linguistic diversity is more like Africa and less like Latin America, and there are significant differences in languages, from Swedish to Russian, from English to Spanish, from Polish to Catalan, from French to German, and with different alphabets. The southern half of Europe can claim some ties and shared roots because so many of their languages emerged out of Latin, a consequence of the Roman Empire's dominance of the region still evident today in such areas as linguistics and legal traditions. Germans spread across a broad width of territory barely or not at all controlled by the Romans, while to the northwest periphery are Nordic (Scandinavian) countries and to the far east Slavonic. The result is that unless a student of the history of computing is a polyglot, doing research on computing's events in such a linguistically and culturally diverse region becomes a very difficult, if nigh impossible, task to accomplish thoroughly. These types of diversities remain some of the largest impediments to a deep understanding of broad European patterns of behavior with respect to computing. Additionally, there is frequently a paucity of data, largely accessible archival records, or national studies to draw upon with which to write a more comprehensive pan-European history of computing.[2] Given the fact that different countries had unique national experiences, what we know of the role of computing points to the fact that their stories are varied. Therefore, one should not be overly critical of historians and economists in how they have approached the topic in the past.

Another challenge reinforced by the linguistic and cultural diversity and lack of extant data is the fact that there are so many countries. Today, Europe is a region with over two dozen nations. Historian Norman Davies reminds us that Europe did not simply consist of the Western half of the continent; rather, it extended from at least Ireland, possibly Iceland, and included Russia all the way to its eastern borders. His study is only one of a plethora of interpretations of what constituted Europe.[3] The existence of the European Union with its over two dozen members is testament to the political and cultural diversity of its participants, which includes such distinctively different nations as Spain, France, Germany, Great Britain, Netherlands, and a host of smaller Central European societies. Europe includes others as well, such as Switzerland, Andorra, and even tiny Liechtenstein, the last two also extensive users of computers (the first in retailing, the second in banking) but hardly known by many people around the world. Many of these nation states have existed for over a half millennium, developed their own business, political, legal, and social practices that affected their use of computing. These realities complicate the story as historians identify issues of multiple patterns of adoption across countries and even within their borders.

But we understand a few things that mitigate the problem. From studies of various uses of computing, telecommunications, and other technologies, we know that certain European nations represented the behavior of others. British, French, and German computing is one collective example, but there are others as well, such as the role of the Nordic nations, despite their small proportion of population and GDP within larger Europe. The economically wealthiest countries with the largest populations were the first to adopt many technologies, not just computing, while less prosperous, and often smaller ones, followed in successions as the costs of a technology dropped and its viability proved out in other nations. In the case of late twentieth century business practices, international corporations introduced new uses of computing to various countries regardless of local languages and culture, economic vitality, or demographic size; recall the case of IBM introducing e-mail into all its offices around the world in over 150 countries in the 1980s and 1990s.

How are historians to deal with the diversities presented by Europe? If one accepts the notion posited in chapter 1 that computing spreads in waves, and that these come in different sizes, often multiple ones simultaneously crashing on our metaphorical shores, then we should be able at least to identify the large national waves and describe their common patterns of behavior under the assumption that other nearby nations would be affected. To that end, Great Britain, France, and Germany represent such a wave of adopters in Western Europe.[4] We will deal with the earliest uses of computing in other West, Central, and Eastern European countries in subsequent chapters. In the west these three are important practical entry points into the history of European computing. Almost every country in Western Europe began tinkering with computing, largely in academic environments in the 1950s and 1960s, with some governments and large corporations using them in the 1960s through the 1980s as their initial wave of significant adoptions, and other nations following at various speeds.

The case for these three countries can be made briefly. First, they represent a large swath of Europe. If we look at the population of Western Europe, using the European Union's construct prior to the addition of the latest 10 nations, most of which are in Central and Eastern Europe (discussed in chapter 6), in 2000 the population of the EU comprised 377.6 million people. At that time, Great Britain, France, and Germany, made up 47 percent of that population (59.8 million, 59.2 million, and 82.3 million, respectively).[5] A similar examination of the EU's total GDP in 2000 shows it amounted to 9.40 trillion dollars, and the three countries accounted collectively for 49 percent of the total (1.44 trillion, 1.31 trillion, and 1.88 trillion, respectively).[6] The proportions of population and size of the economies of each remained relatively similar throughout the second half of the twentieth century. To provide more context the economy of the United States in 2000 amounted to 9.83 trillion, with a population of 281.4 million residents. Europe had more residents, while the American economy was larger by a half trillion dollars.

Additionally, extant information on the use of IT illustrates that understanding the experiences of these three nations account for a significant proportion of Europe's earliest encounters with computing. Because these three were the largest, most extensive, and some of the earliest users of computers in Europe, their experiences affected the understanding about computers in other European countries and, subsequently, their rates and patterns of adoption. In fact, while each nurtured indigenous suppliers of IT knowledge and products, all three, and the rest of the world, also learned and bought IT from many of the same vendors, beginning with Hollerith in the 1890s–1920s and later IBM beginning in the 1910s–1920s, extending to Apple and Dell in most recent years from the US/European outlets of these firms, and from local British, French, and German firms as well. It is the combination of size and sharing of learning and products that make it possible to justify a chapter devoted to an examination of the early experiences of these three countries.

Finally, on proportionality, some European countries are so small that to have treated them as independent cases to be studied would have called for a similar method to be used within the United States, where many states were either the same size as these nations or larger. The state of Wisconsin in the United States and the nation of Norway in Europe are of comparable sizes in terms of population and GDP. The state of California is about the size of Italy, while Andorra is smaller than the state of Rhode Island, the later the tiniest in the United States. Iceland is smaller than hundreds of individual counties in many American states. Size matters when the adoption of a technology is influenced by their cost and because one's scale of market is either big enough (or too small) to warrant selling into (or not) as an important variable. Large nations also have big government agencies and companies and because the earliest systems were complex and expensive, national scale proved important in influencing the rate of adoption of this new technology for many decades. Thus, examining the experiences of the biggest and wealthiest economies often makes very good sense, hence historians and economists gravitate naturally to this approach.[7]

All large European states had been active users of every type of information technology to appear since at least the eighteenth century. In the instances of these three countries, they also had early and extensive experience with punch-card technologies and other forms of mechanical data processing during the first half of the twentieth century, and in the development of scientific understanding of many of the base technologies, such as the physics underpinning semiconductors. Indeed, the recent work of Danish historian Lars Heide on the role of punched-card uses in the United States, Germany, Great Britain, and France clearly demonstrated the extensive uses of these technologies by various government agencies, banks, railroads, and large retailers.[8] Heide's own earlier examination of computing in Denmark also revealed the extent of diffusion of mechanized data processing to other parts of Europe prior to the availability of the computer.[9]

This chapter introduces the experiences of these three nations in what was clearly the first wave of the diffusion of computing, the time when computers were invented and when governments and companies installed this class of technology for the first time. Agencies and companies dominated deployment of information technologies, not individual consumers. The latter came into their own later, beginning with PCs in the 1980s and more significantly with the Internet (along with companies and agencies) in the 1990s, heralding potentially the dawn of a second wave in the diffusion and use of these digital technologies. Here we look at Wave One in these three bellwether nations; the next chapter reviews briefly activities in other parts of Western Europe during Wave One.

GREAT BRITAIN'S INITIAL EXPERIENCE, 1930S—EARLY 1990S

The story of how the British government stimulated development of an indigenous computer industry is normally told from the perspective of the evolution of computer technology, the history of some local vendor (for instance, International Computers Ltd [ICL], J. Lyons & Co.),[10] or, as part of a pan-European defense against the onslaught of American computer companies dominating their indigenous market. While public policy receives attention from historians, economists, and political scientists, they normally characterize it as failed, or in isolation from other actions taken by the British government. While this is not the place to tell the whole story, key elements of the circumstances suggest what public officials were attempting to do, administrating policies not so dissimilar to those in other countries and so we begin our story on the supply side of the ledger.[11] It is an important narrative because many European governments expended enormous efforts between the 1940s and 1990s to foster development and growth of indigenous IT industries, use of the technology by their home companies, and later by citizens of the Internet. Public officials across Europe watched what the British, French, and Germans did.

By the time World War II began in Europe in September, 1939, Britain led the world in the development of technologies that would later be seen as components of the modern computer. During the war, both the Americans and British continued to lead development of these technologies. The British remained ahead of the Americans largely by creating and funding a secret community of engineers, mathematicians, and others who built computers called Colossuses for deciphering German military communications. The project proved highly successful, making German military communications virtually an open book to the Allies, evolving into one of the most remarkable activities of the war, but one that remained secret until the 1970s, when reports of the work done at Bletchley Park began to seep out in bits and pieces.[12] In the 1940s, this project impressed senior British public officials and academics who had been involved in the work to press forward with development of computers in the postwar period.[13]

In the mid-1940s there were no computer companies in Britain, instead, there were vendors of electronic equipment and others that rented or sold business and accounting equipment. Each had an active market, many dating back to before World War I. All wanted to transform from a war-time business to a civilian-based one and to create new products, largely, however, based on noncomputer—hence proven prior technologies—products. Two office appliance firms in particular reflected this pattern (later changed when computers became more marketable): Powers-Samas and British Tabulating Machine Company, better known as BTM.[14] In the short term, they were right to focus on such issues for in the years between 1945 and the end of the

1950s demand for their products grew year-over-year as the British economy transformed. These companies simultaneously built up their export business.[15] In other words, demand for tabulating equipment remained strong, discouraging moves to any technology that had not yet evolved into commercially viable products capable of displacing these earlier ones and the revenues they generated.

The Labor Government elected in 1945 kept a number of wartime economic controls in place, such as restrictions on raw materials, export quotas, and machinery licensing that also affected supply-side data processing. Punch-card suppliers benefited from these policies as they were encouraged by public officials to expand their exports and were treated as members of a favored industry. Both firms were financially in good shape. Thus, they enjoyed preferential treatment in their requests for building licenses and access to raw materials. In the late 1940s, punch-card businesses were exporting some six times the volume of products as they had in the last full year of peace-time trade before the war (1938).[16] As the economy worsened in 1947, if anything, public officials put additional pressure on them to export even more. Devaluation of the British pound in late 1949 made British products more attractive in foreign markets and so the two companies benefited. Sir Stafford Cripps, Chancellor of the Exchequer, called on all industries to take advantage of the devaluation, "let us seize it eagerly and with both hands."[17]

In the 1950s British firms continued to invest in research and development of new technologies, both in tabulating equipment and later in computers, while casting a nervous eye on the growing marketing prowess of such American competitors as Burroughs, NCR, and IBM.[18] Work on what became computers had started earlier with government financial support provided to Cambridge University, Manchester University,[19] and the National Physical Laboratory (NPL), sustaining continued technological leadership by British engineers and scientists and, most notably, construction of the EDSAC (Electronic Delay Storage Automatic Calculator), an important early computer.[20] However, in the late 1940s and early 1950s, office appliance vendors saw the market for computers as small, but one that would grow slowly as the technology evolved into reliable forms.

Government officials were more optimistic about the potential of computers than commercial suppliers and to back up their optimism Harold Wilson, President of the Board of Trade, announced establishment of the National Research Development Corporation (NRDC) in May 1949. Until the early 1960s, the NRDC served as the primary public administrative tool for stimulating development of an indigenous computer industry in Britain. It had three objectives: promote R&D leading to patents, exploit commercially British developments, and recover its expenses by licensing patents. Lord Halsbury, an experienced manager, and socially well connected to senior public officials and business leaders, served as its managing director throughout the

1950s. He spent the decade urging companies to develop and sell computers, urging electronics firms and office appliance companies to merge, and sought funding for development of local computer products.

His proved a tension-filled mission because firms demonstrated more interest in expanding their existing businesses and devoting their R&D energies to bring out new products that leveraged existing technologies—classic path-dependency thinking. Yet Halsbury wanted these companies to shift that strategy to computers and to build small ones that would sell in the market.[21] In the 1950s, 12 British firms entered the computer business, three with direct financial support from NRDC. By the mid-1960s, most had left the computer market, overwhelmed by the faster moving American firms that had developed a variety of less expensive products quicker than the British and were marketed more aggressively, most notably by IBM.[22] During the period from the early 1950s to the early 1960s, the only way NRDC could get local companies to develop computers was largely by granting subsidies for their development, an initiative always under funded, in effect paltry. In fairness to the government and firms, everyone around the world underestimated the massive costs of R&D required to develop commercial-grade computers and peripheral equipment, including all the American manufacturers. It was, after all, a new family of technologies just getting underway.

Nonetheless, NRDC pushed forward its agenda largely by persuasion through Liberal and Conservative administrations, as it was the primary public administrative agency used by the government to promote economic development in high-tech industries. Halsbury used his considerable social prestige to promote his agenda. As one historian of his efforts observed:

> If a firm were to be approached by the NRDC, it would very rarely be through a formal letter on the NRDC's notepaper. Rather it would be through a luncheon at one of the London clubs, arranged informally, either directly by Halsbury or through a mutual friend. Ideas would then be aired and opinions expressed, often quite strongly, but the context was made as personal and unthreatening as possible.[23]

In its roles as regulator, manager, and promoter of the use of patents, the NRDC worked to translate developments at Cambridge, Manchester, Telecommunications Research Establishment (TRE), and the NPL into revenue, even signing an agreement with IBM—a rival of the British industry—to allow it to use computer components developed in Britain in the early 1950s. Ironically, none of the British firms displayed the same enthusiasm for the technology now licensed to IBM.

NRDC faced a number of challenges in the 1950s that extended into the 1960s, a period for both the British and French governments and their computer vendors that one student of the process characterized aptly as filled with a "sense of crisis," yet stimulated policies and actions.[24] With limited funds, which firms should it support? By law it could not favor one over another in bidding, while each company displayed a range of interests in participating in the fledgling computer industry. For example, Ferranti and BTM proved very reluctant to jump into the computer business with the enthusiasm NRDC wanted, yet on the other hand this agency delighted at the interest displayed by the smaller, although weaker, firm Elliott Brothers. Public officials promoted development of the industry and its firms, but could not favor any one vendor when they later sold back to the British government as products what they invented. Complicating matters, the NRDC also had to recover its investments in those firms. NRDC operated in a setting in which its officials experienced a sense of urgency to address the looming problem of American competition in their computer market, a problem identified as early as 1954 but, by 1957, had reached a severe level. NRDC was one of the earliest public agencies in Europe to take on the Americans through support for the development of products to stop them. Even though NRDC failed to energize local firms to deal with the challenge in a timely fashion, while part of its legacy, it demonstrated over many years that the British government was willing to manage aggressively technology policies in support of national economic interests. One normally thinks of the French *Plan Calcul* as the first great defense against the Americans, but that happened in the 1960s, while the British had already started to worry about IBM and other firms a decade earlier. There were private sector initiatives to enter the computer market, so placing too much emphasis on the role of government might distort in part what happened. Leaving aside the computers introduced because of government pushes, such as the introduction of the Ferranti Mark 1 in 1951, often considered the first commercially available electronic computer in Britain, there was Lyons, which, in 1954, brought out LEO through a wholly owned subsidiary, LEO Computers Ltd. It too was subsumed by larger vendors in the international computer industry, most notably by IBM.

Simultaneously, what occurred in the market place? In the greater British economy of the 1950s, some 90 percent of all revenues of British punch-card firms still came from tabulating equipment and as noted earlier, business was good. Their executives saw the 1960s as when they would need to sell computers. Meanwhile, however, the Americans were entering the market, and found ready demand for their smaller systems in the late 1950s and early 1960s. As one economic historian pointed out, the British and Irish customers in particular "took relatively quickly to the computer," because "they had market-based financial systems" and due to "their greater emphasis on general education," which he surmised made workers more able to work with the technology.[25]

British firms were just as early in installing computers as the Americans, unlike on the Continent where deployment came a few years later. One expert on the matter suggested that between 1950 and the end of 1959, 127 systems were installed, some in government agencies, but the majority in the private sector.[26] Yet that number of installed systems may represent an undercount. A detailed contemporary inventory prepared by the highly reliable Auerbach Corporation for American clients in 1961 about computing across Europe, told a different story (see table 3.1). The data collected suggested that 240 British-made systems were installed. Auerbach's list also shows inroads made by IBM both in the number of installed and on order systems. From a user's perspective, over 400 were installed and on order, demonstrating that customers were at least developing a healthy appetite for the technology, perhaps more enthusiastically than indigenous suppliers seemed willing enough to nurture, let alone satisfy.[27]

As in all countries in the 1950s and 1960s where computers were installed in the private sector, these were largely used to reduce the amount and cost of clerical labor, to perform existing routine accounting work, and to manage inventory records.[28] Additionally in manufacturing firms computers were put to work in developing production schedules, monitoring controls, and later computer-aided design (CAD) systems.[29] By 1964, the number of installed systems in the country had risen to just less than 1,000, with 56 in public agencies; the others having gone into large enterprises that could afford these early expensive systems, most notably into manufacturing and banking, also into such nationalized sectors as the railroad industry.[30]

During the 1960s, tabulating markets evaporated and the computer came of age all over Europe, including in Britain. By the end of the decade, the computer market had globalized and large American firms dominated it. These

Table 3.1 INSTALLED AND ON-ORDER DIGITAL COMPUTER SYSTEMS IN GREAT BRITAIN, 1960

Vendor	Number Installed	Number on Order	Total
AEI	1		1
EMI	5	9	14
English Electric	25	5	30
Ferranti	52	26	78
ICT	56	36	82
LEO	11	3	14
Elliott	52	12	74
STANTEC	32	10	42
IBM	18	43	61
Totals	255	151	406

Source: AUERBACH Electronics Corporation, "European Information Technology: A Report On The Industry And The State Of The Art," Technical Report 1048-TR-1, January 15, 1961, p. 28, Blachman Papers, Box 217, Mathematics Division, National Museum of American History, Smithsonian Institution.

included IBM, NCR, GE, RCA, Burroughs, Univac, Honeywell, and CDC, among others. Scores of European computer manufacturers in turn consolidated, merged, dropped out of the market, or went out of business, but the point is that already by the mid-1960s American firms held sway in this new market. In Britain the shift away from tabulating equipment came almost overnight as in the United States, particularly for the country's largest firm, ICT. IBM had introduced a small, effective, well-priced system called the IBM 1401 that competed deftly against the large tabulating machine vendors in the latter's own market—the stronghold of British firms—with the result that between 1961 and the end of 1962, British firms acquired 100 of these, representing a third of their expenditures for computers. The majority of those systems displaced ICT's equipment, causing the firm's revenues to drop by a jarring one-third, leading to all the obvious consequences: shrinking production, declining profits, and layoffs of employees. Now ICT faced the reality that it needed to develop and sell a new generation of computers, and do it fast.[31] In this period, the British suppliers consolidated down to three: ICT (Ferranti, GEC, and EMI into one), English Electric-Leo-Marconi (EELM), and Elliott Automation.

Then in April, 1964, IBM introduced the System 360 family of five new compatible computers and some 150 other products, superseding all earlier generations of computers. It is difficult for historians and others not familiar with the product line or the computer industry to understand how profoundly important this line of computers turned out to be, a point that will be made repeatedly in this book to reinforce its significance all over the world. Over the next several years demand for these systems caused the computer industry to grow at annual ranges of 19–21 percent, led to IBM doubling its revenues within five years, and to move from a mid-sized to a large international high-tech enterprise, a set of events described by one British computer historian as "entirely unprecedented," catching his nation, businesses, and government "largely by surprise."[32] Additionally, IBM's 360 technology and products became the standard of the day, acquired by public agencies and firms all over Europe, including in Britain, while the NRDC still attempted to work with local firms to respond to these developments. But with the return of Harold Wilson to power in 1964, his Labour government changed government's emphasis from less direct support and promotion of a local industry and more to promoting other industries, such as machine tools, telecommunications, and electronics with computing, while still important, playing a supporting role. In short, no longer a magic medicine intended to save Britain's economy, computing settled into the more widely defined role for it in the American and European economies. Meanwhile, newly consolidated ICT struggled along, hamstrung by its various computer systems that were not technologically compatible with each other, driving higher operating costs for the firm while it faced a more fractured market than IBM.

Yet, Harold Wilson's Labour Government did worry about a "technology gap" putting the British economy at risk, a concern also expressed by the French between the 1960s and the early 1990s.[33] This meant promotion of local computing would not be totally abandoned. The government stepped in with loans for R&D channeled through the NRDC to support work to create a response to IBM's products. But the amounts were quite modest, in one instance 5 million pounds in 1965. To put that monetary commitment in perspective, IBM spent several hundred million dollars to develop the System 360.[34] IBM often spent as much just to fix problems in individual software or hardware components subsequent to their introduction to the market in the 1960s as the British government invested in their entire national R&D for computers. However, one should not forget that public officials still considered the existence of a healthy local computer industry strategically important to the nation and even attempted to collaborate with other European countries. But the primary concern was now to diffuse IT into all manner of British industries and the economy at large.

Yet all through the 1960s, 1970s, and 1980s, there were myriad attempts by various European governments to form cross-border alliances to support growth and success of local or pan-European computer industries, which the British government supported.[35] These initiatives should not be confused with the demand side of the story, because European governments and firms were also eager consumers of computers in these decades, often at levels similar to what existed in the United States.

To reinforce his policy commitments to various high-tech industries, such as computers and nuclear energy, but not dominantly computing, Harold Wilson established the Ministry of Technology (Mintech) in 1964 to do much like the NRDC. Mintech faced many similar challenges. For example, it wanted to support development of a large computer, but could not convince key firms to collaborate, let alone concur on what technology standards to support. Additionally, the government established the Industrial Reorganization Corporation (IRC) to add other administrative instruments in support of national economic development. Consolidations continued in the industry as the stress of competition increased, and public officials worked to facilitate mergers. However, they added little, because their proposed financial infusions to facilitate such efforts proved too paltry, they complicated negotiations, and increased the time to complete them. As one British observer noted in 1973, "the British computer is very little better today than it would have been if he [Edward Heath] had left it alone."[36] The government implemented procurement practices to help the local computer industry sell to public agencies, spent on training British managers at the newly established National Computing Centre (NCC), and invested in R&D in such emerging areas of technology as manufacturing computer-aided design software and military systems. But shepherding creation of ICL in 1968 into the largest

British, indeed European, computer manufacturing firm—Britain's national champion—proved to be at least the greatest achievement it could claim, even though ICL (which had grown out of ICT and other merged firms) was not able to respond adequately to the surge of business that went to IBM in the 1970s.

The world experienced its first global recession in the computer industry in 1970–1971, initiated largely by a downturn in the American economy starting in 1969, and that immediately affected capital goods markets. Demand for computers in the United States shrank rapidly by 20 percent; similar declines in demand occurred next in Europe and in Britain. That had the twin effects of lowering revenues for local firms within the British economy and, of course, from exports. Consolidation of computer firms took place all over the world, largely because they were running out of cash with which to conduct R&D, or to compete. GE, for instance, sold its computer business to Honeywell, and as part of that to Machines Bull, the French firm. IBM responded with new products and lowered costs, and entered the small computer market, historically one in which the British firms played well.

The Conservative government of Edward Heath came to office in the middle of this economic crisis in June, 1970, and began quickly implementing a policy of disengagement with industry, retreating from the notion of supporting national champions, for instance. Rather, it favored market forces over aggressive government interventionist policies to stimulate success in the market place. The government dismantled the Ministry of Technology near the end of 1970, replacing it with the Department of Trade and Industry. It stopped financially supporting various computer R&D projects in the private sector, such as ICL's, then the largest computer firm in Britain. ICL wanted help, yet had problems developing new competitive products fast enough. The government did not have a fully developed policy about what to do with such firms, other than to let capitalist market forces work their will. A major public task force in 1971 published a report that led the government to articulate a policy. Called *The Prospects for the United Kingdom Computer Industry in the 1970s*,[37] it advocated favoring procurement of local systems over those of foreign competitors, criticized the government for not having a computer R&D policy, and recommended additional support by government for the industry more in line with what public officials were providing in France, Germany, Japan, and even in the United States. It opined that the amount should annually be on the order of 50 million pounds.[38] The government provided more financial support over the next few years, largely to ICL for developing new products.

The British government and industry faced more than just American competition for control over the research agenda and market demand for products. Other European states were also vying for lead positions and as the report indicated, investing more in the effort. Just using 1973 as an example, a period of heightened diffusion of computing across Western Europe, France and Germany each invested several times as much in computer R&D as did the British.[39] While ICL introduced new products in the mid-1970s, along with other firms across Europe, it could not catch up with IBM and others. Additionally, the market was changing again, from just serving the needs of large computer users to a rapidly growing mid-range one for minicomputers. IBM-compatible vendors were also in the market with lower cost systems.

The arrival to power of Margaret Thatcher's Conservative Government in May 1979 brought with it a sharp break from previous public sector support for the computer industry. Already telecommunications had been deregulated in 1974. That move had proven relatively successful in stimulating product and service innovations and lowered prices, just as occurred almost simultaneously in France. Indeed, historian Martin Campbell-Kelly saw the mid-1970s as the start of a second wave, or era, in British computing. While I do not want to quibble over dates of eras, increasingly we can conclude that sometime in the 1970s, and most probably certainly by the time this new prime minister was able to start changing public policies, a new period in public approaches to computing had begun that extended into the twenty-first century. Adoption of computing had already entered a new phase sometime in the mid-1960s, setting a pattern of diffusion that remained essentially the same until the arrival of the Internet and the wide diffusion of smaller computers and PCs, the former in the 1970s, the latter as early as the late 1980s. Give Prime Minister Thatcher her due in IT's history as she played an important role, regardless of how either political party or the industry felt about her policies. She reaffirmed the Conservative stance of distancing government from directly supporting industries, selling off shares of firms that had been acquired by previous administrations, and dismantling agencies engaged in influencing industries, such as the National Enterprise Board (NEB), bringing to an end a 30-year effort by the British government to promote development of an indigenous computer industry. Local vendors in various parts of the electronics industry faced yet more rounds of global competitive pressures as new products rapidly introduced one after the other, and by constantly lowering prices, all of which put pressure on less agile firms. The NEB experienced mixed success, partially, if not largely, justifying Thatcher's hostility toward it. Strategically, NEB had wisely chosen IT and biotechnology as industries on which to focus its attention, but tactically, proved a dismal failure as it was unable to stick with its champions when these faced difficult economic times or intense competitive pressures. Additionally, critics accused it of not understanding how the markets worked, particularly when it came to software.[40]

The iconic initiative of the 1980s was the Alvey Programme, which ran from 1983 to 1988, and represented the major British research initiative of the decade in information technology. It had many purposes, not the least of which was to continue the attempt to constrain American vendors and now, it would seem, the more dangerous Japanese, both of whom were moving into the European computer market, the latter touting its Fifth Generation computing initiatives.[41] British efforts promoted more formal and close-knit collaboration among university, industrial, and government research, focusing on technologies that had not reached a point of being competitive in the market, with particular attention to software engineering, knowledge-based applications, man–machine interactions, and advanced microelectronics, notably VLSI. The program failed to meet expectations for familiar reasons: inadequate public funding, lack of sufficient commitment by the private sector (particularly funding too), ultimately costing the participants about £350 million (£200 from government, the rest from industry). To a large extent, now with the various industries already operating globally—telecommunications, semiconductors, computers, and software—and "big science" part of how R&D was being done in such countries as the United States and Japan (e.g., Japanese Fifth Generation IT initiatives), it seemed too little too late. It received more publicity at the time than it ultimately deserved. Years later Jon Agar did not even mention it in his history of public sector IT in Britain, while Campbell-Kelly and Richard Coopey mentioned it only in passing.[42]

The thrust of government policy during the late 1980s continued deep into the next decade, promoting civilian use of IT through myriad collaborative research projects. Some of these involved R&D whereby academics and private sector scientists and engineers developed new products and services, and, in other instances, supported basic research on computer science at British universities. As occurred at the same time in France, there was particular emphasis on the role of information and communications in society at large. The combination of initiatives led to spillovers in research on materials technologies, IT of course, and also on biotechnologies and space-related technologies. Along with support for R&D the government encouraged foreign vendors to locate factories in Great Britain, much as the Irish did successfully in the 1990s, the latter using tax incentives and European Community assistance. A third set of initiatives involved government programs supporting expanded computer literacy, particularly in the 1980s, which extended to the early 1990s. In the 1980s, for example, the national government launched a "Computers in Education" program to promote use of computing in classrooms, funded in part by government but also by private industry. But, lest one think the British government was energetically moving with the passion and focus of the 1950s and 1960s, the times had changed, and in the words of one observer of events in the 1980s, officials were "decidedly non-interventionist."[43]

Before returning to the demand side of the British story, we should note that one of the byproducts of any R&D initiative is knowledge transfer, whether intended or not, and in the case of the Alvey initiative some of that occurred, despite any negative conclusions one may draw about its mission. Beyond the actual funds invested is the fact that 109 British firms participated in the initiative, along with 85 schools and universities. While not all projects were about IT, most were and as a consequence knowledge of computing continued to seep into society. At the same time, a European-wide technology collaboration program called ESPRIT also involved some 50 British firms and projects at over 40 schools and universities.[44] That too diffused knowledge of IT. Because some understanding of computing has to come before use of the technology, these kinds of programs must be seen as causing effects far beyond whether or not they pushed the boundaries of technological innovation.

Encouraging development of an indigenous computer industry is only part of the diffusion story. The other is the adoption of computing by government agencies and the private sector in Great Britain. Adopters lived in a near parallel universe alongside government's pump priming, intersecting largely when officials acquired locally made systems for military purposes or universities, the latter largely for experimental engineering and scientific projects. The earliest users of computing, hence the first to diffuse the technology in Britain, were government agencies in the 1950s and 1960s, which evolved into extensive users of IT by the early 1980s. Like the Americans, French, Dutch, Danish, Swedish, and German national government agencies came to computing after a long period of experience with punched-card applications and the use of other, smaller, computational equipment, such as desk calculators and adding machines. Additionally as the Americans, Austrian, and Russian officials, the British had also conducted large and complex operations involving IT, most notably national census counts and the resultant analysis of the data.[45] Government employees had broken German cipher codes and worked on other military applications that used computers during World War II, building a growing base of knowledge about the possibilities offered by the new technology, not the least of which the ability to handle larger amounts of data in an ever-growing complex world for officials at faster rates than with prior machines.[46]

The British government had some 425,000 employees at the dawn of the computer (early 1950s) and so the productivity of these employees became a natural locus for where all manner of IT could be used. Recall that in the late 1940s there were five experimental computer projects underway in Britain that not only improved the technology, but, in the process, led to growing familiarization of computing by professors, engineers, and officials. These kinds of projects serve as unappreciated (indeed unstudied) examples of the diffusion (often support) for new technologies in interesting places. In this and in so many instances in each decade and in many countries, in addition to the scientists and engineers working on a computational project, funders and managers of the institutions in which these existed also became familiar with and sympathetic of the technology. We can assume some of that empathy and support for computing rubbed off on members of one's family and other colleagues in the same institution, other enterprises, and in government agencies. One has only to read the hundreds of obituaries of British, European, American, and Asian computer scientists that appeared for decades in almost every issue of the *IEEE Annals of the History of Computing* to see that these people worked with many organizations and individuals, often back and forth between academic and government agencies and private companies. The early computer seminars discussed in the previous chapter in which Europeans and Americans discussed computing in the 1940s and 1950s became a practice that has continued to the present, again another form of diffusion at work.

Like their American counterparts, the military sought to build air defense systems to counter the potential threats of Soviet attacks, while the Ministry of Supply had, in addition, a keen interest in other possibilities, such as in development of nuclear weapons, much like the French (although the latter in the 1960s). All during the 1950s, each branch of the military embraced computing—a complex story that cannot detain us here, except to say that like their American and Soviet counterparts, they challenged every evolution of the technology to ensure that, in fact, innovations proved reliable. As the Cold War evolved over the years, moving from airplane bombing threats to potential missile-centered attacks by the end of the decade, so too radar and air traffic control requirements changed, providing the basic incentive to acquire continuously transforming computer technologies. In short, the extensive interest displayed by British officials concerning all manner of air traffic control applications in the 1950s and 1960s derived from military concerns.[47] As Agar observed, the acquisition of knowledge and interest in computing had been absorbed from the Americans. He described "the story of British defense computing" as one "of application rather than novelty. Much was borrowed from the United States."[48] Finally, we should note that as the Americans, British officials encouraged and funded various IT projects at universities in order to create new technologies for use by both military and civilian agencies and not just in support of the local computer industry. By the 1970s these activities within universities and the military also assisted in the diffusion of skills about computing.[49]

Almost every historian of British computing tells a dreary tale of failures. These include failures to create a viable local computer industry, failures of firms to compete aggressively, and failures of public policy. Yet, civilian uses of computing by government agencies present a different story. Treasury Organization and Methods (O&M), long influential in matters related to the use of IT, played an active role in investigating and promoting use of computers in government. It had helped many agencies, for example, embrace punched-card equipment; in fact, in 1948 26 departments had or were planning to install such machines. By 1954, some 80 departments did.[50] Note that like the experiences of so many other countries, agencies that operated punched- card equipment as integral parts of the way they did their work were natural candidates for using next generation IT—computers—for the kinds of uses evident in the private sector then and in future years: accounting, inventory control, payroll, personnel tracking, production, and so forth. Early computer installations in the 1950s came next, appearing at the National Physical Laboratory, at the Post Office, and at the Ministry of Supply. By the end of 1954, there were five systems installed and seven one year later. At the time, however, none were installed in the private sector. In the late 1940s and all through the 1950s and 1960s, diffusion came about through the advice and interest of government committees examining the potential of these devices, the work of computer manufacturers (British and American) in promoting their products, and declining costs over time of what otherwise was an expensive technology. In the 1940s and 1950s, many in the public and private sectors had viewed computers as useful only for scientific and engineering applications. Not until it became evident late at the end of the 1950s that these could be employed for other uses did interest grow for such applications as clerical data processing. By the late 1950s O&M personnel had concluded that clerical staff could be displaced by computers, thereby saving on operating costs; so they began promoting new uses of computing. Vendors demonstrated their wares also, particularly IBM more than its British competitors.

By early 1958, seven departments in government had computers, with another 28 digital projects being considered. By 1965, the corner had been turned, with 45 systems installed, and over 250 additional ones contemplated during the 1960s, scattered all over the government doing civil service payroll, production of statistics of various kinds, collection of census data, and managing pensions, for example.[51] Rapid deployment resulted in a population of some 120 installed systems by the end of the decade and right behind them a growing constituency of advocates favoring use of computers in government, namely, IT public servants. That community grew from barely 1,000 in 1967 to over 8,000 in one decade.[52] As happened in France, IBM computers were installed in government at the same time that senior officials were trying to develop a local industry in the 1960s and 1970s. The story of deployment in the 1970s and 1980s proved to be one of further computerization, spread of minicomputers and PCs across both national and local governments, and growing use of IT service providers to perform an expanding variety of data processing.[53]

As mentioned earlier, private sector adoptions of computers began in earnest during the late 1950s and spread during the rest of the century much along the lines that occurred in the United States and across Western Europe. The largest enterprises were first to embrace the technology, largely to lower costs of labor. British manufacturers adopted the same kinds of computing applications as the Americans and West Germans, including development of automation tools, such as numeric controlled (N/C) devices, beginning in the 1950s. By the mid-1960s, and continuing right into the 1980s, adoption and export sales of such tools by the British far surpassed other major European users; only the Americans exceeded in volume. As measured by number of N/C machines per 1,000 employees at the end of the 1960s, the British and Swedes were ahead of all Europeans, 0.3 and 0.4, respectively, versus West Germany at 0.2 and Italy at 0.1. American deployment was three times that of Great Britain (0.9 vs. 0.3).[54] Users in Great Britain mirrored those in other countries, with aerospace firms the most intensive in such places as Britain, Sweden, and the United States. But the applications also were used extensively in mechanical engineering and in motor vehicle design and manufacturing.[55] Alan Booth, who studied British use of computing in manufacturing, concluded that not only were local engineers and their companies aggressive users of digital technology, but also their record of innovation was too.[56] He also documented other uses, which paralleled those in other nations and that came online at the same time, for example, in the 1950s and 1960s with payroll, stock control, cost accounting, sales invoicing, sales statistics, and production control.[57] British firms accelerated their use of computers in the 1980s to levels approaching deployment in the United States. Extant survey data indicated that during the second half of the decade deployment had reached 56 percent of all workers using computers, a jump from 39 percent at mid-decade alone. That was made possible by an increasingly available skilled work force, the motivation of intensified competition, particularly for manufactured goods, much as a result of government policies favoring automation. As Booth correctly pointed out, management teams had become increasingly aware of Japanese, American, and European competition, and learning from their practices pushed forward.[58] In fact, the situation had changed enough that British managers and entrepreneurs were less interested in seeking government support than in earlier decades, while driving down their operating costs by reducing labor content of work. These efforts often required both redesigning work flows (the Japanese approach) and implementation of computerized operations (more an American style).

The results of continued interest in using computers could be tracked through IBM's activities. It was in the private sector that IBM proved most agile against other computer vendors in placing its products. Its machines were seen as more powerful and less expensive than those of indigenous suppliers, beginning with the IBM 650 and IBM 1401, as happened in the United States, often the first computer models installed in any company. Moreover, IBM's products and services were reliable; the firm was around and could support its customers, while British, many American, and European firms came and went. The combined public and private sectors had some 1,000 systems from all vendors installed in Great Britain by the end of 1964, of which roughly 220 were operated by government agencies, a significant increase of governmental installations in just the prior several years. As a whole across the entire economy from the late 1940s to 1965 we can conclude that momentum for diffusion had gone from experimental in the 1940s and 1950s, through government as an early and source of proof-points in the 1950s and early 1960s, to the wider acceptance in the early to mid-1960s by the private sector's largest enterprises.[59]

The experience of the British banking industry provides a window into the diffusion of computing in the private sector. During the 1950s right through to the end of the 1970s, banks that loaned money for homes and other buildings, known in the United States as savings and loan banks, expanded from 1,455 branches to 5,434, while the firms themselves shrank from 795 in 1952 to 287 in 1979. Clearly major shifts took place in how work was done and one way that occurred was through automation of various functions to reduce operating costs but also leveraging technology to generate all the paperwork required by such loans. One byproduct emerging from increased use of computing was management's ability to identify problems quicker, hence resolve them earlier, or before they became unmanageable and too costly to address. To be sure, computing had its fits and starts as software for banking applications had to be written, technologies did not always work as anticipated, deliveries of new systems were often quite late, but in time became highly automated.[60] Bank managers were not shy about acquiring technologies from multiple vendors. Even the Mighty IBM discovered that. The chairman of IBM's World Trade Corporation wrote his brother, Thomas J. Watson, Jr. (CEO of all IBM) in 1967 that competitive pressures were increasing in Great Britain largely "in banking. Burroughs has been particularly successful with their terminal," and that his staff was "trying to come up with an answer" to this problem, and "in the meantime, a specially engineered World Trade terminal is helping to hold the line."[61] In retail banking, the adoption of ATMs also suggested patterns of diffusion. Although first introduced in the late 1960s, during the 1970s networks of ATMs expanded, with nearly 200 networks by the mid-1980s, followed by consolidation resulting in one national ATM network by 2000. Along the way ATMs evolved from simple cash dispensing machines into systems that provided various real-time interactive services. As in the United States, a retail bank could not function by the mid-1990s without being part of an ATM network. Bernardo Batiz-Lazo has studied the British industry more than anyone else, documenting that the public embraced the technology quite well. His research showed that all through the 1970s and right into the next century, the total value of transactions and the number of ATMs grew. As an indicator of diffusion, in 1975 there were some 568 ATMs in the country; in 1985, 8,845; then in 1995, nearly 21,000; and in 2003, 46,461.[62]

The population of installed computers across all industries suggests that the British were not as slow to embrace the use of the technology as one might otherwise conclude. In fact, in both the public and private sectors, adoptions paralleled the timing of similar installations in the United States, albeit fewer in number because the British economy was smaller, hence the number of agencies and companies lower, an issue we discuss further below.

Additionally, as one observer of the British computing scene noted, by 1968—the year ICL came into existence—"government purchases of computers were increasingly not a large enough proportion of sales to significantly shape the manufacturers' strategies"; in short, the market had taken off.[63] Flamm and Campbell-Kelly have explored the history of the British computer industry, a tale of declining number of indigenous participants, with American vendors satisfying and stimulating local demand for these systems. By the middle of the 1960s, half the computers installed in Britain were American, which implies that U.S. vendors had become primary instruments for educating and promoting use and diffusion of computing. They maintained their roughly 50 percent share into the 1970s. When considering the total population of systems, British machines accounted for only 25–30 percent of inventory by 1974, the rest were American and European.[64]

The story of a small local industry at work in Great Britain and its limited abilities to function on a world stage continued to be discussed and lamented in the 1980s. Brian Oakley, who ran the Alvey Programme, noted in the late 1980s, "due to protectionism, each country has its own industry which has fed its own internal requirements, but it has not built up to any significant extent, with the exception of Ericsson, into a world market."[65] He pointed out that if one ranked by revenue generated by the top two dozen firms in the industry, the first British company on such a list would be ICL, ranked twentieth, with revenues in 1984 of $1.22 billion. To put that in context, IBM, ranked number one with revenues of $44.3 billion that year, while the top five firms were all American, leading him to observe, "you can see that we really are pigmies in a giants' game." The challenge for the British industry was to turn its interest in computing and its various developmental projects into "economic successes," otherwise the future was going to be "extremely bleak" for the local industry.[66]

With respect to IBM, as occurred in other countries it was viewed as an outsider, not an indigenous company, even though the firm always hired British nationals and had a long history of investing locally in assets, such as factories and office buildings. In 1951, IBM opened a punch-card manufacturing plant in Harrow and began assembling typewriters in Britain. By the end of 1954, IBM had 1,000 employees in the country and sold its first 650 in Britain in 1957 to Rolls Royce, perhaps the first 650 sold in Europe. Computer development projects in the country took place all through the 1960s, continuing to the end of the century. As an indicator of the expanding local market, one should note that in 1964—the year the S/360 was introduced—the local IBM establishment had 2,350 employees, of which 1,200 were engaged in manufacturing. The manufacturing arm at Hursley built S/360 Model 40s, ultimately one of the most popular 360s in Europe. Indicative of the expanded adoption of computers in Britain and Europe, by the end of 1969 IBM had 10,000 employees in the country; by 1980, some 14,700. In 1983 IBM employed 15,475 people, while revenues from within Great Britain reached

£1.677 billion. All through the 1980s, growing local demand for computing led the company to continue expanding its indigenous base such that by 1990 it employed 18,370 people, with 9,300 of these working in sales and services directly with British users (customers). 1990 is an interesting milestone year as it was just a few years before the massive adoption of the Internet by individuals all over the world, while the PC was seeping slowly into the economy. That year IBM exported from its British factories products valued at £4.324 billion, while it purchased £1.341 billion in goods and services in the local economy, making it a major firm in the country, buying from 3,320 other local firms of all sizes.[67]

IBM was also profitable during these years. In contrast, ICL—Britain's largest computer business—experienced losses in the 1970s and recovery in the 1980s; in its merged form with STC in 1985, it generated combined annual revenues in the second half of the decade of £1 billion in the prior year of 1984 and over £1.3 billion in 1988.[68] IBM played an important role in creating demand and interest in local computing. In short, the combined performance of IBM, ICL, and other vendors in Great Britain demonstrated the increasing diffusion of computing throughout the British Isles in this period.

Agar and others have argued that in the 1980s and 1990s a great deal of public sector computing was outsourced to local companies as a new way of infusing expertise about IT into the operations of the British government, while simultaneously much was done in-house with every new form of IT that came along.[69] Expenditures on IT by national government agencies in the 1990s had grown, now hovering annually at £2.3 billion, and additionally with local governments at roughly half that amount. These volumes reflected a combination of new uses taking advantage of Moore's Law at work in the form of new machines and software, and continued expanded deployment.[70] If that had not happened, the amounts expended would have declined. One final way to measure growth in penetration of computing in the greater United Kingdom is to note what the Organisation for Economic Co-operation and Development (OECD) calculated. Using its very broad definition of "information occupations," which incorporated more than computers, including professions that relied increasingly on IT, as early as 1980 41 percent of the British workforce fit into that category. For context, it placed 45.8 percent of the US workforce in that category, 41 percent of all Western European workforces, 33.5 percent of Germany's, and 32.1 percent in France. The absolute figures are not as terribly important as their relative size.[71] These data suggest British users of computers were a bit less intensive adopters than the Americans (but not by much) and only slightly more than their European cohorts. Providing more specific data about volumes of adoption of specific technologies and uses remains problematic for researchers for the period prior to the late 1990s, but the pattern is clearly evident. It is a subject which had already created angst among observers of the local industry.[72]

A survey of workers in various professions at the end of the century asked about the importance of computing in all manner of their work, which included everything from mainframe-based uses to PCs and the Internet, providing additional evidence of the local extent of deployment and diffusion of IT. While the diffusion proved uneven by type of worker (with higher ranked workers more extensive users of IT, particularly the Internet and PCs, than skilled and unskilled manual labor and other lower paid positions), the range of deployment of even these new technologies went from a high of 80 percent to a low of 41 percent using these at home, with two-thirds characterizing IT as "essential" to doing their work. Managers and senior administrative staff were the most dependent on the technology, while professions hovered in the 60–70 percent range, calling it "essential" and some 13–23 percent only "very important." The point is that as in the United States and certainly across all the Nordic countries, diffusion of IT had reached quite extensive levels and across British society by the end of the century.[73] Individuals had become extensive users of ICT; online transactions were common; business-to-business sales proved extensive; while the British government was implementing a wide range of e-Government applications.[74] In short, all the essential ICT, economic, and institutional infrastructures necessary for a modern information society were in place, suggesting that the British were possibly moving into the early stages of Wave Two computing in the late 1990s.

FRANCE EMBRACES THE COMPUTER AND ITS INDUSTRY

The French experience mirrors many of the same characteristics of the British approach, but also had its distinctive features. Both were keen on developing national capabilities in such high tech areas as aircraft manufacturing, nuclear energy, transportation, and computing, the British beginning to deliver results in the 1950s and the French a few years later. Each supported research and development at universities and national laboratories, funded work in the private sector, and nurtured military–industrial initiatives. Vendors in both countries sought to sell existing products for as long as they could, moved into computers either too slowly or with inadequate funding for product development, and in the 1960s, were swamped by the rapid movement of IBM and other American firms expanding into their markets and by the 1980s, Japanese vendors. Those who have studied both experiences are in general agreement on this comparison.

There were differences in style and intensity as well. French policies tended to be more directly managed by public officials than by the British. Students of events in both countries also generally agree that initiatives undertaken by the French government proved inadequate in addressing the rapidly evolving computer and software technologies of the 1950s–1980s.[75] Yet one American observer of the French and European computer scenes reported in 1961—and before the French computer industry was overtaken by IBM—that "computer activity in France is proceeding at a moderate level, concentrated primarily in a few large, active firms."[76] As in Britain, IBM dominated over 50 percent of the French market for large systems by the mid-1960s, largely on the back of its 1401 and 360 computers combined with aggressive marketing and effective sales, a share that exceeded 60 percent in the 1970s.[77] The extensive similarities with the British case is why we can dispense with as detailed an account of the French experience and instead focus more on the unique features of the later which often proved important.

Motives were one. While both saw the need for national champions in the computer industry for purposes of national prestige and sound economic development, particularly in the 1960s, the French also had an immediate concern different from that faced by the British. In mid-decade the United States embargoed the export to France of state-of-the-art computer equipment needed largely for computing required in the development of nuclear weapons. The Americans had been concerned about computers possibly falling into the hands of the Soviets.[78] This was seen by French officials as a direct affront to their government, which, as much for purposes of national defense and prowess as for pride and economic development, felt they must respond. The story normally told is that President Charles DeGaulle drove the creation of *Plan Calcul*, the French strategy to promote development of an indigenous computer industry that could compete against the Americans in France and across Europe. French accounts of national policy in the 1960s and 1970s, however, make it evident that many cabinet level agencies and corporate leaders participated in the numerous events that led up to the deployment of France's responses.[79]

French historian Pierre Mounier-Kuhn has pointed out in a monumental history of early computing initiatives in France, however, that just discussing French reaction to perceived American affronts to Paris, or glossing over the industry's early years, leaves out many elements of a complex history of information technologies in the country. Many institutes and academics engaged in discussion, research, and focus on mathematics and computing in the years following World War II, yet he bluntly points out that "France is the only industrialized country whose public research was not converted into the construction of a computer during the pioneering period."[80] This happened despite work at various institutes, which he criticized for not focusing more practically on the technology and proving unable to make a significant contribution. He attributed this weakness to various causes: France being cut off from work in other countries during World War II, but most importantly an absence of a prewar academic tradition and heritage of working with calculating technologies in universities. The result was a slow start for France. So, the French had to create in the 1950s and 1960s a new profession, a new national competence, indeed from a French perspective a new science.[81] As a result, early development research on computing in France had to be done "by the private sector which assumed all the risks of innovation in this technology."[82] That circumstance in the 1950s stood in sharp contrast to the experiences of every other country discussed in this book; all others received substantial help in that decade from academics and their governments. In time, that too occurred in France, but hardly in the late 1940s and too anemically in the first half of the 1950s, suggesting why France was late to the computer game, and why the national government had to become insistent and impatient to catch up with its urgent and substantive initiatives in the 1960s. By his count, between 1952 and 1965, 14 academic institutions acquired their first computers (some more than one), while another 11 systems were installed in civilian or military organizations.[83] When compared to developments in other European countries, the list is not so small, just that with the exception one organization (CNRS *Institut d'optique*) in 1952, the rest acquired their first systems between 1957 and as late as 1967. But, his point was made: academics were too theoretical, they focused on mathematics, and they were late to the process when compared to their colleagues in many other countries. With his research, we can now make more sense out of the enormous variety of events that took place in France during the 1960s and 1970s, beginning with those in society who bore the R&D risks he mentioned.

In France, on the industry side it was the Compagnie Machines Bull that had served as the nation's largest tabulating and business machine vendor since the 1930s. It became a supplier of French computers, yet like its British counterparts, proved late to the game and remained undercapitalized for R&D to sustain continuous developments of ever-more current IT products.[84] General Electric purchased part of the company when Bull found itself too undercapitalized to compete. The sale disturbed French officials who saw, in effect, a major French company being taken over by an American enterprise, another example of the *defi americaine*.[85] The shock of IBM's success and the embargo of Control Data's 6600 supercomputers to France cannot be underestimated, as both events led directly to the *Plan Calcul* by the French government in 1966–1967.[86]

The details can quickly be summarized. France established two scientific organizations to help develop new technologies, the *Institut de Recherches en Informatique et Automatique* and the *Laboratorie d'Electronique et de Technologie de l'Informatique*. The industry was reformed and protected with the creation of CII out of two existing firms (*Compagnie European d'Automatisme Electronique* [CAE] and *Société d'Electronique et d'Automatisme* [SEA]). Bull was considered "lost" to the Americans, which unfortunately for the others employed much of the existing local expertise in computing that remained working in France, however, and thus not lost to the nation as a whole. As occurred in most mergers of computer companies around the world since World War II, factions and rivalries developed internally within CII with the result that it could not respond effectively to the rapidly changing market conditions presented by technological changes and IBM's more disciplined approach.[87]

The French government backed financially these structural changes and R&D, but both CII and officials could not invest sufficiently, or fast enough, to gain traction in the market.[88] It was not without trying. In the intense period of economic investments made from 1967 through 1980, the government invested 2.5 billion francs in the plan with about 66 percent going directly to firms in the computer industry. By contrast, the British government invested £47.1 million, of which 77 percent went into the industry itself. In both countries the remaining percentages went to research institutes, academic organizations, and toward education.[89] Part of the reason the French were not able to respond as quickly to the innovations coming from IBM, and even the British, was the lack of sufficient scale and speed of investments in the private sector.[90] Additionally, *Plan Calcul*, to use Richard Coopey's words, "amounted to a Soviet-style 5-year plan," with an implied lack of realistic targets or operational practicalities.[91] At the end of the decade officials introduced a second version of the plan, focused more on the minicomputer market rather than on the mainframe and computer chip markets that had been targeted in the first initiative. It was also developed with greater collaboration on the part of CII. A third *Plan Calcul* appeared in 1976; it proved more modest given that by then officials and CII had recognized the national champion model had not taken off. Bull also continued to suffer due to its chronically insufficient speed to market with products and underfunded R&D. Bull faced criticism in the press, along with the French policy, in what came to be known as *l'affaire Bull*.[92]

So why did the French also fall short of achieving their public policy decisions for computing? They clearly made greater investments in developing a computer industry and involved more agencies than the British. President DeGaulle personally played an important supportive role, and there existed a national security dimension to the story far more intense than experienced by Great Britain. The French market was larger and Machines Bull, in particular, had a sales network in Europe and early successes in introducing computer products. Such a question may either be unfair to public officials or simply difficult to answer. A president of Machines Bull, reflecting on this period, observed that "the slowness of decision making by the state was a handicap, an intrinsic factor recognized by all. The result of numerous people and services led to conflicts in the priorities of various agencies rising to the top of government for decision-making."[93] Conflicts of interest between agencies, compounded by the varying priorities of the companies involved, including his own, led to the failure. In short, he blamed the state.[94] Other French observers, also critical of government policy, faulted Bull too.[95] Those who have looked at the history of *Plan Calcul* lay greater emphasis on the success of IBM's more rapid introduction of new products and success in placing them, even into French government agencies. In those years, it was not uncommon for IBM to generate around 45 percent of its revenues outside the United States, with a substantial portion of that coming from a handful of nations in Western Europe, including the large French market. As in Britain, local companies came to market too slowly, while experiencing divergent priorities both within individual firms and with respect to public policy, what one historian characterized largely as "internecine rivalry."[96]

The French government favored technological innovations that supported military requirements, such as those for atomic weapons. Military uses of computing did not lend themselves to commercial applications, which meant firms working on those technologies would not find it as easy to port over such innovations to commercial markets. Or, they would require time and additional investments, all while IBM and other firms were busily penetrating commercial markets with civilian-centric systems. Thus, as in Britain, French firms relied on public sector customers for sales and support, while American vendors went after the larger civilian market. By relying on public sector markets, which favored local companies, French firms were not able to make the transition to the more competitive larger private sector markets. By the mid-1970s, it was essentially over; the mainframe computer and semiconductor industries had become international and supporting national champions as a national policy itself required a transformation to more specialized computing. This became the direction that the French in particular continued to move toward by the late 1970s and throughout the 1980s, but to be sure, on a less dramatic scale than with the three *Plan Calculs*.

French government enthusiasm for IT came from various sources that should not be dismissed or ignored. One major source was the success the government experienced in forcing the modernization of its telephone system, initially addressed in its Fifth Plan (1966–1970). Without diverting our attention from computers in France, suffice it to point briefly to several results. In 1960 the number of phones in France per 100 inhabitants hovered at 9.1 (in Britain it was 15 and in Germany 11, while in the United States it was 39.5). A decade later, French diffusion had jumped to 17.2, for the British it was 25, the West Germans 22.4, while the Americans were at 58.6. During the 1970s French use of telephones exploded in growth, to 41.5, which compared handsomely with 48.1 in Britain, 43.4 in West Germany, while the Americans were ahead of everyone in the world, with 79.9 phones per hundred inhabitants.[97] The French government had addressed the lack of telephones through a successful national program. In the process a number of innovations came into French society, such as digital telephone exchanges in the 1970s. During the same decade, the number of telephones installed increased by about 300,000 each year. In the 1980s, the government supported deployment of Minitel, which brought online access to French homes long before the Internet. In 1988—the first year for which we have reliable data—French subscribers connected online for 68.8

million hours; usage grew to about 106 million hours by 1997. In the 1990s there were over 6 million terminals installed and on average the French used them about 2.6 times per day.[98]

The project, however, has gone down in history as somewhat of a failure because it relied on existing technologies of the day and, thus, when the Internet came along, proved both primitive in comparison and incompatible. For example, the specialized Minitel terminal did not lend itself to the flexibility of the personal computer. Users simply did not care for it so much, but they used it.[99] Overlooked by most commentators is the fact that in the process millions of French citizens of all ages came to understand the possibilities of online information retrieval and the concept of online transactions before most people anywhere in the world. For them concepts of the Internet and mobile computing was but a short leap in the 1990s and beyond. For policy makers prior to the late 1990s, this telephone initiative served as a continuing source of hope, confidence, and motivation for applying similar practices to computers. Minitel was unique to France and it would affect this nation's appropriation of the Internet, discussed below.

IBM's role is important in understanding both the supply side of the story—French national policy—and the diffusion of computing across public and private sector entities. As in many other European countries, the company and its predecessor firms had had a presence in France dating back at least to 1914, when International Time Recording Company (ITR) established an office in Paris. In 1945, the company employed 870 people in France, of which 500 worked in manufacturing. In the 1950s, IBM opened up sales offices in various cities and expanded its manufacturing and research facilities. Almost as important, IBM moved its European headquarters back to Paris from Geneva where it had been since 1935. In 1955, IBM built its first European 705 computer at its French plant at Essonnes, and in 1956, its first European-built IBM 650. Machines were installed in the 1950s in banks and government agencies. In 1964—the year IBM introduced the S/360 and also in the same period when French officials were already anxious over the issue of IBM's growing presence—the company employed 5,500 people in France, of which roughly 2,000 were in manufacturing. In addition to its European headquarters in Paris, IBM had manufacturing facilities at Essonnes and Vincennes. The company operated 24 sales offices and 12 service bureaus throughout France, and several small R&D centers. In short, it was well entrenched in the country, populating its facilities with French employees. In the early 1960s, approximately 10 firms were building computers in France, making IBM's success impressive because much activity was underway in the development and sale of computers.[100]

While government employees and their management were officially wary of IBM, in practice they acquired its products, even the military that had been one of the early roots of DeGaulle's concern about the Americans. For example, in

1965, the French Department of Defense requested of IBM that it establish a military-centered research capability, which it did, called the Special and Military Systems Division at Essonnes. IBM staffed it with 250 engineers and technicians to design systems for the French army, navy, and air force. Other ministries acquired IBM systems all through the 1960s. Meanwhile, demand for S/360s in Europe grew, thereby enhancing French manufacturing of the system, with satellite production facilities opened in the mid to late 1960s. During 1967, a new facility at Montpellier, built its 200th System/360 Model 40, which went to the Blaise Pascal Institute of Scientific Research in Paris. By 1975, IBM exports out of France had made it a major participant in French foreign trade as it had shipped products valued that year in excess of 2.5 billion francs, all to other European countries.[101] These data demonstrate that IBM had become a major participant in the French computer industry, while aggressively placing its products throughout France, just as it had in Great Britain.

IBM's local footprint continued to grow. In 1980, IBM employed 20,506 people in France, manufacturing each of IBM's major computers, and selling these across the nation. That spring the French subsidiary of IBM announced its revenues for the prior year in France (not counting exports), reaching 6.5 billion francs.[102] IBM's presence in France continued to expand through the 1980s as its French and European markets grew. France had become a major manufacturing hub for IBM for all its major hardware products, for research on its base technologies, and for training IBMers and customers. Its workforce shrank to 14,000 in the early years of the new century, however, as IBM found more competitive labor markets elsewhere in Europe.[103]

The story of diffusion in France cannot be understood without appreciating the role of the French government as well beyond its *Plan Calcul*. All through the 1960s and to the end of the century, officials introduced one plan after another to create a national computer industry and to offer incentives for French institutions to acquire its products. But despite its intentions, these plans failed, particularly when aimed at IBM because by the time they began, IBM already dominated the global mainframe market; Bull was no longer thought of as French, hence the focus on CII as the local champion. All through the 1970s IBM enjoyed half the market share for computers in France, which is why understanding more of the specific details of IBM's presence in France proved important—it provided a window into the diffusion of computing in the country. West Europeans, and not just the French, had begun installing American computers in the 1960s; one estimate placed the number at nearly 20,000 across the continent by 1970, of which 80 percent came from American firms, most of which built these systems in Europe: West Germany, France, Italy, Belgium, Netherlands, Denmark, Norway, Sweden, Spain, and Switzerland.[104]

The French government sought alliances with other European governments to go after IBM and the Americans in the 1970s by establishing interfirm alliances and promoting European technical standards with French leadership; all these failed too in the face of the onslaught of American success in introducing effective technologies and reliable new products. By the mid-1970s IBM dominated the computer markets in every country in Europe. But, there was more to the story than IBM's prowess. It had the wind to its back in France as well because Western Europe's military community was linked to that of the United States through NATO, an extension of Paul N. Edwards' notion of a closed world. That link, even though France stepped out of NATO, led other Western governments to standardize significant amounts of military computing to American models in order to satisfy American military tactical concerns about interoperability. Even the French military had to make sure they could connect technologically into NATO, or put in more prosaic terms, Western defense. There is evidence to suggest that part of the reason American firms did so well in Western Europe emanated from the NATO dependency. Part of the reason for even French companies forming alliances of various sorts with American firms can also be partially explained by this factor. When the French tried to lead a pan-European resistance to American computing in 1968, it failed because other European vendors saw that it was in their best interests to align more with both NATO members and leading lights in the OECD than with the French camp, a situation that did not change through the 1960s and 1970s.[105]

In late 1976, French president Giscard d'Estaing commissioned a study on how computers would impact society, calling on two public servants to prepare the study, Simon Nora and Alain Minc. Published in 1978 as a government report, it quickly became a best seller in France, stimulating discussions about information societies, their features, the future of French culture and nation, and the role of computers in an age of telecommunications linked with computing. In 1980, the report was even published in English, extending the discussion outside of France.[106] Much like the equally sensational discussion in French society about the cultural, political, and economic "American challenge" for France triggered by the publication in 1967 of a book by Jacques Servan-Schreiber,[107] it brought to the attention of the French much information about computers and their role. The book and subsequent dialogue motivated business leaders, for example, to keep expanding their use of computers, regardless of the vendor offering systems and, in IBM's case, brought customers to it curious about its machines.

The report called for a unified national policy with respect to computing and how it should be used, introducing, in the process, a new word into the French language, *télématique*, which blended notions of telecommunications and computing together to describe what eventually occurred, the massive deployment around the world of online systems. In the American edition Daniel Bell, a distinguished sociologist, observed in his guest preface that the French authors thought the technology could reshape a society—the primary reason for the national debate that ensued—but also called attention to "the tone of urgency regarding the role of U.S. domination in computers, in satellite systems, and in telecommunications, and in the need to meet that challenge."[108] The French writers of the report portrayed Americans as posing a threat to France's sovereign independence. After mildly criticizing previous French policy, but nonetheless criticisms in this official government report, the authors took direct aim at IBM:

> It [French government] has to take into account the renewal of the IBM challenge. Once a manufacturer of machines, soon to become a telecommunications administrator, IBM is following a strategy that will enable it to set up a communications network and to control it. When it does, it will encroach upon a traditional sphere of government power, communications. In the absence of a suitable policy, alliances will develop that involve the administrator of the network and the American data banks, to which it will facilitate access.[109]

The authors called for national action to block this development through policy and the formation of a large-enough technological "cartel" to fight IBM, nothing less than creation of a cabinet level department, a Ministry of Communications, to direct the work of the *Direction Générale des Télécommunications Administration* (DGT), *Télé-Diffusion de France* (TDF), and the *Centre National des Estudes Spatiales* (CNES).[110]

That was quite a recommendation aimed at one firm, and to be sure, to the broader needs of society. That recommendation was not implemented. Already, over 250 large firms in France had some 80 percent of all installed computers in the country, making it clear that what the authors proposed would affect directly the activities of many employers, not to mention in the years to come, users of telephone systems. This report cemented two local public myths for many years: that society was becoming overwhelmingly telematic and that IBM was becoming a global behemoth with telematics driven by companies and individuals outside of France, threatening the actual *sovereignty* of the nation.[111] Nowhere else in Western Europe had that level of concern over the international computer industry reached such heights of rhetoric and debate. While public officials spent the 1980s and 1990s deploying computing plans and collaborating through the European Union with other European states, French companies and government agencies continued to install mainframes, minicomputers, and personal computers, beginning in the 1980s and later, during the 1990s, the Internet. Installation of systems progressed slowly and incrementally in the 1960s and the early 1970s, then around 1977 took off. If we include personal computers, the growth became exponential in the 1980s, doubling in number installed just in the first half of the 1980s, continuing its growth right through the decade.[112] The French nation survived. IBM thrived.

Europeans collectively launched what they called Frameworks that stimulated research on IT and its continued diffusion. They provided funding for hundreds of projects also involving hundreds of firms and universities, all aimed at developing a uniquely European technology. The first Framework Program (1984–1987) focused on microprocessors; the Second Framework Program (1987–1991) integrated other features such as social, industrial, agricultural, and environmental factors; the third one (1990–1994) supported the ESPRIT initiative, despite the fact that the European computer industry had been overwhelmed by American and Japanese vendors. Like the earlier ones, this one failed to create an indigenous technological Fortress Europe. But, from a diffusionary perspective, when added to the earlier ones, it drew in thousands of participants into the world of digital computing.[113] ESPRIT faltered and in the process Europe lost some 250,000 manufacturing jobs in the IT industry by the end of the century. The fourth one (1994–1998) acknowledged that the strategy of supporting European champions had failed, and instead, shifted emphasis from developing technologies to diffusion, that is, to the use of all manner of technologies regardless of where they originated.[114]

By the early 1980s diffusion in France paralleled what was going on in Great Britain and across the rest of Western Europe. Accounting, banking, and manufacturing applications stimulated demand for all manner of office appliances, from word processors to large mainframes and as the 1980s progressed, telecommunications for online uses of computers. Sales of software of all kinds started in the 1970s and extended rapidly in subsequent decades. Economists who have looked at the contributions of IT to economic development characterize the French and Germans as in the middle of the pack when comparing IT capital investments and effects on worker productivity. At the high end of IT productivity were national economies such as Britain's, Ireland's and the Scandinavian nations, while at the lower end Spain and Portugal. With respect to sales and use of software, the Nordics led in most years, specifically Denmark, Finland, and Sweden.[115] However, all through the 1970s, French consumption of large mainframe systems continued to build on the diffusionary momentum of the 1960s. One French report showed that at the height of sales of IBM System 370s, for example, across the entire French economy government agencies and companies installed about 26 percent of all systems that West Europeans acquired in 1974 (1,467 in France versus 5,500 in Europe), a level of acquisitions that remained relatively constant for several years.[116]

Both supply and demand sides talked, collaborated, shared information, and experiences, and in the process created a community of IT advocates and experts that facilitated diffusion of computing through many one-on-one interactions, conferences, and by hiring employees from each other. As Alain Beltran has begun to document, these interactions promoted acceptance and use of IT, particularly by end users in the private sector. Mounier-Kuhn also reminded us that, as in the United States, end user communities—clubs and professional associations aligned by type of technologies they used—operated at the nexus of line management, IT organizations, and computer vendors, beginning as early as about 1960.[117] Collectively, they played similar, and important, roles evident in the United States and elsewhere in Europe.

By 1980 France's economy was investing some 2.3 percent of its gross fixed capital formation (GFCF) in office and computer equipment as measured by share of total nonresidential GFCF; a figure that jumped to 3.7 percent by 1985, then settled down to annual ranges from 3.2 to 4.0 for the rest of the century.[118] The government continued to manage formal developmental strategies and policies in support of IT diffusion in this decade.[119] In comparison, the British had invested more in the 1970s so they began the 1980s at a higher rate (3.2), grew to 5.7 by 1985, and continued expanding to 8.9 percent by the end of the century. The West Germans started at a similar rate as the British (3.1), jumped to 6.0 percent in 1985 and ended 1990 at 5.4 percent. Subsequent years showed lower rates for Germany but by then East Germany had folded into a new German state, bringing with it problems with tracking performance and lower investments in IT.[120]

Economists are deliberating the impact of such investments and expenditures on national economic performance, much as the Americans debated the "productivity paradox"; in the European case, they have not reached consensus.[121] They are in general agreement that demand for the use of IT in France, Britain, West Germany, and other national economies continued to grow all through the 1960s to the mid-1990s, when across Europe a slowdown occurred, signaling the end of the Wave One diffusion in most of Western Europe of all manner of IT and the emergence of new dynamics in play that we can begin to think may have been the birthing of Wave Two. But the evidence is not hardened yet. For example, in this late period, economists argued that, in fact, the capital deepening that occurred late in the Wave One, or early Wave Two (my terms not theirs), may have led to too much deployment in the late 1990s, particularly by the banking industry worldwide as it anticipated Y2K problems at the end of the century.[122]

In the 1980s France adopted two forms of online communications that made it possible for the nation to use ICTs in ways that were more widely applied in the United States and earlier than in Europe: the Minitel telephone network, largely by residents in the country, but also by enterprises, and electronic data interchange (EDI) predominantly by businesses as a major channel for commerce. Not until the end of the 1990s did France engage in a rapid adoption of Internet-based communications. Minitel, a videotext system was launched in the early 1980s, with some 84 percent of all businesses using it by 1990, and by a third of the population that same year. By the end of the decade some 35 million users relied on Minitel for a variety of applications in what became known as e-commerce, or B2C (business to customer) transactions. These included accessing databases, relations with customers and vendors, and simple purchase transactions. Small businesses found this a useful tool;

the population did too, especially because the terminals were provided free of charge by the government. People did online banking, made travel reservations, placed online purchases, and conducted information inquiries. By the early 1990s a stable volume of use had settled in, followed by the growing availability of the Internet beginning in the mid-1990s as an option (competitor) to Minitel. By then Minitel had made creation of various online businesses a reality in France.[123]

EDI also made its appearance in the 1980s and was mainly used by large enterprises, most notably the automotive and distribution industries, also sporadically across other industries as well. The most widely used applications involved the management of collaboration among suppliers and other participants in supply chains and to exchange documents with government officials and with banks. The effect of using EDI was the same on the commercial side as Minitel on consumer behavior: both developed a practice of doing online transactions before the arrival of the Internet. France's telecommunications network was digital, modern, and efficient by the early 1990s. That circumstance slowed French adoption of the Internet until the late 1990s. PC adoptions had been slow since people had Minitel terminals with 22 per 100 people with PCs (a prerequisite for using the Internet at the time) as late as 1999 as compared to 25 in EU and 52 per 100 in the United States.[124] Cable networks, useful for high speed transmission of information over the Internet, were underdeveloped. Then in 1997 the government began a major shift in ICTs toward the Internet, which led to national adoption of the Internet taking off. Internet users per 1,000 people in 1995 had been just about 16; 63 in 1998. Then the takeoff began to nearly 222 users per 1,000 in 1999 and 305 in 2000.[125] As one student of the takeoff noted, "Until 1998 the French did not really pay attention to the rise of the Internet," in part because the country was recovering from a recession in 1993 and so could not invest in the technology, but also because "the Internet was seen as a competing technical (and U.S.-dominated) standard."[126] A new government came to power in mid-1997 and concluded that the Internet represented a strategic technology for the nation to embrace and so used its resources and influences to change the nation's direction. Large firms had already started to make the shift, often using concurrently Minitel, EDI, and the Internet. Nearly 98 percent of all large manufacturing firms were users of the Internet by 1999 and about 67 percent of smaller enterprises.[127] Surveys of French and American users of the Internet as the end of the century demonstrated they all used it for the same purposes and essentially to the same extent. Most used applications were for travel and lodging, followed in descending order purchases of PCs and software, books, CDs, and video.[128]

In short, the French had been some of the earliest and most intense users of online computing nationwide in Europe, ahead of the Japanese and nearly comparable to the Americans in extent of use. The same held true for use of

EDI. However, these prior technologies had slowed the move to the Internet when compared to other countries that had not already invested significantly in earlier online transaction technologies, such as Minitel. People and firms continued to use the older technologies even after adopting the Internet, one complementing the other. Telecom and public officials also continued supporting earlier technologies, in part out of habit but also due to inadequate availability of funding to invest in new ICT infrastructures. The result is that by the end of the century household use of the Internet remained quite low, and for the whole nation we can conclude that France was still in Wave One.

But more broadly at the dawn of the new century, what was France's IT situation? A nation now populated with 60 million residents had belatedly fallen in love with the Internet with one survey of global diffusion remarking of 1999–2000 that "the growth of Internet use in France. . . . has been nothing short of phenomenal," attributing that expanded use to experience with the Minitel videotext system, which predisposed them more than many Europeans to go online, energizing another round of IT adoptions.[129] At the start of 2000, nearly two thousand firms sold over the Internet, representing a 100 percent increase over the prior year. But there is more to the story. For one thing, all fixed line services were not digital—that was not the case even in the advanced United States—and the telecommunications industry had been deregulated in 1998, heating up market-driven competition amongst the industry's members, resulting in a growth in sales of services of 12 percent in 1999 just over the prior year. In that important year of 1999, a total of 405 out of the largest 1,500 companies in France had a website and 13 percent of all firms conducted transactions over the Web.[130] Over a half million residents did online banking in 1999 with some 90 banks. By 2000, over 20 percent of households had a PC, most of which had been acquired just in the previous 18 months.[131] In all, while the French were slower than northern Europe or the United States to get online, and their volume of activities too low to declare them as having entered Wave Two fully in 1999–2000, they certainly had within the next several years.

WEST GERMANY ENCOUNTERS THE COMPUTER, 1930S–1990S

During the first wave of the diffusion of computing across Europe, there were two Germanies, and before the end of the century, a third that replaced these two, and with new borders that did not fully reflect the original geographic footprint of the earlier states. In addition, the consequences of both losing World War II followed by the exigencies of the Cold War affected profoundly the German ability to respond to the arrival of computers. The response was as much economic as political. So before discussing the diffusion of computing, a brief review of political and economic factors is in order.

When Germany lost the war, it was broken up into four zones of occupation, each run by a separate Ally: France, Great Britain, United States, and the Soviet Union. The first three began quickly coordinating their policies in what came to be known as West Germany, while the eastern zone, occupied by the Soviets, folded into Moscow's broader pan-Eastern European policies and programs. For that reason, East Germany's involvement with computing took a different path from that of West Germany and, thus, is discussed separately in chapter 6. The three Allied zones were united into a new West German government called the Federal Republic of Germany (FRG) in May 1949; the Soviet zone (normally known as East Germany) acquired its own national government in October of the same year called the German Democratic Republic (GDR). The West Germans allied with the non-Communist countries and integrated rapidly into the greater economy of Western Europe, while the Soviets wove the GDR's economy and government into their broader east European sector that Winston Churchill described as behind the "Iron Curtain." To complete the German political genealogy, in 1990, both German states were reunited into one nation, called the Federal Republic of Germany (FRG), resembling more fully (although not exactly) the footprint it had before World War II when it was known as the German Reich.[132]

All of Germany had suffered enormous physical damage during World War II. Half the homes in all cities had been damaged or destroyed, while some 90 percent of its railroad network was either obliterated or blocked with the wreckage of rolling stock destroyed by the Allies. Germany's industrial production shrank by an amazing 80 percent from prewar levels. More serious, however, hardly anything functioned as before: no telephone, mail, or rail services existed, while basic utilities did not begin to come back online for months, including water, electricity, and gas. Industrial production had, in effect, come to a halt. Indeed, the Americans, British and French had conversations about turning the whole country into one large agricultural sector (known as the Morgenthau Plan), while the French and Russians quickly dismantled factories and moved them to their respective countries. The Americans and British rounded up German scientists and engineers with specific skills relevant to national security, importing them back the United States and Great Britain. Allied officials confiscated billions of dollars in German patents; in fact, no scientific or other R&D was allowed to be conducted in the Allied zones for over two years following the war.[133] Food production dropped by 70 percent, while Germans scrambled for food for two years, with daily caloric consumption hovering at roughly 800 per day for city dwellers, hardly sufficient to sustain adults, let alone a work force. In short, the country was devastated and had come to a halt. To be sure France suffered enormous destructions of its infrastructures too, particularly where fighting actually took place, but less than Germany. The British experienced their greatest damage in central London due to German bombing, but minimal destruction of its industrial base.[134]

At first, because both Germanies had been such powerful industrial sites before the war, all four occupying nations wanted initially to prevent the rebirth of a strong economy in the region for fear that at some future time it might nurture the rise of another totalitarian German empire—a concern that did not go away until the dawn of the 1990s, and even lingered in some circles into the early 2000s.[135] Not until the Cold War came into its own in 1946–1947, did the Allies, in particular, realize that they would have to energize West Germany's economic recovery as a bulwark against the Communist bloc. The introduction of the Marshall Plan in 1948 in Western Europe led to a massive infusion of aid into Europe (later Germany), encouraging European exports of manufactured products and import of American technical knowledge, all the while fueling investment-led recovery and growth across the continent. As one historian observed about the Plan, it "defined the conflict between East and West as a choice between central planning and the market."[136] After two years of rampant inflation, in 1948 the West German government halted the economic crisis by introducing a new currency and lifted price controls. Those two actions led to rapid economic stability and rising employment as the nation went back to work to rebuild what had been destroyed. Those activities ushered in years of low inflation and rising economic growth characterized largely by the return to industrial manufacturing that had made Germany such a powerful force in Europe's economy in the eight decades preceding the start of World War II.

The recovery that came over the next few years proved so extensive that West Germans gave it its own name, *Wirtschaftswunder* (economic miracle). As a result of reforms and infusions of aid in the late 1940s, the Allies knew that Germany would soon become one of the strongest economies in Western Europe. The nation focused on long hours of work with dampened wages, which led to enormous increases in labor productivity. Profits went toward rebuilding the nation's productive capacity rather than to raising salaries. Industries produced goods that were in demand for export all through the 1960s and 1970s. Labor proved plentiful, if for no other reason than until the Berlin Wall went up in 1961, many Germans migrated from East Germany willing to work for wages considered low by West German standards but not compared to what they could obtain in East Germany. West Germany, perhaps more than most European countries, copied many "best practices" of mass production honed in the United States, which included whatever new technologies American industries used, including data processing and market-driven capitalism.

The West German government followed more the practice of encouraging internal and external market competition than protecting local industries in order to make them competitive and prosperous in export markets. These actions worked; in 1950, exports accounted for 9 percent of the nation's income and by the end of 1960, 19 percent.[137] Part of the success can also be attributed to the Korean War (1950–1953), which led to shortages in many countries, hence to rising demand for goods Germany could supply. In 1957, the Saarland, center of German coal mining, was returned to the nation; war reparations paid to the Allies ended in 1971. West Germany was back as an economic powerhouse by the early 1960s, a position it held even after the merger with the more underdeveloped East Germany in 1990. That year Germany had the fourth largest economy in the world behind the United States, Japan, and China. It is against the economic background of recovery, growth, and success in the postwar period that computing came into its own in Germany.

As with France and Great Britain, West Germany's attraction to computing involved various trajectories of activities: prior use of mechanical forms of computing, and other forms of automation and mechanization; role of government; rise and success of an international computer industry. Adoption of new, less expensive technologies over time proved highly effective, first in large enterprises and government agencies, followed progressively by smaller firms and public institutions during the first wave of the World's adoption of digital technologies.[138]

Starting in the second half of the nineteenth century, excellent German research-oriented universities, in combination with an expanding set of science-based industries in such fields as chemicals, pharmaceuticals, electronics, and steel, set off a half century of innovative research in many scientific fields, including electricity and even on the role of semiconductors that led to the development of a transistorlike device at the same time as at AT&T. The difference was that the German telecommunications industry did not see the value of it as quickly as its American counterpart. Indigenous office appliance firms emerged, many of which exported their products to other countries, such as Brunsviga-Maschinenwerke Grimme, Natalis & Co. A.G. (Normally just known as Brunsviga).[139] The emergence of large industrial enterprises, and more importantly by the start of the 1900s, of statistical agencies in the national government, led to the adoption of punched-card equipment and other mechanical calculators and adding machines, such as by officials conducting census studies. By the start of the 1930s, the use of information-handling devices were roughly as prevalent across the German economy as in other industrialized nations.[140] That pattern of adoption did not slow during the 1930s; rather, use of data processing continued to progress, but largely by national government agencies.[141]

The story told by historians of the birth and early years of computing in Germany is of one civil engineer, Konrad Zuse, who, working alone, built three computers between the late 1930s and nearly the end of World War II (Z1, Z2, Z3) and constructed yet a fourth one (Z4) that eventually went into productive use in Switzerland after the war.[142] The history of computing technology has often been distracted by debates of who did something first, and, in the case of Zuse, both he, and most historians of the technology, acknowledged that he had built the first functioning program-controlled electromechanical digital calculator that ran when he "booted it up" on December 5, 1941 (later named the Z3). Working on these machines on his own time, buying parts with his own money, and with a few contributions from friends, he constructed the Z1 and Z2 literally in his parents' large living room in their apartment in Berlin in the late 1930s. Deferred from active military service because of his job at a German aircraft company (Henschel Aircraft Company), Zuse found time to work on his projects during the war and even to persuade his employer to use the Z3 to solve some simultaneous equations and other mathematical problems.[143] That is the story normally found in the history books. However, a careful reading of Konrad Zuse's memoirs, *The Computer—My Life*,[144] reveals a great deal more about the early days of computers in Germany, presenting a very different view of events when compared to those in Great Britain, France, or the United States.

To be sure the story told by historians of his remarkable work is true. His sketches and photographs of the early machines in his parents' home that survived, and the documented memoirs of some of his war-era colleagues are testaments to his many skills, which included drawing. At the time he began his work in the mid-1930s, there were no major computational projects underway as existed, for example, at MIT in Cambridge, Massachusetts. There were no government initiatives; those underway related to information technology focused more on extending the use of punched-card tabulators to increasing number of agencies.[145] Zuse described his support in the late 1930s, when he was only in his mid-20s in age:

> My friends at the university also pitched in. They gave me money, sums which might seem modest today. . . . My sister was also among the first contributors, my parents also helped again and again. Each gave what they could, and in the end several thousand marks had been collected.[146]

Attempts to raise funds through companies failed. In 1937, for example, he reached out to a manufacturer of calculators, with the response that "in the field of computing machines, practically everything has been researched and perfected to the last detail. There's hardly anything left to invent. . . ."[147] Meanwhile, the Allied bombings during the war frequently damaged his equipment, which then had to be repaired, often two or three times a week. He wrote in his memoirs that despite demonstrating the Z3 to various firms, academic departments, and government agencies, "it was never put into everyday operation," furthermore, that "the Z3 was not considered vital."[148] At the same time, the British had initiated their Colossus program, while the Americans were developing the ENIAC to calculate artillery firing tables. He established a firm during the war with the intention of commercializing the Z3, which would have to be rebuilt because it had been destroyed in an air raid. Zuse also commented frequently about how there was hardly anyone else working on any computing issue at the time (1930s and early to mid-1940s). He was cut off from any knowledge about computational projects underway in Britain and the United States.[149] He observed that "an important difference between computer development in the United States and Germany was that, even if American researchers did not work together, they did know of one another and kept in contact with each other," while Zuse and his few colleagues "were not even aware of the existence of our potential German colleague, Dr. Dirks."[150] He also noted that the Americans spent "ten to hundreds times more than the German devices" in the 1940s.[151] The war's defeat cost Zuse and potential other German developers any lead he thought he might have had, while the British and Americans pushed forward.

Zuse went on to rebuild his company after the war, using the Z4 in Switzerland, and renaming his firm ZUSE KG. He held a few conversations with IBM, which only flushed out a mild interest on the part of the large computer firm in acquiring rights to his inventions, not in investing in his projects. Zuse subsequently obtained some research funding from Remington Rand to sustain his early postwar efforts, while making some sales to Swiss customers. German office appliance and electronics firms were not interested in computers in the late 1940s and early to mid-1950s, taking "a wait-and-see attitude," and they "played virtually no-role in the development of electronic computing machines."[152] He recalled years later that "long after electronic computing machines had been accepted, the electrical industry continued to put its money on the new semiconductor technology rather than on the computer itself," and when development of such machines in Germany started, it "began too late."[153] In Germany in the 1950s and early 1960s, when computers were built they were done so by academics in physics and electrical engineering at several universities: Göttingen, Munich, Darmstadt, and Dresden, thus not all even in West Germany. Scientific conferences were held in Germany in the 1950s and 1960s that began disseminating information about computing, but the government maintained a low profile. Meanwhile, Zuse built new models, began placing machines in the optics industry, and a few in other sectors as well. Universities simply could not afford to acquire these and other commercial IT systems before 1958.

To put Zuse's experience in a broader German context, we need to acknowledge that as a very early step in the diffusion of computing in any country is the emergence of a group of technical experts in computing. Two ways to establish that is to follow the example from chapter two of recognizing local conferences amongst interested parties and also by tracking when publications became available in Germany. In the 1940s and 1950s, the Americans had many such conferences, such as the Eastern and Western Joint Computer Conferences, and others hosted by the ACM all through the 1950s, for example. In Germany a few similar conferences were held that led to the creation of a virtual community of experts helping each other. Conferences were held in Aachen (1952), a second at Göttingen (1953) that emerged as an important center for the study of computing (a role it still played in the early 2000s), and another at Darmstadt (1955).[154] As early as 1953, the German Research Corporation (DFG) had formed a committee to look at computer development issues, comprising academics and leaders of existing computer projects in West Germany; DFG received some of its funding from the national government. Much early and important work in the diffusion of computing also came from use of the G1 computer system at Göttingen University where it did calculations for a variety of scientific problems, while staff there built the G2 and did preliminary planning for a G3 in the early 1950s. To sum up the status of projects in the early 1950s, there was the G1, with the G2 almost ready for operation by early 1953; the Zuse Z5 for the Leitz Company in Wetzlar to do mathematical calculations; a system under development for mathematical work at Darmstadt, and another under construction at the *Institute für Elektrische Nachrichtentechnik und Masstechnik* in Munich.[155]

In the 1960s German conferences increased in number. The system in Munich drew the attention of American observers stationed in the American embassy in Germany and experts, who noted the completion of the construction of the PERM (*Programmgesteuerte Elektronenrechenmaschine*, Munich) in 1955. Important for our story is the fact that many German computer engineers came to see it and very quickly made Munich a small hotbed of computing in Germany.[156] Coupled to these projects were publications, which served as effective vehicles for knowledge transfer. While one could assume Germans, Austrians, and Swiss interested in computers in the early years might have been able to read English, they soon had books in their native language, translations of American publications, and others written by local authors.[157]

A problem faced by Zuse and others in the 1950s and 1960s, and that appeared elsewhere in Western Europe and in the United States, was the chronic need for ever larger quantities of funding for R&D. Even in the 1950s and early 1960s, when Zuse was IBM's largest competitor in Germany, he had difficulty accumulating enough funds. His own testament of conditions said a great deal about computing in Germany around 1960:

> Government sponsorship of data processing, which could have helped us, did not yet exist in Germany. We also did not have the opportunity, which many companies in the United States had, of more or less branching off civilian products from comprehensive development contracts from the Ministry of Defense.[158]

Banks did not want to loan money for something that was strange to them. But, like so many other small computer firms in Europe and North America, "a larger capital investment was required," and "since the banks did not want to issue us any more credit," he had to find a buyer for his firm to at least assume existing debt.[159] In 1964 Brown, Boveri and Company (BBC) took over the firm and by Siemens in 1967. At that point, Zuse's role in the early development and diffusion of computing ended. By the time he began trying to sell the company (1962) as the only remaining option available to him to keep the firm going, he had delivered some 200 computers to customers in Europe, with the majority in Germany, Austria, and Switzerland; by 1967, a total of 251 systems had been delivered. He out-produced the three other German manufacturers of computers: Siemens, Telefunken, and Standard Elektrik Lorenz (SEL); in short the German computer industry proved quite small if one omitted IBM from the picture. Yet, this local industry actively developed and sold products too, as one analyst noted in 1961, "despite the country's relatively late start in the field," meaning late commercially to the market not to the interest in the R&D of computing.[160]

But IBM was lodged firmly in West Germany, also across all of Germany since 1910 before the nation was divided after World War II but since 1945 was not present in East Germany in the 45 years following World War II. IBM quickly dominated the West German market, reformulating its prewar presence into IBM Deutschland GmbH, headquartered in Bömlingen. Later headquarters moved to Berlin. When in 1955 prohibitions on the development of computers within the country were fully lifted, it became possible for Siemens, SEL, Telefunken, and Zuse KG to expand operations, as opposed to relying on R&D in neighboring nations. Telefunken built its first system in 1961 and, like Zuse, could not raise enough funds to continue so in 1964 Allgemeine Elektrizitäts Gesellschaft (AEG) acquired these assets and renamed itself Telefunken-AEG. The result of all these various activities and those of others, such as IBM, was that by mid-1962, West Germany had some 450 computers in operation across its economy, and another 32 as research devices, provided by six vendors. To put those numbers in perspective, at the same time the British had 400 installed and another 56 in research, from eight vendors.[161] Italy had 240 installed, another 19 in universities, while the French had 210 in operation and another 20 in research provided by only four vendors. In total in Europe at the time there were 1,695 systems estimated to be in production (research machines excluded), with the British, Germans, French, and Italians accounting for 1,250 of these systems.[162]

A key event occurred in 1967 when the West German government embraced the notion that Germany needed a national computer industry and instituted requirements that official agencies buy local systems. German firms received public subsidies amounting to roughly 50 percent of the R&D budgets companies requested, while various government agencies invested in training workers in the use of computing and word processors to help fill a growing demand by West German firms for such skills. Between 1967 and 1987, the national government invested some 8 billion DMs on R&D in IT as well, with

Siemens and Nixdorf collecting the lion's share, the latter, however, hardly any in its early years. Smaller firms and many universities also received subsidies, all of which helped to diffuse knowledge about computers into the West German economy. In these years new firms came onto the market with mid-range and other IT systems, including Nixdorf, Konstanz, Triumph Adler, Kienzle, Dietz, and Frantz, all of which often survived by offering niche products, rather than to challenge IBM in the large systems market.[163] AEG-Telefunken's own computer entry of the 1970s—the TR 440—while never a mainline product for the firm, and perhaps thus accounting for why it did not do well, was technologically one of the fastest built at the time and were placed in 46 installations, capping a process of computer development in the firm that had started in 1957.[164] However, to provide a balanced view, by the mid-1970s other American vendors besides IBM were active in Germany, including Digital Equipment Corporation (DEC), Data General, Honeywell, and Hewlett-Packard in the minicomputer-market, and at the high end with large mainframes Honeywell (via Bull), Univac, and, of course, IBM, the latter with 62 percent of the German market.[165]

Germany's second IT program, covering the last half of the 1970s, paralleled that of Japan, to build up local IT capacity. It involved a sixfold increase over the funding provided in its first plan. Roughly half of this funding went toward education and basic research (universities and research institutes), the rest went to infrastructure. For some, it seemed like a dual national champion program involving Siemens and Nixdorf, the first one ailing, the second a rising young star.[166] An early student of the process, Kenneth Flamm, argued that German support for these two firms "represented a middle way between the French strategy of placing all bets with a single national producer and the Japanese policy of supporting investment in a common generic research base to be incorporated into the products of a number of highly competitive national producers."[167] Germany launched a third program in 1976, preserving the same levels of financial support for R&D and product development, but reduced extensively funding for education programs. R&D levels of support proved inadequate and, of course, shrinking support for education in IT did not help push forward the government's agenda for expanding the nation's IT capabilities and infrastructure.

Of all these firms, Nixdorf proved the most successful. Founded by Heinz Nixdorf in 1952, and armed with a large grant from RWE, then West Germany's largest supplier of electrical power, he built his first system for doing accounting and census work. Specializing in minicomputers, and well managed, the firm grew all through the 1960s and 1970s. Nixdorf became one of Europe's most successful computer enterprises of its day, with a presence in 16 countries by 1974, sales of over $3.5 billion, and a global payroll of some 25,000 employees.[168] In time Japanese competitors, and Microsoft with its widely accepted technical platforms, caused Nixdorf grief with its proprietary standards, costing it to lose market share, much as Apple experienced in the United States on those occasions when it too stuck to its computer proprietary standards too long in head-to-head competition in the PC market. In 1988, Nixdorf's sales began falling off sharply and in the following year, the firm failed to make a profit. The subsequent story of decline mimicked that of so many other indigenous European companies. In Nixdorf's case, the German government pressured Siemens to rescue the firm, which it did but kept it intact with its own products, rather than integrating both lines of computers. Losses compounded right into the early 1990s, with layoffs of several thousand employees per year at Nixdorf. To end the story with one more sideline, in June, 1999, Japanese computer manufacturer Futjitsu acquired Siemens, moved its headquarters to Amsterdam, making the Japanese firm Europe's second largest electronics firm, not even a German enterprise headquartered in Germany. But, we must return to our narrative to summarize developments in the 1970s.

With respect to public policy, by the 1970s the West German government had embraced the notion of supporting national champions, following the lead of the British and the French, with particular emphasis on electronics and computing. Siemens became Germany's local champion in support it received for developing computers, chip manufacturing, and telecommunications. In the 1980s, support for local suppliers declined as Germany began, as in so many other European countries, in the words of a World Bank report, to "shift in the focus of many industrially-oriented IT programs away from the supply side towards diffusion."[169] The government got out of the business of "picking winners" among firms, a strategy that had failed in almost every instance in every country in Western and Eastern Europe once the computer industry had gone global by the early 1970s.[170]

But what was going on with IBM? Its progress has more to tell us about the diffusion of computers in the country than Zuse's story, which served more as a testimonial about the slow start on the supply side of the story, the early governmental actions (or lack of), and the challenges of funding than about technological prowess. Based on prior experience in prewar Germany, IBM managers knew that once recovered, this country would return as one of its largest national markets, because it would come back as a strong manufacturing, banking, and services economy, ripe for the kinds of goods IBM offered. The firm slowly expanded its presence, employing 2,400 in 1950, with manufacturing and sales facilities. An important laboratory at Boeblingen opened in 1952, which quickly became a company-wide global center for product development.[171] All through the 1950s, IBM expanded its R&D, manufacturing, and sales infrastructure in West Germany. In 1957, IBM even manufactured its popular 650 computer at its Sindelfingen plant, then two years later the IBM 305 RAMAC. By the end of 1963, IBM employed 10,000 people in West Germany. Various manufacturing facilities were built in West Germany in the 1960s, while the Boeblingen laboratory played a crucial role in the development of the IBM System 360 Model 20, variations of which came out of this facility right into the early 1970s. IBM Germany had, as early as 1967, shipped some 2,000 S/360s from its German manufacturing sites, making it one of the largest producers of computing equipment in the World at that time. During the 1950s and 1960s, IBM's computers were installed in manufacturing, process industry, chemical, paper, mining, banking, and retail firms, following the pattern of adoption already evident across Great Britain, most of northern Europe and many parts of southern Europe, with the significant uptick coming in the second half of the 1960s.[172]

By the end of 1980, IBM employed 26,362 people in Germany, more than in any other country in the European Community. The little lab founded in 1952 at Boeblingen had grown to considerable size, employing 1,800 people, making it one of the largest R&D organizations in the IT industry. All through the 1970s and 1980s, IBM Germany developed and produced many of the firm's hardware products that sold across Europe. By the 1970s, Germany had become IBM's largest user of its products in Europe.[173] This company's share of German government orders for computers kept rising in the early 1970s to the point where in 1974 it had about 80 percent of the market share within government. That success led to an order to public agencies to scale back their acquisition of IBM systems to 60 percent within the national administration, largely to benefit Siemens, an initiative that made it possible within a short period of time for the entire local and national public sector market to have half the installed computers coming from German vendors.[174] The American firm's overall success, however, can be explained. IBM was able to leverage its global development of products, and financing, its strong presence in Europe (and specifically in Germany), and the fact that the company's sales force called on large institutions in the 1950s through the 1980s that were acquiring large systems.[175] Western Europe's largest economy was known inside IBM as one of "the majors," along with the United States, Japan, Great Britain, France, and Italy, and, therefore, received considerable attention.

Specific uses of computing in Wave One diffusion proved important in West Germany. Computer-aided-Design (CAD) was a major use of computers in all manner of manufacturing around the world, and especially in advanced manufacturing economies such as Germany, the United States, and Japan. In such industries as automotive, aerospace, industrial equipment, and electronics—Germany's strong suit—this was particularly the case. CAD/CAM systems from all vendors were, like rising waters, lifting all IT boats, and in Germany's case some 300 installations by 1981.[176] Another major area of computer use was in the advanced machine tools industries. Germany was the world's third largest producer of machine tools after the United States and Japan by 1981, and from the mid-1960s to the late 1980s, the world's largest exporter of such products.[177] This was an industry that became an extensive user of all manner of IT, beginning in the 1960s, and by the early 1980s, of robots, which can be seen as yet another variant of industrial computers.[178] Numerical control tools were, most specifically, in great use in this industry. Germany also manufactured these, an example of computing embedded in other equipment, making it difficult to tabulate total computer populations around the world. However, we know a few facts. In 1976, West Germany produced $270 million in N/C tools, out of a global production of $1.35 billion; that consisted of half the production of the United States and half of all Western Europe combined. Over the next decade world production of N/C tools increased four-fold, Germany's 3.5 fold, with all of Western Europe increasing three-fold, but Japan now more than six-fold.[179] In the related field of robotics, West Germany dominated Western Europe with 800 systems in 1980, versus 1,500 for all of Western Europe; Germany doubled production by the end of 1982. To be sure the United States and Japan excelled in volume over Germany in production of robots, but West Germany remained ahead of any individual European country throughout the last decades of the twentieth century.[180]

As economists note frequently, one way to determine the extent of deployment of IT in a country is by examining the amount invested in the technology by a national economy. Beginning with 1980, hard reliable data are available to suggest some patterns. If we look at the percentage of nonresidential gross fixed capital investments, we see some interesting patterns. Between 1980 and the end of the century, Germany went from investing 12.2 percent per annum in 1980 to 13.9 percent in 1990, then to 17 percent by 2001. That was roughly twice the rate of investment by France and Great Britain in 1980 and about the same as its two peers by the early 1990s, with France just slightly less than Britain. However, with unification and new investments called for, the percentages changed by 2001: Germany 17 percent, France 12 percent, and Britain at just over 22 percent. All of Europe had generally been low in the 1970s and 1980s, with these three countries always investing more than the others.[181] 1980's numbers are useful to historians because one does not reach any level of investment without prior buildup or decline. In the case of Germany, investments in IT had been greater in the 1970s than in most other European countries and its experience was consistent with the strength that the national economy experienced in the decade. When broken down into expenditures for IT hardware, communications, and software, with hardware often the most important, because it indicated either new installations of replacement or older equipment, with software an addition to those hardware systems, we can conclude that France and Britain were progressing in their diffusion of IT all through the 1970s and 1980s at almost comparable rates, but in Germany a takeoff occurred which began in the 1970s and continued through the 1980s as its economy became very teledense. In the 1980s alone, West Germany was investing a third more in hardware and telecommunications equipment than the other two.[182]

The uses to which Germans put these computers included accounting, finance, inventory control, and a large number of manufacturing applications, which made much of that country's manufacturing industries current in its practices and uses of IT. All key manufacturing industries participated: automotive, aerospace, electronics, transportation, and heavy equipment to mention the most obvious. CAD/CAM applications arrived early in West Germany and by 1981, there were some 300 installations in the country, mirroring uses in such areas as mechanical engineering, design of electronics, and cartographic work. Germany began using robotic machines in 1974 (130 systems) and by 1985, 7,500, ranking it third in worldwide use of this technology after Japan and the United States.[183]

The biggest national event to occur in Germany since World War II was unification of West and East Germany in 1990. Telling the story of how computing spread in the consolidated Germany, without having reviewed yet in this book East Germany (see chapter 6), can lead to misleading observations. To a large extent, however, the story of the united Germany can be told as part of late Wave One activities in the spread of IT in Europe, involving such technologies as PCs and mobile devices within the context of a growing availability of the Internet. Several activities were underway that set the context. First, with consolidation what used to be the old West German government reformulated itself and established the national capital in Berlin. Those activities led to significant administrative reorganizations of government that occupied the attention of senior leaders for much of the decade. Second, a high priority for what used to be the old West government was to integrate East Germany into what was West Germany across myriad facets to bring together one common structure involving laws, regulations, economic development, and infrastructures of every kind from electricity and water to telecommunications, health, education, police protection, army, and so forth. Simultaneously, corporations that sold in West Germany now either had to expand for the first time into what used to be East Germany or to extend and upgrade their coverage of parts of Germany that operated differently than the western end and that had lower economic standards. Thus, for example, IBM could expect less business from a commercial set of customers in eastern Germany than it received in the western end of the country. Corporations were very concerned about property rights in what used to be a Communist country, security of their investments, and the rule of contract law. Additionally, from an IT perspective, West Germans were using IBM, Dell, and Compac PCs among other brands in 1990. In East Germany there were fewer such systems, many bootlegged or brands not supported by vendors in the west. As this chapter was being written (2012) the process of integration was still underway, funded almost entirely by the Germans themselves as there was no Marshall Plan at work.[184]

The OECD, European Union, and German economists have attempted to collect data on the unified Germany to begin documenting events. It became clear by around 2000 that the rates of investments in IT in the new Germany were coming back to the levels prior to the merger. Thus, for example, when one looks at the Total Gross Fixed Capital Formation in machinery and equipment for 2000, Germany (all industries) was investing heavily in hardware, moderately in communications, and about the same amount on software as other European countries.[185] That data suggest economic development was taking place in the eastern sector of a more traditional type (such as mainframes, centralized computing). But, in time, the reality will probably be shown to be less simple than just described. GDP in the integrated Germany dipped in the early 1990s, then expanded all through the decade as the nation invested in integration. Unemployment for the nation as a whole increased, becoming a chronic problem that spilled over into the twenty-first century as it attempted to find work for East Germans while various European and global economic crises and recessions buffeted the nation along the way.

Yet in aggregate during the 1990s, the German economy grew, as measured by Gross Domestic Product (GDP), by 14 percent between 1991 and 2000. Labor productivity expanded by approximately 20 percent, while per capita income rose from DM 36,700 in 1991 to DM 48,500 in 2000. These events served to establish effective preconditions for the adoption of computers of all kinds, from large mainframes in factories, banks, and government agencies, to PCs, laptops, and cell phones at the individual level. However, the Internet or e-conomy that so influenced events around the world in many countries in the 1990s had yet to arrive across all of Germany. As one German government economist wrote in 2001, "looking at the German economy as a whole, no significant impulses of the Internet economy can be observed."[186] However, one can conclude from various pieces of evidence that Germany as a whole spent a great deal of effort in the 1990s expanding the rate of diffusion of IT in enterprises and in building a modern telecommunications infrastructure in the old East Germany, while upgrading existing systems in the western half, much as other nations were doing as normal activity. Citizens acquired mobile computing fastest in western Germany and, increasingly, at some as-yet-not-understood rate, in the eastern end of the country.

Hard data support this conclusion. By 2001 some 24.2 million residents accessed the Internet, up from 19.9 million just barely a few months earlier. More interesting is that 11 million Germans reported being online every day; e-commerce proved healthy, with the Germans some of the most extensive users of online financial services. The telecommunications market, comprised of 1,800 service providers, was intensely competitive and the infrastructure was improving rapidly in the late 1990s and early 2000s. Nearly all small and medium-to-large enterprises had a website by 2000, so they had gone through the initial phase of institutional adoption of the Internet.[187] The statistics and applications point to a situation similar to what occurred in the United States in the mid-1990s, just as it was entering the earliest phase of Wave Two. Perhaps Germany was also by the early years of the twenty-first century.

CONCLUSIONS

The experiences of these three European countries have much to teach us about the spread of IT across Europe. They shared some common features and yet others that made each unique. Regardless of where they were with respect to computer development and prior uses of data processing equipment in the 1930s, and despite their varied experiences and consequences of World War II, the three ended the twentieth century extensive users of IT across all sectors of their economies. Computers were installed in considerable quantities across all major industries and in virtually every large enterprise (e.g., those with over 1,000 employees each) by the end of the century. Deployment often exceeded in a matter of two decades the extensive adoption of prior basic information technologies of punched-card tabulators and calculators. They were displaced almost universally between the late 1950s and the end of the 1960s, a remarkable achievement because their deployment had been going on for a half century at quite an intense rate of adoption, particularly in the 1920s and right through the depression years of the 1930s. In fact, even the disruptions of World War II failed to reduce demand for these technologies. During the war years government agencies of all types continued to expand their use of such equipment, often making the seizure of an enemy's inventory an important military objective. In the postwar years computing expanded into many organizations that had not yet deployed tabulating equipment. In others, if they had, they expanded the computational capacities of their offices and factories. This expansion was made possible by new uses of computing not practical with earlier technologies. In short, computers far surpassed tabulators in the number of installations and extent of use. This pattern mimicked the same trend evident in the United States across the century.

In addition, citizens in each of these countries became extensive users of IT in their private lives by the end of the 1990s. They learned about the technology at work and school, but also as new generations of small devices affordable by individuals became available, beginning in the late 1970s and by the end of the century had become a massive consumer market around the world that equaled in size that for the more traditional institutional computing of large mainframes, corporate and Internet-based networks, and diffusion of minicomputers.

Briefly summarized, I argued in this chapter that the way computing diffused so quickly through these three economies, hence across half of Europe's population and economy (due to their combined size), was roughly the same. National governments sponsored early development of computing, funded the invigoration of local computer industries, and used the technology within agencies to displace earlier data processing equipment (e.g., tabulators).

Then the private sector began adopting the technologies, usually in those industries with the largest enterprises, such as in manufacturing, finance, and transportation and that were extensive handlers of information. In the process indigenous computer companies emerged, failed either to receive enough government funding, or could not compete against large American computer companies that moved through their markets faster with newer, less expensive products, creating by the early 1970s a global computer industry. All three governments fixated on the rapid expansion of IBM, but even their own agencies acquired this company's products while their national leaders sought to promote the rise of local "national champions." American firms, and most clearly IBM, expanded their presence in Europe. IBM established a very large, significant presence in all three countries with sales forces, manufacturing, and research and development overwhelmingly staffed and managed with local nationals. In all three counties, IBM also became a major exporter of IT. In the 1980s Japanese vendors of all manner of IT hardware and their components did the same. With the passage of time the emphasis an earlier generation of historians placed on military incentives and support in propelling development and adoption of IT needs to be tempered. In these three economies, we see momentum from the private sector—users and vendors—providing more of the drive forward in the diffusion of IT than the military by the late 1970s. That shift in the source of drive behind the momentum adds to our understanding of why concerns about IBM, Japanese vendors, and local firms attracted more attention, hence in this chapter, than they might have if written a decade ago, or if the story ended in the 1960s–1970s instead of at the end of Wave One in the 1990s.

As in the United States, early uses of computing by large organizations were to automate existing processes, often to drive down labor costs or need for workers. Over time, governmental programs led to some of these policies becoming more nuanced, as in the case of France where, by the 1980s, national policies of all kinds aimed at increasing jobs and protecting those already in the economy, rather than in displacing them with computing. Over time new uses appeared as the technology made that possible. For example, in the 1960s organizations began using terminals and online systems, ATMs came into their own in the 1970s, while minicomputers and PCs began appearing in offices in the public and private sectors in the 1970s and most intensively in the 1980s and early 1990s. From the early 1980s, investments in IT expanded so extensively that by the end of the century, expenditures on this class of capital goods ranked in the top two or three in each country, and in many other European nations as well. Economists have argued for decades that part of the incentive for adopting IT was the effect of Moore's Law, that is to say, that the cost of computing kept dropping each year by significant percentages, making investments in the technology more productive than in labor. While true, the actual cost of computers proved quite high, which goes far to explain why in the 1950s through the 1970s only the largest enterprises and agencies could afford the price of admission to modern digital computing and that only after minis and PCs became available were IT products small enough to be affordable by ever smaller organizations.

Diffusion of this technology required three other conditions evident in these three countries. First, the local economy had to be healthy enough to afford the technologies. Despite the inadequate investments in IT by all three sets of national governments to create local computer industries, in all three economies provincial and national government agencies acquired significant amounts of computing for their internal operations, making them each IBM's top customers in Europe, for example. The same applied to the companies operating in these three local economies. Second, knowledge about computing proved a prerequisite for its use, a fact recognized by all three governments and over time, individual companies. As a result, through the period governments promoted training and education of students and workers about the technology. Meanwhile, as R&D continued, those individuals who performed such work seeped into their local economies in various roles as teachers, professors, computer experts, and so forth, facilitating diffusion of the technology in the process. Third, a national economy needed work and enterprises complex enough and technologically advanced such that they needed the functions made possible by computers. All three societies met this requirement, although Germany not until the early 1950s. That explains, for example, why Zuse encountered less interest in his machines in Germany in the late 1940s than in nearby Switzerland, a nation that had escaped the devastation of World War II and had intact the complicated-enough firms and economy to warrant use of such technologies.

Each nation had other differences as well. Great Britain began the period ahead of both the United States and all of Europe with respect to knowledge of computing in the 1930s and early 1940s, a lead it lost to the Americans by the end of the 1940s. France had to develop its base of knowledge of computing following World War II, but did so quickly, demonstrating the power of clustered expertise as it had in Grenoble, for example, where two academic institutions developed computing expertise literally across the street from each other by the mid-1950s, while other sites slowly converted their interests in mathematics or theories of information to the practical considerations of digital computing. These experiences mimicked the grouping of skills resident in Boston, albeit on a smaller scale. In the case of Germany, the case of Zuse demonstrated how few resources were put to the issue until the 1950s and even then, far less than in the United States, Britain, or France. The West Germans managed to deploy the technology as the nation's economy thrived in the 1950s and beyond. The French were quite paranoid about American government policies and the role of IBM in the effects they could possibly have on national defense and culture, concerns brought out in a series of books and reports that began appearing in France in the 1960s. On the other hand, the West Germans accepted the market-driven realities of computing more readily than either the British or the French.

At the most macro-level one can observe shared features of their diffusion of computing. Common across all three included relatively similar public policies that proved successful or failed on the same points (e.g., successful in implementing digital applications in defense, failed in creating "national champions"), existence of common products and practices from the most successful vendors, most notably IBM, and a generally healthy European economy during the second half of the twentieth century. Finally, in all three nations a large body of managers, public officials, and many thoughtful commentators on current events accepted the idea that use of computers was crucial to the economic and social welfare of their nations and individual institutions. This acceptance spread quite intensively in all three nations by the end of the 1960s, in other words, about 15 years from when such technologies became commercially available. That receptivity remains a topic hardly examined by historians, but it was ultimately a more important story than more examined themes, such as national champions or local IT industries. As suggested in the first chapter, the rate of their acceptance and subsequent adoption of IT proved remarkably fast when compared to other technologies, such as the telephone.

The story told in this and in other chapters is that the diffusion of IT involved far more than the development of new, better, and more cost-effective hardware and software.

It necessitated government policies, development of a computer industry to promote the technology as well, often resulting in inefficient but nonetheless collaborative alliances with local governments, existence of large institutions that could afford the technology, and that also meant healthy national economies, and finally a literate workforce that increasingly learned how to use IT. These features of IT diffusion mirrored to a large extent the experiences of other nations, not just the United States, or the three European ones just discussed. All of this suggests the need to step away from characterizing stages of IT evolution from the popular, more narrowly conceived notion of technological "generations" and reach toward a broader, more nuanced conceptualization of patterns of adoption and use, as sociologists have been suggesting for over three decades, most notably Manuel Castells, but earlier also Alvin Toffler and Daniel Bell. The experience of these three European nations offer powerful evidence that as in the United States and in many other countries, there seemed to be a Wave One at work in which development of the technology, its uses, and adoption were driven by large institutions. To be sure, the timing varied for each. Takeoffs in the United States began in the late

1940s, Western Europe in the mid- to late 1950s, Asia in the 1960s, Latin America in the 1970s, and parts of Sub-Saharan and South Africa at the dawn of the new century.

Because this book is centered on the general patterns of diffusion, which resulted as a consequence of both supply and demand activities at work in unique national economic and political circumstances, our discussion has been very limited with respect to subsectors of the IT marketplace, such as about software and semiconductors. Like other products, some nations were better at it than others. Germany's SAP software company is a shining example in one country where software development and sales of such products excelled generally what occurred in France or the United Kingdom.[188] On the other hand, all three attempted to get into the semiconductor business and did poorly, all for the same reasons as with computers. The Americans and then various Asian economies simply were faster than the Europeans to market with ever-improving products. Public officials fretted over strategic implications of their national economies not able to have a competitive local semiconductor presence, especially since this technology served as the building blocks for many other industries and, of course, for their Information Age.[189]

Wave Two was studiously not discussed in this chapter, but it should be noted that at its dawn individuals embraced the technology and adopted new uses of IT at a personal level, such as PCs, digital games, mobile telephony, and, of course, the Internet. For Wave One participants, the Internet served as a technological extension of networks they had relied on since the mid-1960s, which suggests while exploring the role of the Internet specifically is important, it must be situated into a larger mosaic of activities. Minitel taught us that acceptance and use of online systems during Wave One affected it and other adoptions of IT in Wave Two, in which circumstances had less to do with specific technologies than with other issues, such as cost, applications, and inability to migrate its services to the kinds increasingly available over the Internet. Yet in the process, in this particular example, a generation of French citizens learned about the concept of online access to information, a fact that the literature on Minitel has usually failed to acknowledge adequately. Ultimately that exposure may be recognized as the most important consequence of the Minitel's technology in France.

The experiences of these three countries were particularly visible to the rest of the world. In the 1940s, knowledge of computing remained within a small, closed circle of IT experts, office appliance firms and public officials in a few agencies, most so in France and Great Britain. Based on the evidence presented about Germany, only a few German scientists and engineers, along with some office appliance firms, and probably almost no public officials, were familiar with the technologies in the same decade. That hidden circumstance of IT began to change in the 1950s as the number of experts on the topic increased in all three countries, as office appliance and computer vendors began advertising, and as management teams in various organizations acquired these systems. In the 1960s, computing became a public, noisy affair. Academics began writing about knowledge or information societies, French writers wrote widely read books that complained about American companies overcoming their society with foreign technologies. We have the extraordinary case in France of President DeGaulle becoming involved, the highly publicized *Plan Calcul*, and then later Nora's and Minc's report to the nation on the future of French information society. In each decade the volume of conversation, writing, and media exposure about computers expanded in all three countries as it did across much of Europe, North America, and, to a lesser extent, Latin America and parts of Asia.

By the end of the century, terms used to describe modern life routinely included Information Age or Knowledge Society. All three nations discussed in this chapter were ranked and rated by international organizations on their tele-density. On a personal note, I led a team of researchers in IBM that collaborated with the Economist Intelligence Unit ranking scores of nations on their "e-readiness," in the early years of the 2000s. Each year the citizens in the nations we ranked number one read in their largest newspapers headlines like "Denmark Is No. 1; USA Ranks 3." In short, we no longer lived in the Nuclear Age or in the Cold War, now we were told that whole societies lived in the Information Age, now an almost universal cacophony of digital mantras and the *de rigueur* view of our times. The point is that many individuals in these countries viewed themselves through the eyes of information technology, or to cite a German example, *informations-gesellschaft*.[190] For the French, they were now living through *La Révolution Informattionnelle*. It seemed everything was changing because of the widespread use of computers.

Enough economic studies have been done on these three countries to draw the conclusion that the technologies involved (hardware, software, telecommunications, and services) led to increased labor productivity, jobs, and economic improvements, not to mention national competitiveness in various global industries. In the case of Germany, for example, numerical control machines— a competitive advantage for Germans—are highly computerized, thus the technology essential to national competitiveness. The longer computing is used, the better the value-add that develops, suggesting that there is a cumulative result. Between 1980 and 2000 in France, for instance, annual increases in productivity attributable to the use of IT was normally 0.25 percent annually, and since 1995, more in the order of 0.36 percent.[191] Outcomes varied by country, so, for instance, in the case of the United States during the same period, results proved greater because the French were more hesitant to swap out labor for computers than the Americans, and there remained a measurable lag in the speed with which the French used IT when compared to the Americans.[192] The closer a national economy was to the development of a new technology, the more it benefited from it, such as the United States with Silicon Valley, the Irish with their industrial parks outside Dublin, and the Germans with their research universities. The interactions among individuals in close proximity, along with national infrastructures that support innovations, account for this, such as access IT-knowledgeable local venture capitalists.[193] Finally, one should note that economists have observed when they measured the growth in GDP and investments made in IT, that, first, all the major economies of the world, including all Western European ones, invested in IT in large enough amounts to be measurable, and second, by 1990, the Germans were investing as a percentage almost twice as much as the French and British. The United States out-invested all three across the last six decades of the twentieth century.[194]

The rest of Western Europe had similar and differing experiences that need to be explored to offer a fuller picture of the diffusion of computing in Europe as a whole. Italy, for example, while always cataloged as one of the major users of computing had unique features, while the Nordic nations as a group embraced computing more extensively than many leading adopters around the world, a fascinating situation to explore among others. It is these other parts of Europe that we turn to next to add to our understanding of Western Europe's introduction to computers.

CHAPTER 4

Diffusion of Computing in Italy, Netherlands, and Sweden

It is like a gold fever; those who are not moving fast will miss the boat. Those who fail to take a lead, are forever left behind; those who fail to join the race, are forever handicapped.
—Dutch politician, *1982*

Not all European nations mirrored the economies, size, and cultures of France, Germany, or Great Britain, nor were other European countries as concerned in the same way about the invasion of American computer vendors in the 1960s–1970s, or later Japanese suppliers. Western Europe's experiences with modern technologies were conditioned by many circumstances, some of which we have yet to discuss. Just being a large country, as measured by GDP, population, or physical geography, while an important factor in the diffusion of ICT, did not automatically mean that a country would embrace computers early and massively. The experience of Italy, a nation with a large agricultural sector and many small industrial and services firms, has much to teach us about how diverse IT diffusion can be and of its rate of adoption. The example of the Netherlands gives us insights into the role of a small, yet highly productive nation in its use of computers. The four Nordic countries, while just as different amongst themselves as the citizens in the Benelux cluster of nations feel they are within their part of Europe, offer yet another path to computing remarkable in illustrating the role of a nation's long-term investments in educational and scientific assets. To test this last observation, we describe the Swedish experience. Each of the three national cases in this chapter is different, yet all became extensive users of computers by late Wave One. Their experiences shared some common elements with other European nations, but were also shaped by local circumstances.

As a result, there are several questions related to the diffusion of a new digital technology that these countries help us to address. First, what effect did the lack of extensive government involvement when compared to the American or British instances have on the diffusion of IT? That is an important question that the Italian and Dutch cases can help with, as well with a second question: To what extent does the existence of large enterprises and institutions affect absorption of computers? The Italian, Dutch, and Swedish cases reaffirm that during the earliest decades of the existence of computers, large organizations were important because they could afford these new technologies and could actually use them. What industries were most attracted to computing in Wave One? All three teach us that financial, telecommunications, process, and manufacturing industries, along with large government agencies proved essential, agriculture and retail less so, as in the cases of Britain, France, and West Germany. Do specific languages help or hurt diffusion, to the extent that there is evidence with which to judge? It appears in the cases examined in this chapter that languages spoken by only a few people did not retard diffusion of computing, only an appreciation for what they accomplished by those scholars who did not understand these languages (Dutch, Danish, Swedish, and Finnish come immediately to mind). Finally, we continue to explore the influence of the size of an economy on the rate of diffusion in this first era of computing, a period dominated mostly by public institutions and businesses and less by individual users of computers.

THE ITALIAN WAY

The model described so far would have one assume that a country with a large geographic footprint, substantial population, and a hefty national GDP would be an early and extensive adopter of information technologies, an observation one might also want to make of India and China. However, the Italian case suggests that such an assumption would be fraught with problems, as is also the case with Asia's two largest nations. To be sure, Italy was big. As measured by GDP in 2000 the nation had a GDP of roughly one trillion dollars, which was the same as China's, and only $300 billion less than that of France.[1] In 2000, Italy's population hovered at 57.9 million, nearly comparable to France's (59.5 million) and Britain's (59.8 million); only Germany was substantially larger (82.3 million).[2] However, it long experienced a low GDP per capita. For example, in 1950, with a population of some 47 million its per capita GDP was half that of the more prosperous Dutch. While GDP tripled by the end of 1973, and its population grew to nearly 55 million by that year, it lagged behind that of many other European nations; not until the 1980s did Italy enjoy comparable levels of GDP per capita as many of its neighbors.[3] In other words, the size of a nation's per capita GDP mattered. Poor countries took more time to use IT, such as India and China as a whole, even though there are sections of these nations that are teledense, but these are in a minority both in terms of the percentage of population engaged in the use of IT and as a percentage of the per capita GDP of their populations. As also happened in Africa, wealthier nations embraced computing earlier than poorer ones, such as South Africa earlier, and Sub-Sahara more sparsely later.

Italy's economic structure was also quite different than those of the United States, Britain, France, and West Germany. Italy's structural variations influenced profoundly the rate at which computing and other forms of ICT were adopted. The agricultural sector was the largest component of Italy's economy in the first half of the twentieth century, which remained so during the first two decades after World War II. This circumstance gave Italy the distinction of having the largest agricultural sector of any country we have looked at so far. Furthermore, most of its farms were small family enterprises. Agriculture was not an efficient sector by comparison to those of other nations. As late as 1971 this sector still employed over 18 percent of Italy's workforce. For comparative purposes, West Germany employed 8.3 percent of its labor force in agriculture, France 13.2 percent, and the British only 2.7 percent.[4] Agriculture did not become an extensive user of computing anywhere in Europe until the 1980s. Second, as many economists have pointed out, what industrial firms existed in Italy were small privately owned companies. This meant they leveraged largely only capital they raised within the enterprise and family, which also employed very few people. From the 1950s through the 1970s, it was not uncommon for industrial establishments to employ a dozen or fewer employees each. Again for context, other nations supported larger firms; for example, in Britain, the average industrial firm employed 88 people in 1971.[5] Earlier, in the 1950s, the employment profile of Italian firms reflected a pattern closer to 90 percent employing five workers or less.[6] In short, for decades the Italian economy remained less developed in comparison to those of many other countries in both its structure and extent of enterprises ideally suited for using computers in the 1950s–1970s.

Italy had one additional problem relatively unique to itself. Northern Italy proved sufficiently different from the South both culturally and linguistically that the North almost constituted its own economy, with extensive exports (often 40 percent of Italy's total exports were textiles, not industrial products) and home to many of medium- to large-sized enterprises. The South was viewed as "backward" (term often used by Italian public officials) while economists preferred to describe it as "underdeveloped," because it did not have an adequate number of jobs for its many unskilled potential workers. Its highly inefficient agricultural sector was a major problem, contributing to the region's low per capita GDP, and had inadequately provided education. Following World War II, national governments struggled with what to do with this region, while millions of individuals (one out of every five Italians) migrated north to the expanding industrial cities in the 1960s and 1970s. The social changes that occurred were some of the most profound in the Western world: women's access to birth control pills in a Catholic country, legalized divorce, day care for children, and legal rights for women to sign contracts to mention a few.[7]

Italy experienced with the rest of Europe many of the same types of damage caused by World War II. Italy was the locale of much warfare as the Allies fought their way up the Italian boot from the Sicily to Rome, while the Germans methodically destroyed property on their retreat. By the end of the war, agricultural livestock had declined by 25 percent from prewar levels, while agricultural productivity shrunk by 40 percent. Industrial plants declined in number by 20 percent (largely in the South) while the rest of the nation had largely obsolete worn out equipment. In 1945 industrial output only reached 25 percent of what it had been in 1941, and more like that of 1884. One-fourth of all railroad tracks lay destroyed, along with a third of all bridges, and 60 percent of all vehicles and wagons. About a third of all roads were unusable. All ports had been bombed, and maritime traffic amounted to only 15 percent of prewar levels.[8] Economic gains achieved during the Fascist administration of the 1930s evaporated. Like many European countries, postwar recovery involved the familiar story of avoiding starvation, overcoming inflation through the monetary reforms of 1948, and recovering by "repairing wartime damage and putting idle resources back to work."[9]

Like other nations, Italy began modernizing its economy in the 1950s, in both the agricultural and industrial sectors, doing well with exports of small manufactured goods, fashion items, textiles, and agricultural products. Recovery proved slow; in 1950, 44 percent of the workforce still toiled in agriculture, for example.[10] Yet between 1951 and 1971, Italy's economy grew rapidly, with national income rising, stable and highly competitive relative to other European nations. Per capita income increased more rapidly in this twenty year period than in the entire previous period (1861–1950). Leading contributions to Italian growth came from agriculture, industry, manufacturing, and exports. Unemployment dropped and the nation invested extensively in its economic recovery.[11] Movement of inexpensive and plentiful labor from farms to factory meant that the productivity of labor increased because the industrial sector, in particular, had more opportunities to modernize and increase outputs in the decades following World War II, with improvements in scale, scope, and technological prowess, and with minimal pressure for wage increases in the early years.[12] All of these actions occurred in spite of poor educational standards, universities more suited to train lawyers and humanists than engineers and mathematicians, and with some of the lowest school attendance levels in Europe. Firms that did well often exported goods to more prosperous economies, leveraging production of vehicles, chemicals, and textiles, to a lesser extent metallurgy, and exports of agricultural products and minerals. Wage inflation in the 1960s, led by increased union activities, and caused by a diminished supply of cheap labor, put pressure on the profits. Activity slowed but did not stop economic modernization, which continued into the 1980s and beyond.

Industrial policies of the national government in the postwar period concentrated first on recovery from the war, then on expanding exports, promoting modernization of industry, dealing with the continuously inadequate investments in social improvements in the South, and tourism. Only in the late 1970s did officials finally turn their attention to the issue of information technology, albeit for all intents and purposes, giving it only minor attention in comparison to the British, German, and French governments.[13] While it paid little attention to computers in the 1950s and 1960s, the government did invest in development of the nation's telecommunications infrastructure, as did most countries in Europe at the time for building infrastructure and in support of indigenous industry.[14] Key, however, is that Italian diffusion of computing is about a different path taken.

Lest one think that Italy was an economic or technological wasteland for information technologies, it should be pointed out that prior to World War II, businesses and government agencies used small forms of IT, such as typewriters, adding machines, and calculators, not to mention some tabulating machines. IBM, for example, began selling products in Italy in the 1920s, and established a local firm there in 1927, the Società Internazionale Macchine Commerciali (SIMC). Its first customer was the Ministry of Transport, which used IBM's equipment to do warehouse accounting and railway stock control. The following year SIMC opened its first sales office in Milan with 11 employees, followed by a punch-card manufacturing facility in 1931, the same year it established an office in Rome to process the national census data. In 1940, IBM had 309 employees in the country, a small number to be sure, but nonetheless a large payroll to be met. We know little about the history of the firm during World War II in Italy, but like other IT companies, its products were used, probably in a manner similar to what took place elsewhere in Europe. Meanwhile the Olivetti Company, an Italian enterprise founded in 1908 to sell typewriters, entered the local adding machine business in 1940.[15]

Because one way computing became available in a society was through academic projects involving construction and use of experimental digital systems, such episodes represent early markers of digital diffusion. Italy's experience was no exception. Four projects in Italy all began almost simultaneously in the early 1950s, initiating the nation's introduction to the technology, and in the installation of these few systems. These activities made available access to computing to a small number of people in the 1950s, largely academics, but also engineers at a few companies, most notably at Olivetti, at the time Italy's leading indigenous office appliance firm. These projects have been publicized largely as evidence of Italian involvement in early computing. To be sure, these are important early Italian computing, although exceptions to a technological landscape that otherwise remained fairly bereft of digital tools until the early1960s.

The Centro di Calcoli Numerici at Milan Polytechnic became one of these "first" Italian laboratories to install a digital computer, called the CRC 102A. Gino Cassinis, a professor there, had attempted to establish a calculation center since 1940, but the world war made that impossible. In 1951, while rector of Milan Polytechnic, he applied for a computer under the terms of the Marshall Plan and to his surprise, received approval for the project in 1953. He turned to a California manufacturer, Computer Research Corporation (CRC) to supply a system, which it did—indeed its first system—to the Italians in Milan. Luigi Dadda, then a young professor at Milan Polytechnic, traveled to California to observe its construction and learn about its operation. By late October, 1954, the system went into operation at the Polytechnic, and the following year into its permanent home at the Numerical Computations Center. Very quickly various academic and private-sector users gained access to the system to conduct both scientific calculations and industrial work. Users included employees from large Italian firms, such as Pirelli and Edison. In time the staff developed formal classes in computing which relied on this and subsequent systems (including Olivetti products) to train Italian engineers and scientists in the use of digital computing. From the earliest days of computing in Italy, Olivetti—Italy's one major indigenous computer vendors by the late 1950s—engaged in collaborative projects with each institution housing a system.[16]

A second system went into the Istituto Nazionale per le Applicazioni del Calcolo, a center established in Rome in 1932, and thus had a long history of working with various mechanical aids to calculation in mathematics and science. As in the Milano case, local academics decided they too wanted to buy a system and chose to acquire a MARK I from the British Ferranti company for approximately 300 million Liras, funded by the Italian National Research Council (known as CNR) and the Azionola Rilievo Alienazione Resioluati (ARAR), a war surplus disposition organization.[17] The MARK I arrived in Rome in December 1954, whereupon it was renamed the FINAC (based on Ferranti and INAC). With the help of two members of the Ferranti Corporation, the local staff worked on the installation of the machine. The system went "live" on June 22, 1955. As with the CRC 102A, both commercial and public institutions were given access to the system. FINAC remained in use for ten years, used by hundreds of individuals and by such organizations as the Air Ministry and the Treasury Ministry, various institutes located in Rome, Milan, and Turin, largely for scientific, engineering, and mathematical applications. As with Milan's experience, diffusion of knowledge of computing was not limited to experts using the system, but also through teaching about electronic computers by professors of mathematics, physics, and engineering at Rome University and at the

Istituto Superiore delle Poste e Telecomunicazioni. In time, this center collaborated with Olivetti on the construction of a replacement, leading in 1966 to the deployment of the CINAC (Computer—INAC).[18]

A third initiative took place at the University of Pisa, where local municipal support for acquiring a scientific instrument led to its funding. It turned out to be a complicated project. Deciding what kind of electronics to acquire did not center initially on a search for a computer; only after consulting with Nobel Prize physicist Enrico Fermi was the decision made to construct an electronic computer with Pisa's funds. Fermi believed it would be useful "for the students and researchers who will be trained in the use of these new tools for computation," again affirming the intent of the first three organizations installing computers in Italy.[19] Pisa University, other agencies, and Olivetti, collaborated in the funding and subsequent work. In March 1955, Pisa University established the Centro Studi sulle Calcolatrici Elettroniche (CSCE) to house the work effort involved in the construction and use of this system. CEP, as it was known, went "live" in November 1961.[20] During the 1950s and throughout the 1960s, Pisa became a hotbed of computer expertise attracting the likes of Olivetti and IBM both to establish local research laboratories. As with the other systems, CSCE trained and made possible diffusion of knowledge about computing in Italy in the 1960s and 1970s.

The fourth early Italian project involved Olivetti and its ELEA 9003. Based on its experiences in collaborating with the French Bull Company and the various local Italian computational projects, the firm acquired expertise in computing and, more importantly, an appetite to explore further the possibilities of such a product at the time commercial computing was still an undetermined risky business in Europe. Olivetti established its Electronic Research Laboratory in Pisa in 1955 for the express purpose of designing a commercially viable product. Its management hired a team of physicists and engineers and by mid-1958 they had designed a new system, the Elaboratore Eletronico Automatico (ELEA 9003), the company's first digital computer; it was an early European transistor-based device. In time, Olivetti built some 40 of these commercial systems in its factory in Milan, with the first one shipped to a company (Valdagno) in 1960. Olivetti eventually had orders for 200 systems, although it could not raise enough capital to fund the business, or deliver all of these, for various economic and managerial reasons. The company sold its computer business (Divisione Elletronica) to General Electric in 1964. The Italian government had proven unwilling to help fund Olivetti, an action that further encouraged senior management to sell off its product line to the American firm.[21] In a remarkable turn of events, however, the company decided after the sale to reenter the computer field, but this time with a much smaller system called the Programma 101, a product targeted at a different segment of the market for computers than before. The 101 has been characterized as the "first desktop-computer in the world," progenitor of what eventually became better known as personal computers, although it is unclear if its existence even influenced the development of early American desktop computers.[22] Ultimately, Olivetti built some 40,000 of these relatively easy-to-use inexpensive machines, beginning in 1965; the company retired the product line at the start of the 1980s.

These stories might lead one to believe that by the mid-to late 1960s, Italy was awash in computers. One inventory of Italian systems dated January 31, 1961, indicated that there were from all vendors installed in Italy only 20, of which 7 came from IBM (largely 650s), another 2 were French Gammas, while the rest came from various vendors and suppliers, including the ELEA, CEP, and FINAC for a total of 3 Italian made machines. The same list suggested that there were 17 systems on order, of which 11 were with IBM, another one each from Bull and Bendix, and four from Italian sources, the ELEA.[23] Another survey put the population of installed systems in 1960 at around 85 with another 65 on order, with the majority with American vendors. Of the on-order systems, one observer noted "that two-thirds of this equipment is for commercial banks or the manufacturing industries," while "Italian government agencies have followed the commercial companies in becoming interested in computer applications."[24] Additionally, given IBM's considerable presence both in Italy and in Europe, looking at its business results suggests what was happening in the Italian market. In an internal marketing review of sales operations in Europe written at the start of the 1970s, IBM managers in Europe proudly boasted that "IBM Italy is the sixth largest subsidiary of [IBM] World Trade," generating a gross income of $310.4 million in 1970, a sum from combined sales in Italy and exports of locally manufactured products to other European markets. More important, "it [IBM] ranks as the eighth biggest company in Italy." During the 1950s and 1960s, the firm expanded such that by 1971 it had 20 sales offices and 11 data centers in Italy to address needs of its customers, and, in addition, an education center used by customers and employees in Brescia, and three scientific centers. IBM manufactured System/360 Model 20s in Italy, along with two models of its minicomputer, the System/3. As both an indicator of the health of the business in the 1960s and what was to be expected in the 1970s, the firm's local management reported it had 6,864 employees in the country, of which 1,900 were in manufacturing, leaving nearly 5,000 working directly with customers. IBM headquartered its Italian operations in Milan.[25]

Business grew to such an extent in the 1960s that in August, 1971, IBM Italy could report to senior executives in New York that it "owned" 78.5 percent of the computer market. It knew which firms, universities, and government agencies had both IBM and competitive products installed or on order (the kind of information all IBM country organizations and their competitors routinely collected throughout the twentieth century). By late 1971, Italian customers had installed 1,094 systems from IBM's competitors. Doing the calculation to include IBM's, leads to the conclusion that in total there must have been some 4,000 systems installed in Italy, an achievement that allows us to conclude that there were thousands even in the late 1960s.[26] IBM's presence in Italy was overwhelmingly clustered at large industrial and banking complexes and government. In 1952, IBM opened sales offices in Milan and Palermo, in 1953 in Bologna, and in the following decade, offices all over northern and central Italy, fewer in the south. In the 1960s, large banks and government agencies, along with the few large manufacturing firms began acquiring IBM computers, along with typewriters and other smaller office appliances and cards. By 1977, IBM had 9,000 employees in Italy and just three years later, 11,829. Despite difficult economic times in Italy, the company continued to prosper. In the early 1980s the company expanded further its sales and manufacturing staffs, employing over 12,000 people by mid-decade.[27]

What the various indigenous computer projects and IBM's story demonstrate is that Italy's adoption of digital computing was largely a startup process in the second half of the 1950s with the most minimal of governmental support on the supply side of the story, and that the period of the 1960s and 1970s represented the era in which the diffusion of digital computing began in earnest. As in many other European nations, local companies (in this case only Olivetti) did poorly with computing while IBM came quickly to dominate the supply side of the diffusion of computing, coupled with the ancillary activities of a few academic institutions which trained engineers and others so necessary for the spread of computers.[28]

The experience of the 1980s, however, reflects a period in which Italian adoption of computers took on the same characteristics as in many parts of Europe. The national economy had continued its march toward industrial expansion at the expense of agricultural predominance of earlier times, development of larger institutions and firms that became natural consumers of big computers, while the technology itself became available in smaller, less expensive units. Banks, government agencies, and large manufacturers became extensive users of digital technologies. For instance, Italy's machine tool industry kept abreast of technology, making it an important contributor to the nation's export business. Olivetti also played a role as an important provider of numerical control products to the European market, offerings that had embedded in them digital computing. Comau, a Fiat-owned company established in 1977, quickly became known for its robotic products, and for handling assembly and control functions in manufacturing, which incorporated embedded digital computing.[29] Other manufacturing industries infused IT into logistics and telecommunications, beginning in the 1970s and extending through the 1980s, but mostly into the largest enterprises (for example, Fiat with CAD/CAM, logistics, and planning). Smaller firms—and there still were many in Italy—continued to lag in the period, most notably in textiles and automotive supply. To be sure, that situation evolved as networking extended to the smaller suppliers, creating an interdependent web of activities between major companies and smaller firms, much as occurred in American and Japanese automotive industries of the period.[30]

Vendors and industry watchers were normally bullish about the Italian market for computers, beginning in the 1970s and extending to the end of the century, although they recognized that it remained smaller than those in France, Germany, and, of course, the United States. By the end of the 1980s, however, all markets were large. The United Kingdom had some 6.49 million computers installed of all sizes, West Germany 5.01 million, France 3.83 million, and Italy 2.49 million. Interestingly, the human populations in Germany and Italy were roughly the same (61 million in West Germany, 57.6 million in Italy) which suggested that the rate of penetration in Italy lagged or, in market terms, that there was additional opportunity, especially with small systems, such as personal computers. Italy also remained a popular market for the commercial software vendors that now proliferated the world IT market by the end of the 1980s, albeit less so than other national markets.[31]

There has been much discussion in Europe among economists about the extent of investments in IT. Their evidence indicates that in Italy's case, its economy's investment in IT in the last two decades of the century hovered at between 9 and 13 percent of total capital investments, with the greatest amount in the 1980s and surging in the first half of the 1990s, rising to about 16 percent by the end of the century. That level of investment in ICT (includes telecommunications) was roughly a third less than other industrialized nations in the 1980s, but comparable in the 1990s.[32] Asking how much IT investment as a percentage of national GDP indicates that, at the start of the 1980s, it hovered at about 1.5 percent, by 1985 rose to 1.9 percent, grew to 2.3 percent in 1990, then leveled out for the rest of the century. For comparative purposes, it tracked almost exactly the average for the European Union in this period and the measured effects of these rates of investment in Italian IT on national productivity and growth in GDP mirrored that of most other European countries.[33] Data accumulated by the Bank of Italy indicated similarly that in each year during the 1980s the amount invested in IT increased each year as a percentage of total investments, flattening out in the early 1990s during a near-global business recession then continued to rise until 2001 when once again both a national and international global recession slowed investments. Unlike in the United States, where the IT industry also contributed substantially to national productivity, that in Italy had a less positive effect, because it remained small in comparison to those of Japan, United States, and Germany, for example.[34] Economists at the OECD noted that for all of Europe, computers were more influential in national economic gains in such areas as wholesale trade and financial services, than in manufacturing or in indigenous IT industries.[35]

Size also mattered. Already noted throughout Europe's experience, smaller enterprises either did not acquire IT as quickly or waited until they could afford it. In Italy's case, these circumstances prevailed. But to put a sharper point on the matter, many of Italy's small firms and their industries specialized in products and services that simply did not need IT as much or as early as others and that were subject to competition from emerging economies, such as in textiles, clothing, leather, and footwear. Industries that required early and extensive use of IT were not as prevalent in Italy either, such as in chemicals, pharmaceuticals, and consumer electronics. But ultimately it was about size and opportunities for improving productivity. One Italian economist noted that "adoption of ICT technologies is positively correlated with the size of firms" in his country.[36]

Because the existence of an IT industry, and employment in IT, are indicators both of how IT diffused in a nation and its extent of deployment, economists routinely track such information. In the case of Italy, ICT industries contributed 4.4 percent to the nation's GDP by 1990, suggesting that in the 1980s this industry expanded to 4.4 percent, ending the 1990s at about 4.7 percent. For comparative purposes in 1990, ICT industries in Germany contributed 5.4 percent to its total GDP, France 5.0 percent, Britain 5.7 percent, and the United States 6.6 percent. As a percentage of employment, again using 1990 as an indicator of how much had been accomplished in the 1980s, and projecting into the early 1990s, 3.6 percent of Italy's labor force worked in ICT industries. In Germany labor employed in ICT comprised 4.6 percent, in France 3.8 percent, in Britain 4.6 percent, and in the United States 4.5 percent. Economists have had difficulty measuring how many industries and companies in various countries became extensive users of IT, but extant evidence suggests that Italy's experience mirrored that of Germany, Japan, France, Britain, and the United States by 1990, remaining so through the next decade. Where IT went into Italy mirrored that of the rest of the world as well by the 1990s: publishing and printing, chemical, electrical and electronic machinery and equipment, medical and measurement tools, wholesale trade, telecommunications, financial sectors, and select government departments.[37]

In Italy and elsewhere the effects of growing numbers of workers using IT did not reflect the full story of diffusion. Two economists from the OECD concluded that how nations deployed their work forces had a greater impact on a nation's productivity than computers and their use by better-trained employees. These economists identified a relationship between investments in IT and economic results, reporting that

> Most of the countries that experienced an acceleration in GDP per-capita growth also recorded an increase in labor utilization (e.g., the USA, Australia, Ireland, and the Netherlands), while most of those where employment stagnated, or even declined, saw a deterioration in their growth performance (most notably Germany and Italy).[38]

From this pattern of behavior they concluded that in Germany and Italy, "labor productivity growth has not been able to offset the negative contribution to growth coming from poor employment performance," while more specifically in Italy improving skills of workers proved inadequate.[39] Nonetheless, in each OECD country, as the cost of technology declined, investments in IT increased. While we can quibble over the data itself, Italy's experience in the 1970s–1990s with respect to labor productivity had less to do with how much IT was installed than in the classic Wave One problem of firms and agencies not yet fully reorganizing how they did their work and organized their enterprises. The evidence presented of earlier adopters indicated that once such a process of transformation occurred then labor and national productivity (including the yield from IT investments) increased sharply, most documented by the experiences of the United States and Japan early on, and South Korea later.[40]

What role did the Italian government play in the diffusion of IT from the late 1970s to the end of the 1990s, which for Italy was its Wave One era? In addition to being a consumer of IT for its internal operations, about which we know little other than it was a major customer of IBM, Olivetti, and even Hitachi, Sharp, Fujitsu, and other vendors,[41] there remained the issue of the role of public policy and how it influenced national diffusion of IT. The Italian government focused its initial economic development energies on reestablishing its core industries damaged by World War II, then built on these during the 1960s through the 1980s, from textiles to fashion, tourism to chemicals, and agriculture, of course. Like other European countries it published formal economic development plans every few years. In the late 1970s government officials began paying serious attention to IT matters, funding R&D work, for example, at the University of Milan.[42] Olivetti was normally starved for capital, and so R&D in ICT by this firm and nation as a whole (computing and telecommunications) remained at lower levels than in most other West European countries. Investments in local champions in IT were virtually nonexistent. Nonetheless, the Italian experience shows that capitalism worked; every major IT vendor in the world attempted to sell products in Italy between the 1970s and the end of the century. This applied to software of all types, mainframes, personal computers, and earlier, to minicomputers. That is why, for example, a company like IBM generated healthy volumes of revenues in the country. With the value of the local labor and land costs lower than in some other parts of Europe, it could establish a base there for producing products for export to the rest of Europe. So the lack of a heavy-handed public policy role for government did not inhibit the diffusion of IT as much as one might otherwise have thought.

Observers of the Italian scene thought otherwise, commenting on the paucity of public support, which did not translate into slower diffusion, as much as did the structure of industries and sizes of companies. The widely accepted mantra of the period held that for Italy to become a modern society it needed to participate more aggressively in creating an indigenous IT industry and to use the technology more extensively in the nation's work. Observers suggested legislative initiatives, infusion of capital into local suppliers, more expenditures on R&D (especially in those industries that were users of IT, such as machine tools), and in training and education.[43] The government supported some initiatives, albeit late and less than in other nations such as Germany and Britain. For example, it invested in superconductivity research, beginning in 1978, and the following year in information technology. Like other nations it also had other investments to support, such as air traffic control, lasers, chemicals, metallurgy, transport, and energy.[44] It also invested in the use of IT as an economic development initiative, a slow process that extended into the twenty-first century.[45]

Why did the Italian government do less than others? While more research is required to answer confidently this question, the answer might actually be quite obvious. Given the size of Italian enterprises, and the mix of industries, the nation had an economy that did not necessarily need (or could use) expensive computers in the early decades of computing to the extent as could more industrialized economies which had more highly concentrated, even oligopolistically dominated industries. Second, the nation had a workforce and level of GDP lower than more technologically advanced countries. But a third reason, rarely acknowledged, is that whatever IT was required was amply supplied by IT vendors, most notably IBM, also by Bull, Olivetti with smaller systems, and a plethora of largely American PC suppliers, such as Compaq and later Dell.[46] With respect to local computer chip manufacturing, an IT subindustry different from the computer and software IT industries in both products and customers, the government did express more interest and concern. The market for such components lay in the manufacture of current complex products, such as for automobiles, trucks, and machine tools. Even in this market, international suppliers, such as Intel and various Asian suppliers, proved aggressive in finding Italian customers.

In this submarket of IT, government initiatives proved too little too late, particularly by the 1980s. Nonetheless, officials made small investments in the local microelectronics industry throughout the 1980s and 1990s.[47] Universities added courses about computer technology in these last two decades of the century, mimicking what had started in the United States and Great Britain in the 1950s and in France and West Germany in the 1960s. Therefore, the process of diffusing knowledge about IT in Italy was not left just to large computer vendors to achieve through their normal product-oriented training programs, although extant evidence suggests they did more such training than public educational institutions.

In the late stages of diffusion of computing in Wave One, evidence of widespread public discussion exists about the role of technology in shaping modern societies. It began in the early 1950s in the United States with discussions about "Giant Brains" and "automation" in the Netherlands, in France in the 1960s and 1970s with concerns about American computing affecting the local economy and society, and by the early 1980s, with other debates in West Germany about its economy and society. Evidence of such discussions are normally public policy pronouncements (such as with Simon Nora's and Allain Minc's report on France in 1978),[48] newspaper and television coverage, and ultimately in book-length analyses of local IT events. In the case of Italy, book-length discussions began in the late 1980s, but not until the 1990s and later did studies appear about technology and information societies. Most focused on the effects of computing in American, British, French, and Japanese societies. There was a near complete absence of any analysis of the Italian situation, or discussion about how technologies were beginning to cause fundamental changes in how corporations did their work in Italy.[49] By the mid-1990s, however, academics began suggesting more aggressive national policies to promote the transformation of Italian society and economy into some information age paradigm. For example, publishing in 1996, two authors, a political scientist and a software engineer, described the elements of an information society, dedicating a chapter to suggesting Italian public policies. They argued that IT was waning in Italy and that government officials needed to "stop the decline," with more aggressive policies.[50] They advocated strategies for developing a local IT industry along the lines followed by other European countries.[51] The conversation seemed late by West European standards.

Some scattered data on utilization of IT reinforces this case. First time users of the Internet came online late in the 1990s, but then did it with a rush with about 12 million users in late 2000. Individual subscriptions went from roughly a half million in 1998 to an excess of 3.7 million the following year, ending 2000 with some 4.5 million; but keep in mind multiple people were tied to one account, such as a family with one household subscription. At the dawn of the century roughly 32–33 percent of all households had a PC, still a low rate of penetration certainly by northern European standards and a limiting factor in Internet use because in those years one could access the Internet only by using PCs. Online transactions have been described as "marginal," while business-to-business transactions barely reached $195 million in 1999, although it climbed to some $350 million the following year. So the results are clearly those of a late Wave One society. The rapid growth in usage that came at the end of the century was made possible by improved access to the Internet and ICT in general, to free access to the Internet, and to favorable government programs designed to promote usage generating results. These results included new consumer protection legislation aimed at online transactions, which led to additional expanded deployment of the Internet in subsequent years. In some categories of uses, volumes lagged, for example in online sales, which remained relatively low, always a lagging use of online services.[52]

NETHERLANDS—A CASE OF A SMALL YET DYNAMIC ECONOMY

Fabled for its international trade and prosperity in the seventeenth century and its iconic windmills, themselves clever technology, the Netherlands has long enjoyed a prosperous, dynamic economy. Situated to the west of Germany, south of the Nordics, and north of Latin Europe, it served as an economic and cultural crossroads. A densely populated, well-educated society, with most citizens able to converse in two or more languages (particularly those expert in IT), the Dutch had a population of nearly 16 million in 2000, qualifying it as a small European country, organized into 12 provinces. However, its per capita GDP was impressive, nearly twice Italy's already by 1950 ($5,850) and double in 1973 ($12,700), and which grew additionally to $16,900 by the end of 1992.[53] Its economy comprised a few very large enterprises, such as Philips the electronics firm, a large usually highly successful agricultural sector, and two ports through which a great deal of goods and raw materials flowed. Rotterdam is Europe's largest port or entry into Germany's economy from the West, and Amsterdam, which for centuries has been Europe's access point to the Dutch colony of Indonesia, and that served much trade in the Atlantic. While a late arrival to the Second Industrial Revolution (1890s), the Netherlands became an active participant during the twentieth century. By the end of the 1960s, it was also comparable to Scandinavian countries in its social welfare spending, as measured by share of social services in GDP. Labor unions wielded enormous influence on public policy. Several features of this society important to the adoption of computers included generally lower inflation rates than Europe for most of the twentieth century, higher labor productivity compared to neighboring countries, a stable sociopolitical system of society and government, and open to trade. While the nation's tradition of financial and monetary stability changed in the 1960s and 1970s, with the implementation of expansionary policies, nonetheless, the economy remained stable. That stability ensured a strong currency, all resulting in rapid growth in GDP.[54]

Throughout this book I have argued that having large enterprises and governmental agencies facilitated rapid acceptance of computing. Did these kinds of institutions exist in the Netherlands? They did and came to dominate their sectors in chemicals, electricity, machine building, oil refining, banking, insurance, and consumer electronics, to mention a few. In classic Chandlerian fashion they invested in R&D, production, distribution, and service of their offerings, leading to the development of multidivisional firms. To put a fine point on this topic, if we look at the value of all assets owned by the top 100 firms in various countries, we see the Dutch highly concentrated. In 1950, the top 100 Dutch companies controlled 62 percent of all assets, and for comparative purposes, in the United States that figure was 18 percent, 22 percent in Britain, and 16 percent in West Germany. If one leaves out Royal Dutch Shell, then the nation's figure is 34 percent, still very high. In 1973, at the midpoint in Wave One deployment of IT, the Dutch percentage had risen to 88 percent (52 percent if we leave out Shell), which still outpaced other countries: 30 percent in the United States and 18 percent in Britain. Moving to 1990, the Dutch percentage remained at a high 77 percent (42 percent without Shell), which compares to the United States at 35 percent, to Britain at 38 percent, and to the unified Germany of 31 percent.[55] When combined with the growth of government within the economy, one has a nearly ideal institutional environment suitable for computing.

In the 1960s and 1970s, when compared to its economic rivals in West Germany, France, Britain, and elsewhere for the all crucial export markets within an increasingly pan-European economy, to compete effectively Dutch firms in banking, insurance, and manufacturing had to expand and take advantage of economies of scale, and stem declines in profit, hence the fundamental attraction of computers to assist automating activities to help drive down operating costs. As computers became less expensive and smaller in the 1970s and 1980s, smaller enterprises could also leverage technology to become active players. They became a major source for innovative products and services in the Dutch economy in the 1980s and 1990s as well. When the total number of large and small firms is combined, they actually increased in number during the 1980s and 1990s. In industry, mining, transportation, and telecommunications, their number grew by over a third, while in business services they doubled.[56]

The Netherlands enjoyed one of the fastest recoveries from World War II experienced by any European nation. The period from 1950 to 1973 is often referred to as the "Golden Years" by most historians and economists, one characterized by a liberalization of the economy contributing to stable economic growth. The global oil crisis of 1973, which so drove up the costs of energy—the Dutch were some of the most extensive users of all manner of fuel—ended the era. From the oil crisis to the mid-1990s, GDP grew at about 2 percent per annum, down from the more normal 5 percent experienced in the previous quarter century. In the 1970s growth still exceeded that of most European countries, but lagged behind in the 1980s, then after 1987 the economy began a long period of growth relative to European and OECD countries in general. Why all these activities turned out the way they did is still subject to much debate by economists, which we will not address here, other than to intuit that technological innovations of all kinds (not just computing) affected the performance of organizations in combination with other activities. For example, during the depression of the 1980s, labor relations were restructured to dampen inflation in wages then underway, many firms retooled with new machinery (not just IT), and public policies aimed at economic stimulus. It all worked, and most largely by the success of restraining the growth in salaries than through increases resulting from the use of computers or exports, since even in Europe as a whole the entire region's export trade stagnated during the second half of the 1970s and first half of the 1980s.

As in most European countries, in the Netherlands use of punch-card tabulating equipment and other office appliances by large organizations in the decades prior to the arrival of the computer had been widespread. Large data-handling enterprises such as insurance companies, banks, and government agencies reflected this relatively normal practice. For example, the Centraal Bureau voor de Statistiek (Netherlands's government main statistical agency) had used Hollerith equipment since the dawn of the twentieth century to create statistics, and all manner of contemporary IT by the start of World War II. To leverage effectively such technologies, many of its operations were already centralized by the time computers became available. Thus, it became institutionally receptive to the new technology, albeit with the usual skeptical assessments in the beginning about the efficiencies and value of the new technologies, of course, evident all over the world in the 1950s and early 1960s.[57]

Mimicking the interest in the new technology in the 1950s in the United States and Britain in particular, a number of early projects got underway in the Netherlands, numerous given the fact that the nation was relatively smaller than others which also launched digital initiatives.[58] These can be summarized quickly as they mirror patterns of early adoption already described for other countries. The Mathematisch Centrum (MC), established in 1946 in Amsterdam, could not afford to buy a computer, but wanted to experiment with the technology to learn about it and of its possible use in mathematics.[59] It chose to build one, starting work in August 1947, even though its staff had little access to extant literature on computers, due largely to the world war interrupting the normal flow of scientific and engineering information. They attempted to build a digital computer (named ARRA) between 1948 and 1952, but failed to get a workable device, largely for technical reasons. The MC then hired Gerrit A. Blaauw, a Dutch electrical engineering graduate who had learned about computers at Harvard under Howard Aiken, applying what he learned at the MC. Thirteen months after starting, the ARRA II ran its first program in December 1953, and the MC used the system for three years. While a primitive computer, it afforded many engineers the opportunity to learn about the technology. These individuals went on to create the Dutch computer industry in the late 1950s and early 1960s.[60] One way IT knowledge transfer continued occurred with the MC building two additional machines between 1953 and 1956, called the FERTA (Fokker Elektronische Rekenmachine Type ARRA) for the then legendary Dutch airplane manufacturing firm, Fokker, to conduct flutter calculations in the design of its airplanes, and the second machine was the ARMAC (Automatische Rekenmachine Mathematisch Centrum) for use by the MC. The ARMAC held a certain celebrity status among historians because one of the most important early European computer scientists specializing in software and programming, Edsger W. Dijkstra, worked on its construction.[61]

With commercial availability of computers now a reality, the original purpose of building machines in the MC dissipated and so the computer builders at the MC, with the financial backing of the Dutch insurance firm Nillmij, established the first commercial computer firm in the Netherlands, called Electrologica, in 1956. By 1958 45 members of the MC had moved to the new firm and built their first commercial computer, the X-1, completing the work in August 1958. A contemporary American observer noted that "the X-1 computer has been well received in Europe" and that as of mid-year, 1960, 14 were on order of which six were destined for Dutch organizations and eight for others in West Germany.[62] In time they sold 40 of these systems. The X-1 was fast, was constructed using transistors, and proved very advanced technologically for its day. Other machines followed, such as the X-8, introduced in 1965, but it turned out to be a commercial failure in the face of IBM's more reliable products and effective sales operations. One official at the Dutch firm acknowledged later that "discipline in their [IBM's] factory was better than ours."[63] The firm sold machines only in Holland, Germany, France, and Switzerland, possibly one in Italy. Following an all-too-familiar pattern evident in other European countries, this local firm accumulated too much debt, proved grossly under capitalized, and had to be sold to salvage some value. In 1968 Philips acquired the firm.

A second line of diffusion of computing took place at the Dutch Post, Telephone and Telegraph Company, best known as PTT, originating in its mathematical and research departments, where the need for aids to mathematical analysis of telephonic network behavior mimicked the kinds of concerns expressed at AT&T in the United States, particularly with respect to modeling traffic. Dutch engineers were added to the staff, including W.L. van der Poel, who, like Konrad Zuse, studied and tinkered with computing on his own during World War II, designing with telephone stepping switches and, like Zuse, had no funding before joining PTT. In the early 1950s, he and others constructed PTERA (PTT Elektronische Reken-Automaat), completing their work in September, 1953. Next, they built a commercial machine, called ZEBRA (Zeer Eenvoudig Binair Reken-Apparaat), with initial delivery of the system in 1958. It remained in production to 1964; in total PTT sold 55 systems around the world. As with IBM's products, a small group of ZEBRA experts in the Netherlands and elsewhere promoted use of digital computing. Elsewhere in the firm various other computers were installed for normal accounting and operational automation in the late 1950s and throughout the 1960s, largely sourced from IBM.[64]

The third early Dutch initiative involved one of the world's largest electronics firms, Philips, which launched a computer project at The Philips Nat Lab. The company, much like Siemens in Germany or General Electric in the United States, had adequate technical staffs, funding, and facilities to mount an initiative to get into the commercial computing business in the 1950s. It was large, wealthy, and well linked to other firms, including IBM. It launched its first construction project in 1953 and by 1960 had put into operation three systems for its internal use to conduct scientific and engineering calculations, and in support of administrative and accounting operations. Senior management decided initially not to enter the commercial computer market, turning down offers to collaborate with other Dutch enterprises on such a project, most notably PTT and the MC. Management believed that Philips' core competence resided in the construction of electronic components, not electronic machines, including computers. Furthermore, the company signed an agreement with IBM to sell components to the American firm in exchange for not entering the computer market. This arrangement proved to be a strategic blunder made by Philips, sadly the first of several made regarding IT in the years to come, because in time IBM made its own components and came to dominate both the Dutch and European computer markets before Philips's management realized their mistake. Furthermore, the agreement to produce components for the American firm terminated in 1963 at which time IBM chose not to negotiate a new arrangement. Philips then entered the computer industry later that year, accessing the market by acquiring Electrologica as its way of making a fast entrance into the field. Philips introduced its first computer in 1968, but by then it was far too late; the market had moved on and it was largely IBM's.[65] Philips became a cautionary tale for many vendors from which they learned that entering a high-tech market required one to be quick, early, and decisive, with emphasis on early speed to market.

On a more positive note, the Dutch nation's experiences with and use of computers in the 1950s and early 1960s matched the capabilities of the technology quite well, focusing on scientific, engineering, and mathematical applications to manage telephone networks, analyze stress of materials, perform telephone accounting, and insurance actuarial analysis. In short, computers became tools quickly useful in response to the growing complexity of mathematical and statistical work. Various organizations designed about ten different types of computers in the Netherlands before 1960, placing the Netherlands in comparable volume of engineering creativity underway in the United States and Great Britain, and built machines that were normally state-of-the-art for small systems; not for large ones, however.[66]

A reasonably reliable inventory of installed systems in the Netherlands suggests that between 1952 and the end of 1960, some 30 systems were installed in the country in the agencies described above, but also in other large enterprises, such as at Shell, Fokker, Nillmij Insurance Company, Unilever, Royal Dutch Airlines (KLM), at the Central Bureau of Statistics, Twentsche Bank, and at various universities and government agencies. All were large organizations involved in diverse industries. Some applied the systems quite effectively. An American employee of the U.S. Embassy stationed in London prepared a detailed survey of West European computer installations in 1957 in which he commented about Shell Laboratories' Mark I (installed in 1955) that "their utilization record is one of the best the writer has ever encountered."[67] IBM 650s proved popular (5) along with several other IBM computers, while locally built systems dominated with 18 machines.[68] Another inventory of installed systems covering the years 1962 through 1971, and also reasonably accurate, reported that 69 systems were in use in 1962, 553 in 1967, and 1544 by the end of 1971, which would make the Dutch intensive users of computers at the time. This inventory determined that IBM supplied about half the systems, Dutch suppliers roughly 10 percent, and the rest divided among other American and European suppliers.[69]

A tight labor market in the 1950s and 1960s provided economic justification for these early and subsequent installations, especially by such large employers as Nillmij and PTT. Government agencies and Shell, on the other hand, were less interested in improving the efficiency of labor through investments in IT than in technical and mathematical uses made possible by the new technology. A more urgent personnel issue was the inadequate supply of people expert in the use of computers in this era, a paucity of skilled IT professionals sufficient to serve as a partial brake on the rate of deployment since one needed operators for machines; programmers and systems analysts to design, write, and maintain software; and management knowledgeable about the technology to run data-processing organizations. It was often a global problem not limited to the Netherlands.

American vendors came into the Dutch market in the 1960s and 1970s, such as IBM of course, but also Bull/GE, Sperry Rand, CDC, and DEC. By 1967, IBM dominated the local market, with about two-thirds of all locally installed systems bearing its name.[70] Part of what the firm had to do to increase demand for its products was to offer service bureau services to allow smaller enterprises to start using computing and to train users about data processing in general and more specifically about its products and ways of using IT. IBM offered these two services regularly in all its markets for the majority of the twentieth century. Those twin activities increased the pool of users, hence advocates for computing within enterprises and agencies, a group whose largest body of knowledge was how to use only IBM equipment. This is not the place to discuss features of that community, but suffice to say that as IBM technology spread, it was in the best interests of technical staffs to be expert on this vendor's system if they were either to progress in their careers, wherever they were employed, or if they wanted to move to another enterprise, one that probably also used IBM equipment.

The availability of IBM's 360 systems in the mid-1960s raised expectations for advanced computing, and while IBM experienced many technological difficulties in making its new system work in mid-decade, Dutch users like others around the world were keen to move their programs from the IBM 1401 and 650 to the larger 360s, to have more capacity and to save on the costs of hardware.

Given that the Dutch had become extensive users of computing by mid-decade, the challenges for IBM were greater than in many other countries. IBM expanded its operations in the country in the 1960s and, by 1969, had a combination of nine sales offices and manufacturing facilities in the country.[71] As it experienced in so many other European countries, IBM first began doing business in the Netherlands in 1920, working with a local subsidiary. By 1940 it had 100 employees in the country, most of whom were able to work with tabulating equipment in Holland during the war. In 1950 there were 272 employees there. Throughout the 1950s manufacturing and sales offices were established or expanded, with sales offices opened in rapid succession in The Hague (1953), Amsterdam (1955), and Utrecht (1956). By 1964, the company had 2,000 employees in the country, just as the S/360 came to market. The factory in Amsterdam was one of IBM's largest manufacturing sites for typewriters and a major locale for the production of other products. Between the early 1950s and 1972, IBM had manufactured a half million typewriters in Holland. By 1974, the company had almost 6,000 employees in the country, further evidence of the IT-density of that nation's economy.[72] The key observation to draw from the various data about IBM is that the firm invested in enough staff not only to meet growing need for computers, but also to stimulate that demand beyond what other American, European, and Dutch suppliers could do, particularly through the expansion of its sales network.

Results were palpable. In a study commissioned by the OECD in the late 1960s to assess the extent of computer uses, economists took a measure of the number of computers per million workers in a dozen countries. The data are fascinating (displayed in table 4.1) because it suggests that the Dutch had already begun using computers more intensely than the British, Italians, Japanese, French, or their neighbors, the Belgians. Only the Germans, Canadians, Swiss,

Table 4.1 COMPUTER INSTALLATIONS, SELECT COUNTRIES, 1966

Country	Number of Systems per million workers
Belgium	79.6
Canada	132.9
France	76.4
Germany	100.9
Italy	57.9
Japan	42.9
Netherlands	89.3
Sweden	91.1
Switzerland	138.7
United Kingdom	64.8
United States	361.2

Source: Adapted from OECD, *Gaps in Technology: Electronic Computers* (Paris: OECD, 1969): 11.

and the Americans outpaced this country.[73] The Americans accounted for about 95 percent of the world's production of digital computers in the 1960s, the majority of which were installed in Western Europe and in North America.

The same report noted that in Western Europe the annual increase in computer installations in the period 1962–1966 averaged 45 percent. It also points out what economic historians and technologists have always found remarkable about digital technologies, and so it is worth quoting one of these scholars:

> This high figure, however, greatly underestimates the effective growth of computing power: prices have been falling rapidly, computer systems have become considerably more powerful, more reliable and easier to handle; moreover, the range of applications has expanded considerably and they have become more sophisticated. This evolution has been made possible by great improvements in the computer hardware . . . by the development of new types of software . . . and by a more general phenomenon consecutive to the diffusion of computers throughout the economy: computer users have become considerably more experienced, and have been developing many new applications which the manufacturers themselves had initially not thought of.[74]

The quote mirrors the Dutch experience, as much as that of any other nation.

The aggressive, yet effective, push by American vendors contributed substantially to adoption as well. In 1962, 55 percent of all computers in the Netherlands were American supplied; that installed share grew to 67 percent in 1967. For comparative purposes, in France the percentages went from approximately 51 percent to 65.5 percent, in West Germany from 75.5 percent to 78.3 percent reflecting a de facto maximum possible saturation (since there were multiple vendors in the market), while the Italians were at approximately the same range. Only in the United Kingdom were the percentages dramatically different, reflecting the existence of an indigenous computer industry, 20.5 percent in 1962 and 55.4 percent five years later.[75] OECD reported that in the Netherlands, Dutch suppliers only satisfied 12 percent of their local market in 1962 and that their share fell by half by the end of 1967 to 6 percent. French and German suppliers were able to satisfy just over 20 percent of their respective local markets, so even in those nations, we can see that diffusion depended a great deal on American suppliers just at the moment in time when Western Europeans were experiencing their initial great take-off in their adoption of computers.[76] To define the scope of the market, by 1970 nearly 20,000 mainframe systems were installed all over Western Europe, over 80 percent of which came from American vendors, although many of these U.S. systems were actually manufactured in Europe by European employees, as in the case of IBM.

During the 1970s the variety of systems expanded, as in other countries, to include smaller mainframe systems that were less expensive, minicomputers and by the early 1980s, PCs. In the 1970s, the community of Dutch users transformed and expanded as elsewhere, with the technologists in the centralized data centers that had emerged from scientific and engineering communities of the 1950s and early 1960s now sharing the stage with end users in manufacturing, offices, public government agencies, and banks.[77] Dirk de Wit, in his highly detailed case studies of the diffusion of IT in various Dutch agencies and firms made repeated references to this growing community of users who also played a crucial role in adopting, resisting, or shaping the use and role of computers in the 1970s and 1980s. They comprised office workers, academics, and engineers in many industries. His case studies mirrored the experiences of thousands of large firms and agencies all over Europe and North America in these years. Wit included new vendors in these communities, such as Philips for example, in what he correctly called "social groups," and the "technological community," members who often set their technological worldview upon the design and operational principles of IBM's S/360 family of computers, giving these people a sense of community, shared language and practice.[78]

His study of these communities uncovered the same reasons evident in other countries for using computers, often sharing findings and strategies with each other. These included desires to improve efficiencies, to expand services and offerings, and to reduce operating costs. Managers accepted the "dialectics of progress," but also were restrained by organizational structures and interest groups that resisted the kinds of changes computers made possible.[79] The technology helped some. For example, throughout the 1960s–1980s, PTT's institutional desire to centralize operations was reinforced by early computer systems that encouraged consolidation of computing into large data centers stocked with centralized databases. On the other hand, the Centraal Bureau voor de Statistiek aspired to decentralize its operations and so the smaller systems and networking that became available in the 1970s made that possible.[80]

By the mid-1970s, Philips had become an important European supplier of computers, despite American dominance, providing jobs to Dutch citizens who in turn learned more about the technology, building small and medium-sized computers. By mid-decade Philips had about 18 percent of the European market for computers, which made it a force to be reckoned with. But competitive pressures proved too intense. By the end of the decade Philips was no longer a major player on the European computer scene and retired from the market.[81] The Dutch government joined with those of France, Germany, Britain, and other European nations in the late 1970s, and continued through the 1980s investing in digital R&D to reduce their dependence on the Americans.[82] Philips had eagerly advocated Dutch government involvement, and participated in various European initiatives in the 1970s and 1980s, for example, in Unidata.

Meanwhile an increasing number of Dutch firms and agencies installed computers, often at a faster rate than in many other European countries, a reflection of the configuration of their economy, as in West Germany, for example, especially by large enterprises. The Dutch government proved ambivalent in policy making with respect to IT, mirroring more the German experience, and possibly even Italy's, than that of the British or French. Its behavior led to slower decision-making and inadequate funding to make a significant difference, especially with such indigenous institutions as Philips and its local universities. Reports with recommendations appeared all through the 1970s and 1980s, as such pronouncements had in most European countries, however, which could lead historians to think there were great investments underway.[83] Dutch officials recognized the need for action, particularly with respect to semiconductors, even if it came slowly, as the epigram starting this chapter suggested was the danger at hand. The biggest of these pan-European initiatives, ESPRIT, launched in the 1980s, but proved marginally effective given the fact that so many firms and agencies were already actively and extensively deploying IT across Europe, using American and Japanese designed and manufactured computer chips and other IT.[84]

A few more observations about Dutch public policy shed light on the diffusionary patterns evident in the country. Public policy prior to the early 1980s explicitly supported the one major potential national computer champion: Philips. Agencies were required to seek bids from Philips when they had a requirement for computers, and later in the 1970s, for software. There was no other overt policy with respect to the promotion in the use of computers prior to the 1980s; instead, policy encouraged the local supplier through procurement practices. By the early 1980s, as in many other countries in Europe, it had become clear that IT would be a major force in modern society and that pan-European collaboration would be necessary. Philips's management eagerly embraced such initiatives as ESPRIT in recognition that both their home government and national market was too small to help Philips scale up to what had become a highly competitive international market. Dutch officials, like so many colleagues in other countries, were implementing IT stimulus projects for the first time and thus had to feel their way forward in the 1980s. Unfortunately, initial efforts concentrated on mainframes—a market already captured by the Americans and Japanese—then switched to components (a core strength of Philips), but missed the opportunity to invest in emerging technologies, such as PCs, and even the Internet. When initiatives in those areas were launched, they were routinely too little too late.[85] On the other hand, Dutch officials simultaneously in the 1980s also promoted demand for the use of computing, with a significant shift in emphasis to this side of the diffusion equation discernable by 1987–1988. By the early 1990s Dutch officials concentrated largely on meeting the IT needs of the private sector through a combination of development, R&D, and training programs, among others, although some economists thought at the time that there still remained too much of a focus on supporting the local IT industry.[86]

The private sector continued to install computers, along with most government agencies at all levels. As measured by share of capital investments made in ICT (which includes communications so important to IT by the 1980s), by 1985, the Dutch were investing about 12 percent of their capital, comparable to the Portuguese, Italians, Irish, and slightly less than the West Germans and substantially below the Americans (over 20 percent) and Belgium (20 percent). By 1990, while the percentage of investments had risen worldwide, the Dutch only increased marginally their rate of investment to about 13 percent. This rate remained ahead of most European countries, however, and reflected the fact that the Dutch economy had already invested considerably in earlier years, thus requiring less of a catch-up as for others. The world's investments in ICT jumped from an average of just about 12 percent in 1985 to 20 percent by 2002. The Dutch increased their investments at roughly that rate, reflecting the worldwide implementation of broadband, other forms of telecommunications (such as mobile phones), expanded use of personal computers, and just the start in use of the Internet.[87]

IBM kept chasing and shaping the expanding deployment of IT in the Netherlands, despite difficult economic times in the nation during the late 1970s and early 1980s. It began the decade with some 6,000 employees in the Netherlands, an employee base, however, that did not grow in the 1980s. Yet IBM expanded manufacturing facilities, introduced the IBM PC into the local market in 1983, and by 1988 was generating $1.6 billion in gross revenues.[88]

As a result of a pan-European discussion amongst local economists on a perceived lag in European adoption of ICTs, when compared to the Americans in the years from the early 1980s to the turn of the century, they did much work to explain what happened in Europe. They concentrated their studies on the countries discussed in this book, and across all members of the OECD. We have already noted the Dutch performance with respect to investments in ICT. Emerging evidence about the 1970s through the 1990s suggest that how organizations were structured and used IT had a great effect on the economic returns from investments in ICT, essentially the conclusion reached by Dirk de Wit in his study on the use of computers by several large Dutch organizations. Computers did make Dutch process and manufacturing firms productive and facilitated their competitiveness, although even as late as the end of the century the opportunity to enjoy significant economic benefits had yet to be harvested by the services sector. Dutch economists and national government officials were quite concerned about the whole issue of productivity and automation because in the 1990s national productivity had slowed to well below 1 percent per annum, which stood in sharp contrast to the 3 percent annual growth rates enjoyed for many decades following World War II. This slow-down presented a severe problem since growth in GDP can only come if there is expanded productivity and outputs as well, such as more goods and services. Computers were already understood to be useful inputs into an economy that wanted to grow its GDP. During the 1990s, investments expanded worldwide, raising the table stakes for any nation's economy wishing to compete successfully in the modern economy.

Part of the explanation for the Dutch and European experience involved one of the inputs into the economy that always remained higher and more expensive than in Asia and North America: labor. As one Dutch economist explained about his own nation: "Following the recovery from the serious economic crisis during the early 1980s, economic policy in the Netherlands has been strongly focused on the creation of more employment by raising the labour force participation ratio," a policy that proved successful.[89] In the Dutch case local manufacturing and services sectors were extensive users of ICT, and when compared to other countries somewhat more so than the European Union's 15 members, but less than the United States. The effects of ICT on the Dutch economy were positive and similar in form to that of the United States, suggesting that so much IT had been diffused in the economy that macro effects could be measured. These were positive for both ICT-producing and ICT-using parts of the economy, although in both instances less than that of the United States, the nation viewed by all European states as the benchmark economy with which to compare. Beginning in 1995, the ICT-producing part of the economy began declining, suggesting that Asian competitors were proving more productive, although the local electronics industry was normally blamed for pulling down the nation's performance (that is to say, Philips and its cohorts).[90] In the case of the Dutch, significant improvements in the adoption of ICT only came after 1995 as increased use of ICT in distribution, transport, communication, and knowledge-intensive businesses finally had accumulated sufficiently to provide a return on productivity. That also occurred as employment actually increased in the nation, because more workers meant greater incomes per household, supporting the sale of digital consumer products, such as PCs, cell phones, and digital cameras.[91]

Because telecommunications was an important component of the nation's ICT environment, facilitating the diffusion of online uses of computing both in work-related activities and in personal activities at home, the role of communications in Dutch Wave One history is essential to keep in mind. Like other small nations such as Denmark and Norway, the Dutch had created universal access to voice telephony early in the post-World War II era. During the early decades of the twentieth century a mixture of state and municipally owned telephone service provided built the initial infrastructure; then they were consolidated into the one PTT by the German occupiers in 1940. The system was privatized through a series of actions, beginning in the 1980s. By then Dutch citizens considered telephony a public good, a necessity, no longer viewed as a luxury. By the end of the 1980s, the telephonic and television infrastructure of the nation was one of the most advanced in the world; telephony was completely digitized by the end of the century. All this made possible the adoption of the Internet, such that by the end of 1997, there were 17.5 Internet hosts per 1,000 residents, which compares to 14.9 for OECD nations on average.[92] The high-quality networks, prior diffusion of telephony, and affordability bode well for the spread of the Internet in the Netherlands as was also occurring across Europe as other nations liberalized their telecommunications markets and upgraded their infrastructures in the 1980s and 1990s.

The first wave of IT adoptions by the Dutch ended sometime in the early 1990s. Its Wave Two had clearly started by 1995–1996 when ICT's direct share of the Dutch economic growth averaged 17 percent (between 1996 and 1998), at a rate that kept growing right into the new century, and substantially greater than the 7 percent more normal in the mid-1980s. Part of the reason economists give for this is the *C* part of ICT, because in the mid-1990s, the telecommunications and telephone industries were liberalized in the Netherlands, as had happened in the United States in the 1980s, and the results proved similar: introduction of new products, expanded competition, and wide adoption of new technologies by businesses and individuals alike, technologies that depended extensively on digital technologies, computers, and software.[93] If Dutch public policy was to encourage the nation to acquire and use IT, it could claim success because after officials reduced their proactive efforts of the 1970s to protect national suppliers and provided insufficient quantities of R&D funds to make a difference, by as early as the mid-1980s government influence on the computer market proved nil, thus diffusion was driven more by normal market activities than by public programs.[94] The extent of the public's use of ICT made it evident that government policies had less to do with adoption than the influence of existing infrastructures, economics, and public attitudes. By 2000, roughly eight million citizens regularly used the Internet at work and at home and experienced one of the highest levels of PC usage in Europe (between 10 and 12 million users). They also used nine million mobile phones, making them additionally one of the most connected European nations at the time. In business-to-business commerce, about $2.5 billion in transactions were done electronically as well, a way of conducting business that in subsequent good economic times continued to increase.[95]

SWEDEN—COMPUTING THE NORDIC WAY

The Nordics, or Scandinavia, as a group of northern European countries are called, constitute a sub-ecosystem in Europe's culture and economy distinctive in its own right, a region in Europe that enjoyed high levels of literacy and standards of living during most of the twentieth century. Depending on whose definition is used, it normally consists of Denmark, Norway, and Sweden as the core group, with Finland and tiny Iceland included. They have different languages from those of other parts of Europe and have a long history of sharing information, trading, and otherwise interacting with each other. The primary reason for selecting Sweden as a case study of diffusion in the upper-left-hand corner of Europe is its size. As of 2008–2009, Sweden had the largest population of the group, with 9.3 million residents, while Denmark consisted of 5.5 million people, Norway, 4.8 million, and the other two combined 5.4 million. Since GDP is important to our story of IT diffusion, in the same period Sweden's hovered at $343 billion, Denmark's at $204 billion, Norway at $257 billion, and the other two combined at $203 billion. Clearly all were prosperous and productive during the time in which computers and telecommunications existed, with the partial exception of Iceland. Each has a long history of using IT extensively, with, again, the exception of tiny Iceland (population 320,000 in 2008).[96] We could have picked any of the mainland European Nordic countries to describe the diffusion of IT, and more precisely ICT, in this part of Europe, because of their collective long history of using office appliances and subsequently computers. One reason for selecting Sweden is that we know less about this country's experiences with computing outside the country than about Denmark, Finland, and Norway, allowing us to fill in some gaps in our knowledge of the region.[97] It is also the largest of the Nordic countries.

The economy and society of Sweden is widely characterized as unique, with such labels as the "Nordic model," or "Welfare State," typical.[98] But examining the data suggests what characteristics affected directly the diffusion of IT into this corner of Europe. GDP per capita in Sweden in 1950 hovered at $6,738, grew substantially to $13,494 in 1973—midway through our first global wave of IT diffusion—and ended 1992 at $16,927. It outpaced GDP with Norway and Denmark all through the period, although by 1992—a period of enormous economic stress in the region—the Norwegians had a per capita GDP of $17,543 and the Danes $18,293. For the majority of the period, Swedes enjoyed a per capita GDP higher than the average for all of Europe by about 12 percent, a statistically significant amount. Only Denmark and Switzerland routinely reached comparable levels through most of the second half of the twentieth century.[99]

The "Swedish model" is, however, what observers want to turn to first, rather than to some dry economic statistic, going to a discussion of this "modern welfare state." One student of the region defined this society as "marked by the use of big, centralized institutions and large-scale transfers, commonly provided on a universal basis," in order to reduce inequality and poverty, and to minimize social risks.[100] Taxes on individuals were high throughout the second half of the twentieth century; collaboration and interactions between the state and the private sector intense and compared to many other European countries quite effective; regulatory activity was pervasive, while wages were frequently set through centralized bargaining mechanisms. OECD's assessment of Sweden's "tax wedge" (combination of personal taxes and social security contributions) placed it second after Germany in percent of one's income (just over 50 percent), with the rest of Western Europe's averages ranging between 30 percent (Switzerland) and the other Nordics in the high 40s percentiles by the late 1990s; the percentages remained relatively the same in the early 2000s.[101] Government revenues, as a percentage of GDP, made up of a combination of taxes, fees, capital income, and other sources of income, ranged between about 56 percent in 1980 to a high of just over 60 percent at the end of the decade, followed by a return to levels closer to 1980 all through the 1990s.[102] Finally—and important for our story—the government owned many enterprises.

The economic strengths of the nation over the past seven to eight decades included effective public administration, extensive and good support for education and health services, provision for high rates of employment year-over-year, social peace, and, by the late 1980s, a solid stake in the high-technology sectors of modern economies. One historian described Sweden's economic circumstance as "a mixed economy, partially capitalistic and market-oriented, but with a public sector which at the same time controlled the allocation of resources."[103] Sometimes, all was not well, however. State expenditures expanded all through the 1960s–1980s as public support for expanding welfare programs grew, creating a crisis in national debt in the early 1990s that led to a reverse in direction for a while, and steps toward expanding the role of the private sector. Yet for the entire period in which computers were available, the role of government in the private sector remained visible and overt. This was largely because the public sector operated monopolies in telecommunication, the post office, energy, forest management, the retail sale of alcohol and medicine, and funded the use of computing by various governmental agencies.[104]

Sweden's modern era began in the 1930s, following considerable turbulence in the prior decades as it evolved from just a highly agriculturally dominated economy to one that included industrialization and expansion of the financial sector both coupled to a growing urban-based labor force. It remained neutral during World War II, thus suffered no physical damage or the destruction of its social and political institutions. This circumstance also allowed its population to thrive and grow. A new generation of Swedes went about the business of creating families in the 1940s. One byproduct of its century of economic growth was a slowed average personal standard of living during the second half of the century relative to other European nations as more resources were channeled into the state to maintain a more evenly balanced level of services for its growing population. This circumstance stood in contrast to developments in other nations where the private sector operated in a less fettered environment. Like the Dutch, Swedish industries relied heavily on energy for its work (especially in forestry, pulp, and automobile industries) and so the oil crises of the 1970s hit the nation hard, as it did others, most notably the Dutch and the Americans. The era of extensive growth and prosperity slowed severely enough in the early 1990s that one can conveniently label the start of the last decade of the century as a contributing factor in declaring as over Sweden's Wave One period in its use of IT. Both its national and personal GDP and GNP fell below the average of OECD.[105] It is enough to simply call out the problem rather than engage in a discussion of the various possible public policy mistakes made in the 1970s and 1980s that led to this situation; however, the inability of the state to sustain its levels of debt and services after about 1993 have not been reversed. But we are getting ahead of our story in describing the economic landscape.

While economic growth around the world was common from the end of the 1940s to at least the beginning of the 1970s, in the case of Sweden, GDP grew faster than in other European countries (3.3 percent in 1951–1954, 3.4 percent, 1956–1960, 5.2 percent in 1961–1965, and 4.1 percent from 1966–1970).[106] In the 1970s it was closer to 2 percent. Standards of living improved, workers migrated to urban centers from the countryside, and the local capitalist market economy thrived. For centuries exports had proven important because the country was so small. Since the late nineteenth century for a large firm to thrive it had to extend its sales to other countries and that gave the economy and nation an Atlantic orientation, in contrast, for example, to Finland which had focused its economic, political, and military attention eastward toward Russia, largely for reasons of national security and survival as an independent nation. Part of Sweden's prosperity of the 1950s and the 1960s came from extensive capital investments in modern equipment and new plants, leading to continuous improvements in economic and labor productivity. Swedish government agencies and businesses invested extensively in all the modern technologies that came along and also became producers of some of these, such as telecommunications and automobiles.[107] Electronics and chemical industries expanded in the same period, with engineering jobs (particularly mechanical) and industries doing the best all through the second half of the century; these became extensive users of all manner of ICT. After the 1960s more traditional First Industrial Revolution industries and extractive ones declined in their relative share of GDP, such as forestry and mining, as high-tech industries came to the fore.

Swedish industries as a group were highly concentrated in fewer, larger firms, representing some of the highest levels in Europe. By the early 1960s, for example, the 100 largest firms in the country accounted for 46 percent of the value added in the industrial sector.[108] One of Sweden's most distinguished economists, Lars Magnusson, argued that this situation made it possible for firms and, by extension, the economy, to be far more stable than those in other nations, but also because they had become leaders in their respective industries early in their corporate lives, such as SAAB in aircraft (later automobiles) and Ericsson in telephony. Their high-tech competency is frequently acknowledged as he does too.[109] Magnusson also pointed out that many of Sweden's industries lent themselves to adoption of the mass production processes and organizational forms already implemented in the United States during earlier decades, particularly in such areas as in the production of iron and steel, paper and pulp, and in the manufacture of automobiles, telephones, and other high-tech products sold in quantity. Automation expanded all through the period, leading to the expansion of an educated white collar workforce equipped over time with technical skills suitable for a modern industrial-services rooted economy.[110] Extensive use of regulations supported, indeed facilitated, job security for a large portion of the nation's workforce, and a stable financial sector so crucial in funneling capital into new industrial investments. The key word he used to describe the period to the end of the 1980s in Sweden's economic history is *stable*. While the economy had its issues and problems in the 1960s through the 1980s, not until the arrival of a severe recession at the start of the 1990s was the situation upended severely enough to bring to an end one period and to signal the start of another, as yet unclear at the time to users of ICT. Magnusson argued that there were problems as early as the 1960s, followed by the negative effects of oil crises in the 1970s, and subsequent drops in Swedish industrial capacities. Partial deregulations of banks' ability to make loans in the 1980s and growing labor pressure for increases in salaries added churn, contributing to Sweden's declining industrial performance in that decade, while encouraging a round of inflation. The "overheated economy" (his words) coupled with an international recession which dampened Sweden's export performance put a brake on the economy that spread to all sectors.[111] With parallels to the banking crisis around the world that began in 2007–2008, Sweden's banking industry experienced severe erosion in its capital stock. Unemployment increased too, creating a combination of circumstances that dominated Swedish affairs and concerns in the 1990s.

Against this economic backdrop one can appreciate more clearly Sweden's experience with computers. By the end of the twentieth century the nation had become extensively IT-intensive in its use, a situation driven largely by the twin developments of an IT-thick industrial base of manufacturers of computing equipment and software (supply side of the story) and as users (demand side). Government officials also embraced computing early and promoted its adoption, particularly in the 1960s and 1970s. At the risk of drawing too close a parallel with another nation's experience, one could see some significant similarities to the experience of the United States, albeit on a far smaller scale: government funding of research then support for use of the technology within public and private sectors, and public support of education, collaboration with large enterprises which were the natural early users of computing during Wave One. Unlike the British example, where public policy also proved important, but wavered inconsistently over the decades, the Swedes actively supported diffusion of IT to a very reasonable and effective extent during the past half century. While officials flirted with the notion of national champions, as did all of Western Europe, ultimately in practice Sweden as a whole focused more attention on how best to use the technology, not merely on who should make and sell it abroad.

Sweden's story began in a manner remarkably similar to that of the United States on the supply side of the equation with government involvement in its first two decades. Following the conclusion of World War II, and as a result of Swedish engineers and diplomats having become familiar with American computational developments, there was interest in acquiring such technology, or at least components and blueprints. The Swedish Naval Forces and the Defense Research Agency expressed interest much along the lines that the U.S. Army had a few years earlier in using computers to calculate firing tables and later, to do mathematical calculations related to missile guidance issues, an interest that unfolded as the Cold War came into prominence. Swedes had talked to Vannevar Bush, the professor at the Massachusetts Institute of Technology (MIT) who had worked on early analog computational machines in the 1920s and 1930s and maintained excellent access to U.S. government computing projects.[112] Other contacts with other American academics also made it possible for Swedish engineers to visit various projects in the United States in the late 1940s and early 1950s, thereby transferring knowledge of these emerging technologies to the Swedish government and to others in their home country.[113] As with the Americans, concerns regarding warfare provided much of the initial interest in IT by public and private institutions.[114]

The first major action taken as a result of this growing awareness came in early 1947 with the establishment of a study committee by the government to determine Sweden's needs; in turn, that committee's work led to its work at the National Committee on Mathematical Machines.[115] In 1948 the committee recommended the purchase of this technology from the Americans, so the government established the Swedish Board for Computing Machinery to handle the transaction. However, by then that approach proved impossible to implement as the U.S. government had begun to limit the export of technologies that it deemed might get into the hands of the Soviets. The Cold War was now influencing events. The Swedes concluded that they had to build their own machine for their military, and for other purposes, such as to acquire knowledge of the emerging technology.[116] Beginning in 1949 and extending into the early 1950s, using American blueprints and components that could still be imported, work began. The first device, called BARK, was a relay-based system constructed largely as an interim machine until a more advanced one could be build.[117] The construction of BARK mimicked the nearly ubiquitous pattern of the late 1940s and early 1950s that helps in part to explain why so many such projects could be done in various countries in Europe and Asia, and often at far less cost than experienced by the American and British computer developers. As a contemporary described it, the strategy was to use "standard telephone components, and by borrowing designs and principles from other machines, particularly American machines." The people doing this work comprised a "group of some fifteen people; including engineers, mathematicians and technicians" while keeping "in close touch with American developers."[118]

In November 1953, Sweden's initial digital system went live, called the BESK (Binary Electronic Sequence Calculator).[119] Designed based on American ideas obtained from such places as RCA and the Institute for Advanced Study at Princeton, New Jersey, for a while it proved to be one of the fastest and most advanced systems in the world. The project was open and so other firms in the country were able to watch developments and express interest, most notably the office appliance firm Facit and Saab, the aircraft manufacturer. One observer of the European computer scene at the U.S. Department of State commented in an internal report (1953) that Sweden had "one of the most outstanding computers groups in Europe," while another expert made a similar comment three years later, opining that the Swedes "must be ranked as among the three or four" most outstanding computer developers in Europe.[120] By the early 1960s academic interest had grown. Professors began to offer classes at local universities and initiated research on IT themes. These various activities led to the expansion in number a cadre of computer savvy technologists that has continued to flower to the present, serving academia, industry, and government agencies.[121]

In this early period (1950s) the driving force for computing shifted from just the government to these two firms (Facit, Saab). A distinguished supplier of office equipment since the 1920s, Facit's management recognized that computers might represent a new opportunity for the firm, particularly because it was an export-oriented business in Europe. Management concluded that since American vendors were too busy supplying their own domestic market that a window of opportunity existed to enter the European market for computers. To move quickly, it hired the bulk of the engineers who had built the BESK, who then acquired the nickname carried down in history as the "BESK Boys." It was able to do this by offering higher salaries while taking advantage of morale issues, because the engineers did not feel the government was sufficiently supporting their efforts.[122] Facit's management recognized that since other firms wanted to borrow or rent time on BESK machines there might be a domestic market for this technology as well. Additionally, like office appliance giants of the day such as IBM, Burroughs, NCR, and others, its senior managers concluded that the future lay with computers. Like those firms, they believed Facit could use its existing network of sales and support to reach the logical customers for this new technology, paralleling, for example, IBM's unfolding strategy in the same decade of the 1950s. A few years later, but not much behind Facit, Saab became interested in computing for its own internal calculating needs. By the end of 1956 it had built its own BESK (called SARA—Saabs Räkneautomat).[123]

Unlike Facit, however, at the time Saab had as its primary customer the Swedish government and so identified more with the strategy of the American defense contractors who served the U.S. military in the same period. While these two firms were engaging with computers, the Swedish government relinquished leadership in the development of this technology to them, but supported growth of a local competitive market for computing by beginning to acquire systems in an open bid process. Demand in the country began to develop; Facit built and sold 11 BESK systems, the majority to local firms, universities, and government agencies. IBM's market analysts and management noticed the growing market and, since by the end of the 1940s the American government was now allowing its firms to export some high-tech products, the time to expand into Sweden's market seemed propitious. In 1956, IBM placed copies of its 650 system with Kungliga Flygförvaltningen, Thule, and Folksam. As in other Nordic countries, these were used for both engineering and data processing applications. Even the Swedish government acquired systems from IBM, such as a 7090 in 1961 for military research, and a series of 1620s in the early 1960s. For IBM and other foreign vendors, the market had taken shape in Sweden by the end of 1959 when further export restrictions had been lifted by the government of the United States and the largest customer in Sweden (the national government) had established a record of open bids for acquisitions of computing.[124] The government was quite open in announcing in the mid-1950s its intention to acquire systems, which it was prepared to spend millions of kronors on them, and let bids for such devices and software.[125]

The emergence of a market for computers ushered in a new phase of diffusion by the end of the 1950s, which extended to the start of the 1970s. Saab emerged as the local informal national champion driving demand. In the late 1960s, it introduced its own products into the market. Facit, meanwhile, evolved into a subcontractor supplying Saab, while IBM became an important foreign supplier (but by the 1970s not as dominant in the public sector as it was in other countries). The Swedish government evolved into a major customer for these systems. Additionally, Saab received considerable support from the government in the form of funding for R&D and, of course, as a primary customer of such products, selling up to 25 percent of its systems to government agencies. To quote Tom Petersson, historian of Swedish technical developments, "the development of the Swedish computer industry from World War II to 1970 can be characterized as a state-led or state-propelled process as in the U.S. case, at least in the initial stages."[126] As with American vendors, the two local computer companies could not collaborate in the 1960s, despite efforts to do so. They simply had roles to play too different to foster significant collaboration. Facit oriented its business toward export and battling competition from other office appliance vendors, while Saab concentrated increasingly on being the key supplier of the Swedish government, hence protected in part from having to compete as extensively as either IBM or Facit for sales of computing products. Saab could get the Swedish government to fund its R&D and then buy its results, such as its popular D21 computer, introduced in 1961, brought out by Saab's computer organization, Datasaab.[127]

In the mid-1960s a lengthy benchmarking exercise pitting IBM's hardware against Saab's resulted in orders for 12 IBM 360/Model 30 and 8 D21s. A subsequent assessment of the performance of all these systems concluded that the D21s had proven more reliable. This finding resulted in the government ordering all IBM systems to be replaced with D21s by the end of August, 1970.[128] This turn of events proved to be a major hindrance to IBM's efforts to penetrate the important public sector market in Sweden, although it had customers in the private sector. One consequence was that Saab continued to expand both the number of employees knowledgeable about computers along with another cohort of experts who worked in government agencies using its products. Thereby, Saab continued to serve as a conduit for the diffusion of knowledge and expertise about IT throughout the local economy, much as IBM did in so many other countries. Expertise spilled into the other Nordic countries almost from the earliest days, as it did from the United States at the same time as Swedes were learning from the same Americans about the new electronics.[129]

What was demand like for IT systems? In 1953 there was a handful in Sweden. Demand increased and by the end of 1963 there were some 120 systems installed. Since most computers in use in the early years had a life of between 4 and 8 years, most of those installed in the early 1960s were of relatively new vintage, with government leading the way in acquiring these.[130] Table 4.2 shows the value of acquisitions made by the Swedish government to 1981, to demonstrate its important role as a user of this equipment.[131] IBM's share of this sub-market peaked at about 72 percent in 1969 within public sector; then declined by about 5 percent a year until 1979 when it bottomed out at 25 percent, still an impressive market share. Interestingly, in the same period (1969–1981), Saab grew from roughly 10 percent to 25 percent, before losing market share in the Swedish public sector. It only maintained that amount largely because it had formed a partnership with Univac. Other major suppliers were Honeywell Bull, Philips, and various others with market shares of less than 10 percent each.[132] In the private sector, IBM enjoyed about 70 percent of market share.[133]

Some interpretations of IBM's role in Sweden could leave one with the impression that it played a diminished role in the emerging Swedish digital landscape.[134] Since a central finding in this book is that vendors played a crucial role in diffusing IT around the world, and that IBM played a *primis* role in that process, we need to understand more precisely its role, one that proved

Table 4.2 PROCUREMENT VALUE OF ALL COMPUTER EQUIPMENT
ACQUIRED BY THE SWEDISH GOVERNMENT, 1962–1981
(MILLIONS OF KRONORS)

1962	50
1967	185
1972	450
1977	860
1981	1,500

Source: Statskontoret, *Statliga Dataorer 1981: State-Managed Computers* (Stockholm: Statskontoret, 1981): 6.

extensive throughout the Nordics across most of the century, and with Sweden more the normative case than the exception. IBM's first foray into Sweden dated to 1911. IBM appointed its first local agent in 1920, and established a Swedish subsidiary company in 1928. By the mid-1940s IBM sales staff in Sweden was actively renting out punch-card equipment to Swedish enterprises, manufactured punch cards, and soon after typewriters for the pan-European market, which meant they had, like Facit, a healthy market network for accessing those who might buy computers. By the mid-1950s the company was selling electronic calculators and in 1956 installed the first 650 computer system in the country at a local insurance company, Livforsakringsanstalten Folket. Over the next few years IBM installed EDP systems in banks, the Swedish State Railways, and in manufacturing firms. By the end of the decade, it had eight sales offices in the country, employing in total just over 600 people, the majority of whom who were Swedish. In 1960, it established an R&D laboratory in Stockholm, and the following year opened a data center there to demonstrate its products and to sell computer use on these.[135]

A careful reading of the marketing reports from IBM Sweden held in the company's archives reveals that the firm proved active, profitable, and thrived all through the 1950s and 1960s. Its revenues from data processing products grew from the late 1950s right through the next decade, doubling in fact. When the System 360 was introduced in 1964—the system that historians of computing agree overwhelmingly expanded fundamentally diffusion of IT across dozens of countries—the directors of the local company noted in a quarterly report to IBM World Trade Corporation that business "has developed satisfactorily during 1964," and that with the 360, "several are scheduled for customer installation in the fall of 1965."[136] An internal market assessment of IBM's presence in Europe showed that it was generating revenue from its entire product line (not just computers) in Sweden in every key local industry: process, manufacturing, utilities, government, universities, transportation, communications, commerce, insurance, and banking at rates comparable to what was happening in other countries.[137]

To be sure, IBM had to earn its business because, as one manager pointed out in the mid-1960s, "competition is harder but in no way alarming."[138] The assessment proved realistic, in that with the System 360, within two months of the product's introduction, IBM had orders for 18 in Sweden. IBM had also grown its data processing services business, reaping between 25 and 30 percent of the Swedish market before even installing its first 360. During the second half of the 1960s, 75 percent of all computers installed in Sweden were IBM machines.[139] In short, IBM's position in the Swedish market was as broad and dense as in other countries, a situation that continued through the 1970s and into the 1980s with mainframe systems.

So what were Swedish enterprises and agencies doing with all these IBM systems, and those of its competitors? As always, having suppliers of the technology, while crucial to its diffusion, was not enough; one needed good economic incentives and uses to justify these early expensive systems. Let one Swedish veteran describe an early milestone:

> A very spectacular event was the procurement of computers for the Government tax system in the 60's. The Government had created a fund from which all Government agencies could be reimbursed for their computerizations. A report on the use of the fund was issued annually and it is therefore easy to see the development of computer use in the public sector.[140]

This observer concluded that the 230 systems installed by 1981 became possible through this funding mechanism, what he called "a driving force for the early use of computers in Sweden."[141] The data presented in table 4.2 reflect the results of this initiative. As to the use of many systems, "There was thus a central procurement of computers in the 24 tax regions and 5 companies provided offers." It was this procurement that also led to all the government benchmarks of the 1960s that ultimately made it possible for Saab to sell 32 D21s, followed by orders for 71 D22s, a faster operating replacement product for the D21s.[142] During the 1960s and 1970s government uses mimicked those of other public institutions across Europe and North America.

A second group of users were academics in local universities. These institutions came to computing later than in some other countries, essentially beginning in the early 1960s when the national government proposed funding various installations. Lund became an earlier user of computers, as it had its own BESK system. Both universities in Stockholm and Upsala gained access to the government's BESK and Facit systems. The problem was overuse as demand accelerated beyond the capacity of these systems to handle the volume of work required; furthermore, these three systems were aged by the mid-1960s. Chalmers University of Technology in Gothenburg relied on minis (Alwac III E devices) and on one D21, largely for scientific applications. The Government's funding led to important developments between 1965 and

1970, making it possible for universities to acquire largely American systems, most notably IBM 360s. End users were funded to use these systems all of which were housed in regional centers. By 1970, combined these centers employed about 200 technical staff. Users included academics and other individuals from industry, but largely faculty and students, ranging from physicists and chemists to medical and social science faculty. By the early 1970s several hundred people relied upon these facilities every day, serving as an indirect channel for further diffusing knowledge about IT into the greater society throughout the 1970s and into the 1980s, although the centers were dismantled in the late 1970s in favor of decentralization of computing into the various universities. In fact, by the early 1980s, with various departments acquiring their own mini computers, perhaps half the computing used by any given campus were lodged in departments and laboratories, not in centralized data centers. Diffusion came about for similar reasons evident in both public and private sectors around the world: minis were often less expensive to own and operate, usually easer to maintain and use, and provided more flexible access to their owners.[143]

Courses were offered to students in computing, beginning in the mid-1960s, involving thousands of students over the next four decades. The University of Stockholm and the Royal Institute of Technology (KTH), in particular, also ran training programs for local and national government employees, including adult education courses, also starting in the late 1960s. Demand for training and academic education increased in the early 1960s because by mid-decade alone, there were more than 200 computer systems installed in the country just in the public sector, with at least an equal number already in use within the private sector. At the time Sweden employed between 8,000 and 10,000 individuals in computer-related jobs in programming, systems analysis, machine operations, and in data processing and managerial roles.[144] This last set of statistics demonstrates that the Swedish economy had a large number of skilled people in the field. That population expanded in the 1970s and 1980s such that by the mid-1990s Sweden's economy and society were generally considered some of the most IT-intensive in Europe. Public officials recognized in the 1970s that even pre-university students needed to learn about computers and went so far as to start designing computing into their curriculum and in the 1980s even funded the manufacture of a PC designed just for students, called the Compis. The school initiative, however, enjoyed limited success against expectations, mirroring what had happened in elementary and secondary schools in that decade in many other countries.[145] But, on balance, across all sectors of training and education, a transformation was underway that led to increased IT skills in the workforce, starting in the early 1960s thanks largely to public funding.

The private sector followed a trajectory in its use similar to that of other corporations in such countries as the Netherlands, Great Britain, France, and the United States. Suppliers of goods and services to the military were early users, such as Saab, often collaborating with their customers, in this case the Swedish Defense Research Agency (FOA) through military contracts.[146] IBM, Facit, and others also worked with data-intensive customers who already used other forms of mechanical aids to calculation and data processing, such as insurance companies and those with large numbers of employees. As one veteran of the 1960s recalled, "the IBM school educated far more people than the meager government system," and that the company, "had an extensive base of contacts from its punched cards business and a cash flow from its rental system that was unassailable."[147]

Uses reflected other common patterns evident in similar industries in other countries. For example, in many mid-sized to large enterprises in both process and manufacturing industries, Computer-Aided-Design (CAD) systems were installed, beginning in Sweden in the early 1970s, about a decade after they were first used in the United States, and at a time when CAD software and hardware systems had become reasonably reliable and cost-effective. The first companies in Sweden to use these systems were largely electronics firms, but then these applications spread to other companies, such as to Kockums Shipyards, SAAB-Scania, and Volvo, all in the 1970s. One survey of the number of CAD systems in Sweden estimated 60 in use in 1979, with 40 deployed in tandem with numerical controlled (N/C) machine tools and robotic devices in manufacturing. These included such large firms as L.M. Ericsson, ASEA, Sandvik, and Electrolux, in addition to the ones cited above. Small firms supporting these and other industries also became users in the 1980s, such as architectural companies. Another inventory of users conducted in the early 1980s placed the total number of systems in the country at 205 used for a variety of applications: mechanical, electronics, architecture, civil engineering, construction, and by service bureaus (22 systems), which meant other firms also had access to these systems. For an historian use of CAD systems serve as an early indicator for gauging how much computing was spreading into a modern industrializing economy, because long before these were installed traditional accounting and financial systems always made it into an enterprise.[148] So, once CAD systems came to be used one could usually find accounting and financial systems already in use in the organizations or industry, inferring the diffusion of computing out of the "glass house" into user departments. In Sweden this happened at the same time as minicomputers also spread within the private sector, followed soon after by yet another wave of computing, the personal computer in the 1980s and 1990s. In a few cases in Europe, as that pattern of diffusion expanded, users decided that they too could manufacture computers, as occurred with Saab, beginning in 1971.

As early as 1978, various officials in the national government began to notice the uptick in computing. To explore the possibilities of expanding use, they established the Computers and Electronics Commission (DEK). One of its first set of recommendations was to suggest steps be taken to expand the use of CAD/CAM systems in the early 1980s, with training sourced from five technical universities.[149]

Another activity that drove forward the diffusion of IT in Sweden was the manufacture of high-tech products themselves and this was no more profound or illustrative than the examples of the two telecommunications firms, Ericksson and Telia (local telecommunications monopoly). Ericksson became the largest ICT company in Sweden, with an extensive export business. By the end of the century it had become the world's largest manufacturer of mobile networks, and ranked third in global production of telephone handsets. Both firms invested earlier and more extensively in mobile networks than cohort firms in other countries, with market and user penetration rates at the end of the century comparable to those of the United States. In fact, work on the first mobile network began in Sweden at the dawn of the 1970s, called the Nordic Mobile Telephony, launched in 1981 across all the Nordics.[150] The two firms collaborated on the introduction of the Global System for Mobile Communications (GSM) in 1992, one of the world's first global mobile telecommunications standards, and the dominant one in Europe. Part of the success of GSM grew out of the government's deregulation of telecommunications before other European countries, which stimulated competition, led to its *de facto* dominance in Sweden, hence served as a springboard to exports. In the 1980s and 1990s, the national government provided subsidies for its citizens to lease computers (for instance, PCs) from their employers to use at home with communication hookups and economic incentives for firms also to use the technology. A byproduct of these initiatives was the continued education of a highly skilled work force in the ways of ICT, which not only helped companies use the technology, but led to the injection of that technology into all manner of high-tech products.[151]

Writing in 1995, three economists at the World Bank reminded us of a characteristic of the Swedish economy mentioned early in our discussion about this country, namely, that it was export-driven, hence concerned about being competitive in the world market. The World Bank observers noted that "20 of the Fortune-500 non-US corporations are Swedish-owned," and that they, and other Swedish firms, were extensive users "of robotics and other advanced manufacturing technologies such as CNC [Computerized Numerical Control] and FMS [Flexible Manufacturing System]," indeed, they were "among the most intensive in the world."[152] They also pointed out another cumulative result of the ramp up in diffusion that had occurred between the 1970s and the 1990s, and that was the competitive strength of Sweden's telecommunications domestic industry, centered most notably around LM Ericsson, and to a lesser degree with defense electronics. While the story told in this chapter gives credit to the Swedish government for playing a facilitative role in encouraging the use of computing all through the second half of the twentieth century, the World Bank's observers in the mid-1990s thought that the government did more discussing than supporting of ICT developments than already evident in the private sector—again more evidence to the effect that in Sweden, as in the United States, the bulk of the innovations and adoptions of ICT had clearly shifted to the private sector. As in the United States, however, the national government did not abandon economic developmental policy or support. Indeed, the government had an active agency with this mission called the National Board for Industrial and Technological Development (Nutek), nestled within the Ministry of Industry and Commerce.[153]

My relatively positive account of the role of government has to be balanced by some of the realities of the 1970s and 1980s that help explain the shift of initiative from the government to the private sector. Writing in the early 1980s, one observer characterized government commitment to all manner of industrial support as stogy, because over 70 percent of its investment funds went to such traditional industries as shipbuilding and steel rather than to more high-tech ones; yet its support for promoting export business in the 1970s and 1980s remained strong. A witness noted at the time, "the government pours vast amounts of money into R&D," which did lead to the development of a substantial national electronics capability, for example, to over 80 percent of all production of IT equipment in the mid-1970s being exported.[154] The bigger operational problem grew out of the fact that some 20 government agencies were involved in supporting the development and use of IT, thus creating their own pattern of diffused decision-making.[155] As of the early 1980s, one European expert (not Swedish) on public administration noted, "If Sweden has little experience of government telling big firms what to do, it has an equally indifferent record of purposeful government procurement policy, outside of the aerospace sector and the remarkably self-sufficient area of defense."[156] Charges of incompetency seem misplaced given the economic successes of local firms in using IT and who thrived in the increasingly competitive global economy of the 1970s and 1980s. When the government did get proactive, it was often in highly targeted areas, most notably, for example, in the use of CAD/CAM and N/C systems in the 1970s and 1980s. In short, like so many other European nations, it had a strategy to support diffusion and like other governments, published its plan for all to read who cared to understand it.[157]

In the private sector, expenditures in IT in the 1960s and largely into the 1990s rested fundamentally on capital investments, because of the substantial expense of the hardware in computers and manufacturing and processing equipment. Clearly capital stock, as measured per employee, grew through the period across all basic sectors of the economy: manufacturing, paper and pulp, basic metals and metal products, and mining. Collectively, on average, they invested 50 million kronors per employee in 1963. Ignoring the effects of inflation for our purposes, since investments increased far more dramatically than inflation, the per capita investments continued to increase steadily until the start of the financial crisis of the early 1990s, peaking at over 250 million kronors, important evidence of the extensive substitution of capital (e.g., IT) for labor.[158]

Another measure of the accumulated diffusion of IT is to look at the volume of electronic fund transfers that took place across the entire economy, since this is a type of transaction that can only be done using computers. By the early 1990s, about 3 percent of Sweden's GDP went through this application of ICT. For setting context, at the time, using the same measure, for Belgium it was 5.4 percent and France 6.4 percent, exceeding Sweden; however, that country used this application more than the Netherlands (2.3 percent), Switzerland (1.1 percent), and in North America Canada (1.2 percent) and the United States (0.2 percent), and Italy (0.4 percent). Clearly the Swiss had embraced this important form of computing far more extensively than many other nations, in all probability due to the extensive use of computers by the banking industry and government for financial transactions.[159]

As in other countries, diffusion of IT built up its own economic and operational momentum in an incremental fashion over time to such a point that its role became visible and dramatic, adding onto prior knowledge of the subject. Growth in dependence on the technology increased once operations had changed to such an extent that computing had been integrated deeply into work (as with CAD and electronic funds transfers), and as a result of government incentives combined with the twin efforts of suppliers of the technology in developing useful products and then selling them. Data collected by economists documented what the investment was in computing, hence a measure of deployment, but to appreciate these kinds of statistics one should keep in mind all those activities that had to occur incrementally over many years to make possible whatever measure is taken in a particular year. A favored measure of economists is to look at the investment in ICT as a percentage of GDP, and for Europe and the United States we have such data for the period 1980 to the present. What is striking is that for the 1980s—hence a report card on the trends of the 1970s—Sweden invested 1.6 percent of its GDP in ICT, placing it in second place, after Belgium (1.7 percent), tied at the same rate of investment with the Netherlands, and slightly ahead of Italy and Denmark (each at 1.5 percent). West Germany hovered at 1.3 percent, United Kingdom at 0.8 percent, and the French at 1 percent. For comparative purposes, the United States came in at 2.5 percent. So the Swedes were heavily invested in the technology, coming out of an aggressive diffusion process in the 1970s. Data for 1990 reflected what occurred in the 1980s, demonstrating that Sweden had increased the rate of its investments to 2.7 percent of GDP, just behind Denmark (2.9 percent). The EU-15 on average was at 2.2 percent (versus 1.1 percent in 1980). Many countries had increased their rate of investments between 50 and 100 percent, a pace that actually picked up further in the last decade of the century, with Sweden absolutely the most invested by 2001 at 4.7 percent. What these data tell us is that diffusion across Europe, and particularly in Sweden, had increased in rate and that it accelerated over time.[160]

It has been argued quite sensibly that a major driver of adoption, and the increased rate of that adoption of ICT, was due to the simultaneous rising cost of labor experienced by Europe and the United States. While there raged a debate among economists in Europe about why European labor productivity rates were below those of the United States, it is quite possible that the simple answer is that in general for the entire period in question IT cost more in Europe than in the United States caused by remaining barriers to competition and imports slowing adoption below what might otherwise have been possible. It was not uncommon to see IT products cost as much as 20 percent more in Europe than in the United States.[161] But, I conclude that a more important factor was Europe's high cost of labor. Economists for decades outside of Europe have criticized the national practices evident in so many European countries of restricting the ability of enterprises to reduce their labor or alter work practices, suggesting that from the perspective of improving productivity of ICT, the kinds of changes in organization and work processes necessary to optimize the economic effects of ICT could not be implemented. It remains an open issue for historians of IT to explore further. The case of Sweden remains interesting, however, since its rate of adoption (hence investment) of IT continued right into the new century, often more than by other European states, only surpassed by the Americans.[162]

Before moving beyond Sweden's experience with IT during Wave One, we need to address an imbalance in the accounts given of the nation's early experiences with the development of an indigenous computer industry, because it distorts the conversation about the pattern of diffusion that occurred. As happened in so many other European countries, and with those local observers who recorded their memoirs and accounts of their nation's experiences with ICT, the Swedish government, local companies, and experts in ICT discussed the feasibility and desirability of creating a local computer industry in the 1960s and 1970s. Exploring that possibility and the potential economic benefits of such a development made sense, of course, and it was a conversation held in almost every Western European country during those two decades. As with such nations as Great Britain, France, both Germanies, and Italy, the debates led to direct programmatic actions on the part of national governments, which the historical record demonstrates clearly played important roles in both the debate and through activities tipped in favor of nurturing local computer and semiconductor industries. Later we discuss the effectiveness of such initiatives, but with respect to the Swedish case, the dialogue continued to the present, differentiated largely from that discussion in other countries by the fact that many Swedish individuals who lived through the earlier decades are still discussing the issue, often describing the role of their government as flawed.

Many of the accounts of Sweden's computer industry characterize events much as other Europeans as well, expressing hopefulness in the 1950s and 1960s about the economic possibilities presented by the technology, but then displayed a sharp and noisy turn to cries about the aggressive and dominating behavior of the American computer companies, most notably IBM. As late as in two large conferences held in the early 2000s on Nordic computing, there were many discussions and papers presented on the "tragedy" that Sweden could not counter the Americans, and to a lesser extent in the 1980s, the Japanese.[163] The emphasis of the debates centered largely on the supply side of the diffusion issue, about the possible development of local champions (such as Saab) and fighting off IBM. As one witness to the debate recalled later, "Of course, technical and commercial advances crushed any dreams about national and independent computer industries," acknowledging the large size of the American economy and the support funneled to the local computer industry by the U.S. government through defense contracts. That action made it impossible for any European nation to compete, so the argument went, and even within the American economy for many local computer companies to succeed against IBM. The logic, therefore, was that IBM found itself in the enviable situation where it "kept its position until the introduction of personal computers," which "changed the rules of the game."[164] By then (1980s) the possibility of a vibrant local company competing in the world economy had largely bypassed Sweden as it had so many other European states.

When the view of Swedish events is turned around by looking at what users of computers accomplished, a different narrative emerged. One sees a nation that embraced computing in all industries and in every large enterprise, using equipment and software from all over Europe and the United States in the 1960s and 1970s, and an economy not shy about acquiring Japanese products in the 1980s and 1990s. A collection of over 150 oral histories of end users conducted in the early 2000s in Sweden clearly demonstrate that this was a nation whose users of computing were timely and astute in understanding the potential benefits of the technology, clearly able to install and use it, and in the process transformed how they did their work, particularly in process, manufacturing, banking industries, and in public administration.[165] This experience leads to the conclusion that the economists at the World Bank embraced when they observed that Sweden had become one of the most tele-dense societies in the world, along with their Nordic neighbors. Thus, the debates about Swedish failures on the supply side had focused too much attention on an issue that diminished in relevance over time.[166]

So how did Sweden end the century? To what extend had it started to move into Wave Two? The data suggest it was in its early stage. About 60 percent of the population had at least one mobile phone, which meant in effect, every adult, while over 70 percent of all households had a PC, one of the highest percentages in the world. Internet usage by the population ages 12–79 was close to that same statistic (59 percent). Nearly 20 percent of all Internet users made purchases online by 2000, although only 0.7 percent of all retail sales in the country were conducted online; still it was not a low volume, as it was second to the world's leader in online sales—the United States—and far ahead of the rest of Europe. Commerce and customers intersected electronically elsewhere too, for example in banking, where there existed about 1.6 million Internet bank accounts. The nation had the second largest number of online brokerage accounts in Europe. Swedes were early adopters of various Internet-based services too. Business-to-business digital transactions were far more extensive than at the consumer level.[167] Based on our embryonic calculus of what constitutes features of a Wave Two society, it appears Sweden was on its way there.

CONCLUSIONS

Italy, Netherlands, and Sweden offer insights into Europe's appropriation of computing. In some ways, these three reflected patterns evident in most, if not all West European countries, but also local variants. Each had individuals ensconced in universities, institutes, or in relatively high-tech firms who learned about digital computer developments, largely occurring in the United States in all periods, but also in Great Britain in the 1940s and early 1950s, and sought ways to acquire or build similar ones in their home countries. While the dates of these early, almost individual initiatives began at slightly different times (Netherlands and Sweden late 1940s, Italy mid-1950s) they varied by only a few years, so one can declare with some confidence that initial forays into computing occurred within the first decade following the end of World War II. In parts of Europe, and specifically in these three countries, these early efforts were humble affairs, lacking the massive financial and institutional backing evident in the United States and Great Britain in the 1940s and 1950s. They were similar as well in that local national governments supported them, beginning in the mid-to late 1950s in most European countries, although Italy was an exception until years later. Government support proved modest in these three countries, but not as different as in other nations.

In time in all three nations a few people, then groups, came to the discussion about the feasibility of developing a local computer industry (1950s–1960s), relying on existing indigenous high-tech firms, typically electronics firms, large manufacturers, or telecommunications utilities, with some aspirations for pan-corporate collaborations, and invariably involved local universities. In all three cases, as in some other European countries, including the other Nordics, the events were modest investments in comparison to those made by the

American, British, and French governments by the 1960s and by West Germany by the early 1970s. In the three countries discussed in this chapter, investments proved inadequate to the task. What officials and management in various companies did not realize was the extent to which they had to invest in order to get to the same level of activity already evident in the United States, which really was the benchmark for what it would take to compete by the mid-1960s. Earlier, the levels of investments were simply not clear and even now just becoming more so to historians. By the end of the 1970s, it was evident to all the players that an initiative to promote national champions was a lost cause and instead turned their attention to pan-European collaborations to form a European industry, an initiative that also failed, the reasons for which are discussed in the next chapter.

Looking at the national economies themselves, the three countries provide a great deal of insight on how Europeans accepted computers. If we leave aside the geographic size of Italy, and to a considerable extent the great landmass of Sweden (mostly sparsely settled), and also exclude from our discussion for the moment the size of populations, in each country the economy consisted of a large percentage of small enterprises sprinkled with a few very large firms that dominated their respective industries. Each also had an important agricultural sector, which barely used much computing until the 1990s. Small enterprises only began using computers in any quantity in the late 1970s relying on minis and in the 1980s PCs and telecommunications when they could be used in tandem. So, while Italy was geographically and demographically a big country, it was also a nation with many small enterprises, just as in so many other nations. One could argue that Sweden fit more the profile of West Germany, and to a lesser extent the Netherlands, because it had a few large enterprises, organizations perfectly suited to use computers in the 1960s and beyond.

Also relevant of firms in all three countries is that they had an export orientation that caused them to strive to be competitive in a globalizing market. Put in more tactical terms, they had constantly to apply any tool or technique that could optimize their performance, and for all, that meant using computing. In fact, this seemed more the case than with the British, although the French and West Germans proved keen to serve at least a pan-West European market. The Swedes, Italians, and Dutch did too, but also had a trans-Atlantic view of possible markets. These orientations made adoption of computing easier at a number of levels: through academic and business trips to the United States to learn what the Americans were doing with computing, through collaborations among firms and with American counterparts in computing, and through national policy commitments to constrain the increased domination of local computer markets by the Americans, most notably IBM. But, in each instance, most activities and interactions were not coordinated among countries in an effective way during the half century, while individual scientists, engineers, and firms proceeded often without necessarily knowing what others were doing; this was especially true of engineers and academics. Often, their awareness of what others did outside their country occurred when they visited the United States and in later decades through their attendance at IT conferences, by reading trade journals, and in their deliberations with suppliers of ICT.

On the demand side, how individual institutions used computers in these three countries mimicked practices evident in every other European and North American nation. The uses—applications—began with accounting and finance, next moved to manufacturing and distribution, and subsequently evolved in transforming finance and retail. Organizations acquired every new form of IT that came along and when telecommunications linked to computers in the late 1960–1970s, ICT. Rates of adoption of a use proved similar within any particular industry and largely so around the world, evidence that international competition kept export-oriented firms abreast of a wide range of industry-wide trends, a pattern of behavior that has been documented in considerable detail of the American experience.[168]

Public policy played both a profound role, yet on occasion, a nearly irrelevant one. In the case of Italy, it was clear that not until the end of the 1970s did the Italian officials weigh in with significant public policies and actions, and even then still largely as just another consumer of digital technologies. Italy's lesson is that a national appetite and consumption of IT was not dependent on pump priming on the part of a government. This is a lesson that one could also extract from the American experience if we left aside the massive infusion of public investments in R&D, because even in the United States the local computer industry relied less on government support to continue its trajectory of expanded use of IT in the local economy by the 1980s. The Dutch example parallels very closely those of Great Britain, France, and West Germany, where local firms dined too frequently at the public trough at their peril as it turned out, since it made them less competitive than they needed to be in a rapidly emerging global market. It proved to be a business environment that not only extended beyond their local shores but also dragged along their largest companies, because they too were enmeshed in the thick of international markets for their own goods and services.

With the Swedes, we face a delicate case, because it is quite normal for observers outside Scandinavia to lump them in with all the other Nordics. While this is not the place to review in detail the experiences of Norway, Denmark, and Finland, which indeed proved quite divergent, nonetheless the clumping together happens all too frequently. All of these nations are routinely characterized as different from the rest of all Europe, that each (and collectively) enjoyed an exceptionalism with respect to culture, role of public administration, and the functioning of their economies and businesses. What the story of Sweden's experience with computers demonstrates is that there is no significant exceptionalism at work. The Swedes ran a market-driven economy with enough flexibility for its firms to compete around the world; the same was true for Denmark all through the period, and increasingly so for Finland. Local historians have demonstrated that the Nordics integrated themselves into both the Atlantic and pan-European markets for centuries, often with business partners in other countries, such as England and Holland, running their business and economics in similar fashion to other market-driven European economies.[169]

What perhaps made the Nordics unique from everyone else was their justifiable reputation for having very high personal taxes, social stability, and governments that involved themselves in more aspects of their nation's private and economic life than was evident in other countries. But when it came to the diffusion of computing, that happened for reasons evident in other countries: availability of the technology, a skilled workforce able to use it, an effective collection of vendors who could invent and sell hardware and software all through the period, and, after the mid-1960s, telecommunications as well, and ultimately high-enough national and per capita GDPs to afford the technology in the first place. There is one important piece of evidence in support of some exceptionalism to acknowledge, namely, that the Nordics maintained higher levels of education for a greater percentage of their populations useful for the introduction and continued use of computers than all the countries around the Mediterranean rim, including France. These levels of education—not merely literacy and the ability to write—paralleled, and sometimes exceeded, similar levels of training evident in all Anglo-Saxon nations without exception, and that also proved comparable to the role of education in the Netherlands in the twentieth century, more than in the Benelux nations.

Finally with the Dutch, historians examining use of ICT by its public and private institutions observed that these proved attractive to install when they reinforced existing ways of working and organizing activities—a lesson that applies to all the countries discussed in this book and a characteristic of early Wave One. Bank of Italy economists reached the same conclusion about Italian firms: the more they changed their operations and improved the quality of their workforces, the more productive IT proved to be.[170] The Dutch and Italian experience is a reminder that at least half the period of Wave One was characterized by users furiously bending the technology to existing organizational and operational practices.

Clearly by the time firms, and to a lesser extent governments and educational institutions above the secondary level, began moving out of Wave One, leading private sector enterprises with high rates of productivity attributable to the use of IT had started to change the structure of their organizations and how work was done to optimize use of computing and communications. These were lessons learned, first, by companies in the United States and Japan, and that the international firms in these two nations then exported to their subsidiaries and business partners in Europe in the late 1980s and throughout the 1990s.[171] This occurred to such an extent that one can argue that a feature of Wave Two computing was the transformation of organizations and work flows.[172] Thus, we can come back to the Dutch experience to point out that here, as in other countries, this transformation was underway. The comment quoted earlier about IBM's factories being more disciplined than Dutch ones presaged the emergence of this new way of working.

Looking at the broader pattern of European diffusion of IT, these three nations, and essentially all others in both developing and advanced industrial economies, participated in the Wave One spread of computers in what has been demonstrated as a recurring pattern of adoption as we moved from one chapter to another. They started their adoptions in varies years, which were not so greatly dispersed from each other (largely mid-to late 1950s to early 1960s) when viewed across events of a half century. When they completed their Wave One also varied, although the chronological ends are less defined since some are just exiting it (European countries) and others are still there so far in Latin America, Russia, and wide swaths of Asia. It appears the Italians remained in Wave One until the end of the new century, with tentative forays into Wave Two in the late 1990s. The Dutch began to evolve out of Wave One by the mid-1990s, much as the United States and for all intents and purposes, the western part of the newly unified Germany, and certainly all the Nordics. Sweden, along with its regional neighbors, may have become a source of early features of Wave Two because of its strong international involvement in the C part of ICT, along with Finland, for instance, with the latter's Nokia, the mobile telephone manufacturer. Because of the interconnections among vendors racing to introduce and sell products as quickly as possible in as many countries as they could, and the focus within the private sector to leverage whatever technologies, skills, and managerial practices that were identified in a highly literate, well-served environment by academics, associations and publications, to achieve business success both at home and in a rapidly expanding global economy, all contributed to the pace of adoption which picked up speed over time. All the thousands of trend line charts on volumes of installations invariably looked the same, with lines pictured moving higher from lower-left-hand corners to upper-right-hand corners, with compound rates of expansion in many years at double digit percentile rates, from the 1950s to the end of the century.

Even looking at the supply side of the equation, one can see that in each country Wave Two has a technological demarcation, such as the time when the number of IT enterprises dependent on the Internet to do their core functions increases. In the case of Sweden, for which there is excellent data on opening and closings of enterprises by industry, one can clearly discern a divide that began, beginning around 1990–1991. The number of Internet-based IT firms founded (but not necessarily survived) increased in number, doubling by

2004, from 6,122 to 12,068. In the entire decade of the 1980s, only some 5,000 were established. Some grew out of existing telecommunications enterprises, while others were complete green-field establishments; the former tended to survive more frequently, probably due to access to capital and customers from the mother enterprises which new starts would not have necessarily had. The point is, things changed, with new start-ups taking advantage of the pre-existing cadre of experts in information and communications technologies that had developed over many decades.[173] As in Germany, Latvia, and Estonia, for example, the widespread introduction of telecommunications and later the Internet with government support, created a fertile environment for a sharp increase in a new breed of supply side diffusionary activities in the nation.

While the story of how diffusion in these and other countries are perforce told one after another to arrive at the specifics of what happened, unlike many earlier technologies, ICT itself helped to facilitate the diffusion and to affect the rates of adoption. That self-facilitative feature may ultimately result in historians attributing to digital technologies a feature of its emergence and evolution unique in the annals of humankind's love affair with all manner of technologies. But that is a discussion for another time. To facilitate it, however, we must step back and aggregate our findings about what happened in the West European community. In aggregate it was the world's largest collection of modern economies in the second half of the twentieth century, before exploring what happened in the Soviet sphere. The latter was also a very large swath of the world's economy and population, but one that functioned with fundamental differences. The Communist region tests our notions of how computing spread dependent on innovation and free market economies, while further defining features of Wave One around the world.

CHAPTER 5

How Western Europe Embraced Information Technologies

I arrived in Trondheim on a very cold January day in 1963, straight from Seattle. . . . The computer was Danish. It had 1K words of core, a 12,800 word drum, no tape or printer; and it had a wooden door. I thought it was the broom closet.
—Norman Sanders[1]

Western Europe's experience with computers and telecommunications is a remarkable story of individual pioneers building machines out of parts, of vendors developing what ultimately became a large market for the new technology from all over the world, and of governments in almost every country nurturing development of both local computer vendors and ICT industries and users. These events made it possible for West Europeans to use these technologies. It is an epic tale that remains highly fragmented, told on a country-by-country basis in hundreds of articles, government reports, dissertations, and books in over a dozen languages, many that rarely seemed to circulate outside the national markets in which they were published.

This chapter continues the process of providing a more composite view of Western Europe's experience before moving to a description of events in Eastern Europe. This is a limited discussion because, if the individual case studies in early chapters have taught us anything, it is that once we pass through grand generalizations about patterns of adoption, the detailed circumstances and events in each country proved unique and important to understand on their own merits. It is why this chapter follows those with the more detailed discussions instead of before them. With this chapter it becomes possible to describe some of the shaping forces at work that scholars might consider exploring as they dive further into the specific histories of European nations not covered in this book, but that are worthy of study, such as Switzerland, Austria, Denmark, Norway, Finland, Czechoslovakia (if we want to consider it as part of the West), Spain, Greece, Belgium, and Portugal. So, there is much work to be done. Here we provide an overview of many key patterns, describing major sources of influence and impetus for diffusion across Western Europe and, in the process, reinforce themes developed in earlier chapters.

The overall finding presented in this book is that the world has—and is—going through two waves, or phases, of adoption. The initial one, which I dubbed Wave One, involved largely the invention, acceptance, and use of large mainframes, minicomputers, and, to some extent, PCs connected via telecommunications (overwhelmingly pre-Internet) by the end of this first wave of deployment. The primary users were initially large institutions and toward the end of Wave One, additionally smaller enterprises and organizations, while individuals were beginning to use PCs and other devices connected to networks. The major users were corporations in all industries, including extensive adopters in national and state government agencies, and universities. As the technology became more modular and less expensive, those West Europeans who had little or no prior experience in IT began to embrace the new IT, just as had North Americans. Wave One represented the time when the technology first came into existence and matured to a great extent. The key diffusionary point to make is that the technology evolved to such a degree that it was routinely and extensively used for all manner of data processing and calculations across all industries in almost every aspect of modern work life. All West European countries have passed through Wave One, but not all nations around the world. Europeans accomplished this in roughly a half century, a remarkably fast diffusion for any technology, let alone those as complicated as ICT. It was an achievement that started from nearly zero, as some of the cases demonstrated in this book, with only experience using office data processing equipment (such as adding machines, punch-card tabulators and calculators), knowledge of electrical engineering, and both mathematics and physics[2] to prepare them for the arrival of the computer. They had limited knowledge and resources they could apply to the modern electronics that underpin early computing technology. The British were the sole European exception to the latter observation because they led the world in understanding the base technologies and had the capability to invest resources to exploit these in the early 1940s. The epigram at the start of this chapter is a testament to what the definition of starting from zero meant for some computer pioneers who lived through the process.

The start of Wave One occurred with the construction of the machines, creation of software, and initial uses, while the end of Wave One came—or was recognizable as concluding—when two new circumstances surfaced: when in addition to Wave One uses wide diffusion of digital consumer electronics and use of the Internet occurred beyond a small community of academics, public officials, and individuals in the private sector, with emphasis on the word *small*. Wave Two was in full bloom when consumer use of IT proved demonstrably widespread along with their interactions with the Internet.

Other factors differentiated the waves discussed earlier in this book. In the United States Wave Two dawned when individuals began their extensive use of PCs connected to various forms of telecommunications, not just the Internet (latter happening in the late 1980s–early 1990s), while in Europe it started often with mobile phones followed soon after by wide use of PCs connected to networks, including the Internet (typically mid-1990s), and earliest in Europe north of France.

There appeared to be no clear cut end to Wave One and start of Wave Two. They overlapped, characterized by a transition from one to the other that often lasted a decade or more, which in itself is a short period in historical terms. Additionally, all the activities that went on during Wave One continued subsumed into Wave Two. One big difference for Wave One diffusion and its use of IT is that when mixed in with Wave Two activities, the first wave's efforts evolved, such as occurred when companies using large centralized systems in Germany and Great Britain gave their customers access to services over the Internet in the late 1990s and early 2000s, and a few years later in the United States, through "apps" on Apple iPhones and other "smart" phones.

The timing of when Wave One began and ended varied, reflecting a pattern of measurable diffusion out of the laboratories and experimental projects also in evidence across many parts of Asia and all of Latin America. The North Americans began in the late 1940s, at the latest by 1950–1951; the British almost at the same time, perhaps off by only one to two years; the French commenced in the mid-1950s, while Germany began later that decade, along with most of the German-speaking parts of Europe and the Nordics. The Mediterranean basin engaged at the end of the 1950s, yet more frequently in the 1960s. The European case suggests that the amount of time a nation spent in Wave One before starting the process of transitioning into the initial phase of Wave Two also varied. Most Europeans were recognizably in the earliest stage of Wave Two by the late 1990s, although it appears the Nordics arrived there earlier (circa mid-1990s and beyond), while the French could argue that their nation's Minitel gave them the dignity of early-entrant status. What is remarkable about the West European experience with these cycles is that there existed a wide diversity of starting times for entering Wave One (see table 5.1), but remarkably little variation in when they began evolving into Wave Two. While the study of Wave Two behavior in Europe awaits its historian, the fact that Europe as a whole embraced mobile phones at roughly the same time as the Americans accessed the Internet and their (American and West European) adoption of myriad consumer electronics with digital features, all signaled a nearly synchronized changed circumstance from just IT use to ICT diffusions of all kinds. Global enterprises introduced new online functions to both their institutional and consumer customers in multiple countries in a nearly simultaneous fashion which also led to the synchronization of the transition. An example from IBM occurred when it introduced e-mail to its European employees: it happened

Table 5.1 EARLY WEST EUROPEAN IT DIFFUSION MILESTONES WITH
U.S. COMPARISONS, 1930S–1960S

Country	Year Began Work on Computers	First Working System	Initial Local Commercialization
USA	Early 1940s	Mid-1940s	Early 1950s
U.K.	Early 1940s	Early 1940s	Early-mid-1950s
Germany	Mid-1930s	Early 1940s	Mid-1950s
France	Mid-1940s	Early 1950s	Mid-late 1950s
Netherlands	Early 1950s	Mid-1950s	Late 1950s
Italy	Mid-1950s	Mid-1950s	Early 1960s
Sweden	Late 1940s	Early 1950s	Mid-1960s
Denmark	Mid-1940s	Late 1950s	Mid-1960s
Austria	Early 1950s	Late 1950s	Early-mid-1960s
Switzerland	Start 1950s	Mid-1960s	Early-mid-1960s
Average Range	Late 1940s-Early 1950s	Mid-Late 1950s	Late 1950s–mid1960s

There is much controversy in the historical literature about "firsts" caused by multiple projects in any given country. Therefore, a composite view by country is given here in the form of ranges of time.
Source: Adapted from OECD, *Electronic Computers: Gaps in Technology* (Paris: OECD, 1969): 33.

across Europe almost at the exact same moment. Another is when Google and e-Bay came into existence, before they hosted national nodes in local languages; the world could access Google's website, housed in North America, as long as one had an Internet connection and, in the beginning, a PC.

To put a fine point on the matter, adoption of specific machines and software, telecommunications, and applications were so concurrently underway in Europe that it raises the obvious question of how could that be? The case studies in this book suggest that the reason for that near simultaneous adoption was the combination of factors discussed in the rest of this chapter: wide availability of information about IT, pan-European products from vendors like IBM and Bull, later various American PC vendors and Swedish and Finnish telephone manufacturers, emergence of a class of technologists able to build then later use computers, among other causes, and the European Union's push for liberalized telecommunications markets and pan-European technical standards, beginning in the 1980s. Differences in rates of diffusion and timing of adoptions can be better explained as more frequently the result of varying levels of structural economic features of a society that lent themselves to early adoption (for example Great Britain and Netherlands with a core set of large enterprises and government agencies) or delayed it (as in the cases of Italy and Portugal which had high percentages of small firms). The extent to which governments backed financially the myriad investments necessary in training, R&D, cash outlays for equipment, software, and required staffing (such as in Sweden and France, but less so in Spain) affected rates of diffusion too. Costs always captured the attention of all participants in the diffusion of IT. It is the first factor influencing European diffusion that we explore.

ROLE OF COSTS AND ECONOMICS

Costs and economics are two distinct subjects. The first speaks to the issue of expenses for machines, software, and people to run them, subtracted from the measurable potential savings in operating expenses (largely the financial rationale for IT in the 1950s–1970s) to arrive at some discernable economic benefits. Cost is also the cash outlay for IT that permits new functions to be done that have financial value, less the expense of IT to derive such a benefit, often called the "business case" for IT or the "financial" rationale for some IT initiative (increasingly also a reason for IT in the 1970s to the present). Costs also can be a budgetary conversation, that is to say, "my IT department has a budget of $5 million and next year possibly $5.3 million or $5.8 million if I can make the case for it." Users of computing viewed IT through the cost lens when they weighed decisions about adopting IT. The rate at which they could acquire IT was profoundly influenced by costs, but also by their understanding of the benefits of using IT; the reliability and functions of the technology; and the extent to which they had the necessary skills with which to operate the equipment and use of the software. That is why this chapter discusses various facets of diffusion. But, costs are so important that economists and business writers have long commented extensively on this topic, beginning in the earliest days of computing and in every capitalist economy across Europe and North America.[3] To a certain extent price was discussed in the first chapter, but more needs to be said with respect to Europe's experience.

Since price and economics are interrelated, there is that issue of economics to integrate into the discussion. The thrust of any economic assessment of diffusion of IT moves rather quickly above the firm level, from where costing considerations dwell, to matters related to the impact of IT on employment, labor, capital productivity, and to growth of GDP for an industry, nation, or continent. Price becomes an input into the conversation. Costs are of intense interest to users and their enterprises while economic considerations are more the distant preserve of economists and public officials. Economic issues remained very important affecting the rate and nature of the diffusion of IT in any nation. Economic considerations influenced profoundly the policies of governments, as we see in each European case and with the United States, and observe in dramatic form with China and Japan in our global history. Economists keep score on the prosperity of nations and their rate of national productivity. They describe the source of a country's activities and incomes based on such inputs as the extent of investments in IT versus labor or some other tools that offer the potential of financial or economic benefits.

Economists are nothing if not prolific in commenting about the economic effects of IT. The infusion of IT into the world's economy provided a new surfeit of activity and growth for economic discourse. There is nothing more attractive to economists than seeking out new ways to define an economy once somebody dubs it as the New Economy or the Information Age.[4] Indeed, measuring the old Second Industrial Economy had stabilized globally to such an extent before the arrival of the computer that international bodies were established in the confidence that people knew so much about how an economy worked that these national economies could be commanded to do things in a predictable fashion. That mindset facilitated creation of the International Monetary Fund (IMF), the World Bank, Organization for Economic Cooperation and Development (OECD), American foreign aid economic development programs, regional deployment around the world, and various agencies within the United Nations. So, economics is an important variable in the diffusion equation, and why it receives so much attention in this book as a factor to consider. This observation may appear to the casual reader as obvious, but in the diffusion literature, it lags behind such other considerations as the nature of the technology itself and its uses.[5] While this is not the place to engage in a discourse about the matter, suffice it to say that by the start of the 1990s, no serious discussion could be had about the economy of an "advanced" nation without engaging on the role of IT.

Every vendor to which historians had access to archival records illustrated that they discussed, fretted over, and carefully managed pricing of their offerings. Every user of such technology to which historians could examine their decision-making archival records demonstrated that they too worried about and haggled internally, and with vendors, over budgets and prices. The accumulated results of literally millions of individual decisions across all of Western Europe aggregated into the economic results to which economists began paying attention in the 1970s, and that in turn influenced public policy. These results affected even national debates about the evolving shape of modern society, as we saw with the French. In fact, the debate took place all over Europe.

From the earliest days of commercially available computers, customers discovered a pattern in pricing that continues to the present, namely, the decline in cost per transactions, while quality and capacity increased on a continuous basis—the phenomenon introduced in chapter one as a factor that influenced the attractiveness (and uncertainty) of timing one's acquisition of computing. From the beginning pricing dynamics were always dramatic and volatile. For example, if one looked at the cost of an IBM S/360, the system that fundamentally compelled so many enterprises to adopt computers in general, and not just IBM's, the pricing data is very impressive. Table 5.2 displays the cost of 100,000 multiplications of three generations of IBM's computers. There are two dynamics to observe in this data. First, note the dramatic decline in cost

Table 5.2 PRICE COMPARISONS OF THREE GENERATIONS OF IBM
COMPUTERS, 1954–1965

Type	Year Shipped	Cost/100,000 Multiplications
IBM 704	1954	$1.38
IBM 7090	1958	0.25
IBM 360	1965	0.035

Source: Adapted from Organization for Economic Co-Operation and Development, *Gaps in Technology: Electronic Computers* (Paris: OECD, 1969): 73.

over time, and that is just in one decade; declining costs of transactions continued to the present even though salaries for users kept rising. Each generation of new computers brought into the fold waves of first-time users and
encouraged those already using systems either to add more into their operations or to swap out older, more expensive machines to rent and operate,
because with each new generation of products, operating costs frequently
declined as well, but not always. The second dynamic is that these machines
and their software were introduced so quickly, one after the other, that it
proved very difficult for the vast majority of rivals to keep pace with IBM. It
seemed IBM always had a new machine with which to compete against an
"older generation" product from someone else, whose price was set to the
prior generation's higher costs.[6] The same occurred with PCs and now is happening with tablets and mobile phones involving thousands of IT vendors
worldwide.

During the second half of the century, it was customary for a computer
vendor to introduce a new system at between 9 and 18 months before initially
delivering these to their customers, much as happened in the 2000s with the
introduction of a new commercial aircraft. When a vendor unveiled a new generation of computers, users proved reluctant to acquire anybody's current
generation technology (now viewed by them as "old") because they could see
that by waiting for a reasonably reliable delivery date, they could acquire a less
expensive, better, and bigger system. This happened all over the world and
given that for many reasons computers from most vendors cost roughly 20
percent more in Western Europe than in North America, waiting made even
more sense than, say, in the United States. This behavior became so set by the
late 1960s that economists were generalizing about how long a generation
lasted. For example, in 1969, economists at the OECD were telling their
readers, "that the market share of a given generation begins to fall within approximately six years, after three years of very rapid growth and three years of
relative stability."[7] If companies were looking for a machine and they were late
in the six year cycle they quickly placed an order for the new generation literally on the day it was introduced; there was even a name for this action, a "first

day order," which would be time stamped in a sales office and placed in a sequential queue for assignment of a delivery date. The whole process of announcing and ordering computers this way represented a crude form of digital applied economics at the firm and industry levels for decades, proving to be an important gating factor affecting the rate of diffusion of computers. The announce-to-order received cycle affected how both early and late adopters thought through their adoptions strategies.[8] It is a behavior yet to be explored by historians.

ROLE OF ENGINEERS AND ACADEMICS

Across all the European countries studied, and others not discussed, small groups of electrical engineers in high technology firms were often some of the earliest individuals to become involved in the construction of one-of-a-kind computers in the 1940s and 1950s. They often evolved into groups of employees in companies that went on to manufacture early systems in the 1950s through the 1970s. In addition to the countries examined in this book, there were such groups of skilled workers in Norway, Denmark, Austria, and Switzerland; however, there is currently insufficient evidence to extend this generalization to Portugal, Spain, and Greece.[9] An extended group of like-interested individuals also surfaced from other disciplines, particularly mathematics, physics, and other branches of engineering. Many were employees of automotive companies, telecommunications firms, large technical agencies in government, some banks and insurance firms, and, of course, universities. Academic cohorts represented a parallel community that expressed interest in computers at the same time as the first group. In short time they found each other and collaborated in the early decades by sharing information, raising funds for projects, building machines and writing software. To a great extent, the technical accomplishments of these communities have been the best documented so far, largely because (a) many of the pioneers wrote about their experiences (still do) and (b) historians have studied (and continue) to examine their early projects. As this book was being written in 2010–2012, one of the largest historical projects underway in computer circles in Europe was the documentation of the development of a European programming language of the 1950s and 1960s called Algol, while in another initiative Nordic computer pioneers gathered information on their experiences with IT, and held conferences to share their recollections and observations. The journal of record for the history of IT, *IEEE Annals of the History of Computing*, published dozens of well-researched articles on the work of European engineers and scientists. Indeed, our base of knowledge about early computing in Western and Eastern Europe relies largely on the materials published in the *Annals*.

Discussed in both chapter two on the United States and in subsequent ones on Western Europe, these early communities often interacted with each together across borders, sharing information.[10] This practice remained constant and open all through the 1940s and 1950s, slowing only as the computer industry emerged, when vendors reduced the amount (but not all) of sharing of data to protect their competitive positions and patents. All through Wave One, however, academics continued holding conferences all over the world; the largest number in the United States. Perhaps the most critical period of sharing occurred in the beginning, when a few engineers could almost single-handedly introduce computers into their countries, as happened in Sweden and Italy, for example. This model of introduction stood in sharp contrast to the larger, more institutionalized approaches taken by the Americans, British, and French.

Herman H. Goldstine, an American scientist working in Princeton, New Jersey, at the site of important pioneering work on computers at the start of the 1950s, recounted in his memoirs-history of early computing numerous visits by Europeans from all over the Continent to where he and others worked. At the Institute for Advanced Study in Princeton, Goldstine received British visitors in 1946–1947, Swedes in 1947–1948, Norwegians in 1950–1951 and again in 1952–1953, and Swiss engineers in 1948–1949, to mention a few.[11] The overwhelming number of these visitors came from newly established institutes created after World War II to explore mathematics, engineering, and other advanced technologies. By the end of the 1960s, there were almost no countries without such quasi-academic-developmental laboratory types of organizations operating across all of Western and Eastern Europe, working in parallel with the industrial development labs in the private sectors, such as those run by IBM, Bull, ICL, and others. In every instance I could find, these industrial laboratories hired staff from other institutes and academia, while frequently employees in industrial labs moved back and forth to academic positions, even across national borders, as occurred at the German IBM labs, which had employees from Austria and Switzerland as well.

By the late 1970s, it took fewer computer scientists and engineers to build computers and write software than before, as the process for manufacturing equipment had been sufficiently systematized that lower-skilled workers could assemble machines from parts manufactured all over the world, software written also largely by individuals who did not have to be technically trained to the degree, say, as an electrical engineer. Programming, while still a craft, indeed many called it an art then (and now), did not require a Ph.D. in anything. Rather, specialized training of a slightly higher grade than vocational skills proved effective, with mathematical capabilities an advantage in the 1950s–1970s, because of the availability of what became known as "higher-level languages," which were easier to use. However, development of new technologies and training people skilled in their use did require computer scientists, engineers of all types, mathematicians, and other skilled individuals, who remained in growing demand all through the half century across Europe (not just the West), North America, and East Asia.

These communities—resident in companies, universities, and government agencies—had twin roles to play: to invent and use new IT and to train users, such as programmers. Around the world they succeeded spectacularly in the former, as evidenced by the continued evolution of the technologies still transforming after seven decades of development, but who failed to train a sufficient number of end-users, particularly programmer and systems analysts. In any given decade, there were complaints about the lack of a sufficient number of computer scientists, programmers, women in computing, and engineers skilled in the use of this technology in embedded industrial equipment, even for designing electronic systems in automobiles in the 1970s and later. In a pioneering work on the role of automation of Swiss railroad systems between 1955 and 2005, historian Gisela Hürlimann illustrated the dependence end users had on the availability of IT skills with which to transform their rail system, so important to the integration and functioning of the daily activities of Swiss society.[12] The cases reviewed earlier of Dutch organizations adopting computers across the half century demonstrated this same pattern.

There is no reliable information yet about how many people in total there were in Europe during Wave One developing and building computer systems, let alone operated them in the early years. One can reasonably speculate, based on anecdotal evidence that this ran into the hundreds at most in the 1940s, a few thousands until the mid-1950s, after which the number climbed in the 1960s, before shifting massively higher to the present.[13] The OECD attempted to study a related, broader question about how many knowledge workers, or individuals primarily making devices or using information there were and their impact on national economies. It began the study in the 1970s, as part of a growing recognition in the advanced economies that the services sector was increasing as a percentage of a nation's GDP, at the expense of the agricultural sector, and possibly even of manufacturing. OECD reported that while early users of computers embedded these systems into process and manufacturing operations largely in the 1950s and 1960s, by the end of the 1960s computing and telecommunications were now converging for new uses. It forecasted that the next wave of adoptions would be in offices, which is exactly what happened. Of most European nations, OECD observed that from 1950 to the late 1970s, "a progressive shift has occurred in these countries towards occupations primarily concerned with the creation and handling of information," which it characterized as "a profound change," quantifying the rate of change as "3 per cent in each five-year period," leading to one-third of all workers in the OECD now called "information workers."[14] What made this transformation crucially relevant to diffusion of IT was that their work was "vulnerable to substitution by information technology."[15]

In the mid-1980s, OECD came back to the issue of information workers and did further analysis on the number of information occupations, slightly lowering its estimates of how fast the workforce had transformed from the 3 percent reported earlier to 2.6 percent—still an impressive number. For all of Western Europe, the percent of labor engaged in information tasks at the dawn of the computer's arrival (approximately 1951) varied from 10 percent of a nation's workforce (Denmark, Finland)[16] to as high as 26–27 percent (UK, Sweden), while that of the United States was just north of 30 percent. By 1981–1982, however, Denmark and Finland had caught up with where the United States had been in 1951, through continuous and sharp progress to that level all through the previous 30 years. Sweden and West Germany had not rested on their laurels; rather, they extended their transformation to percentiles in the range of 32–35 percent, although more gradually since they had started with a higher base in 1951. Finally, the British were now at 40 percent while the United States closer to 47 percent.[17]

The reason for reviewing these data is to suggest that across Europe the diffusion of computing came into professions that lent themselves to automation, indicating that the nature of work was a direct cause for how IT spread so quickly. Lest there be confusion about carts before horses, with OECD's early analysis of information jobs at the dawn of the 1950s we have the indication that prior to the arrival of the computer there already were large swaths of European economies with jobs (hence work) that lent themselves to automation, indeed potentially 10 percent of all labor. So the ground was already fertile, and when computers made their appearance in various forms, these systems had a place to go within the broad spectrum of Europe's economy. Thus, credit for adoption of computing must be shared by engineers, scientists, and vendors with users.

ROLE OF PUBLICATIONS

The American experience in publishing and reading articles and books about IT proved instrumental in the diffusion of IT in the United States, especially for end-user communities. Publications educated engineers, customers, and computer scientists as well about the subject. Numerous bibliographies appeared.[18] There were relatively comprehensive bibliographies published of European publications that suggested interest on the part of Europeans; yet, no serious attempt been made to study the extant literature on IT and its influence on the key actors. Anecdotal evidence abounds that across all of Europe (East and West) there existed an appetite for this kind of material. In the process of conducting research for each of the country case studies for earlier chapters, I examined publications in local languages that demonstrated their presence. Early on libraries began to collect materials. For example, all of IBM's European laboratories maintained technical libraries. Its laboratory in Zurich, Switzerland, in the early 2000s maintained a large collection of thousands of publications in many languages, including bound copies of various

European and American technical journals dating back to the 1950s. Perhaps the most important hard evidence of the extent of publications available in Europe was a German bibliography funded by the German Science Foundation (Deutsche Forschungsgemeinschaft—DFG) which published 73 reports on the "new" literature from 1954 to 1966 in order to "make up ground" in the field of computing, in the words of the DFG. Given that each report published between 200 and 400 citations, the total list cataloged for European technical audiences ranged between 14,600 and some 20,000 titles.[19] The majority of the titles appeared in English and German, many were articles. Materials also existed in other European languages, including Russian, Polish, Swedish, and French. The Dutch Studiecentrum began collecting materials in all languages in the 1950s and in 1971 published a two-volume bibliography of its collection of thousands of titles.[20] A Belgian bibliography, published in the early 2000s, ran to nearly 270 pages, citing materials available in that country.[21]

As new subfields of IT emerged, so too did publications in Europe. For example, in the mid-1950s the topic of cybernetics (artificial intelligence) took off in the United States, or so the story goes. While research and debate on the topic centered the most in the United States, a detailed study of cybernetics written in German published in 1965 makes it manifest that Europeans were also engaged in the subject, and through bibliographic citations in the book, one could see that they had been for a number of years.[22] In short, cybernetics was an international topic, bringing into question the current U.S.-centric views of the subject and the nature of its influence on both computer sciences and how societies viewed themselves in recent decades. The conversation about cybernetics was more central to the history of the evolution of IT around the world than we currently give it credit for, especially across Western and Eastern Europe at a minimum in the 1950s through the 1970s.

By the early 2000s, online Internet-based searches of library catalogs of European universities and research institutes became possible. Queries of these repositories made it evident that the body of literature published in Europe on IT had been extensive since the early 1950s. In countless trips to Europe over the past 40 years I saw for sale many of the key American publications on IT either in English or in foreign language translations, in addition to locally written volumes.[23] Interestingly, books and articles originally published in European languages were rarely translated into English and sold in the United States. The single most obvious exception was, of course, the French government's report on computing in France released in 1978, the subject of extensive discussion in chapter 3.

Of course, publications alone did not stimulate use of IT; rather they had to be used in combination with other activities. Some of the most important, if not central players in diffusion of IT, were vendors of computing products, because they compelled potential users to make decisions to acquire, use, and become reliant on the technology; publications rarely forced action in such an explicit way.

ROLE OF VENDORS

European vendors in the computer industry worked all over the continent. They were tracked by many interested observers. These industry watchers concentrated more on describing the histories—and failures—of these firms than on the role they played in the process of facilitating diffusion of IT in Europe. Without vendors diffusion would have progressed much slower, despite the giant role IBM played, because it took many suppliers to educate users, develop technologies, install and maintain equipment and software, and engage in national debates about the "Information Society." They energized an enormous amount of governmental actions leading to funding national champions, sponsoring R&D situated in firms and local universities, educating citizens on the use of IT, and in conditioning their societies for an age in which use of computing would be extensive. Even the broad liberalization of the telecommunications industry evident all over Europe in the 1980s resulted from trends in computing often articulated by local vendors. So, just discussing them as collections of failed European enterprises, which one can argue was the case in classical market terms, is too limiting an analysis and misleads us about diffusion of IT in Western Europe. Yet, discussing their inability to scale up to the needs of the global market also offers lessons for the ongoing evolution of the IT industry and possibly of other future technologies, especially for Europe.

Finally, positioning their fate as a story of falling before mighty IBM is too narrow because, even by IBM's admission, these firms occupied a good 20–25 percent of the market in Europe. And that market was never small. One data point late in Wave One makes the case when supposedly all these firms had been devastated. In 1988, the global market for information technology amounted to about $350 billion dollars. The United States and Canada accounted for an estimated 43 percent of that number, Japan 18 percent, and Western Europe 29 percent ($102 billion)—not a small amount by any stretch of the imagination. To be sure, Europe's overall global share declined in the 1980s, not because Europeans were acquiring computers in fewer quantities, rather because the rest of the world was buying more, as we see in the subsequent chapters discussing Asia's adoption of information technologies. Asia in 1983 accounted for 22 percent of the world market and by 1988, for 29 percent; even the United States combined with Canada lost share, moving from 45 percent in 1983 to approximately 41 percent in 1988. Yet, overall the dollar value of business increased during the decade.[24] So, one can conclude that European firms had an expanding field of play, even selling products in the United States.

Against that background, we can quickly summarize the European IT industry's pattern of evolution. The first phase, evident in the late 1940s, 1950s, and early 1960s involved initial injection of knowledge about computing into academic settings in many European nations, followed by early forays into the commercialization of computing as a product by three types of enterprises: existing electronics firms familiar with some of the component building blocks of the new technology (e.g., GE in the United States, Philips in Holland, and Ericksson in Sweden); old office appliance firms familiar with data processing using precomputer technologies coupled with their access to a set of customers who were logical candidates for the new technologies (e.g., Olivetti of Italy, Britain's BTM, Facit in Sweden), and a variety of either start-ups (e.g., Zuse in Germany) or high-tech firms which saw an opportunity to sell the technology they were using themselves (e.g., Saab in the Nordics). There were scores of such firms all over Europe. During the late 1950s and continuing through the 1970s, they sought to develop and sell products, often grossly underestimating the costs involved and complexity of R&D. But when they did, they sought financial assistance from their local governments. The latter often responded too slowly and too inadequately to assist, a few not at all. European firms moved slower to introduce competitive products than the Americans in the 1950s–1970s and the Japanese in the 1980s. A third phase overlapping the 1980s and the 1990s reflected the dominance in all product lines by American and Japanese firms. Some Europeans participated too, often as subsidiaries owned by American and Japanese firms, such as GE and Honeywell as early as the 1960s, and Futjitsu in the 1980s and 1990s.[25] The largest of the European firms in the last half of the century included ICL (British), Bull (French), Olivetti (Italian), and Siemens (German), which evolved into Siemens–Nixdorf, the latter the most successful long-term of the lot. But there were other smaller players, such as Sweden's Saab.

We have reviewed the role of these firms in earlier chapters, so we can dispense with details and focus, instead, on broad patterns. The big shock for all was not IBM's introduction of the 650 in the mid-1950s as an earlier generation of scholars used to think, but rather the arrival of the System 360 in the mid-1960s and at the end of the decade, the System 370. Governments in London, Paris, and Bonn reacted by seeing these events as massive threats to indigenous industries and pushed through funding and other forms of support to their local champions, but too little as the momentum of development, pricing, and marketing had already shifted to IBM at a speed that no European firm proved able to match. Sales for all these firms "collapsed," the term used by business historian Alfred D. Chandler, Jr.[26] Independent European vendors imploded rapidly. Great Britain's ICL became fully part of Futjitsu in July 1990; Bull a subsidiary of GE in 1964, then in 1970 of Honeywell, which eventually sold products made by the Japanese firm, NEC; while in Italy Olivetti's computer business became part of GE the same year as Bull. The latter manufactured PC's for the American firm AT&T in the mid-1980s, but in 1996 was sold to a non-Italian financial consortium.

Siemens stood alone as Europe's single integrated computer company, thanks in part to the support of the West German government, but also due to an internal technical capability in electronics as good as Philips and better than that of most other European firms. For a time it collaborated with RCA, French Cii, and Dutch Philips in the Unidata mainframe initiative until that project failed in the face of American and Japanese successes.[27] In relying on its knowledge of microprocessors, Siemens progressed. In 1990 it acquired the financially troubled minicomputer manufacturer, Nixdorf; but even Siemens had to form alliances with American and Japanese firms to survive, such as with Fujitsu. Meanwhile, Japan's market share for mainframe computer systems in Western Europe grew as did that for American PC vendors throughout the 1980s.

That is the essence of the trajectory of the supply side of Europe's experience. On the demand side, as described in earlier chapters, these firms, and others across Europe, installed computers in local government agencies, facilitating movement of knowledge about the technology into agencies hitherto unfamiliar with it. They hired and trained workers in IT who, over the course of their careers migrated to other suppliers and to their customers, following a similar pattern noted with the Whirlwind alumni in the United States. They interacted with computer scientists in universities in almost every European country throughout the period, creating several generations of computer scientists, programmers, and systems analysts—all crucial to the acceptance of computing in Europe. This demand side of the history of IT diffusion in Western Europe has not yet been examined to any great extent by historians, sociologists, or economists. Related was the diffusion of knowledge that occurred even within the industry as, for example, when IBM's managers recruited people from these firms into their laboratories and national sales forces, or conversely, IBM's employees were lured away to jobs in competitive enterprises. Everyone followed or paralleled the same pattern evident in the United States through Wave One.

European vendors, along with their American and Japanese rivals, and often in collaboration with computer scientists in universities, government laboratories, and telecommunications companies, facilitated diffusion in some fundamental ways. The most obvious involved development and manufacture of computational products (both hardware and software) that (a) became increasingly reliable over time, hence making it possible for users to rely on the technology with which to conduct ever-increasing amounts of work critical to their missions and success; (b) products that could do ever-more diverse forms of work, initially scientific and mathematical calculations then business data processing and transactions; and (c) at lower costs per transaction over time (the Moore's Law phenomenon) made possible by use of ever-improving, smaller components and their accompanying mass manufacture, and more diverse and flexible, easer-to-use software. Simultaneous with these technological and application developments were others in the telecommunications industries that, by the late 1960s, made it possible to link computers together so that they could transmit data back and forth. That capability led to online processing, a major step forward in how computers were used. If we include modularization of systems so that one could use mainframes, minis, and microcomputers, then embedded systems in industrial and consumer goods, one begins to realize the enormous impact these firms had in making digital technologies attractive to use, beginning in the 1960s.

Linking these technologies to prior patterns of using IT, as in the Dutch examples, to a similar extent in Sweden, and discussed as part of the American experience, demonstrated that office appliance vendors were able to tap into a set of customers already in need of greater "horsepower" or who knew enough to appreciate the value of the new IT.[28] Old data entry equipment and formats were used with the new systems, such as punched cards and their associated input/output equipment, with which they were familiar and had used earlier to create massive files, and that they could integrate with the new equipment. This compatibility goes far to explain the success of early computing from both the supply and demand sides of the story. It clarifies, for example, why office appliance firms did better than electronics companies in the industry; the former had customers about whose data-processing operations they knew quite well and with whom they had relationships that informed the elements of their digital product offerings. The electronics firms, which clearly had greater skills in the components that went into computers, and thus could be expected to develop great products, did not have as in-depth an understanding of the potential users of the technology nor even access to them. These points suggest that in discussing the evaluation and adoption of a modern technology one has to pay close attention to the critical role of knowledge diffusion, who knows about it, how information is shared among vendors with customers, customers with each other, and even suppliers with competitors with whom they must collaborate or are aligned with as business partners at times, but who are also competitors in other situations.

ROLE OF IBM

What the publicly-available contemporary evidence and the internal records of IBM clearly demonstrate is that the firm played a large role in the diffusion of computing in Western Europe. In fact, if one were simply to measure market share for the installation of systems in the 1950s through the 1980s, IBM "owned" a bigger share of the market in Europe than it did in the United States. In Europe, its share varied between 50 and 80 percent, depending on which decade and country one looked at, with over 70 percent more normally the case. In the United States the figures were closer to 50–60 percent.[29] IBM's revenues grew faster in Europe than even in the United States between the 1950s and the mid-1980s, a testament not only to IBM's success but more importantly to Europe's appetite for IT. Also accounting for IBM's greater success in Europe was the more intense competition it experienced in its home country of the United States that demonstrably proved greater than against its weaker, smaller European rivals. Either set of market share statistics is impressive, but more so in Europe, and with that kind of dominance, we must ask what made this so possible, because without that kind of penetration it would be quite feasible to conclude that diffusion of computing in Europe might have progressed more slowly than it did. The reason we can make that assertion is because in each case examined for this book, IBM's competitors always seemed undercapitalized for what they needed, if for no other reason than to compete against each other, and they clearly did not invest in adequate sales coverage of their various markets, with most more often inclined to devote the bulk of their marketing and sales energies within their individual national borders, as we saw with Saab. To be sure, exceptions perdured, such as Siemens and Bull, but even those companies were usually undercapitalized for R&D and manufacturing and underfunded their investments in sales infrastructures. So, IBM did something different. What was it?

The orthodox explanation embraced over many decades by IBM and proffered by many of its admirers is that the company had great products and an outstanding sales force.[30] We even have Thomas J. Watson, Jr. admitting in his autobiography that in the 1950s and 1960s IBM did not have the most advanced (read, best) products at any specific time, yet its sales force "outsold" its competition by relating benefits of a particular system to the needs of individual customers. To be sure, the skills of its salesmen in justifying a sale to a customer had a long track record of success.[31] Second, its products were normally quite good for the markets they served, were well priced, and made available earlier than those of its rivals. All observers of the company are in agreement on these points, even those who called out the fact that IBM experienced severe software problems with its System 360 for several years in the 1960s, because eventually even those performance issues were resolved. At the height of the software problem, Arthur K. Watson (head of WTC) wrote his brother, Thomas J. Watson, Jr., that "market share has declined overall in World Trade from 64 percent to 63 percent this year," suggesting various problems, such as competitive inroads into the British market, but he also had problems on the continent.[32] Soon after, the 360 product line became the best seller in the industry's history, at least certainly until the late 1980s when PCs began anticipating the impending arrival of Wave Two.[33] To an important extent, these twin explanations are accurate, but, the European experience suggests that there was more to the story which goes farther to explain why the company was able to place so many computers into the hands of experienced and new users.

Recognizing that Europe represented a growth market following World War II, and for reasons related to how the Watson family wanted to divide leadership responsibilities for itself within the company, IBM established the IBM World Trade Corporation (WTO) in October, 1949, as a subsidiary to handle all business outside the United States. It consolidated its various disparate national operations already in place into a more organized, integrated collection of local firms reporting to WTC, as it was best known. Headquarters for the WTC sat in Paris, in one of the countries with the largest potential markets at that time. For the next several decades, Europe's contribution in revenues and profits to WTC hovered at between 65 and 70 percent. The history of WTC's enormous success has been the subject of many piecemeal discussions, which we do not need to revisit here, although we still lack a formal history of this aspect of IBM's operations.[34] With Europe, IBM's senior executives (mainly Thomas Watson, Sr.) had made the fundamental decision to treat all of Western Europe as one large market as early as the 1920s. For legal and cultural reasons it had to establish independent companies in every country, each with its own board of directors. IBM employed local citizens in these firms while their boards were always populated with distinguished nationals as well. Yet all took direction from WTC in Paris, which, in turn, received extensive strategic and oversight direction from American headquarters in New York City. IBM centralized and coordinated management of product development and production across the world, and most specifically in Europe. Sales strategies and execution were left to the individual country organizations to implement, tailored to what made sense on a country-by-country basis. But a pan-European view of sales and marketing completed the management process.[35] Thomas Watson, Jr. noted in his memoirs that his father made all the plants dependent on each other to develop and manufacture products, which allowed the company to optimize various national tariff and tax realities, and to shift work from one facility to another when needed or advantageous to do so. He explained the key benefit of such an arrangement: "This trading around allowed us to operate on a much larger scale, and far more efficiently, than any company that was bound to a single country."[36]

Underpinning this pan-European approach was the investment made by IBM across the continent. As in the individual country case studies, the company built laboratories and manufacturing facilities across Europe, insisting that they collaborate with each other and with others in the United States to gain benefits from diverse thinking, depth of specializations required in high-tech innovations, and synergies of costs and productivity. As described in earlier chapters, some of these facilities and national organizations employed thousands of local and foreign national employees, and had profound economic impacts in their respective regions, although most only housed several hundred people each during most of Wave One. By late Wave One, IBM had some 100,000 employees in Europe. Beginning in the 1950s and extending to the end of the century, the number and size of these facilities grew. Expansion began early enough such that by the mid- to late 1960s—when IBM had become, in effect, the dominant European computer vendor—it maintained a substantial footprint in Europe.

Table 5.3 documents two facets of that investment: number of laboratories that worked on development of software and hardware (and sometimes conducted some training of customers), and manufacturing plants. The data are as of 1969, the year in which sales of 360s were some of the highest. Obviously these figures reflected the accumulated investments in facilities and people that had occurred over many prior years. Note the geographic spread of both laboratories and manufacturing facilities and their number: 6 laboratories and 13 plants for a total 19 sites in 6 nations, a cluster of localities that grew in number in subsequent years. No other vendor had that level of investment in Europe. Note further that labs and plants were often housed on the same campus, just as in the United States, where experiences gained as early as the 1920s and 1930s taught senior management that collaborative synergies resulted by locating these facilities together, since these pushed forward innovative product development and quality and cost effective production.

When we turn to IBM's sales penetration, the firm's "tip of the arrow," where its employees worked who actually made acceptance and installation of so many computer systems possible in Europe, one customer and one sale at a time, one sees a similar pattern of having offices staffed by trained sales and support personal recruited in the local markets of two types: distinguished citizens (often poor members of the nobility and other local elites) and well-educated young engineers and business graduates of prestigious local universities. Table 5.4 uses the same year 1969 as for plants and labs to illustrate what happened. At the height of the popularity of the System 360, IBM had 157 sales offices and related locations (usually for repair staff or centers to teach customers about IT) in 19 countries in Western Europe. These facilities existed in all key industrial, commercial, and public sector centers in Western Europe, as we saw, for example, with the establishment of sales offices in greater numbers in industrialized northern Italy than in the more agricultural southern half of the country. The list is ranked by the number of offices as yet another, if indirect, way of suggesting where the largest concentration of users worked. These locations were urban, large, and near enough to each other to reinforce their decisions to acquire IBM equipment and software to be sure, but also to exchange information about how best to use data processing, regardless of whose equipment they acquired.

While some of these facilities were small, with less than a dozen sales personnel in the 1950s, most were many times that size by the end of the 1970s, expanding in response to business and market demand. Additionally, because they all reported directly to their national headquarters, then into the one European office, it was possible to channel specialized resources from one country to another in response to customers' needs or to concentrate them to leverage economies of scale. Typically, for instance, a customer who needed to

Table 5.3 NUMBER OF IBM LABORATORIES AND MANUFACTURING
SITES IN WESTERN EUROPE, 1969

Laboratories	Manufacturing Plants
Sweden (2)	West Germany (5)
France (1)	France (3)
West Germany (1)	Great Britain (2)
Netherlands (1)	Italy (1)
Great Britain (1)	Netherlands (1)
	Sweden (1)

Source: IBM World Trade Corporation, *Annual Report 1969* (Paris: IBM World Trade Corporation, 1970): 24–25, IBM Corporate Archives, Somers, N.Y.

Table 5.4 NUMBER OF IBM FIELD OFFICES IN WESTERN EUROPE, 1969

West Germany (35)	Sweden (5)
France (30)	Denmark (4)
United Kingdom (13)	Finland (3)
Spain (11)	Portugal (2)
Italy (10)	Ireland (2)
Netherlands (9)	Cyprus (1)
Switzerland (8)	Greece (1)
Belgium (8)	Luxembourg (1)
Austria (7)	Malta (1)
Norway (6)	

Source: IBM World Trade Corporation, *Annual Report 1969* (Paris: IBM World Trade Corporation, 1970): 24–25, IBM Corporate Archives, Somers, N.Y.

learn more about a product than his local salesmen could provide would be taken to one of IBM's laboratories or manufacturing sites in Europe that specialized in his area of interest for education, briefings, and tours of the facilities. Executives in Europe and the United States also called on customer peers in various countries to broaden the personal nature of relations between users of its products and IBMers. Customers received a steady flow of published marketing and technical materials about the company's products, services, and views of data processing and IT matters, a practice that continues. Thus, when one combines good sales execution, a pan-European market coverage strategy—and using 1969 as a sample year—over 175 IBM facilities just in Western Europe, one has to conclude that IBM had followed the classic Chandlerian thesis of investing enough in development, production, and service to dominate its chosen market.[37]

How successful was IBM's strategy in Western Europe? Put another way, how attractive were IBM's products and services to European organizations? Outside the United States, West Europeans installed more of its products than any other part of the world. In the 1950s, WTC grew faster than even the American home company and closed out the decade of the 1960s with gross revenues of $360 million (20 percent of the entire company's revenues). By that year, Europe accounted for 67 percent of the WTC's revenues. It was a remarkable achievement because of the fragmented nature of the European market.[38] One student of the process described European realities of the years between the late 1940s through the 1980s as, "the patch-work quilt of customs regulations, national standards, and taxes that governed commerce among the many independent countries made [sic] this task particularly challenging."[39] One biographer of the Watsons summed up the results of IBM's strategy in glowing terms from the perspective of its customers: "The combination of local management and the stability of an American corporate giant proved irresistible in the war's aftermath," adding that it did not hurt that "IBM filled World Trade with some of the most talented businessmen in Europe."[40] It is because of IBM's success in serving the European market that we can acknowledge its contribution to Europe's acceptance of computing, ranking its influence during Wave One above all other vendors and in some countries greater than that of local national governments, as evidenced by Italy's example, if less so in Great Britain and France. Ultimately, its influence permeated every country in Western Europe.[41]

ROLE OF GOVERNMENTS

IBM's influence occurred simultaneously to when the public sector's activism in IT swelled. There existed three facets to the story of the role of governments in assisting in diffusing computing in Europe. First, there was the effort to get things started in a country by funding local research and development projects. Second, myriad initiatives were launched to train local working populations in the use of this technology, in encouraging adoption through financial and other incentives, and for deploying these in various agencies. These actions had to be taken largely within the context of modernizing economies to make them competitive on the world stage. Yet a third class of related initiatives was the promotion of a national or pan-European computer industry, reviewed below more thoroughly. Clearly, the European experience, more than that of the United States after the 1960s, represented a more integrated initiative evident in almost every West European country; so this circumstance should be thought of as one influencing the others.

The breadth of tools available to governments with which to engage in the complex issue of IT diffusion proved remarkable, compounded by the fact that officials deployed these in various ways since the 1940s. Indeed, they continue to be applied around the world by governments, using IT as a high priority economic development initiative.[42] The orthodox economic and public administrative thinking holds that officials implemented Keynesian practices on the back of intense involvement by governments in the economic and technological affairs of the world; that this thinking emerged during the Great Depression of the 1930s, particularly in Europe, in most nations participating in World War II, and among the protagonists of the Cold War. However true that is, the Swedish case demonstrated that governmental activism probably took place involving officials who probably were not even aware of Maynard Keynes' thoughts on the matter, because all the Nordics had a history of public economic initiatives dating back to at least the eighteenth century, long before Keynes was even born. Nonetheless, the modern prototypical European model for how to leverage various tools available to public officials in support of IT diffusion emerged out of Great Britain, largely because it was the first national government to apply them so extensively to computing. In fact, the World Bank used the British experience with which to inform developing countries on how they should stimulate diffusion of IT in their societies. As recently as 1995, its economists still saw the British example as instructive. Table 5.5 catalogs these various British policy instruments that the World Bank described as a "matrix" of tools. These were used in various combinations in different decades, depending on the objectives strived for by officials. The list is rather impressive in number and variety, illustrating the growing role of governments around the world during the second half of the twentieth century.

However, even this list was incomplete because just like the Americans, European officials paid attention to patent and copyright practices as well,

Table 5.5 VARIETY OF BRITISH PUBLIC POLICY INSTRUMENTS USED TO STIMULATE IT DIFFUSION, CIRCA MID-1990S

R&D support aimed at specific companies
Policies stimulating collective R&D in IT
Funding and other support for basic research in IT
Inward investments
IT Industry sector support policies
Governmental defense and civil procurement policies
Formulation of national and governmental IT strategies
Investment in government and national IT investments (also for telecommunications)
Implementation of IT and telecommunications technical standards
Regulatory and legislative activities
Development and implementation of industrial and technological policies
Non IT-specific diffusion policies
IT-related technology transfer policies
Educational and training policies in support of IT

Source: Nagy Hanna, Ken Guy, and Erik Arnold, *The Diffusion of Information Technology: Experience of Industrial Countries and Lessons for Developing Countries* (Washington, D.C.: World Bank, 1995): 41.

although with software more by the late 1980s. Protecting patents and copy-rights became increasingly more important to governments and vendors of hardware, and especially for software, during late Wave One. World Bank economists also left off their list two important levers: manipulation of both import–export regulations (the issue that triggered heated French response to the Americans in the mid-1960s, and that slowed the flow of American technology to the Soviet Union) and applying monetary policy, particularly the relative value of national currencies as did Great Britain in the late 1940s to make its exports less expensive on the world market and China in the 1990s and early 2000s. Nonetheless, these economists correctly cataloged most of the practices in evidence across Europe. They also pointed out that in action policy frameworks concerned, first, generating new knowledge, technology, and products, second, applying the bulk of the policy instruments to construct bridges between creation, and, third, an aspect of policy: actual diffusion of the technology into a society.[43]

The World Bank's economists also documented many programs—the actual activities of governments translated into action to implement policies (strategies)—that they saw in evidence, and which our own research indicated were used in one fashion or another across all of Western Europe. These are briefly summarized in table 5.6, cataloging what governments actually did to create local technological capabilities, diffuse them, and create bridges to facilitate diffusion. If one wanted to find the skeleton of how public officials in Europe quietly (or sometimes not so silently) went about stimulating development of modern economies and "Information Societies," this is as good a list as exists emblematic of activities in the late 1900s. It certainly reflected actions underway in our country cases from Wave One and, one might add, continued during Wave Two. While these tables seem clean, confident, even definitive, reality proved otherwise, and when implemented by European nations failed to create national champions that could substantively take on the Americans and later the Japanese. Examples abounded of the limitations of policy and programs: governments constantly tried to have firms collaborate that had different cultures and missions, such as Facit and Saab, or the various British office and computer companies; the French had to create whole new enterprises to attempt to get around the impediment described by several observers as the problem of being "in bed with a stranger," in which collaboration proved impossible for all practical purposes.[44]

Some called implementation of these various national policy programs and policies an IT "race," one that continued over the past seven decades.[45] To be sure, there was always a sense of urgency across Europe, either to catch up with some new generation of computers from IBM or to create an economy fit to thrive in the Information Society all envisioned enjoining whole nations and continents into some new World Order that the Spanish sociologist Manuel Castells began calling the "Informational City" by the early 1990s. Indeed,

Table 5.6 TYPES OF PUBLIC SECTOR IT DIFFUSION PROGRAMS

In Support of Creating Technology Capability
Supporting specific R&D projects
Linking institutes and universities
Disseminating technical information
Actions Taken to Diffuse Technology
Implementing technology bootstrap programs
Conducting technology audits
Advising manufacturing firms and industries
Establishing and supporting technology centers
Supporting technology demonstration and awareness programs
When Focusing on Bridging
Promoting, implementing and funding education and training in IT
Coordinating among IT constituents, such as vendors, universities and users
Developing government informatics
Establishing science parks

Source: Adapted from Nagy Hanna, Ken Guy, and Erik Arnold, *The Diffusion of Information Technology: Experience of Industrial Countries and Lessons for Developing Countries* (Washington, D.C.: World Bank, 1995): 75.

Castells had begun to argue his case in the 1980s and extended his notion to that of "The Information Age" by the mid-1990s.[46]

Trade data demonstrated that Europe had a problem as a whole; not just a few nations. By the time the IT industry had gone global (1970), Europe combined was producing a fraction of the products and systems when compared to what came out of the United States. Over the next decade, while European production went from a few hundred million dollars, the Americans expanded output by fourfold over its output of 1970. The balance of trade proved equally out of balance for the Europeans, with IT trade negative for France, Germany, and United Kingdom in the 1970s. Both the United States and Japan enjoyed positive trade balances. Meanwhile, on the critical battlefront of integrated circuits, which all European governments recognized was the key battlefield for the future of the Information Society, by the second half of the 1970s even the Americans lagged behind the Japanese, while the United States surged in the newly emerging field of software products.[47]

The negative trade balances in IT of the 1970s all over Western Europe reaffirmed a key finding of earlier chapters, namely, that the Europeans had become massively important customers for all manner of IT. In fact, they accounted for about a third of the world's demand for computing. While public officials fretted over Japanese's dominance of ICs, and the prowess of American and Japanese vendors in selling computers by the end of the 1970s, they were simultaneously successfully in encouraging use of these technologies in all manner of economic and technological activities, so the economic development mechanism cataloged by the World Bank's economists proved more effective when applied in stimulating the demand side of the diffusion equation. Yet officials remained largely wedded to applying such measures to the supply side of the story. They often decided that microelectronics, and more generally IT, represented "the single most important sector [of the economy] for the remaining years of this century,"[48] providing "increases in productivity." This "circumstance lead us toward an economy in which the major portion of jobs and activity will be linked to information."[49] Policy makers sought to close "gaps" in the use of technology; that Europe's "weakness" related to its slow embrace of IT as an industry and as users, concerns evident in almost every European country, and that remained an intense area of focus in the twenty-first century.[50] To a large extent these gaps were seen first as the inability of Europeans to compete effectively enough against the Americans in the IT industry during the 1960s and 1970s, then in the 1980s, as concerns mounted about whether their societies were technologically invested enough. This second issue explains why so many training programs and investments in embedding IT into all manner of work and products continued to expand, spawning a massive dialogue and extensive literature in its wake for over a decade.[51]

National champion programs failed to achieve their objectives of building vibrant indigenous IT industries to challenge larger global rivals and to grow into substantial export businesses to help their nation's trade balances generating local prosperity. Although to set the record straight, European IT manufacturers engaged in many collaborative efforts and alliances, particularly in the early 1980s and later as part of the Esprit program.[52] While discussion of these alliances rest largely outside the scope of this book, such collaborative efforts, nonetheless, provided evidence that firms understood the need to scale to even have a chance at being successful.[53] Exceptions to failed attempts existed, as with Olivetti in its role as a major international supplier of PCs and Siemens, for instance. In general, however, local companies did not want to collaborate, governments either micromanaged them (as in Britain), did not fund them adequately to scale up (as in France), or made decisions too slowly in response to the constantly transforming market (vendors in all countries complained about this issue).

Another problem for Europe was that national champions cancelled each other out as individual firms competed with other champions across Europe, each creating R&D, manufacturing, and sales infrastructures that could not be realistically sustained by their home markets and thus had to compete in other European arenas. Compounding this problem was the policy followed by all governments giving their national champions favorite status when bidding for local business, limiting the ability of a champion from another country to compete for business locally. In short, governments did not want to cast aside their national champions for the greater good of pan-Europe.[54] A combination of slow moving business development, highly fragmented and sub-optimized operations, indeed small national markets, and inability of officials to understand fully (or accept) the dynamics of a globally transforming IT industry compelled some leaders in both the European Union and many public and private organizations to conclude things had to change.

By the end of the 1970s, after much angst, European governments began switching to a new strategy, a process that took all of Europe nearly a decade to adopt. They sought to fund and create pan-European responses to the Americans and Japanese, while aspiring to see European IT industries stimulate economic renaissance. European firms had spent the previous two decades trying to compete in the mainframe and computer chip markets long after firms in the United States and Japan had "won" in those arenas, instead of finding new niches that had not yet attracted the attention of the Americans or Japanese. Little did they realize that by the end of the 1990s it would also be obvious that if they could not compel companies within a country to collaborate, that they could not marshal whole countries to collaborate effectively either. But we are getting ahead of ourselves in explaining the unfolding pan-European effort to coordinate development and diffusion of IT.

It was an old dream, first articulated as early as the 1960s.[55] While the history of pan-European IT initiatives of the 1960s and 1980s awaits its historian, some key activities suggest what happened. An important shift came in the late 1970s, when the first major European initiative to compete against IBM with the Unidata in the 1970s launched. It failed to achieve its objectives.[56] IBM dominated the European computer market. European policy makers and local IT firms still viewed IBM as American, although as described earlier in this chapter, it was also European, a point overlooked by the vast majority of Europeans concerned with IT on their continent and by IBM which failed to convince the region of its European identity. Had Europeans acknowledged IBM's European features, they might have realized sooner that their problem lay more with the targets of their strategies and priorities. It was easy to blame the Americans, and they could never separate IBM from such U.S. vendors as RCA, GE, and AT&T in Wave One, and Compaq and Dell among others during both Wave One and Wave Two.[57] American PC vendors came into the European markets in the 1980s and dominated that business too, while the Japanese expanded enormously their share of the mainframe market. But that was all in the future at the start of the 1980s. Unidata's ultimate contribution was probably more in informing European policy makers about the dynamics of existing IT realities. One student of the Unidata experience concluded that

> Unidata embodied the characteristics of a policy approach that was built on an uneasy balance between nationalism and internationalism. Its goal was internationalist in spirit, but the measures taken to achieve this goal were all designed to preserve national sovereignty in data processing. Not until the fundamental political reorganization of the EC in the 1980s did a different kind of European IT policy emerge—and with less nationalistic forms of technological cooperation.[58]

The shift came finally toward a European-wide industrial policy, advocated by the European Communities (better known as the EEC). The initial challenge was how to get past the inability of the 12 largest IT firms to compete. These were ICL, GEC, and Plessey of Great Britain; Siemens, AEG, and Nixdorf in Germany; Thompson, Bull, and CGE in France; Olivetti and STET of Italy; and the Dutch Philips, collectively called the "Big Twelve." The pan-European EEC developed new ideas for a replacement strategy to displace the national champion model in the early 1980s, covering key IT areas from microelectronics to computing, software of course, and additionally integrated computer systems and office applications. It intended to continue working with the largest firms because they had the technical abilities and capacity to sustain pan-European programs, while smaller enterprises were seen as not to having that capability. The 12 firms agreed to collaborate, advocated a European-wide set of technical standards to which they could adhere in their development of products, while funding would come from the Commission, largely for developing mainframes and memory chips.[59] Meanwhile on the broader front of European economic policies, member nations of the European Union agreed to develop a single European market, as articulated in the Single European Act (1986). That commitment called for the European Union to implement programs that strengthened Europe's scientific and engineering prowess across multiple fields, not just IT, and also led to more consistent policies regarding telecommunications.[60]

Beginning in 1984 and extending into the next century, the European Union launched and funded a series of what came to be known as "Framework Programmes," often simply referred to as Esprit, but really extended to initiatives in other areas of technology as well. Funded R&D projects and results would be made available to any member of the Union; the EU's intent was to make European technology competitive on the world market, hence its focus largely on the supply side of the IT diffusion issue. The CEO of Olivetti from 1978 to 1996, Carlo De Benedetti, observed in the early 1980s that IT company executives and public officials came to "understand that Europe's weaknesses are a result of excessive market fragmentation, national protectionism, inefficient and costly public intervention, and a reduced capacity to compete in the new international cycle."[61] The first Framework Programme (1984–1987) focused on creating new microprocessors, the second (1987–1991) continued this priority, while the third (1990–1994) shifted attention to such issues as multimedia, miniaturization, and mobile technologies among others, all at a time when indigenous hardware industries were virtually a thing of the past. Despite years of inadequate returns on the extensive investments funded by the European Union, its authorization to continue was reaffirmed with the signing of the Maastricht Treaty in 1993.

The Fourth Framework Programme (1994–1998) finally made the fundamental shift to the demand side of the diffusion story by emphasizing use of all manner of ICTs. The Big Twelve, which had not performed well enough in the 1980s and early 1990s, were now joined or displaced by smaller firms better able to present project proposals that concentrated on applications and diffusion. In the middle years of the 1990s, EU funding for all manner of IT initiatives continued to increase as well, in some categories of investments nearly doubling over prior programmatic levels. It was a good shift, because European IT industries had shrunk all through the 1980s and 1990s. In fact, Europe lost some quarter of a million jobs in IT manufacturing alone, although the number of IT professionals continued to increase during the last two decades of the century, as in other nations that had become extensive users of computing.[62]

The story told here left out one other pattern of public policy that national governments continued to implement quietly all through the 1980s and 1990s, specifically, support for their national champions. Simultaneously they endorsed pan-European initiatives of the European Union, as evidenced, for example, in the Lisbon Strategy first promulgated in 2005 and reaffirmed several times in subsequent years. But, they also upheld local initiatives in support of national economies, job creation, helping local users, and, in general, activities that made their societies competitive in the emerging Information Age. This practice has continued to the present. On the one hand, the European Union attempted to develop continental technological capabilities aimed at the supply side of diffusion until the mid-1990s, which one can easily characterize as a feature of Wave One public policy in Europe. We can also argue that after the mid-1990s, with the shift, although not totally so, from an industry to a user focus that the EU and European governments began moving slowly into Wave Two. There still existed national champions supported in albeit smaller, less obvious fashion, with local public policies characteristically Wave One in form. National shifts in public initiatives to Wave Two remained too tentative for one to conclude that they were forceful enough to declare the earlier wave over. As of this writing (2012) public policies at the national and pan-European levels were still in transition, with policy feet planted solidly—and often indecisively—in both eras.

INFLUENCE OF EUROPEAN INDUSTRIAL STRUCTURES

In every case examined of the pan-Atlantic community of Europe and North America, we paid close attention to the size of firms and to the industries that comprised each national economy for good reason. The more a country had large enterprises and government agencies, the greater the penchant of those national economies to embrace early and more extensively all manner of IT. Behind that behavior is the underlying cause for IT being so attractive: uses of computing that optimized work productivity, cost of labor, and capabilities of an institution to do more or different work and, if in manufacturing, to produce different, often more complex products. The most important industries for Europe for IT diffusion included banking, automotive, aerospace, industrial equipment, machine tools, large social services public agencies, and military establishments. Industries that used IT less than, say, those in the United States, included education at all levels, including universities, insurance companies (probably because there were fewer and smaller ones in Europe), and in the early decades of Wave One, retail and wholesale enterprises, the latter because they tended to be small when compared to their American counterparts. These generalizations varied widely, of course, depending on which country one considers and changed over time; however, anecdotal evidence suggests that, in general, uses of computing were more or less adopted for the same reasons and usually within a couple of years of each other across an industry within a nation and typically also across the same industry in multiple nations. In short, IT vendors called on everyone at the same time about the same technologies and applications, while their customers were talking to each other across borders about similar issues. Yet that said, uses of computers in Europe still await their historians; the American experience, however, suggests such research will demonstrate uses of computers had a far more profound impact on Europe than currently appreciated.

Because of the large industrial base with which the Nordics, British, and West Germans began with during the 1950s–1970s, before losing share to Japan and others, manufacturing served as an important stimulus for adoption of computing. A thread running through the industrial sector, both in the west and behind the Iron Curtain, is defense contractors. In Sweden, the military aerospace and defense manufacturers became early adopters, especially using CAD applications, just as happened in the United States. Microelectronics for weapons systems were developed in France and Great Britain in the West, for example, but also in East Germany and in the Soviet Union for similar customers on their side of the Cold War. The dynamic need for more advanced and continuously changing military hardware and aircraft, then rockets, created a massive ongoing demand for new, more complex products that could only be designed and built with the aid of computers. As the cost of computing dropped and, more important, new functions became available, additional computing went into the military industrial complex of Europe, just as in the United States.[63] One observer of the European scene, Margaret Sharp, described some of the activities underway in the 1960s–1970s with respect to CAD applications:

> Computer aided design (CAD) is an example of a new activity which has emerged as a new product of microelectronics. It originally developed as a highly specialist activity associated with the aerospace industry, with the United States leading the way because of major defence contracts, but with France and Britain

both developing appreciable specialist facilities in the 1960s. Popularization of the techniques waited on the fall in computing costs that came in the early 1970s, and the small, new American firms in particularly forged ahead on the basis of 'packaging' the techniques for drawing offices and design shops. These small firms, established at this time as spin-offs from the American defence contracting projects, rapidly made their mark. . . .[64]

She commented, however, that lower vocational skills among workers in an economy made it more difficult to implement such new uses of IT as CAD applications, which suggests that when viewing diffusion of computing that we take into account the constraining role played by inadequate skilled labor. To quote from her assessment, "Countries such as the United Kingdom, which have a generally lower level of skill training, often find it more difficult to introduce new technologies because the work-force both lack the minimum requisite skills to make good use of them, and tend to be more suspicious of the new techniques."[65] Historians may ultimately conclude that how computers were used proved more influential on the rate and nature of diffusion than public policies to which I give considerable attention because of how the historical record reflects events so far.

DIALECTICS OF THE INFORMATION SOCIETY

A remarkable feature of the world view shared by Europeans during Wave One, and in evidence in Wave Two, is an effervescent positive expectation of the benefits to be derived when society finally arrives at the Information Age. One would be hard pressed to find any contrarian statements in public pronouncements, in the documentation of scores of national economic programs involving ICT, in the commentary of social and economic observers, or in the expressed views of other European elites. Vendors and academics touted the same mantra of progress. It is a remarkably sustained, indeed deeply seated view shared across all of Western Europe, to say the least. To be sure, there was a dark tinge to this enthusiasm as concerns about the privacy of one's data, and security of the same, bubbled, beginning in the late 1980s and still in evidence, but even that hardly slowed enthusiasm for the impending start of the New Age. Critics of national economic and IT development abounded, of course, beginning in the 1960s, but their criticism focused largely on programs not getting their societies to that promised land of the Information Society; they did not question whether Europeans should go there, just the means of transportation, and often the duration of the journey. Implicit in all the discussions wafted a sense of progress. Society moved from agriculture to industrialization, and that had improved the quality of life, as would the next phase in humankind's evolution into the "Information Society" with its "New

Economy," which would continue the historic progress of people. One astute observer, writing in the late 1990s, commented that "the theme that we are entering a new industrial revolution is becoming more and more common-place," when old factory jobs would disappear while new ones in services took their place.[66] In every national case study examined for this book, it was the same, even in other European nations not reviewed in detail.

While the subject deserves its own historian, and already has many sociol-ogists, economists, public administrators, journalists, and experts on media joined in the task, for our purposes, it is enough to acknowledge its existence. For it provides a fertile background, a world view, that officials and member of the IT community could present as their case for the use of IT. For politicians and other public officials, it made political and economic sense: Who could criticize an official for wanting to help implement this new world? IT vendors were selling the new technology for modern management. Sociologists spoke of networked societies, journalists did too; it all seemed so clean and proper. That view of the world became so permissive of using IT, later ICT, that one cannot dismiss this broad-based attitude as not being a facilitative engine moving forward diffusion of computing into European society. The links were there, and in time, no doubt, historians will document them more fully. At the point of acquisition—businesses, government agencies, and universities—this rhetoric, indeed dialectic of an emerging information society, permeated business cases presented in private conference rooms, in proposals for funding thousands of projects by the European Union, and earlier by national govern-ments. Read the titles of articles and books in the endnotes to this book and you will see the dialectic in evidence.[67]

The same can be said of the world views of North Americans, who were as prolific in their declaration of the New Society as the Europeans, almost to such a fervor that one can think of it as a new digital mantra of the old Amer-ican idea of Manifest Destiny, the notion that the nation would some day in-evitably dominate North America and that this would be a wonderful event. People on both sides of the Atlantic shared similar ways of thinking about this new world, and often the best of their writings were translated into multiple languages, as happened with the French report written by Simon Nora and Alain Minc as far back as 1978 and the writings of Alvin Toffler, best known for his book *Future Shock* (1970) in the United States. While these views were also expressed in Japan and other parts of Eastern Europe very late in Wave One, and to a lesser extent in Latin America, it was not so evident in Eastern Europe until after the fall of Communism. Was that a consequence of a different view, or simply one not so openly expressed as in the West, where unfettered expres-sions were more visible and encouraged?

No attempt has been made to document the effect of this new dialectic on the diffusion of IT in the West; much work would have to be done in archival mate-rials documenting minutes of meetings of officials, reviewing presentations by vendors, and so forth. However, the few available case studies hinted that this was a subtext for decision-making, as we saw with the various Dutch organizations that became early users of computers. The issue is extensive enough that no history of the diffusion of IT should ignore the topic anymore, especially with respect to Wave Two, when the ability of individuals to express their views expanded and opinion surveys about technology abounded, such as those done routinely by the Pew Foundation in the United States and the Economist Intelligence Unit in Europe.

Outside the community of opinion makers, officials, and vendors, the voice of the European public has rarely been heard. While national opinion surveys began to be administered in the 1980s, they became more available in the 1990s as diffusion of consumer electronics and IT had made significant inroads into societies. However, a fairly large international survey conducted in the mid-1980s gives us insight into the public's perception of the digital dialectic. Conducted on behalf of the Atlantic Institute for International Affairs, it demonstrated that the more people had used IT, the greater the likelihood they believed IT was good for society. Over 25 percent of British and French workers had experience with computing of some sort, while West Germans and Japanese citizens had lower levels of experience. Those societies with the highest levels of experience were the most interested in using computers and word processors; conversely, interest among West German and Japanese citizens registered quite low. The one exception in Germany and Japan were professionals who had experience with computers and, thus, like individuals in other countries, expressed interest in continuing to use the technology. In Italy and Germany blue collar workers disinterest in computers exceeded those willing to use them, a pattern not as evident in the rest of Europe where interest proved much higher. Half the surveyed Europeans from all countries believed computers would/could cause job layoffs, except in Spain and Great Britain where two-thirds believed that would not be the case. Those with the greatest experience with computers were individuals who felt the least threatened by computers taking away their jobs; blue collar workers in general feared loss of employment the most of all groups that could result from automation and IT, while in the United States all groups believed computers would create new jobs. On the value of IT and word processors as useful tools, West Europeans were generally positively inclined (between 63 and 75 percent agreeing) while the West Germans were quite negative about the value of the technology (38 percent approving) along with the Japanese (39 percent approving); Americans had the highest positive views (77 percent) followed by Norway (74 percent) and Spain (75 percent).[68]

American and French workers were the most willing to be retrained, while those most hostile to the notion were West Germans (39 percent saying no) and Japanese (63 percent); once again, the United States was the most willing (65 percent saying yes, 31 percent no). When asked what inhibited development and use of IT, they largely said popular prejudices and lack of education were the most significant. The report writers thought people did not know enough to respond thoughtfully to the issue's questions and that they were too concerned about potential loss of jobs caused by automation. Two-thirds of Americans—the society most familiar with computing—saw inadequate education as the primary inhibitor to innovation and adoption. The surveyors reported that attitudes varied enormously by country, conditioned largely by what opinion makers and other elites had to say about the general subject of IT. They could not explain why Germans and Japanese—two societies with extensive use of IT—proved so hostile to IT, using the word "inexplicable" to characterize the problem, one intensified by the extreme negativity of the responses. They surmised that the German case could be a result of elites in the country not having as uniform an opinion as those in other European countries did. The report warned that people feared for their jobs, a concern that could be overcome if IT led to the creation of jobs.[69] Clearly, the issues and attitudes were some that had existed in the early stages of IT's deployment in the United States in the 1950s–1970s, suggesting that the public had not moved quite as far through Wave One as had management, public officials, and other opinion makers.

WAS THERE A WEST EUROPEAN WAY?

The case studies suggested a tentative answer of yes there was a West European way of embracing IT that had a style of its own, but one that also reflected practices evident across the world during Wave One. The risk of suggesting a European exceptionalism exists when asking if there was a West European Way, but it is one worth taking, because the central question asked in this book is how did the world embrace computing? As already suggested in the way chapter two described the experience of the United States, and how chapter six explored Communist Europe, and subsequent chapters on Asia, there occurred in Europe some behaviors different from those of other regions. These were often characterized less by the actions taken by key players, such as IBM or national governments, than by the extent of the behavior. These speak as much to the way Europeans diffuse any modern technology as to how they responded to IT and to traditional telephony separately, and then to ICT before and after the availability of the Internet.

As in the United States, academics and their universities played important roles in the embryonic stages of computing, often, however, in ad hoc relatively underfunded ways when compared to what occurred in both the United States and Japan. Once a minimal critical mass of interest and activity existed, national governments began investing in these academic initiatives, although never to the extent as the Americans, Russians, or the Chinese.

European academic research on IT remained far less in all decades than in the United States, with the Americans always investing the most during most of the second half of the twentieth century. Even in countries with extensive R&D programs, there were just a handful of academic institutions involved, such as in West Germany, Sweden, and Great Britain, while in some American states there were more working on computers than whole countries in Europe, such as in California and Massachusetts. So volumes of money invested and number of institutions engaged were fewer in Europe.

A second aspect of the European way involved the role of national governments. Americans funded R&D and national champions in the 1940s and 1950s, but by the end of the 1960s, the private sector's investments in IT exceeded that of the U.S. Government, resulting even in the latter lagging the private sector in its use of computers by the 1980s. In sharp contrast, Europeans channeled much of their investments and pushed deployment through the public sector right into the twenty-first century. Put another way, the most aggressive investor and promoter of IT in Europe was normally its public sector. To be sure, there were exceptions, such as Italy, Spain to a certain extent, and Greece. Otherwise, all other countries had active national programs, in fact so much so that some national champions became so dependent on their public sector customers and supporters that it made them insufficiently competitive in the private sector market, as occurred to a certain extent with Saab, ICL, and French vendors. The one dramatic example of the opposite being the case was Olivetti, which received minimal support from its home government, and even it suffered. Public support came through national governments implementing country-wide strategies to transform whole industries and economies from the 1950s through the 1980s. Simultaneously, however, beginning in the 1980s the European Union took the reigns of developing, funding, and promoting pan-European initiatives aimed at (a) restraining the local success of IBM and other U.S. and Japanese vendors (1950s–1980s), while (b) creating a European-wide market that would operate in an integrated fashion as Fortress Europe (1960s–2000s). The fact that Europe failed to achieve both objectives does not allow us to ignore the vast sums spent on the effort, the other commitments of time and resources engaged, or the consequent changes in how Europe digitized. A great deal of Europe's supply side story was tied up in that publicly-driven set of activities. Put in other words, Europe's diffusion of IT involved the public sector more than in the United States. In the process of aggressively supporting the supply side of IT's diffusion, demand side activities increased as a byproduct of various government policies originally aimed at supporting indigenous IT industries.

Third, there is the role of vendors. The Americans and Japanese essentially operated in Europe as elsewhere around the world, blocking rivals, acquiring them, or forming alliances, but always pushing to sell products without relying on European governments for help. Local vendors typically starved for capital, found it difficult to establish meaningful and profitable pan-European alliances, and generally relied too much on their fragmented national markets, which were individually tiny in the global scheme of things. Examples to the contrary are just that, exceptions to the pattern. All rhetoric aside, the European way was to sell first in one's national market and to rely too much on the help of local government. Most other vendors in Asia and North America, however, operated in a global economy encompassing over 100 national economies by the start of the 1990s.

Fourth, European users of computing reacted remarkably like their cohorts around the world. Big firms became early adopters of the technologies and optimized the innovations in IT fairly quickly, that is to say, at about the same speed evident around the world give or take a few years. Liberalization of the telecommunications industry sparked integration of IT with communications into ICT with remarkable speed, which affirms that regulatory practices related to the telecommunications industry of any country is one of the factors historians should weigh in any assessment of IT activities, certainly by the 1970s at least about Western Europe and worldwide for the 1980s. International firms, like Philips, IBM, Brown-Boveri, and others, implemented uses (applications) across their enterprises around the world at the same time, as IBM did with e-mail and online company-wide business practices spreading across its offices worldwide within a matter of a few years. At the individual level, PCs came to Europe later than in the United States, despite the Italians claiming a "first" with their machine in 1968; the market began in the United States in the late 1970s, spread around the world in the 1980s, and included Western Europe.

Fifth, language used to describe computing's role in Europe varied from that of the Americans. Words are important as they carry with them special meanings, ideas, and implied values. Europeans spoke about the "information society," Americans thought in terms of the "Third Industrial Age"; Europeans thought about "informatics," "tele-density," and "telematics." Americans never used those terms, preferring instead such notions as "networked enterprises," "networked economy," and to a lesser extent, "networked societies," while "Post-Industrial" remained a perennial favorite moniker for decades. While mathematician Norbert Wiener at MIT in Cambridge (Boston), Massachusetts, invented the term "cybernetics" in the 1940s, Europeans liked it more than the Americans who preferred, instead, terms like "artificial intelligence" and later as well, "expert systems." American economists and business commentators in the 1950s developed and promulgated concepts behind "knowledge workers" and "knowledge management," such as Fritz Machlup, the economist, and Peter Drucker, the business management consultant; but Europeans embraced these ideas more than the Americans. John Diebold, the American consultant, introduced the world to the word "automation," which both Americans and Europeans latched onto in the 1950s and used for decades, but it was the former who then invented the concept of "process reengineering," which heralded the reduction of labor content in work in the 1990s, but left the Europeans nervous since they were attempting to do the exact opposite in preserving and creating jobs as a byproduct of their New Economy.

Other features could be added to our list, but the case studies provided in this book support those listed to such an extent that we can conclude they were shared to one degree or another across Western Europe. Timings of their adoption varied, as well as the intensity with which they were used, to be sure. The French protected their language, culture, and economy against the Americans, the Swedes and Germans embraced international trade with energy, while the Mediterranean Basin trailed in deployment and its rhetoric. Hence, one should caution that it can be misleading to overstate the case for either a pan-European way or a distinct exceptionalism for any individual country. That sense of exceptionalism, that is to say, "my country is unique," when coupled to language barriers has, for instance, made it difficult for European historians to create pan-European histories of IT, leaving the field largely to journalists and a tiny cohort of Americans.[70]

WHAT WERE THE RESULTS?

The answer to this question will be limited to a few remarks about economics for two reasons. First we are too close to the events of Wave One to have sufficient perspective to deal with such issues as political, social, cultural, educational, business, and other impacts. Second, the only available data to help answer that question are preliminary economic productivity and GDP information; even those are not comprehensive enough to give us absolute confidence. But, since Europeans embraced overwhelmingly information technologies to further their business and economic interests, it is time to begin an assessment. A second benchmark could be to evaluate the impact of ICT on national defense since, as occurred with France, Britain, and to a certain extent Sweden, realities of the Cold War provided incentives and a crucial rationale for the development and deployment of IT, as happened with the Americans and Russians from the earliest days of the computer. It continues to be a source of interest and use now by many nations around the world. For our purposes, discussion is limited to a few economic data points that suggest possible lines of future research.

The best documented national case in the pan-Atlantic community is that of the United States, in part because it began using IT a few years earlier than many nations and thus its economists were some of the first to explore the effects of the technology on this country. In addition to the results already described in chapter two, they taught us several other things. First, investments in IT have a cumulative effect, that is to say, it is the accumulated investments in IT and subsequent changes in how work was done and enterprises organized that led ultimately to productivity gains of a sufficient quantity that they could be measured and their effects at a national level described. Elsewhere I have explained that process of accumulated effects of IT in some detail.[71] In the European case, the patterns of adoption of IT were similar. To seek answers to our question, we need to look beyond the time frame of Wave One, in the case of Western Europe deep into the 1990s. Economists in most of Europe's central banks engaged in this kind of an exercise not only in the countries discussed earlier, but across all of Europe. Interest in the "productivity paradox" so evident in the United States in the late 1990s–early 2000s, was shared by economists and policy makers in scores of countries. No central bank in Europe ignored the issue. In fact, often the conversation centered on the question of why the Americans enjoyed more IT-productivity than the Europeans, again illustrating the phenomenon of Europeans looking constantly over their shoulders to compare their IT performance to that of the Americans. It was a nervous fixation because while, on the one hand, Europeans were worried about the prowess of U.S. IT vendors, on the other hand, they understood that they could learn a great deal about how to be competitive economically and how best to use the new technologies from the Americans.

What is known about Europe in general? Based on measures of economic performance we know IT had measurable positive effects on the national economies of Western Europe throughout the 1990s. Indeed, gains in economic productivity in many countries actually doubled in the years 1995–2000 over what they had been from 1990 to 1995, evidence of the cumulative pattern of productivity increases so characteristic of investments in IT, indeed ICT. This finding applies also to the more advanced economies in Asia and to a lesser extent in Latin America. One of the ways economists measured productivity was to examine returns on capital investments made in IT versus in non-IT capital goods. In some instances they also compared results of investments in labor to that in capital goods in general (non-IT) or more specifically in IT. Investments in IT actually increased as the 1990s rolled by, seen as an indicator that managers were making decisions about IT that they perceived to be beneficial to their individual enterprises and government agencies that had started long before the 1990s. Western Europeans invested roughly 32 percent of their capital in IT in the first five years of the 1990s and some 45 percent in the second half of the decade. One economist wrote in 2000 in the dry language of his profession that "the contribution of ICT capital was positive and substantially increasing."[72] One would expect that results varied from one nation to another, and to be sure, such was the case in Europe as we saw in previous chapters; but largely, Western Europeans enjoyed greater returns than Eastern Europeans, Latin America, and many part of Asia. Those European states late to the IT game had fewer results to show in the early 1990s and later than earlier adopters, but in the second half of the decade made significant strides in their accumulated results, for example, Italy and especially Sweden. This pattern of economic results reflected the well-understood phenomenon of late entrants leapfrogging in adopting a technology already in use elsewhere.[73]

What effect did ICT have on economic growth? Again the West Europeans experienced growth in the 1990s attributable to investments and use of ICT, ahead of Latin America, Asia, and Eastern Europe. The factors that most influenced rates and results of adoption of ICT around the world, and in Western Europe as well, included costs of the technology, extent of education, openness in the flow of information (knowledge transfer), and quality of operations within an enterprise.[74] Those firms that experienced the highest level of incomes and productivity found it easiest to invest in IT, leading economists to recognize that income level was a major factor affecting rates of diffusion by both organizations and individuals. The reasons are fairly straightforward. Better education meant workers who could learn how to operate the new technologies.[75] Elevated levels of productivity encouraged more adoptions, particularly in those cases where high cost of labor was a factor. Using a PC, for example, required a minimal level of learning capability, such as being able to read instructions. Europeans also had an open-enough trading environment that made it possible for the Americans, Japanese, and internally within Europe other European states, to sell their IT goods and services. Western Europeans could access a large variety of options. Thus, as in the French case, France's military establishment decided it was sometimes more attractive to acquire IBM System 360s than locally made products. In the language of the economist, they had better quality investments that they could make.

The concept that Europeans had better run enterprises than in some other country(ies) was a function of increased certainty that an investment in IT would return a profit or some anticipated benefit to the managers in exchange for making such a commitment. This is an enormously important point that seems to have been overlooked by most, if not all, historians who have examined adoption of IT. Influences of management's confidence of success and their perceptions of risk of failure in decisions they made to acquire IT (products, service, and reliance on the technology) permeated managerial affairs across all of Western Europe, just as in other countries. Those who worked in market-driven economies, both in the private sector and in democratic governments, shared similar forms of confidence and risk, suggesting that decision-making became yet another shared activity of the Atlantic community.[76]

Combine all these elements together and what happened in Western Europe over the course of the 1990s was that ICT had cumulatively increased its share of economic growth within various national economies. That investment had a greater positive effect on a country in the more developed nations than elsewhere, placing Western Europe and the United States at the head of the line, with Japan and South Korea close behind. As one economist calculated for any highly developed economy, such as Western Europe, "a 10 percent increase in the ICT capital stock adds (*sic*) 0.8 percentage points to output growth."[77]

There are various explanations one could put forth to explain why in Western Europe companies invested more in IT. A critical reason was firms that used a great deal of IT tended to generate more output per unit of input invested, which is an elegant way of saying they obtained more work from employees and made more products per manufacturing plant than they would otherwise. The primary macroeconomic finding of many economists looking at IT investments around the world from 1980 to roughly the mid-1990s is that net returns on IT investments proved greater than for non-IT projects, which demonstrated strongly that commitments to IT spurred growth in efficiencies at both the firm level and more broadly within a national economy. Evidence of this outcome is now global, not limited just to the United States or to Western Europe. It is visible in both advanced and developing economies.[78] The broadest conclusion one can reach about the economic effects of IT on Western Europe during Wave One, and that intensified in the early years of Wave Two, is that ICT had a profound influence on economic growth and vitality in excess to those investments of a non-ICT nature.

Because there has been extensive discussion around the world in recent years about whether or not there is a New Economy, what can one conclude from the experience of Western Europe? First, European nations embraced IT at different rates over time for all the reasons cited in earlier chapters: nature of the local enterprises, what industries they were in, availability of funding through the state or private capital markets, quality of their work forces, and access to international markets, among others. The Americans grew economically at almost twice the rate of Western Europe in the 1990s, evidence that the variables counted in rates of diffusion and effects. During Wave One, that is to say from about 1960 to the early 1990s, Western Europeans and North Americans experienced about the same rate of increase in investments in IT, while in the 1990s, the Americans expanded their investments as the Europeans pulled back. These events suggested yet another reason why Wave One in Western Europe ended some time in most countries in the1990s. A productivity cycle in the same period also had its beginning and end. In the case of Western Europe, it actually did better than the Americans right into the 1980s because it was rebuilding from the damage of World War II, allowing it to gain ground near parity with the Americans. But, beginning in the last decade of the century, the reverse occurred when European productivity declined while that of the United States soared. Given the importance of the role of ICT, the changed economic pattern should be viewed as another marker that one wave was possibly ending and a new one beginning.[79] That is the macro economic view.

At the microeconomic level the role of labor proved important. One of the fundamental reasons why firms and agencies all over the world wanted to use office appliances and later computers was to contain the cost of labor. In the United States and Western Europe that cost of labor was high all through the twentieth century. Since the expense of labor can often amount to as much as 80 percent of the cost of running an organization, the economics of labor are important. In those economies where enterprises could hire and fire employees at will, with minimal regulatory restrictions in comparison to competitors in other countries, the more attractive IT investments became. In the case of Western Europe in comparison to the United States, the former had more restrictions in place than the latter, with the result that potential improvements in productivity made possible by displacing investments in labor with others in automation surfaced slower in Europe. That result accounted for some of the delta between the positive returns the Americans experienced when compared to those of their European cohorts. Even within Europe variations existed, however, such as in Great Britain and the Netherlands, and to a certain extent in Denmark (although not discussed in this book) which reduced regulations affecting hiring and firing of workers to a greater extent than did Germany or France. These variables resulted in greater returns for IT for the British, Danes, and Dutch than for the Germans and French. This issue remained constant, affecting rates of diffusion and consequences, but it also existed as one of the least studied topics in the history of ICT.

Western Europe's entry into some New Economy construct did not occur during Wave One. Economists believed in the early 2000s that it would do this much later than the Americans who were already seen as entering this new economic world and, indeed, were already deeply into what I am calling Wave Two, although they did not use my notion of two waves. The limiting factors affecting returns on investments in ICT (hence incentives to adopt the technology) for Europe were the same as during Wave One: more regulated product development, labor markets, and less mature financial markets that could support venture investments and sufficient funding for R&D than in East Asia and North America. As two economists noted in 2000, "EU as a whole is not yet a sufficiently open economy," with the result that "the foundations are not in place for the efficiency gains and reorganization of production factors that have generated the high rates of growth seen in the U.S."[80] As a consequence, "there are, as yet, few signs of the new economy emerging in Europe," one important feature of an economy that would serve as a useful marker that a nation had moved from Wave One to Wave Two diffusion of ICT.

Primis above all discussions about the economics of IT diffusion and public policies related to this topic was the pervasive influence of the Cold War—the single most important worldwide circumstance affecting societies on every continent in the second half of the twentieth century, even in the decade after the public believed it was over. Both the United States and the Soviet Union, and their political allies and satellites (which accounted for the vast majority of all nations at one time or another, including colonies), viewed the use of IT as critical to national security. Their concern translated into extensive investments in the development of various forms of information technology, a focus that did not relax until at least one decade after the end of the Cold War. For Europe, that meant in Western Europe whole nations would align with the Americans and their experiences with IT would be influenced by the realities of the Cold War. The same held true for all European states behind the Iron Curtain, and for the same reasons. Both superpowers needed their allies and neighbors to develop IT, to apply these to various forms of military weapons systems, and to have economies and societies that could continue to create and operate such technologies. Both sides had pan-European military alliances into which they injected all manner of IT, and that led to the establishment of manufacturing facilities on either side of the Iron Curtain. In the West it was the North Atlantic Treaty Organization (NATO) and on the other side the Warsaw Pact.

Military innovations funded and developed by both super powers were exported to Europe, giving European allies of the two sides from an economic point of view, and in the words of an economic historian, a "free ride on the security system provided" by their respective superally. In the case of Western Europe, "less defense spending allowed" these countries "to devote more government revenues and investment to private ends," which goes far to explain why countries like France, Germany, and Sweden could afford to fund national champions quite generously in some instances and invest in peaceful infrastructures, such as public transportation systems, while the major powers had to spend their funds on rockets and bombs.[81] In the process, defense contractors thrived in Western Europe, acquired and applied modern IT, such as military aircraft manufacturing in Sweden, while generations of young military draftees were exposed to the new technologies.

While the role of IT in the Cold War activities of the United States has been extensively studied Western Europe's experience awaits its historians. What they will find is that the British and the French wanted computing to develop nuclear bombs, the Nordics to have air defense systems that kept them out of harm's way from either side, the West Germans effective deterrents against a ground invasion from the East, and the Mediterranean community the foreign aid and rental revenues from allowing the Americans to have facilities for their air force and navy, particularly in Spain.[82] At the height of the Cold War in the late 1980s, countries on both sides of the Iron Curtain were extensively dependent on the arms race for their prosperity and capability to embrace IT, a reality that influenced how quickly Europeans adopted computers and after the Cold War civilian IT applications, including the Internet, GPS, and mobile communications.

A PAN-ATLANTIC DIFFUSION

To a large extent, the world view on all matters related to computers during Wave One and largely, so far, in Wave Two reflected a pan-Atlantic diffusion of all manner of information technologies over the past seven decades. The trails of bread crumbs lead us to this conclusion. Computers, and before them office appliances, were invented and were most thoroughly deployed first in both North America and in Europe.[83] As earlier chapters illustrated, there occurred a trans-Atlantic dialogue about computing, sharing of information on its form and uses, and a conjoined discussion about the emerging new societies of the future so variously described as the Information Age, Post-Industrial Society, a Networked Era, and by other labels. Articles and books published on one side of the Atlantic made their way to the other side. Early pioneers in computing shuttled back and forth across the Atlantic, and later vendors. It was no accident that IBM directed much of its work out of New York across the Continent, but also gave European executives corporate positions in America, as happened with French citizen Jacques Maisonrouge, for example. The same technologies (systems, products) developed on one side of the Atlantic were used on the other, from American computers of the 1960s to Microsoft's operating systems in the 1990s. Not all the traffic flowed west to east; European software made its way west too, such as Germany's SAP products in the 1990s, and the Internet's World Wide Web (WWW) from Switzerland. It was an exchange that had a long history. When the British led the world in the advancement of computers during World War II, Americans went to Great Britain to learn about the technology, just as they had about cryptographic cipher machines in the 1930s.[84] The bonds were strengthened with the emergence of English as the preferred language of exchange between Europe and North America regarding IT. Military and business mattered. American dominance of Western Europe's computer industry reinforced and reflected yet another manifestation of a connection bigger than many of the events discussed in this book.

People, companies, vendors, and governments cajoled and influenced each other to embrace the kind of information technologies developed on both sides of the Atlantic, as well as for what they were to be used. Models of how corporations and government agencies, armies, and schools that evolved since World War II reinforced their bonds, with IT often serving as an informational infrastructure supporting the transatlantic connections. To be sure, shared heritages in how to do science and engineering, both products of hundreds of years of practices originating in Western Europe yet over time transformed collectively by Americans and Europeans, provided the intellectual, scientific, and philosophical underpinnings that helped bond transatlantic diffusion of many technologies, not just IT. Developments in IT and its diffusion in Asia grew out of the output and practices of the transatlantic world. To be sure, as Asian countries moved into Wave Two, they began to contribute, but even as this book was being written, some of the most exciting consumer electronics were coming out of Apple, a non-Asian firm headquartered in California. The rest of the world followed the transatlantic nations in what they diffused, how they did it, and even when. Thus, we must conclude that the shape and scope of how IT diffused around the world during Wave One was largely shaped by the transatlantic community. This view stands in sharp contrast to the received orthodoxy which holds that the United States almost single-handedly drove forward diffusion of IT in the twentieth century. As the country cases demonstrate, the American-centric perspective is too narrowly focused, indeed flawed, or at least no longer viable given what scholars have learned over the past decade, because it leaves out a section of the world that had collectively as large a GDP as the United States, a bigger population, and economies and governments similar enough to leverage effectively a pan-Atlantic body of technologies and products. To appreciate more fully the diffusion of IT, one must take into account the roles of such nations as the Nordics, Mediterranean countries, Canada, and the surrogate Anglo-Saxon countries that physically sat outside of Europe but were socially, economically, and politically the children of the pan-Atlantic world, most notably Australia and New Zealand. Emerging histories of earlier forms of information technologies, such as office appliances and telephony are illustrating the same pan-Atlantic patterns of diffusion evident with computers, calling into question notions that IT's diffusion in Europe was a highly Balkanized affair. In short, the story of diffusing the most advanced technologies of the nineteenth and twentieth centuries require a much broader framing than has hitherto been done. As important as American leadership was in the process, the United States did not operate in a vacuum.

If the Atlantic community was the epicenter of the diffusion of information technologies, there had to be peripheries beyond its pale. Indeed, during Wave One these existed, and based on what is normally assumed, participated in the diffusion of IT as distant followers and as less intensive adopters, particularly in Eastern Europe, but also in other parts of the World. That received orthodoxy is challenged in subsequent chapters, questioning old assumptions, beginning with an analysis of events on the other side of the Iron Curtain. Regions around the world demonstrated diversities of practices in their use of IT, but often also demonstrated an ability to leverage quickly knowledge and best practices that emerged first in the pan-Atlantic world. That is why, for example, eastern Asian diffusion of ICT occurred so rapidly in both their Wave One and now is happening in its Wave Two. But, first we turn to the fascinating experience of Eastern Europe.

CHAPTER 6

Limits of Diffusion

Computing in the Soviet Union, German Democratic

Republic and Eastern Europe

This is not the kind of technology that is easily nurtured by the Soviet system and, for a long time, the leadership did not perceive much need to go to a lot of trouble to develop a full range of capabilities and promote widespread use.
—Seymour E. Goodman, *American economics professor, 1979*[1]

The nations behind the Cold War's "Iron Curtain" present a collective case of limits to diffusion of information technologies. Limits unfolded in three ways. First, the volume of computers installed in this part of the world were far less than in the pan-Atlantic community, measured by known populations of computers, number of systems per capita, or any other metric conventionally used to quantify the amount of IT in any economy. Second, the extent to which IT-based automation—normally referred to as data processing in this era—became integrated into the fabric of work in Communist Europe remained far less than in the West, a situation that had not relatively changed in a substantial manner by the early 2000s. Third, the degree to which the self-perception of society east of the Iron Curtain as moving into some post-Fordist, or Information Age, was the least pronounced of any European and East Asian society. Only nations largely in Sub-Saharan Africa and in the most economically underdeveloped parts of Asia and Latin America discussed these issues less.

The first five chapters of this book told essentially a positive story about computers flowering, spreading across many nations, all the while with local economies and standards of living improving. The collective experience of the pan-Atlantic community led many observers, public officials, business managers and their employees to conclude that computers had a profound and normally supportive effect on the evolution of their societies. That proved not to be the case in Eastern Europe; in fact, the story told in this chapter is one of slow and anemic adoption, even significant criticism and resistance to IT's adoption, despite public support displayed by senior political leaders. The situation proved starkly more negative, and so the tone of this chapter is decidedly jarring. That dramatic contrast teaches us a great deal about the limits, or brakes, on the diffusion of information technologies in the late twentieth century. The experience of this region provides insights on possible effects on other emerging technologies, such as biotechnology, and graphical software in video games and movies, which are now spreading world-wide. In short, the history of Communist Europe's IT helps to refine our understanding of the forces that propel, constrain, and shape use of a modern complex technology. Such insights can assist in identifying what influences speed in deployment, also at other forces that influence diffusion. For example, more than in the West, Marxist–Leninist ideology framed much thinking and action with respect to the use and diffusion of IT, as did the fundamental structure of Communist societies in which centralized government agencies adjudicated who could use computers and when. This pattern of diffusion stood in sharp contrast to the experiences of market-driven economies where many more institutions made decisions about uses of computing without depending on approvals by a collection of complex bureaucracies in government.

To someone reading about the history of Eastern Europe nearly a quarter of a century after the demise of communist regimes in Europe may seem odd, particularly to historians who began their adult work after the Cold War. However, during the Cold War—the period covered largely by this chapter—the term was used widely to mean that portion of the continent behind the Iron Curtain and under the sway, if not control, of the Soviet Union. I use the phrase Eastern Europe to mean that and consciously sidestep discussions about Central Europe and other designations of the region which, to be factual, was as large as Western Europe, the latter designation referring to those parts of Europe that did not live under Communist control. For most of the second half of the twentieth century Europe lived split in half, despite interactions among the two discussed below. That reality influenced profoundly all that took place even after the demise of Communism in the eastern stretches of the European continent.

Scholars in the West who have looked at the role of computers in Eastern Europe learned to accommodate an alternative reality. Those interested in the global diffusion of IT must do so as well. The names of agencies and industries were different, along with titles of key players. Decisions to acquire and use IT often differed one from the other (i.e., acquisition vs. uses), as well as who made them. Attitudes proved equally varied from the West and, as demonstrated by the experience of the pan-Atlantic community, in which worldviews often played a more decisive role in the diffusion of a technology than engineering features of the devices themselves. The Soviet experience illustrates that computers several to 10–15 years behind technologically those in the West did not slow adoption. Rather, more influential on rates and extent of diffusion were political, ideological, and rhetorical considerations, as Slava Gerovitch illustrated in his study of the rhetoric of Soviet computing. He noted that "while American academics preferred precise, unambiguous wording, Russians often valued more intricate and vague formulations open to multiple interpretations," and that way of thinking affected attitudes toward deployment.[2] The structure for development, manufacture, and deployment of computers— institutional organization of the state—was even more influential in setting the pace of diffusion of IT in Communist Europe. One early finding about the study of Communist computing is that scholars need to exert extra effort to understand the idiomatic peculiarities and idiosyncratic institutions of this part of the world before they attempt to study the role of IT in these nations. While that requirement applies to Asian and pan-Atlantic communities as well, most people studying the history of computing already have a deeper knowledge of capitalist societies and how they function than about Communist ones.

The requirement to spend a great deal more time understanding the politics, economics, and expressions of these societies is compounded by the fact that for decades hard data on the role of computing rarely became available to scholars. There is no other part of the industrialized and industrializing world for which we have less useful, reliable information about the role of computing than for Communist Europe and China, although there is much useful data about the technologies themselves. While we encounter this paucity of reliable data all through this chapter, in comparison a wealth of data exists about the pan-Atlantic community. Part of the reason for the poverty of reliable information derived from institutional practices, as explained by an American economics professor in the late 1970s:

> Standard statistical information is withheld, not merely from foreigners, but from the Soviet public as well. This traditional secrecy lowers the effectiveness with which domestic economic decisions are made, since on any specific matter only a handful of people are well-informed. An advanced economy needs accurate widely-available economic information, and Soviet authorities hamstring their own efforts through continuing their secretiveness.[3]

After reading this chapter, one can be less sympathetic to those economists and historians in the West who complain about lack of data for such countries as the United States, West Germany, Italy, or Spain. A brief tour through the history of Soviet, East German, or Bulgarian computing should disabuse them of their *angst*, and instead cause them to see Western data as the kind one should seek to pull together about Communist Europe much as various international associations have successfully started to do documenting economic activities in post-Communist Eastern Europe, such as the United Nations and OECD. When examining extant materials on Communist computing prepared largely by Westerners, we see in nearly every instance a disclaimer that there is inadequate, let alone accurate, information. In over forty years of doing historical research, I have never encountered such a poverty of reliable information. There are data, of course, and discussions about these but often these debates reflected more guesswork than one encounters for many other countries on the part of some economist, CIA analyst, Soviet computer scientist, or public official. In other cases, a rosier picture presented the capabilities, use, and deployment of some technology, or computer, to position a politico or agency in a better light and, in the process, too frequently purposefully obfuscated information.[4]

Archival sources were closed to researchers interested in Communist Europe since World War II, and in the case of the Soviet Union, since 1917. Not until after the collapse of the various Communist states, beginning in 1989, and in the case of the Soviet Union the end of 1991, did archival sources slowly become available. Now researchers have a monumental task before them to reconstruct the modern history of nearly two dozen European nations and published results of their investigations are only now just appearing, including a few monographic morsels on computing. Understandably, most of the earliest studies focus on the political history of the Cold War. One historian looking at archival sources, Vladislav M. Zubok, commented in 2007—16 years after the collapse of the Soviet Union—"that the opening of archives in Russia and other countries of the onetime Communist bloc provides fascinating opportunities to write about the Soviet past," and spoke about "the abundance of sources" as being "astounding," all paralleling the kinds of materials routinely available to scholars in the West.[5] At least the role, motivations, and actions of individuals and institutions can be potentially described and analyzed with greater certainty than before, portending a positive future for historians of computing in the old Communist bloc.

In an ideal world, then, it would be better to write this chapter 10 or 20 years from now. The experience of the Communist world displays before us today an opaque view of events and details underpinning the definition and extent of trends. It is opaque in that we see shadows and images of information technology machines (peripheral equipment too, not just computers), even less about software, deployment, activities of individuals, and a bit of the landscape where computing was used. Broad general outlines of behavior are visible—a perspective that has not changed among Western and Eastern observers over the past three decades—but some details that sharpen our image of Communist computing are becoming more discernable. However, all the statistical data presented in this chapter should be considered suspect in their specificity, useful only as general indicators of trends. In other words, if we state that a particular Russian reported 281 machines installed in an agency, we can only accept with confidence that computers were being used by that agency, and that the number in all probability is either higher or lower than 281. To make matters more obtuse and frustrating, we do not know if in our example 281 refers to 281 systems (computers with their peripheral attachments, such as printers and data storage units), or if such data catalogs specifically all peripherals attached to a system, along with the computer being one component of the system in which case the number of computers (systems) in this hypothetical example would be closer to 10.[6]

Another circumstance to keep in mind is that unlike in the West, public officials implemented a concerted and relatively effective effort to slow or block the flow of information about all manner of things and activities in their societies for ideological purposes, reasons of national security, and political concerns. By contrast in the West over four centuries of myriad experiences reinforced practices in support of open flows of information. However, the example of the role of computing in the Soviet bloc is beginning to illustrate that more information about computers moved around the bloc and from the West into the Communist world than thought previously. In fact, the history of how knowledge entered the region about computing makes it appear that the Iron Curtain proved less a solid barrier to information flows than a chain link fence through which one could observe what was happening on either side and through which information could continuously pass in small amounts. More information about computing flowed west to east than east to west. Contrary to our image of an impenetrable Iron Curtain, Soviet and East European computer engineers and officials attended technical conferences in Western Europe, even visited the United States, and routinely acquired Western publications (which they often translated into Russian and other East European languages). One would expect cross-curtain visits to occur before the Cold War solidified around 1950, and indeed some trips were made. But, even during the most intense years of the Cold War, when Premier Nikita Khrushchev told the West "We will bury you," scientists and engineers shared information, often quite openly, for instance in the late 1950s.

Several examples demonstrate the nature of the exchange of information. In August, 1958, a team of American computer scientists visited Soviet scientists and engineers to learn about what Russian computing projects were underway and shared with the Soviets developments in the United States.[7] Then in April, 1959, a Soviet delegation of computer scientists visited the United States, followed by another American group to the Soviet Union in May, 1959. The Soviets toured about a dozen American factories, installations, research centers, and other facilities in the United States. The Americans visited all the key computer development and manufacturing centers mentioned in this chapter, met almost all the leading lights of Soviet computing, and had very open and candid exchanges about the architecture, design, construction, use, and performance of Soviet and American computing. The Americans reported that the Soviets were very familiar with the American literature on computing. For instance, they learned that about 95 percent of all Soviet computing at the time focused on scientific and engineering calculations (just as was the case in the United States earlier during the 1940s and very early 1950s); and that they had the All-Union Institute of Scientific Information (VINITI) that translated and disseminated American articles on all manner of scientific issues drawn from about 2,000 publications.[8] The only substantive knowledge about how Soviets manufactured computers in the 1950s that we have today came largely from observations made by this team of Americans from their visits to the Penza Computing Machine Factory. Located about 350 miles slightly south of Moscow, this facility employed some 4,000 workers, and at the time built the new URAL-I, which the visitors concluded "looked quite modern." The Americans reported that 120 had been built so far, and that 24 were on the floor in various stages of construction.[9]

Soviets also bought and smuggled an unknown quantity of Western technologies despite restrictions placed on the export of American systems to the Soviet Union deemed most advanced by the U.S. Government.[10] In fact, the IBM S/360s–S/370s became the foundation technology for some two decades of Soviet computing and the source of controversy about Communist computing between Moscow and some of its East European satellites. More is said later about this important strategic dependency on Western technology and the effect it had on deployment of Communist computing.

Before discussing Communist IT, we need to recognize that the region of the world in which IT was used represented a large land mass. Figure 6.1 is a map of all Europe from the perspective of the eastern side of the curtain. Note the size of the footprint on both sides and it becomes quickly evident that combined Eastern Europe and Russia are massive. The map shows over a dozen European countries that were either part of the USSR or buffer states, while the USSR also had non-European components as well at the eastern and southeastern end of its empire that extended to the Pacific and to the warmer waters of the Black Sea.[11] Within the USSR Russia had the largest land mass of any country in the world. Among the Western satellite Soviet republics Poland, Georgia, and the Ukraine ranked as big as the largest states in Western Europe by any measure of land mass or population. At the time of the demise of the USSR in late 1991, it had a population of some 290 million people—about the same as the United States, while the population of all Communist Europe hovered at about 350 million.[12] Eight states comprised what came to be known as the Soviet bloc: Poland, Hungary, Czechoslovakia, East Germany (German Democratic Republic, or GDR), Romania, Bulgaria, Yugoslavia, and Albania, although the latter two nations broke with Moscow. Most joined the Warsaw Pact, the Communist equivalent to the West's defensive alliance, NATO; the exceptions were Albania which pulled out of the alliance in 1968 and Yugoslavia which never joined.

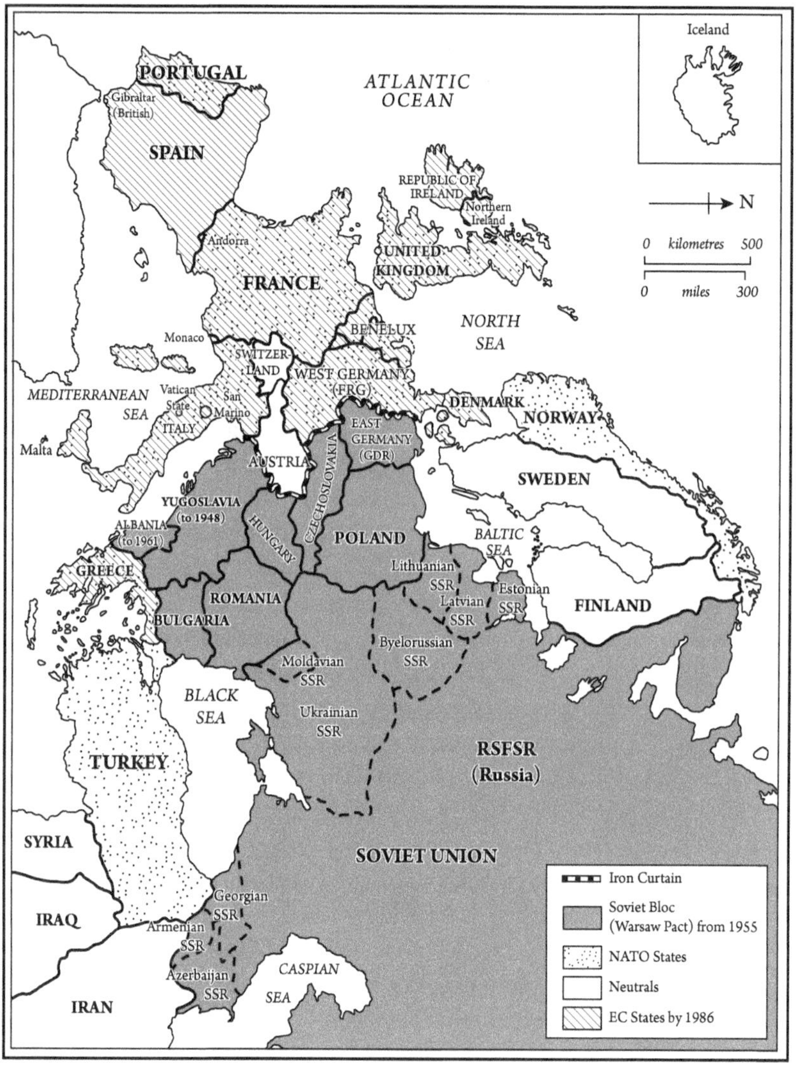

Figure 6.1
Europe from the perspective of Eastern Europe, 1949–1989.
Source: Norman Davies, *Europe: A History* (New York: Oxford University Press, 1996): 1056. Reprinted with permission.

Before the Cold War, nations in eastern and central Europe were more heavily agricultural than the West, with the exceptions of East Germany and Czechoslovakia which had extensive industrial bases prior to World War II. Immediately following the end of World War II, Communist Europe's economy did exactly as Western Europe's: it rebuilt war-damaged buildings and physical, economic, and social infrastructures, with physical destruction far more extensive than experienced by the West in some areas (for example, 50 percent more in Poland) and in the process achieved national income levels comparable to prewar levels by 1950. That was all made possible less because of some Communist ideology than by the need to rebuild so much infrastructure. The Communist bloc came out of the war with an economically underdeveloped labor market and financial infrastructure. Its political system evolved rapidly into Communist dictatorships. This political path was unlike that of the West which either continued or re-established democratic institutions and market-driven economies.[13]

This chapter examines the diffusion of IT in the Soviet Union (primarily Russia), East Germany, and to a lesser extent in the rest of Eastern Europe.[14] The Soviet Union dominated the Communist bloc, managed both its structure and influenced profoundly activities of all nations behind the curtain, becoming a major influence on IT diffusion in the region. However, each country also managed its IT destiny to a larger extent than supposed previously—a process explained in some detail here. The path to Communist computing lay through the north of Central Europe via East Germany and from the east through the Soviet Union, while the rest of Eastern Europe embraced computing in similar if variegated ways. The two detailed country cases allow us to normalize our understanding of trends across the entire region. East Germany is particularly interesting because it emerged as a major user of IT, playing a crucial role in shaping applications and deployment of the technology across the entire region ever in the shadow of the even larger Soviet Union. Every country behind the Iron Curtain used computers, albeit to lesser extents; almost all also began the era of the Cold War with individual embryonic one-off computer projects, just as occurred in the West at various institutes and academic settings.[15]

HOW COMMUNIST EUROPE DIFFERED FROM CAPITALIST WESTERN EUROPE

Before describing diffusion of IT in Communist Europe, we need to appreciate the environment in which users and officials imposed this technology, because IT's story turned more on political and institutional imperatives than on technological developments. Government agencies made decisions affecting diffusion across their entire national economies. Normally, deliberations of officials were more affected by non-technical considerations, such as political agendas, personal careers, and ideological perspectives.

The central source of authority in Communist countries resided in the local Communist Party, with the Soviet Communist Party serving as *primes*. Governments worked on behalf of the Party, which directed the political and economic activities of a Communist society. The Party operated with three fundamental elements which can best be thought of as ways of conducting its work that proved so distinctive that we can think of them as a Communist *style*. This consisted of a political philosophy and world-view of what a perfect society should look like couched in Marxist–Leninist terms, which its members strived to develop, that incorporated protection of national security (largely against the Capitalist West), and that fostered economic development. Supreme authority rested with leaders with authoritarian power who led both the Party and the state, normally holding simultaneously both positions.[16] No other recent ruler of the USSR did this more assertively than Josef Stalin, who had led the Soviet Union since the 1920s, and ruled with a brutal iron hand until his death in 1953. He determined much of the destiny of the Soviet Union and its Communist neighbors even after his death, by leaving behind a legacy of centralized public administration still lingering today, albeit far differently than before the collapse of the Soviet state. By keeping decisions respecting political, social, and economic policies highly centralized at the national level in the various Communist nations, these countries became directly beholding to Moscow's authority. Much of this decision-making mechanism emerged in the Soviet Union in the 1930s as Stalin implemented rapid industrialization, then extended in the period 1948–1949 when he exported this style of public administration to the newly occupied East European states. Centralized decision-making by government officials covered all manner of activities: education, scientific research, economic investments, prices of goods and services, censorship, and diffusion of computing. The Soviet's communist party made many decisions about the strategic role of IT from the late 1940s to the end of the USSR in 1991, while a large collection of planning organizations and ministries translated strategic intentions into more precise plans and tangible targets. Indeed, any understanding of the role of IT in Communist Europe entails appreciating the role of planning activities and the politics associated with those actions.

Planning agencies and their superiors in the Communist Party initially concerned themselves with those IT matters related to needs of the military in response to the rapidly evolving—indeed escalating—nature of high-tech armaments that erupted at the start of the Cold War in Russia, beginning with nuclear weapons followed by missiles and various weapons guidance systems. Second, they and all Communist states also focused on requirements to handle ever-increasing complexities of running national economies which individuals using desk top calculators could no longer perform and, thus, needed to turn increasingly to computers for assistance. Third, they had to satisfy the need to manufacture and make available goods to its citizens and for trade; later to improve the productivity of such economic activities.[17]

Communist economies are often described as "command economies," run largely by public officials who direct the activities of producers of goods and the actions of consumers through the medium of administrative bodies of national and regional governments. These institutions made decisions about the total amounts of goods to be made by a particular industry, in other words, what and how much to be manufactured. The state owned all means of production, since there was no private property as known in the West. A manufacturer of computers in Russia or Hungary, for example, was an agency of a ministry, and all its employees, government workers. The state had a monopoly on all forms of economic activity; there was no free market in which individuals bought and sold goods and services as in a capitalist system, based on supply and demand, competition, and price. In practice, there always existed a small legal, often also "black market," in which private enterprise existed, even for the illegal sale of small IT items, Western music and American blue jean pants in the 1970s, vegetables and fruits, and PCs in the 1980s. In the West the working economic assumption relied on the notion that the interests of the individual firm or consumer dictated behavior and priorities of actions, while in the East the state's interests replaced those of the consumer and privately held firms. With the state owning property and means of production on behalf of the community at large it controlled all assets of the nation, making it possible, indeed necessary, for the government to direct the economy almost as explicitly as a company would its own operations in the West.

The system of planning and economic administration evolved into the heart of what drove rates and nature of diffusion of IT in Communist European states. It became a complicated process affecting all manner of goods, not just computers, peripheral IT equipment, and software. For instance in the Soviet Union in the 1980s, central planners routinely set prices on as many as nine million items, and since centralized planning functions resided in multiple agencies, the process of even coordinating costs and quantities of available goods proved byzantine and highly inefficient.[18] Planners dictated to factory managers and farmers—themselves public employees—how many and what to produce, and told them what raw materials and components they would receive, how many people to employ, what salaries to pay them, and at what prices to sell (think distribute) their goods to other government agencies or consumers. Management in factories received bonuses for achieving their targets, which normally were not set in a manner designed to encourage innovation, to meet market demands, or to increase productivity. This is an important feature of Communist management to keep in mind as we look at the history of computing in the Soviet empire, because in the West a great deal of the reasoning behind adoption of computing emanated precisely to promote innovation, to offer what customers wanted, and to optimize economic productivity. Since the Soviets dominated eastern and central Europe and Russia to the Pacific, a few examples of the process at work introduces key players into the story of IT's diffusion.

The most important planning agency was GOSPLAN (Soviet acronym for State Planning Commission), which reported to the Council of Ministers, which in turn answered to the leader of the country, such as Josef Stalin or Nikita Khrushchev. GOSPLAN developed five-year economic plans and their high-level targets. Its employees worked with other Communist European national planning agencies as well, because the latter's plans also included statements about what economic assistance would (should) be provided to satellite states. The ministries then developed operational one-year plans supporting these longer-term ones. Ministries also set specific targets for the various organizations that produced, sold, and used goods and services. These producers and users were called enterprises, and served the same analogous role as private corporations in capitalist societies. All enterprises of a particular type were collectivized under the authority of a specific ministry, such as all automotive manufacturing enterprises that reported to one ministry responsible for coordinating the activities of all automotive factories to meet goals set in the five-year plans. Multiple industries, hence numerous ministries, wishing to operate in some relative mode of independence from each other in order to make their objectives sometimes developed, built, and used their own computers and software. They frequently made them technologically incompatible with another ministry's on purpose. Incompatibility made it impossible for GOSPLAN, for example, to build a national integrated (meaning, automated) computerized planning and feedback model to optimize goal setting and monitoring of activities that measured attainments of targets across what in the West decades later one might refer to as either a "value chain" or "supply chain."[19]

Up to 1958, sectoral and regional ministries covered all of the Soviet Union, called All-Union ministries. Then some of their activities transferred to more regionalized ministries. In 1965 the Soviet Union reorganized these into All-Union and Union-Republican ministries. The former coordinated all activities of a particular sector across the entire Soviet Union while the latter served as planning and operating agencies within individual Soviet republics—a structure that also existed in a somewhat similar form in the satellite states. By 1984 in the Soviet Union 39 All-Union ministries existed. They were responsible for such areas of economic activity as defense manufacturing, electronics, transportation, steel, machine manufacturing, machine tools, shipbuilding, and so forth. There also existed 46 Union-Republican ministries responsible for the all important agricultural industry, but also for mining, construction materials, fisheries, food production and distribution, among others. One credible estimate from the early 1980s suggested that the All-Union ministries controlled about 54 percent of economic production in the Soviet Union, the rest by the Union-Republican and other local republican ministries.[20]

In addition to GOSPLAN, the State Committee for Material-Technical Supply (GOSSNAB) had responsibility for ensuring the availability of all materials needed by various enterprises to fulfill their assigned numerically expressed objectives. GOSSNAB failed often to accomplish its fundamental purpose over the years, resulting in shortages, inadequate supplies of various types of information technology, and disrupted activities of enterprises and end users of computing devices. Yet a third agency at the top of the government was GOSTEK-HIMKA (State Committee for Science and Technology), which changed names and mission from time-to-time, but had primary responsibility for developing, manufacturing, and using computers to improve the national economy. All three—GOSPLAN, GOSSNAB, and GOSTEM-HIMKA—played important, often overlapping and confusing (some contradictory) roles in the diffusion of IT in both the Soviet Union and through their various counterparts in Communist Europe.

Activities at the enterprise level are other key areas in the economy that historians of IT need to examine because they played a crucial role as well in the deployment of computing into the Soviet economy. David Lane, a well-informed student of Soviet economics, described these in the mid-1980s:

> Once plans have been drawn up by Gosplan after consultation with the ministry, ministries in turn discuss projected plans with the directors of production enterprises. Enterprises' plans stipulate specific quantitative inputs and outputs, and the enterprise is legally obliged to fulfill these directives. That is, with so much resources of materials and labour, so many units of cars or tables must be produced. All the prices of inputs and outputs are defined by the plan.[21]

Enterprises were on their own to recruit and train workers; but employees could also switch jobs at will, which they did frequently. That behavior proved disruptive to the functioning of many organizations. Manufacturing facilities were often larger than those in the West, and while good data on this point are hard to come by, it was not uncommon for many manufacturing sites to employ from a few hundred to several thousand employees, particularly in Russia, complicating management of their activities, often without the help of computers.

Enterprises and their economic ministries shared as almost their only measure of success fulfillment of their assigned plans. The process was complicated by the practice of setting next year's targets as a larger one of that achieved in the prior year, in other words, doing more normally with the same amount of resources, and routinely with few considerations for the level or changed circumstances of demand for a particular product. Senior political leaders focused on quantities of output, rather than on their price or demand for specific products, thereby making it possible for a government to favor one sector over another for strategic or political reasons. In the postwar decades across all of Communist Europe officials skewed resources and production to heavy industry (capital goods) along with the raw materials these required (such as coal, steel, electricity), at the expense of investments in agricultural improvements or in provision of consumer goods. One consequence of this skew in favor of heavy industry was that light manufacturing and consumer goods had a smaller presence in Communist economies than in the West. In hindsight, it was the latter two areas of economic activity that so profoundly stimulated growth of Western economies, not heavy industry. Yet it was heavy industry that the various European Communist regimes favored the most. A consequence of this economic system was under-investment in infrastructures, such as roads, housing, urban transportation, even for provision of electricity and heating of homes.

Despite these limitations, the various Communist economies thrived in the 1950s, slowed in the early 1960s, recovered in the late 1960s, then, in the words of one economic historian, "staggered through the 1970s," and declined in the 1980s.[22] Across all of Communist Europe, for the entire half century the standard of living lagged behind that of the West; industrial production proved far less efficient, and there existed an almost universal chronic problem with poor quality of goods.

Flows of goods, the basis of an economy's performance prior to the expansion of the service sector in advanced economies by the early 2000s remains a useful path for following the economic behavior of the Communist economies. It is best understood by observing activities at the enterprise level. In Communist economies, enterprises had captive markets. Someone in the planning apparatus would identify who were their customers and yet another organization set prices at which they sold their goods. Enterprises were not allowed to take initiative, for example, in developing new products without permission of a designated planning agency. There were almost no incentives built into the economic system either to increase the variety of goods produced or to improve their quality, endemic problems in all Communist European states during most of the Communist era. Senior government officials attempted to implement various reforms all through the half century in each country to address such issues, but to little avail.[23] There always existed a desire to balance efficiency and political control, a feature so different than in the West, because in Communist economies leaders valued more the notion of meeting needs of the proletariat than in optimizing economic efficiency (the profit mantra of the West), the latter which ideologically driven advocates and practical public officials considered contrary to the "social good" of society. So reforms were at the margins; not at the core of the economic structure of communist societies.

Party and government officials always faced problems with the centralized economic planning model, probably the most crucial structural issue that historians encounter when trying to understand the political activities and management of the economies of Communist Europe. Imbalances in resources, inaccurate alignment of need (demand) with supplies of goods and parts, the far more extensive bureaucracies evident than in the Western economies all proved to be serious problems. Others also existed, not the least of which included quality problems, making it difficult to improve efficiencies or even to earn hard currency in the world market already accumulating vast quantities of high-quality goods throughout the second half of the century; and inadequate investments in infrastructure, training and education, health care, and R&D. All these issues were byproducts of what Western observers noted during the period when central planning and management dominated as the operative model of behavior in communist economies.[24] One Westerner concluded in about 1980—when it was evident on both sides of the Iron Curtain that the Soviet economy was not well—that the problem "with the old economic structure is that it gave maximal encouragement to decision makers to favor established products and processes and to discriminate against innovations," further "that the central government agencies bear the major responsibility for providing the enterprises with their inputs and disposing of their outputs; in contrast to a market economy, for example, in which the enterprises bear the responsibility for carrying out these two functions."[25] That was the environment in which Communist computing had to emerge and diffuse.

INFORMATION TECHNOLOGIES IN THE SOVIET UNION

Soviet computing was inextricably tied to the military, political, and economic evolution of the nation. IT was subject to the realities of the war's devastation in the mid-1940s, then to the necessity of supporting development of advanced weapons systems in the 1950s and 1960s, followed by economic and administrative (most notably planning) uses from the 1960s to the end of the century. Finally, IT played a slowly expanding role in data processing for operations in enterprises and government agencies from the 1960s to the present. To a lesser—yet important—extent the quality and availability of the technology also influenced how it was adopted and its rate of diffusion. In the late 1940s, the level of advancement in computing in the Soviet Union was just slightly behind that of the Americans, but then lagged all through the 1950s to the demise of the Soviet Union. However, lags did not mean lack of computing, no improvements in both Soviet technologies, or of their use.

Observers over many years often estimated this technological backwardness and implementation lag in the range of 10–20 years. However, Soviets had some access to Western technologies so using older generations of equipment itself was not the primary reason for lagging use of IT. Older generations of hardware and software had been used in the West successfully, both when they were state-of-the-art and after new generations of equipment and software became available. The condition of the machines and software per se, while a factor, was thus not the dominant cause limiting use. Nor did the cost of computing appear to be as sensitive a factor influencing adoption as in the pan-Atlantic community or in Asia, since decisions to install and use IT were not made on the basis of cost or on some cost–benefit analysis, despite guidelines for such efforts. For an explanation of the lag in use of both state-of-the-art systems and the extent to which Soviets used IT one has to turn to non-technical realities, which proved more influential. A group of scholars looking at Soviet software, and most concerned about the technology itself, made the same point: "the lack of horizontal ties between developers and commercial enterprises, the absence of a national market in intellectual property and of incentives for innovation meant that basic advances were not translated into commercial success and diffused through the economy."[26] One must add meat to the bones to understand the limits of diffusion across the entire pan-Communist ecosystem.

But first, the early introduction of computing in the USSR needs to be appreciated, a story that can be summarized quickly. As in the West, in the late 1940s there existed isolated pockets of interest in computing, although not yet formally embraced by officials. To the best of what we know, there were two projects underway, based on Western technologies, with initial designs of computer architectures recognized by Soviet officials in 1948. Then in 1951, a team of engineers built the MESM, most probably the first stored program digital computer built on the European continent, constructing it just two years after the British had built the EDSAC. The MESM was an important milestone for Soviet computing since it influenced future work on local systems, not because it was a "first" in European technology.[27] In that same year another group of engineers at the Special Engineering Bureau SKB-245 tested their own machine, called Strela. They built their first production model in 1954, which was used for calculations by the Institute of Applied Mathematics of the Academy of Sciences (IAM AS) of the Soviet Union. Many of the calculations concerned nuclear explosions, an application mathematicians used with the system over the next several years. Best estimates suggest that seven copies of Strela were built between 1954 and 1958, representing the first Soviet production system; all were used for internal state applications.[28]

The key technologist at the time, Sergei A. Lebedev, then built one of the first early important Soviet systems, known as BESM-1, at the Institute for Precise Mechanics and Computer Engineers (IPM CE), putting it into limited production in 1953. Lebedev and his system went on to influence Soviet IT all through the 1950s and 1960s.[29] In 1958 he introduced the M-20, also a first generation system built as a collaborative project by the IPM CE and SKB-245. While others have described its technical features, suffice it to note that it was a high-speed computer that led to development of a family of computers in the 1960s which spread throughout the economy, albeit on a limited basis, known as the M-220 and M-222.[30] The amount of information about computing, particularly Western publications available to a closed circle of experts in the Soviet Union, proved considerable and essential for Soviet IT developments. We are led, therefore, to the conclusion that the issue of technical capability of potential builders of machines was not a serious gating factor affecting (or retarding) development and diffusion of computing, compelling us, instead, to examine other considerations.[31] It is to these we need to turn to now.

As in the United States in the early 1950s, Soviet computing focused on scientific, engineering and military applications, not on data processing, accounting, and other civilian uses in the Soviet Union. In the United States embracing civilian applications had started by 1950–1952. As one observer noted in the 1970s in the case of the USSR, "the technology was not available for such [civilian, ed. note] applications" and also because "there was serious ideological suppression of the use of economic theory and quantitative methods in economic planning," resistance to the use of IT in a way that Soviets did years later.[32] From the early 1950s to 1961, Soviet engineers designed and built some two dozen different digital computer models, all used for scientific and engineering applications, most for the military. There was no office appliance industry with the likes of IBM or Burroughs pressing potential users to embrace computing; nor was there a large base of users of pre-existing punch-card tabulators who would have naturally found computers attractive innovations and who could have pressed developers to create even better systems faster, as occurred in the West, most notably in the United States. Public rhetoric and debate about computing—specifically cybernetics—was a debate about the pros and cons of computing, which began in the 1950s and extended to the end of the 1980s, making it clear that officials, engineers, and scientists discussed the possibilities and limitations of using computers.[33]

The military consumed the vast majority of computing in the 1950s and early 1960s for three initiatives: development, construction, and operation of nuclear weapons, ballistic missiles, and antimissile defense systems. IT communities kept secret as much as possible information about these military technologies and uses, unlike in the United States where knowledge of the technology spilled out, leading to the development of commercial variants of the machines for use by companies, universities, and non-military government agencies. As in the United States, pilot projects became production machines when the military threw its support and resources behind the technology, which occurred slowly in the case of the Soviets during the 1950s, a few years after the Americans had started to do, but still largely for use within military circles. MESM got them started, while Strela became a mainstream production system for the military. Because the government controlled all production of those computers almost all these early systems went to the military.

Lebedev and other experts on computers began telling their political superiors that the Soviets were falling behind the West as early as 1952. By 1958 various Soviet officials began thinking about using computers to facilitate state economic planning. Yet civilian applications remained limited, as one student of the period explained:

> Civilian computer applications were excluded not only by the heavy militarization of computing, the scarcity of computer time, and the ideological controversy

around cybernetics, but even more effectively by the wall of silence and the barriers of clearance requirements built around the early Soviet computers. In the paranoid atmosphere of the Cold War, the cloud of secrecy surrounding military computing not only concealed Soviet computers from the enemy, but also created serious internal obstacles for the development of Soviet computing.[34]

All through the 1960s a second generation of computers emerged in the Soviet Union, beginning in about 1962 and concluding in 1975. This second generation included greater capacities, faster processing, yet still aimed largely to support mathematical, engineering, and scientific processing.[35] During this period initial economic planning applications first appeared. The technological "star" of this period was the BESM-6 which appeared in 1967, built by the IPM CE under the direction of Lebedev. It was seen as a supercomputer with new software and hardware architectures and functions. The BESM-6 influenced the forms and features of subsequent Soviet computers as well and remained in use until the late 1980s. A third generation of computers began to appear in the 1970s as well. At the beginning of the decade some 20 different systems were either under development or in production in the Soviet Union, with various architectures (hence incompatible with each other), mimicking the diversity of approaches so evident in the United States prior to the introduction of the IBM S/360 family of computers in the mid-1960s.[36] Some were designed specifically for narrow military applications, others for satellites and space programs. A similar dialogue about the value of having modular, compatible systems that occurred in the West in the 1950s and early 1960s took place in the USSR in the 1960s and early 1970s.[37]

Meanwhile, Soviet officials became interested in using IT to support economic planning. In November, 1962, deputy chairman of the Soviet Council of Ministers, Aleksei Kosygin (later premier of the Soviet Union, 1964–1980), called for such an application of computers, making it official when the Party and the government issued a resolution to that effect in May, 1963. Various government agencies were ordered to build and use computers for such applications.[38] Plans were developed for massive diffusion, but intense debates and disagreements also took place about how to go about this, an issue that has been described elsewhere and need not detain us here.[39] What is important to note is that the use of computers threatened the power of some agencies. For example, automated collection of economic data using a network of computers would take work away from the Central Statistical Administration (CSA), or so it thought, and its leaders resisted such use of computing.

By 1965, Premier Khrushchev began decentralizing economic planning, and so the issue of the role of computers elevated as an aspect of new methods for managing the Soviet economy. The worried CSA was given responsibility for developing and running a confluence of computers. It wanted regional data centers, but management at GOSPLAN insisted that use of IT be organized along the same lines as the agency, by groups of industries. The dispute led to inaction on implementation of these systems. Nonetheless, ad hoc installations of IT systems proceeded all through the mid- to late 1960s, with one estimate suggesting just over 400 were installed.[40] Another calculation from a Soviet source stated that there were "about two thousand" computer centers in the country at the start of 1973; either number is quite small when compared to what existed in many other countries. At least one Russian commentator was quite candid about why deployment had been slow: "it still has to prove itself. First, plants are not interested in computers for economic reasons. Second, there is a shortage of trained personnel. Third, the quality of the equipment itself is inadequate. Moreover, existing computers are used poorly with regard to the correct distribution among their customers."[41] While debates on use of IT continued, the military upgraded their systems to a new generation of technologies not suitable for use in economic planning. There appeared to be no sharing of technologies between those responsible for developing military systems and others charged with providing computers and software for economic planning during the 1960s and 1970s. Nonetheless, it was also in these years that some enterprises began to experience the effects of computing in their work streams. Concurrently, the slowly growing community of computer experts pressed on with their activities.

Simultaneously, as part of the third generation wave of computers, which had as a feature upward compatibility of systems, the Soviets launched an important initiative to create a unified system that could be used by all Communist countries and that would involve such nations as East Germany and Hungary in its development. There were several reasons for involving so many countries: to leverage good thinking and available resources scattered about these societies, to solidify economic and military interdependencies among all the Communist societies in the face of the Cold War, and to stimulate economic development. The decision to engage all proved more strategic than practical because the Soviets had already demonstrated that they could build their own systems without engaging their satellites, but so too could some satellites, notably Czechoslovakia, Hungary, and Romania. In other words, the "Unified System" emerged as both a political and technological initiative.

Known as ES-Ryad (Russian acronym), it began to emerge in 1968, largely at the Research Center of Computer Technology, followed in 1969 with the creation of satellite agreements and enterprises to participate in Bulgaria, Hungary, East Germany, Poland, Romania, and Czechoslovakia. In time it involved in excess of 46,000 scientists and other engineers and experts, and possibly as many as 300,000 workers in over 100 organizations.[42] If future historians can learn more about military computers in the USSR, we may be able to conclude that the ES initiative was indeed the largest in Soviet history; but for the moment, we can only speculate that this probably was the case. All of these people worked from 1970 to 1985 on software, peripheral equipment, and computers with the objective of making them compatible. As we see in the discussion below about East Germany, many of the participating scientists and their nations had various opinions about what to incorporate into ES, leading to intense debate about the extent to which the Soviets and others should simply copy IBM's S/360 versus creating a whole new architecture. The Soviets decided to copy IBM's S/360 architecture and later the company's S/370. Three series of systems were developed in various sizes with names beginning with the term ES followed by a model number (e.g., ES 1020, ES 1035), imitating naming conventions for computers that had been applied in a *de facto* manner (a name or family numeric followed by a specific model number) around the world by the early 1950s for production systems. Ultimately, 32 models were built, along with some 200 types of peripheral equipment, 12 versions of its operating systems, and an undetermined, but possibly thousands, of application software programs, not including those available in the West from IBM and IBM-compatible vendors.[43]

The project was fraught with difficulties, much as IBM had experienced earlier in developing the S/360. A few key technologies simply were not even available in the Soviet Union, of which the most important involved disk drives until 1973, over 15 years after they became available in the West. Non-Communist Europe dominated development of online systems which could only be done using disk drives to access data directly, unlike the earlier approach of using cards and tape which had to be read sequentially to access information. That earlier approach remained the common mode for civilian uses of IT in the Soviet Union at the time, but not for military applications, which relied on the more advanced online methods.

To manage the work of large groups of national participants, the Soviets created the Council for Economic Mutual Assistance (CEMA). Central Europeans recognized the project as a Soviet initiative, less a pan-Communist one. CEMA quickly encountered resistance to this Soviet initiative. Hungary, Bulgaria, and East Germany at first signed on willingly. Polish officials wanted the project to be more like the British ICL systems they were using, while the Czechs had their own approach which they championed. Thus, many Eastern European experts and officials had simply not embraced enough to the merits the Soviet's Unified System, as it became known, to ensure unanimous support. The most problematic participant turned out to be Romania, which looked to the West for its computing styles, most notably to French models. Military uses of computing were, of course, subject to Warsaw Pact requirements just as in the West these folded into the needs of NATO, hence to American computing technologies, technical standards, and products. While computing activities in the satellite states were smaller than those in the Soviet Union, they were often more advanced, because they relied on technologies developed in the West. The German Democratic Republic (GDR) advocated using IBM's architectures, making Ryads (also known as Riads) compatible with S/360s, a path it had already taken, although based largely on an earlier IBM technological schema (IBM 1401). Nonetheless, each nation contributed resources and participants to this pan-Communist project, one of several lines of computer development in the Soviet Union.

Western students of the project observed that the systems ultimately were not fully compatible with IBM's. They also lacked time sharing, and reliability proved to be a continuing problem.[44] Production of these machines remained low; one estimate held that only 5,000 Ryad-1s were built in comparison to 35,000 S/360s in the same period.[45] While developers had access to all of IBM's S/360 manuals, many of which were translated into local languages, they did not have IBM's service support infrastructure to address problems at various data centers. Severe shortages of peripheral equipment held back installations of systems and their use. This problem was often the result of multiple ministries making peripherals different than those manufacturing and distributing computers, or of systems shipped with the minimal—often inadequate—number of peripheral devices, including tape drives, printers, and later disk storage.[46]

Observers concluded in the late 1970s, however, that the new systems had "brought some real progress," with more and better quality systems than before. The Communist Party supported people using computers with the old cybernetic debates about whether or not the technology was compatible with Marxist–Leninist thinking now a thing of the past.[47] Yet, a knowledgeable source at the time concluded that "Soviet institutional structures tends to inhibit the customer-oriented design, development, and diffusion of software," which in the West influenced more the rate of technological adoption than innovations in hardware.[48] Very high level public officials managed the project. They touted the new computing extensively prior to its initial deployment, before many technologists and public officials had begun to encounter the number of technical problems that would plague their work for years. In short, many of the problems users encountered were as political as they were technical, with careers at risk and the public face of Party performance potentially liable to criticism. We can conclude that the Unified System approach, while commendable, devolved largely—but not completely—into a strategy to reverse engineer existing Western technology, one that took longer to implement than developments in the West. That meant the Soviet's approach guaranteed its base technologies would always lag those of the West and that the technological gap would grow over time.

Western observers knew that as probably did many Soviet scientists and engineers; those in East Germany and Romania certainly understood the implications of the pan-Communist IT strategy. But, it was the path of least resistance and greatest speed the Soviets were capable of at the time. The reason is not hard to find. With this strategy they did not have to invent from scratch whole new computer architectures, which would have taken too long to develop and at some undetermined very high cost in funds and with technical risks of failure. Soviet officials and computer scientists had witnessed IBM's very public array of difficulties in birthing the S/360, particularly with its operating system, and did not want to risk the same experience. If IBM had difficulty and was recognized as an efficient technological powerhouse, so the logic went, the Soviets would probably experience a similar, if not greater, set of problems if they went on their own path. Furthermore, an incompatible system with IBM would automatically shut the Soviets out of using the rapidly growing library of software applications from the West that were written to run on IBM's S/360 and later, S/370.[49] Despite lingering debates about the decision, given what the Soviets were dealing with at the time, it was probably the correct managerial one to make, even though not the most exciting one for some Soviet computer scientists. As in the West, often technological strategies were determined by political and line management, those who paid the salaries of the technologists, not the other way around.

The Ryad systems deserve less obscurity in the historical literature and their developers more credit for what they accomplished. The systems were in use for over 20 years, and during the course of those years, developers improved them to reflect changes in computer science in general and not just in response to whatever changes IBM implemented. Perhaps the most knowledgeable Westerner about these systems, Richard W. Judy, got it right when he pointed out that when the designers decided to mimic IBM, "they ended up not only with some of the beauty of the 360/370 systems but also with many IBM warts sprinkled across the face of Riad."[50] But they also learned a great deal about building computers in the next two decades, insights not there in the beginning, making it possible to depart from IBM's designs from time to time, acting more like "other IBM-compatible manufacturers such as Amdahl, NAS, Fujitsu, and Hitachi" in improving upon earlier designs.[51] The rate of introduction of new systems lagged IBM's rate, but so too did many of IBM's competitors. In the Soviet case with this class of systems, the lag rates were lowest in the 1970s and 1980s, and closer to five to seven years, unlike in the 1950s and 1960s when they were in the range of 10–20 years. Give Judy the final word on these systems: "Not all is perfect; far from it. But from the shambles of the 'sixties has arisen an IBM 'plug compatible' computer manufacturing entity of impressive scope and capability," with each generation of Ryads improving "hugely on its predecessor."[52] The most impressive work occurred with the central processors and the worst with peripheral storage and input/output units.

This family of systems should be recognized for more than just its technical features, which seemed "good enough" to be used as the Soviet workhorse for many years. From the perspective of IT diffusion, it was a technical channel for diffusing knowledge about IT, both Soviet and Western (e.g., IBM). While Western observers overwhelmingly discounted the prowess of these systems, from a Soviet perspective they proved more valuable, in spite of technological problems, particularly when compared to what they had come from in the 1950s and 1960s.

The Soviets had problems with users, an issue never resolved because these involved the fundamental system of society then in place. Utilization of the technology proved slow and limited when compared to how the West appropriated IT. A couple of observers of the issue from the 1970s offer a possible explanation of the problem:

> Soviet managers tend to be older and more inhibited than their American coun-
> terparts. The system in which they work stresses straightforward production
> rather than innovation and marketing decisions. Soviet economic modeling and
> simulation activities stress the necessity for reaching a 'correct socialist solution,'
> and are not oriented towards being alert for general and unexpected possibilities
> in a problem situation. . . . What does one do with a computer system for the 'of-
> ficial' operational management of an enterprise when actual practice is different?
> Does one dare use the computer to help manage 'expediter' slush funds, under-
> the-counter deals with other firms, etc.?[53]

By the late 1970s, as happened by the end of the 1960s for IBM with the S/360, many of the earlier Soviet's technical problems had been worked out good enough. Reliability increased, along with availability (up time) of systems, although never with enough peripheral equipment. By the end of the 1970s, of course, on a world-wide basis this Soviet technology was not competitive. IBM and Japanese vendors were already engaged in a titanic battle for the global market for mainframes, each equipped with generation of products superior to that of the Soviets, and with a sales and support infrastructures that far surpassed that of any Communist country.

During the 1970s, the Soviet "computer industry," if one can call it that, became public enough to describe. In addition to the network of institutes, academies, and economic planning agencies, and the still little understood mil-itary/defense computer world, other agencies existed responsible for the pro-duction of these systems. The Party and government assigned manufacturing of computers largely to two ministries. The Ministry of the Radio Industry (MRP) was the primary source for general purpose and large-scale scientific computer systems, such as the Ryad computers. The Ministry of Instrument Construction, Means of Automation and Control Systems (better known as Minpribor) developed and manufactured systems for such uses as industrial control, production, planning, and management. Machines built by both over-lapped, thus, on occasion, competed with each other as early as the 1960s. MRP seemed "the winner" as the primary provider of large systems in the Soviet Union by the late 1970s, while Minpribor became the key supplier of minicomputers, such as its own ASVT M-4030 system. Other ministries, how-ever, participated in the Communist computer world as well. Keep in mind that ministries need to be viewed by Western readers as analogous to what one called industries in capitalist economies. In the Soviet instance, the Ministry of the Electronics Industry (MEP) had responsibility for developing and manufacturing components used by Minpribor and MRP, among others (including sometimes the military), and which it used to manufacture some computer equipment for its own internal uses and special projects.

In addition, the Ministry of Communications Equipment Industry (MPSS) made telecommunications equipment and wrote software. Some ministries with strong ties to the military also manufactured equipment and developed software, such as aviation, shipbuilding, and various "machine-building" ministries.[54] Still other ministries supplied components, material, and equipment to those manufacturing computers, such as computer punch cards, paper tape, printer paper and forms, magnetic tape, magnetic disk packs for storing digital data, and even air conditioning equipment for data centers. The State Committee on Science and Technology (GKNT) worked with, or in competition with (not yet clear) GOSPLAN, the Central Statistical Administration (CSA), Military-Industrial Commission (VPK), the USSR and Republic Academies of Sciences (AN) which played such a crucial role in the introduction of Soviet computers in the 1950s, and with the Ministry of Higher and Secondary Specialized Education.[55] Other agencies were involved too, the Party, and the most senior public officials of the nation, all working through such institutions as the Central Committee Secretariat of the Communist Party of the Soviet Union (CPSU) and the Committee for State Security, better known as the KGB.[56] To what extent the military depended on these agencies for their computing remains unclear, but these civilian ministries must have played a significant role because they were so large and numerous and manufactured a substantial number of machines.

Despite the dearth of reliable Soviet statistics, some data suggest there clearly existed momentum across parts of the economy in using computers. Between the mid-1960s and 1970, the percentage of first generation vacuum tube systems in production declined from possibly two-thirds of all machines built to about 15 percent, while the percentage of second generation systems (which used transistors) rose correspondingly from a third of all systems built in the mid-1960s to over 80 percent by the mid-1970s; third generation computers, relying on integrated circuits, only came into production, in tiny numbers in the early 1970s. These data demonstrate that the Soviets made a concerted effort to modernize (by their standards not Western) their installed base of computers, beginning in the late 1960s, although these percentages are based on a small population of computers. One creditable observer speculated that by the late 1970s, the Soviets were building just over 1,500 Ryad computers per year.[57] To put that number in some perspective, IBM manufactured about 1,200–1,400 computers per year in that decade just in its American plant in Poughkeepsie, New York.[58] In an internal market analysis prepared by IBM in

1974, an employee noted that the USSR had "only one computer per 35,000 inhabitants. U.S.: one computer per 2,000 inhabitants," adding that "their technology [was] well behind that of the west."[59] An analyst at the U.S. Central Intelligence Agency estimated in 1979 that while production of computers expanded very rapidly between 1965 and 1975, in 1977 there were only about 20,000 computers installed in the Soviet Union, as compared to 325,000 in the United States, despite the fact that "the number of computers needed to flesh out all of the Soviet schemes is staggering."[60] There is currently no credible data on the amount of peripheral equipment made in the period 1960–1985. One observer noted in 1979, however, "that there are serious shortages of every-thing."[61] For software, historians face the problem that many data centers wrote programs specific to their needs, hence not practical to share with other enterprises or ministries. So these were probably never counted.

In order of diffusion in the Soviet Union following, first, military applications, second, uses of computers for economic planning, came what in the pan-Atlantic community was better known as commercial applications. In the West this class of uses of computing did more to infuse computing into all manner of human activities than any other type of IT. Manufacturing, retail, banking, and other financial industries led the way in embedding computers into the fabric of their work, beginning in the 1950s, with manufacturers and banks ahead of other industries.[62] In the Soviet Union, these kinds of applications of IT were generally called Automated Enterprise Management Systems, better known simply as ASUPs. When discussing these uses, Western and Soviet commentators referred to computer systems that ran such applications not by the name of the hardware or software but by this term, ASUP. When these kinds of uses were first initiated in the 1960s, it was not uncommon for the industrial sector to run one or very few ASUP applications on a system, which makes it understandable why one might use the term ASUP, or something like "an ASUP," or when cataloging the number of localities having data centers or computer systems installed to speak about the number of ASUPs. As in the West, beginning in the 1970s with the availability of new database software, programming languages, and even some software packages, and hardware that could handle larger programs and bigger data files, multiple ASUP applications could be linked together in a more integrated fashion, leading to many ASUP programs running in a system; but the term ASUP continued to be used to describe a system as if it only ran one application.

The reason it is important to understand this use of language lies in the fact that while we may have confidence in describing an application in one system in the 1960s and early 1970s and, thus, by counting the number of computers installed in factories begin to measure the extent of deployment, that language gets in the way by the early 1980s, because it masks the extent of reliance on computing to do the work of, say, a manufacturing site. The term leaves us only with information about the number of machines installed and not about the extent of diffusion and number of applications, hence degree of reliance upon computers. It is not enough to know how many machines were used to arrive at a true understanding of the diffusion of IT in any society. Ultimately, historians need to discuss uses, and our ability to do that for the Soviet Union and all Eastern Europe remains constrained. Another frustration is that the implementation of ASUPs became the most important Soviet initiative in computing in the 1970s, extending deep into the 1980s, really to the end of the USSR, beginning in manufacturing and eventually seeping into planning ministries.

There is broad agreement among Soviet watchers and historians that ASUP-like uses of computers originated largely with a decree of March, 1966, issued by the Central Committee and the Council of Ministers making various organizations responsible for developing IT systems for planning and auto- mated management systems to be used at the enterprise level. This action followed many years of discussions within the Soviet IT community and with economic planners about the "pros" and "cons" of using computers at the enterprise level.[63] ASUPs were intended to extend state-level economic plan- ning down to the enterprise level and so should be seen as part of a larger process in Soviet economic management. At first little was accomplished, but with the decision made in the same decade to build systems based on IBM's S/360, then subsequently these systems becoming available at the enterprise level, along with funding for them, ASUPs began to be installed, essentially beginning in the early years 1970s.

What were ASUPs? One report in the West from 1986 asserted that 75 per- cent of the processing done by these systems were "for accounting/statistical functions; 20–24 percent are [sic] for planning; and 1–5 percent are for optimi- zation," largely for "functional departments and lower level management."[64] Following Soviet practice, planning ministries dictated who were to install such systems, rather than wait for enterprises to come forward requesting such applications. Often applications were limited to producing few or even one report; therefore, these cannot be seen as comprehensive in the tasks they performed when compared to pan-Atlantic and East Asian ways of using com- puters. All through the 1970s and early 1980s, most of these applications were installed on second generation computers which were capable of processing such small jobs. Some software came from Minpribor, probably the majority, while others were locally developed by an enterprise. Observers of the process noted that this software was often badly written by programmers who were equally poorly trained, frequently creating software that proved quite unreli- able and incompatible with other software applications.[65] Table 6.1 lists some of the most widely deployed applications in evidence by the 1970s.

What kind of enterprises implemented these ASUPs? Extant evidence from various sources allows us to build a preliminary, yet reasonably confident pro- file of that community. Soviet surveys of end user communities conducted at

Table 6.1 ASUP APPLICATIONS IN THE SOVIET UNION, CIRCA 1970S

Soviet Name	Western Equivalent
Material–technical Supply	Materials Requirement Planning, Inventory
Operational Management of Basic Production	Operational Management, Production Scheduling
Auxiliary Production	Various applications in manufacturing
Accounting	Accounting and Finance
Technical-economic Planning	Production Planning
Sales	Sales and Distribution
Technical Preparation of Production	Production and Product Engineering
Cadres	Personnel
Quality	Quality Control
Finance	Financial
Normative Base	Database of Norms and Standards

Sources: Largely adapted from William Keith McHenry, "The Absorption of Computerized Management Information Systems in Soviet Enterprises," (Unpublished Ph.D. dissertation, University of Arizona, 1985): 7, but modified based on experience of American and European manufacturing practices, James W. Cortada, *The Digital Hand: How Computers Changed the Work of American Manufacturing, Transportation, and Retail Industries* (New York: Oxford University Press, 2004): 89–127.

the start of the 1970s suggested that the largest enterprises became the earliest users, as occurred in the pan-Atlantic economies. Since some Soviet factories were far larger than Western equivalents, the term *largest* means that in the Soviet Union these often concerned facilities with more than 10,000 employees.[66] By the mid-1970s, these applications were most widely installed in three types of industries: machine building, extraction, and processing, with an undetermined number in various non-industrial settings.[67] After 1975, as ASUPs were installed on Unified System (ES) computers—the S/360-Ryad styled systems—ASUPs spread to other manufacturing, process, and distribution industries.[68] An earlier attempt to identify ASUP user industries, circa 1970s, provided more granular information that suggested, as a percentage of all users of ASUPs, that one-third came from heavy, power, transport, and machine-tool manufacturing; almost another third from instrument manufacturing, automation systems manufacturing, and electrochemical; just over 20 percent of systems were in petroleum and chemical machine manufacturing and automotive production; while nearly 10 percent existed in tractor and other agricultural machine manufacturing; the rest spread across light machine manufacturing, food industries, construction, highway, and other manufacturing enterprises.[69]

As to how many installations there were census data vary and given our caveat that Soviet numbers are not to be taken too literally, extant data suggest that when compared to the West, diffusion was quite limited. Martin Cave published data in 1980 based on Soviet studies, suggesting that some

700 systems were in place in 1967, providing services to 12,000 organizations, and that the number of systems installed may have grown to about 1,000 in 1968.[70] Cave estimated that by the end of 1977, 1,200 locations had systems with another 200 installed in various ministries.[71] Another study, published in 1986 by American observers, suggested that about one-third of all Soviet enterprises with more than 500 workers had were using ASUP systems, with installations of new systems occurring at about 200 per year, a rate so slow that it led the authors of the report to predict that "only a small minority of all Soviet industrial enterprises will have their own systems by the year 2000."[72] One of the co-authors of that report, William K. McHenry, testified before a U.S. Congressional committee in 1987 that he thought some 3,600 ASUPs had been installed by the end of 1985, largely in heavy industries. Per Soviet expectations this was a disappointing performance. McHenry observed, "Of the approximately 44,000 industrial enterprises now in the USSR, only 8.4 percent" have a computer system and that "the number of ASUPs appears minuscule when compared to the approximately 580,000 enterprises, organizations, and institutions that the Soviets say have a need for computing in management applications."[73] In his testimony McHenry repeated his earlier comment about rate of anticipated diffusion, this time ranging from 200 to 300 per year.[74] His is the most authoritative account we have so far for rate of deployment, which he compiled for his doctoral dissertation in 1985.[75]

Before leaving this class of uses of digital computing, we need to acknowledge that the planning agencies also embraced ASUPs, and their attendant computers, beginning by the early 1970s, albeit slowly. Specific examples of ASUPs in planning included a system for collecting and processing data for planning and controlling the entire Soviet economy; GOSPLAN's system for automating planning calculations; another for tracking scientific and technical progress by the State Committee for Science and Technology (GKNT); an automated state statistics system by the Central Statistical Administration (CSA); for standardization tracking at the State Committee for Standards (GOSSTANDART); and at the State Committee for Prices (GOSKOMTSEN) a system to help set prices.

What is remarkable about these systems is how broadly and massively they were conceived, even though for the most part they failed to work as planned, let alone be fully deployed.[76] What the Soviets did not understand at the time was how complex these applications were, exceeding all capabilities of any IT in the world, let alone computer science to deal with—indeed not addressable in any realistic manner even in the West until the 1980s, and then only in a qualified manner. It was a case of political world-views uninformed at the top of government about technological capabilities clashing with the limits of existing technologies. The most perfect managerial system of the day in the world would not have been able to overcome the technical realities in existence to make these systems work well. Observers in the West did not realize these technological limitations as much as we did so many decades later, at the time preferring instead to focus on the resistance of Soviet line management to automation and on the lack of adequate technically skilled people to implement such systems, which, nonetheless, were realistic limits to implementation. The technological limits were masked and otherwise compounded by more visible issues. Most notably, the Soviets suffered all through the second half of the century with an inadequate IT ecosystem for computer scientists, suppliers, and users necessary to implement nationally deployed and connected IT systems.[77]

Why did deployment prove so tepid and slow? Why was the prerequisite IT ecosystem not in place? We have touched on the issue already regarding all manner of IT diffusion, and that applies specifically as well to ASUPs. Cave considered the general inhibitors evident across all computing to be the same with ASUPs. He found that Soviet observers were particularly frustrated with the inadequacies of Soviet computing technology, especially in the 1960s and 1970s. However, he also observed that ministries and enterprises consistently, indeed routinely, failed to meet deadlines for installing and using the technology, which he attributed to the lack of clear directions given to the various parties involved detailing what they specifically had to do. Building systems that were unique to various locations, rather than relying on standard software and hardware, also impeded installations.[78] McHenry and Goodman probably got closer to the heart of the problem in their assessment:

> The great difficulties that have been encountered in absorbing ASUPs are not due to any one overriding cause, but represent a confluence of organizational, economic, and political constraints on the part of users, service suppliers, and higher level organizations.[79]

They concluded that "the ASUP programs shows [sic] the limits of 'reform from above' when implemented via a complex technology. Even in what might be regarded as 'the world's largest corporation,' centrally formulated mandates and standards" were not enough to "ensure their successful diffusion and absorption."[80] Table 6.2 catalogs a list of problems McHenry identified that placed less emphasis on the larger ecosystem and more on the technology. While this list would seem to narrow the assessment he and Goodman made, it rather suggests another level of detail more in line with what some end user manager would have encountered and, therefore, should be acknowledged. In presenting this list to a U.S. Congressional committee, they supported his contention that "computerizing enterprise management is risky for managers not only because of inadequate computer services, [the finding we can extract from table 6.2] but also because it threatens some of the fundamental ways that the enterprise does business under Soviet conditions."[81] Others who

---

Table 6.2 PROBLEMS WITH INSTALLING SOVIET USES OF
COMPUTERS, 1970S–1980

---

1. A reduced scope of applications that could be implemented and longer development times because of slow, unreliable hardware, small main memory sizes, and small disk sizes
2. The inability to completely rely on the computer because of hardware failures and difficulty of obtaining service
3. Increased costs due to the necessity to maintain hardware and software locally, and to fill in the gaps left by the infrastructure
4. The difficulty of obtaining new machines and help in migrating from old ones, leading to a tendency to hang onto old systems longer than necessary
5. The inability to procure packaged software, leading to incorrect specifications and the delivery of unusable products
6. Poor user training and difficult-to-use systems which alienated users.

Source: Text is a direct quote, William K. McHenry, "The Integration of Management Information Systems in Soviet Enterprises," in U.S. Congress, *Gorbachev's Economic Plans*, vol 2, *Study Papers Submitted to the Joint Economic Committee, Congress of the United States* (Washington, D.C.: U.S. Government Printing Office, 1987): 187.

examined the flow of work from planning to execution in manufacturing industries generally shared this perspective.[82]

Soviet experience with IT during the 1980s proved just as contentious, problematic, and frustrating as in earlier decades. Much of the arc of these activities extended that of the 1970s, particularly deployment. However, there were new facets to the story as well, such as the arrival of personal computers. The history of computing in the Soviet Union in this decade needs to be seen against the backdrop of major economic problems, changes in political leadership, and a heightened interest in using all manner of technologies to invigorate a declining economy. In short, Wave One, Soviet style, showed no signs of ending. By the mid-1980s, economic decline was accompanied by growing public deficits; initial efforts to energize the economy and lower the deficit through a combination of *perestroika* and centralized economic development with the complete and enthusiastic support of Mikhail Gorbachev and the Politburo were not working, indeed one historian wrote that they "lay in ruins."[83] Beginning in 1985, the Soviet Union began spending more hard currency than it took in, which made the option of importing Western technologies of all kinds more difficult to accomplish. Students of Soviet history now look back on the second half of the 1980s—Gorbachev's period of rule—as a major turn toward new directions improving relations with the West and the United States, in particular, and attempts by this Soviet leader to modernize his nation's economy, leverage greater use of science and technology, and a reformed political system.[84] The early phase of *perestroika* began with cautious economic reforms intended to boost growth, followed by other more aggressive initiatives when milder ones did not prove effective enough in the late 1980s, such as expanding private ownership of small businesses and agriculture. Coupled to *perestroika* was the policy of *glasnost*—the opening of dialogue and knowledge—leading to a period in the late 1980s of information flows about all manner of subjects, including technology. A student of the period, Anders Åslund, quoted one Russian official from about 1987 who observed that "from the very beginning this whole system [meaning, Soviet economy] was characterized by economic romanticism, tightly linked to economic illiteracy."[85]

In early 1987, industrial production dropped precipitously by 6 percent, particularly in those industries that had been users of computers in the 1970s and 1980s. That decline led to shrinking tax revenues that might otherwise have been used to acquire more computers. Foreign aid and military budgets already were consuming some 40 percent of the Soviet government's budget, limiting availability of economic wherewithal to prime economic growth. Nonetheless, the government allocated increased amounts of funds to heavy manufacturing industries in the late 1980s in an effort to resuscitate them. Attempts to right the economy were mixed, demonstrating that the Politburo had little grasp of macroeconomics.[86] In effect, the state had run out of economically and politically acceptable options, and so the old Iron Curtain fell apart. The political will to hold things together no longer existed. As one historian reported, based on archival evidence recently made available:

> The chairman of the Council of Ministers of Hungary, Miklos Nemeth, informed Gorbachev [March 3, 1989] of the decision 'to completely remove the electronic and technological protection from the Western and Southern borders of Hungary. We have outlived the need for it, and now it serves only for catching citizens of Romania and the GDR who try to allegedly escape to the West through Hungary.'[87]

The Soviets did nothing to stop this; 20 years earlier they would have sent tanks into the country. The fall of the Berlin Wall on November 9, 1989, proved to be the culmination of Eastern Europe breaking from the Soviets, even if it did catch everyone by surprise. Eastern Europe turned its back on the Soviets, and the latter could do nothing to stop it. One student of the period, commenting subsequently on Gorbachev's rule wrote about "his chronic inability to choose a consistent course of economic and financial reform," destroying Soviet finances, increasing debts, and bringing the USSR to near default. In short, "the domestic trade and distribution system ceased to function."[88]

This decline, while partially obvious to Western observers at the time, came to a head much faster than they or the Soviets anticipated. Writing in 1987, one noted that while Gorbachev's economic reforms were intended to modernize his economy, he proceeded in part by renewing emphasis on using many types of technologies, not just IT. That strategy was not new, "since the end of the Stalinist era, Soviet policymakers have sought ways to redirect the

economy from an extensive to an intensive pattern of economic growth,"
although "his activist pursuit of the goals of modernization sets him apart
from his predecessors."[89] The Soviets were increasingly concerned about the
technological lag with the West, but could do little to reduce it.[90] IT could
hardly be used to help energize the economy since roughly only 8 percent of all
industrial enterprises even had one computer![91] The problems plaguing deployment in the 1970s persisted through the 1980s: rewards for fulfilling
targets set by central planners with none for risk-taking and innovation; conflicts between the rigid financial planning process and the need for flexibility
and agility; emphasis on modernizing existing facilities instead of building
state-of-the-art factories; and myriad problems with Soviet computing involving poor quality products, services, and training. Soviet IT innovation was
in response to national objectives not to market demands; servicing priority
industries, such as military, government agencies, and manufacturing sectors,
all with a continuing strong commitment to centralized planning and management.[92] Reforms proved insufficient and not comprehensive enough; indeed,
needed Soviet reforms were so great and fundamental that they could not be
realistically accomplished, let alone in such a short period of time as required
by the urgencies of economic, fiscal, and political problems facing the nation.

One American economics professor deeply steeped in Soviet computing,
Seymour E. Goodman, wrote in 1987:

> Three serious problems arise from the surrounding environment. First, the whole
> planning and control mechanism is a highly political process. Despite efforts to
> computerize Gosplan since the early 60s, most of the automation simply replaces
> the calculators of yesterday without changing the methods used to balance the
> plan. Are Soviet planners really ready to allow a computer to make decisions for
> them when their decisionmaking power is their most valuable possession?
>
> Second, computerization does not substantially change the nature of the
> data which is being collected, nor does it address the problem of collecting data
> in machine readable form.
>
> Finally, there is the problem of planning from the achieved level, on which
> much of the incentive system is based.[93]

This text could have been just as easily written in the 1960s, an assessment
shared by other observers.[94] On a more positive note, the U.S. CIA noted that
in the mid-1980s senior managers in the military were being appointed to
civilian positions in an attempt to bring to the civilian side of government
the economic skills perceived to be better within the military. The CIA identified 15 individuals by name transferred just in 1985.[95] Table 6.3 lists the
objectives and strategies these officials were expected to implement, an
achievement they barely had time to accomplish before the collapse of the
Soviet Union.

Table 6.3 SOVIET MEASURES TO APPLY DEFENSE INDUSTRY BEST PRACTICES IN CIVILIAN ECONOMY, CIRCA 1985–87

• Improve effectiveness of bureaucratic levers—Party's sponsorship and oversight of new technology development and strong centralized management
• Strengthen role of long-range scientific forecasting and technology assessment in economic planning—Use of management tools used by the military for decades
• Create big, goal-oriented projects to accelerate development of key technologies (lasers, computers, robotics, biotechnology)—Apply methods and organizations as used in Soviet nuclear and missile programs
• Task defense industries to help develop and apply new technology for critical civil sectors
• Organize new superagencies at Council of Ministers, led by deputy premiers—Pattern on Military-Industrial Commission to oversee and coordinate work of related ministries
• Introduce military-style quality control inspections at most important nondefense industrial enterprises
• Move top defense executives with experience in managing high technology into critical civil jobs

Source: Report prepared at the U.S. Central Intelligence Agency by Paul Cocks, "Soviet Science and Technology Strategy: Borrowing From the Defense Sector," in U.S. Congress, Joint Economic Committee, *Gorbachev's Economic Plans*, vol. 2, *Study Papers Submitted to the Joint Economic Committee Congress of the United States* (Washington, D.C.: U.S. Government Printing Office, 1987):152, and for the list of agencies staffed with military executives, 152–153.

Looking back 20 years after these events at developments of the late 1980s and early 1990s, one European economist concluded that key economic problems—which we surmise affected the rate of diffusion of IT in these years—included liberalization of the economy to such an extent that managers of enterprises became increasingly focused on optimizing personal profits and wealth without accountability for their behavior due to the lack of adequate regulations, let alone their enforcement. For instance, the Soviets had an unregulated banking industry in the 1990s—the only such unregulated financial sector in the developed world—emerging as the Communist Party collapsed and as the peripheral republics and Communist nations of Eastern Europe were breaking away from the USSR. Problems were not limited to one or few industries. Economically by 1991 the Soviets were experiencing a rapidly expanding budget deficit, a "hot house" of wealth accumulated by a few individuals without accountability to the state or a market, and reduced tax revenues flowing to Moscow from the republics as local populism and nationalism surfaced.[96] The economic apple cart had been upended, complicating the manufacture, distribution, and use of computing beyond levels achieved prior to 1985–1986.

Against this background politicos and public officials on the right (pro-Stalinist or advocates of the traditional *status quo*) and others more on the left (advocates of *perestroika* and *glasnost*) had to go about their work and politics, including making decisions about IT. One can see that with the emergence of new technologies. Harking back to a debate about computers, Marxist–Leninist political rhetoric and beliefs, and Cold War security concerns of the 1950s and

1960s, similar old issues bubbled to the surface in the new context of Gorbachev's era with the appearance of personal computers. These issues illustrated once again the chronic difficulty Soviets faced in accepting technological change. They were, of course, aware of the existence of PCs in the 1980s and moved slowly to incorporate these into their deployment of IT, but not without some pain. In the mid-1980s, the government began authorizing their use, with the Politburo endorsing a program to install up to 5 million in secondary and vocational–technical schools over the next 15 years; that did not happen, but the plans and permissions were established. A CIA analyst in 1987 reported that there were political issues related to such a decision, unique to the Soviets:

> The increased use of PCs is a potentially serious threat to party control. Equipped with word-processing software and a printer, a PC could revolutionize the *samizdat* (Soviet underground publication) process. Moreover, PCs provide plant managers with a sophisticated tool that could be used to challenge production quotas and supply figures set by the State Planning Committee and the ministries.[97]

So, it should be of little surprise that the computer literacy program approved by the Politburo was slowed by officials who viewed "widespread use of PCs as a threat to the traditional state monopoly of information in the USSR," just as Gorbachev was initiating *glasnost*.[98]

On a more practical level, regardless of political considerations, the supply of PCs was quite low, so the number diffusing into Soviet society in the second half of the 1980s and very early 1990s proved minimal. From a more strategic perspective, use of PCs in the enterprises represented a shift from centralized management and processing to decentralized operations, as was occurring with use of that technology in the pan-Atlantic community. As early as 1984, Western observers were already calling out this political issue at a time when even having a fax machine or copier was prohibited, paying less attention to the lack of supply as a more substantive issue. Yet political considerations dominated Soviet views of PCs. Soviet officials recalled the typed *samizdat* documents dissidents prepared in the 1960s; PCs could do the job of typewriters faster and produce more copies than this older technology. Polish dissidents had been doing that already. Many senior Soviet officials concluded that the risk of Soviet citizens owning PCs was too dangerous, and so the fact that supplies were limited was viewed as a good circumstance.[99] The flipside to this issue, of course, was that by limiting accessibility to PCs, the Soviet Union slowed diffusion of computer literacy in its society, which officials had long recognized they had difficulty in improving, in turn slowing economic productivity that could be improved by further use of information technologies.[100]

Again, the greater issue of how much information should flow through society came up in discussions, thanks to the arrival of PCs.[101] Officials had long constrained access to telecommunications and printed materials, making it, in the apt words of one Western observer, Loren R. Graham, "the most secretive industrial power in the world."[102] Checking accounts were still almost unheard of in the 1980s; the telephone system one of the worst in Europe; schools hardly used any information technologies, rarely even teaching basic typing skills; and university IT instruction focused more on theoretical than practical user operational matters. In combination with the obfuscation of statistics and other information about conditions in the Soviet Union, and sloppy record keeping in general, quality of information of all kinds proved far less reliable and accurate than in many other parts of the world.

These issues surfaced quickly in other ways. For example, it was not long before it became evident to the Soviets that PCs could be linked together via telecommunications, such as dial-up telephone lines, as was occurring very rapidly in the West by 1984–1985, most extensively in the United States. The Soviet telecommunications network was of such poor quality that even if officials had encouraged wide diffusion of modern PCs, it is doubtful their citizens could have taken advantage of that technology to begin their evolution into a telematic or information society. One observer in the West who studied Soviet telecommunications in the early to mid-1980s labeled it "technologically antiquated and inefficient" and its switched networks "too thin even to tie the economy and the country together."[103] Extant—and questionable— data suggest that in the early 1980s the USSR might have had 29 million telephones, as compared to 180 million in the United States, of which just over 50 percent were in homes, as compared to 84 percent in America. Conditions did not improve much in the last years of the Soviet Union. The Soviets had in their limited telecommunications infrastructure a bottleneck that few officials proved willing or able to modernize.[104]

After the breakup of the Soviet Union in the early 1990s, IT took off in Russia and in other ex-Communist countries with a decidedly market-oriented approach that led to some amelioration of the problems posed by the lack of PCs and connections to networks. Drivers of innovation in the late 1990s and early 2000s were now private companies, generating great demands for such goods and services by both individuals and the private sector that paralleled practices evident in the West. As with new technologies in the West, early adopters were those who could afford them the most, including government agencies by the early 2000s, suggesting that it was at the dawn of the new century that Russia was achieving late Wave One IT usage.[105]

Before leaving the era of the Soviets, there is the unanswered question of how diffusion occurred in the 1990s after the fall of the old Soviet Union, because the change in political and economic structures of society did not include the conclusion of Wave One IT. In fact, despite the economic and political churn, indeed economic revolution of that decade, events stimulated adoption of IT in ways that increasingly mirrored practices long in evidence in the pan-Atlantic community. Adoption of IT occurred slower and, since we are so close to the events, in a manner not so clear to students of the process.

While space and scope limit discussion about the political and economic history of the post-1991 breakup of the Soviet Union into various republics, including the emergence of Russia as an identifiable state independent of the USSR umbrella, several historic events need to be acknowledged. As the economic problems of the Gorbachev years spilled over into political ones at the end of the 1980s, leading to a relatively moderate set of political reforms up to 1991, the public came to realize that further changes would be needed because the old Soviet regime was proving itself moribund, ungovernable and insolvent.[106]

Boris Yeltsin seized power from Gorbachev in the autumn of 1991. He promptly dissolved the Soviet Union, and initiated a series of radical economic reforms that quickly transformed statist economic practices into more market-driven ones. These took root in the 1990s as managers of enterprises and other public officials saw opportunities to acquire wealth at a time when normal capitalist controls of the state over economic behavior were essentially absent. One economist living in Russia at the time opined later, "the main achievements of Russia's capitalist revolution were the peaceful dissolution of the Soviet Union, the building of market economic institutions, and privatization."[107] There were relatively clear ideas of what needed to be done and Yeltsin was prepared to take more decisive action than his predecessor. The problems the Russians wanted to focus on during the 1990s were more political; in fact, most economic issues were viewed by senior public officials through a political lens. After Gorbachev's attempts at opening up the political process in the late 1980s, he and others lacked the same clarity in policy development and implementation as on the economic front. *Glasnost*, while it expanded dialogue and debate, did not lead to necessary political reforms, despite Yeltsin's embrace of some democratic reforms, reinforced by his winning three elections in a row. His contributions to reforms and changed circumstances were essentially completed by 1993, but he remained in power until 2000. In the interim a form of parliamentary sovereignty flourished, which proved ineffective.

When Yeltsin stepped down from his position in 2000, he appointed Vladimir Putin as his replacement. Putin spent the first decade of the new century recentralizing political and legal authority in the national government, and extending his political influence through war, trade, and diplomacy into neighboring states. He continued to support the market economy, which flourished throughout that decade, but he also tolerated a high level of corruption and severe censorship, features of this emerging authoritarian structure. Several aspects of the Russian economy during the 1990s, however, carried over to the 2000s: creation of privately held enterprises; reduced governmental interference in the day-to-day operations of a firm, and, by the early 2000s, a relentless national emphasis on economic growth. This latter objective led to annual growth rates in GDP often in excess of 7 percent, fueled by rising prices for oil and natural gas (of which Russia had a great deal), expanding markets for consumer goods, and implementing macroeconomic policies in support of a more capitalist economy.[108]

Our understanding of the role of IT in Russia in the post-Soviet years remains fragmented and suspect. Nonetheless, R&D in computing continued far below world standards, as it had in earlier years, although it had always been strong—sometimes even world-class—with respect to space and military innovations, for instance, just inadequate for consumer electronics and civilian IT.[109] After the demise of the old Soviet Union, enterprises that had largely supported the military began converting to civilian forms, recognizing in the early years of the 1990s that they needed to discard many of the managerial practices of earlier times in favor of new relatively not well understood capitalist ones. They had to improve employee productivity, pay attention to profitable initiatives, and replace old incentives for meeting targets with others that improved efficiencies and responded to consumer tastes. It was a painful learning curve for many managers and their enterprises.[110] The world of IT was not immune from these tensions. The "old" lingered on, largely the state-run sector even as it collapsed in the early 1990s, with many of its leaders "privatizing" state entities. The Academy of Science, which had played such an important role in the development of Soviet computing since the late 1940s, lost its influence on technological issues as republic-level academies ascended. On the other hand, there rapidly emerged a mixed new sector consisting of private, state, foreign, and black-market IT enterprises, the latter most obvious in the sale of small consumer IT products. IT service providers emerged as some of the newest small private sector firms, filling a desperate need for technical support for existing data centers.[111] Western enterprises were encouraged to enter local markets as well. For example, IBM established a wholly owned subsidiary called IBM USSR Ltd. in Moscow in 1991, and began manufacturing PCs in the country in 1993. In the next few years IBM sold a few large systems to various state agencies.[112]

Little is known about the extent of deployment of IT in the 1990s. As Russians acquired PCs with modems,[113] they began accessing the Internet, particularly in the late 1990s. However, private networks had flourished in a limited fashion in large urban centers since the 1980s, in such cities as Moscow and Leningrad, within agencies and large enterprises. Some of the earliest available data describing Russian networks (circa 1998) suggested there was one computer connected to the Internet for every 2,189 residents. Observers speculated variously that there were between 600,000 and 2.5 million Russian users of the internet. Russians had 4 percent of the world's Internet domains. That last statistic would suggest that the Russians ranked 23rd in its use of the Internet at the time. Extant data on number of hosts demonstrated that use of the Internet took off as in most countries in 1994–1995, as measured by the number of hosts connected to the Internet.[114] This development is quite remarkable given the fact that the old Soviet telecommunications infrastructure was, in the language of a UN report, "among the least developed of any country" with "penetration rates for telephony . . . at Third World levels."[115] Almost half the telephone lines in Russia were connected to internal networks within enterprises and agencies (approximately 45 percent), the rest largely in homes and other organizations through more public systems and largely in urban centers. The poor quality and extent of the network served as a break on deployment of PC-based Internet use in the 1990s and early 2000s, with most users accessing online data largely through off-line e-mail and Usenet groups, not through reliable, full-time online access. By 1998 urban concentration of users had expanded quite extensively, however. Some 70 percent of all nodes were concentrated in just three cities: Moscow, St. Petersburg, and Ekatarinburg; the capital city alone accounted for 40 percent; vast tracks of Russian territory did not have IT of any kind. Given where the population lived, only some 40 percent could even hope to have access to the Internet at the turn of the century.[116]

The timing of the early adoption of the Internet in Russia coincided with two other developments. The first was the emergence of technical features of the Web that made it accessible to people around the world, most notably user-friendly browsers. Second, controlling forces of the Soviet state evaporated, creating for a brief period of time a government and circumstance in which the political leadership did not have an appetite or capability to limit access to the Internet. That confluence of the two concurrent developments contributed mightily to the general diffusion of the Internet in large Russian urban centers. This occurred as many privatizing enterprises were also installing modern Western computers and software. It was a period we can think of as a techno-perestroika backstopped by a *glasnost* political environment.

As Putin clamped down on traditional media (print, radio, TV) and other sources of expression in the 2000s, the flow of information through digital means increased, although hard data to document the extent are not available. For those with access, all was not as bleak as circumstances might suggest. One student of Soviet telecommunications suggested that

> The Russian Net—built upon the cultural tradition of personal *blat* networks—
> served to extend and empower those social networks by "routing around" the
> hierarchical dominance of the institutional order, while providing a mechanism
> for the exchange of much-coveted private information. In this sense, the virtual
> space that the Net created—cyberspace—acted as a kind of surrogate civil society,
> a space that allowed for the unfettered pursuit of personal contacts and group
> interests outside the strictures of the Soviet institutional order.[117]

After the coup that led to the demise of the Soviet Union in late 1991, the importance of the Internet had increased, of course, along with more traditional telephone networks becoming more obvious. For instance, they were used by some activists to organize political opposition to various regimes and programs during the halcyon days of the Duma's power in the 1990s. Notions of Internet-based e-business and e-government, however, did not appear to a great extent in the thinking of many Russian citizens and managers in the late 1990s; those developed in nescient form in the 2000s.

The Y2K concerns bedeviling IT worldwide in the late 1990s existed in Russia, where work to mitigate its potential effects began quite tardily, following a formal decree of the national government addressing the problem in May, 1998. In fairness to the Russians, we should note that some public and private officials had begun to consider the issue as early as 1996. For our purposes, the Y2K issue resulted in some statistics to surface about Russian computing, such as the fact that there were well over 3,000 national government systems that might have problems; to be sure, a small population of mainframes, but nonetheless evidence that computers had continued to be installed all through the 1980s and 1990s. One American expert estimated that the Russian national state administration had only 25 percent as many mainframes as did the U.S. Government. Evidence presented to the Russian Duma regarding the number of legacy systems suggested that between 1967 and 1991, some 15,000 ES systems (IBM S/360–370 styled) had been produced and of these between 3,000 and 4,000 were still in use, the majority by the military–industrial sector. As part of the Y2K debate, one learned that the Soviet airline, Aeroflot, was an extensive user of computers for the same applications as other airlines around the world, along with the ministry that ran the train systems (MPS). The 29 nuclear reactors operated with analog computers, and thus not subject to Y2K problems to the extent as digital systems. There was, however, much concern because most of these systems were second and third generation Soviet computers of questionable reliability. Banks had also installed computers. Between 300,000 and 500,000 national government employees used Novell Netware software, with at least a third (some estimated as much as 80 percent) pirated copies, thus not on any official inventory to be remediated as part of the Y2K initiative.[118]

Finally, one should recognize the increasing role of software sales and use as a source and cause of diffusion of IT in post-Soviet Russia. The Soviet Union had a reputation within the global IT community of producing excellent programmers, a accolade dating back to the 1960s, which diminished in the 1970s and 1980s as programmers did their work largely on systems one generation behind those available in the West. With the flood of Western PCs into Russia in the early 1990s, local programmers became current on the technology to such an extent that most programmers switched away from mainframes to PCs. In fact, by 1997, they and other users were acquiring approximately 1.4 million systems from all sources, both local and international. From that community sprang a large number of small software development and support companies, expanding until 1998 when a severe financial crisis for the nation led Russian officials to default on its government's debts. That event precipitated a sharp decline in the value of the ruble, resulting in a massive increase in the cost of imported PCs while disrupting sales of software and services to the West. Following the crisis, the software and commercial IT industries came back, and grew.

By 2001, one estimate had it that for every 100 citizens 5 had PCs, hence also modern Western or Russian software. To put that deployment in context, the comparable American statistic for that year was 62 for every 100 people. More important to note, however, was that many businesses and other organizations relied more on PCs now than on old Soviet mainframes.[119] Diffusion came from the expanded number of employees of these firms and training of students. In 2000, there were between 50,000 and 80,000 programmers. Student data suggest that in 1996 some 8,000 university students were studying computing and 25,000 in 2000. Regardless of the precise accuracy of the statistics, the trend is obvious—knowledge of computing diffusing through the economy, much in the manner evident in the West.[120] Members of the IT industry had created their firms and industry from scratch in the 1990s, as it had not existed before. Their private sector for-profit industry kept growing in the 2000s, largely situated in major cities, most notably Moscow, but also in St. Petersburg and Novosibirsk, creating in those urban centers critical mass much as existed in Silicon Valley near San Francisco, California, albeit as smaller regions. These were also much smaller firms than the old Soviet enterprises, with 50 to 250 employees the norm.

It is tempting to declare that with the arrival of all these programmers and their PCs, and the emergence of a market-driven economy in the 1990s, that Russia was now grandly marching into Wave Two IT deployment. We have to suspend that judgment until further evidence surfaces for the case. For one thing, all through the 1990s and 2000s, software piracy was as rampant as in the Soviet period, and so obscured statistics on volumes installed and uses to which IT was put. Telecommunications infrastructure had yet to approach the quality had by Wave One nations. The legal and regulatory infrastructures remained problematic, more equivalent to what existed in many underdeveloped countries. Not until the early 2000s did the national government begin implementing policies nurturing growth of the software and IT services industries. This was still a time when national investment in IT as a percentage of GDP lagged that of central and east European countries.[121] In short, the Soviet/Russian road through Wave One IT was a long one, extending over at least five, possibly six, decades, as compared to the pan-Atlantic community, which accomplished that feat variously between four and five decades, and the most advanced Asian economies, largely in five.[122]

INFORMATION TECHNOLOGIES IN THE GERMAN
DEMOCRATIC REPUBLIC (GDR)

The diffusion of IT within the German Democratic Republic (GDR) paralleled many of the experiences of the Soviets, even a few common to the West. Others were unique to the nation. The first can be understood as the consequence of Soviet dominance of Central European political, military, and economic affairs.[123] The second can be attributed to the interactions between East and West through our metaphorical chain link fence. The third—unique circumstances in the GDR—can be explained by the fact that the Soviets allowed its neighboring governments to set local policies and run their own nations and economies, so long as nothing was done that conflicted with international, national, military, economic, and social policies and political beliefs of the U.S.S.R. When they did, the Soviets did not hesitate to send in their tanks and soldiers as they did in East Germany (1953), Hungary (1956), and Czechoslovakia (1968). In the case of the GDR, information technologies were used as an important element in the national government's strategy for economic and political survival and vitality, beginning in the 1960s, and that had become central to its actions in the 1970s and 1980s. In short, IT played a more visible, significant role in the GDR than in the Soviet Union. Yet, the extent of deployment remained less than, for example, in its neighbor, West Germany. Central to any appreciation of events in the GDR is an appreciation of the tensions and political dynamism that existed between the Soviets to the east and the Federal Republic of Germany (FRG) to its west.

In the past decade there has been a renewed interest in explaining the role of IT in the German Democratic Republic, often also known in the United States and Western Europe as East Germany during the Cold War, with major collections of proceedings having been published. Collectively they provide another layer of detail beyond what was known even just a few years ago, although the role of IT in the country has not profoundly been described differently, yet. Perhaps the most important finding of these newer studies is that there was a great deal more activity underway than previously known, particularly in universities and in other public institutions.[124]

Following the practice of the time, and largely by non-Communist pan-Atlantic citizens, writers, and officials, the GDR was frequently and conveniently called East Germany, since the names of both Germanies were confusing with the communist one using the word "democracy" in its title and the non-communist government the words "federal" and "republic," which one might have thought should have been reversed with the communist country using the West's name. FRG and GDR were unclear then except to those concerned professionally with the difference. So, just as the term Eastern Europe is used in this chapter, so too East Germany synonymously with GDR, following the same practice of the period 1950s–1980s. That use helps to distinguish clearly differing patterns of behavior between the two German states and also with other governments in the Soviet orbit.

To a lesser, but not insignificant extent, one can say the GDR—East Germany—had technological issues and relations with the other members of the Soviet sphere. As part of Stalin's response to the Marshall Plan he organized the Council for Mutual Economic Assistance in 1949, better known as Comecon that became the stage for many discussions and economic interactions within the Soviet sphere, including the GDR. It served as the institutional vehicle for coordinating much economic—mostly trade—relations within the Soviet orbit. Comecon served as the context, the field upon within which much occurred with respect to the diffusion of IT in Central Europe until the end of the Soviet era.[125] Comecon was not a trivial enterprise. By the early 1970s, one-tenth of the World's population lived in its sphere and it functioned on 20 percent of the world's land mass. It generated 20 percent of the World's national income and a third of its industrial output. Comecon's members represented collectively a larger entity than all of Western Europe, albeit a loose economic confederation at times. It comprised a potpourri of different nations and economies.

With respect to technology, as early as the start of the 1960s, the GDR had both modern and very backward industries, efficient and inefficient enterprises, and shades of variation in the operation of government reflecting essentially the Soviet model of centralized management of local economies. The economic underperformances described earlier of the USSR—the biggest member of Comecon—occurred in all its other member states as well. Public policies determined the interactions among science, technology, production, and political objectives in all its member states. The Soviet's used Comecon to further its ends, including in the GDR. As in the Soviet Union, East German public officials were bent on putting science and technology on an organized and solid footing just as the Soviets had established their State Committee for Science and Technology. In GDR's case the local government established the Ministry of Science and Technology to tie together in a more organized manner earlier fragmented attempts to coordinate national science and technology policies and programs.[126] The results proved disappointing for GDR and its neighbors for largely the same reasons noted earlier regarding the Soviets, albeit with local variations. Policies in one were mimicked in the others to address performance. Thus, for example, when the Soviets decided in the second half of the 1960s to begin decentralizing management of its economy, so did a few other Comecon countries, which led to some improvement in economic performance. The GDR, however, opted to maintain tight centralized control of its economy.[127]

With respect to West Germany, there were dramatic differences. The FRG was a medium-sized nation with a large-enough internal market to support many industries, including coal. The GDR was just slightly more than 25 percent of the size of FRG in population, with some coal and uranium for energy. Prior to Soviet occupation, East Germany had a sophisticated and relatively strong industrial base, although much of the corporate management of these firms were headquartered in West Germany, thus, lost to the GDR in its bid to rebuild its local economy after the war. Much of its industrial might centered around machine manufacturing and machine tools. Indeed, some 80 percent of the old office-machinery business of pre-War Germany resided in what was now East Germany.[128] Many of the technologists and scientists who had lived in the area, and could normally have been expected to revive R&D, scientific research and technological innovations, largely located in the industrialized Berlin metropolitan area, were in the Western zone and thus not available to the GDR, or had been moved to the Soviet Union. Ultimately, the biggest difference was, of course, that FRG became a democratic state with a market-driven economy, discussed in chapter three, while the GDR went down the Communist path.[129]

As in so many countries, devastation of war influenced activities of East Germans in the mid-1940s. Russian occupation had been hard and brutal, far more so than that in West Germany by the French, British, and Americans. In addition to suppressing citizens, the Soviets extracted reparations in their zone that local Germans had to pay in the form of assets and goods until the second half of the 1950s, imposing a severe drag on local economic development. Soviets moved about 15 percent of East Germany's industrial base to Russia. Direct war damage to industrial capacity hovered at about 22 percent. Both statistics were comparable to the experience of West Germany. More buildings and homes than factories had been damaged by the war. That is why the Soviets could move entire plants to Russia and put them to use quickly. Some industries lost about 25 percent of their capacity, largely due to war damage, including machine building, automotive, and electrical industries. Soviets dismantled facilities in many high-technology industries, including metallurgy, machine building, vehicle manufacturing, electrical goods, mechanics and optics, and soda chemicals the most, leaving behind about 25 percent of the local capacity to manufacture that had existed during the war.[130] One highly influential consequence of the division of the country into zones was that economic and technological relations between East German enterprises and those in West Germany were disrupted. When coupled to the fact that so many senior leaders and scientists were living in the Western zone, while the Soviets occupied the Eastern zone, a fundamentally different situation existed in what became the GDR, exacerbated by the movement of technology, facilities and technical labor out of Germany into the Soviet Union.

Expropriations were central to the Soviet's strategy for its own economic recovery, but at great cost to the tiny East German zone. The largest transfers occurred in 1946 with best estimates suggesting it involved over 3,000 individuals and their families from all forms of technology and science; however, most returned to Germany between 1952 and 1956.[131] One student of East Germany's postwar science and technology concluded that "the loss of so much technological talent was especially damaging to the Soviet zone of occupation and the GDR, where the need for reconstruction and technological modernization was so great."[132]

Not until after the GDR was established in 1949, allowing Germans to begin running their own country within the Soviet Stalinist orbit, could this part of Germany begin shaping its own technological destiny, hence also its response to the concurrent emergence of digital technologies. Early initiatives centered on adopting Soviet-styled centralized planning and allocation of resources for R&D and manufacturing tied to national priorities and the political priorities of local Communist leadership. Others have written about the period of startup and crisis in science and technology in the GDR from the late 1940s through the 1950s that need not detain us.[133] However, in this period, Soviet-styled organization of science and technology took time to set up, all the while many East Germans migrated to West Germany, perhaps as many as 3.5 million before the Berlin wall was erected in 1961 to stop this outflow of people. Many of the migrants were young, educated, and experienced in scientific and technological matters.

The story of computing in East Germany began essentially in the early 1950s at the Carl Zeiss optical firm, located in Jena, an enterprise with deep skills in fine mechanics. It built the first computer in the GDR, which, albeit slow and primitive, became operational in 1955, called the Oprema (Optische Rechenmaschine). The computer was constructed in about nine months using some 25,000 component relays and other existing parts. However, engineers at Zeiss had been designing a computer since at least 1952, aware of developments in digital computing occurring in the West.[134] Over the next several years, the engineers who built this first machine, which proved successful in performing mathematical calculations, began work on a second system to succeed it, called the Zeiss-Rechenautomat (ZRA 1); it became operational in 1958. Management intended it to be used by its staff, but that could also be mass produced. In time Zeiss manufactured 32 copies.[135] Simultaneously, some preliminary work on a computer took place at the Technical University of Dresden, culminating in the construction of the D 1 in 1956. Over the next several years it became the basis for the construction of other systems, including the D 1–2 and D 4.[136] East German scientists and engineers were aware of various computer projects underway in West Germany, and had visited the industrial fair held in 1959 in Hannover, where they observed that they were behind. This conclusion was reinforced by visits to other West and East German industrial fairs in the early to mid-1960s where they learned more about the technology from Western exhibitors, bringing home information and ideas.[137] However, after the construction of the Berlin Wall, exhibitors of computer equipment at East German industrial fairs did not display their newest most advanced products, rather typically those one generation behind—Cold War politics reduced information sharing among technologists.

Senior political leaders in the GDR realized in the 1950s that they would have to ramp up modernization efforts in their industrial base in order to compete politically and economically with the Soviets to the east, and with the FRG to whom they were constantly compared. They responded with a strategy of building self-sustaining capacities to make the nation able to avoid relying on either neighbor for economic and political survival. GDR's strategy evolved toward technological autarkic approaches and actions taken in the 1960s in such areas as petrochemicals, semiconductors, and computers.

As in so many other countries, officials saw development and manufacture of semiconductors less of a way to learn about computers and more as a path to competitive electronics industries and modernization of local economies. GDR moved quickly on the heels of the Americans. In 1952 engineers at the VEB Factory of Electrical Components for the Communications Technology (VEB Werk für Bauelemente der Nachrichtentechnik "Carl von Ossietzky," or WBN), near Berlin, began work on semiconductors. WBN was renamed the Institute for Semiconductor Technology, staffed with 74 technologists and a total of 625 employees by the start of the 1960s, initially giving them enough critical mass to keep up with developments in the West. WBN had produced small quantities of semiconductors in the 1950s when compared to the West (by several orders of magnitude with the United States, for instance), largely because of few resources to manufacture and as one student of its work suggested, inadequate appreciation of the technology's complexity and what it would take to be competitive, or to scale up production.[138] Nonetheless, all through the late 1950s and 1960s, WBN expanded operations and its work became part of various central economic plans. To improve, engineers simultaneously did three things: conducted R&D, acquired semiconductors and information from the Soviet Union, and bought technology from the West, most notably from the FRG, United States, and Great Britain, all with the intention of staying current and acquiring enough supply without becoming too reliant on either East or the West—early evidence of GDR developing an autarkic strategy.[139] As in so many other countries, engineers had access to Western literature, but lacked hands on experience in the manufacture of semiconductors.

The Soviets proved less forthcoming in collaboration than the East Germans wanted, probably the result of nervousness about sharing too much information relevant to its military–industrial sector because of concern that East German scientists might take such information to the West. Until the Berlin Wall went up in 1961, the West served as the most important source of information for the GDR, particularly West Germany. While the Americans discouraged and outlawed the transfer of its most advanced technologies to Europe, including to the GDR through its COCOM restrictions,[140] GDR turned more aggressively to the British for their technology in 1959–1960, sending a delegation to view

British developments and then acquired specialized manufacturing equipment. Others still came from West Germany, but did not acquire any turnkey plants from any nation.[141] Statistics on volumes of manufactured devices, although not fully complete or reliable, did lead students of the effort to conclude that as of the early 1960s, semiconductors had had minimal impact on GDR's economy, let alone on any industry.[142] Construction of the Berlin Wall slowed access to Western technology, reluctantly forcing the GDR to increase its reliance on Soviet methods and technologies, despite Russian reluctance to collaborate much. At the time, the Soviets had an impressive record of results: their launch of the Sputnik satellite, atomic bombs, and ballistic missiles, all proof points that Soviet methods of doing technological development perhaps were not as backward as one might otherwise think, if quite different from that of the West. Long term it proved an unfortunate decision because the Soviets were not able to sustain this level of performance; however, given the few options available to the East Germans, they really had no alternative partners.

In 1961, GDR established a formal and extensive research and development plan for various technologies, called the Plan Neue Technik (New Technology Plan), mimicking Soviet-styled planning from top down through enterprises (called combines in GDR).[143] Such planning helped focus the attention of officials on the reality that it was becoming more difficult to acquire knowledge and technology in advanced electronics. They also knew that between 1958 and 1961 some 1,600 scientists had defected to the West, a large number given the size of the GDR. The erection of the Berlin Wall proved profoundly influential in the direction of GDR's development and use of semiconductors and computers. One student of the situation explained the turn to the East in development as a new experiment in technological development to meet the needs of the nation:

> Although it changed character over time, the experiment undertaken in the GDR in the 1960s was driven primarily by ideological commitment and by political expediency. Notions of technological excellence, of the power of scientific planning of technological change, and of the capacity of the GDR to reform its system from within underpinned many of the programs developed in the 1960s.[144]

GDR became increasingly isolated from the West and partially held back by the Soviets, which complicated execution of its intentions. Industrial enterprises wanted access to computers and the latest technologies. Political leaders recognized the need to supply them without having to use too much of their limited supplies of foreign currencies with which to buy these; hence, growing pressure to expand internal development of supplies. The Soviets were also becoming increasingly concerned for reasons of security with the rapidly expanding use of ever-more advanced IT in weapons by the West, which reinforced their penchant for not wanting to share too much information or devices with its member Comecon partners. In this context, the aging

General Secretary, Walter Ulbricht, launched his New Economic System (NES) in 1963, an initiative that continued to the late 1960s. NES led to large increases in the manufacture of semiconductors. Meanwhile German computers were installed to do economic planning and to help manage large industrial complexes.[145]

Collaboration with the Soviets deteriorated in the mid-1960s, while the GDR attempted to create internal capabilities to manufacture all manner of computing (mainframes, peripheral equipment, software). As the evidence from experience in the 1960s and 1970s revealed in hindsight, the country was too small to launch such a major comprehensive initiative. By the early 1970s, there were five semiconductor and computer manufacturing and R&D facilities in the country. To put that number of industrial facilities in context, in all of Comecon there were 36, of which an estimated 15 were known to have been in the USSR (probably too low a number), 6 in Czechoslovakia, 4 in Poland, 3 in Hungary, 2 in Bulgaria, and 1 in Romania. There were at least as many facilities making semiconductors, other subassemblies, parts, and peripheral equipment. At the time across the entire Comecon, some 30 different lines of computers (systems) were in production, with about 130 different models within those systems. At least 35 were in "mass" production.[146]

It was in this period that the debate over IBM's S/360 took place, resulting in Comecon making the decision in 1969 to use IBM's technical standards, then negotiated arrangements with its members to develop portions of the Ryad system based on the IBM technology. GDR received the assignment to build the R-40s (otherwise known as the ES-1040, introduced in 1973) with Poland and the USSR making ES-1030—resulting in all three making medium-sized models of Ryad computers; production began in 1972.[147] But leading up to the implementation of the plan was much debate behind closed doors as each nation's IT decision makers determined what they wanted. For instance, Hungary, Bulgaria, and the GDR favored relying on IBM S/360's approach, while the Poles promoted its ODRA approach, which relied on the Britain's ICL architecture. The Czechs had their own computer development initiative which they promoted, while the Romanians were the most resistant to the Soviet plans, trying to stay wedded to the computing technologies coming from the French. In 1969, a pan-Comecon committee in Moscow began integrating the work of all these nations, with East German support. The process went well enough such that later in the 1970s, a similar strategy for a pan-Comecon minicomputer was launched called the System of Mini-computers (System der Kleinrechner, or SKR), which were based on DEC (VAX) architectures. In both instances, within IT circles these systems were called "IBM analogues" for mainframes and for the minis, "DEC analogues." In the 1980s, the strategy was implemented a third time for PCs, based on IBM's approach.[148]

To appreciate GDR's predilection for favoring Western technology, in these instances American approaches, one cannot simply point to the chain link fence phenomena of being physically closest to the pan-Atlantic community, because already East German officials were formulating their strategy of maintaining independence from West and East. Rather, one should look to the considerable experience East Germans had with computing's technical features when compared to their Comecon neighbors. While a comparison with their experience to that of the West is not impressive, within their pan-Comecon world it was and their technologists understood the value of the Western approaches which they knew could be taken, retro-engineered, and adapted to their needs. From experience in the 1960s, reinforced by those of the 1970s, they could see that adopting Western approaches was quicker than embracing some as yet undeveloped, new untried Soviet architecture. It was a delicate task not always achieved because, on the one hand, the East Germans admired and wanted to rely on American and West European technologies and knowledge but, on the other hand, had to distance themselves from the political and social systems that generated them. As German historian Simon Doing has wisely pointed out, "technology could not remain a neutral artifact in the bipolar Cold War world."[149]

By the end of 1970, one estimate of the number of installed computers in all of European Comecon—6,450—amounted to about 6 percent of the World's installed supply. While these numbers are suspect, if for no other reason than Soviet military systems are not included in these statistics, table 6.4 at least suggests the proportions of installations compared to other countries. GDR's share put it ahead of all its neighbors, including the Czechs who were always considered quite advanced by East European standards yet understandably behind the far larger Soviet Union. When compared to the West, the entire region was far behind in extent of diffusion of IT.[150]

The same compiler of these data calculated the number of computers per million people by country as yet another way to measure deployment. By that method, for the Soviets were at 23 and for the GDR 21 systems/million residents; the Czechs were at 16, Hungary 8, and Poland at 6; all the other Comecon nations had lower rates of diffusion. The global average was 31.[151] Soviet diffusion was 15 times lower than the Americans and, because most of these systems were smaller or less advanced than those of the West, the amount of processing capacity proved much lower than the table otherwise suggests was the case.

The story emerging from such data for Comecon nations as a whole, and specifically for the GDR of the 1960s and early 1970s, paralleled the Soviet Union's experience of a slower rate of diffusion of computers into local economies. The compiler of the data in table 6.4 visited Comecon countries in 1972, discussed diffusion of IT with various officials, engineers, and users, reporting that "the Comecon countries had a late start in the production of computers and their progress was slow," producing only 6 percent of all computers in the World yet one third of its industrial output. Such slow IT diffusion meant that

Table 6.4 ESTIMATED NUMBER OF INSTALLED COMPUTERS, 1970

Comecon Members: 6,450 Systems

USSR	5,500	Hungary	85
GDR	360	Romania	50
Czechoslovakia	235	Bulgaria	40
Poland	180		
Major Other Adopters: 106,520			
USA	70,480	France	4,570
FRG	6,710	Italy	2,580
Japan	5,790	Canada	2,280
United Kingdom	5,070		

Sources: Data collected in *Electronic Data Processing in the Soviet Union and Other East European Countries* (Brussels: East-West, 1972): 10, 29 along with other data presented in, J. Wilczynski, *Technology in Comecon: Acceleration of Technological Progress through Economic Planning and the Market* (London: Macmillan, 1974): 114. GDR's inventory is accepted by Erich Sobeslavsky and Nikolaus Joachim Lehmann, *Zur Geschichte von Rechentechnik und Datenverarbeitung in de DDR 1946–1968* (Dresden: Hannah-Arendt-Institut für Totalitarismusforschung, 1996): 74.

a time lag in adoption emerged, ranging from 10 to 16 years behind the United States by 1970, and "at least" 6–12 years behind Western Europe. In the case of the GDR, that lag behind Western Europe (of concern to GDR's leaders) was at least 6–7 years, of less importance the lag behind the United States, which was 10–12 years. These slower rates of adoption applied to Czechoslovakia and the USSR as well, while for Hungary, Poland, Bulgaria, and Romania, the situation proved demonstrably worse, with an average lag behind Western Europe more in the range of 10–12 years and 15–16 with the United States.[152]

As in the Soviet Union, the longer a lag lasted, the older the stock of installed equipment that continued to be used, which meant the fewer newer or additional applications that could be installed. Reports of poor quality and inadequate support and maintenance mimicked complaints in the Soviet Union, with responsibility for these scattered across multiple agencies and organizations. When enterprises reluctantly adopted computing, they did so for similar reasons as well. The results were the same too. Demand for systems built in the country remained low through the 1960s–1980s. This even occurred in industries that were almost perfect fits for this technology and ideal uses, such as the relatively simple calculations of the banking industry, which remained highly manual in its operations all through the 1970s and 1980s.[153] As one student of IT in the GDR argued, "computing hardware was very difficult to come by, particularly as Robotron exported the largest share of its production to the Soviet Union," compounded by poor quality software for what machines there were, and none available for start-up entrepreneurs.[154] That lack of interest was in part state driven. For example, for reasons not fully clear, the government did not place a high priority on investments in computerized numerical control (CNC) technology, yet given the nature of heavy industry production all over the Comecon it appears in hindsight as an obvious candidate. That would require focusing resources on narrower projects, such as CNC, at a time when the government wanted to implement full autarky across all of IT. Meanwhile, the state moved forward to produce S/360-like systems per the Comecon plan in the 1970s, despite the fact that some engineers had expressed a short-lived interest in developing their own technologies, or simply acquiring Western products. Ultimately, the East Germans favored the technology developed by IBM in its Soviet variant.[155]

There is some question as to whether interest in computing by GDR's users languished in the early 1970s. The decline in support of IT by the government in the early 1970s was probably due to a combination of factors, such as the rise of a consumer society then underway and the simultaneous expansion of the welfare state during the Honecker era. Officials reduced their investments in IT to support the provision of such consumer goods and services as housing, health care, and such foods as fruit and coffee.[156] One could argue that a new surge of interest came when, in 1977, the Central Committee of the East German Communist Party (SED) increased investments by billions of marks. It shifted R&D labor and production to almost every kind of IT one could produce in those years. That support for IT continued to the end of the GDR in 1989. Officials wanted to create technologies that East Germans would use, to simplify central economic planning, lead to improved—read internationally competitive—goods, and to enhance the GDR's trading capabilities with other Comecon nations. At most, GDR made only nominal progress against each objective. Given its small market, both at home and within Comecon, it could only operate small production runs of any IT device, which limited the opportunity to leverage economies of scale, while simultaneously constraining its ability to improve a few focused products as opposed to a broad range of devices. Relying on existing and older technologies also meant GDR's IT industry was condemned always to be behind the West and, in some instances, even behind the Soviets.

The institutions making decisions operated increasingly in isolation from realities faced by users and manufacturers of IT. By the early 1970s, and extending to the end, the amount of information that either was presented to senior public officials or that could get to them proved so limited that they did not necessarily understand the practical problems of implementing their four objectives. One observer of the process noted that as a result, their view "remained stagnant or, at best, were subject to infrequent lurches when a top decision maker had a change of views."[157] Industry experts understood that IT was a rapidly evolving set of technologies, and they occasionally wrote about that circumstance in public, but no evidence exists yet that senior officials took note of their observations, let alone acted upon them.[158] More than in any other Comecon country in the late 1970s, East German officials viewed development of IT as central to the welfare of their nation.[159] Their concern about the nation's dependence on foreign raw materials and energy and the poor competitiveness of the GDR's economy in its ability to raise hard currency with its products to pay for energy and raw materials first expressed in the 1960s continued to bother officials in the 1970s and 1980s. Their thinking remained locked on the notion that IT could be used to make its products more advanced and competitive. Even as late as 1986, Erich Honecker, still GDR's General Secretary after 15 years at the helm, spoke about the need for "expanding the GDR's maneuvering room in trade policy and assuring that the country cannot be threatened economically."[160]

All through its history, GDR never seemed to have enough hard currency with which to buy adequate supplies of raw materials and energy to meet the needs of its priority enterprises, let alone to improve the productive capacity of the economy, or the standard of living desired by its citizens who increasingly questioned their situation when compared to that in West Germany. Although recent historical research suggests that commercially oriented enterprises managed to develop, albeit at a lesser pace than in the West.[161] GDR continued to build up its internal IT capabilities in the 1970s and 1980s, despite problems, even selling some of its IT products at a loss in exchange for hard currency, needed especially for supplies of oil from the Soviet Union which usually wanted to be paid in hard currency. The cost of energy increased in the second half of the 1970s and remained high in the 1980s as world prices rose, putting further strains on GDR's balance-of-payments and demands on its supply of hard currencies.[162]

Official positions on autarky and the critical role of IT remained relatively unchanged in the 1980s and, if anything, actually hardened into a more determined stance as the economy deteriorated along with internal public support for the GDR. Odd consequences of this increasingly arthritic stance surfaced. For example, the Carl Zeiss Jena (CJZ) combine was forced to produce ever-more current computer chips at enormous cost to the nation, even if it made it the first socialist enterprise to make a full range of microchips. The expensive to develop 1 megabit DRAM (made in mid-1980s) was rushed into production so that Honecker could present one to Gorbachev as evidence of the modernity and prowess of the GDR. The German Robotron combine also emerged as the leading manufacturer of computers in the GDR by the late 1970s, hence an important exporter of 16-bit computers. Comecon customers wanted more advanced systems than those, illustrating yet again the kinds of disconnects that often existed in many industries in Communist countries between supply and demand not benefiting from the kind of precise synchronization characteristic of market-driven economies. In the 1980s, the GDR became an important supplier of personal computers as well. GDR also produced most of its own software, often copied from Western products, and most notoriously, IBM's and, of course, without the firm's permission.[163]

By the early 1980s, there existed four major combines, each with a number of enterprises, producing products for internal and external use: Carl Zeiss Jena (CRJ) with 19 enterprises; Mikroelektronik Erfurt, with 20 enterprises; Elektronische Bauelemente Teltow (EBT), with 11 enterprises; and the Keramische Werke Hermsdorf (KWH), with 18 enterprises. All reported to the Ministry for Electro-technology and Electronics which in turn served the Council of Ministers; the latter the Politburo. All combines produced microelectronics and most many other products—a substantial number of organizations for a country not much different in size than the Netherlands.

Extant data exist on PC diffusion that suggests the broad diffusionary patterns evident in the 1980s. In the years 1980–1985, it appears the GDR manufactured about 12,700 PCs, exporting about 45 percent of these. That meant only 7,000 of these became available for local use, a tiny number by Western standards. In 1986 and 1987, nearly two-thirds of production was exported with 30,800 made in 1986 and over 49,000 the following year. In 1988 a similar percentage were exported on a production run of 57,400 and in 1989 almost the same percent, leaving less than 30,000 for local use.[164]

Diffusion data for mainframes demonstrate even more emphatically the limited and late start of this technology's appearance and diffusion, illustrated by table 6.4. The absolute numbers are not as important as their relative size one to another. The proportions coincide with what is known about Communist computing and that of the West. Despite the limited number of systems in the GDR, and the discussions in this chapter about its diminutive use of computing, when compared to some of its neighbors, the East Germans were ahead in its early deployment of such systems. Poland always is acknowledged as a large economy, hence its numbers suggested a relatively slow start too; Hungary is normally seen as aggressive in its use of digital technology, but again the data suggest that maybe such was not quite the case in the beginning but more so later. When compared to the West, there is no contest—Communist Europe began slowly with the exception of the USSR.

Between 1972 and 1978, a total of some 380 mainframes were fabricated in the GDR and of these only 203 were installed within the nation. In the years 1978–1982, only 150 were manufactured, of which less than 25 percent remained in the GDR. The final set of data for these comes from 1982–1983 when 172 systems were built, of which about 90 percent were exported—all to the USSR.[165] Some mainframes came from the Soviet Union to the GDR, but not many. In the years 1965 through 1970, the Soviets sent to GDR probably 27 computers; between 1973 and 1976, 51 ECs (Ryad class systems), another 100 between 1976 and 1980; 79 EC computers between 1980 and 1985; and 24 more in 1986. Across the period 1979–1986, the Soviets exported to the GDR a total of an additional 119 other types of computers.[166] In short, for many years, GDR exported about two thirds of various IT products to the Soviet Union and imported very few.

One can opine on whether the combined numbers of machines made by either GDR or the Soviets is extensive or not, but even these data are not complete because there were other systems that were bought, some smuggled, from the West as well of an undetermined number. There was no extensive program or strategy for importing Western components and finished products into the GDR as it wanted to avoid dependence on the West. The GDR was also constrained in its purchases by the fact that it had insufficient supplies of hard currency to spare. Honecker prohibited his IT enterprises from forming joint ventures with Western firms, in contrast to Hungary and Poland which did.[167] Furthermore, as occurred in the Soviet Union, just because a system was delivered to a factory or government agency did not mean it was used. As in the USSR, if one did not have software, an adequate number of peripheral devices, supplies (e.g., printer paper), or maintenance and training, a system was going to remain in its shipping boxes or silent in a data center. So at most the numbers just cited represent an expansive citation of possibly installed systems. One can conclude, however, that local demand within the GDR was not met for the totality of products needed in order to support an industrial economy of its size. Hard currency once again proved of greater value to the government than improved productivity derived from internal use of computers.

However, the number of people who acquired knowledge about IT certainly was impressive. Extant data on the number of employees on average in an electronics enterprise suggests that at a minimum a growth trend existed. First, these enterprises expanded from an average of less than 400 employees per factory in the 1960s to over 600 in the 1970s, to over 1,000 by 1980, and in excess of 1,400 by the late 1980s. For the microelectronics industry, data from 1989 showed that there were four conglomerates: CZJ with 69,000 employees of which 13,000 were involved in R&D; KME with 59,000; Kombinat Elektronische Bauelemente Teltow (EBT) with 28,000; and at the Kombinat Keramische Werke Hermsdorf, another 24,000. The one combine devoted to computers, Kombinat Robotron, employed 68,000 across 20 enterprises, of which some 8,500 focused on R&D.[168] That group brings all employment in 1989 to a total of 248,000 workers, not an insubstantial commitment on the part of the GDR.

Second, such extrapolations do not account for those who may have worked in the industry then left for other firms or careers.[169] One estimate, made after the fall of the GDR, posited that 20,000 scientists and engineers worked in IT and related fields, impressive but not excessive when compared to similar cohorts in other countries.[170] One would also need to add individuals also familiar with the technology, most of whom lived and worked in cities with IT enterprises and local universities closely linked to these, most notably Jena, home of CZJ, and Dresden, and with facilities run by CZJ and Robotron, for example. Few had PCs at home for the same reasons as in the Soviet Union—fears of information sharing and that might support political agitation. Probably more important, these were expensive and supplies always inadequate to meet demand.[171]

Looking back on the period of the 1960s through the 1980s, several trends can be identified and conclusions drawn from the East German experience. More than in any other Comecon country GDR's rulers were more committed to developing a digital capability than their neighbors, most notably in the manufacture of semiconductors. In no other Communist country was an individual technology so closely aligned to national intentions, public policy, and, to use a marketing term, a nation's brand by its political leaders. Development of just the 1-Mb chip absorbed approximately 14 billion East German marks, or 20 percent of GDR's total R&D budget during the years of its development. Between 1986 and 1990, GDR spent up to 30 billion East German marks in microelectronics and computers. That sum also translated into two-thirds of GDR's 85,000 researchers in R&D in just this one technology.[172] Yet the results were poor, most notably technology that not as advanced as what already existed in the West or was needed by indigenous industries, such as in machine tool manufacturing. It was also a technology still evolving rapidly beyond the nation's ability to keep up with.

Second, they single mindedly practiced policies of autarky to avoid over-dependence on the West for technology, to demonstrate their legitimacy vis-à-vis West Germany, and as a means of creating a source of hard currency through sales of high-technology products to neighboring Comecon countries. The GDR sent over half its supply of locally manufactured semiconductors and computer equipment to the Soviet Union both at the insistence of Moscow and as a way to help pay for supplies of energy. The remaining roughly one-third proved too few to meet the needs of the internal market within the GDR, and, as in other Comecon countries, the quality was often questioned. Supplies of peripheral equipment and software proved inadequate as well for reasons similar to those experienced by Soviet users. The challenges faced by GDR in developing advanced technologies were shared with other Comecon nations, which all experienced enormous difficulties across multiple industries and technologies in generating innovations that could compete on the world stage.[173] A combination of lack of incentives to innovate, poor-quality equipment, and inadequate supplies and services also led East Europeans enterprises to resist automation, with the lack of supply of adequate systems probably the trump card in explaining the slow rate of diffusion.

A further word on incentives is needed to round out our understanding of events. As in other Communist countries during the years of the Cold War, a worker's bonus and salary were most frequently tied to the amount of physical labor one did, so there were few incentives for individuals to promote and support the use of automation, such as provided by computers. Even modest automation threatened overtime salary payments, providing an additional reason to suspect the benefits of computing. Reducing the complexity of work through automation resulted in lower salary levels too, particularly in GDR's factories, and having a ripple effect on future pension payments. Management also shared these negative consequences in addition to those risks that come whenever a new process or technology was implemented, particularly if they posed a threat to someone being able to produce as many goods or tasks performed as required by government administrators. This is the similar problem that affected use of computing in other Comecon countries, such as in Poland and in Russia. The extent of these issues has yet to be clarified, but they were widespread enough in discussions about the value and use of computing over the years to not be dismissed as trivial, particularly during the 1950s when the whole issue of computing was a new topic for discussion.

Meanwhile, public officials engaged in the use of computing. One of the most interesting findings of Gary Lee Geipel's study of GDR's technology was how few people set policy for the entire nation. Unlike in the Soviet Union, where many participated in various decisions regarding IT, he demonstrated convincingly that the combination of two leaders dominating East German affairs for so many decades—Walter Ulbricht (1950–1971) and Erich Honecker (1971–1989)—along with a tiny cadre of public officials who remained in office, hence in positions of autocratic authority for so many years, led to a hardening of positions on policy, most notably by Günter Mittag, who dominated economic policies in the GDR. There was no East German *perestroika* or *glasnost*, as in the Soviet Union and in other parts of Central Europe. The Germans failed to make an effective 1-megabyte chip, saw their machine-tool industry deteriorate in the face of more modern products developed in the West and in Japan, and proved unable to manufacture computer systems competitive with those of Western suppliers.[174]

Critics of this small group's emphatic reliance on an unchanging IT strategy from one decade to another were silenced. In the Soviet Union the influence of senior leaders ebbed and flowed on IT policies as they came in and out of office. In no other country in Central Europe did senior leaders so depend on IT as the symbol of their country's strategy and portrayal of national pride and modernity than in the GDR.[175] Ultimately, their failure to effectively use the technology, in other words, "to make good" on computers, contributed to discrediting their government in the late 1980s. No other country in the world did the lack of IT performance contribute so mightily to the downfall of a government, at least until the Arab Spring in 2011 when the use of social media IT facilitated the collapse of several regimes.

But why did this small group not manage the situation better? To be sure, allowing the public to see IT as a policy tool, an apparatus of the political system, turned out to be a mistake because decisions about how to develop and deploy it were largely decoupled from the cadence of the technologies's evolution underway within the Comecon ecosystem and around the world.

That led to the state not leveraging effectively, or in sufficient amount, the technology to solve social, economic, and operational problems, let alone to increase efficiencies. These problems were highly visible to workers in factories that could have used the technology, to scientists and engineers working with semiconductors and computers, in both development and operation, and to mid-level government *apparatchiks*. Experts knew that technology per se could not solve social or political problems, as one wrote at the time: "There is no such thing as a pure technological fix," and "that to turn to technology as an unencumbered way to solve a political problem is a delusive goal."[176] In short, technology was not a political silver bullet; sound public and business administration was called for and not provided. Combined with a failure to develop a systemic IT sector in the country, one had the makings of a failed model of diffusion that in turn resulted in the public losing confidence in its national government. While it would be an exaggeration to argue that inadequate IT diffusion led to the fall of the Berlin Wall and of the state, it would not be unreasonable to conclude that failed IT diffusion contributed mightily to the state's weakened viability at the end of its life.

In defense of the regime one should acknowledge the majority of industrial workers proved generally hostile to technological innovations throughout GDR's life, which did not help; but the state had promised its 2 million workers a relatively easy time of it in the work place in exchange for political support. Some 120,000 scientific–technological intelligentsia would rather have seen more innovations, although many gave up trying to advocate for these. Media and artists were generally critical, or at least suspicious, of technologically based modernization programs. One student of the role of IT in GDR concluded years later, that as a result "by the late 1980s, most East Germans reacted to the regime's promotion of IT with cynicism, not with respect or confidence."[177] One historian argued recently that as the state touted IT as progress it was put in a position that "consigned the GDR, more than ever, to eternally trumpeting its achievements in essentially outmoded technologies," and everyone knew it.[178]

The East German case of IT diffusion presented an example of enthusiastic senior public officials in support of deployment that was hampered by their misunderstanding of the internal and external larger technological ecosystem and the influence its players around the world had on events. Coupled to the kinds of institutional and structural nontechnical problems that also existed in the Soviet Union and in other Comecon countries that made development, delivery, and use of IT unattractive to too many potential users and unachievable exacerbated problems. While historians criticized GDR's technological affairs, and especially its misplaced exuberance for science and technology, the country shared many of the same weaknesses and problems evident in the larger pan-Communist world in sharp contrast to developments in the pan-Atlantic world.

Finally, we might ask what about the end users of computing? The new historiography and earlier studies of computing in East Germany suggest strongly that the applications were similar to those in evidence in the West: accounting, shop floor data collection, logistics management, and the same kinds of uses seen with engineers, mathematicians, and scientists. In fact, the latter's use of computing has increasingly been documented in the past decade, particularly by Christine Pieper, who demonstrated that more technical applications were in evidence than suggested by earlier studies.[179] One can conclude that the combination of similar applications, on the one hand, and considerable activity to develop locally designed technologies, on the other hand, goes far to explain why user communities admired Western innovations, wanted to use computers more than their economy could provide, and contributed specifically to the discontent exhibited against public policies that attempted to satisfy these wishes.[180] But there were also stop-and-go issues that mitigated more successful implementations as well, such as the start, then reduction, of work on flexible manufacturing systems in the 1970s when the West was beginning to implement these, and that was not reversed in the GDR until the 1980s, by which time it was too late.

INFORMATION TECHNOLOGY IN THE REST OF COMECON EUROPE

The rest of Comecon Europe comprised Czechoslovakia, Poland, Hungary, Romania, and Bulgaria.[181] Individually they had unique cultures, languages, history, and varied economies. Each, however, were required by the Soviets to operate with Communist political philosophies and a strong centralized state that owned, managed, and developed their national economies, particularly heavy industry and collectivized agriculture. Soviet leaders expected these governments to align politically, diplomatically, and militarily with the USSR and, if they did not (or could not on occasion due to local unrest), were punished (or helped as the case may be) through a myriad of political, military, and economic means. Political and economic dissent was to be controlled by local national governments, and if not, then there were Soviet troops and tanks at the ready, most emphatically in the first couple of decades following World War II, and benignly by the mid-1980s. Many of the economic problems and experiences they had were similar to each other and to those of the Soviet Union for the two most obvious of reasons: their economies were managed essentially in the same way and linked to each other more than to the global economy. To be sure, individual countries had their own circumstances, as evidenced by the story told about the GDR, but their structural and operational similarities outpaced their differences.[182] One could see this pendulum of similarity and local circumstances swing back and forth as well with IT.

The numbers of computers built and used varied in these countries due to the degree to which a nation's government and economy proved able to develop, acquire, and deploy IT. However, one should also keep in mind that the size of these nations also varied enormously, skewing comparisons. The easiest way to demonstrate differences in size is to look at populations and landmasses. We have already discussed the GDR which had the second largest population in 1950 of all the Comecon nations (18 million) after Poland with 25 million. As to the others in 1950, Hungary had nearly 10 million, Bulgaria 7.3 million, and Czechoslovakia 13 million. By the mid-1980s, all these countries had grown in population with the exception of the GDR, which had shrunk by about 1.3 million. Poland possessed the largest landmass (nearly 313 sq.km), followed by Romania with 238 sq.km, next the Czechs at 128 sq.km, GDR with 108 sq.km, and Bulgaria quite similarly with 111 sq. km, Hungary at 93 sq.km, and Albania at nearly 29 sq.km. Density per square kilometers varied, with the more density indicating greater urbanization and that statistic correlated with more industrialized economies by 1985: East Germans at 155, followed in descending order by the Czechs at 121, Poland 119, Hungary 114, Albania 103, and Romania 96.[183]

As in the West, Soviet Union, and GDR, diffusion of IT was influenced by the extent to which a government supported its development and use, and made available the technology. Industrial profiles were also important, as in the West, so a society with heavy, large or technologically complex industries would be expected to be a more extensive user of IT, assuming that its government played the role just described, while a more agricultural society would be inclined to be a lesser user. This paradynamic model helps to explain why Poland and Hungary gravitated to computing faster than, say, Bulgaria. Other factors influenced events as well, such as technical and scientific education. The Soviet Union, Poland, and the GDR invested extensively in the education of engineers and scientists all through the postwar period; each had done so throughout the twentieth century before becoming Communist. Finally, as argued at the start of this chapter, these nations had a similar public sector infrastructure to manage their economies, hence, also diffusion of technologies of all types. Often IT remained lower on the list of priorities, particularly before the 1970s or 1980s. Later the pendulum swung toward favoring its importance. In all cases, in the post-Communist period of the 1990s and 2000s a surge in deployment of IT of all kinds occurred. The Internet proved to be one of the least important in the early 1990s, since all had to upgrade their telecommunications infrastructures in a period when official attempts to constrain the flow of information and data no longer perdured and access to the technology from sources from many parts of the world became available. The shared pattern of adoption of IT occurred in an environment where the main players—governments—proved so similar, and so tied to Soviet supervision, that one can generalize sufficiently about the rest of the Comecon community's experience with IT. To be sure, the more closely one looks at the role of IT in any particular nation, the easier it becomes to identify nuances in behavior and circumstances that make it more difficult to generalize as glibly as we can today, currently made possible by the lack of more details, data, case studies, and memoirs.

Initial forays into computing can quickly be summarized. Eastern European projects surfaced during the 1950s in various countries, initially in institutes and academies interested in mathematics and engineering, as occurred even in the West, most notably in France, also in the East Asia and within the Soviet Union. In the case of Poland, early work took place at the Zakład Aparatów Matematycznych (Institute of Mathematical Machines), a part of the Polish Academy of Sciences, located in Warsaw. Employing several hundred individuals, work began in 1957 on what came to be known as the XYZ1 digital computer. In 1958, academics and students began using it to do mathematical calculations. A second system was quickly built to support work at the Institute of Aircraft, and still a third called the SKRZAT 1, which was characterized at the time as a control computer, but functioned more like a flight simulator.[184] The Romanians began at the Institut de Fizică Atomică near Bucharest, starting actually earlier than the Poles in 1953. In quick order the institute built three systems: CIFA-1, CIFA-2, and CIFA-3, and a variant of these for use by the University of Bucharest in 1957. CIFA-2 and CIFA-3 became operational in the second half of the decade. Finally, Czechoslovakia quickly and earliest began at the Výzkumný Ústav Matematických Strojů (Research Institute of Mathematical Machines) located within the Ministry of Precision Engineering in Prague, constructing one of the best known computers of Central Europe, the SAPO (Samocinny Pocitac). Soon after, other institutes began building computers in the country.[185] The other members of Comecon had their own projects as well.[186]

In the 1950s, Western observers noted that many Central European machines were constructed based on locally grown knowledge, although as asserted earlier in this chapter, the amount of Western technical literature available locally was impressive and goes far to explain how the necessary knowledge about computer technologies and their relative similarity one to another was made possible. Yet in the West comments appeared in the technical press that spoke about slow starts and primitive systems, even though the Czechs and Poles had a long and distinguished national core competency in mathematics and engineering. For example, in the then important Western publication, *Computers and Automation*, in 1955 an article on Communist systems opined that "in contrast to the Russians, who seemed completely informed on developments here, were the Czechs, who seem to have been almost completely uninformed about developments either here or in Russia," adding that "cut off from all outside information, the Czechs had built at Prague a small magnetic drum computer called SAPO," going on to describe a computer that would have been quite recognizable to Western and other Eastern European, and American, computer builders.[187] Such comments did not always make sense in hindsight, such as about the ignorance of Western computing but building systems recognizable by the West. While more research on the history of indigenous IT is needed, it is clear from extant evidence that Central Europeans were familiar with technological developments in many countries. In time, the Czechs, Poles, Romanians, and East

Germans either had technologies as contemporary as those of the Soviets, or even more advanced, the result of more borrowing of ideas from the West.

Development of computers remained a prerequisite for almost every European and East Asian country to install and use computers, especially in the 1950s and 1960s; nowhere was this more so than in Communist countries. These early systems made it possible for mathematicians, engineers, military manufacturers, and scientists to use the technology, but just as important, to learn much about how to develop and later manufacture it. When coupled to the promotion of their use by public officials, one can begin to piece together the story of their diffusion. Table 6.4 illustrates this point. Every country on the list "invented" computers of some sort or another, mostly built at institutes and academic institutions during the same decade of the 1950s, and roughly the same number in all—a handful of systems. But note the wide diversity of installations just a decade and a half later, especially in Eastern Europe. Both Western and Eastern observers thought the Soviets, Czechs, East Germans, and Romanians had the most advanced technologies, yet we know today that the Hungarians were probably as advanced as any of the others on the list. Within a few years (certainly by mid-1970s), the number of manufacturing facilities, and the distribution of their output, became a more important factor in explaining diffusion than the stage of sophistication of a nation's technology. Across all Comecon countries around 1975, of the thirty-six known manufacturers half were in Eastern Europe.[188]

We can explore further the start-up of computing in the Central Europe, and its maturation and deployment by looking briefly at Hungary's experience, just as we did with GDR's. Like the East Germans and other Comecon states, after the Stalinist period, local leaders assumed more initiative in governing their nations, crafting their own economic and technological plans so long as these were not at odds with the Soviet's initiatives or style of public administration. Hungarian scientists and officials always kept an eye open for Western practices all through the second half of the twentieth century. As one observer of the scene noted in 1981, "as East European leaders and planners commit their economies to qualitative improvement, modernization becomes increasingly synonymous with emulating Western technology and management," especially in Hungary, opening "more of their [*sic*] economic sectors to world market competition," and "developing economic mechanisms more compatible with Western free enterprise economies."[189] It is against this background of being able to observe what was going on to the West and East, and in the context of having more freedom of action than might otherwise have been assumed, that computing came to Hungary.

In 1955, a small group within the Institute for Measurement and Instruments began examining mathematical machines; then in 1957 the Research Group for Cybernetics within the Academy of Sciences (KKCS) came into being, thereby creating an initial path for computers by taking some of the very first steps in development. At the Budapest Technical University another group built the MESZ-I computer, putting it into operation in 1958; at least one other university also built a system at the time (Szeged), and another group constructed the EDLA system. In the 1960s the Hungarian state created a number of research institutes and launched economic reforms under the label of New Economic Mechanism, which called for better control of the economy and its modernization, including greater tolerance of some private sector initiatives and additional work on computers. In 1969 it launched the Hungarian Central Development Program for Computers (SZKFP) to develop and produce systems and to educate and train people in their use, combining basic and applied research and deployment. This initiative became an important source of IT knowledge and work in the 1970s and 1980s. By the late 1980s, some 700 people designed and constructed computers, including software for use in agriculture, food, energy production and distribution, traffic control, and manufacturing. A manufacturing facility for production of computers produced equipment in the 1970s and 1980s.[190]

Interesting to note is that development and manufacturing occurred on the same site, roughly the same model used by IBM. Because of limited access to Western products for sale, the Hungarians began to develop their own local IT expertise, beginning largely in the 1970s. Along the way they adopted, adapted, and learned from such American technologies as those in DEC's PDP and VAX-11 systems, making their own called the TPA. This family of machines proved effective and popular, leading to some 600 installed by the late 1980s in such diverse facilities as at three power plants, in several liquid and gas hydrocarbon processing centers, at transport and storage depot firms, the Post Office, and in various factories. Some Western systems came into the country as well, providing instructive insights to other technologies, such as those from IBM.[191]

A Hungarian computer scientist looking back on her country's experience wrote that "by the 1980s, the computing culture had become well-established in Hungary, owing to the consequent state support and enterprise interests," adding that "perhaps the most important reason for this growth was the constant and generous investments given to IT training and education," received by over 100,000 individuals at local universities.[192] It is an important point to make because in the Soviet case we noted inadequate training. Further learning came through proliferation of personal computers in the country, beginning in the early 1980s with what one observer called "a mass influx" of such devices, while mainframe computers were installed in universities, research institutes, various ministries, and companies. The normal range of applications was in evidence: data processing, accounting, manufacturing, economic development, and CAD/CAM being some of the most obvious by the early 1980s. In 1985 the national government reorganized the SZKFP into the Central Economy Development Program for "Electronization" (EGP) to further promote diffusion of the technology along with other electrical devices, such as automation and telecommunications. This restructuring led to a new round of national initiatives in 1987, which were soon swept into the vortex of the collapsing then birthing of new governments all over Eastern Europe.[193]

Hungary shared problems with other Comecon countries, such as its inability to modernize sufficiently to meet local expectations in the 1970s and 1980s. In this local leaders were not alone; we have GDR's challenges while the Poles could not even get some of its local computer products deployed in any substantive manner, such as the Odra system developed in the 1970s.[194] There existed differences as well. For example in Poland, the chronic lack of venture capital, let alone budgets for acquiring systems, constrained both innovation in IT and implementation of existing technologies. Acquiring Western imports proved difficult too as a result of inadequate supplies of hard currencies, existing supplies of which went to a large number of products, making competition for these funds to use for IT difficult. Polish exports could not generate enough foreign currencies, due to their lack of features making them sufficiently competitive in the pan-Atlantic, let alone pan-communist markets.[195] However, in the Polish instance, by 1990 computer products, office automation, telecommunications, and other "high-tech" products were coming into the country in substantive quantities, far more than into most other Comecon economies.

While IT skills were increasing in the GDR, Soviet Union, and Hungary, the rapidly growing reliance on Western technologies in Poland as the local economy opened to the West led to declines in locally skilled workers in IT, since these machines were increasingly coming from other Comecon and Western suppliers. With the wide opening of Eastern Europe to the global economy, beginning in 1989, the problem became worse, with Polish employment in computer equipment declining from about 9,500 employees that year to 4,800 in 1992, and in telecommunications from 19,000 to 8,500—all representing loss of skills available in any focused manner to the nation. In all Comecon countries, scientists, engineers, and other IT experts, such as programmers, migrated by the tens of thousands out of the region into Western Europe in the 1990s or took jobs that did not always leverage their IT expertise.[196] Accurate numbers are difficult to come by for all of these countries, but one report, again using Poland as the example, dated the loss back into the 1980s: "10 percent of scientists left Poland," and of these "about 30 percent of scientists of such specialization as mathematics, information technologies and engineers."[197] The Polish situation demonstrated what was happening in such other places as Czechoslovakia and Hungary since in the 1980s as Polish investments in local IT declined, foreign involvement increased, filling in some of the slack, albeit in a tiny manner with what were called "Polonia" companies. These provided such services as importing hardware and software, being final assemblers of components and machines, and performing some software development.[198]

Table 6.5 NUMBER OF TELEPHONES PER 100 INHABITANTS IN
EASTERN AND WESTERN EUROPE, 1990–91

Eastern Europe, 1991		Western Europe, Select Group, 1990			
Bulgaria	22.2	Denmark	57	Belgium	39
Yugoslavia	15.1	France	50	Greece	39
Czechs	14.3	Luxembourg	48	Italy	39
Baltics	13.0	Germany	47	Spain	32
Romania	9.2	Netherlands	46	Ireland	28
Hungary	8.6	UK	44	Portugal	24
Poland	8.2	EC 12	41		

Sources: OECD, ITU, Paul Gannon, *Trojan Horses and National Champions: The Crisis in Europe's Computing and Telecommunications Industry* (London: Apt-Amatic Books, 1997): 278–280.

As in the Soviet Union and GDR, the fall of Communism was more than a political watershed for all of Eastern Europe. It was also a milestone event for IT,
initiating the twin trends of privatized enterprises installing or selling IT, and in
increased use of more modern systems typical of Western societies. A third trend
important to IT involved upgrading telecommunications in the 1990s and 2000s,
which made it possible for these nations to start using the Internet widely. Since
the border between Wave One and Wave Two diffusion is importantly defined, in
part, as including a surge forward toward consumer-based use of IT, and specifically Internet-based applications, it is important to be aware of installations of
telephones. Table 6.5 provides information on the extent of the diffusion of regular point-to-point voice telephone service's diffusion in a number of countries at
the dawn of the post-Comecon era. Specifically, the evidence about the state of a
technology prerequisite to the more complex transmission capacity, technology,
and speed of data transfer necessary for using the Internet is essential to understand.[199] The data make obvious that the number of telephone lines was lower in
Eastern Europe than elsewhere, indicators that diffusion of online computing
would be lesser than in the West and limited to those few with personal computers
to access the Internet. Another more serious problem of growing concern involved
aging national systems in dire need of technological updating in the 1980s and
1990s. To be sure, governments across the entire region made efforts to improve
their local telephone networks, in many cases importing Western technologies as
early as the 1970s and continuing through the 1990s and beyond.[200]

DIFFUSION OF INFORMATION TECHNOLOGIES
IN POST-COMMUNIST EASTERN EUROPE

We now turn attention more fully to the post-Communist era, one largely
characterized by Wave One diffusionary activities. The fundamental historic
change that came to Eastern Europe at the end of the 1980s was the demise of

the old Soviet empire, and the authoritarian regimes across the region. Tied to that move away from authoritarian government and to democracy was an extensive, quick alignment with Western Europe across a wide spectrum of a nation's activities: culture, jobs, politics, and economics. In subsequent years various Eastern European countries applied for membership in the European Union, NATO, WTO and other international associations. While the history of that historic shift has yet to be understood adequately, we know that it occurred quickly and broadly, representing a near wholesale desertion of Soviet connections, with the most notable exception of still being dependent on Russian sales of energy, such as natural gas to Poland and Germany. To be sure, all did not go smoothly, such as the conversion from a state controlled economy to capitalism, a transition that twenty years later was still unfolding.

As the region began transforming so too did the area's experience with information technologies. The evidence of change appeared all over the region in many ways. PC sales rose as the cost of products declined and quality rose, and access to Western software improved. Western firms came into the region setting up offices, as did IBM.[201] Countries that had specialized in manufacturing IT before the breakup transformed enough to experience increases in the percentage of their GDP earned from this product set, for example Hungary.[202] Using the same country to illustrate the pattern, adoption of computing across the economy rose in the 1990s, reaching levels equivalent to those in such places as South Korea, Finland, and Ireland at the time, in the range of just over 5 percent of GDP. The share of GDP generated by the manufacture of such technologies within the industrial sector rose too, from 2.69 percent in 1995 to over 15 percent by 2002.[203]

Because Hungary was one of the more technologically endowed countries in the region, its experience with diffusion should not be ignored. So, in addition to diffusion underway at the institutional—that is to say, enterprise and agency—levels, adoption of IT by individuals became substantial. Extant data indicate that in 1995, 6 out of every 100 households had a PC, a machine population that rose to 10 out of a 100 in 1998, and to 15 per 100 in 2000, in short more than doubling in less than a half decade.[204] Computer installations lagged in schools, at 48 percent equipped with at least one in 1998, in contrast to more than 70 percent on average in OECD's member states, and that was much higher in the United States where the Clinton administration was extensively involved in bringing the "Information Highway" to all schools.[205] By the early 2000s, as in what used to be the old GDR, economic productivity in Hungary began to rise as a result of modernizing industries and transforming economic behavior.[206]

Because Poland was such a large economy, indeed bigger than GDR's or Hungary's, we should ask, what happened there? Leaving aside the lack of precision in economic data, but accepting general trends, economists who looked at Poland's economy concluded that spending on ICT (including telecommunications) as a percentage of total GDP rose all through the 1990s from 2.06 percent in 1993 to 5.95 percent in 2001, albeit lower than for the entire region on average.[207] Table 6.6 provides early evidence of spending levels in Central Europe, demonstrating that a surge in the diffusion of IT occurred across the region. Reasons for Poland's surge are not hard to find: pent-up demand for IT being the most obvious as a consequence of underinvestment and insufficient supply in the 1970s and 1980s, and related to that, the need to become more competitive in trade with Western Europe. The availability of better quality, lower-cost ICT from European and Asian sources played a major role, not only for Poland but also for all of Eastern Europe. As the data in table 6.6 indicate, the rate of investment in ICT increased as the 1990s progressed; reflecting a pattern historians of IT often take for granted and so rarely highlight. Namely, as a firm, industry, or national economy begins to invest in IT, its appetite—think dependence—increases too, forcing, if you will, someone now dependent on the technology to do their work to continue expanding and upgrading their deployment and use of IT.[208]

To close the Polish story, the fact that world-wide the rates of spending on ICT increased (17.1 percent in the EU and 29.6 percent in United States by the end of the century), with Poland going from some 6 percent investments in 1992 to 14.2 percent in 2001, should be seen as part of the dramatic expansion in the use of IT there as occurred in so many other countries.[209] Furthermore, as in other Eastern European countries, these investments were in modern IT, so we no longer needed to keep in mind that diffusion often meant increased uses of older technologies, some 10–20 years behind those of the West, as we had to do when discussing Comecon's computing of the 1960s through the 1980s nor that these systems were smaller, not capable of doing as much work as Western systems. What were the economic results? In the case of Poland, IT's contribution to GDP growth ranked it in the middle of a multinational sample that included Western Europe and the United States in the second half of the 1990s, largely in profit-maximizing enterprises.[210]

Table 6.6 ICT SPENDING IN EX-COMECON EUROPEAN COUNTRIES, 1993–2001, SELECT YEARS (AS PERCENTAGE OF GDP)

Country	1993	1995	1997	1999	2001
Bulgaria	2.2	2.3	3.0	3.6	4.2
Czech Republic	5.6	6.0	6.4	7.9	8.7
Hungary	4.2	3.9	4.5	8.2	10.0
Poland	2.1	2.2	2.6	5.4	6.0
Romania	1.1	0.9	1.3	2.1	2.4
Russia	4.0	1.8	2.0	4.1	3.2
Slovakia	4.2	4.0	3.9	6.8	8.8
Slovenia	3.0	2.9	3.4	4.4	4.7

Source: Data from modified table in Marcin Piatkowski, "The Output and Productivity Growth Effects of the Use of Information and Communication Technologies: The Case of Poland," in Kezysztof Piech and Slavo Rzadosevic (eds.), *The Knowledge-based Economy in Central and Eastern Europe: Countries and Industries in a Process of Change* (London: Palgrave, 2006): 82, based on information collected by WITSA, *Digital Planbet 2002/2000: The Global Information Economy* (Vienna: World Information Technology and Services Alliance, 2002). WISA looked at 51 countries worldwide and the averages from the larger group were: 1993: 4.5%; 1995: 4.5%; 1997: 5.0%; 1999: 6.2%; 2001: 7.3%.

As in other East European countries, in the 1990s many of these investments were made by small, often new enterprises, rather than just by the
longer established large state-run companies. Expenditures were driven by
the availability of the technology and significant drops in their costs. By
2004, citing Poland's case, and thus not one of the most advanced economies in its use of ICT in Central Europe, over 90 percent of small and
medium-sized firms reported they used computers, usually PCs and the
Internet, with word processing, spreadsheets, Internet browsing, and
e-mail their top four applications in descending order of use.[211] To put that
performance into a broader diffusionary perspective, in the same year
(2004), the public as a whole had limited broadband use (0.6 percent),
which translated into 3.8 percent of all households with the capacity to access the Internet if they chose to do so. That year, however, a subset of the
nation's population (22.4 percent) appeared to be regular Internet users.
On a global basis that usage put them in the bottom quartile. Nearly 29
percent of businesses had access to the Internet, and it is within this group
that the small and medium-sized firms dominated.[212] The data suggest that
Poland, like so many of its neighbors, had entered late Wave One diffusion
of IT, albeit later than the West, but nonetheless had caught up very quickly
from where it had been in the 1980s as compared to many other European
nations.

Economists and historians have long known that catching up economically
and technologically is often faster to accomplish than developing a new class
of economic behavior or technology.[213] For Central Europe to embed existing
IT into its processes and copying best practices already implemented elsewhere was much faster than inventing them. One economist who looked at
the Polish situation made the same point: "The relatively large contribution of
ICT capital to output growth is due to an extraordinary acceleration in real ICT
investments, which were growing between 1995 and 2000 at an average rate
of more than 20 percent a year," accounting in a substantive manner for
Poland's diffusion and economic results that led it to the middle of the pack
when compared to other countries.[214]

One factor, however, requiring further study is the actual effect(s)
beyond the numbers because, as in the case of Poland, even its ICT producing sector was so tiny to begin with (certainly no more than 5 percent as
of 2002) that its impact on the nation as a whole was probably also small
when compared to that in other countries. In nations such as Sweden, Germany, Netherlands, South Korea, Japan, and the United States, large internal ICT manufacturing and services sectors sped up local diffusion of digital
technologies as knowledge of these spread through the economy. Even as
those sectors themselves used their own technologies, the role of internal
ICT industries became an important contributor to IT spreading so quickly
in the second half of the twentieth century. It appears that the role of these

kinds of IT industries in Central Europe was far less and thus should not be seen as a significant explanation for how IT spread in the region so quickly in the 1990s or, for that matter, before, even in such nations as Hungary and the GDR which had indigenous IT industries. The answer for speed rests largely on the demand side as occurred across the entire pan-Atlantic community.

CONCLUSIONS

In 1987, Boris Naumov, director of the Institute of Informatics Problems, commented on IT at the start of Gorbachev's *glasnost*:

> Certainly we have some problems, and these aren't simple ones. The biggest problem is that we do not have enough computers. It's not the design of the computer that is the main obstacle, but organizing the production. It's a problem of developing a modern industry in computers which can provide what the users want.[215]

In this short quote one reads an important part of the answer to the question of what influenced the rate of diffusion in the Soviet sphere. The problems faced were complex, supply always inadequate, and the organization of distribution and use clearly the largest challenges facing the entire region, not just the Soviet Union. It was not for want of trying. With the exception of civilian interests during the Stalinist years, public officials sought to find ways to make computing available for many of the same reasons evident around the world. But ultimately, their ability came down to the extent that they could, or could not, implement their plans. The centralized approach to managing technology and the economy simply did not work as effectively as the more decentralized ways common in market-driven economies of the pan-Atlantic community and other non-Communist states. During the Putin period, however, a resurgence of the centralized approach appeared once again. In 2010, Russian officials made it known that they wanted to create a Silicon Valley-like haven for local entrepreneurs and tech-savvy experts so they could apply their talents and skills. In the West questions were raised if this could be done since the Russians had never been able to convert IT experts into business enterprises through state initiatives. Rather, as occurred in the 1990s, it happened on its own regardless of the state and even then on a limited basis. In part Silicon Valley worked because of absolute freedom to find capital, to use it wisely, and unwisely, and to move quickly from one initiative to another. When asked in April, 2010, about that freedom of action, a senior Russian official snapped back with "consolidated power in Russia is the instrument of modernization. I would even insist it is the only one."[216] It appears that old ways of diffusing technology remain difficult to change.

Another influence on the rate of diffusion and on the nature of IT's role involved the attitude of all Comecon nations toward the free flow of information. It is a complicated issue. As with the debate on cybernetics in the 1950s and 1960s, it was tied up in the region's Marxist–Leninist world view. Yet there were no absolutes, because information about IT flowed rather freely within technical circles in these Comecon countries, influencing views of experts. There were many exchanges between Western European, American, and East European computer scientists and engineers, even during the harsh years of the Stalinist regime. A small coterie of individuals worked with access to Western technical literature. The bigger issue was the role of information outside that community, specifically in society at large. At the risk of picking on the Russians too much, but making clear that the issue of freedom of access to information extended to all Comecon nations, on the one hand, officials were adamant about minimizing availability of technologies that would allow the diffusion of information, hence their many bans against printers, copiers, fax machines, and later PCs. On the other hand, they wanted to harness the power of computing for many of the same reasons as in the West and in parts of Asia. How they wrestled with these issues determined the speed with which these information technologies could seep into a nation's economy and society.

Loren R. Graham, a Western observer of East European affairs, recalled that in the 1960s, when the Soviets were implementing their first major post-Stalinist reforms, an important gating factor for availability and use of information technologies revolved precisely around this issue of open flows of information. Yet Graham observed that in the early 1960s while the state struggled with the issue, "the formerly closed society of the Soviet Union was perforated by hundreds of thousands of holes, through which outside information streamed inward and inside information streamed outward."[217] It was not a static situation because over time information flowed more freely with or without the blessings of the government. Let Graham explain the contrast: "in 1960 . . . if a scientist was arrested, it might be months before we had reliable information in the West on the event," yet in the late 1990s, "such an arrest would be known in the West within hours. New information technologies are not the only reasons for this dramatic change, but they are crucial for understanding it."[218] Communist leaders all over the region had to loosen the knots on information, and their technologies, or risk condemning their nations to what Graham characterized as "hopeless obsolescence."[219] Communist exceptionalism gave way before the need to compete in a global economy and to adopt Western technologies, beginning with the kinds of decisions made to use IBM's S/360 technology, and letting information flow through society.

The trilogy of an ecosystem that proved unable to collaborate effectively to the extent that occurred in the West, reluctance in allowing free flow of information and knowledge, and reliance on continuously uncompetitive technologies, put the entire region on a path to slower diffusion of IT. Indeed, those three factors ensured that no Comecon nation would use whatever IT they had in ways that mirrored the effectiveness of computing in either the pan-Atlantic world or in East Asia. Not until Communist regimes and economies gave way, information flowed, and more modern less expensive information technologies became available, did all these nations begin to catch up and use IT more extensively, rapidly, and effectively. As the Polish statistics demonstrated, it did not take much time for diffusion to begin mirroring patterns evident with earlier adopters of IT in the European Union.

By the early 2000s, economic data began demonstrating that the catch up was well underway. One simple example illustrates the point. In 2008, OECD reported that ICT workforces were expanding in number around the world, over what their populations had been in 1995. As evidenced from the experiences of the pan-Atlantic community, a nation with ICT experts is a society that has woven into its fabric individuals who, when economically and culturally permitted to ply their knowledge, represent a major force for diffusion of IT. So, having more of these people in an economy open to innovation and competition is an important cause for adoption of IT, and as the American experience demonstrated, rapid diffusion. In 2008, OECD reported that in all the ex-Comecon nations ICT workers made up nearly 3 percent of the local workforces. That extent of diffusion of ICT skilled workers throughout an economy proved comparable to those in such other countries as France, Portugal, Italy, Spain, Belgium, and Austria. Only the Nordics, Switzerland, North Americans, and Anglo-Saxon countries exceeded these percentages. For all of the EU 15, the average was just barely above 3 percent; in short, not much above the old Comecon countries and even some Western countries were not up to the same levels (Turkey, Greece, Ireland).[220]

We are left with the conclusion that the entire Comecon region of Europe remained in Wave One until at least the dawn of the new century, taking longer than the pan-Atlantic community and parts of Asia to pass through this phase to Wave Two. As in the cases of Western Europe and the United States, prerequisites for Wave Two began appearing all over the region in the late 1990s: better telecommunications capable of supporting Internet use, wireless phones, citizens equipped with PCs able to access the Internet, more modern computing across larger numbers of enterprises and government agencies and IT optimized new work processes, albeit by every measure far less so than in the West. To be in Wave Two one would want to see in society a circumstance in which the majority of its citizens had access to consumer electronics (such as digital cameras, cell phones, and printers attached to PCs) and the Internet. In the ex-Comecon countries extant evidence suggests that even around 2000, less than a third of a nation's citizens used the Internet; in the West it was already twice that percentage and rising rapidly. While Eastern Europe was finally on a path to Wave Two, its rate of travel remained slower than the others to that new phase in its technological evolution.

After arguing that making the flow of knowledge within a society essential for its modern evolution, two East European experts on IT in the region acknowledged the interdependence of networks, computers (largely PCs) but also "that to achieve such network developments is neither automatic nor simple for either foreign or local firms, and other support (for example, from government) may be required."[221] From a practical point of view, economic development in all these countries in the late 1990s and early 2000s was also uneven, as it had been in Communist times, and thus the diffusion of IT, and its industry, appeared to be a condition that would continue to affect the rate of adoption, the speed with which these countries could evolve into Wave Two societies.[222]

The next three chapters examine the intense diffusion of IT across large swaths of Asia, from Japan that set as a national policy to become an Information Society, to South Korea, which became the most extensive user of broadband in the world, to China, which rode a massive tsunami wave of technology hurling it into the twenty-first century with observers wondering if it would be China's Century. If Wave One diffusion seemed orderly and paced in the pan-Atlantic and Comecon communities, it was anything but that in Asia.

CHAPTER 7

Computing Comes to Japan

It seems fairly certain that during the next decade the computer industry will develop to a point where it will exercise a powerful influence on the national economy and the social structure of every industrialized country in the world.

—Shohei Kurita, *1973*[1]

No nation generated more interest and controversy about its use of information technologies than Japan. The Japanese underwent one of the most dramatic, rapid, and extensive transformations in their economic and political makeup experienced by most industrialized societies in the shortest period of time. The size of its economy expanded from being ranked lower than other major industrialized economies in the world in 1950 to the second largest by the early 1990s. Although devastated in World War II, by the early 1960s it had returned to being Asia's most industrialized economy. A decade later Japan had become a major exporter of steel, automobiles, semiconductors, telecommunications equipment, consumer electronics (such as TV receivers and radios) and by the 1980s, computers to other Asian nations, the United States, and Western Europe. Its relatively closed economy and aggressive export practices set off pan-Atlantic diplomatic protests, trade tensions, and public policy countermeasures that smoldered for years.[2] Academics, public officials, and other commentators heralded Japan's enormous economic growth of the 1960s–1980s as bordering on ideal approach. Many touted it as the model of national economic performance that other firms, industries, and nations should emulate. Observers of the IT experience were emphatic on the point. Tom Forester, an influential IT writer working in Australia, reflected the conventional views of the 1970s and 1980s: "Japan is about to overtake the US to become No. 1 in information technology, the key strategic technology of our times."[3] Even after the economic crises of the early 1990s were well underway, Westerners were still being admonished to learn from the Japanese. Two distinguished Japanese scholars, Ikujiro Nonaka and Hirotaka Takeuchi, typical of many writing about quality management practices and knowledge management in the 1990s, presented managerial advice to the West, "a universal model of how a company should be managed based on the converging of management practices found in Japan and the West." Even in the face of economic hard times, they declared Japanese companies would "emerge stronger from the current recession, since the seeds for continuous innovation have already been sewn."[4] By the late 1990s, many observers were dismayed with the economic troubles faced by Japan during that decade, a malaise that continued into the next one.

The ever-shifting performance of the Japanese economy had other features, however, that evolved more slowly, or, seemed quite stable. Major Japanese telecommunications and electronics firms became the primary providers of IT, for instance. These were companies that had been in existence for decades, and continued to be major participants in the economy during the second half of the twentieth century. The close triad relationships and collaborations among leaders of firms, large national banks, and government agencies—often referred to as the Iron Triangle—remained a signature feature of Japanese economic activity during good and bad times. In fact, consensus among non-Japanese observers of Japan's economy held that the triad's behaviors made it very difficult for these three communities, and most specifically high-tech industries and banks, to respond quickly enough to changing economic and technological innovations with new products and business strategies. Instead, they sustained long standing business practices when on the face of things these should have changed. Practices requiring more rapid transformation included opening up earlier, and to a greater extent, more of the economy to world trade after an embryonic industry's need to stand on its own became obvious to government agencies, in particular the Bank of Japan and the Ministry of International Trade and Industry (MITI). Other needed changes included reducing the negative aspects of lifetime employment practices, allowing more wide-spread competition among entrepreneurs in technology-oriented industries, and make possible—and accessible—Schumpeterian gales of "creative destruction" to become a more common feature of the local economy.[5]

The Japanese story displays contradictions and dichotomies; positive and negative economic performances; optimization of public policies and investments that enhanced the early surge in development of local IT industries, but also failures to keep up with global technological innovations; outstanding and innovative business practices (such as Total Quality Management (TQM) and just-in-time manufacturing), yet others too that made it difficult for firms to change their business practices to remain competitive in the world economy. The list of contradictions is long and important. Core to that story is the role of information technologies, their vendors, and users.

As a result of these difficult, various, and serious conflicting currents of activities, it should be of no surprise that our understanding of Japanese events evolved substantively over time. Swings in views and understanding had many phases of causes and effects. Sometimes these were the result of specific events, such as World War II, later imposition of a new constitution and national government on Japan by the American occupation forces, followed by economic recovery and expansion, then recession. In its wake Japanese activities spawned a massive body of publications in Asia, United States, and Western Europe almost in the same quantity as about the North Americans. Two byproducts of that commentary have been the widely conflicting perspectives on Japanese events and, second, about its culture and functioning of its economy and business. In the process this discussion generated a great deal of misinformation and confusion about the country at large, often reflected in studies published by Western media and in the rhetoric of politicians, policy makers, and business executives in the pan-Atlantic community. The history of Japanese information technology (IT) was not immune from inclusion in this conflicting analysis; indeed, it was often at the center of it, because computing was beginning to play a crucial role in Japan's economic affairs by the end of the 1960s. During the 1970s and 1980s its export practices for semiconductors, computers, and robotics heightened diplomatic and trade tensions in ways reminiscent of similar problems in the 1960s and 1970s with textiles, consumer electronics (specifically radios and television sets), and steel, largely with the United States. American firms accused the Japanese of "dumping" onto the U.S. market products at below costs, or, at least for sale below what American companies could charge, in order to "buy" market shares in violation of U.S. laws and trade practices.[6]

The reasons for so much Interest in Japan are thus not difficult to identify. The most obvious was the size and characteristics of its economy. Mentioned earlier, it boasted the second largest national economy after that of the United States for roughly the last three decades of the twentieth century and the first in the next century.[7] Some of its industries became highly advanced in manufacturing, telecommunications, semiconductors, and computing, which spun off myriad business practices viewed as innovative around the world, particularly in the manufacture of automobiles, consumer electronics, and processing by small steel mills. Its personnel procedures were often admired and to varying degrees emulated, such as team-oriented work groups, continuous quality improvement; but, others rejected, most notably and categorically its life-time employment practices, criticized largely by Americans as too rigid and limiting managerial flexibilities. At the same time other Japanese industries remained complex, inefficient, and ineffective, such as its telephone network, wholesale and retail industries, and, to a considerable extent, its financial sector. Japan's practice of discouraging imports of non-Japanese goods into its economy, while basing a large portion of its own economic development on extensive exports to the West, posed an enormous problem for dozens of countries all over the world, often swaying many of the authors of academic studies and "think tank" policy statements, for instance.[8]

As Asia's most advanced economy, other emerging regional ones looked to Japan as a role model to emulate (1970s–1980s); became hosts to Japanese manufacturing plants (1970s–1990s), which helped diffuse know-how about high-tech manufacturing and use to other nations in the area; or evolved into economic rivals (1990s–2000s). Japan exported many of its most technologically advanced products into Asian consumer and industrial markets, although most of the attention outside Japan focused on pan-Atlantic–Japanese trade, not on Japanese activities in other Asian markets. A long-time student of Japanese industrial practices and resident in Japan, Gene Gregory, summed up the situation in his well-informed study of this nation's role in electronics with the blunt statement that "by 1970, East Asia had become the epicenter of the world consumer electronics industry, with Japan in undisputed leadership."[9] The core skills underpinning consumer electronics were evident in IT as well within the next decade.

Japan's style of industrial organizations (*keiretsu* described later) in collaboration with government's managed economic development and highly protectionist trade practices stamped the country's economy with some unique ways of functioning. However, while many commentators argued the case for Japanese exceptionalism, largely with respect to its business practices, these were no more mysterious or so differentiating than those evident in many other nations. Yet, many particular Japanese practices were featured in some exaggerated style by commentators.[10] As Japan's history with IT demonstrates, many of its activities reflected common patterns of behavior and motives evident elsewhere around the world, such as use of IT to reduce labor content of work, competition of computer hardware vendors for profitable sales, and a healthy appetite for export business. Reading memoirs by Japanese executives serve as a useful, yet partial, antidote to the arguments in favor of Japanese exceptionalism.[11] Just as the pan-Atlantic community began worrying about Japan's rapid expansion into their economies, Ezra F. Vogel, a leading expert on the Japanese, reminded Westerns that "I became convinced that Japanese success has less to do with traditional character traits than with specific organizational structures, policy programs, and conscious planning."[12] Vogel's observation is supported by the fact that these were the same categories of influences evident in other countries with respect to economic performance in general, and more specifically to the business practices of its IT suppliers and their local customers.

Japan had many of the prerequisites evident in other countries which also embraced digital technology that proved so essential to their rapid diffusion.

Japan had a highly educated workforce, although like many advanced econ-omies, more often than not it seemed to need more engineers and programmers of all kinds than it had.[13] Japan built a large and productive industrial sector, beginning in the late 1860s, over a half century before most other nations in Asia. Its timing mimicked approximately that of the United States and West Europe's Second Industrial Revolution, although to be more exact the pan-Atlantic community entered the Second Industrial Revolution by the 1840s. It was the demonstration of the results of that economic and technological trans-formation that led senior public officials to embrace what was the West's Second Industrial Revolution, but Japan's first. Eight decades later the Japanese began rebuilding and modernizing many factories and whole industries almost from the ground up as a result of the nation's devastation during World War II. This reconstruction meant Japan could embrace the latest American and European technologies and managerial practices with less effort than required of experi-encing the slower, more difficult refurbishing of existing systems, factories, and firms as occurred in many American industries in the postwar period. William Chapman, observing Japan's economic growth in the postwar period did not mince words: "Virtually all of Japan's modern miracle can be explained by events that took place on that empty landscape left by war and occupation. The ele-ments of success were formed by the special circumstances of the times, by decisions men made, by the interactions of people and social forces, and some-times by the sheer luck of the draw." More to our point, "New institutions were developed and new relationships between them were constructed, not because of any cultural predisposition but because certain choices were made and cer-tain understandings took root," including the "cooperative labor-management system" which "was almost entirely newborn."[14]

Japan's workforce has long been admired for its work ethic and discipline, particularly after its lifetime employment practices were affirmed during the 1950s. Yet, as in Western Europe, its workers were no more disciplined; indeed, Japan experienced considerable labor unrest in the 1950s while the Europeans did not. Both had excess labor in the 1950s so salaries remained lower than in subsequent years when their labor forces proved insufficient in quantity to meet demand. Per capita GDP rose along with the cost of living. Each region achieved a common *de facto* commitment among government, management, and labor to rebuild their nation as a national objective. Like Europe, the Japanese recovered economically by the mid-1950s. Japanese firms proved willing and able to learn about other nations' technologies and business practices, adopting them quickly and effectively, as its history of patent agreements and industrial partnerships with American and European firms made quite clear.[15] A CEO of NEC, one of Japan's leading IT suppliers, Koji Kobayashi, devoted an entire chapter in his memoirs to how his company learned from others around the World, citing specific examples of knowledge transfers that came from General Motors in the United States, of required readings by his managers about foreign business practices, and of visits to other countries and their firms.[16] Each of these types of activities contributed substantially to Japan's ability to learn about IT and to participate in its subsequent development and adoption, particularly between the 1950s and the end of the 1980s.[17]

This chapter presents evidence, however, that Japanese adoption and use of IT is not a tale of consistently stellar performance. Some industries used a great deal of computing, others were bereft of knowledge about what the technology could do. The image of Japanese as an economic powerhouse for many decades proved partially illusionary for in many ways its eco-political environment was slow to adapt to global economic and technological changes. IT did not contribute to the nation's productivity to the extent evident in many pan-Atlantic nations. The evidence suggests problems evident in centralized national economic policies in Communist Europe and China appeared similar, if fewer, in Japan as well. Over time the ability of the nation's economy to innovate its manufacture and use of IT slowed, certainly by the mid- to late 1980s, while simultaneously a consumer electronics renaissance flourished, continuing to the present, fueled by products that had embedded in them highly advanced electronic technologies.

Our narrative, therefore, offers a partial antidote to the widespread image of Japan as one of the most advanced users of computers, qualifying that image with a more nuanced account of its diffusion. While the vast majority of historical and economic literature on Japan focused on its export business, and largely on the supply side of the story, addressed as well here, attention is needed on the role of IT within the nation to provide a more rounded, albeit brief, view of Japan's involvement with IT. The bulk of its experiences with IT continued to sit squarely in Wave One, but with increasing evidence of consumer uses of digital technologies suggesting a march into Wave Two underway by individuals, if not also by some enterprises by the end of the 1990s. That tension separating and simultaneously overlapping the two waves needs to be set in context because consumer electronics has an enormous presence across Asia, Latin America, and Africa with more significance for the way life and work unfolds than in the pan-Atlantic community. Japanese products dominated many niches in this industry world-wide. I want to emphasize more than earlier observers that Japan's experience with IT served as a propellant pushing this technology into the rest of Asia. Put in counterfactual terms, if Japan had not been an advanced industrial economy that made and used IT, and if the Japanese had not embraced export and outsourcing strategies for growing their economy, adoption of computing by the rest of Asia would, in all likelihood, have occurred at much slower speeds and probably to a lesser extent, perhaps more as occurred in Latin America, South Africa, and parts of the Middle East where other national less technocratic economic role models proved more influential.[18]

JAPANESE SOCIETY, ECONOMY, AND INDUSTRIAL STRUCTURE

Because local economic and political circumstances conditioned the course and rate of diffusion—a key theme of my global study of IT diffusion—Japan's circumstances must be explained. In 1945, Japan had a population of just over 76 million residents, two-thirds more than either Italy or Germany and almost half as many as the United States. Japan's population was young then and within twenty years expanded to nearly 99 million; in 2000 it exceeded 127 million.[19] In 1950, 60 percent were under the age of 40, prime candidates for the local workforce. Japan's highly mountainous terrain forced the majority of its population to live on the coasts, resulting in an urbanized society for more than half its residents.[20] This demographic feature is an important attribute because the diffusion of knowledge and technology concentrated largely in urban regions, as in other nations, such as in Silicon Valley in California and Boston in the United States, or around London and Amsterdam in Europe.[21]

Table 7.1 documents Japan's fast growing and large national Gross Domestic product (GDP), an economic circumstance essential, indeed prerequisite, for IT diffusion in any country, since the technology was expensive and only economies with complex manufacturing and services industries could afford to use these tools for the entire period of Wave One. Tables 7.2 and 7.3 demonstrate that Japan also enjoyed a high per capita GDP and economic growth, also indicators of affordability for IT in both Wave One and Wave Two. Rapid growth is part of Japan's IT experience. Per capita GDP data in particular provide a snapshot of the obvious trend during its best recent decades to illustrate how quickly Japan had become an economic powerhouse, impressive to many nations around the World, not the least of which its Asian neighbors. In subsequent chapters, when we examine economic performance after the later takeoff of IT in those nations, such as in China and India, essentially in the late 1980s, 1990s, and beyond, how Japanese connected to their achievements is described.

Economists increasingly view Japan's economic evolution as having passed through four to five periods, each important to our story. The first, beginning with the end of World War II and the nation's occupation by the United States, covers the years 1945 to 1950 at which time the Korean War erupted. During this initial period, Japan rebuilt much of the damage it had suffered during the global conflagration and implemented an American-style constitution drafted largely by the American occupation forces and approved by the Japanese Diet. The Japanese government brought its economy under control by redesigning the Bank of Japan, dampening inflation, and launching reindustrialization, largely on the backs of prewar enterprises, such as those in telecommunications, which would prove so crucial to the introduction of computers into the country.

Table 7.1 JAPAN'S GROSS DOMESTIC PRODUCT (GDP),
1945–2008 (GEARY-KHAMIS DOLLARS—INTERNATIONAL
DOLLAR—IN MILLIONS)

1945	102.6
1955	248.9
1965	586.7
1975	1266.0
1985	1851.0
1990	2321.0
2000	2628.0
2008	2904.0

Source: www://ggdc/net/MADDISON/orindex.htlm (last accessed 11/11/2011).

Table 7.2 GDP PER CAPITA, JAPAN COMPARED TO SELECT
OTHER ASIAN ECONOMIES, 1950–2008, SELECT YEARS
(1990 INTERNATIONAL DOLLARS)

	1950	1973	1993	2008
Japan	1,921	11,434	19,478	22,816
South Korea	854	2,824	10,232	19,614
Taiwan	916	3,448	11,929	20,926
Thailand	817	1,874	5,666	8,750
China	448	838	2,342	6,725
India	619	853	1,390	2,975
Indonesia	803	1,490	2,994	4,428
New Zealand	8,456	12,424	14,031	18,653
Australia	7,412	12,878	17,853	18,653

Source: Angus Maddison, *Monitoring the World Economy, 1820–1992* (Paris: OECD, 1995): 250; www.gglc.net/Maddison (last accessed 10/1/2010).

Table 7.3 JAPANESE GDP GROWTH RATES, 1961–2008,
SELECT YEARS (PERCENTS)

1961–70	10.2
1971–80	4.5
1981–90	4.0
1991–03	1.2
2003–08	1.0

Source: OECD Economic Survey, various years.

The start of the Korean War in June 1950 signaled a second phase. The war led to a rapid and massive increase in demand for heavy industrial goods by the Americans, most notably steel, which helped speed up Japan's economic recovery and launched this second era normally viewed as extending to 1970. This war prompted a nationwide economic surge as it did in the United States and in parts of Western Europe by creating demand for many products needed for the conflict, such as steel, but also outputs from many industries, in large part for Japan because it was so near Korea. This surge in business proved sufficient to make computing affordable and practical, raising the nation's GDP and causing complex organizations to grow large enough to serve as logical candidates for using computers. Unemployment declined and sale of existing inventories soared in many industries, creating demand for new products and materials. Observers in Japan reported that U.S. demand for Japan's goods "proved to be powerful enough to reanimate Japan's economy, which, due to the disinflationary policy, had been in a coma."[22] As a consequence, in the two subsequent decades (1950–1970) Japan reindustrialized massively to such an extent that these years assumed their own name—High Growth Period—a time characterized by many annual growth rates in real GDP of about 10 percent, while inflation hovered at 5 percent. Japanese prosperity proved tangible. Political stability usually prevailed and after labor disquiet in the 1950s subsided unions became stable as well.[23] As in so many other countries with advanced capitalist economies, Japan enjoyed a brief period of near idyllic economic performance.

A third period extended neatly across the decade of the 1970s, ending in 1980 when Japan's relative stability ran headlong into numerous political, economic, and technological head winds, including many from IT innovations that caused managers to begin questioning the continued practices of existing ways of doing business. It is crucial to keep in mind that as Japan's economy expanded in the 1960s its indigenous IT companies ran out of local customers and, thus, in conjunction with government economic development policies, began expanding sales through aggressive export trade. Almost every segment of the IT world participated: semiconductors, consumer products, computers, IT peripheral equipment, telecommunications, and even the most primitive robotics and other industrial equipment. Software proved to be an exception to an otherwise impressive array of exports, uncompetitive for a complex set of reasons discussed later. Prior to 1980, the government had already implemented a variety of measures that protected Japanese industries it considered strategic to the nation's interests from being overrun by foreign competitors. The strategy was to protect these Japanese firms until they matured and grew to sufficient strength to compete effectively in world markets. All IT industries were deemed worthy of this attention and protection by government, most specifically the Ministry of International Trade and Industry (MITI).[24] During this period the Japanese government implemented various measures to constrain IBM's success in the national computer industry, while funding R&D and other initiatives to help local computer suppliers mature and expand sales with competitive products at home, against IBM, and abroad.[25]

The fourth period began after 1980, a time when many developing problems came to a head, characterized by demand of governments around the world that Japan open its markets to free trade, which it liberalized to a limited extent.[26] Changes in domestic political leadership also took place. Rapid introduction of new technologies became available worldwide to which Japanese IT and telecommunications firms responded much too slowly, such as distributed data processing brought about by minicomputers and PCs, and later the Internet. This fourth era extended right into the new century and because of our proximity in time to this period, it is not yet clear that one can declare definitively that a fifth one now exists.[27] But historic change was in the air. As two observers of this time noted in the 1980s, "Japan seemed invincible . . . whereas in the 1990s and the beginning of the new century, Japan seemed unable to do anything right."[28]

But much is clearer about events of earlier years. From the early to mid-1950s through the 1970s, Japanese businesses and government agencies sorted out how they would work with each other, how competition would occur. They funded and exploited R&D, promoted foreign trade, and implemented monetary policies to protect the nation which saw itself as alone against the World and handicapped by expensive energy, most notably oil and electricity. These events have been the subject of hundreds of books and thousands of articles. In the 1970s and 1980s observers wanted to explain why the Japanese economy performed better than seemingly everyone else's. Most sounded the alarm that "Japan, Inc." was going to take over such advanced industries as consumer electronics, semiconductors, and automobile sales worldwide, especially after it had taken the lead in consumer electronics and was well on its way to doing the same in various classes of semiconductors.[29]

Beginning in the late 1990s, observers wanted to explain why Japan's economy failed in the latter 1980s and did so miserably throughout the decade and beyond in sharp contrast to the spectacular period of growth in the 1970s and 1980s. To achieve their explanatory objectives, economists, political scientists, and even Japanese business executives advanced various concepts, also applied to the role of IT. These Japanese, American, and Westerner commentators used such descriptive notions as *corporatism, bureaucratic, neomercantilist,*[30] *political economy,* and outright *protectionist* or *autarkic.*[31] Most focused on similar themes: close collaboration among private enterprises and with large government agencies, the powerful role of banks in funding companies designated by the national government as worthy of protection, and national trade practices.[32]

The dialogue proved complicated because different nations had varying economic models that became the lens through which their observers judged Japanese activities. The Anglo-American form of economic behavior, normally characterized as "liberal capitalism" did not reflect, for example, what France, Germany, South Korea, and many other nations practiced, which could be more precisely described as "non-liberal capitalism." In the latter group's approach, the state played a more proactive role, placing less trust in Schumpeterian free market forces than did the Americans, for example. It is into the non-liberal camp that Japan fit and it is this dichotomy in how economies and public policies unfolded that helps explain many of the IT trade battles that flared up, beginning in the late 1950s and extending variously to the present. Ultimately, the most important motivation for many to look at Japanese business practices and public policies resulted from Japan's highly successful foreign trade results.[33]

One of the most useful of these descriptions of Japanese eco-political behavior for understanding IT's story comes from Marie Anchordoguy, who has been examining the role of IT and other technologies from both economic and political perspectives for over a quarter of a century. Anchordoguy describes the Japanese ecosystem as "communitarian capitalism." In her words, "Communitarian capitalism is an economic system characterized by an activist state and a number of private-sector organizations that manage markets to promote development and national autonomy in the context of the broader goals of social stability, predictability, and order."[34] Most confusing to many Westerners, therefore, was its byproduct, "elaborate social conventions about how the state, firms, and individuals should behave in given situations. The rigidity of these customs binds community members into a strong collective identity."[35] In the realm of IT trade when Japanese faced transformations in the global change in technology it proved unable to respond as quickly to these as could other economically advanced nations. This is a problem discussed at length in this chapter.

From an end user's (consumers of IT) perspective in Japan adoption of computing for competitive reasons did not necessary require speed or as elaborate forms of cost justification as in many other national economies. Social stability proved generally more important than maximizing short-term profits at an individual firm, particularly in large enterprises; management valued full employment, providing it resulted in an increasingly productive workforce.[36] Following World War II, Japanese society desired security, stability, a way out of the humiliation and horrors brought on by the war, and return to international legitimacy and economic well-being. In particular, public officials valued national and industry-wide successes more than those of individual firms. One government official stated clearly the objective: "Japan's system prefers no big winners and no big losers."[37] Another declared that "all companies should survive."[38] These objectives varied slightly over the past six decades with the greatest shifts coming only in the late 1990s and early 2000s, when deteriorating economic conditions, transformations in technology, and rising unemployment forced all players to change their behavior. Companies, however, sought growth, protection, and independence from too much collaboration and sharing of patents and R&D results among their competitors. Elsewhere, even protective West European regulators were prepared to let some firms die, while the Americans considered the birth and death of enterprises positively essential for a healthy capitalist economy.

The national banks served as the key stockholders, behaving more like stakeholders. Individuals played a far lesser role in Japan than they did in pan-Atlantic economies. Banks funded private enterprises with an expectation of longer term, not quarterly or even in some instances, annual returns on investments. Often financial officials and retired government regulators served on the boards of these firms, creating the close symbiotic relationships at the heart of the Iron Triangle. In short, banks were the primary owners of firms across multiple industries. Companies aligned with a particular bank formed an enclosed market, an ecosystem, where they overwhelmingly bought from and sold to each other. A company that made computers, such as Fujitsu, sold its products to other members of its bank's invested firms, and to a lesser extent to others outside its circle, and, conversely, if a corporation wanted a machine and Fujitsu was part of its ecosystem it would tend to acquire its products. Customers of IT were routinely discouraged—but not necessarily prohibited—by government officials from selecting a foreign made system, such as those sold by Americans, even by IBM Japan, even though it manufactured locally and employed Japanese nationals. The concept of the enclosed community was encapsulated in a Japanese word, *keiretsu*, which loosely translated referred to industrial groups, a core concept to keep in mind when thinking about how one bought and sold IT in Japan during the second half of the twentieth century.[39]

If Japanese banks ran into difficulties with their investments, government agencies would backstop them, much like an insurance agency. Public officials declared which industries needed protection and aid for the good of the nation, thereby putting the state in the business of managing markets, literally down to the level of determining which companies and products should thrive (or not), and in what product categories.[40] Within such controlled limits, firms were expected to compete, for example, on excellence of products, although they had to share risks, even negotiate joint solutions to problems, with the result that they often had similar goods priced almost the same. In such an environment, it was very important to hold onto existing customers.

In IT that meant most vendors and their customers adhered to IBM's S/360 standards and notions of centralized computing, but with enough proprietary software (or changes to IBM's) to prevent (restrict) a customer's ability to move easily to another vendor. This strategy proved increasingly ineffective when decentralized data processing came to the fore in the late 1970s and spread in the 1980s. Changes in technologies highlighted that incumbent vendors were inadequately prepared to respond quickly enough to these disruptions to their prior centralized computing, stable ecosystem, let alone allow the demise of a company unable to react to changes adequately in time. The process of creative destruction that flushed out insufficiently responsive vendors unable to adapt had "been blunted," in Anchordoguy's words.[41]

There were several primary government players in this system, including the Bank of Japan, of course, but also MITI with its enormous legal power dating from the late 1940s and not challenged successfully by rivals, foreign suppliers, customers, or even local economists until the late 1980s. Its authority did not diminish until the 1990s and early 2000s. In its heyday (1940s-early 1980s) MITI dictated who received import and export licenses. It funded R&D and other initiatives, persuaded firms what they could or could not do (euphemistically known as giving "administrative guidance"), and prioritized industries its officials believed most essential to the nation's economic welfare. For example in the 1950s, it favored heavy industries, such as steel and ship building, giving these economic and financial help, while sheltering them from global competitors. After the oil shock of 1973–74 with its resultant higher costs of oil, MITI's officials concluded that they needed to promote industries that used less fuel, such as electronics, and to reduce support for others that consumed a great deal, such as steel manufacturing and shipbuilding. MITI used its economic power, legal authority, and influence to get a chosen industry access to foreign technologies so that it could quickly produce and sell these both within Japan and on the world market. MITI's facilitative role exerted a profound influence on activities of local IT suppliers, IBM, and other American IT firms all through the period.[42]

By favoring some firms at the expense of others, and in collaboration with Japanese banks, MITI discouraged new enterprises from forming, because they would not be able to obtain venture capital. This practice left existing vendors with fewer incentives to change than those in other more competitive economies where new rivals emerged constantly, challenging incumbent companies in industries experiencing the kind of technological turbulence that occurred in telecommunications, semiconductors, computing, software, robotics, consumer electronics, and later, biotechnology—all industries that MITI chose to support with the exception of consumer electronics. Between the 1970s and mid-1980s, IT, telecommunications, and other firms acquired access to technologies and produced goods off the backs of these. They and MITI expanded trade outside Japan, while government officials blocked competition at home from non-Japanese firms through trade barriers created to shelter indigenous companies, and with technology licensing agreements designed to give local firms access to non-Japanese technologies. Japanese enterprises, however, were also often forced by government to collaborate in sharing patents and in co-developing technologies and products. This practice resulted in costs of transactions (internal expenses) and technological collaborations to remain relatively low when compared to those in pan-Atlantic nations, although the expense of products to their customers remained higher, since there was less domestic competition over prices. Rather, intense domestic rivalry centered on quality of products and service.[43]

Between 1950 and 1980, MITI and other government agencies negotiated over 25,000 licensing agreements with non-Japanese companies around the world worth some $6 billion, including with most key U.S. IT and West European IT vendors. Those arrangements reduced and, in some instances, eliminated the requirement of Japanese firms having to invest time, resources, and energy to develop similar technologies and products. This strategy of rapid adoption of existing technologies, combined with fast turnaround in producing products based on these, often with some incremental innovations, when coupled with protection from foreign competitors at home, went far to account for the rapid speed with which Japan's economy became so modern and up-to-date in its high technology product offerings by the mid-1970s.[44] Some Japanese firms refused to play by MITI's rules, however, and so were denied as much access to funds, permits, and development projects as made available to those that collaborated with government initiatives or that were deemed to be in an important industry. Through the 1960s, in addition to technology transfer agreements and licenses to sell in Japan or export, MITI also applied currency controls to supply low cost capital to targeted industries (such as those in IT) and to exercise benign neglect to those it was not interested in helping. To be sure, MITI was a large agency in which intra-agency and interagency priorities conflicted, but MITI's achievements with respect to effects on the IT world were as it generally desired, particularly in the startup decades following World War II. Japan built from scratch competitive industries in semiconductors, IT peripherals, computers, and, despite its lack of interest, consumer electronics flourished too.[45]

Particular policies proved enormously problematic for many global vendors of IT eager to compete in Japan. Their imports were subjected to high tariffs, which helped local vendors be more competitive in home markets. Just as important, MITI did not allow foreign companies to invest directly in local firms to get around tariffs. That meant, for example, American companies had to license their technologies and products to local firms at low prices, which reduced the yield they might otherwise have obtained if they could have invested directly in Japan. MITI's officials often confronted individual Japanese managers who requested permission to acquire a foreign computer to convince them to buy from indigenous vendors, or delayed issuance of requisite permissions. MITI implemented all these strategies in the 1950s and continued to use these to one degree or another to the end of the century.[46]

Simultaneously, however, what one would characterize as normal capitalist rivalry existed in Japan that on occasion became part of the Japanese IT diffusion story. If not initiated into MITI's inner circle of select industries or companies, and into the *keiretsu* ecosystem, firms competed with rivals as anywhere else in the World. Yet, they usually experienced enormous difficulty convincing local banks to invest venture capital in them as these were some of the most conservative financial institutions in the industrialized world. Banks insisted on tangible collateral, such as land, unlike venture capitalists willing to invest on the basis of a patent or good business plan, as so frequently occurred in the United States with emerging technologies and practices.

The perfect example is Sony, a firm established after World War II that sold the world's first transistor radio. MITI's officials thought radios were trivial, overlooking the fact that Sony's founders acquired a license from Western Electric in 1953 to use transistors—soon a core component of computers—to embed in its initial products. Sony acquired knowledge of electronics from a combination of staff educated in Japan before World War II and later by establishing a U.S. subsidiary with a laboratory in New Jersey, home to many high-tech consumer and computer operations, most notable RCA, a giant in American consumer electronics in the 1940s and 1950s. Sony's engineers added to their knowledge of electronics, and specifically about transistors; one member of the original team forming Sony, Leo Esaki, received recognition for his work with a Nobel Prize in 1973. The firm succeeded, often having to obtain funding for expansion and R&D outside of Japan. It went on to introduce innovative products, such as the CD-ROM (compact disk-read only memory), which gave computers video and audio capabilities, the Walkman in 1979 of which it sold over 200 million units over the next 30 years, the DVD, and the Sony PlayStation. In 1982 it entered the personal computer market where it remained competitive, offering new products over the next three decades. Like Toyota, Honda, Canon, and Kyocera, Sony was an outlier in the story told in this chapter of what computer vendors and MITI did, following its own strategy and finding its own way to success.[47] More telling, Sony could also respond to new opportunities and changes in the market *quickly enough* to become a world-class competitor in its chosen markets, demonstrating, like other exceptions, that firms in Japan could be managed effectively, avoiding some of the problems local vendors experienced with their more lethargic strategies and tactics.

MITI's and Bank of Japan's efforts, the cumulative results of the *keiretsu* system, along with such other developments as stable political regimes, culminated in the high water mark of Japanese economic success in the period 1980–1985. The period stood in sharp contrast to the difficulties faced by many other nations at the time. Japan avoided inflation stimulated by governments in other nations, largely in response to the second oil price increases of 1979–1980 and subsequent bouts of inflation. The banking system was carefully managed, the internal economy isolated from global effects, and exports expanded, as discussed below for IT products. But in 1985, the Bank of Japan made some errors too, driven by the government's decision to assuage complaints from the United States and European countries that Japan had closed off its domestic markets from their firms for such products as semiconductors, computers, and automobiles, signing the Plaza Accord in September 1985.[48] As the dollar simultaneously declined in value, and to make it possible for foreign vendors to sell in Japan per the accord, the Bank of Japan took steps to slow appreciation of the yen. The bank expanded the money supply and credit that fall (1985). Expanded cheap money and flawed policies designed to liberalize finances led property values to rise rapidly—the real estate bubble economists talked about afterwards—diverting investments from technology and to those that led to the bubble which burst in 1989–91. The economy began to seize up, investments in capital goods (such as computers) slowed, while the public reduced its expenditures on such things as new homes and consumer electronics, instead increasing their savings.[49]

Central to our story is that Japan's national government continued to nurture development of indigenous suppliers of IT available in its national market that competed effectively around the world. At the dawn of the new century, four out of the World's ten largest manufacturers of computers were Japanese: Fujitsu, NEC, Hitachi, and Toshiba. In telecommunications (another protected industry) NEC and Fujitsu made the global top ten list; but, the greatest success came with semiconductors where the Japanese held six out of the top ten positions: NEC, Toshiba, Hitachi, Fujitsu, Mitsubishi Electric, and Matsushita. Martin Fransman, a leading authority on Japan's high-technology industries, cautioned, however, "while Japan's computer and communications companies have dominated global markets in areas such as memory semiconductors, optoelectronic semiconductors, microcontrollers, and liquid crystal displays, they have been significantly less successful outside Japan in crucial markets such as mainframe computers, workstations, servers, personal computers, microprocessors, packaged software, and complex telecommunications equipment."[50]

Success in some sectors of the IT market resulted from being able to ride the wave of overall economic expansion of Japan's economy which could use its products, where local competition proved effective, and demand lubricated with government backed initiatives, especially when they responded to the specific needs of customers in the local economy. But, effectiveness and support at home did not necessarily mean that they would be as successful outside Japan against other competitors. For example, while the Japanese were enormously effective at home selling mainframe computers, where demand for this technology remained strong, they did not develop as successfully product lines in minicomputers and PCs, let alone software—all new classes of technologies which grew in importance outside Japan in the 1980s. If the local market, which shaped the products and work of indigenous IT vendors, caused these firms to introduce goods attractive at home but not abroad, then these enterprises did not do as well worldwide.[51]

Path dependencies, or "learnings," acquired by Japanese vendors from their customers played an enormous role in shaping their ability to compete, regardless of MITI's or any other government agency's policies, suggesting a limit to the influence of governmental and banking actions. As in the case of Communist European government policies, constraints existed, diminishing the ability of officials to pick "winners" and "losers" or to define realistically features of technologies needed in both domestic and foreign markets. So, was there evidence of an underlying private sector strategy?

Economist William K. Tabb best summed up the role—indeed contribution—that the various IT industries and their users played in Japan's economy in the past half century when he argued that "Japan's pattern of rapid growth was one of moving continuously upscale to more sophisticated process technologies and product mix."[52] It is the short answer to the complicated question asked by almost every observer of the Japanese: How did Japan go from being such a devastated nation at the end of World War II to being the second largest economy in the World? As with so many other nations, IT played an integral role in the economic evolution of whole industries and economies. The Japanese were not exceptional, rather generally more like the others.

ORIGINS OF JAPANESE COMPUTING, 1950S–LATE 1980S

While many aspects of Japanese IT diffusion paralleled events in other countries, as with those other nations, however, it had its own unique features as well. Specifically, the origins of computing in Japan are rooted in different traditions from those of the pan-Atlantic world. For one thing, motives for developing and using these technologies and, consequently, their funding, did not spring from military requirements of the Cold War so evident in the R&D activities of the Soviet bloc, the United States, Great Britain, France, Sweden, and a few other nations. In Japan's case, incentives to create IT competences grew out of civilian economic needs for foreign trade and national economic development. Therefore, sources of funding, incentives for developers of computers, and practices of local vendors often displayed a pattern of their own, just as in every country there were specific local features of adoption and evolution. Second, and largely unique to Japan, the majority of computer development came out of the pre-existing telecommunications industry, particularly during the 1950s through the 1970s, but as in many other countries, also included work done at government agencies and universities. Paths of learning by which Japanese engineers came to acquire knowledge about the technology, and found support for their work, originated less from universities and, of course, not at all from the nearly nonexistent military–industrial complex. This stood in sharp contrast to the experiences in many other nations where local electronics firms, office appliance suppliers, and military laboratories collaborated in the development and early diffusion of IT, most notably between the 1940s and the end of the 1960s. Although a few universities participated in the early decades of computing in Japan, theirs' was a diminutive role when compared to those of Western academic institutions.[53]

Third, those Japanese companies engaged in developing and diffusing computing did so across a broader set of digital technologies than their cohorts in the West. A firm built computers, manufactured semiconductors, wrote software, sold telecommunications technologies which had embedded in them digital technologies, and later diffused all manner of ITs into such spill-over products as consumer electronics and robotics. In the West, these various classes of digital technologies (lines of business and products) emerged from companies that specialized in one or few of these. Intel made semiconductors, not semiconductors *and* computers; Zuse built computers and avoided offering robotics, or, becoming a software powerhouse, and so forth. A few limited exceptions existed, of course, such as IBM which for many decades manufactured semiconductors for its own use in its products, did basic research on all manner of IT, introduced computers and software, but rarely robotic devices and never consumer electronics with the one notable exception—the PC. The Japanese case presents a situation in which the transfer of knowledge about IT across an entire supplier's enterprise occurred quickly, often effectively, facilitating development of computing.

Yet in the beginning, Japanese computing briefly mimicked Western experiences, starting with early activities at universities and public research laboratories in the 1950s. The first relay computer was designed and built at the Electrotechnical Laboratory (ETL), an agency within MITI, in 1952, a remarkably early date given that the Japanese economy was still recovering from World War II and the positive economic effects of the Korean War were just starting to be felt. A machine built at the University of Tokyo in 1958 came into use as well, using parametric oscillation of magnetic ferrite cores, called the Parametron (PC-1). The Electrical Communication Laboratory (ECL), a department within the Japan Telegraph and Telephone Public Corporation (M-1) constructed a similar device. Eventually these experimental machines led to commercial versions developed by Hitachi, Nippon Electric Company, Fuji Communications Manufacturing Company, Oki-Electric Industry Company, and other firms. The following year scientists at the University of Tokyo had a vacuum tube computer, a project that had taken them six years to complete. Japan's first digital computer appeared in 1956. Additional early design and construction of computing equipment took place at the University of Osaka. In the private sector Fuji Photo Film Company put a system into operation a system in August 1956, named Fujic, considered the first electronic computer built in Japan for production work. The first Japanese computer to use a CRT and vacuum tube technology was built at Tokyo's university in collaboration with the Tokyo-Shibaura Electric Company, called TAC; it went into use in early 1959. Table 7.4 catalogs various systems built in the 1950s, demonstrating that quite early in the history of computing numerous projects were underway in Japan involving a small but diverse group of academic, government, and private sector firms.[54]

Several patterns are observable from the data in this table. First, a variety of organizations and companies became involved quite early in constructing machines. Second, they built them for each other, borrowing knowledge from collaborators and European and American sources (note the use of the British-styled Mark series). Third, there existed a considerable number of early systems. If we had extended the table to include machines that went into use in 1959, over 15 additional systems would be listed coming online at the rate of one to two per month.[55] Additionally, most of the work occurred in the Tokyo area where knowledge sharing occurred quickly from one organization to another, a point not lost on observers of the Japanese scene trying to understand why computing spread rapidly at the time.[56]

Behind the production process evident in the table, American computers began to come into Japan which local producers learned from, beginning in 1954. Simultaneously in that same year five local firms were already manufacturing transistors under licenses from American firms: NEC, Fujitsu,

Table 7.4 EARLY COMPUTERS IN JAPAN, 1952–1958

Developer	User	Machine	Operation Date
ETL	ETL	ETL Mark I	December 1952
ETL	ETL	ETL Mark II	November 1955
ETL	ETL	ETL Mark III	July 1956
Fuji Film	Fuji Film	FUJIC	August 1956
ECL	ECL	M-1	March 1957
Hitachi	Hitachi	HIPAC-1	August 1957
FCM	Sakura Film	FACOM 138	September 1957
ETL	ETL	ETK Mark IV	November 1957
NEC	NEC	NEAC-1101	January 1958
Tokyo Univ.	Tokyo Univ.	PC-1	March 1958
FCM	Canon Camera	FACOM 128 B	April 1958
HEW	HEW	H-1	October 1958
NEC	JEIDA	NEAC-2201	October 1958
NEC	Tohoku Univ.	SENAC-1	November 1958

Source: Japan Electronic Development Association, J.E.I.D.A. and Its Computer Center (Tokyo: JEIDA, undated [circa 1959], CBI 714, Box 5, Charles Babbage Institute, University of Minnesota, Minneapolis.

Hitachi, Matsushita, and Toshiba, each endowed with extensive skills in electronics of all kinds accumulated over many decades. The first three became the major suppliers of commercial computers in Japan; each was also expert in the development and manufacture of telecommunications. As the table also demonstrates, nonprofit organizations were some of the first to experiment with computers since it was not clear to local telecommunications and electronics firms that there was a commercial market for such devices until about 1955, and so, many chose to proceed cautiously into the new technology. The two most important early players were MITI's ETL, and Nippon Telegraph and Telephone (NTT)'s Electrical Communications Laboratories (ECL), which transferred their work on computers to NEC, Fujitsu, Hitachi, Toshiba, Mitsubishi Electric, and Oki, beginning in the late 1950s.

With initial imports of American computers starting in 1954, along with licensed production of transistors in Japan that same year, Fujitsu and other firms approached MITI with the recommendation that it foster development of an indigenous computer industry; a suggestion put into action with the passage of enabling legislation in 1957 (Electronics Development Provisional Act). This law authorized funding of R&D subsidies to these companies, loans for the acquisition of these new products, accelerated depreciation of manufacturing plants and production of such machines. In the first five years (1957–1961), however, financial aid barely reached $1 million, so the process of assistance proved slow to arrive in the beginning.[57] Of all the vendors, the one which most enthusiastically committed itself to the new market was Fujitsu, leading to its early production of new systems (see table 7.4).[58] At Fujitsu a young generation of engineers wanted to pursue computers, including a future chairman of the board, Taiyu Kobayashi. He approached senior management to commit the firm, despite the old guard's reluctance. The transitions by these various companies to computers mimicked much of the same process of transformation into this industry that IBM underwent internally with Thomas J. Watson Jr. during the early to mid-1950s, leveraging a new opportunity at a time when—in the case of Japan—other communications firms had a stronger position in the locally known and less risky telecommunications market on which they were more focused. At IBM the issue concerned tabulating equipment versus computers. In each instance, entrenched interests advocated the status quo, others favored emerging new products and markets, and both sides competed internally for support and, ultimately, for control over the destiny of their enterprises' futures.

NEC was in a strong position in the telecommunications market, yet also saw the opportunity to sell computers by the late 1950s. Management decided to use that technology to complement its sales of telecommunications products, developing what famously became known as its "C&C" strategy, involving early online computing and communications.[59] Hitachi, weaker than Fujitsu or NEC in the telecommunications market, had a stronger presence in consumer electrical and electronic goods markets, and looked to similar firms in the United States for its business models, most notably RCA and General Electric (GE). Like the other two Japanese firms, it entered the market for computers in the mid-1950s, as did Toshiba and the smallest of the firms, Oki.[60] Meanwhile MITI established an Electronic Industries Section and the Electronic Industry Deliberation Council to coordinate its response to this new market. Membership on the Council included vice-ministers at MITI and from the Ministry of Finance, presidents of various electronics firms, president of the Japan Electronic Computer Company (JECC beginning in 1961 after its establishment), and academics, for a total of some 40 individuals who, over the years, coordinated many IT-related activities and built consensus around national strategies and collaboration by individual firms.[61] MITI was clearly organizing to support local firms.

Takeoff on the supply and demand sides in Japanese computing occurred in the 1960s, although not obvious at the start of the decade. Japanese firms entered the decade with experience building small, expensive computers that were technologically several years behind those manufactured in the United States. Demand for these systems remained quite limited, but interest was growing in government, banking, telecommunications, large machine manufacturing, and automotive firms. In 1962 and 1963 combined all Japanese vendors built 5 large computers, while an additional 44 came into the nation from foreign vendors, largely American. However, medium-sized systems (in the same range as IBM's 1401s and 650s) proved more popular, with 197 manufactured between 1960 and the end of 1963, although in this class of machines, foreign products also came into the country (122 systems). During these same years 128 locally built small computers were installed, while another 130 came from outside the country.[62] This surge in demand began in 1960 with mid-sized systems, an interest in computing not lost on local firms, or MITI. Fujitsu set up a separate computer division in 1961, while MITI's experts in the industry began sorting through what actions to take. MITI wanted to expand the capabilities and capacities of local firms, and chose to negotiate licensing of American and European technologies to reduce the technological innovation gap and bring quickly into Japan IT expertise and ability to make systems. In quick order agreements were signed: Hitachi with RCA (May 1961), Mitsubishi Electric with TRW (February 1962), NEC with Honeywell (July 1962), Oki with Sperry Rand (September 1963), and Toshiba with General Electric (October 1964). As Tasiyu Kobauashi, the senior executive at Fujitsu recalled years later, "it was, of course, the easiest method of obtaining such technology."[63]

As in so many other countries, there was also the issue of IBM's role. IBM's Japanese presence began in 1925 when a local agent for the company rented the first products. IBM established a local company in 1937 (Watson Business Machines Company of Japan, Ltd.). In 1949 IBM regained control over its local properties confiscated during World War II. It installed its first 650 in Japan during 1958 at the Atomic Research Institute and the first 704 in the country the next year at the Government Meteorological Agency. In short, it participated in the embryonic Japanese computer market from nearly the beginning. The domestic history of IBM in Japan mirrors that in many European countries with plants established or expanded all through the 1960s and 1970s, along with opening of sales offices in major cities.[64] IBM could sell directly in Japan because it had been present there before laws passed in the 1940s prevented foreign companies from doing that; other American suppliers of computers had to participate through technology-sharing agreements. The major events for IBM, however, were the ongoing negotiations between IBM and MITI on how to sell its new products in the late 1950s. IBM refused to establish an operation dominated by Japanese firms, as proposed by Fujitsu and later MITI, leading to protracted negotiations. IBM's lead negotiator, James W. Birkenstock, recalled that in late 1960, during an impasse, MITI's lead negotiator, Akasawa-san, attempted "to bully me into acquiescence," stating "that unless IBM accepted MITI's terms, MITI was prepared to impose severe sanctions on IBM Japan, crippling its current operations and clouding its future."[65] That threat did not deter Birkenstock. MITI eventually gave IBM permission to operate in Japan, while the company agreed to cross-license patents with five Japanese firms for five years.

In the 1960s, demand for IBM's systems remained strong, because their quality proved superior to that of local firms at the time and there existed availability of supply; but, MITI pressured buyers to buy local first. IBM's corporate records regarding its Japanese operations are full of letters complaining about the constraints put on its ability to sell computers, such as the government requirement for permission to open sales offices and for users to obtain authorization to acquire IBM's imported or even locally manufactured products. An internal report, dated December 20, 1967, noted that the only two countries in which IBM's dominance of local mainframe markets hovered below 50 percent were Great Britain and Japan (46–50 percent in GB, 47 percent in Japan). The author of the internal report noted "IBM, however, has been and still is the single dominant factor in the Japanese marketplace, and government policies and administrative measures appear to be directed specifically at curtailing the importance and growth potential of IBM Japan." The same report commented that "demand for computing in Japan is far in excess of local supply capability," going far to explain the desire for IBM's machines, while opining that "all six major local manufacturers must be characterized as being weak."[66] At the time, local firms only participated in 25 percent of the Japanese market, so the issue of what to do with various American and European firms was serious to MITI and Japanese providers.[67]

IBM's introduction of the System 360 in April, 1964, precipitated as much of a shock to local firms as in other countries. Hitachi, NEC, and Toshiba collaborated with their American partners to upgrade their product offerings. Fujitsu, which did not have an American ally, went it alone, updating its product line by introducing the FACOM 230-60 in 1968; the first installation of this system took place at Kyoto University. Japan's most important telecommunications company, NTT, installed Fujitsu systems after concluding that this supplier had finally acquired the necessary capability and market presence to be a competitive supplier of digital products. MITI launched R&D projects to help local firms upgrade their products, most notably the Very High Speed Computer System Project (VHSCS), conducted between 1966 and 1972, specifically to bring all locals up to technological snuff. The term used to describe such efforts was to "catch up." It was through this project that MITI cajoled local firms to specialize in parts of the market to optimize each company's limited resources and capabilities. Hitachi, Fujitsu, and NEC worked on time-sharing computing; the weaker Toshiba and Oki on peripheral equipment for mainframes. NEC specialized in memories, while the technologically most advanced of the group, Hitachi, focused on high-speed logic devices.[68] But, local firms could only offer new products as fast as their American partners could develop new devices, since they were dependent on the latter's technological innovations. Meanwhile local users were urgently demanding S/360s for the same reasons as their cohorts in the United States and Western Europe.

Hardly had local suppliers begun to introduce new products when IBM brought out its S/370 line of computers in 1970. That event put into high relief the continuing slow reaction by Japanese technology firms to evolving digital technologies, which still was deployed while a measured response was underway with respect to the S/360, first shipped to customers by IBM a half decade earlier. Historian Alfred Chandler, Jr. used the word "stunned" to describe how officials at MITI reacted to this product's introduction, because it once again made clear that Japanese vendors were technologically behind and within those companies how badly so.[69] MITI called for mergers of local firms, as did Europeans in their own home markets, but Japanese vendors refused. MITI then proposed a "New Series Project" to stimulate collaboration among its companies to which they reluctantly and slowly acquiesced in the early 1970s.[70] Simultaneously, Fujitsu negotiated an arrangement with American computer engineer Gene Amdahl and his new enterprise to provide him funding he desperately needed to build computers based on the S/360–370 technology that he could sell, some of which he had earlier developed while at IBM. This arrangement provided Fujitsu with a rapid infusion of detailed technological information about IBM's systems which the firm applied to its products. During the 1970s Fujitsu continued to increase its share of ownership in Amdahl's firm.

At the same time Japan's exports of computers and peripheral equipment did well. In 1974, for example, 20 percent of all Japanese exports went to the

United States, grew to 31 percent the following year, jumped to 75 percent in 1976, and then began declining to the 63–66 percentile range for the rest of the decade. The surge of sales in the United States reflected the rapidly growing use of computers in the United States versus Europe, and favorable monetary exchange rates. Dollar flows quantify the results: $16.7 million in 1974 to the United States, up to $99.7 million in 1976, similarly the following year ($96.9), and in 1978 to $218.1 million, all convincing evidence of the continuing growth in demand for computers in the United States. However, demand also grew worldwide as total global sales increased in dollar volumes: $83.2 million in 1974 and increased steadily over the next few years to $331.1 million in 1978, with a third of that year's sales going to Europe and Asia combined. Sales kept expanding over the next several years at similar rates of compound growth. One final set of statistics makes our understanding of the results more explicit: number of systems installed. As the dollar volumes went up, so too did the number of computers sold. Amdahl led the pack with a worldwide total of 189 between 1975 and 1978, of which another 100 went into the U.S., and in the following year 200 went into the United States. Data on Hitachi indicates it did not enter the export business for computers until 1978, when it sold 31 systems and a similar number the following year. Most of these machines would have been sold to existing users of computers seeking to find a less costly alternative to their own indigenous suppliers.[71]

Meanwhile, IBM continued to do well in Japan, notwithstanding public pronouncements by the firm and the American government, overcoming a concerted effort by Japanese public officials to dampen its activities. By 1977–1978, at the height of IBM's success with S/370s and follow-on products around the world, business volumes demonstrated both growing demand for computers in Japan from this firm and indigenous ones. Using 1978's data to illustrate events, IBM Japan sold $1.4 billion worth of products, roughly the same amount as did Fujitsu, and more than Hitachi ($900M), NEC ($800M), or combined Toshiba, Mitsubishi, and OKI ($600M). Because Japanese firms were in many lines of business, their data processing revenues cited above only represented a portion of their total revenues. Fujitsu's business was largely in data processing (69 percent), although published accounts suggest it was higher; only 12 percent for Hitachi, 27 percent for NEC, 23 percent for Toshiba, 29 percent for Mitsubishi, and 35 percent for Oki, while for IBM Japan it was 96 percent.[72] Clearly, attention from Japanese vendors had to be spread across a much broader set of issues than those concerning computers to such divergent ones as markets for appliances, electrical equipment, consumer electronics, and, of course, electrical components.

Because I argue that having effective sales and other field personnel in the computer business exhorting customers to install and use computers contributed importantly to the rapid diffusion of IT, knowing the number of such IT protagonists is useful, if nearly impossible to find. IBM's internal market analysts routinely collected such information and, while probably not always 100 percent accurate, were, nonetheless quite reliable. Using data from the late 1970s, for which we have evidence, for the seven firms (including IBM) there were in 1978 a combined 30,000 sales, sales support, and sales agents. Fujitsu had 9,200 (two to three times more on average than the others), while IBM had 5,075.[73] Sales is a labor intensive activity so these numbers suggest a very active, dense market for IT products in industries that routinely used computers, such as banking, manufacturing, and government.

MITI was uncomfortable with Japanese companies being dependent on IBM's technology and that of other American vendors, who themselves were increasingly reliant on IBM's standards and, thus, also had to react to the pace of product introductions by the large American company, rather than chart their own courses. MITI's New Series Project, however, worked well to strengthen the technological capabilities of local firms to offer products that competed locally and abroad with IBM on functionality, cost, and reliability with the result that by the early 1980s, Japan had a vibrant computer industry. In effect, it had caught up to the Americans. By the early 1980s Japanese firms were exporting machines to Europe, which they marketed through local vendors, such as Siemens, Machines Bull, and Olivetti, which is why Europeans reacted in panic as they saw a Pan-European Trojan horse attack on their local industries by what appeared to them to be a highly coordinated "Japan, Inc."[74] Japanese sales in the United States also occurred in the same decade.

Figures 7.1 and 7.2, created by internal IBM marketing analysts in the late 1970s, illustrate the thick network of relationships two major computer vendors had created to diffuse its products around the world. No major geographic IT market was overlooked. Japanese exports were conducted with an organized infrastructure that grew early in its ability to sell and manufacture wherever it made economic sense to do so. This kind of information would have been known to officials in MITI and, of course, to other Japanese vendors and to their funding banks. Indeed, the evidence from these graphics suggest Japanese suppliers did this earlier and more effectively than the Americans, and far beyond what the Europeans proved able to do since the latter barely could break out of their domestic markets, let alone outside of Western European ones.

Table 7.5 briefly tells us a great deal about trade in computers. Leaving aside specific numbers and just looking at the broad pattern of behavior, several things become obvious. First, both imports and exports rose through the period, with the one exception—and a significant one at that—occurring briefly in the early 1990s when exports declined by nearly 50 percent between 1992 and 1995. Exports rebounded and continued their historic growth, with smaller declines occurring in 1998–1999 and again in the early 2000s. Second, a similar pattern manifested itself with imports. Third, the data tell us that

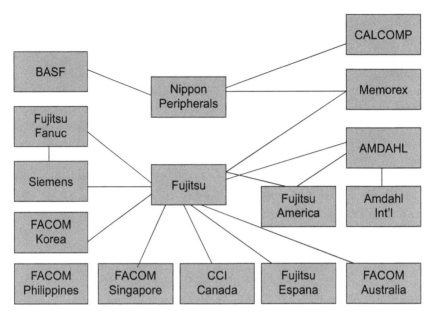

Figure 7.1
Fujitsu overseas relations, 1979
Source: "JCM & IPS Status in Japan & U.S.," RG 5, Box 46, folder 3/4 IBM Corporate Archives, Somers, N.Y.

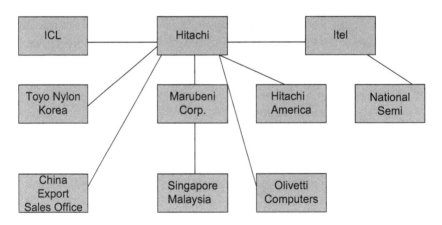

Figure 7.2
Hitachi overseas relationships, 1979
Source: "JCM & IPS Status in Japan & U.S.," RG 5, Box 46, folder3/4 IBM Corporate Archives, Somers, N.Y.

regardless of MITI's attempts to slow imports, they occurred and grew through the period. Fourth, MITI's efforts, and those of indigenous companies, to export were enormously successful, particularly in the 1980s and then in sustaining high volumes through good and bad economic times during the 1990s.[75] Anchordoguy concluded, however, that both imports and exports could have been better had the local industry been more effective, particularly in its software business.[76] Nonetheless, the data documents an expanding adoption of IT in Japan and a strong export business over the course of some 35 years. Finally, table 7.6 documents the growth in the presence of locally made computers over the early "catch up" time, indicating both the growing strength of Japanese vendors and results of MITI's strategies. Besides a steady march toward reducing the rate of foreign participation, which happened relatively quickly (although never shut off), keep in mind that the overall market kept growing, as documented later in this chapter. So, the actual number of domestic and foreign systems increased during the last three decades of the century.

As to specifics underlying the data, Fujitsu, NEC, and Hitachi developed complete lines of computer products from mainframes to personal computers in the 1970s, the latter product line introduced in the early 1980s.[77]

Table 7.5 JAPANESE COMPUTER TRADE, 1975–2000, SELECT YEARS
(¥ MILLIONS)

Year	Imports	Exports
1975	50,000	5,000
1980	75,000	20,000
1985	105,000	220,000
1990	210,000	390,000
1995	600,000	310,000
2000	780,000	590,000

Source: Data consolidated from Marie Anchordoguy, *Reprogramming Japan: The High Tech Crisis under Communitarian Capitalism* (Ithaca, N.Y.: Cornell University Press, 2005): 144–146.

Table 7.6 DOMESTIC AND FOREIGN COMPUTER INDUSTRY MARKET
SHARES IN JAPAN, 1958–1982, SELECT YEARS (EXPRESSED AS
PERCENTS = 100%/YEAR)

Year	Domestic	Foreign
1958	6.9	93.1
1959	21.5	78.5
1962	33.2	66.8
1967	47.2	52.8
1972	53.2	46.8
1977	66.4	33.6
1982	74.0	26.0

Source: Marie Anchordoguym *Computers Inc.: Japan's Challenge to IBM* (Cambridge, Mass.: Harvard Council on East Asian Studies, 1989): 168–171.

Vendors and customers still had to adhere to IBM's standards, relying on IBM's software (operating systems) and derivative programs (such as programming language compilers and applications written to order using these tools). Mitsubishi Electric and Oki Electric became technologically weaker vendors in this schema, which MITI's officials concluded needed to play lesser roles. MITI assigned them the mission to build mid-sized IBM-compatible computer systems, rather than the largest ones, with the former focused on mainframes, the latter on peripheral equipment for these systems. After 1978, OKI narrowed further its focus to the manufacture and sale of banking automated teller machines (ATMs).

Non-Japanese observers of the local computer industry agree overwhelmingly that MITI and the government played an extensive role in protecting and funding development of an export-oriented computer industry in Japan. Japanese commentators sidestepped the issue.[78] Circumstances changed over time as MITI's power surged in the 1960s and 1970s, then began waning in the 1980s. The West's perspectives in the 1980s and early 1990s remained rooted, however, in perceptions about MITI and Japan set in earlier decades.[79] By the early 2000s, even Japanese commentators were criticizing some of MITI's earlier efforts, suggesting that it had compromised Japan's ability to transform and compete as technologies and global trade evolved.[80] Business historian Alfred D. Chandler, Jr. suggested that the key Japanese firms benefitted from over seven decades of experience in electronics, including their ability to manufacture transistors and later semiconductors, minimizing more than others MITI's influence.[81] Yet others, with a variation of his observation, pointed to the Japanese telecommunications expertise, combined with their strategy of developing various types of products within the same firm, such as mainframes, peripheral equipment, semiconductors, and digital telecommunications switching devices.[82] Anchordoguy argued that the broader strategy of an ecosystem of customers, suppliers, MITI and other government agencies, in combination optimized cooperation, leveraged competition at functional levels, and worked within the framework of communitarian capitalism to emerge successfully.[83]

The evidence suggests all of these views were correct to one degree or another and so debates about emphasizing one mitigating influence or another is less important for our purposes than recognizing that each affected the nature and rate of diffusion of IT in both Japan and around the world. MITI's protectionism and funding of R&D took cost and risk out of the process of innovation, giving the industry time to flower. Clearly, learning and expertise about electronics proved crucial to the effort. Funding/ownership mechanisms of the banks and the *keiretsu*-oriented market also helped. An aggressive export orientation expanded the market to a size that made it possible for Japanese firms to achieve economies of scale necessary to compete locally and abroad against larger American suppliers. Below we discuss MITI's role in encouraging local users to adopt Japanese computing, a story which helps fill in the picture of the local digital eco-system.

Before turning to adoptions, there were four components of the Japanese supply side of the story that need various degrees of attention—semiconductors, PCs, software, and robotics—all of which demonstrated the contours and limits to success in the world of Japanese computing. Combined they contributed to the digitization of large swaths of Japan's economy and exports. Semiconductors became the central concern around the world regarding Japan's export strategies, PCs a late Wave One event, software the major gating factor after hardware affecting adoptions, and robotics, which promised to be a core competency of the Japanese economy in Wave Two computing.

ROLE OF JAPANESE SEMICONDUCTORS IN JAPAN

Many scholars and journalists have looked at the global semiconductor industry's spread and, in particular, the role of Japanese semiconductors. That export success on the part of Japanese electronics companies, beginning in the 1960s and extending nearly to the end of the century created enormous alarm in the United States and in Western Europe because governments and local vendors believed that losing domestic capabilities to produce and sell semiconductors posed threats to national military security (a U.S. Cold War concern), loss of potential trade, and ability to nurture indigenous computer industries (West European perspective). Such a turn of events represented the loss of initiative in building an Information Society (a shared pan-Atlantic perspective), with the height of concern occurring in the late 1970s and 1980s.[84] The Japanese proved most successful in dominating world trade in this class of digital technologies over all other types, such as computers or peripheral equipment. While the subject is interesting, we need not retell that story here. The issue to address here is how diffusion of semiconductors supported the broader spread of computers. To begin with, one cannot build a computer without ICs, nor can one manufacture almost any major industrial machine or vehicle without this technology—it had become nearly ubiquitous by the mid-1980s with Japan supplying a large quantity of the world's needs for such technologies.

Able to copy, cross-license, and manufacture various types at highly competitive costs, Japanese suppliers clearly facilitated diffusion of ever-less expensive machinery and IT around the world in the 1970s and 1980s through its success in selling various semiconductor products. To be sure, the Japanese lost ground in this trade by the late 1980s to other Asian rivals. As with computers, in semiconductors Japan's manufacturers pursued similar strategies with predictable results: all chasing better, faster, cheaper, similar products and marketing results, including similar niches for memory chips, logic chips, and microprocessors, products which over time evolved faster than Japanese vendors could keep up with. That technological evolutionary cycle led to extending marginalization of Japanese suppliers in the 1990s and subsequent decline in their global shares in sales. Along the way similar events occurred with computers, replete with American and European accusations of Japan of dumping (1970s–1980s), concerns of pan-Atlantic firms that specialized in the development and sale of semiconductors (for instance in the United States, Intel and Texas Instruments), and international agreements to address these issues (1980s and 1990s).[85] By 1988, Japanese suppliers had hit their high water mark in exports of semiconductors (51.2 percent) then declined steadily to 28.1 percent of global market share in 2001.[86]

From a statistical perspective one can see why concern existed about Japan. Between 1970 and 1974, 95 percent of all 1Kb dram (dynamic random access memory, most widely used in PCs and terminals) chips were produced by the pan-Atlantic community, Japan only 5 percent. The next generation of 4Kb drams (1974–78) demonstrated a shift with 85 percent made by Americans and Europeans, while the Japanese now provided 15 percent. For 16Kb ICs (1978–81) the Japanese produced 40 percent of the total and in the early 1980s—when the storm over exports peaked—64Kb product shipments indicated that Japan now owned 70 percent of the market, the Pan-Europeans 30 percent. By any measure that was an impressive growth performance for Japan, made even more impressive with every passing generation of ICs which were more complicated to manufacture.[87] Meanwhile, other nations had gotten into the game, most notably South Korea with its drams, which cost less for it to make and sell than Japan, contributing to the further decline of Japanese success.

Before leaving the supply side of this story, it should be noted that the same level of concentration that existed with computer vendors did too in semiconductors. Depending on which specific semiconductors (or ICs) one measured, roughly six companies dominated between 66 and 70 percent of production in the 1970s and 1980s: Fujitsu, Hitachi, Matsushita, Mitsubishi, NEC, and Toshiba. The same vendors dominated manufacturing of computers at similar percentages (60–75 percent depending on what kind of computers one looks at). A similar story can be told about advanced manufacturing for such devices as computerized machine-tools (80 percent by a handful of firms) and by the mid-1980s, industrial robots (74 percent).[88]

Use at home was impressive—a circumstance often overlooked by students of semiconductor diffusion. Electronics firms early on understood the importance of semiconductors, certainly by the 1940s, and when transistors (later ICs) appeared in the United States, they negotiated patent usage rights and were able to create their own versions all through the second half of the twentieth century. As demand for chips increased in the 1970s and 1980s, both the number of Japanese workers familiar with the technology and manufacturing capabilities expanded, thereby further diffusing use of IT in both that industry and subsequently into other devices, such as various consumer and industrial products. As with computers, however, the Japanese lagged the Americans in the 1950s and 1960s as they relied on the latter to set the pace of innovation. Nonetheless, by 1966, Japanese firms were producing nearly 300,000 integrated circuits.[89] By the late 1960s, manufacture of ICs took off in a substantial manner, stimulated by local then global demand for these to embed into electronic calculators and in larger IT devices, such as peripheral IT equipment and computers. Production of discrete IC devices jumped from 177,647 in 1970 to over 420,000 in 1983, at a 6.9 percent annual growth rate.[90]

Domestic consumption in this take-off period demonstrates that a healthy appetite existed in Japan for such items. In 1970, nearly 70 percent of all production went to local customers. One could argue that this was because the quality was insufficient to compete on the global market and thus had to be sold within the sheltered one, but the argument can also be made that exporting capability for such products was just being developed too. Regardless of explanation, local consumption increased all through the 1970s. Japanese customers also imported chips from other countries—an action often overlooked in complaints about Japan's "dumping" activities. By 1983, the internal Japanese market was consuming a total of about 125 percent of the volume of chips manufactured at home; in other words, local demand grew faster than domestic vendors could meet. Imports outpaced consumption of locally made devices until the second half of the 1970s when local productive capacity and quality began catching up with domestic demand, a need Japanese semiconductor manufacturers were able to serve in the late 1980s.[91]

Where did domestically consumed ICs and other semiconductors go in the 1970s and 1980s? The question is important as it sheds some light on the diffusion of IT in Japan. By the start of the 1980s, 58 percent were going into consumer electronics, such as video tape recorders, audio equipment, calculators, watches, and television sets. The other 42 percent were deployed in industrial products: computers (16 percent), telecommunications (10 percent), office equipment (4 percent), automobiles (2 percent), and 10 percent into various devices, such as manufacturing tools.[92] That diffusion meant individuals and firms in other product lines were becoming more reliant on IT with which to do their work and in their products. Local demand reflected unique differences among various markets. For example, using 1982, by which time Japanese IT use was in full swing, 51 percent of all demand for ICs in Japan came from consumer electronics, while for that same category U.S. demand was for 11 percent of its own consumption, and in Western Europe 25 percent of its own needs. In the United States, demand for these was most for computers (40 percent vs. 22 percent in Japan and 25 percent in Western Europe). Both the United States and Western Europe allocated about 20 percent of their supplies of semiconductors to telecommunications. Industrial equipment reflected Western Europe's large appetite, particularly in West Germany: 11 percent in the United States, 25 percent in Europe, and at the time 17 percent in Japan. We see the Cold War's effects on U.S. demand that year, which amounted to 17 percent of all American needs for ICs, as compared to 5 percent in Western Europe, and in Japan, zero demand since it was not a major player in Cold War military affairs.[93]

By tracing where semiconductors and other types of integrated circuits went into Japan, we can observe diffusion of digital technologies across specific industries and sectors of the Japanese economy. This diffusion was slow in the 1960s, picked up speed in the 1970s comparable to rates of consumption in Western Europe, and in the 1980s proved extensive, particularly by manufacturers of consumer electronics, industrial equipment used in other Japanese manufacturing industries, and, of course, by suppliers of computers to each other and to members of their respective *keiretsu* ecosystem.

JAPANESE PERSONAL COMPUTERS

Of all the modern economies, that is to say those which were the earliest adopters of computing, Japan was one of the latest to use widely personal computers. Even three decades after development of this class of computing, it had not diffused as extensively as in so many other countries, including in other Asian countries. A brief recounting of Japan's experience suggests that local circumstances do affect deployment of technologies, in this instance, computing. PCs are important components of late Wave One computing and so we need to understand its role in Japan. Japanese computer vendors acknowledged early the arrival of PCs. By the late 1970s NEC, Fujitsu, and Toshiba were manufacturing these; in fact, NEC claimed 5 percent of the U.S. market as early as 1980, before IBM even had a product. NEC brought out its first Japanese language IBM-clone PC in 1982, the first such product introduced in Japan (PC9800), aimed at the business market. It quickly dominated 80 percent of Japan's tiny market, although NEC had been selling kits people could put together since 1973.[94] Hitachi had introduced such a product in 1978, Sharp in 1979. NEC did well by creating retail distribution channels, rather than just selling to business customers. By 1994, approximately 70 percent of its PC sales went to individuals through 900 branches of Micom Shop and over 7,000 other retail outlets. NEC recognized early that sales would occur if software application packages existed in Japanese, so it created and commissioned such products sooner than its rivals. In 1986, NEC had acquired, written, or was selling over 3,400 software packages; two years later some 6,700, and in 1990, nearly 9,600.[95]

Japanese PCs experienced limits that keyboards for online terminals had as early as the 1960s, namely complexity of the Japanese language which had thousands of characters that obviously could not be accommodated by keyboards. As occurred later in China, which had the same problem with Chinese (progenitor of the Japanese language), a shorter set of characters were introduced for keyboards and online systems in the 1970s and 1980s. In the second decade (1980s), software packages using the language were ported over to PCs, albeit late in that period. When IBM introduced a Japanese language PC at the start of the new decade (1991 with the DOS/V operating system), all vendors quickly began using this software, which also made it possible for non-Japanese companies to introduce products for sale in Japan, such as Microsoft. Fujitsu, and Toshiba quickly licensed IBM's software. Apple appeared with a Japanese version of Macintosh soon after and in 1993 Microsoft came with Windows 3; others followed quickly, including Compaq. Much of the PC story in Japan is one of much activity in the 1990s, with firms scrambling to protect their local market share from the Americans, to provide Japanese language software and hardware, and to export, particularly newer forms, such as laptops (Toshiba beginning in 1994). Fujitsu demonstrated Japan's successful penchant to export after building a domestic base of business, becoming the third largest global producer of PCs by 1996. While there is a chicken-and-the-egg debate about who wrote the necessary facilitative software, the unique features of the Japanese language constrained deployment of online systems in the 1960s–1980s, and, later, PCs. For some early adopters, they had to rely on English-language-based PCs.[96]

SPECIAL ISSUES OF SOFTWARE IN JAPAN

The Japanese PC story hinted of a profoundly important technological issue looming in the background—the role of software. One cannot operate computers without software and that of most manufacturers outside of Japan used English or some European language, relying largely on IBM's technical architectures and centralized style of computing. Equally important, however, is the role software began to play in influencing the rate and nature of IT adoptions in the country. This is not to diminish the story of supply side computing in Japan, which proved essential to understanding acceptance of computing in Japan in the 1950s through the early 1980s, the period in which Japan caught up with the rest of the global computer industry. The types of software relied upon by end users slowed Japan's transformation through Wave One computing, keeping many organizations locked into centralized mainframe-oriented uses of IT longer than their cohorts in most other advanced economies. As a result, I believe the delay may have extended Japan's dependence on Wave One computing for at least one decade, with a few notable exceptions discussed below. At a minimum we must conclude that software in Japan remained a crucial element of that nation's experience with computers not to be sidelined by hardware's immanence in Japan's history of IT.

Data on the smaller extent of adoption of software in Japan and small volumes of exports of local software products out of the country stood in sharp contrast to the enormous success the Japanese experienced with hardware adoptions and exports—there is no disagreement on these points by experts. Much of the debate, instead, focused on the question of explaining why. The complexity of the Japanese language has been cited by some, while other experts pointed to inadequate education in computer science and programming, still others to the fact that the closed world of the banking industry's support for and dominance of a half dozen hardware firms constrained innovation.[97] Clearly, each of these circumstances played a role. Of all the effects on the spread of software through the Japanese economy, the most influential was evidently the construct of the computer industry itself, in other words, the supply side of the story of diffusion. When combined with an education system that encouraged conformity rather than innovation and the practice of vendors promoting closed technical standards, impediments to adoption appeared.

In the 1950s and 1960s, hardware providers included software in the rental prices of their machines, such as operating systems and utilities. They wrote to order application software for customers, all of whom were large and thus justified tailored systems—the practice evident in the West at the same time. Beginning in the 1960s software packages emerged in the West that were more reliable and less expensive to acquire and maintain, so sales took off. Meanwhile in Japan individualized software continued to be written and expensively maintained. Each Japanese hardware vendor effectively nurtured and defended its proprietary systems all through the 1950s–1980s to limit the ability of their customers to move to other suppliers. Locally developed systems were designed for the Japanese market, were unique to their customers, and were written in Japanese (for instance, data on a screen appeared in Japanese). No large market existed for Japanese-specific products overseas. They were limited in Japan by the size of the domestic market. Not until the 1990s did Japanese firms and individuals for their PCs begin extensive acquisitions of packaged software, most of it imported. But, we must return to size, because it is an important issue. In China, where similar problems with language existed, the size of the population guaranteed a market for software larger than that of many other nations, ensuring there existed economies of scale and commercial incentives to service domestic demand.

Japan and its language combined was a far smaller market. In 2000, for example, of the top 20 global software firms, only Hitachi made the list, simply because of its large volume of revenues earned domestically.[98] Yet, all through the 1960s–early 2000s domestic sales of software continued to grow, reflecting expanded use of computing in general across the nation. Even in the challenging economic times of the 1990s, domestic sales went from about $9 billion in 1996 to over $17 billion in 2002.[99] This closed system worked within

Japan. As for imports, in the 1960s and early 1970s, these were barely measurable, took off gradually in the second half of the 1970s when the adoption of computing expanded enormously, then doubled roughly every four years, beginning in 1984 and continuing to the end of the century.[100]

The sheltered, fragmented nature of the market centered around standards and products of a customer's preferred provider (byproduct of *keiretsu* or communitarian practices), which meant that the centralized approach to IBM computing upon which local vendors relied not only translated into dependence on American standards, but also meant the rate of change would be dictated by whatever IBM did with its own operating system (OS)—the foundation upon which mainframe programming rested. This occurred despite efforts by MITI to break this dependence. When IBM announced in 1969 that it would offer software apart from hardware (unbundling), and began sharing technical standards, an explosion of software innovation unleashed all over the Western world. MITI chose not to allow its firms to follow suit, because Japan's largest hardware providers resisted taking the risk of losing their customers in such an open market that inevitably would emerge as it had already in the United States. Coupled to government inaction on unbundling with user loyalty to their suppliers made clear that innovation would be constrained, and competition in software more restricted. Not until the government was forced to open its IT markets more fully in the late 1970s and early 1980s did the situation change slowly, as the import data on software suggests.[101]

MITI, however, understood the problem in its broader context—the need to break IBM's technological standard's hold on Japan that clearly had been locked in with the introduction of the S/360. So, as part of its campaign to fund and promote development of hardware responses to S/360 through its Super High-Performance Computer Project (1966–1971), MITI promoted development of software. Officials used the vehicle of their newly established Japan Software Company (JSC), a joint venture among Fujitsu, NEC, and Hitachi funded by the Industrial Bank of Japan, for the purpose of developing a local OS and common technical standards for software. Because the hardware firms were reportedly losing money with such sales, they were reluctant to converge on standards, desperately seeking ways not to share and collaborate, seeking, rather, to protect their customer sets from migrating either to IBM's very attractive products or to other local rivals. IBM's standards were not displaced with innovative systems. Instead, Japanese vendors chose to improve the efficiency of writing software—the software factories of the 1980s that received so much attention all over the world—which focused on economies of scale, much as in routine (repeatable) manufacturing processes, and less on what really drove software adoptions around the world, namely, better code (continuously innovative, hence changing and dynamic) that did useful things.[102]

MITI's efforts failed; indeed JSC collapsed in 1972. IBM's unbundling made it possible for local vendors to use the American firm's operating system as the platform of choice upon which to situate their proprietary software and tailored customer applications through their closed standards. Since most software came bundled with hardware, a vibrant local software market did not really exist, let alone flourish, in Japan in the late 1970s–early 1980s. MITI tried other measures in addition to the failing JSC, such as establishing the Information Processing Promotion Association (IPA) in 1970 to support small independent software firms, but this too failed due to inadequate funding, insufficient protection of software copyrights, and, of course, resistance of the main hardware providers to open their markets.[103] With little venture capital, inadequate training of programmers, lack of an entrepreneurial spirit in a society that resisted business risk and avoided failures meant that the best and the brightest programmers and systems analysts preferred to work for firms that either sold hardware, or companies that used existing software and standards. This behavior largely preserved the construct of the existing market for IT in Japan long after technological innovations around the world had led to more specialized and diverse providers of IT. Customized software—always very expensive and difficult to upgrade—remained the form of new adoptions of applications in the 1970s and 1980s. It was made affordable because the local economy was booming.

Then in 1982 employees from Hitachi and Mitsubishi were snared in a U.S. Federal Bureau of Investigation (FBI)–IBM sting in the United States trying to steal IBM's design specifications for its operating systems. Hitachi, like so many others reliant on IBM's standards had felt the pressure to acquire quickly access to the company's technical specifications so that it could build products for its customers before they agitated for permission to buy from IBM. To cut to the end of the story, both companies and the Japanese nation were humiliated by the incident. Hitachi and Mitsubishi were compelled to pay IBM for access to its technologies, for prior uses of its software, and fines, all substantial. All other Japanese vendors using IBM's operating system too had to pay to use IBM's software. Fujitsu quickly signed an agreement with IBM for permission to use the latter's software. Hitachi's settlement cost it about $46 million and it now had to pay similar amounts going forward for legal use of these. In the case of Fujitsu between 5 and 10 percent of its annual profits now went to IBM.[104] Regardless of affordability for these firms, which they could sustain, the national humiliation once again called attention to Japan's structural problems with software. R&D projects funded by the government proliferated in the late 1970s through the 1980s to develop local software and technical standards, but these too enjoyed limited successes, because Japanese hardware vendors continued to resist unbundling, hurting customers in the process who might have wanted to step out of *keiretsu* practices. Even development of a local operating system called TRON (Real-Time Operating System

Nucleus) for PCs largely failed in the 1980s (it still exists), because microcomputers standardized quickly on Microsoft and IBM PC technological strictures. Again, the Japanese butted against IBM's influence in the world market.

Increasingly in the 1980s, MITI and local users of IT became concerned about the closed standards defended by local vendors, while around the world technological shifts occurred as markets and users moved rapidly from just using centralized mainframe computing to additionally distributed processing with minicomputers, PCs, and networks. All of these new forms of computing required different software, hardware, programming, and management of IT. Yet, one can criticize Japanese users too for continuing to rely on the increasingly backward computing because they were too reluctant to force their suppliers to change their ways, such as by demanding unbundling of software.[105] Right into the 1990s, Japanese vendors and customers remained stuck in early Wave One computing, which increasingly meant relying on more expensive, inferior digital tools, often software that inhibited transition of customers into new practices required by changing global market dynamics in the 1990s and beyond, such as to distributed processing. This occurred despite a few changes in the 1980s that allowed foreign software firms to enter the local market, albeit with tepid results. Marie Anchordoguy pointed out that a specific consequence for the Japanese was their late realization of the importance of the PC as a tool for organizations, not just individuals.[106]

While changes came in the 1990s that began to mitigate the situation, discussed below, there were two areas of software development that did prosper, and so necessary to discuss briefly: software for industrial machines (such as machine tools and robotics) and game software. These two classes of computing applications diffused so extensively in Japan that one must conclude they served as additional domestic paths of diffusion for IT, although not in the form of computers, rather as devices that contained computing and software. As demand grew in manufacturing to produce smaller lots of products and to change quickly what was produced to meet ever-growing demands for various goods in the 1960s and 1970s, Japanese manufacturers developed new processes, called flexible manufacturing systems (FMS).[107] Additionally, automation held out the real possibility of reducing expensive labor content by as much as 95 percent in some instances, leading to subsequent speculations about the "unmanned factory," an aspiration never achieved. However, numerical control tools and robotics developed rapidly in the 1960s and 1970s, each with software that was stable; that is to say, caused machines to do repeatable tasks and that could be embedded in these with fewer changes than required of software ensconced in PCs, for example. By the early 1980s, just over half of all machine tools installed in Japanese factories were programmable, in other words, numerically controlled. That capability led Japanese manufacturers of N/C tools and industrial robots to have products they could also export, reinforcing further development and deployment of such technologies and, in the process, fostering further use of IT in manufacturing industries.[108] Over time relatively simple two-dimensional functions of these devices went into automotive, aerospace, semiconductor, and other manufacturing plants around the world, not just in Japan. CAD/CAM applications spread rapidly, beginning in the 1960s and extended to the end of the century to such an extent that such applications became ubiquitous.[109] As with other IT initiatives, government policies fostered export trade in such software-driven technologies, and influenced the nature of software R&D projects in the 1970s and 1980s, reinforcing such notions as "software factories" and "Fifth Generation" computing.[110]

While such developments were extensions of patterns of IT diffusion evident with computers and systems in large enterprises, video software, also called games software, took a different path. Briefly, Japanese vendors entered the global market in the early 1980s in a substantive way.[111] Nintendo introduced an interactive software-based videogame console and game in 1983, building on the firm's prior experience with arcade games. It proved popular in Japan, resulting in domestic sales of some 2.5 million consoles and 15 million cartridges.[112] As in other IT sectors, Nintendo began successfully exporting its products to other countries, making it possible to build up resources and capabilities at home in IT, both in the hardware and software development and manufacturing. A second vendor, Sega, also enjoyed similar success with its products in the 1980s, but Nintendo controlled the world and Japanese markets for video games. By the mid-1990s, Nintendo's standard dominated over 80 percent of the world's supply of games, mimicking IBM's over mainframe standards in the 1960s–1990s. Sony became another major participant in the highly competitive video game business, a market less regulated or influenced by public policies than such earlier sectors as mainframes and semiconductors. Sony has dominated domestic and global markets with its PlayStation products since the mid-1990s. Between 1995 and late 1998 it sold 40 million PlayStations worldwide.[113]

This success stood clearly in sharp contrast to the malaise evident in other segments of the Japanese software industry. Why? Anchordoguy has convincingly argued that "as new players in a new industry, the game makers were relatively insulated from the institutions and practices of communitarian capitalism."[114] Her explanation extends to customers as well, in this case individual Japanese citizens who purchased this software and its related hardware, not a firm tied to a *keiretsu* network or subject to government pressures. Individuals bought products that were priced competitively, offered attractive functions and reliable performance, and were new forms of entertainment. As they became used to such forms of amusement, they demanded more innovative products, happily served up by firms that increased their use and abilities in software development all through the 1990s and beyond. Video games became the first ubiquitous use of IT across large swaths of Japanese society, presaging the arrival of many Wave Two uses of computing, including extensive use of cell phones (which also offered games).

To set the historical record more thoroughly complete one should also acknowledge that the availability of digital hand calculators in the 1970s began the process of introducing the Japanese public to the use of digitally based goods, albeit in a small way; nonetheless, it created an appetite for digital products that would become a hallmark of the Wave Two environment. In 1965, Japanese vendors manufactured some 4,000 units; a decade later production had climbed to 30,000. By 1985, annual volumes had exceeded 65,000 units; thereafter, production took place in many other countries in Asia. However, about two-thirds of Japan's annual production of these products went into its export market, while less than a couple thousand entered the local market each year by the late 1970s.[115]

ROLE OF ROBOTICS IN JAPANESE IT

Japan is an important source of robotics and other closely related innovative products.[116] Its reputation as the source of these rests on the early diffusion of this technology in industrial applications both at home and as export products. In the 1990s it introduced consumer robotic devices as well, such as mechanical pets and humanoids, all cute and friendly looking, attracting enormous amounts of media coverage around the world. Before looking at Japanese robotics, however, recognize that research, manufacture, and diffusion of robotic applications occurred in other countries as well: Germany in the manufacture and use of these devices, in the United States with important research done over the past four decades at the Massachusetts Institute of Technology (MIT), and the manufacture and sales globally by various American vendors, beginning in the early 1960s. However, one can conclude that robotics played an important role in Wave One diffusion of IT in Japan at an impressive speed if we are to gauge diffusion and reliance on this technology by the number of robotic units installed.[117]

The history of robotics in Japan has a long and complex history, which for our purposes, can be briefly summarized. Industrial robots were the first type of modern robotic devices invented, and that happened in the United States at the start of the 1960s. In 1967, the first use of robots in Japan occurred with imported American devices, which in subsequent years went into use in various manufacturing industries. They were also soon manufactured in Japan, beginning in 1968. By 1980, annual production had approached 19,400 units (see table 7.7).[118] By 1980, there were some 70 Japanese manufacturers and, to put that number in perspective, at the same time there were 26 in the United States, 33 in West Germany, and 18 in Great Britain (see table 7.8). Other countries also had a handful of producers, typically less than 10, many only one. Smaller producers included most West European countries and Canada.

Table 7.7 PRODUCTION OF JAPANESE ROBOTS, 1968–1980,
SELECT YEARS

1968	200
1970	1,700
1972	1,700
1974	4,200
1976	7,200
1978	10,200
1980	19,400

Source: Kuni Sadamoto (ed.), *Robots in the Japanese Economy: Facts About Robots and Their Significance* (Tokyo: Survey Japan, 1981): 131.

Table 7.8 NUMBER OF ROBOTS INSTALLED IN JAPAN,
1979–1990, SELECT YEARS

1979	9,000
1981	21,000
1983	47,000
1985	93,000
1986	116,000
1990	240,000

Source: Frederik L. Schodt, *Inside the Robot Kingdom: Japan, Mechatronics, and the Coming Robotopia* (Tokyo: Kodansha International, 1990): 115.

The number of local installations of robots grew extensively. Early applications in Japan centered on labor-saving tasks in manufacturing, improving versatility of performing repeated actions (movements) on a production line, for instance, manufacturing semiconductors in collaboration with N/C tools.[119] By the late 1970s users in Japan included electrical machinery manufacturers (accounting for a third of all deployed robots in the country), followed by the automotive industry (slightly less than a third of all installations), then a string of additional industries: plastics molding and processing, general machinery manufacturing, metal working, textiles, chemical, steelmaking, and shipbuilding, amongst others. Most robots manufactured in Japan listed in table 7.8 were installed domestically, being either of insufficient quality to compete on the world market, or because local manufacturers had yet to generate sales overseas.[120]

In the 1980s, diffusion of robots in Japan increased rapidly and extensively. By the end of 1986, there were 116,000 installed in Japan and for comparative purposes, 25,000 in the United States. At the time Japan probably had installed about 60 percent of the world's inventory in its homeland, making this country the "Robot Kingdom," to use a phrase widely embraced by Japanese media at the time.[121] To be sure, robots in the 1970s and 1980s were fairly low-tech devices, but they increasingly used IT components (such as semiconductors) linked to computers (N/C machines, PCs, and other intelligent machines), largely in manufacturing. Robots became one of the earliest IT-infused devices that widely captured the public's attention in Japan, leading to widely attended "robot fairs," for example, through the 1980s and beyond.[122]

Japan's experience with robots demonstrates the local style of learning and deployment of IT. Rather than simply buy American or European robots the Japanese negotiated licenses to manufacture these machines for which there unfolded rapidly a growing demand, and for a device they later improved upon. Such circumstances accounted largely for why there were so many companies in Japan making these devices. Kawasaki, for example, licensed with firms in Norway (Tralfa) and West Germany (VFW). American companies specialized in robotics while in Japan many were large multiproduct enterprises, just as in such IT goods as semiconductors and computers, including Kawasaki, Yashawa Electric, and Kobe Steel. All could invest capital and personnel in the manufacture and sale of these devices. Their products were expensive, suggesting that had these been less so and functionally richer that perhaps more would have been installed in Japan. They improved in performance and reliability, particularly as electronic components shrank in size and IT memories increased in capacity during the 1970s and 1980s. The rationale for customers to acquire this technology mimicked reasons for installing computers: to lower labor costs, due to increased quality, and to perform more tasks not possible by humans.[123]

Robotics took off in Japan because the country had a well-trained pool of mechanical design engineers and others familiar with electronics working in companies experienced in these fields of knowledge and skilled in the manufacture of electronic devices, such as industrial equipment and consumer appliances. There existed a natural domestic market for these new devices, including automotive and other transport manufacturers, consumer electronics firms, and other enterprises that were sophisticated consumers of technologies. These uses were motivated by growing shortages of skilled workers in the last three decades of the century, not just because they were increasingly expensive. Early market acceptance led to increased innovations that fostered a foreign trade market too. By 1987, over 300 manufacturers of robotic devices operated in Japan, spawning intense competition. In turn that led to more innovations, also to more users of these devices, enhancing technical capabilities of the industry as a whole. Encouragement in the development of robotics in Japan came from local expertise and markets (global too) for Japanese numeric control devices, motors, optical scanners; sensors and, of course, myriad digital technologies. One segment fed off the other in a mutually supportive cycle of innovation, competition, and consumption.[124]

HOW JAPANESE USED COMPUTERS

Who used computers in Japan, and how, demonstrate clearly that the Japanese acted very much like their cohorts in North America and Western Europe at roughly the same time. For example, large organizations became early adopters, used batch punched-card data processing in the 1950s and 1960s, and relied on the same types of systems as did the others. In the 1960s and 1970s they embraced online computing, and concurrently distributed processing, albeit often later than in the pan-Atlantic world. In short, users and applications did not reflect a Japanese exceptionalism; rather, their experience offered additional evidence that there were common patterns of use—hence diffusion—of IT from one region and nation to another for similar reasons, although with degrees of difference. For instance, reducing labor content of work in Japan during the 1950s and early 1960s was less of an issue than in the United States since in the former labor costs were low and paternalistic employment practices discouraged displacing workers far more than in the United States. Yet, both focused on automating human work. By the early 1970s, even this difference between the two communities regarding labor's role in the use of computing was diminishing.[125] The same factors that either propelled or constrained diffusion in Japan included availability and cost of hardware, supply of qualified software programmers and systems operators, and functionality of hardware–software–telecommunications systems (for example, batch versus online computing, centralized and decentralized systems). The one major exception—already discussed—was the Japanese propensity to write their own application software rather than rely on packaged software products, at least until the second half of the 1980s when demand for pre-written code increased, mimicking rapidly patterns of adoption already underway in the pan-Atlantic community.

Users represent as important a group as the suppliers in extending reliance upon information technologies in Japan. IT vendors were, of course, also users of computers for the same reasons as banks, insurance companies, other manufacturers, and government agencies. The role of users—think customers of IT—is quite opaque in the extant literature on Japanese computing, which overwhelmingly focuses on the supply side of the story and foreign trade.[126] However, as everywhere else, without customers there would be no diffusion. The rate at which they could absorb the technology determined the extent and speed of diffusion of IT. Thus, their experience is the flip side complementary to the traditional story told of Japanese supply-side computing. While users await their historians, various industry associations documented some patterns of adoption and activities which suggest who used computers in Japan and how.

Population counts of installed computers, reviewed in more detail later, make it clear that use of this technology outside of experimental cases commenced in the late 1950s, indeed, in 1959 if we rely on extant inventory counts. In that year the private sector began installing such equipment, including banks, securities firms, electric utility companies, manufacturers of large consumer products (such as refrigerators), and transportation, iron, and steel manufacturers in support of routine work. Some of the largest enterprises in these industries were initial users, but by the late 1960s computers began to appear in ever smaller firms. In addition to the industries already listed, power and gas utilities, local public services, machinery manufacturers, cooperatives, and textile manufacturers also embraced computing. By the mid-1960s they used computers at service bureaus or their own installed systems. Among mid-sized firms were users in foodstuffs, precision instrument manufacturing, glass, cement, nonferrous metals, construction, mining, paper, and pulp industries. Ranked in extent of use, as measured by the amount of money spent on these systems, by 1968 securities firms had become the most extensive users, followed by power and gas utilities, next insurance, and in fourth place, other financial companies. Government agencies came in fifth, not surprisingly, since they did not have the large military establishment which led to more extensive governmental uses of computing as in the Soviet Union, France, Great Britain, and the United States. Behind government were users in electrical appliance, machinery, iron, and steel.[127] In short, Japanese adoption reflected a similar pattern of participants evident in other industrialized countries.

Extensive users of data were early adopters of computers, such as the Meteorological Agency, Japan Telephone and Telegram Corporation, and Japan National Railway—all of which began using these in the late 1950s. Surveys of IT adopters in the late 1950s and 1960s reported that early applications of IT involved processing routine office transactions, although, as one Japanese report at the time noted, at a level of utilization "lower than that of the U.S."[128] By the mid-1960s, newly installed systems tended increasingly to be online versions rather than batch processing, while optical and magnetic scanning surged ahead in the financial, electrical power, and gas industries, partially as a workaround for the complex Japanese language. A detailed national survey was conducted in September, 1968. It evidenced that over 74 percent of firms used computers largely to perform clerical computations, which concluded "indicates that they are intending to overcome the increased burden of office work by mechanization," but also that overall the extent of deployment across the entire economy remained low.[129] Clerical computation took place most extensively in personnel and labor management, sales, finance, accounting, and inventory control. Other areas included production, R&D, and engineering. When computers were being used for more advanced applications, such as analysis, planning, and forecasting—areas of far less deployment (the remaining 25 percent)—sales ranked the highest, followed in descending order by R&D and engineering, production, planning research, and inventory management.[130] The same study concluded that "only a small number of companies have reached the stage where computers are used for judgment work or management decision-making."[131]

Migration from batch to online systems occurred in the largest firms, particularly those with most geographically dispersed workforces. It was, thus, no surprise that the Nippon Telegraph & Telephone Public Corporation (better known as NTT and in Japanese Nippon Denshin Denwa Kosha) would be an early pioneer in the use of such systems, a case watched by many other Japanese organizations. By the end of the 1960s it had 70 online systems, leveraging its internal telecommunications skills and infrastructure, making it possible for NTT to apply its existing knowledge to rely rapidly on I, C, and T. One of the largest and earliest online services was offered by NTT to banks, linking 62 banks to their respective branch offices, creating a new source of revenue for itself. In turn, that success stimulated development of new online systems at NTT in the late 1960s and early 1970s in mathematical programming, process control, office management systems, and in IT operations in general.[132]

Because many companies were reluctant to invest capital or to long-term rental agreements in the expensive technology with their initial applications, did not have the skills to manage such new tools, or wanted to experiment before making extensive financial commitments, many turned to time-sharing options. An undetermined number of enterprises opted for this approach in the 1960s, finding universities had the most extensive facilities they could access, such as at Osaka University with its NEC 2200-500 system and at Tokyo University which had installed an HITAC 5020 system. Each was equipped with MULTICS and MIT's Project MAC, both American-built software systems. Unlike in the United States, where access to computing systems was typically online using CRTs, in Japan most users of online processing relied on teletype printer/terminals which could print Katakana (Japanese script) characters. For more data-intensive or high-volume transactions, even large enterprises had to pool resources to afford computing, encouraging their reliance on systems at universities. NTT was the only firm in the 1960s outside of government able to mount its own large computing initiatives.[133]

There is broad consensus across dozens of studies of Japanese computing that an important agent facilitating diffusion of computing was the work of the Japan Electronic Computer Company, better known as JECC. Established in August, 1961, with a capitalization of $3 million raised equally from each of the Japanese computer companies, it was essentially a hardware leasing firm. Since no single vendor could afford to bear the burden of R&D for new systems, MITI wanted them to collaborate, yet compete. One way to do that was to buy a computer and associated peripheral equipment from a vendor once a customer had committed to acquiring these it then lease back the system to the customer and pay the vendor up front for the cost of the equipment. This immediately moved cash back to the IT manufacturer, obviating the need for the vendor to recoup the cost of the system over the life of a lease. It could use the money sooner to build others or to invest in R&D. That process reduced risks of declining values of equipment as new generations of less expensive systems appeared. JECC's strategy proved to be a stunning success. In the 1960s alone, vendors' sales to JECC went from $3 million (1961) to $230 million (1969). As one commentator reflected on its success, he noted that JECC was "more than a computer leasing company to the Japanese," it represented "a spirit of cooperation, a spirit of solidarity which links all Japanese companies into an invisible union against outside encroachments."[134]

More to the point, the government funneled, or facilitated, loans and subsidies for JECC to lubricate the process of inventing, manufacturing, selling, and buying computers by making funds available at each stage of diffusion while tamping down financial risks for vendors and customers. If a customer returned a machine to JECC, the agency sold the machine back to the original manufacturer at a discounted rate of depreciation, thereby taking a considerable amount of financial risk off the vendors while making it possible for users to change systems from time-to-time. JECC could affect the rate of diffusion by its terms and conditions for leases as well. Its successful and extensive influence on the ebb and flow of supply and demand served as a generous conduit of funds, indeed some $2 billion by 1981, largely generated with low-interest loans backed by the national government. Its influence can be quickly measured by the fact that it funded the majority of installed systems: 65 percent of all domestically installed ones in the 1960s, 30 percent of those in the 1970s, and over 10 percent in the 1980s.[135]

But, just as important, JECC in Wave One facilitated preservation of an orderly competitive environment "by managing a price cartel."[136] Its dominance and positive influence supporting diffusion declined after the catch-up phase with the West had been achieved by the mid-1980s when different economic dynamics made its services less necessary to both vendors and customers. By then, funding came from many sources, unlike in the beginning when banks, for example, proved reluctant to invest in the new technology and customers were too wary to buy the systems outright, let alone acquire funding for leases. In short, JECC was a successful catalyst in getting the Japanese to start adopting computers.

Financing was only one important part of the diffusion story. As salaries rose in the 1960s and early 1970s, largely due to shortages of qualified labor for various jobs, not just in IT, automating manual functions increased. Evidence from the growth in IT populations suggests the growing diffusion across multiple industries. Between 1967 and 1973, for example, the number of card-punching service bureaus increase threefold, part-time IT service providers fourfold, software development firms threefold, and business data processing service bureaus also threefold.[137] IT staffs were widely diffused across many industries by 1973. One survey suggested that 143,217 IT professionals had diffused across a sample of 3,867 companies and government agencies, working with 6,770 computers.[138] In this particular study, 17.1 percent of all users were banks, surpassing electric and electronic equipment manufacturers themselves (12.8 percent of total), which included computer vendors. Whole-sale, retail, and other commercial establishments accounted for 9.6 percent of the group; government agencies 6.7 percent, and data-processing service bureaus also 6.7 percent.[139] NTT still laid claim to the largest network of online systems in the country. Yet online telecommunications systems had appeared as early as 1964, used by various companies, most notably and earliest the Japan National Railway and Japan Airlines, both firms for managing reservation systems. By law, NTT dominated provision of networks until 1971, when other telecommunications competitors were allowed to offer networking services to meet growing demand. Table 7.9 documents the extent of adoption of online systems using telecommunications by all manner of industries, firms, and agencies.

Industries that relied on computers proliferated at a large rate in the late 1960s, setting the base for extended use of IT for ever-increasing numbers of applications by firms all through the 1970s. That pattern can be quickly illus-trated with table 7.10. Of 35 categories of industries tracked by the Japanese government at the time, seven had no systems installed in 1965; in 1970 only one did not (foresting/hunting). The table lists the top 10 users in 1965 and their number of systems installed that year and also in 1970. The total for all systems was 1,683 in 1965 and 7,933 in 1970, a goodly increase, which does not account for the users in addition to those who relied on service bureaus to do much of their mundane data processing. User organizations reported use of IT across a broad spectrum of applications: "personnel and labor" (22 percent), "sales" (18.9 percent), "accounting and finance" (17.6 percent), "stock control" (inventory) (15.7 percent). Nearly 70 percent characterized their applications as clerical, while additionally 31 percent also for forecasting and planning.[140]

By the start of the 1970s, evidence of the benefits of using IT began circu-lating among organizations. Table 7.11 summarizes their observations as of 1971. These data track reasonably well with American and West European ex-periences with the exceptions of reduced inventory costs and reduction of clerical costs, which U.S. firms routinely ranked as highly significant benefits of automation in the 1950s and 1960s. Unquantifiable indirect effects (bene-fits) reported at the time included the following responses: "easier to grasp business situation" (number one), "improved company image" (ranked second),

Table 7.9 JAPANESE TELECOMMUNICATIONS SYSTEMS, 1965–1973

1965	6	1970	139
1966	12	1971	213
1967	22	1972	330
1968	40	1973	490
1969	84		

Source: Japan Electronic Computer Co., Ltd., *Progress of Computer Industry in Japan* (Tokyo: Japan Elec-tronic Computer Co., Ltd., undated [circa 1974]): 29, CBI 32, Box 631, folder 22, Charles Babbage Institute, University of Minnesota, Minneapolis.

Table 7.10 SAMPLE INDUSTRIES USING COMPUTERS, 1965 AND 1970
(NUMBERS OF INSTALLED SYSTEMS BY INDUSTRY)

Industry	1965	1970
Electrical machinery	168	718
Chemical & petroleum	134	436
Transportation machinery	110	356
Finance	109	620
Universities	99	285
Textile	88	173
Service bureaus	85	614
Iron & steel	82	299
Insurance	74	130
Government (national)	71	202

Source: Japan Computer Usage Development Institute, *Computer White Paper* (Tokyo: Japan Computer Usage Development Institute, 1971): 46–47, CBI 32, Box 551, folder 2, Charles Babbage Institute, University of Minnesota, Minneapolis.

and "better information flow in firm" (third), with "greater accuracy and speed in judgment and decision making" a close fourth.[141]

Because government users in the pan-Atlantic community were early users of computing, what can we say in comparison about the Japanese experience? As in other industrialized nations, public officials used computers, and did not simply promote their use through MITI and other administrative mechanisms. Some of the earliest evidence dates to the start of the 1970s, and can be reasonably relied upon to reflect activities of the 1960s, since it took years to build up an inventory of systems and applications. Use of computing by large national agencies, cities, and prefectures (local governments) proved extensive. In 1960, the city of Osaka became the first municipality to start using a computer, followed soon after by Nishinomiya, Sapporo, and other large urban centers. In 1962, Kanagawa Prefecture became the first local government to do the same, followed by Tokyo Metropolis, Osaka Prefecture, Aichi Prefecture and others, spilling over into towns and villages by the early 1970s. To be precise, as of early 1973, a total of 1,874 local governments used computers in support of administrative functions, including 38 out of 47 prefectures, although all prefectures relied on computing, some through service bureaus. In short, the evidence indicates that just over half of all local governments appropriated one form of computing or another by early 1973. Because of their use, people all over Japan became familiar with the technology. At the time over 4,500 IT government workers served in such work as agents of diffusion. Prefectures represented the fullest adopters of computing which one report of the time "attributed to the fact that small-scale computer systems" were "so inexpensive and easy to purchase."[142]

Table 7.11 MEASUREABLE BENEFITS/EFFECTS OF COMPUTER USE
REPORTED BY JAPANESE ORGANIZATIONS, 1969–1970 (RANKED BY
NUMBER OF FIRMS REPORTING THE BENEFIT AND PERCENT)

Benefits/Effects	Percent Reporting Experiencing this Effect
Accuracy and speed in processing	34.4 (740 firms)
Reduction in personnel expenses	24.0 (123 firms)
Better customer service	12.0 (257 firms)
Simplification of file management	7.5 (162 firms)
Reduction in inventory	6.1 (132 firms)
Reduction in non-personnel expenses	5.7 (123 firms)
Reduction in delivery period	4.7 (101 firms)
Efficient use of funds	3.9 (83 firms)

Source: Japan Computer Usage Development Institute, *Computer White Paper* (Tokyo: Japan Computer Usage Development Institute, 1971): 57, CBI 32, Box 551, folder 2, Charles Babbage Institute, University of Minnesota, Minneapolis.

In addition to administrative computing, local governments also used small systems for monitoring environmental pollution, in support of medical facilities, and to conduct scientific tests and research, applications not normally in evidence across the pan-Atlantic community at the local level.[143] Less understood, however, is the extent to which local and state governments were in the grip of the *keiretsu*-communitarian ecosystem so evident with respect to the national government, across the private sector with large firms, and smaller ones operating in the shadow of their major customers.

As extensive as early adoptions were, particularly in the 1960s, a similar concern existed in Japan as in the United States across all industries about the lack of an adequate supply of qualified IT professionals, particularly programmers. This was an important, indeed acute, issue for Japanese users since they preferred software systems designed for their own organizations. A group of American visitors to Japanese facilities in 1973 called attention to the lack of sufficient programmers as a reason why even more systems were not deployed. Indeed, "the greatest shortage seems to be in the supply of application programmers," with the brightest choosing to work elsewhere.[144] Systems programmers too—those who wrote operating systems for computers—for instance, were largely employed in the more attractive career of working in large enterprises such as Hitachi and Fujitsu. Yet, the dearth of programmers did not hold back increased reliance on computing in the 1970s as members of the various industries cited in the last several tables continued to add applications, many of them online. Some systems were now national in scope, each deploying thousands of terminals. These included applications for banking, motor vehicle registration, seat reservations for railroads, and even horse race betting systems deployed by the Japan Racing Association.[145]

By the late 1970s both computer vendors and their customers had built up programming staffs across dozens of industries, making it possible to write and implement IT systems across the economy. The market for packaged software, however, remained small because, as one observer at the time noted, "Japanese users are not yet aware of the value of software, therefore they are not willing to pay separately for software."[146] But, the braking forces on software's diffusion were complicated by causes that made it appear less valuable than the hardware itself. A contemporary explanation suggests the circumstance faced by users: "As software costs are rising and although many computer users have large dp divisions, they cannot keep pace with the growing demands."[147] Hardware vendors sought to speed up development of systems, hence their interest in such concepts as software factories and structured programming. One survey of programming languages showed Japanese relied almost entirely on American-developed compilers already proven to be relatively productive: Cobol (by 60 percent of users), followed in descending order by Assembler (most widely used in the United States in the 1950s, less so in the 1960s, and only with a few legacy systems in the 1970s), Fortran (largely by engineers and scientists), and just starting, PL/1.[148] MITI's own assessment of IT results of the 1970s noted that manufacturing firms enjoyed the most progress in using computers, which facilitated rapid diffusion of IT to its suppliers and customers. MITI officials believed systems had evolved to such a point that office work could be done as "common use" with computers, while large retailers with multiple branches began using computers (much as their American and British counterparts).[149] MITI attributed IT's diffusion in the 1970s to three circumstances, which historical evidence supports: advances in semiconductors and computers, availability of highly educated and "favorable labor management relations," and "appropriate government measures to promote technological development."[150]

The pattern of applications used, and by whom, did not change substantively in the 1980s. Momentum toward greater reliance on computing for daily work spread from the largest organizations to ever-smaller ones as accessibility increased and costs of computing declined, and joined the rest of the world in managing problems of adoption, as observed across the pan-Atlantic community. For example, in 1987 users were still complaining about the lack of an adequate supply of programmers, now known as "software professionals," leading to more collaborative efforts with other firms, joint development projects and subcontracting programs to overcome the problem. As in the West, the percent of IT budget for software rose while unit (hardware) or expense of a transaction done digitally declined. The fastest growth rates in IT expenditures occurred in banking, securities, and insurance industries.[151] The locked-in, mainframe-centered *keiretsu*-styled patterns of adoption and use of computing continued to characterize Japanese computing all through the 1980s and deep into the 1990s as well.

Besides affecting types of uses and tools used by corporations and government agencies, this enclosed approach to computing extended to the adoption of private networks in the 1980s and 1990s. Use of private networks and software inhibited the adoption of open systems and multiple technical standards during the second half of the 1980s and early 1990s. As one student of the process noted at the time, "in the early 1990s Japan had numerous islands of computer networks that could not communicate with each other."[152] That circumstance slowed adoption of interconnected supply and value chains already being used rapidly by other Asian, American, and European industries, indeed whole economies.

Given the enormous economic problems faced by Japan in the 1990s and the simultaneous spread of personal computing and other types of digital consumer electronics, what did users do in that decade? How did external forces affect their work, such as deteriorating economic conditions? Did they change their style of computing or uses? The biggest technological event of the decade was, of course, the rapid diffusion of the Internet around the World, adding to the mix of questions and issues. Worldwide, Japanese competitiveness in IT declined as vendors in the United States and elsewhere introduced new Open Systems products, others for use with the Internet, and customers embraced these. Japan was not immune from the process, just slower to respond. By the end of 1993, the adoption of open architectures in PCs had helped to lower the costs of this technology, stimulating acquisitions by both organizations and individuals. Simultaneously, adoption of more open computer networks picked up, increasingly reliant on the Internet, largely starting in 1995 (see table 7.12). This tripling of adoption meant Japan's global share of Internet hosts expanded from 2.9 percent to 3.8 percent in that critical period of takeoff.[153] Table 7.13 documents applications of both private and Internet-based networks by organizations, supplying evidence that the evolution in usage mimicked global patterns. Use of networks expanded across almost all industries, with participation ranging from 45/50 percent to 60/65 percent of firms and agencies.[154]

We are too close to the events of the 1990s to understand with confidence if the lag in adoption of such forms of computing as PCs and the Internet were due to adoption cultures described by Anchordoguy, Fransman, and myself, or to the harsh economic realities faced by the Japanese. But clearly, customers of IT slowed their acquisitions, both at institutional and individual levels. One study on activities in 1995 looking at global adoption rates of PCs documented Japan's rate per 1,000 people, ranking it eleventh in the world (United States first, Australia second, and Canada third). Yet, it also noted that sales had been rising in the previous several years.[155] That slower pattern can be largely explained by the fact that Japan embraced PC-based applications later than other countries and had an expensive telecommunications market making online computing less affordable than in many other countries in the late 1980s and early 1990s. The lag of roughly 2 years in adoption of the Internet in Japan during the first half of the decade reflected those realities. It was not for

Table 7.12 NUMBER OF INTERNET HOSTS IN JAPAN, 1995–1996

November 1995	225,000
May 1996	400,000
November 1996	700,000

Source: JPNIC.

Table 7.13 APPLICATIONS OF JAPANESE NETWORKS,
CIRCA MID-1990S

Within Inter-firm Networks
Ordering systems
Reservation systems
Technology information management
Distribution applications
Financial transactions
Across Intra-firm Networks
Production control
Sales and inventory control
Accounting
Human resource management
Customer information management

lack of trying.[156] Internet usage essentially arrived in Japan in 1993, followed by the establishment of hundreds of Internet access providers, most small enterprises, which continued to grow in numbers and revenue during the decade.

There were structural supply side problems with telecommunications, however, that influenced extensively events in Japan and that mimicked many of the behaviors and issues evident in the 1960s–1970s with computing in the country. During the 1980s and 1990s Japan maintained a telecommunications environment that served as a drag on IT adoptions, a situation made obvious by what happened in so many other countries where telecommunications were modernized and reorganized to make access better and more affordable, as in South Korea, North America, and Western Europe. NTT, the prime supplier of telecommunications access in Japan, maintained very high costs, slowing development of ICT infrastructures, including new goods and services, a pattern evident in the United States with AT&T, for example, prior to its breakup in the early 1980s. NTT also embraced technical standards which constrained further use of the Internet in the 1990s. Selecting the wrong technologies, rigidity of management practices, and inadequate efforts by government to restructure the telecommunications industries were all sources of problems that in hindsight are increasingly becoming obvious.[157] The Japanese only began addressing these problems in the late 1990s, after a few half measures in the 1970s and 1980s failed to have significant positive effects.[158] As one Japanese expert wrote in 2006, not until the end of the prior century had it become "clear to all that Japan's domestic telecom sector was well behind cutting-edge developments in information technology," and that "the government was becoming increasingly aware that the existing regulatory structure and institutional environment hindered firms from using new technologies and pursuing new business strategies." This problem cost "Japanese firms dearly in global markets," particularly in the fast moving growth market for wireless phones accessing the Internet.[159] After changes were made, utilization of the Internet surged, making it more evident that computing and all forms of communications were now moving into what possibly was the dawn of Wave Two IT.[160]

With those telecommunications issues in mind, we can better understand what market data from the first half of the 1990s indicate. It begins to shed light on how users (institutions and individuals) were applying computers by tracking what they purchased. Table 7.14 shows that the percent of expenditures devoted to traditional mainframe computing versus smaller, more distributed (indeed in many instances network dependent computing) grew incrementally. Large enterprises began to use distributed processing by creating networks over which they could fax documents, supported by national packet-switching networks. As early as 1984, a survey of some 1,700 firms determined that a third already used digital networks linked to computers and other companies. Firms could also conduct teleconferences that year for the first time among sites located in Tokyo, Osaka, Kobe, and Nagoya, continuing the diffusion of information about digital possibilities in yet new ways.[161] The trends are obvious and predictable. Users began moving work from mainframes to distributed processing in increasing amounts at the same time as consumers were entering the market for digital products. To be sure, the demand for computing declined as the recession extended, from a high of $42 billion in 1991 to a low of $34 billion in 1994, before increasing incrementally again during the second half of the decade.[162]

Future discussions about who used computers and for what purposes during the 1990s and beyond will have to take into account the role of individual users, as is now the case when discussing the role of IT in North America and elsewhere. Table 7.15 demonstrates that by 1994, consumers were entering the IT world as individuals, a sign that Japan was moving from predominantly Wave One computing into the initial phase of Wave Two. These data suggest cell phones had become ubiquitous within a decade of their availability and that PC ownership had leveled off.[163] By the early 2000s, many functions available before only via a PC were now accessible using cell phones, including banking transactions, e-mail, and games. The evidence also suggests

Table 7.14 PERCENT OF MARKET SHIFTS FOR VARIOUS CLASSES OF JAPANESE COMPUTING, 1991–1996 (EACH YEAR TOTALS 100 PERCENT)

Year	Mainframes	Minis	Office Computers	Workstations	PCs
1991	51.0	3.7	14.8	8.7	21.8
1992	48.1	4.2	13.8	9.2	24.7
1993	45.0	4.0	11.9	10.9	28.2
1994	39.2	3.6	9.9	10.9	36.3
1995	32.9	3.1	9.0	11.0	44.0
1996	29.1	2.9	8.9	11.9	47.2

Source: Data extracted from Norris Parker Smith, "Computing in Japan: From Cocoon to Competition," *Computer* (March 1997): 30.

Table 7.15 OWNERSHIP OF DIGITAL PRODUCTS BY JAPANESE HOUSEHOLDS, SAMPLE YEARS 1995–2004 (PERCENT OF HOUSEHOLDS)

Year	Personal Computers	Cellular Phones
1995	16.3	10.6
1997	28.8	46.0
1999	37.7	64.2
2001	58.0	75.6
2003	78.2	93.9
2004	77.5	91.1

Source: Information and Communications Policy Bureau, Japanese Ministry of Internal Affairs and Communications.

that the relative cost of computing had declined sufficiently to make it affordable to consumers, despite the persistence of economic recession.[164]

While events in the 2000s lie outside the scope of our study, because they represent increasingly Wave Two activities, they cannot be ignored entirely as they built on behavior evident in the 1980s and 1990s. Most specifically there was the Internet, so we must discuss this issue further, describing the arc of its adoption into the 2000s. Table 7.16 shows diffusion of Internet use in Japan. While starting later than in the United States, its growth rates were high in households, enterprises, and other establishments even during difficult economic times.[165] Over 60 percent of households and establishments of all types used the Internet by 2001 in one way or another, ranking Japan second after the United States in raw number of users and, interestingly, followed by China at third with 33.70 million users.[166] By mid-decade (2005) just over 68.5 percent

Table 7.16 INTERNET ADOPTION IN JAPAN, 1997–2010

Year	Number of Users (millions)	Population (percent)	Penetration of Households (percent)
1997	1.55	9.2	6.4
1998	16.94	13.6	11.0
1999	27.06	21.4	19.1
2000	47.08	37.1	34.0
2001	55.93	44.0	60.5
2005	75.05	60.9	n/a
2007	87.54	68.0	n/a
2009	95.98	73.8	n/a
2010	99.14	78.2	n/a

Source: Ministry of Public Management, Home Affairs, Posts and Telecommunications, Japan, *Stirring of the IT-Prevent Society* (Tokyo: General Policy Division, Information and Communications Policy Bureau, Ministry of Public Management, Home Affairs, Posts and Telecommunications, Japan, 2002): 5; "Internet World Stats," http://www.internetworldstats.com/asia/jp/htm (last accessed 7/31/2010).

of the population had access to the Internet, accounting for 87.5 million people.[167] In short, Japan had become a densely ICT-networked society, or, to use the words of two witnesses, "the country is now a sea of video games, computers, instant messaging, and camera phones."[168] While historians will eventually present reasons for extended use of the Internet in Japan, one can point to several pieces evidence: expanded and affordable high-speed broadband, adoption of state-of-the-art mobile phones and other digital consumer products, and declining costs of all technologies, collectively making use of IT relatively more affordable to people than in the 1980s and early 1990s.[169] Finally, we must acknowledge that technological innovations in the Internet's base technologies spurred demand, such as the availability of web browsers by the mid-1990s and the expanding supply of content over the web that has continued to the present, including online games which are so popular in Japan.[170]

PATTERNS IN DIFFUSION OF IT IN JAPAN BY THE NUMBERS,1950S–1990S

The use of information technology in Japan demonstrates a pattern of diffusion similar to those of other industrialized nations: first came the vendors along with government initiatives to support their efforts, next early adoption by the largest firms and agencies, and then as the technologies improved, dropping in cost, and enhanced with software applications, seeped further into society. An educated workforce employed in industries that increasingly used IT themselves became advocates and users of additional computing, first at work, then in their private lives. We have argued in this book that the most discernable forensic evidence at the scene of the action documenting diffusion is the trail of computer installations, in particular, mainframes, but later such distributed processors as personal computers during Wave One, and in private life cell phones and other digital products in late Wave One and early Wave Two. Japan, like other nations, left a trail of spreading installations of computers across its economy.

The pattern of additions was a familiar one. A handful of systems were built and used in the 1950s, followed by a spike as second generation mainframes came into use in the early 1960s. The big spurt began after third generation hardware became available, driving all the computer population numbers up sharply in the 1960s. The spike extended into the 1970s. By the 1980s tracking populations of installed systems became increasingly difficult since computing had spread to so many formats: mainframes, distributed processors, minicomputers, personal computers, other IT embedded in consumer and industrial equipment, even in automobiles and all other modes of motorized transportation. Table 7.17, however, suggests the rate of diffusion at takeoff of diffusion. Various inventories were kept through the 1950s–1980s with only slightly differing "nose" counts for how many were installed.[171] The great takeoff in installations which began in the mid-1960s can also be seen by the amount of money spent on these. Net domestic consumption of computing cost $359 million in 1967 and increased each year to where it reached $2.21 billion in 1974.[172] These data demonstrated that the nation invested extensively and rapidly in the technology.

Industry observers commented about adoptions in the 1970s and 1980s that confirmed a long term pattern of adoption had remained relatively intact, despite difficulties along the way. This adoption validated comments made by Anchordoguy, Fransman, and others, about the enduring IT ecosystem right into the recessionary period of the 1990s. For example, an Arthur D. Little

Table 7.17 NUMBER OF COMPUTERS INSTALLED IN JAPAN, 1959–1973, SELECT YEARS

1959	50
1961	130
1963	300
1965	581
1967	1004
1969	2057
1971	3787
1973	5025

Source: Various lists prepared by MITI and JECC, including Japan Productivity Center, *Computer Utilization Management in Japan* (Tokyo: Japan Productivity Center, July 1969): 11, CBI 32, Box 471, folder 3; Japan Electronic Computer Co., *Progress of Computer Industry in Japan* (Tokyo: JECC, 1974): 4, CBI 32, Box 631, folder 22, both at Charles Babbage Institute, University of Minnesota, Minneapolis.

commentator wrote in 1978 that the market was being influenced by "the con-tinuing enhancement of technologies and the resulting improved economies," citing recent product announcements by IBM that once again "forced a reaction by the Japanese domestic manufacturers" to lower their prices to domestic customers. The same analyst acknowledged that the uncertain eco-nomic conditions faced by Japan in the late 1970s might dampen investments (we now know it did but only momentarily).[173] The Japanese computer market grew to $4.3 billion in revenues in 1981, and expanded to over $6 billion in mid-decade. By the early 1980s, the domestic market had expanded to become the second-largest single national one after that of the United States and just less than half the total size of all Western Europe's; in short, it had expanded rapidly.[174] At the other extreme of mainframes—PCs—by 1989 the Japanese had just over 5 million scattered across their economy, representing an invest-ment of nearly $19.8 million.[175] Put into an international perspective, diffu-sion of this class of computing remained relatively low when compared to such other nations as the United States, United Kingdom, and Germany, compa-rable to that of France and South Korea, and ahead of Italy in the 1990s, yet always less than the average for the 27 OECD nations in that decade.[176]

The economic crisis that began in 1989–1990 signaled the start of a slow-down in the diffusion of IT in Japan, but not a profound pause. In the mid-1990s, the economy rebounded slightly, leading to several years of extended IT expenditures in Japan, before deteriorating at the dawn of the new century as happened across the entire pan-Atlantic community as well.[177] The ICT community—industry and its users—however, remained important elements of the Japanese economy in the 1990s. In fact, the ICT industry contributed 7 percent of the nation's GDP in 1990 and grew to 7.3 in 1997.[178] By the end of the decade, some 80 percent of large companies were using the Internet in one fashion or another, 70 percent of all government agencies, but only about a third of small companies. Consumers, as noted earlier, had become very active; in the same year (1997) they frequented nearly 12,500 retail outlets for Internet-related products and services, a 20 percent growth in retail outlets since the partial economic recovery in 1995–96.[179] IT investments and uses picked up again after 2002, driven this time, largely by Wave Two considerations, most notably demand for broadband, a story for a future historian to tell.[180]

ECONOMIC EFFECTS OF IT DIFFUSION

As observable in other nations that relied extensively on the use of IT since at least the 1970s, measurable results suggest extent of diffusion and speed of adoption of computing, contributing further to our understanding of pat-terns of behavior. Japan proved no exception *in general*; indeed, while partic-ular facts vary from one nation to another, the broad patterns of Japanese economic effects generated by IT mirror those of the pan-Atlantic community, including, in particular, the United States, with some local variations, such as lower labor productivity than the Americans. As in the pan-Atlantic community, Japan's experience has no similarity to the Comecon experience in Eastern Europe. In subsequent chapters, we compare Japan's experience to those of other Asian economies. The economic data on Japan's use of IT in the 1950s and 1960s is virtually non-existent largely because the role of computing was, to put it bluntly, miniscule, despite the quick ramp up in investments in IT during the 1960s. In large part, this is due to the fact that at a national level it takes 10 to 20 years from the start of such investments before results are large enough to document substantially in economic statistics dealing with such measures as labor productivity and effects on GDP. Although, one can discern and catalog results in exports and the value of national currencies as they occur almost immediately. As we do with other nations, exploring the economic impact of IT offers insights also on the shape of adoption of computing, leaving to economists to debate methods of data collection and various effects—a dialogue underway regarding Japan since the 1990s. The long recession in Japan intensified interest in the economic impact of IT on its economy, and it is from that literature one can draw some lessons for the history of Japanese uses of computing.

Yen investments in ICT—computing and telecommunications—grew steadily from 1970 to 2005, with a noticeable flattening, beginning in the early 1990s, but with some subsequent growth. For the entire period 1970–2005, the annual rate of growth in investment averaged 8.6 percent. Clearly the amount of investments made in the period to the end of the 1980s turned out to be less than the volume of expenditures since 1995, as measured in 2000 constant prices. Put in more precise language, the Japanese invested more in ICT between 1996 and 2005 than in the years from 1970 to the mid-1990s. The amount spent on communications, while rising in the late 1990s was not appreciably significant; the growth came in computer hardware and software. In 2005, ICT expenditures accounted for 18 percent of all investments made by Japan. In that year this class of ICT (computers) accounted for just over half the total investments in ICT, much as in the previous several years.[181] During the mid-1990s investments remained flat, now down over prior years. These picked up quickly during the second half of the 1990s at rates comparable to those in other industrialized economies.[182]

The influence attributable to ICT on economic productivity has been sizeable and the subject of considerable discussion among economists. Its effects were similar to those seen in other Wave One economies, such as those comprising the G-7 countries within OECD. These ranged between 0.1 and 0.4 of 1 percent per year during the 1980s and early 1990s. Japan's investments, hence participation of ICT in its economy, proved lower than that of the United States, but comparable to France, Germany and Italy, caused by less IT installed and used in the economy than to some other factor.[183] Economist Dale W. Jorgenson, in looking at the same issues concluded that "during the 1980s productivity played a minor role as a source of growth for the G7 countries except Japan," but proved more important for G7 countries after 1995, except for Germany and Italy. Japan's investment in various forms of ICT led it to have some of the highest per capita economic output of most industrialized nations with the exception of the period beginning after the mid-1990s, when its economic malaise was on full display in results gained from use of ICT. Japan lagged while all other G7 nations enjoyed growth in ICT-driven productivity, beginning in 1995. Jorgenson noted that Japan's quality of investments in ICT lagged those of other G7 nations over the entire period with the exception of the 1980s, when it did well. Although its per capita investment in IT went from being the fourth highest of the G7 in 1980—an impressive amount—it dropped to third highest in 2001. It invested substantially in the 1970s and 1980s, making it often the leader in growth rates, but again the process slowed during the recessionary years 1989–1995. Jorgenson's observation about the contribution of capital investments in IT could not be clearer, its "contribution of IT capital comparable to that of the US during the 1980s, followed by a decline in contribution from 1989–1995, reflecting the sharp downturn in Japanese economic growth."[184] Non-IT investments rose in the recessionary years having greater influence on overall performance of the economy.

However, he also observed, as in the United States, that the portion of the economy producing IT products and services generally made a positive contribution economically from 1980 through 2001. But it was not enough:

> For Japan the dramatic upward leap in the impact of IT investment after 1995 was insufficient to overcome downward pressures from deficient growth in aggregate demand. This manifests itself in declining contributions of Non-IT capital and labor inputs.[185]

In other words, as in France, Germany, and Great Britain, IT alone could not stir the national economy into growth and prosperity. As for the other G7 nations, investments in such tangible assets as IT was one of the most important sources of growth for these nations from 1980 to the end of the century. In fact, about two-thirds of Japan's output growth since 1995 to 2003 was a result of IT production (supply side of the economy) from both hardware and software. That data reflected consequences of accumulated and substantial investments made in IT over many years, a rebound that stood in sharp contrast to the decline in output of the first half of the 1990s. In other words, the influence of IT on Japan's economy continued to expand in general over the entire period 1970 to the early 2000s, despite the slowdown in the early 1990s.[186]

In the same period, Japanese observers complained that slow management decision-making and *keiretsu*-style practices were holding back use of IT, most notably "more open, downsized computer network systems." With a clearly 1990s twist, one student of the process described the situation this way: "With the traditional Japanese management philosophies that emphasized rich communications among geographically concentrated, closed groups of individuals and firms, open computer networks are neither necessary nor useful." In the 1980s and beyond even "PCs were irrelevant." "Enclosing customers" remained in force in the 1990s as it had in earlier decades.[187] So leaping to conclusions that Wave Two was unfolding with Wave One now in the past must await stronger evidence since Japan was enmeshed in a transitory period, moving from the pre-recession era to as yet new one in the processing of emerging.

Based on both the evidence presented in this chapter, combined with what economists observed, it appears that changes in diffusion of IT and influence on the economy had led to changes by late 1980s or early 1990s. Furthermore, between the mid-1990s and the early 2000s, investments in IT obviously increased, as in the United States, but, unlike in the United States, labor and total factor productivity (TFP) did not improve in Japan. The closer one approaches the present, the more unclear even that kind of data becomes. For example, for much of the second half of the twentieth century, users of IT were large enterprises, as in other countries. There is now a tiny, yet growing body of evidence suggesting that as IT came in less expensive units that ever smaller enterprises embraced them, as occurred in the United States and in various parts of Western Europe, such as in Italy. Recall that Japan's economy in a way it mimicked Italy's, because it consisted overwhelmingly of small and medium-sized companies, in fact well over 90 percent of all enterprises. Leaving aside definitions of small and medium-sized enterprises, they clearly were not in the same league as NTT, Sony, Hitachi, or Nissan.

The always present issue of exports enters the discussion here and for clarity we have the excellent analysis done by Adam S. Posen who got to the heart of its role and relationship to IT: "the bulk of the Japanese economy, in fact practically the entire economy outside of the export-oriented manufacturing sectors, is beset by very low productivity, extreme inflexibility, and long-term stagnation," except, of course, where government stepped in with economic stimulative programs. Armed with a raft of data, Posen concluded that "there has been a complete lack of diffusion of either technical progress or labor productivity from the high-tech sector to the rest of the Japanese economy in the last forty years."[188] While his last statement may have to be moderated somewhat, as our evidence suggests, he is, nonetheless, correct that when comparing Japanese results to those in so many other countries, it came up shorter on results. Yet to give the Japanese their due, Posen also concluded that when they changed (improved) products, they did this very well, with a deep bow of respect to the contributions made to those technologies with which they proved so successful in foreign trade. Nearly a decade later, Japanese economists looking at productivity levels of the 1980s and 1990s reached similar, if less harsh, conclusions. IT had not stimulated productivity to the extent it had in other countries, despite the highly evident increased adoption of IT across the economy in the 1990s. In the tactful language so many Japanese scholars used when discussing the problem, they argue, "our results suggest that Japan's comparative advantage in areas such as production technology and management techniques may have been sharply eroded by advances in IT," which became rapidly available all over the world, from PCs and new software to deployment of quality management practices.[189]

But, of course, at the time users did not know to what extent they would affect national productivity and GDP by their individual decisions. They continued to do in the 1990s what they felt appropriate in the best interests of their organizations. In a survey conducted in 1998 with over 5,000 participants across major Japanese industries, researchers observed that firms of all sizes were using IT: 23 percent with 4 or less employees, 58 percent with 5 to 19 employees, 86 percent with 20 to 49 employees, and over 93 percent by firms with more than 50 employees (97 percent those with 100 or more). The same survey reported firms using IT also produced more innovative goods and services than those that did not. Additionally, companies using IT were more profitable than those which did not, suggesting greater flexibility in creating products, services, and business models among users, or at least more creative management in these firms. More precisely, over 30 percent of the users were profitable as against only 15 percent of non-users. The surveyors did not explain why profitability was so low overall, although small firms were more likely to be profitable and users of IT than medium-sized companies.[190]

Japanese economists have documented the increased expenditures and use of IT since the 1990s along two tracks: supply side (that is to say, by IT manufacturing and service vendors) and on the demand side by users of their products. The combination of the two played an important role in improving productivity in Japan in the 2000s. But to be more precise, "the IT-using sector has relatively slow growth in the 2000s, compared to the IT producing sector, because the recent cyclical expansions have been mainly driven not by an increase in domestic demand but by an increase in exports abroad." This brings us back to an issue that has masked the extent of IT diffusion in Japan since the 1960s: the role of foreign trade in driving production of IT products, hence in causing economic statistics to make Japan appear a greater and more productive user of computing than it was.[191] Additionally, resurgence in product innovations and investments in R&D in this period, often supported by government programs, proved essential to the process.

Japanese economists recognize, as do their American and European counterparts, that while IT producing industries appear quickly in the national statistics, productivity increases as a result of investments in IT by user industries lag by many years, in Japan's case by 10 years or more. Japanese economists, therefore, did not expect to see significant effects in the post-1995 pickup in investments in IT until closer to 2003–2005. The delay in the evidence would become apparent later because about two-thirds of all capital investments made by users of information technologies in all manner of capital went into IT since 2000. As two Japanese economists noted in 2008, "the post-2000 resurgence (in Japan's economy, my comment) in technology has been an IT-centered story, with increases in the rate of technology change for both IT-producing and IT-using sectors."[192] One has to question whether this was good news. Just a few years earlier a thoughtful team of business professors examining events in Japan concluded that investments in IT alone were not enough to improve the nation's economy: "Japanese targeting did not work. Government policy should instead focus on removing constraints to productivity."[193] In other words, the institutional infrastructure described at the start of this chapter continued to put a brake on what otherwise was the implementation of IT in ways evident in other countries.

JAPAN'S IT DIFFUSIONARY ROLE IN ASIA

All through the history of IT's diffusion around the world one encountered Japan. In the pan-Atlantic community it established alliances with local firms to sell its products, beginning in the 1970s. It exported manufactured goods made at home and in other countries to the area. The same occurred in the United States and Canada and, of course, across Asia. Export economics has been a global half century long Japanese strategy for economic development, cutting across many industries, not just IT producing ones. Given its geographic, historic, and cultural ties to Asia, what patterns were evident that we can expect to encounter in subsequent chapters on the region? In a word, they were similar. Exports began with such relatively high-tech items as refrigerators, radios, and television sets, but soon extended to transistors, semiconductors, and other electronic components, and in time to such higher order products as telecommunications and computer machines. However, Asian markets were not as developed in the 1950s and 1960s in a way suitable enough to have customers afford such IT items to the extent in the United States and in Western Europe. As these nearby economies evolved and became more prosperous, Japan's products proved more relevant to them.

Producing in other Asian countries preceded selling IT to these neighbors. As demand for semiconductors, for example, increased worldwide, Japanese electronics firms found labor costs lower in other parts of Asia and so began investing in productive capacities there. The rising value of the yen in the 1970s and 1980s provided additional incentive to move work to less expensive countries. These actions drove down production costs while creating additional capacity to keep up with growing global demand in the 1970s and 1980s. Hitachi made foreign direct investments (FDI) in semiconductors in Malaysia in 1973, NEC in Singapore and Malaysia in 1976, Toshiba in Malaysia in 1975 but in South Korea as early as 1969; a decade later Matsushita did the same in Singapore, and Tokyo Sanyo in South Korea in 1973 and Taiwan the next year.[194] These same companies also invested in semiconductor manufacturing in the pan-Atlantic community, such as NEC in the United States in 1978, Hitachi in West Germany in 1980, Toshiba in the United States in 1980. In the case of Asia, all investments went into nations that were rapidly developing into industrialized economies.

Our narrative should be balanced with an acknowledgment of circumstances that also slowed the process of manufacturing overseas because of how production took place in Japan. Let a student of the process explain:

> The just-in-time delivery system, and their close relations with subcontractors did not lend itself to overseas production. Also, Japanese productivity was based in significant measure on the close working relation between engineering and the constant improvement developed on the line. This depended on attitudes and work habits of Japanese labor and were not easily developed abroad.[195]

The Japanese overcame these impediments, exporting their manufacturing practices, as we saw earlier to the United States, initially through the automotive industry but later to all manner of production practices. By the early 1990s, only about nine percent of Japan's manufacturing capabilities across all industries were located outside the country, so the process described below was slow.[196] That percentage still represented a large volume of transactions, because the overall size of the Japanese economy was so extensive that even a statistically small amount of activity was large within the context of many other Asian economies.

By the late 1970s, buying and selling of ICs in Asia had become important elements of Japan's high-technology import/export business. It's most important Asian trading partners were South Korea, Taiwan, Hong Kong, Philippines, Thailand, Singapore, and Malaysia. Table 7.18 catalogs this trade. Note that where imports were small or nonexistent, Japanese investments were essentially to leverage lower labor costs and often products manufactured in such economies were shipped on to customer in the pan-Atlantic markets.

Table 7.18 TRADE IN INTEGRATED CIRCUITS BETWEEN JAPAN AND
ASIA, 1978 (MILLIONS OF U.S. DOLLARS)

	Imports into Japan	Exports out of Japan
South Korea	$20.0	$20.00
Taiwan	0.5	33.00
Hong Kong	0.00	33.00
Philippines	13.0	0.00
Thailand	0.30	0.00
Malaysia	9.00	0.00

In the case of Singapore, trade was funneled to it through South Korea, Philippines, Thailand and Malaysia.
Source: Gene Gregory, *Japanese Electronics Technology: Enterprise and Innovation* (Tokyo: The Japan Times, Inc., 1985): 402–404.

As in the case of the United States, Japanese electronics manufacturers concentrated their Asian manufacturing facilities primarily in Taiwan, South Korea, and Singapore in the 1970s, expanding these as local skilled labor and world demand grew in the 1980s.

Japan did not implement such import/export strategies on its own. In each instance, Asian national governments welcomed Japanese investments and subsequent trade, a key point to keep in mind as one explores the broader question of how computing and other electronics became such a rapidly defining feature of so many Asian economies by the 1990s. These governments had concluded economic development required active leadership of private enterprises and just as the Japanese had borrowed and learned about technologies from the United States in the 1950s–70s, so too they would from Japan, beginning in the 1960s. They liberalized trade, welcomed Japanese and American investors, and learned from them. This strategy worked for them as it had for Japan in propelling diffusion of IT into their economies. In the process IT raised their per capita GDP during the last three decades of the century.

This process was not small. Just between 1960 and the end of 1979, for instance, Japanese companies established 208 wholly-owned or joint ventures in Asia covering many electronic components and products, including IT.[197] An observer of the process in the 1980s described the emerged ecosystem as "an intricately interactive system" of three tiers of countries, Japan, the four most industrializing Asian nations (see table 7.18) and another four developing economies, such as the Philippines and Malaysia. Each complimented the other in the global market: "As comparative advantage shifts within the region, export-oriented manufacturers have no alternative but to rapidly abandon production which is no longer competitive abroad, developing new products and technology suitable to the changing costs of production."[198] The increasingly expensive economies in turn became consumers of IT products. The system was dynamic and worked fast enough to be globally competitive. Japan sat squarely in the middle of this cycle of activity in the years 1960-late 1980s, bringing in technology, know-how, investments, and in the process created demand for its products and both directly and indirectly, diffusion of IT in the region.[199] This strategy worked well with components for mainframes, but in the 1980s and later even better for PCs because its manufacture and sale optimized manufacturing and sales more globally than mainframes. The latter's manufacture optimized more frequently at the nation level, which is why Japanese production of larger products stayed at home than smaller devices, such as peripherals and PCs.[200] In time the process caused the emergence of local IT companies that could stand on their own, competing against the Japanese and the pan-Atlantic community, beginning in the 1980s and in full flower in the 1990s.[201]

Asian results were impressive. By the mid-1990s, Japan was only one of many computer users of all types in Asia. Relying on the measure of number of installed computers per 1000 residents in 1995 as an indicator of diffusion and use, the United States was at 365 systems per 1,000, Japan 145, Singapore already at 189 and Hong Kong at 171. Countries with larger populations across diverse income levels and industries were also doing well: Taiwan at 98 and South Korea at 78. Growth rates in investments in IT between 1985 and 1995 reaffirm the momentum that had developed: South Korea with growth rates in its investments at 17.8 percent, followed in descending order of investments Hong Kong at 17.1 percent, Singapore at 12.2 percent, and Taiwan at 8.3 percent. Both Japan and the United States actually invested at slower rates, 9.5 and 9.6 percent, respectively, which we can account for by the fact that each had already invested substantially prior to 1985 while the other states were playing catch-up.[202]

Japan stuck to its combined foreign investment/export strategies through the 1990s, despite difficult economic times at home. As late as 1998, worldwide Japan controlled 1,274 manufacturing facilities producing all manner of electronics, with more than two-thirds of these situated in Asia, twenty percent of the global total in China.[203] By 2010, its foreign manufacturing had only climbed to 30 percent of its national capacity across all industries, but its exports around the region were affected by another round of growing strength of the yen in the face of Chinese and other Asian competition now fully capable of taking on the Japanese at their own export game. Yet all Japanese industries had foreign factories that were productive and competent, allowing Japanese firms to continue competing around the world, while their domestic firms declined in productivity and skills, in the words of *The Economist*, "great for Japanese firms but troubling for Japan."[204]

CONCLUSIONS

Japan was the largest and earliest user of information technologies in Asia between the 1950s and the mid-1990s. It was also the most influential source of diffusion of all manner of digital devices across Asia, a role it continued to play into the next century. Just as it was necessary to understand the role of the United States in the diffusion of IT in Western Europe, so too was it essential to appreciate Japan's actions to begin the process of comprehending the role of computing around the Pacific Rim. Since events in the diffusion of IT in both the United States and Western Europe were tied to activities in Japan, and Japan's embrace of computing to diffusionary actions in the United States, additional linkages have to be made to begin building a global-wide view of both the extent and nature of the diffusion of IT. This also had to be done to explain the speed with which this happened. Every nation had a role, but some more important parts to play. If one had to rank in importance the national players, clearly on the short list of a handful of nations, we would have to include Japan. More than the rationale for why I wrote such a long chapter about Japan, this nation's role was, simply put, extensive for over one hundred million Japanese and for over one billion Asians. As argued in this chapter, Japan's way of embracing computing, when combined with its export strategies, became consequential and a model for many Asian countries.

The story told is that Japan relied on a nationally endowed collection of skills and firms in the electronics and telecommunications industries which combined served as the core delivery mechanism for injecting computing into its society. Coordinated, tightly linked public policies to these companies created the economic, political, and disciplined actions that made diffusion possible within the context of Japanese culture and the economic realities it faced across the half century. Regardless of the assessments of various experts about the speed and effectiveness of diffusion, and peculiarities of Japanese business practices or culture, by the 1990s Japan had become an extensive user of computing and influenced measurably by it. A feature distinguishing Japan's experience from that of the United States was the relatively disciplined, indeed effective, collaboration between the private and public sectors, providing protection for the home team but also room for competition. It was the national champion strategy applied effectively, the effort West Europeans failed to deploy as well as the Japanese. The Japanese approach kept Schumpeterian realities at bay, while in the United States government created the environment for success and early investments, then let the market work its will. Eastern Europe's and China's experiences stood in such contrast to that of the West and East that attempting comparisons may not be a useful exercise, particularly since the experiences of one had little influence on those of the others.

Japan stayed in the early phases of Wave One far longer than it should have, wedded to mainframe-styled computing in both what it made and used when the rest of those countries also using IT were moving rapidly into new forms of computing. This resulted in Japan having to play another round of catch-up with the West, beginning in the late 1980s and extending through the 1990s as it embraced distributed processing and the Internet. This constituted a process still underway while this chapter was being written. The evidence presented supports the notion that Japan's *keiretsu*-style communitarianism played a profound role in determining the rate of diffusion into many industries. It reminds us that in every nation diffusion of IT was shaped and influenced by local national culture. I mean that statement in its broadest form: language, education, way of thinking, forms of government, how public administration is done, existing institutions, companies, personalities of individuals, and geographic and demographic realities. Conversely, on a more homogeneous note, the Japanese had access to all the same technical knowledge as any of the most "advanced" economies of the world and again, here the statement is also meant in its broadest form: knowledge about all components of a technology, the underpinning science, manufacturing expertise, entry to distribution channels, legal access to patents and copyrights, and insights on innovation.

Japan's experiences with information technologies were also situated within an economic and business context that it could not choose to opt out of, or to escape if trapped in it, despite heroic efforts to do so. This economic/political context was the environment Schumpeterian rules which described how economies, industries, and firms operated. The economy of any nation, industry, and firm was affected by all the elements of our generalizations made above. All were profoundly influenced by these global activities during the entire second half of the twentieth century. This behavior can be characterized as capitalist-centric, influencing the actions of the pan-Atlantic community, certainly even Communist European countries coupled to political and military vagaries of Cold War. Increasingly in the 1990s and early 2000s, managers and public officials came to realize that many of the behaviors of an increasingly techno-scientifically influenced set of global commercial markets were those described initially by Joseph Schumpeter in the 1930s and 1940s. The Japanese attempted to avoid the churn he described as essential to capitalism's vitality. MITI and other Japanese institutions blunted some of the harshest effects he said were essential, and which manifested themselves in many economies, including those of neighbors in the Pacific Rim. Their blunting efforts proved insufficient to stop the "gales of creative destruction" he is so best known for. This insufficiency cost the Japanese precious time in a moment in history when industries and economies were rapidly becoming linked up to a far greater extent than had been the case since before World War I. Future historians will decide if its being out of phase with the world's economic evolution disadvantaged the nation's welfare. I believe it did.

Our generalization about the role of local cultures in the context of international economic and political actions also applies to a shared ethos about technology. Recall how Japanese electronics firms went to the Americans to learn about technology at AT&T, Bell Labs, and RCA. Those experiences at a personal level influenced the values Japanese engineers applied at home in the development of transistors, semiconductors, computers, and, yes, the humble transistorized radio at Sony. More than the Europeans, who also learned from the Americans, the Japanese more than others embraced viscerally notions of innovation as essential—indeed core—to their success in both product development and manufacture.[205] The subject of the extent of Japan's acceptance or rejection of a managerial and technological value system evident in other countries will remain a subject of debate among historians. But, the message to them cannot be any clearer: Japan was an active participant in the flow of information about the "what," "how," and "why" concerning all aspects of the story of IT's diffusion. It never operated in isolation, and when it attempted to (as in its repeated attempts to cordon off its domestic economy from foreign competitors) it did it overtly. We are too close to many of the events discussed in this chapter; but, it appears that whenever it isolated itself, it paid a high price, repeating a pattern of prior attempts to distance the nation from the World's events in earlier centuries.

Before leaving Japan, one final, brief view of how much this nation invested in IT helps put its IT adoption activities into the context of what so many other nations did. The OECD tracked such investments, providing snapshots as of 1980, 1990, and 2001 across select nations, all of which are discussed in one form or another in this book. In the arcane language of the economists, as a percentage of non-residential gross fixed-capital formation, in 1980 Japan was investing more in IT than most nations, outpaced only by Canada, Netherlands, and the United States at annual rates of about 7–8 percent of total capital expenditures. By 1990, its rate of investments had climbed to a range hovering around 9 percent per annum, outpaced only by Denmark, Australia, Sweden (just slightly), Canada, Great Britain, and the United States, with the United States leading by spending over 20 percent per annum. So, Japan started quickly—reflecting the catch-up phenomenon of the 1970s and early 1980s at work—but manifestly slowed in the second half of the decade with other nations spending higher percentages on IT and in some cases significantly greater proportions, such as firms in the United States. The snapshot for 2001 demonstrates that in the late 1990s like so many other nations, Japan dramatically ramped up its investments (another catch-up at work), with annual expenditures now over 17 percent. The number of countries out-investing Japan in IT had climbed to nine, although 8 were investing just slightly more than the Japanese, while the United States remained the wide leader over all others with rates approaching 28–30 percent. The major global trend between 1990 and 2001 was that all nations surveyed by OECD had stepped up their expenditures as a percent of total capital investments, and quite dramatically over rates of the 1980s. In short, "everyone" was deploying IT in ever greater amounts and faster, with Japan more or less in the middle of the pack.[206]

Having situated Japan in the broader story of global diffusion of IT, we have now entered the vast area called Asia, which for our purposes may also and, more precisely, be termed the Pacific Rim. It is to this large area of the world we turn, a region that entered Wave One later than Japan and the pan-Atlantic community, but which seems to have simultaneously journeyed through it even faster than Japan, entered through some doorways into Wave Two piecemeal, and yet clearly not left the first wave behind it. It is to its experience to which we now turn our attention.

CHAPTER 8

IT Tigers of Asia: South Korea, Taiwan, and Singapore

> We . . . will further strive to become the world's most advanced IT power.
> —Dr. Seung-talk Yang,
>
> Minister of Information and Communication,
> Republic of Korea, *2001*[1]

South Korea, Taiwan, and Singapore represented the next group of digital adopters that surfaced in Asia after Japan. Just as Japan served as a model and an important conduit for initial efforts to use computing in these three very different Asian societies, these too also participated in the global interactions that fostered diffusion of IT by way of knowledge transfers into their home markets, with business alliances through global sales, and as a consequence of government initiatives. All three practices unfolded within the context of local culture and economic and political realities. Each nation developed its own path to *computopia*, a term that increasingly became popular in Asia and elsewhere by the early 2000s. Each also embraced many similar deployment strategies. These three nations became models and agents of diffusion for other parts of Asia to learn from as they initiated their own paths to computing, such as Indonesia, Malaysia, and Thailand. Institutions drove much of the national diffusion of IT: government agencies, vendors, and customers, and only in the latter stages of Wave One, consumers. Universities played a less important role than in Japan, the United States, and Western Europe, largely in supplying graduates in computer science and engineering. As the epigraph opening this chapter suggested, these three nations relied increasingly on IT to facilitate the modernization of their economies and to improve their standards of living.[2] All three countries approached their tasks with more concerted effort and discipline than any pan-Atlantic nation and with a greater sense of optimism about the positive results they could achieve. Future historians will probably moderate this observation as they uncover more evidence of what happened and respond to the continued unfolding of new events over the next several decades. Nonetheless, we already know that these three nations accomplished an impressive technological "catch-up" with the most advanced users of IT in a short period of time. That insight informed the story told here.

In these three cases unique local circumstances differentiated as well adoption of IT. First, each embraced computing later than Japan or the pan-Atlantic community, experiencing isolated encounters with IT in the 1960s, but with their central IT experience commencing in the 1970s. By the early 2000s, these three Asian nations were as extensively (or more) engaged in computing as many pan-Atlantic nations across most dimensions: as manufacturers of digital industrial and consumer products and components, as leaders in technological evolution, as entrepots for regional and global trading, as reliant on the Internet, and users of mobile communications. Most importantly in some instances, they were more advanced in the effects ICT had in shaping their societies than most pan-Atlantic countries, such as in the attraction of online gaming and avatars in South Korea.

Second, and related directly to our first point, is the high speed of IT adoptions—less than four decades—in contrast to the pan-Atlantic community which took at least a half century to reach similar levels of use, and hardly to the same level of gaming and use of social media software. If one counted the number of digital users, speed actually becomes a more emphatic feature of the Korean experience, because wide-spread use of IT realistically took off in the 1980s, compressing the time it took to arrive at late Wave One computing to three decades. That is a remarkable development not fully appreciated in the West. Still, much of the dialogue about computing in these three nations seemed too exuberant, because while adoptions quickened and diffused the closer to 2000 one looked, it was not always as extensive as we are often led to believe. Paper-based bureaucracies still existed in many banks and government agencies late in the last century, for example, so one needs to be more circumspect and precise when describing deployment of IT in these nations.[3]

Third, once the Internet and digitized consumer electronics come into wide use, telling the story of a society's diffusion largely based on the histories of suppliers and institutional users potentially misrepresents local patterns of adoption, because individuals who used computers became agents of adoption too; therefore, their roles must be described as well. Individuals and small organizations became earlier adopters sooner than in the West because their nations spent very little time relying solely on mainframe computing (although they all used it extensively) to the extent evident in Europe and North America.

Instead, they embraced quickly and widely those technologies available to them at the time of their deployment of computing, which often meant PCs and minicomputers. In other words, their time of adoption of PCs and affordable easy-to-use telecommunications coincided when these technologies became available and that did not require extensive technical expertise to use. To be sure, these nations manufactured inexpensive semiconductors in the 1970s, governments and universities installed mainframes (even occasionally in the 1960s), and relied on large systems to operate their telecommunications networks and complex applications in manufacturing and public administration. But the presence of individuals acquiring consumer electronics acting as agents of change and adoption appears greater in these Asian countries than in Japan and the rest of the world.

The implications can be striking. For one thing, the concept of Wave One computing emerging from our understanding of pan-Atlantic events is rooted in the experiences of institutions and communities of users, managers, and interested parties, with individual protagonists portrayed as lesser players, or at least they step onto the digital stage late in the play. Not so in these Asian nations once the process of adoption got underway. The role of individual users—consumers, students, for example—raises interesting questions about these nations, such as: Are they still in Wave One or are they entering Wave Two in some form or another which we have yet to describe? Where are the borders between the two waves? Is it where the types of technologies change, or does reliance on the Internet increase that change to some level of adoption which makes the circumstance so different? Or, does the wave continue until the digital tools fundamentally affect how society functions and thinks of itself, or does its work and play? In other words, while the French discussed the impending arrival of the Information Age as clearly distinctive from the world they lived in during the 1970s and 1980s, but that had not arrived by the early 2000s in France, had the South Koreans entered that new digital era by the early 2000s in Seoul? What do the experiences of these three countries tell us about the shaping of modern societies going forward around the world, and most specific about the Asia-Pacific region?

The challenge in addressing all of these questions using the tools of the historian is that many of the events occurred in these and other Asian nations within very recent years. For example, wide diffusion of computing in financial industries began in the 1980s. For another, public acceptance of the Internet occurred in the 1990s. So, the story told in this chapter feels less like history and more like contemporary reportage. Ironically, there exists a great deal of information about events, thanks to the technologies themselves, such as websites, and extensive data collection by government officials, so the problem is not necessarily (or usually) paucity of relevant facts. Rather, our challenge is the lack of perspective that distance in time from the events offer to the analytical process, and, of course, access to information about the experiences of individual users and their organizations. While archival materials are sparse, memoirs and other texts written by key CEOs of IT firms and by academics are not, and some users are vocal as well. Reliable and useful statistical data on the deployment of digital ephemera while plentiful for Singapore remain problematic for South Korea and for all Asian countries for the period before 1980. This is the last chapter where a large body of data can be used to build a detailed narrative about Asian computing. For all other Asian countries, which were later adopters than these three (four if we include Hong Kong), the paucity of information becomes quite evident, severe, and inadequate, with the exception of Australia.

The Asian experience with IT was profoundly influenced by the activist— one might say paternalistic or neo-mercantilist statist—role of national governments, as in Japan. In each Asian country, institutional practices were—and still are—more directly responsible for making them agents of initial and continued deployment than in any other region of the world. Tensions existed between public and private sectors as the influence of free market forces ebbed and flowed over the years. But, clearly one characteristic evident in Asia was the proactive, often relatively disciplined and effective role of public agencies, indeed even of presidents and cabinet officials in stimulating personally the adoption of IT. The specific example of computing was not an isolated one either, as their pattern of involving the state had emerged earlier in telecommunications, textiles, shipbuilding, steel-making, automotive, and biotechnology.

Historians and economists examining late twentieth century economic events in Asia have often studied this region through the lens of institutional agents at work, writing about the "political economy." The state is seen as the primary catalyst at work. Large private firms are presented as closely collaborating with the public sector in initiatives often funded or supported through governmental control of banks, credits, foreign exchange, and trade protectionist regulations, among others, all the while with some conflicting interests and various levels of local competition. Debates among the experts often focus on the shifting relations, power, and influence of these supply-side agents of IT's use over time, and less on the effects of technological innovations, or the increasing effects of user/consumer interests and actions. Governments implemented policies and programs that encouraged users also to embrace computing and to be able to use the technology so they too must be included in the global narrative.[4]

The story here acknowledges the fundamental interest and importance of institutional agency at work, because the evidence demonstrates that without such involvement, instead of making and using semiconductors, for example, often called "the rice of the information era," Asians would probably have continued to grow more rice than digitized economies.[5] They would have had little choice but to continue doing what they had long done and in such groups as family-run farms and small businesses at a scale that only a family could finance and staff. The case of South Korea, and to a far lesser extent Taiwan, illustrates the process of large institutional actors playing instrumental roles in the diffusion of IT, even more than occurred in the equally impressive case of Japan.[6] In short, institutional involvement made it possible to aggregate sufficient resources—people, capital, expertise—to scale up sufficiently to embrace computing. The social and political disciplines displayed by these societies made possible the speed—rate of adoption—of information technologies.

The public–private nexus of institutional activism regarding economic development would appear to stand on its head some core elements of the Chandlerian model of economic growth as there were few large private enterprises that could set the pace of economic evolution in which the state would play a minimal role. These Asian nations teach us that rules of thumb, or theories of economic and business behavior, are increasingly difficult to adhere to as the details of various localized business and economic activities are understood and compared to each other. But, as the examples of these three nations reconfirm, elements of prior thinking about the role of large enterprises, individual executives and public officials, and governmental agencies remain valid, because no Asian nation embraced IT in a substantive way without large organizations providing the necessary quantity of capital, labor, manufacturing, distribution, and after-sale support networks that were prominent in Alfred D. Chandler Jr.,'s worldview.[7]

Finally, the experience of these three nations in their adoption of IT has a similar feel to the progress and positivism displayed by West European nations and the United States as occurred from the 1950s at least through the 1970s, and again after the wide public adoption of the Internet and use of digital consumer goods, beginning in the mid-1990s in the West. I intentionally do not discuss Hong Kong's experience with IT for two reasons. Telling its story would have made this chapter far too lengthy, but more importantly, the lessons about the diffusion of IT in Asia can be learned by examining the three states chosen for our attention.

SLOW STARTER BUT FAST MOVER: SOUTH KOREA AND THE IT REVOLUTION

South Korea's experience reminds us that the use of information technology is tightly correlated to a nation's extent of industrialization and development of a modern services sector economy. Prior to the 1960s, South Korea's economy was largely agricultural, thus not preconditioned for a natural adoption of computing. To be more precise, using data from 1962, 63.4 percent of the employed labored in the agriculture, forestry, fishing, and mining industries; only 7.5 percent worked in manufacturing and another 29.1 percent in services. In the latter two sectors, little data processing existed that even used such primitive tools as electronic and manual desktop adding machines, calculators, or punch-card tabulators. On the other hand, about 70 percent of the population was literate, and over the next several decades it became highly urbanized—two features of Korean society that facilitated rapid adoption of computing.[8]

Political and military considerations also conditioned the environment into which computers eventually functioned. The Korean Peninsula had been a Japanese colony between 1895 and 1945, with the most senior managerial positions held by Japanese nationals in agriculture, industry, and government until the end of World War II. In 1945 the victorious Allies divided Korea into two occupation zones with the northern part controlled by the Soviets, the southern portion by the Americans. In the late 1940s two separate governments were established, one in each zone, and in 1950 the two Koreas started a civil war that quickly drew in Chinese and American (actually United Nations) forces into a 3 year fierce conflagration that killed off nearly 10 percent of the population, led to massive internal migrations into what is today South Korea, and to the division of the nation that has continued to the present. Our discussion focuses largely on South Korea, which industrialized, beginning in the 1960s, and that transformed rapidly into a post-agrarian, in the majority manufacturing-based, economy, managed by quasi-military dictatorships for over three decades. Public administration began transforming into a democratic government by the early 1990s. Meanwhile, South Korea's population expanded from 27 million residents in 1962 to some 48 million in 2005.

Almost every observer of South Korea's economy has used words like "miraculous" and "rapid" to describe the economic transformation that occurred since the 1960s, in large part because of the extent of the changes that took place. For example, in 1962 per capita gross national product (GNP) hovered at $87, very low by any nation's standards. That level of income led many observers to characterize half the population as living in poverty. Then, the nation's GNP took off and by the end of 2005 per capita GNP had reached $16,413, making its citizens some of the most prosperous in the world. National GNP went from $2.3 billion in 1962 to $790.1 billion in 2005.[9] Table 8.1 illustrates, indeed strongly suggests, why observers of the South Korean experience routinely use notions of speed as both adjective and framework when describing what happened in this country.

Korea's (unless otherwise specified, this term refers only to South Korea) economy was extensively financed in the 1950s with U.S. foreign and military aid, then beginning in the early 1960s, fueled by exports and high rates of savings, mirroring Japan's approach. Early industrialization was designed by government to reduce the need for imports by supplying locally made products and in the 1960s to feed an export strategy serving as a key instrument for rapid economic development. Through a series of 3- and 5-year plans, which were generally well conceived and implemented, the Koreans began their transformation by manufacturing relatively low-tech items, such as textiles. As local capabilities and knowledge increased, industrial production moved up the scale of complexity—the value chain—throughout the rest of the century (see tables 8.2 and 8.3).

During the 1960s and 1970s, the manufacturing sector was able to assemble electronic components designed in other countries (most notably by Japanese and American electronics firms). Components suitable for use in digital products

Table 8.1 KEY ECONOMIC FEATURES OF SOUTH KOREA, 1962–2005

	1962	1972	1982	1992	2005
Population (millions)	26.5	33.5	39.3	43.7	48.1
Absolute poverty (%)	48.3	23.4	9.8	7.6	6.4
GNP (US$ billions)	2.3	10.7	74.4	329.3	790.1
GNP per capita (US$)	87	320	1,893	7,527	16,413
Industrial value added (%)					
Agricultural, forestry, fishing, mining	37.0	28.7	15.9	7.7	3.4
Manufacturing	7.5	14.1	21.9	26.5	18.6
Services	29.1	35.4	46.1	59.5	73.5
Share of capital goods exports (%)	4.9	9.8	25.2	37.5	43.9
Share of capital goods imports (%)	16.5	29.9	25.7	37.7	34.7
Literacy rate (%)	70.6	87.6	92.8	95.9	97.8
University graduates (thousands)	20.5	29.5	62.7	178.6	268.8
Graduates in science & engineering (%)	34.6	45.7	46.4	40.9	39.4
Patents (per million residents)	10.0	6.5	66.3	240.1	1,527.0

Source: data drawn from tables and charts in Joonghae Suh, "Overview of Korea's Development Strategies," in Joonghae Suh and Derek H.C. Chen (eds.), *Korea as a Knowledge Economy: Evolutionary Process and Lessons Learned* (Washington, D.C.: World Bank, 2007): 24–27.

Table 8.2 SOUTH KOREAN ECONOMIC DEVELOPMENT
PROGRAMS, 1960S-2000S

Decade	Goals	Policies
1960s	Build manufacturing capacity Export orientation	Promote light industries Mobilize local and foreign capital Protect local industries from imports
1970s	Build self-reliant base	Upgrade industrial capabilities Apply strategically capital and loans
1980s	Expand technologically-Intensive industries	Increase industrial rationalization Decrease export subsidies Expand imports liberalization
1990s	Promote high-tech innovation	Support technology development Build national info-infrastructure
2000s	Transform into knowledge-based Economy & society	Support market development Promote venture businesses

(such as semiconductors), and, ultimately, advanced telecommunications equipment and digital products (such as flat screens, printers, PCs, and storage devices) became goods made by Korean firms, beginning in the late 1970s. Like the Japanese, for similar reasons, software proved problematic.

Even more so than in Japan with its *keiretsus*, yet in similar fashion, business conglomerates known as *chaebols* dominated Korea's private sector. The national government supported their development to accelerate economic growth and industrial evolution.[10] *Chaebols* attracted the best educated in Korea's workforce. Officials optimized these workers' knowledge by promoting the import of foreign capital and technical knowledge (mostly from the United States and Japan). *Chaebols* were able to transfer capital rapidly from one business to another within their own conglomerates in compliance with national strategies, emerging opportunities, and the availability of capital provided by the government-controlled banks. These conglomerates managed operations in a highly centralized manner. More so than with the Japanese case, *chaebols* remained tightly coupled to government agencies. This circumstance created inevitable occasions of corruption and inefficiencies, on the one hand, but, on the other hand, rapid implementation by the private sector of public policies. These organizations worked best when they mass-produced stable products—as in Japan

Table 8.3 SOUTH KOREAN WORKFORCE DEVELOPMENT
AND SCIENTIFIC/ENGINEERING PRACTICES, 1960S–2000S

Decade	People Development	Technology
1960s	Increase literacy	Create scientific and engineering Institutions Create initial government programs
1970s	Increase vocational training	Expand technical infrastructures
	Improve schools Expand higher education	Build Daeduck Science Town
1980s	Expand higher education	Promote R&D initiatives
1990s	Develop strategic skills in IT, biotechnology, telecommunications	Lead with programs to "catch up" with other countries
2000s	Increase R&D productivity	Build national and regional R&D capabilities
	Improve quality of higher education Implement regional economic development	

and Taiwan—and so also proved often too slow to respond to rapid innovations that required smaller manufacturing runs and changed products, as occurred across all ICT industries and products by the end of the 1980s. This changed circumstance in IT continued to be a chronic problem for Korean manufacturers in the 1990s.[11]

If officials made a mistake, it was to allow the *chaebols* to dominate the economy too much, because their influence came at the expense of the emergence of smaller entrepreneurial firms that could respond more quickly to changing market dynamics, as occurred with small companies in the United States and Taiwan, for example. *Chaebols* operated a "top-down" "command-and-control" style of management often described as militaristic, in large part reflecting the fact that with universal military service and a constant state of potential war with North Korea, a highly disciplined style of work emerged. These firms evolved into a form that proved effective in the early decades of industrialization (1950s–1980s).[12] By the early 1980s, when *chaebols* began to play a significant role in the manufacture, export, and internal use of IT, they dominated major sectors. For example, the top 50 *chaebols* controlled 45 percent of sales in all manufacturing firms, 66 percent in construction projects, and 23 percent in transport and storage companies. Only banking did not participate, because these institutions were owned by the government until late in the century when they were finally privatized. By 1983 the largest 50 *chaebols* comprised 552 firms.[13]

The national government facilitated development of more domestic commercial competition than Japan. It also encouraged greater imports of foreign knowledge and capital into the nation, which in turn stimulated technological innovations, particularly from the 1970s forward.[14] Yet, there remained limits to competition. As one observer noted, *chaebols* did "not cooperate well with other Korean businesses, large or small, and they form export-oriented, vertically integrated networks that attempt to create economic self-sufficiency."[15] Japanese *keiretsus* were more likely to collaborate and even sell products to each other, less so the Koreans. Nonetheless, they competed at home and abroad across a broad range of goods, not just in electronics and IT. For example, looking at 1983 when Korea was becoming engaged in all manner of electronics, Samsung, had one electrical firm, 2 each in finance and insurance, construction, textiles, food, beverages, and trade, 3 in machinery and others in various industries for a total of 29 companies. Lucky-GS, also in electronics, owned 24 firms in diverse industries mimicking Samsung (although with 5 in electronics).[16] Deep into the 1980s companies within the *chaebols* remained small by Western standards, so, these were not always logical candidates for computing until the 1980s or later. Much like the Italian model, as late as 1984 smaller enterprises (those not part of a *chaebol*) accounted for a third of the industrial production of the nation.

By having an authoritarian state, officials were, in the words of one political scientist, "able to insulate the economic decision-making machinery from contending social and political pressures, resulting in the resolution of collective-action dilemmas as well as the formulation of efficient, coherent, and consistent economic policies."[17] This circumstance stood in sharp contrast to that of most pan-Atlantic nations, and even to the Japanese. Government could more easily apply the usual tools of policy makers: taxes, financial credits, regulations, laws, monetary policies, trade protectionist practices, and administrative interventions to help these small entities within the *chaebols* grow and compete globally. Such measures were combined with a reasonably effective bureaucracy. Public agencies experienced less internal infighting than evident in Japan between MITI and other public Japanese agencies in the late 1980s and 1990s.

When Michael E. Porter studied the Korean economy in the 1980s, he readily acknowledged that South Korea had few natural resources, but called attention to its large workforce (17 million) who could be developed. Porter attributed the development of these workers' skills as a primary reason why Korea did well, "a principle reason why Korea has been able to upgrade its economy," arguing that the root cause of that ability was the nation's deep "commitment" to education.[18] When put to work in what he called the "cutthroat rivalry that characterizes every successful Korean industry," one begins to understand the positive results that came to Korea. For instance in semiconductors there were 21 firms in the late 1980s, 200 making printed circuit boards, and 31 making PCs and other small computers all competing against each other.[19]

When we combine all manner of small and large IT firms from these early years, one can see that a startup process was underway. In 1980, there were 4 IT firms, by the end of 1982 there were 26 such companies, and two years later 57. The following year (1985), during difficult economic times, seven disappeared, but after the crisis, the number of companies increased again. These IT companies generated $9 million of products in 1980, $47.4 million in 1982 and in 1984, $428.4 million. While the number of firms declined in 1985, the value of products produced continued to climb to $519.3 million, proof that extant firms also grew in size.[20]

Two broad conclusions can be drawn from the Korean circumstance. First, as the nation traveled closer to the end of the twentieth century, IT played a growing role in the economic affairs of the country, the story about diffusion discussed in the pages ahead. Second, the economy as a whole did well, even in the face of periodic hardships, such as those financial and recessionary problems endured in the late 1970s, mid-1980s, and the late1990s–early 2000s. Many explanations have been offered to account for the success of the economy. Young-Iob Chung, a long-time watcher of local economic developments identified seven reasons that can be quickly summarized. First, the regime wanted to modernize the economy and so made that a priority. Second, the local authoritarian government used its power to do this quickly. Third, officials used Japan and the United States "as their development model." Fourth, a sense of extended family values led individuals to strive for success for the betterment of the clan (family and state). Fifth, old traditional leadership styles and institutions were swept away, replaced with new forms and leaders. Sixth, there developed "a high-quality and adaptive labor force that worked for low wages." Seventh, public officials proved proactive and effective in doing such things as promoting capital formation and economic development.[21] A combination of these seven activities occurred in each decade, with some more influential in one period or another, but all present nonetheless.

The origin of computing in South Korea remains a fuzzy story due to a lack of information.[22] However, we know that the story began in the 1960s when the first computers were installed in Korean organizations in 1967. A few details on some of computing's beginnings are emerging largely from IBM's internal records. The U.S. military brought IBM punched-card equipment to Korea in the early 1950s to help manage accounting, budgetary, and personnel activities. Local use of IT by the American military has continued to the present, with the military upgrading to more modern computing over the years. IBM maintained equipment for the Eighth U.S. Army in the 1950s, and when needed, brought in help from IBM Japan. In the early 1960s, the U.S. Army acquired an IBM 7010, one 1460, and an 1130 which IBM installed at Daegu, Bupyung, and Yongsan, respectively. IBM's records indicate it did data processing work for the South

Korean government for the first time in 1961, when its employees analyzed data from two census, one in agriculture, the other for a national population count. The Korean Bureau of Statistics within the Department of Interior installed tabulating equipment—not computers—in March, 1961, to analyze these various census data.[23]

In 1967, IBM established a small sales office in the capital city of Seoul, which expanded quickly its services and sales from punch-card equipment and typewriters to data processing and computing. Data on sales revenues from the local IBM office suggest that demand for computing took off. Because the company measured revenues as incremental sales over prior years, rather than by the simple expedient of all cash taken in, we have to extrapolate. I estimate conservatively that revenues by 1970 amounted to between $150,000 and $175,000 annually, and that by the end of 1977 IBM had taken in some $10 million in revenues that year. This information on revenues indicated solid growth in demand for computing. The installed IBM equipment was manufactured outside of Korea, largely in Japan. By the late 1970s, IBM's customers included the Bank of Seoul, Honam Oil, Ministry of National Defense, both the Air Force and Navy, Hyndai International, Cheil Wool, and the Citizen National Bank—all large organizations fitting the global profile of early adopters and typical IBM customers. By the end of the decade, other large enterprises were using computing in such industries as life insurance, steel, electronics, and pharmaceuticals. These firms and government agencies relied on IBM's current generation of products, such as System 370s models 125 and 135, even an early minicomputer (S/3).[24] But clearly, the Korean economy remained largely paper-based in the early 1970s.

In 1969, for the first time, public officials had turned their attention to the broader issue of electronics, enacting new legislation and launching an economic plan to promote the broader electronics market. Known as the Electronics Industry Promotion Law, it established a development fund, policies, training, and other programs. This law became the genesis of programs implemented all through the 1970s intended to foster domestic and export-oriented electronics manufacturing and that in the 1980s led to the more current production of semiconductors, a story reviewed below.[25]

Koreans focused initially on the supply side of diffusion for the same reasons as elsewhere in Asia. Asian regimes chose IT as a primary path to economic growth and prosperity. Indeed, more East Asian countries settled on IT as their central area of focus for economic development than any collections of nations elsewhere in the world. In Korea's case this priority on IT led to a relatively long period of intensified actions from the end of the 1960s into the new century. That focus was often directed personally by the president or a senior cabinet official; this style of engaged leadership remained substantially the same to the present. Because domestic markets were definitely too small in the beginning, and all through the second half of the century, exports were seen as essential both to economic growth and to supplying the nation with ICT tools and capabilities. The supply side began with simple consumer electronics and components (1960s–1970s), such as assembling radios and television receivers, expanded into telecommunications almost simultaneously, evolved into production of parts for IT equipment (such as semiconductors) in the 1970s, then into small devices such as PCs and minis (1980s–1990s), and into digitally-based consumer electronics (1980s–present).

Government programs were designed to reduce Korea's dependency on Japanese technologies and patents, which help to explain public support for local R&D, while an assortment of tax credits, and later procurement practices favoring Korean vendors, were implemented, all measures aimed at bolstering the supply side. In the elegant language of economics, these measures were designed to foster "import substitutions," that is to say, make South Korea more self-sufficient technologically while saving its foreign currencies for more strategic investments in manufacturing capabilities, patent rights, and "know-how." In the next five year plan in 1982, for the first time the government focused on promoting the manufacture and export of microcomputers and semiconductors. To protect local suppliers from foreign competition, it banned importation of PCs and various IT peripherals. The prohibition was not lifted until 1987 as part of a broader strategy to make the economy more competitive and to assuage American complaints of protectionist practices. Table 8.4 summarizes a clearly articulated explanation of the various stages of public policy described by Korean professor of public administration, Sung Gul Hong, with respect to semiconductors. His explanation applied to other digital items and uses. He argued that the story of Korea's public policy involved a lengthy, complex, and evolving interplay between public and private institutions with the influence of the former waning as results and strengths of the latter grew. Phases and time lines for semiconductors applied reasonably well to other products, particularly PCs and flat screens in later stages. The government led in the beginning and the private sector took greater charge later of events that facilitated product innovation, sales, and

Table 8.4 ECONOMIC DEVELOPMENTAL PHASES IN SOUTH KOREAN ECONOMY: SEMICONDUCTOR CASE, 1965–1995

First Phase	1965–72	Influx of foreign investments for assembling and packaging Semiconductors; promoted exports
Second Phase	1973–79	Semiconductors/electronics designated strategic industry for government attention
Third Phase	1980–87	Develop local semiconductor fabrication capabilities
Fourth Phase	1988–95	Promote all manner of IT R&D, manufacture, exports

Source: Phases based on Sung Gul Hong, *The Political Economy of Industrial Policy in East Asia: The Semiconductor Industry in Taiwan and South Korea* (Cheltenham, U.K.: Edward Elgar, 1997): 80–126.

exports as these firms built up capabilities, all the while driving exports more vigorously than on internal consumption for sales.

The 1980s proved to be an important time in Korea's IT history on almost all fronts: government programs, launch, or expansion of multiple types of export-oriented IT production, and domestic uses of computing. By the end of 1982, Koreans had installed some 800 computer systems across the nation, many using online applications with 10,000 terminals attached to these systems. The profile of leading users mirrored that evident at the time in dozens of countries: large financial institutions (not just banks), the nation's most substantial manufacturers, suppliers of telecommunications equipment and services, and government agencies. Telecommunications was beginning to take off too. Customers leased approximately 3,000 networks, while some began using digital offerings from DACOM Corporation.[26] Networks were augmented during the decade with other value added network (VAN) providers as well, indeed 140 of them by 1993. These vendors offered access to online databases, electronic mail, and electronic data exchange, all uses deployed in large organizations.[27]

The bigger problem was not just the United States accusing Korea of dumping products or keeping its domestic market closed, rather lower-cost producers of components in Taiwan, who were taking market share away from the Koreans. Local rivals led to additional protectionist actions, beginning in 1988, such as the imposition of import duties on PC components (largely motherboards). Officials designed these actions to blunt the effects of expanding Taiwanese IT manufacturing capabilities by promoting competitive quality and by affecting relative costs of Korean–Taiwanese products.[28] Dumping and constraint of trade were thus not issues in Korea. The Korean government encouraged mainframe vendors to bring their products into the country, a practice that stood in sharp contrast to the Japanese one in their home market. By the end of the 1980s key foreign vendors were thriving in Korea. IBM's computers accounted for nearly 70 percent of the installed population of large and mid-sized computers, a market share quite normal for that company whenever it operated in a competitive market at the time. Control Data and Unisys—two other American vendors—enjoyed a combined additional 16 percent of installed systems, while combined Japanese Hitachi (HDS), NEC and Fujitsu (Amdahl) provided another 12 percent of systems. The remaining less than 3 percent came largely from Europe. All major players not mentioned already were also in the Korean market: Digital, H-P, Tandem, Data General, Prime, Pyramid, NEC, NCR, Bull, Sratus, and Wang, each selling products in the mid-range (mini) computer market.[29]

In 1987, the government launched an initiative in support of the development of a locally made minicomputer in support of its broader initiative to increase use of computers in government (National Administrative Information System). Its effort involved the four major electronics enterprises, Samsung,

Goldstar, Daewoo, and Hyundai, all collaborating with a key government agency, the Electronics and Communications Research Institute, better known as ETRI, to build a system named TICOM 1. The system became available in 1988. It relied largely on American technology from Tolerant Corporation. A second, more powerful model also went into production (TICOM 2), which, after Tolerant's demise, was redesigned to work with AT&T's Unix operating system. TICOM 3's production began in 1992, using Intel's Pentium technology. This family of three systems enjoyed modest success, with 503 sold by the end of 1994. Two-thirds of these went into government agencies, nearly another 10 percent into government-owned telecommunications companies, and just over 4 percent into various financial firms; 96 systems went into the private sector, with the majority into firms controlled by the four *chaebols* involved in the development and manufacture of the TICOM systems. As observers of the rollout of TICOM noted, however, its pattern of adoption suggested TICOMs were not viable commercial products.[30] One can speculate on why, but clearly relying on technologies of a defunct firm (Tolerant), which was already in financial trouble when the Koreans embraced its technologies, was never a good strategy for an end user to accept if given the option to use that of a thriving supplier who could support continued evolution and operation of its products. Neither was it wise to embrace systems whose performance did not match best-in-class available at the time, nor if their price and the expense of maintaining and using these were not competitive with those of other vendors.[31]

In the early 1990s, public officials and Unisys (a U.S. computer manufacturer) agreed to collaborate in development of massively parallel processing systems. The government also brokered the sale of AT&T's GIS (ex-NCR) System 3600 in Korea. At the time, however, the world demand for mainframes was declining, while distributed processing was growing, leading to rapid growth in the use of networked PCs and minicomputers. More successful was an initiative launched in 1992 that resulted in the manufacture of a local multimedia workstation developed by ETRI, Goldstar, Daewoo, Samsung, and Hyundai.[32]

In every country in which IT diffused widely, and normally early in Wave One, there existed an indigenous semiconductor manufacturing capability, "the rice of the digital age." In Asian economies manufacturing these kinds of electronics components were aimed at the export market, both in the region (such as to Japan) and around the world. Yet, as in pan-Atlantic societies, those building blocks of computers were also subsequently embedded in myriad other products, from consumer electronics to transportation and telecommunications systems, even toys. People knowledgeable about semiconductors also worked in many industries and personally relied on IT with which to do their work.

Asian economies had two features that quickly made them the primary global choice for manufacturing computer chips. First, their cost of labor was far lower than that in the pan-Atlantic community, a characteristic of these Asian economies that made it attractive for American designers of semiconductors to export work to the region, moving it from one nation to another to take advantage of the lowest cost of labor. Americans and Europeans began this practice of sending work to Japan (1960s), shifted to South Korea, then to Taiwan and Singapore (1970s), and next to even less expensive economies in the 1980s and 1990s, such as Vietnam, Malaysia, Indonesia, and the Philippines, to mention a few. Governments in most Asian nations welcomed the influx of technology, training, capital, and revenue, using myriad financial and regulatory tools to build up local capabilities and economic attractions as a way of developing their economies. South Korea was part of that regional process. Only Japan moved quickly away from extensive dependence on foreign know-how, equipment, and capital.

Second, there existed enough of a local electrical manufacturing infrastructure and skill base that Asian workers could be trained on the fabrication and assembly of semiconductors, then later to absorb knowledge on how to design, build, and subsequently compete against the best that the Americans and Europeans had to offer, a capability that emerged over the course of the past half century. Those capabilities involved in the beginning the low-skill tasks of putting parts together, next in managing whole processes for the assembly or manufacture of parts, subsequently how to run high-tech plants, followed by the management of entire businesses, and finally in creating, making, and selling profitably competitive products in the world market—all while both the technologies involved and the business organizations created to invent, build and sell these were continuously and simultaneously changing. And so it was with South Korea, and why the story of semiconductors in this country is a crucial facet of the history of this country's adoption of computers.

In the 1960s, government economic programs began encouraging expansion of the electronics industry through financial support and encouragement of foreign enterprises to invest knowledge, production equipment, and capital in local manufacturing. Initial forays into electronics were low tech, such as manufacturing small consumer electronics, but that changed over time. This effort proved so successful that for several decades, descriptions of the "miracle" of Korean economic growth were largely shaped by what happened in electronics. Close examination of events demonstrated, however, that significant transformative actions occurred in a number of industries, not just in electronics, discussed below after describing the role of semiconductor products in Korea.[33] By 1994, exports of semiconductors had become Korea's most exported items, particularly DRAM chips, making it the second largest global producer after Japan, and ranked third in all manner of semiconductor production, following Japan and the United States. Local capacity and output grew constantly throughout the 1970s, 1980s, and 1990s, often tenfold per decade.

As table 8.5 demonstrates, however, while production went up sharply, internal consumption lagged until the 1980s when diffusion of semiconductors into other Korean products (some for subsequent export) provided evidence of growing reliance on the technology in other parts of the Korean economy, even though limited use had begun in the 1970s. The data do not reflect the importation of semiconductors into Korea, which did occur, but is intended to demonstrate that an internal market for Korean-made components had developed.

Production of semiconductors has always been a very capital intensive, expensive process, requiring mass production to cost justify, as happened in Japan. These investments had to be repeatedly remade as newer generations of semiconductors came out in shorter periods of time, a feature of the business that became increasingly expensive and complicated to manage by the mid-1980s. It should be of no surprise, then, that their manufacture was highly concentrated, indeed within three very large *chaebols*—Samsung, Hyundai, and LG. By 1994, in world ranks as providers of such products, Samsung came in 7th, LG 20th, and Hyundai, 21st. Their global market shares were 15.1 percent, 6.4 percent, and 6.2 percent, respectively.[34] Local manufacturing began in the 1960s when Signetics, Fairchild, Motorola, Control Data, AMI, and Toshiba began assembling IC devices in South Korea. In the 1970s, the government stepped up the nation's technical knowhow through local R&D, leveraging the newly established Korea Advanced Institute of Science and Technology (KAIST), Korean Institute for Industrial Economics and Trade (KIET), and capabilities already available within local electronics firms. The story of how semiconductors spread subsequently has been told by various observers, and need not detain us here, just an outline of their role in Korea to place these products into the broader context of Korea's computing experience.[35]

The broad outlines were clear: Koreans manufactured every new class of semiconductors that emerged in the 1980s and 1990s, with the three major players central to the process, leveraging foreign technologies, Koreans trained in Japan and the United States, and through the purchase of American firms. Linsu Kim has extensively studied this process, observing that the lag

Table 8.5 ESTIMATED EXPORTS AND LOCAL CONSUMPTION OF SEMICONDUCTORS IN SOUTH KOREA, 1970–1994 (US$ MILLIONS)

	1970	1975	1980	1985	1990	1994
Local Production	32	231	424	1,155	5,104	14,800
Exports	32	178	415	1,062	4,541	11,720
Local Consumption	0	53	9	93	563	3,080

Source: Data derived from Linsu Kim, *Imitation to Innovation: The Dynamics of Korea's Technological Learning* (Boston, Mass.: Harvard Business School Press, 1997): 150.

time from when a new computer chip was developed, say in the United States, and when the Koreans manufactured it shrank with every new generation of semiconductors, particularly as the Koreans learned to develop next generation products themselves. For example, the 64K DRAM first appeared in the United States in 1979, in Korea in 1983, a 4-year lag. With the 256K chip, the lag shrank to 2 years, from initial U.S. introduction in 1982 to 1984 in Korea. The 1M DRAM lagged by one year (1985 vs. 1986), and the 4M DRAM to 6 months (1987–1988). In late 1992, it simultaneously brought out the 64M DRAM, hence no lag in development. Shipment times shrank from 3.5 years for the 64K DRAM in the early 1980s to no lag in the late 1980s for the 4M DRAM. With the 64M DRAM in 1994, Korea became the first nation in the world to ship (sell) this product.[36]

As had the Japanese early in their forays into advanced technologies, the Koreans relied on foreign knowledge and alliances—a point made earlier—that continued to prove important to South Korea even in the 1990s. Samsung, for example, established strategic alliances with 12 firms between 1992 and 1995, of which 7 were Japanese, the rest American. To be sure, while that kind of activity ensured a rapid flow of technical know-how into the country and channels for distributing products, it continued the century-long dependence on the Japanese by the Koreans, which the government would have preferred be more limited, indeed even independent completely of Japan's technological knowledge base and technologies. The aspiration to be self-sufficient was an important reason why officials also encouraged local R&D in semiconductors and on related components and processes to reduce dependence in general on foreign sources of knowledge. The number of Korean patent applications in the late 1980s through the mid-1990s suggests some progress occurred toward that end. In the area of chip design, Koreans filed 130 patent applications in 1989 and by the end of 1994, had cumulatively in the years 1989–1994 filed 978; for manufacturing processes they acquired a total of 11,335 patents. Combined, the Koreans accounted for 63 percent of all patents filed in semiconductors, with LG the leader among the Korean firms with 3,182 filings. Samsung weighed in with 2,445 patents and Hyundai 2,059. The data about patents make clear that the Koreans became highly disciplined, aggressive developers of local knowledge about the technology, insights it could apply across many classes of products, from cell phones to automobiles, from industrial equipment to PCs and flat panel displays, and in particular, to the art of their manufacture. Korean vendors moved up the value chain from making low profit components and performing cheap assembly to higher profit design and manufacturing.[37]

These various initiatives were made possible and executed quicker than in most countries for several reasons all at work simultaneously. First, the *chaebols* could invest in such efforts by shifting money from other businesses they owned into this highly expensive industry, just as the Japanese *keiretsus* had done.[38] Additionally, between 1965 and 1973 alone, well-known foreign manufacturers invested in Korea, such as Fairchild and Motorola from the United States, and from Japan Sanyo and Rohm, thereby adding to the local pool of funding, a great deal of which went into firms owned or dependent on the *chaebols*. Second, they learned from other companies around the world how to be in this business.[39] Third, they followed a common process of creating local know-how, initially by hiring experienced technologists, then by studying extant technical literature, upgrading intrafirm knowledge, setting short-term urgent deadlines for development, and building on the knowledge and experience they gained. Fourth, experts moved from one enterprise to another within and outside of *chaebols*, spreading insights and skills into various parts of the economy. Fifth, a point made by Kim, and crucial to the process, these firms had learned enough, and had sufficient state-of-the-art capabilities, to become attractive partners for equally advanced manufacturers and users of semiconductors in such countries as the United States and Japan.[40]

One of the outcomes Korean officials and business executives wanted was success in exporting semiconductors. For our larger story of diffusion of IT around the world, Korean exports adds further evidence that the process of global diffusion was a trans-national affair in which information and products flowed through dozens of countries. That flow depended on collaboration and dependencies among vendors and users in multiple countries. In short, a nation's IT diffusion depended in part on the activities of other countries. Korea was not immune to the effects of this process. For Korea its digital role started quickly. While Korea's largest trading partners for semiconductors were Japan and the United States over the past four decades, like the Taiwanese, however, Korean exports also went to Hong Kong, West European countries, and a smaller portion to other Southeast Asian markets. This pattern of exports had stabilized into the way it worked by the end of the 1960s and early 1970s for the Japanese and Americans, and for many other countries by the mid-1980s.[41] The special competitive role played by Taiwan in Korea's efforts is discussed later.

Software received very little attention between the 1970s and the early 1990s for reasons similar to Japan: it was perceived as lacking value; there were inadequate intellectual property rights protections; writing software was not as fashionable for IT professionals; there existed limited demand for Korean language products outside Korea; and computer firms were unable to move quickly enough to develop products in short runs with rapidly introduced new releases. That situation began to change by 1993. Officials introduced a Basic Plan of National Strategy for Software Industry Promotion with much fanfare and publicly stated commitments to invest in this area of computing.[42] With online gaming a few years later, Korea entered the software business in an impressive way (reviewed below).

Officials also supported the supply side of the economy through various initiatives intended to promote domestic use of all manner of IT, most notably by its own public agencies and by creation of a data communications network. Because Korea is known today for its extensive use of networks and broadband, one can conclude that of the two initiatives, the second proved historically the more important. As late as the 1970s, Korean telecommunications was primitive. A large unfulfilled demand existed for landlines desired by consumers and all manner of large and mid-sized agencies and businesses. The nation was saddled with an aging technological infrastructure and expensive services—all typical circumstances evident in developing economies at that time.[43] In the 1980s, policy makers addressed these problems through local expenditures, by attracting foreign capital investments, and by increasing service and usage fees to pay for upgrades to the nation's telecommunications infrastructure. The Ministry of Communications (MOC) managed much of the government's initiatives. Some services were privatized to meet local demand through more local competition. In addition to various changes in regulations, in 1982 the government created the Korean Telecommunications Authority (KTA), which it owned 100 percent. It also designated which firms were to provide services. Simultaneously, the national government established the Data Communications Corporation of Korea (Dacom) as a private firm to serve as yet another common carrier. In 1985, restrictions on the creation and use of private data bases and data processing services were lifted to further stimulate a local market. Additional liberalization of the communications market came. In 1989 the Telecommunications Business Act was revised to reduce monopolies in telecommunications, following a pattern of deregulation evident worldwide at the time.[44]

Special mention has to be made, however, of the Plan for National Computerization, introduced in 1988, which combined telecommunications and computing with the objective of creating a National Basic Information System (NBIS). Initially conceived as a way of improving the public and private sectors' productivity through the use of ICT applications, it became a signature feature of the economy. The first stage (completed in 1991) involved writing software and installing the National Administrative Information System (NAIS) and the Financial Information System. The first system cost the government some $200 million (150 billion won) while the second cost the bankers about $650 million (500 billion won). These two uses led to the installation of 160 mainframe systems in various government agencies.[45] A second phase in deployment followed, extending across the 1990s to link together government networks and also to existing private networks. Progress on the second phase proved slower than the first due to various budget cuts in the mid-to late 1990s. The importance of this two-phased initiative is increasingly becoming evident because it fostered deployment of IT faster in Korea than would otherwise have occurred by funding implementation and by providing a market for local products, such as TICOMs, a pillar of the local domestic market for ICT.

The government also facilitated development and diffusion of IT skills within the local economy which paid handsome dividends in the 1990s and in the early 2000s as Korea became one of the most networked societies in the world. Government strategies in the 1980s and 1990s worked relatively well in this regard by combining various initiatives implemented simultaneously in the development (education and training) of people, in the creation of tele-communications networks, and in the establishment and funding of various R&D activities. All of these actions made it possible for Korea to have a strong, quite modern, and practical ICT infrastructure by the mid-1990s. For example, by 1995, Korea's approximately 296,000 software engineers were comparable on per-capita terms to similar populations in the United States and Japan, and ahead of other Asian and most European nations. Many Koreans with advanced scientific and engineering degrees who had studied in the United States and elsewhere brought home new knowledge about ICT.[46]

While it is easy to focus on public initiatives intended to promote use of IT in Korea, providers themselves were also active. They experienced startup growth pains in the 1970s, enjoyed a great boom in the 1980s (as did Japan), then went through both a slowdown in sales with mixed results in the 1990s and later more recession and prosperity. Massive exports spurred successes in the 1980s, but poor quality of these goods encouraged local users to prefer foreign-made PCs, beginning in about 1989, a choice which contributed to some of the mixed results of the 1990s. The vast majority of locally-made computing involved PCs (not mainframes) and a few minicomputers. As late as 1991, combined values of locally-made computers (but not PCs) amounted to only $39 million and barely $44 million the following year, with TICOM claiming the lion's share in these years.[47] Taiwanese and Singaporean rivals bested them in the manufacture and sale of components and such peripheral products as disk drives, PC motherboards, sound cards, scanners, and other digital devices in both local and global markets.

Table 8.6 summarizes the production, consumption, and exports of these classes of products by Korea. While the data may not be 100 percent accurate, these are quite comprehensive, demonstrating that exports drove manufacture and revenues more than domestic consumption, following patterns evident in Taiwan, Singapore, and Japan at the same time. Because PCs became one of the most widely used computer devices in Korea in the 1980s and 1990s, it should be noted that with the lifting of the ban on imported PCs in 1987, local consumption for this class of hardware expanded dramatically (by 100 percent) in less than two years, and by another 100 percent within three years. The key market segments that acquired these machines ranked from greatest to least throughout the 1980s and early 1990s remained unchanged. First, the public sector was the largest consumer, followed in descending volumes of

Table 8.6 KOREAN PRODUCTION, EXPORTS, AND LOCAL
CONSUMPTION OF PCS AND PERIPHERALS, 1982–1992, SELECT YEARS

(US dollars million)

	1982	1984	1986	1988	1990	1992
Total Production	42	391	830	2298	3159	3630
Exports	37	314	843*	1942	2163	2703
Local Consumption	5	77	n/a	356	996	927

* Includes products made in prior years
Sources: Data extracted and re-assembled from tables in EIAK, *Electronic Industry Yearbook* (English version) (Seoul: EIAK, 1993); John Dedrik, Kenneth L. Kraemer, and Dae-Won Choi, "Korean Technology Policy at a Crossroads: The Case of Computers," (Center for Research on Information Technology and Organizations, University of California-Irvine, 1995): 32.

adoption finance, distribution, manufacturing, and home users. If we use 1992 as a snapshot of where these PCs were distributed, home users spent nearly $200 million on IT that year. Public and financial sectors each accounted for over $800 million in expenditures.[48]

Software sales were tied to these machines, much of it imported. One study of the local market demonstrated that in the late 1980s and early 1990s, software and services accounted for only 15 percent of the local IT market. Put in monetary terms, software sales in 1992, for example, amounted to $240 million (probably an inflated figure, however). Exports of software were a miniscule $15 million.[49] Services consisted largely of custom programming and maintenance, much as occurred in Japan in the 1970s and 1980s.

Can we create a composite view of users of computing in Korea prior to the great surge that occurred after the mid-1990s? The question highlights the limited amount of data available to answer it but also that computing came late to South Korea. We leave aside the U.S. military's use of IT in the 1950s and later in Korea, because it was not a significant part of indigenous uses of computing, except by the Korean military by the 1980s. The story of domestic Korean diffusion begins in the 1970s, at roughly the same time as a local IT industry was starting up. At the time (1960s–early 1970s), there only existed sporadic awareness and use in academic and public agencies. As the national manufacturing base shifted from simple to more high-tech complex products the need for data processing, and later computers, grew too. For example, when manufacturing semiconductors and other complex electronic components in the 1970s and 1980s, these manufacturers themselves needed to use computers and computer-driven industrial equipment to do many of the relevant assembling tasks. As brokerage firms began to operate in a global financial network in the 1970s they too needed to use computers linked to telecommunications. Any major firm conducting trade overseas increasingly needed to "plug in" to the procurement and supply chains of their customers, clearly a requirement to participate in trade in evidence around the world by the mid-1980s. Korean organizations were the earliest adopters of computers for the same reasons as in dozens of countries and industries.

The greater the number of large enterprises and government agencies a nation had, the more likely they were to be early adopters of computing. The smaller a firm or an agency the later they came to rely on computers to assist in their work. Many were too small to be practical users of large mainframe systems, but if they were, these were often already early adopters. About two-thirds of Korea's agencies and businesses were of the size that most paralleled users of minis and PCs around the world. So, until these two classes of machines became widely available in the 1970s and 1980s, such organizations were not logical candidates to embrace computers. Once available, however, their take up became evident as the various tables and data presented in this chapter suggest.

The industrial demographics of Korea did not change during the era of large, expensive, and complex mainframe computing. As late as 1992, for example, the number of firms with 330 or more employees constituted only 1.4 percent of the nation's total number of manufacturing companies; although this tiny number accounted for 55.2 percent of Korea's share of shipments of all manner of exports and 54.2 percent of the nation's value added. To put those numbers in perspective, Japan's largest enterprises accounted for 48.2 and 43.8 percent, respectively, so Korean economic concentrations were still considerably greater. The 1.4 percent demographic constituted the ideal candidates for early and continually expanding use of telecommunications and computing. IBM's own lists of Korean customers of the 1980s and 1990s confirm this pattern of consumption. These large enterprises were also members of the highly centrally managed *chaebols*. As these conglomerates diversified across multiple industries, managerial practices in one were brought to others, sharing knowledge and investment support within these conglomerates. These shared practices included using computers in accounting and process work most evident in manufacturing and financial firms.[50]

Missing from our story are the individually documented case studies so available about the pan-Atlantic community, largely because the number of industry publications that normally would have documented such events were late comers to Korea's publishing ecosystem and corporate archival records have yet to be studied widely by historians. Extant stories however, make it clear that large organizations were adopters, initially in the 1970s and 1980s. The government examples reported above represent one collective data point. Corporate histories of the electronics firms demonstrate that they not only used the technology themselves, but also sold it to others in Korea.[51] Occasional examples appeared in the published literature, such as that of the large Korean shipbuilder Hyundai Construction Company (HHI), which employed many people, had to track thousands of parts per ship, and could no longer do this manually in a cost effective way when the price of ships dropped in the mid-to late 1970s. HHI needed a cost-accounting system to deal with the problem—the same IT solution used by manufacturers of capital goods in other economies—and so expanded its data processing operations. In the 1970s its IT staff grew to 300 souls. They introduced incrementally the new cost accounting system to the organization, following the time honored tradition in the private sector of enhancing (modifying) existing systems with improvements.[52]

The Pohang Iron and Steel Company, Ltd. (POSCO), on the other hand, offered a glimpse into the limits of IT adoption. The use of process control techniques in the firm to improve productivity did not work as expected. POSCO's management concluded that computing would constrain the institutional learnings needed to improve operations and, thus, chose to do process engineering work without recourse to data processing technologies, instead doing it manually so that people could learn the processes and be more familiar with what the data indicated. Its employees collected and analyzed data using manual methods. Not until 1975 did POSCO install its first process control computer and even then, only for use by a limited number of locations. Business computing had only arrived at this company the previous year. Computer-based applications spread slowly within POSCO in the 1970s, while its global rivals had already become extensive users of various digital technologies. This company of some 60,000 employees took its time, but trained extensively all its staff about the firm's activities. Not until the early 1980s did this training routinely include informing employees about computers; in 1984; approximately 1000 people learned about digital applications. By then, employees had become extensive users of Japanese management and operating practices, most notably Total Quality Management (TQM), which necessitated collecting large quantities of data in support of such campaigns as POSCO's Zero Defect initiative. Ultimately, the benefits of using computers for such activities accrued to POSCO in the same ways in evidence around the world in other manufacturing companies.[53]

Government agencies decreed, and waxed eloquently and extensively about the value and need to use computers in Korea's economy, beginning at the end of the 1960s and continuing into the early 1970s. These pronouncements and early actions came simultaneously with their arguments about the need to develop domestic capabilities to build and use such systems. However, actual progress proved quite slow. As of late 1980—during the height of the global mainframe era—only 475 computers had been installed in Korea in government, private enterprises, educational institutions, and research centers (U.S. military users not included). I use the term "only" as a relative term. Given the fact that Korea had only started seriously to industrialize a decade earlier, perhaps employing 475 expensive systems was reasonable to expect, but nonetheless, Korea's adoption of digital tools was only just starting. In fact, government forecasts from the period held that the number of installations would expand dramatically through the 1980s and indeed that happened, especially after the arrival of minis and PCs.[54] Servicing these 475 systems was, however, a large collection of workers trained in IT, a population approaching 16,000, big enough to sustain additional installations. Meanwhile universities and the private sector had training programs sponsored by government agencies to train thousands of additional IT professionals.[55]

As recently as the early 1990s, however, local knowledgeable observers still noted that use of computers by the largest enterprises remained less in evidence than in other economically advanced countries. Banks, which for many decades had been owned by the government, and only privatized in spurts, beginning in the early 1980s, were often cited as an example of slow adoption of IT in contrast to faster deployment in large manufacturing enterprises. A Korean critique of banks described the situation as of the early 1990s:

> Currently, Korean institutions are making strenuous efforts to fully computerize. However, they lag far behind those of developed nations both in amounts invested in their information systems and levels of computerization. Furthermore, their computerization activities are confined mostly to internal business activities in contrast to the advanced financial institutions, which are very quickly becoming integrated information businesses.[56]

The same observer, Hyung-Koo Lee, an ex-cabinet finance minister and Governor of the Korea Development Bank, also lamented the small size of so many financial institutions, which made it difficult for them to invest in the kinds of IT needed to make them competitive in a global financial market.[57]

The automotive industry, however, used computing as extensively in the manufacture of its products as firms in other countries. Hyundai, for example, began using digital systems in the 1970s and over time, computer-aided design/computer-aided manufacturing (CAD/CAM) systems. Hyundai also installed various assembly line control processors over the next quarter century. In some instances it borrowed Japanese practices and technologies, others from the United States. As part of the process of implementing a full line of computer uses from design to distribution, Hyundai acquired the skills to design its own automotive products independent of the Japanese and Americans. Its knowledge of automotive IT applications also made it possible to diffuse knowledge on digital matters to other enterprises, such as suppliers and business partners, as occurred in this industry in other countries at the same time.[58]

In 1982 government officials started to fund installation of 5,000 PCs in public schools. Soon after, they launched the National Administration Information System (NAIS) with an initial $179 million to support it. 80 mainframe systems and some 5,000 workstations for public employees were installed as part of NAIS. Additional IT projects launched by the government in the 1980s included initiatives in the postal system, others automating tax management, an education research network, and military applications. Government agencies established technical standards for the use of IT. By setting technical standards, funding their adoption, and managing national initiatives using project management techniques, officials were able to speed up deployment and the use of IT all through the 1980s and 1990s more than might otherwise have happened.

The national government continued to fund various IT research initiatives, such as those run by the Korea Institute of Electronic Technology (KIET), which in the late 1980s it privatized. By then KIET had developed sufficient skills and momentum to continue doing R&D on its own. A key element of the government's promotion of the use of IT in the 1980s was a dual strategy of seeding initiatives and supporting a combined public/private campaign in R&D.[59] The private side took the campaign serious. Samsung Electronics, for instance, expanded its R&D initiatives from $8.5 million in 1980 to $905 million in 1994. It was the leader in South Korea in spending on R&D. Samsung invested at a rate comparable to IBM in the same period. To speed up results, the government encouraged, and firms took up, the idea of acquiring companies in other countries to license their patents and share knowledge, such as in Silicon Valley in California.[60] As one of many examples, in 1995 Samsung Electronics acquired a controlling interest in AST Research, a very large American manufacturer of PCs. This acquisition made Samsung one of the top five PC producers in the world. Similar patterns of R&D, partnerships, and acquisitions encouraged by government officials were taken up in such areas as flat panel displays and other electronics components.[61]

Beginning in the mid-1990s, South Korea entered a period of enormous economic and technological volatility that may someday lead historians of Korean computing to designate events before and after as belonging to independent eras, perhaps even in different waves. Conventional Korean thinking at the moment, however, designates 1988—the year of the Olympics in Korea—as the crucial historical date because in preparation for that event and by that date the nation had (a) to enhance its telecommunications network, from satellites to telephones to support the global media, (b) to prepare and install the IT needed to run the complicated operations of such an event, and (c) to improve many parts of the nation's infrastructure for the anticipated arrival of several million visitors, from transportation to housing, television and satellite facilities, to roads and airports. The Olympics was to be South Korea's "coming out" party that announced to the world that it had evolved into a modern industrial economy. It seemed everyone in Korea prepared for what turned out to be a successful Olympics. IBM, the major provider of IT support to this event, installed a large 3090 system, two smaller 4381 systems, 47 System/36s, and more than 600 workstations all to run over 70 applications in 87 locations within the country for the games.[62] Korea's major firms

participated; Samsung became an official sponsor with money and technologies provided to the games, representing the electronics sector. One result of
its involvement was that Samsung served as a proud sponsor of subsequent
Olympics around the world.

Before turning to the enormous number of changes in the use of technology that occurred beginning in the 1990s, coupled to a severe national economic crisis, we should ask how extensively had IT diffused in Korea? Using
inventories of installed equipment as a surrogate answer to the question,
table 8.7 presents a snapshot, circa 1989–90. The proportion of large mainframe systems to medium-sized mainframe and mini-computers reflects patterns of adoption similar to those in nations with a few large organizations
and mid-sized firms. However, many large private and public organizations
had multiple large and mid-sized systems too. Deployment across a string of
industries indicates typical Wave One sources of interest in computing and
the degree of implementation by industry evident in many other nations at
that time. Koreans installed over 2,000 systems, making the 1980s a very
good decade for IBM, Japanese computer firms, and for end users seeking to
leverage information technologies in the operations of their organizations.

The formal end of military rule in 1993 was not a turning point for the use
of IT as it was for other activities of the nation, since democratization and the
evolution toward greater capitalism had preceded this event; nor the arrival of

Table 8.7 INSTALLED COMPUTERS IN SOUTH KOREA,
CIRCA 1989–1990

(Number of systems)

Industries	Large	Medium-Sized	Total
Manufacturing	62	605	667
Nonbanking/finance & Insurance	21	227	248
Government	25	208	233
Higher and K-12 Education	6	189	195
Wholesale/Retail	21	151	172
Banks	41	113	154
Communications & Utilities	12	113	125
Business Services	6	76	82
Health	2	57	59
Transportation	6	19	25
Other	4	132	136
Totals	206	1,890	2,096

Source: Derived from data in International Data Corporation, *Korea Multiuser Market Review and Forecast,
1989–1994* (Framingham, Mass.: International Data Corporation, 1990): 9, CBI 55, Box 103, folder 12,
Charles Babbage Institute, University of Minnesota, Minneapolis, Minnesota.

the World Wide Web at nearly the same time. Rather, a string of intermittent, yet related, events—call them environmental changes—coupled to a major national economic crisis in 1997–98, cumulatively facilitated a surge in both deployment of IT in Korea and a change in the public's acceptance of the view that Korea was emerging as some form of an information society. That all of this occurred during the living memory of a 60 year-old person is remarkable, that it happened largely within the life time of a 40 year old who could remember a time of even limited uses of computing, even more arresting. It is the accelerated adoption of IT since the mid-1990s that led so many observers to speak once again about Korea's transformation as "miraculous" and "fast."

Although we are very close to the events of the mid-1990s, already it is evident that digitization's effects on South Korean society intensified during the second half of the decade. Korea began 1996 in seemingly good shape; most notably the nation's per capita GDP had surpassed $10,000. Korea had also joined the Organization for Economic Cooperation and Development (OECD), a prestigious event in its own right, because only the most "advanced" economies were members. Even scholars benefited from Korea's membership in OECD, because historians and economists would begin to have access to information about Korea compared to similar categories of performance of the other 25 members, all of whom were extensive users of ICT.

But then Korea's economy ran into trouble. The economic prosperity of the 1970s and 1980s had rested largely on the growth of revenue rather than profits, resulting in declining productivity of the *chaebols* that had so dominated the economy. Their debts accumulated. In the summer of 1997 Thailand's economy collapsed, leading to economic crisis elsewhere in the region, most notably in Indonesia, Hong Kong, Laos, Malaysia, Philippines, and South Korea. In November-December foreign creditors began extricating themselves from Korea's economy, fearing similar problems as they had witnessed in Thailand and elsewhere in Asia. Their pull-back resulted in the sharp decline in the value of the South Korean currency (*won*) at a time when Korean firms had been borrowing heavily from foreign banks, debts which now grew in volume as the *won* depreciated. Many Korean companies increasingly could not meet their debt obligations, so they defaulted on loans. These defaults saddled local banks with their debts since these financial institutions had served as intermediaries in the foreign loan process. Inevitably, these banks began to fail. To cut to the end of the story, the government had to turn to the International Monetary Fund (IMF), the Asian Development Bank, the World Bank, and to other governments, such as that of the United States, for financial support. The IMF led the international response. It insisted on painful reforms in exchange for loans that resulted in the local population objecting, feeling humiliated, and that compromised the credibility of both business and government leaders. The South Koreans accepted the terms, and IMF helped the nation. But the *won* had declined in value by 50 percent, the local stock market by 75 percent. *Chaebols* laid off workers, driving up unemployment to 6.8 percent. Ultimately, 40 percent of these conglomerates were proven to be insolvent. Some of the more prominent conglomerates went bankrupt in 1999, including the Daewoo Group. Korea's per capita GDP shrank from over $10,000 to $6,000, an enormous decline by any measure.[63]

Koreans from all walks of life viewed this severe and rapid economic crisis as the nation's failure tantamount to a military defeat. Opinion polls in 1999 unearthed a national and personal sense of shame. Respondents indicated that as a nation they had failed as "industrial warriors."[64] The task ahead was to implement quickly the IMF's demands, actions that changed the environment substantially in which diffusion of IT would subsequently occur. The IMF called for an increased focus on profits over growth in revenues, on the reduction in the power of the *Chaebols*, and on transformation toward a more neo-liberal economic system. The IMF also insisted on reforms of various banking regulations and practices. The IMF wanted more transparency in business dealings, market discipline within corporations, increased competition, more effective business practices, elimination of weak banks, and an economy more open to foreign competition. To various degrees these reforms were implemented over the next decade.

Critical to our story, Samsung laid off nearly 50,000 employees at the height of the crisis (some 34 percent of its workforce). It began to abandon its long-standing strategy of reverse-engineering high tech products at the low end of the electronics market, deciding to move up the value chain to increase profits. That strategy called for more innovations and less imitation across multiple product lines. With respect to IT that approach required immediate expansion of R&D and the introduction of new products, a process of renewal and innovation which the firm intensified in 1998–99. Samsung saw rapid results. 1999's net income reached $2.2 billion (1998's had been only $201 million). In 2004 it earned $11 billion in profits. In 2007 its global sales amounted to $100 billion, which was comparable in size to IBM's or H-P's.[65]

Not all large enterprises did so well, yet overall, small and medium-sized firms now had more flexibility to operate in the economy, leading to innovation in many industries, including IT. Between 1997 and 2002 the number of companies seeking venture capital went from just over 9,000 to nearly 22,000. The IT industry recovered from the crisis to such an extent that in 2002 it constituted 14.9 percent of the nation's GNP. To put that result in comparative terms, that of the United States' IT industry reached 8.3 percent of GNP in 2002.[66]

Meanwhile, following four decades of practice, government officials crafted recovery plans for the nation. Part of that effort involved leading Korea more comprehensively into a digitally-intensified way of life, reflecting some of the rhetoric evident in Japan, but putting more operational substance behind policy. This proactive effort to move closer to some notion of an information society made sense since increased use of IT had continued all through the 1990s: more adoptions, greater use of new products and services, and formation of additional IT firms. As early as 1995, the national government had introduced the Informationalization Promotion Framework Act in support of the IT industry. Even earlier in 1987, public officials had begun changing the role of telecommunications from being a tool to control society to one that could serve as an engine of economic prosperity—a crucial step in the development of a modern economy and society. They combined this approach with additional public investments digitizing networks in the 1990s. The concept of the Information Revolution appealed to the public as a strategy for moving out of their economic malaise, and so they supported new government initiatives.

Then in 1999, the government announced Cybor Korea 21, designed to further the nation's digital transformation with three programs. First, in education it provided 158,000 PCs, along with local area networks to 2,500 high schools, and training for 85,000 teachers in the sue of IT. Second, it funded e-commerce strategy consulting. Third, officials began to implement use of electronic documents as the routine way to communicate among government agencies. Simultaneously, the government published its next five year economic plan which was designed to promote further use of digital technologies across the economy.[67]

By and large the government implemented its announced initiatives, as it had earlier ones. By the end of 1997, for example, fiber optics now reached almost all business and residential buildings across South Korea. Access to broadband services increased, with a target to connect 80 percent of all households by 2005; it reached 75 percent by that date.[68] Access to the Internet became a priority, resulting in a major shift from less than 1 percent of all Koreans using the Internet in 1995 to 39.6 percent in 2000, and to 63.3 percent in 2005.[69] Public officials implemented e-government applications as well.[70] Urban rates of adoption were higher in all manner of ICT, but not that far ahead of rural communities and farms. In 2002 Korea Telecom privatized and officials admonished the telephone company to extend broadband to rural areas. Meanwhile, at the same time the national government pumped $24 billion into rewiring its own buildings and other public institutions.[71]

Korean officials looking back on the 1990s concluded that progress in the growth of IT-centered sales and deployment was "achieved through both deregulation and competition," an acknowledgment of the various actions taken to make the economic environment fertile for IT diffusion.[72] Officials could generally be quite pleased for good reason. The emergence of the Internet and new products helped, but they also saw the IT industry grow at an annual rate of some 22.5 percent just between 1995 and the end of 2000. Employment in IT remained flat at between 20,000 to 21,000 workers in the IT industry in the same years—a result of increased productivity in the industry because of improved "value-added" activities of employees in manufacturing. The only negative productivity trend in the IT industry was in the services business. In the trade arena by the end of 2000, the IT industry's exports amounted to 22.1 percent of all items sent out of the country. Imports reached 22.1 percent of the nation's total, reflecting in part components coming in that were embedded in higher value IT products, which were then exported. Within government circles, optimism on the future remained quite high: "Korea's IT industry is expected to grow stronger with continued efforts in IT research and development. Korea is thus expected to join the world's leader countries in innovative technology."[73] The economic crisis made it clear that not having the economy more open hurt the nation and "as a result, the Korean government adopted a globalization strategy" to enable local firms to integrate into to the global economy.[74]

Before describing results, one might ask, how did the events of the late 1990s feel on the ground within the IT industry itself? In 1998, IBM management in South Korea prepared a briefing for the chairman of the board, Louis V. Gerstner. It noted that "the sudden collapse of the Korean economy in late 1997 was due to a structural weakness combined with a series of unforeseen events . . . due to a fundamental weakness in Korea's financial and corporate sectors." It opined that the government took too long to acknowledge, let alone address, these problems (IMF's position too about the government's role), and predicted that the national economy would continue to contract until the end of 1998. "As such, IT demand has contracted drastically this year [1997], causing severe price competition," creating financial crises for many IT firms. That crisis led some vendors to seek financial support from outside the country, mainly capital through joint ventures or by selling assets. The Korean IBMers anticipated a large increase in the demand for process reengineering work from many industries as firms restructured themselves to improve profits and productivity.[75] More specifically about IT in South Korea, IBM estimated that the IT market it served represented 1.9 percent of GDP in 1996, and that by the end of 1998 would have contracted by 28 percent to 1.3 percent of GDP. Sixty percent of the market comprised of IT hardware, driven largely by the sale of PCs, which made up 52 percent of the overall IT market; services accounted for only 25 percent. They calculated that the market amounted to $6.591 billion in 1998 (down 28.4 percent from 1997) and that IBM held 10 percent of the total in 1997 but would increase its share to 13 percent in 1998. Small and medium size businesses accounted for 46 percent of the total consumption of IT, with manufacturing another 10 percent, followed in descending order in share finance, government/health (each 6 percent), education (3 percent), and telecom/media, process/petroleum, and distribution firms (each at 2 percent), with utilities, insurance, travel and transportation trailing with 1 percent each.[76] With PC sales so crucial to the industry, not just to IBM, the chairman was told that the "PC market is frozen and is expected to decline by almost 40 percent." Small and medium sized businesses were expected to consume about 47 percent of all IT over the next five years.[77]

Local IBM managers reported noticing changed behavior with respect to how computers were being used. First, for example, *chaebols* were restructuring in 1997, exiting some external businesses and outsourcing select IT functions to local Korean firms (a problem for IBM, H-P and other foreign suppliers). Second, this shift in who did IT for large enterprises was occurring as local vendors of PCs were attempting to sell off portions of their businesses, notably Sambo and Daewoo, creating some confusion among consumers about where to acquire such products. Third, demand for IT services as part of restructuring activities was causing many IT vendors to increase their capacities to offer such support and as a way of backfilling declines in sale of hardware with new sources of revenues. But big firms still dominated deployment of IT. Half of IBM's Korean revenues in 1997 had come from 10 sets of customers, all *chaebols*. Finally, with companies and governments all over the world now focusing increased attention on Y2K remediation,[78] IBM's assessment of the local situation was not positive: "Overall, Korea is not well prepared for Year 2000 problems, with little improvement due to the economic difficulties facing Korea since the fourth-quarter of 1997." Local IBMers thought, however, that "the finance industries are well positioned," as in most advanced economies. IBM's assessment of conditions on the ground was sober and telling.

One final reflection on local IT priorities can be seen through how IBM had deployed its employees. Relying on data from 1995 in Korea (because the firm usually looked out more than one year up to three years when positioning staff in a market), IBM had 1474 employees, 89 percent in field and sales positions, of which half were in IT services, 39 percent in sales and distribution, and the remaining 11 percent in infrastructure. This distribution of employees suggested where future business would come from since the firm normally positioned staff to capture current business and to be situated to do the same over the next two to three years in each of its markets.[79]

1999 turned out to be a year of economic recovery for South Korea with GDP, however, growing at 10.5 percent. Driving an important part of this recovery were the information and telecommunications industries, beginning in 1997, and that extended their successes through 2002. Did IBM simply assess the situation incorrectly? IBM's local analysis focused on those parts of the conomy it served: big institutions and for large, expensive IT products and services, such as mainframes and Y2K remediation projects. Yet Korean IT industries grew in markets not largely served by IBM, such as consumer communications, which contributed to the nation's growing stock of I-C-and T.

Koreans increasingly saw IT, and particularly advanced telecommunications, as a way of enhancing their society and economy. As one commentator put it, "Wireless for [South Korea] is the same as the space mission to the moon was for the U.S. Everyone is behind it."[80] Wireless and broadband consumption increasingly symbolized modern South Korea. The number of cell phone subscribers expanded from 6.91 million in 1997 to 23.44 million in 1999, and to 32.34 million in 2002. In 1997, 27 percent of all households had a PC; in 2002 60.1 percent. All these increased number of users was more impressive as one observer noted because "these booms took place during a financial disaster that placed many citizens out of work and limited domestic spending power."[81] While access to the Internet was less expensive to a user once they had the necessary hardware, use of it too signaled the extent of diffusion of IT in a Wave One society. In South Korea's case, the number of users expanded from 1.63 million in 1997 to 26.27 million in 20002, ranking Korea sixth in the world in Internet availability, even though its population was the 25th largest.

The broadband effect became evident too as the number of users of high-speed Internet went from some 13,000 to 10.4 million, which is why Korea came to be ranked the number one user of broadband in the world. Broadband services had a speed triple that of even the most advanced systems in the world (18 Mbit/s) on the way to much higher speeds in the next several years.[82] By the end of 2008, 95 percent of all South Korean households connected to broadband services (still the highest percentage in the world) and with some of the lowest overall costs for these services too.[83]

Historians and scholars in other disciplines are just now beginning to explore the implications of what used to be known as far back as the 1980s as "sub-second response time." Access to the technology facilitated online commerce.[84] However, perhaps the one most immediate result of high speed computing involved the rapid expansion in online gaming. The more one explores the role of the Internet and computing facilitated by high speed broadband the more it becomes evident that we may be at the frontier where Wave One begins intersecting with or transforming into Wave Two. Even deployment and use of the Internet with high speed broadband in both the workplace and in private lives is very much a part of the Wave One story; but perhaps less so what has been unfolding in South Korea toward the end of the first decade of the new century.

The first circumstance that strikes one after seeing how extensively the development of broadband and Internet had become is the amount spent by Koreans on electronic/digital games, and increasingly the numbers of people who played these online via the Internet over high speed broadband networks. Just between the two years from 2007 and 2009, sales of online games in Korea jumped from $2.24 billion (a great deal for a small country) to $3.49 billion, in other words, by 56 percent.[85] This data hides the fact that the industry was indigenous and the extent to which the youth of the nation had become users of these software products was substantial. Online gaming dominated all other form of digital games, such as console-based varieties so popular in Japan and in the pan-Atlantic community. The leading student of Korean gaming, Dal Yong Jin, concluded that "online gaming has become a massive cultural phenomenon," not just around the world but especially in Korea, a nation that "has developed and pioneered several areas of online gaming, from software development to eSports" which this expert considers to be the next stage in the development of online gaming.[86] Another observer noted that broadband and Internet access resulted in "the lives of Korean citizens" moving "increasingly online, creating a dual existence as real-world citizens and cyberspace netizens," and not just among teenagers and young adults, but across large swaths of society and age groups.[87] Jin's analysis of how this adoption happened followed very much the story told earlier about IT diffusion in general: funding and deployment of Internet and broadband services by the government, promotion of competition among telecommunications firms, and decline in costs for these services in the early 2000s from an average of $40/month in 1999 to $20/month in 2006. Introducing new uses of IT into that preexisting infrastructure proved far less complicated and expensive to do, such as online gaming or other digital services.

When unemployment increased at the end of the century, many of the unemployed and young Koreans began visiting Internet cafes, called *PC bangs* (*bangs* means room in Korean), where many Koreans were first exposed to the attractions of high-speed Internet. For many recently laid off workers, it was also a business they could get into without extensive capital investments. They had essentially only to rent a hall, subscribe to broadband, and acquire PCs for their customers to use.[88] A second application, hence source of involvement, came from online stock brokerage transactions which became very popular after the financial crisis of 1997. By 2001, slightly over 8 percent of Korea's population engaged in online stock trading, holding a third of all shares traded. Individual investors accounted for 97.2 percent of all online trading done by 2002, which amounted to two-thirds of all stock trading conducted online by all people and financial institutions. As recently as 1998, instead of two-thirds, the volume had been closer to 10 percent.[89]

Jin attributed growing desire for entertainment and online gaming as sources for increased demand for broadband services (access, speed, reduced costs). He cited a 2003 survey that indicated some 47 percent of all broadband subscribers used the service largely to play games, with viewing movies and television second (37.4 percent) in popularity. Jin observed that Koreans were less familiar with monthly payments for services than Europeans and Americans. They were more comfortable paying for use by the hour, and thus often preferred to play games at *PC bangs*.[90] One individual familiar with Seoul—ground zero for the Korean information age—noted that "within the city it is rare for a person to walk the streets for two blocks without running into a computer room, or a PC *bang*." At the turn of the century (2002), over 25,000 such salons were already operating in South Korea, most of them 24 hours a day. While the young may have been at these Internet gaming cafes, a third of their elders reported blogging, leading the same observer, when assessing all modes of usage, to conclude that "nearly every aspect of the society has moved online, and more than a few aspects of it exist only within cyberspace."[91]

Another application, similar to Myspace or Facebook in the West, was Cyworld (begun 1999) in which in excess of 21 million Koreans maintained a home page, called Cyworld *minihompys*. Visitors could be invited to *minirooms* to interact with *minims* (avatars). In the words of one observer very familiar with such systems, "with nearly half of South Korea's real-world population active within the space, virtual society and real society blend together to form a dual community experience, producing an Information Age society where real/virtual exists in tandem rather than dichotomy."[92] To fulfill the needs of all these avatars and minirooms, some 30,000 businesses came into existence in the early 2000s, selling over a half million "virtual merchandise" and services using a virtual currency called *dotori* purchased with real money through cell phones. These virtual products and services were largely focused on creating identities for avatars, in short, customized Internet personas. The market for such services had reached $114 million in 2004.[93] Cyworld had become the repository of over 100,000 videos loaded daily into the web site by the early 2000s, making it a busier place than YouTube, even though the latter drew its users from around the world while Cyworld was overwhelmingly Korean. This same service sold more music online than any other, with the exception of Apple's iTunes, with this Korean vendor often selling on average over 200,000 full-track MP3s daily.[94]

Jin joined a growing number of observers of modern Korean life who commented on the role of speed, "The primary Korean cultural characteristic that expedited the growth of high-speed Internet was a demand for quick change," reflective of a feature of recent vintage in which "Korean society is well known for its impulsiveness . . . its desire for quick communication, quick contact, and quick results."[95] This meant early adopters were also plentiful (helpful to the rapid emergence of gaming). An enthusiasm for online entertainment and education both contributed to acceptance of the new digital application.

A few statistics help define the rate of adoption and the economic infrastructure required to deliver it. The Korean online game market in 2001 amounted to $191 million in revenue. The online gaming industry then took off massively, doubling its revenues in 2002, and nearly doubled it again in 2003. This performance was followed by a slower, yet very impressive cycle of growth: just over a billion dollars in 2004; then $1.4 billion the next year, followed by $1.8 billion in 2006, and ending 2008 with a record $2.76 billion. Exports did well too, rising from $130 million in 2001 to $898 million in 2008, with the only decline in sales occurring in 2007 as many national economies slipped into recession. On the supply side, most products came from many small firms: 1,381 in 2001 and expanding each year to 3,317 enterprises in 2008. We know they were small because employment in these in 2001 amounted to only 23,594 people, grew to 60,669 in 2005, then settled down to 42,730 in 2008.[96] Normally, one would consider these people important agents of diffusion of IT, as were software developers in other countries. While that certainly was true in Korea, their numbers were so small when compared to the volume of users that we can safely conclude they played a minor role in introducing people to IT *per se*, playing instead a most profound role in satisfying a demanding, already extensively IT-savvy society.[97] In short, a PC *bang* operator was more an agent of IT diffusion than a software game programmer. That is a different effect of IT experts than normally observed in most countries.

If there was anything additionally new that game writers and players caused it was the rise of the professional full-time online gamer, the best of whom became national celebrities. The number of full time players amounted to less than 200 in 2001 but closer to a thousand by 2009. They played in leagues, competed, and had all the sports and occupational infrastructures one would expect to see with full-time expert players in other sports, such as European soccor or American professional football.[98] Major Korean corporations began sponsoring professional online gaming teams, including Samsung Electronics, KTF and SK Telecom, much as Nike in the U.S. supported professional basketball players and STP NASCAR teams. Each of the Korean armed services did too as both a public relations initiative and recruiting mechanism. In short, e-sports had a solid foothold in Korea.

Along with gaming and other online services came activities that, while similar and evident in all Wave One societies, appeared more extreme in South Korea. Cyber crimes reported to Korean police by 2004 exceeded 200,000 incidents annually, making digital crime appear much like a high growth industry, expanding by nearly 100 percent in two years. It included such offenses as chat room insults and cyberstalking, better known by the Korean term "cyberviolence," along with such traditional Wave One digital crimes and misbehaviors as online advertising for prostitution, drugs, and adulterous activities.[99] Gamers increasingly have been identified as addicted to continuous playing, leading the medical community to consider the problem serious and worthy of research, diagnosis, and treatment by 2006. The phenomenon has been recognized by the government as "internet addiction," a form of compulsive psychological disorder such as encountered in gambling or alcoholism. The only other countries to recognize this digital phenomenon early were Israel and China. A survey conducted in 2005 by the government indicated Korea had 546,000 Korean addicts between the ages of nine and thirty-nine, while another study suggested as many as 30 percent of Korean teenagers displayed symptoms.[100] As in prior decades when dealing with an IT issue, the Korean national government took action, establishing the Center for Internet Addiction Prevention and Counseling in 2002. Simultaneously, NCSoft Corporation funded forty counseling centers to deal with this e-medical problem.

The effects of IT spilled over into other areas of private and public life as well. The outcomes of the presidential election of 2002 were directly affected by the use of the Internet. The winning candidate, Roh Moo-hyun, used the Internet to reach many voters through an organization of 80,000 members called Nosamo. They raised over $7 million through the Internet, a strategy implemented in the United States by then Senator Barack Obama in 2008. Texting brought people to thousands of rallies in South Korea, while voting via the Internet was tested for the first time. Initial experimentation in Korean Internet politics had taken place as early as 2000, however, giving politicos ideas that led to the more extensive use of the technology in 2002. Rumors of misbehavior by legislators that spread over the Internet resulted in their losing elections, suggesting for the first time the power of this new technology. By approximately 2005 lobbyists and other interest groups were routinely trying to affect legislative outcomes through online petitioning and campaigning.[101]

To close out our discussion of IT diffusion in South Korea, one must account for more than just young Koreans using broadband gaming and the Internet now as part of Korean political behavior. Diffusion is always facilitated by the assertive activities of IT vendors generating demand for and use of their products. Our analysis of Korea's experience ends with a brief review of that existing supply side circumstance so essential to diffusion. The three parts of Korea's ICT market—telecommunication services, information communication equipment, and software—accounted for 13 percent of South Korea's GDP in 2000, up from the 8.5 percent in 1997. The rate of growth in demand for all of Korea's ICT products and services varied widely from 37 percent in 1997 to only 6 percent the following year (as also reported by IBM above), back up to 40 percent in 1999, then down to "only" 29 percent in 2000 and, in the recessionary year of 2001, 7 percent. Yet during these years, it was not uncommon for ICT to contribute up to 50 percent of the nation's growth in GDP, and a third of the nation's total exports. By any measure of international comparisons, IT's presence in the economy proved substantial. The ICT eco-system employed 1.3 million workers in 2000, a population that grew slowly over the next few years.[102]

As use of broadband and the Internet increased, consumers and users in business, education, and government pressed for more services and applications; online gaming was only one of these. The telecommunications market was now about as open and competitive as any in the world, a circumstance which facilitated continued competition, generated ever-higher speed broadband services, led to lower prices, and to the introduction of more functions. Because of growing demands for such improvements, paybacks (ROI) for providers shrank. Unlike in Europe and North America, where economic returns on investments were uncertain, unmeasured, or often took many years to realize, by the early 2000s, some telecommunications providers in Korea were reporting breakeven or payback periods for incremental infrastructural improvements of closer to a year.[103] Historians of Korean telecommunications may someday conclude that the uptake in demand for broadband facilitated the rapid deployment of this new class of technology, resulting by 2002 in the Koreans using the Internet at the same rates as Sweden, United States, and Netherlands—the last three taking over 15 years to reach the same rates of penetration and use that the Koreans did in less than a decade.[104]

Finally, we should acknowledge the Koreans' use of cell phones as one of their preferred digital platforms for accessing computing and communications. Between 1984 and 1994, subscriptions expanded slowly with two subscribers for every 100 residents. Services were provided by a monopoly (Korea Telecom). Subscriptions were few by Asian-Pacific standards of the day. Service was analogue, proved slow, remained too expensive, and was unreliable. Between 1995 and 2000, the market opened substantively to competition. Four providers competed and offered digital services (Shinsegi Telecom, Korea Telecom Freetel (KTF), LG Telecom, and Hansol, later named M.Com). Use increased so fast that Korea surpassed other extensive regional markets, such as those in Japan, New Zealand, and Australia. After 2001, growth in use continued, although the rate of adoption slowed, partly attributable to the economic crisis in Korea but also to the fact that so many users already had services. Within a few years cell phones were ubiquitous in South Korea, and just as important, some of the richest in function of any national phone systems in the world.[105] These phones increasingly made it possible for Koreans to access video on demand, music, streaming TV, in-car navigation systems, shopping online, interacting with government agencies, and playing games. All of these users raised Korea's collective rate of mobile data usage to the highest in Asia as early as 2001 (nearly two-thirds of regional mobile Internet users).

As in prior decades, the national government continued to stimulate rapid adoption of IT through its own use of ICT and digital interactions with citizens. Space does not permit a detailed discussion of its activities but a few are worth ticking off. By the dawn of the new century almost all government ministries were online using a shared national backbone network. Some 80 percent or more of government documents and forms had been digitized and citizens were interacting electronically with these documents for half of their transactions with public agencies. Major online applications relevant to the population at large included a real estate registration system with over 200 million property citations, with issuance of titles on demand; online birth, death, marriage, and divorce information accessible online (beginning in 1993), making it one of the first such applications in the world. Leading many countries in recognizing the economic development benefits of IT, the government installed these and other applications as part of its overall campaign to digitize portions of Korean life, an effort that had its roots in the 1980s and early 1990s. When combined with sustained commitment from one regime to another, and lubricated with state funding, public officials facilitated diffusion of such technologies. Even at the height of the nation's economic woes between 1998 and 2001, the national government doubled its expenditures on IT, from $544 million to $1.1 billion, while many other nations' governments cut back their spending.[106]

As suggested earlier in this chapter, the government developed plans (see table 8.8), backing up its rhetoric with meaningful actions that generated many desired results. The boldness of the government's intentions and role made initiatives underway in the pan-Atlantic community appear somewhat anemic, even paltry, in comparison, although most European and American governments aspired to implement similar intentions. They simply executed their plans for implementation less effectively, or were more casual (less focused) in approaches. While speculation, I believe that with the number of key public and private leaders involved in transforming Korea into a modern economy rather few, when compared to the volume of entities that would have to be engaged in the United States or Germany, for example, when coupled to common experiences (such as military service), probably gave the Koreans a subtle advantage that made implementing their plans easier to achieve.

However, all was not always well in government agencies when it came to managing centralized economic planning. As the last several pages would suggest, the government's role seemed to diminish in effectiveness and extent as the private sector and citizens took off in their use of IT in ways that made sense to them. In 2004, for example, the government called on the nation to focus on twenty different ICT technologies and industries that would define the future of Korea. By 2010 it had become evident that almost all of these had failed, even though the government had used its normal budgeting, regulatory, and R&D muscles to help foster them. These failures included digital mobile television (few people wanted subscription services to watch shows on 2-inch screens while traveling), and WiBro wireless data standard unique to Korea (too expensive for subscribers while "good-enough" less expensive service from WiMax already existed and was used in over 50 countries). As two commentators on the government's inability to predict future champion technologies to

Table 8.8 SOUTH KOREA'S NATIONAL GOVERNMENT
ICT PLANS, 1987–2006

1987–1985	Various initiatives to promote development of local IT industry
1987–1996	National Basic Information System
1995–2005	Korea Information Infrastructure Initiative
1996–2002	National Framework Plan for Information Promotion
1999–2002	CYBER KOREA 21
2002–2006	K-Korea Vision 2006

support wrote, "the Korean government proved to be absolutely awful at picking winners in emerging technologies," a weakness of public administration similarly evident in many economies around the world.[107]

Nonetheless, large swaths of Korean society seemed engaged in IT. 100 percent of all students at all grades had access to the Internet by 2001. The vast majority of employees in most industries did too, reporting continuously being active users. In that new century's first year, there were more credit cards in use in South Korea than there were citizens; sixty percent of all businesses had a presence on the Internet, and nearly two thirds of all Koreans used the Internet on a regular basis. As one report on deployment concluded at the time, Korea had the potential of "becoming an electronic commerce paradise."[108] Finally, one should note that South Koreans had begun to see themselves as participants, if not leaders, in the development of an information society.

THE SPECIAL CASE OF NORTH KOREA

All discussions about substantive issues concerning South Korea from the second half of the twentieth century to the present acknowledged the influence of North Korea, the single most important and unsolved problem hanging over the entire Korean peninsula. It is universally assumed by academics, business leaders, public officials around the world, certainly by South Korean citizens, and many in neighboring countries that the two Koreas will someday be reunified. Each country prefers that the other join it, but the consensus is that the smaller North Korea will ultimately reunite with the larger, wealthier South unless China blocks such a move. South Korean officials have long expected that they would have direct responsibility for integrating the North Koreans into their nation, much as the West Germans had to assume responsibility for integrating East Germany into the larger West Germany.

The North Korean situation presents a case study of what happens when a national government works extensively to block the creation, use, and flow of information in a society. One is so used to assuming that information swirls around in a society that as a topic it does not receive much attention in discussions about the diffusion and use of computers.[109] North Korea's experience suggests that information technologies, including computers, telephones, and telecommunications infrastructures are not terribly useful unless a society is fundamentally committed to the use and flow of information in the course of its mundane activities. An authority on North Korea, Nicholas Eberstadt, has studied the issue of the lack of information in this country. While writing an economic analysis on North Korea, he could not find any local reliable basic data, even to state accurately the size of the population, although it is believed to have been around 22 million in about 2000 (versus South Korea's at 47.3 million). So, even basic statistics were notoriously unreliable, or not available. He described the government's "clumsy but obsessive effort to control all information entering, circulating within, and leaving the country, as a major activity of the regime."[110]

It is universally accepted as a nearly failed state, characterized by a low per capita GDP, massive expenditures for a standing army, tension-ridden nuclear weapons capability, and an economy unable to feed its own citizens, let alone make possible for them to earn a reasonable quality of life. All of those circumstances suggest, however, that other than for the military (about which little is publicly known) the desire for computing remains minimal.[111] Scientific and technological innovations are few, other than for nuclear weapons systems, and with almost no international exchange of information, particularly scientific. As recently as the early 1990s, it appeared that North Korea was the only country in Northeast Asia unable to manufacture microchips—a core capability the majority of countries around the world considered essential for the functioning of a modern economy.[112] Finally, it should be noted that the population was widely dispersed in small urban communities and in the countryside, while by the early 1990s some 80 percent of all South Koreans lived in towns and cities.

Extant data on telecommunications suggests how little technology existed in North Korea. As of 1992, a credible estimate held that teledensity in the north was 3.5 telephones per hundred people, while in the south it was 36.3. What service it had was one of the worst in the world and antiquated. In 1992, the North Koreans had queried Japanese telecommunications officials about acquiring telecommunications switching equipment, inquiring about pre-digital hardware no longer used in Japan.[113] An analysis done in 2001 of the North Korean telephone system concluded that there were no cell phone services, but a tiny number of Internet users worked in government. As late as 2004, there still was no cell phone service in the nation and, as one report put it, "it is not believed there are any web sites in the country."[114] In 2012 it had one million. Regardless of how accurate these observations might have been, it is clear that North Korea has used fewer computers than most nations, and with the majority in use controlled by military organizations working with aircraft, naval ships, missiles, and nuclear energy.

The more interesting aspect of this lacuna in the use of all manner of ICT is how the South Koreans propose to unify the two nations. Regardless of how that comes about—regime collapse in the north, civil war, foreign occupation—officials in the south recognize that one strategy will have to be to link the two through communications, initially via television and radio, followed by deploying modern telecommunications throughout the north, probably wireless. In other words, as part of the short list of initial essential activities that include the supply of food, the establishment of public administration, law and order, disarming of the military, and job creation, will be the extension to the north of South Korea's information society infrastructure and culture, leapfrogging the many phases of implementation experienced by South Korea and the rest of the world.[115]

TAIWAN: IT SUPPLIER TO THE WORLD

Taiwan is located 75 miles off the southeast coast of mainland China and to its east is nearby Japan. Taiwan is also known as the Republic of China (ROC). It is a tiny country (just under 14,000 square miles) with a population of some 23 million people (2010). When it began industrializing in 1960, it had a population of just over 11 million inhabitants.[116] The reason Taiwan is important to the story of the global diffusion of IT is not just because in this nation one had an additional 20-plus million people eventually using computers; rather it produced so many of the components and digital devices used around the world. Economists have marveled at its rapid transformation from an agricultural economy to a high-tech manufacturing one, and even now also having evolved into a services one. The long-term historical interest in Taiwan's IT experience will rest on the role it played in supplying many of the digital objects that made diffusion of IT a reality. Like South Korea and Singapore, Taiwan entered the digital era as part of the second Asian surge of adoption after Japan.

Like the others, including Hong Kong, Taiwan became quickly a major manufacturer of parts and products. Table 8.9 suggests the remarkable results achieved by 1999, a year after the severe Asian financial crisis had largely spent itself, but before the global IT investment bubble and industry-wide recession of 2001. Note the global rankings this small country enjoyed in what was by then a globalized, highly-integrated IT industry, one of many worldwide manufacturing supply chains that characterized the production and the flow of goods in many high-tech industries by the end of the century.[117] For many components that went into finished IT products it was not uncommon for Taiwan to have produced 20 to 60 percent of the World's supply of parts by the late 1980s-early 2000s. Its components were used in products sold by such firms as Fujitsu, H-P, Gateway, Dell, and IBM. Of the nations reviewed in this book, local use of IT was of less significance than the role it played as supplier (exporter) to the World.

Taiwan shared with other nations in the region the experience of being a Japanese colony early in the twentieth century, of being caught up in World War II, and then following the loss of mainland China in a long-festering civil war to the Communist Party of China, retreat to this island of 1.5 million Mainlanders. These transplants to Taiwan quickly gained dominance over the local government and controlled it over the next half century. Under Japanese rule, management of public and private institutions had followed models similar to

Table 8.9 TAIWAN'S WORLD RANKINGS OR SHARES
AS IT SUPPLIER, CIRCA 1999

Component or Product	Global Ranking by Size of Contribution
Integrated circuits (ICs)	3rd
Image scanners	91%
PC Housings	75%
Power supplies	70%
IC Foundaries	65%
Notebooks	49%—3rd
PC motherboards	64%
Analog modems	57%
CD-Roms	34%
Graphic cards	31%
IT Industry	3rd

Source: Various tables and charts in Chun-Yen Chang and Po-Lung Yu (eds.), *Made by Taiwan: Booming in the Information Technology Era* (Singapore: World Scientific Publishing Co., 2001); Teresa Shuk-Ching Poon, *Competition and Cooperation in Taiwan's Information Technology Industry: Inter-Firm Networks and Industrial Upgrading* (Westport, Conn.: Quorum Books, 2002): 58.

those in evidence in both Koreas and the Philippines, and that lingered in the form of business alliances and trading partners during the second half of the twentieth century in many Asian societies. Public administration remained largely authoritarian through half of the era in which IT became important. The government implemented protectionist, neo-mercantilist public policies that sheltered local industries, much as was done in such other regional economies as Japan and South Korea.[118] Mainland Communist China—known as the People's Republic of China (established in 1949)—claimed to be the successor of the ROC, hence owner of Taiwan. That disputed status hovered over all military, diplomatic, and many economic issues over the next six decades, just as similar "unfinished business" existed with North-South Korea.

Since Mainlanders controlled the government of Taiwan public officials had little familial or professional ties to leaders in local businesses for most of the past six decades. Taiwanese families dominated ownership and the staffing of local businesses and interacted with each other through familial and business ties. Relations between the Mainlanders and the locals were not always smooth, and were made more difficult by the fact that as in mainland China, and in varying degrees throughout Asia, family ties and extended social/familial networks dictated staffing and trading practices of those firms that came to be such critical participants in the global IT supply chain. An example of the process at work involves extended families in which a member might complete their education in engineering or computer science in the United States, then work for a major Silicon Valley firm or semiconductor manufacturer (such as Texas Instruments), trade with the home firm, or eventually come back to Taiwan to take up a senior leadership position in the family-owned business. Conversely, such an individual might move to Silicon Valley or to Europe and establish a local branch of the business or trading house.

An important feature of Taiwanese culture was its profound reverence for education, which in turn facilitated the rapid and timely transfer of IT knowledge to Taiwan from more technologically informed countries, beginning in the 1970s and continuing to the present. As two local observers noted in the mid-1990s, "the high number of trained professionals has been a primary factor for Taiwan's high-growth of information industries."[119] By the mid-1990s, over 18 percent of all Taiwanese over the age of 25 had some form of post-secondary education. In Taiwan's equivalent to Silicon Valley (Hsinchu Science-Based Industrial Park—HSIP) over 60 percent of the nearly 83,000 people working there in 1999 had post secondary education, nearly 20 percent had masters and Ph.D. degrees. The overwhelmingly majority of these advanced degrees were in the hard sciences.[120] A large number of these workers were trained in the United States, but many also in Japan. 2,800 returned from the United States in 1991 and 5,700 in 1995, encouraged by an economic recession in the U.S. in the early 1990s, a time of economic boom in Taiwan.[121] Coupled to a strong work ethic was the need to succeed in the eyes of fellow family members, the discipline and style of operating gained from the experience of compulsory military service, and the willingness to embrace Japanese and American managerial practices. These attributes, education, and work experiences combined contributed to Taiwanese business behavior that cannot be ignored as one considers the macro-economic policies of the government that facilitated the flowering of a strong IT-centered manufacturing base.

Use of data processing in the years following the Chinese civil war began in the mid-1950s when National Cash Register (NCR) and IBM established operations in Taiwan. In IBM's case the firm set up an office in Taipei in 1956 with a staff of three. In 1960 IBM established a local service bureau which processed Taiwan's agricultural census the following year. IBM's first installation of a computer came in 1962 with a 650 system at the Institute of Electronics at the National Chiao Tung University. The National Taiwan University ordered a 650 the following year. Tracing IBM's experience, one can see the start of other computing projects in the country during these early years. The university-based computer systems were used to solve engineering questions related to national infrastructures, such as to forecast demand for electricity, help requested by the Taiwan Power Company, a state run utility. In 1964 the Council for International Economic Cooperation and Development (CIECD) used an IBM 1620 at the National Chiao Tung University to do economic planning. A great deal of the early uses of punch-card equipment in the 1950s and first academic uses of computing were funded and introduced in large part through the foreign aid provided by the United States. In short, Taiwan's earliest introduction to data processing was also made possible in part by the global politics of the Cold War.[122]

Taiwan Sugar Corporation ordered a S/360 in 1964 and in the same year installed an IBM 1440 for materials accounting. That same year the Bank of Taiwan acquired an IBM bank proofing system, while National Chiao Tung University replaced its 650 with an IBM 1620. Within several years additional systems were installed: System 360's at IBM's local data center (1967), Taiwan Power Company (1967, another in 1970), and several IBM 1130s (1967). The Ministry of Economics ordered a S/360 (1968), a large S/360 went to the Taiwan Sugar Company (1970). Cathy Life Insurance Company installed a S/370 (1973). IBM acquired a S/370 Model 145 in 1974.[123] Table 8.10 displays IBM's revenues from all sources (computers, punched card equipment, 80-column cards, data processing services), documenting the rapid increase in demand for IT goods and services that occurred during this early stage of Taiwanese diffusion.

During the 1960s, other high tech manufacturers began investing in local manufacture of electronics, including General Instruments, Texas Instruments, Sanyo, Matsushita, RCA, Philips, and also IBM. All brought into the country uses of computing that they also deployed in their home factories. Most were assembling products out of electronic components manufactured in other countries, taking advantage of low labor costs in Taiwan. However, no computer components were assembled in Taiwan until the late 1970s. While government interest in implementing economic policies in support of IT came late in the 1970s, it had earlier established the Electronic Research and Service Organization (ERSO) in 1974. In time ERSO played an important role in nurturing local use of computing and development of an indigenous manufacturing capability, including facilitating the introduction of foreign companies' alliances with local firms.[124]

Between 1979 and 1982 local Taiwanese firms began assembling parts and subassemblies for PCs, including clones of Apple II computers without permission from the American company. That illegal manufacture ended by 1984

Table 8.10 IBM REVENUES IN TAIWAN, 1967–1974, SELECT YEARS

(U.S. dollars)

1967	363,664
1969	1,156,331
1971	1,658,708
1973	2,600,283
1974	3,999,690

Source: Profit and Loss Statement, undated [late 1974], Legal Records/Taiwan, Finance, 1971–1975, IBM Corporate Archives, Somers, New York.

after both Apple Inc. and the American government pressured Taiwanese officials to stop the practice; a request that could not be ignored because of the
large amount of American foreign aid coming into the country. Local manufacturers started to assemble IBM clones under contract to American firms,
leading U.S. manufacturers of components also to start operations there, such
as Digital.[125] In 1979, the government established the Institute for Information Industry to spread information about computing, the use of this technology, and the development of an indigenous IT industry. The following year
the government took the very important step of establishing Taiwan's first
industrial park at Shinchu to recreate the sharing of information and technical know how already in evidence in Silicon Valley in California. The prior
year, ERSO had developed a master plan for the development of a computer
industry and strategy for training workers in IT for this new industry; the establishment of the science park was part of that plan. In the 1980s, American
firms largely displaced Japanese companies as major providers of investment
funds to local electronics firms, leading to a rapid shift from manufacturing
low-tech electronics, such as television sets and radios, to higher-tech components for computers and IT peripheral equipment, especially PC sub-assemblies
and their peripheral devices. These included power cords, monitors, graphic
cards, mice, and keyboards. Most of the firms were quite small but two founded
in these years evolved from being other-equipment-manufacturers (OEMs)
into manufacturers of branded products: Acer Computers (1981) and Mitac
International Corporation (1983).[126]

Before explaining in greater detail the takeoff in computer usage that began
in the 1980s, attention should be paid to what probably turned out to be tens
of thousands of largely young Taiwanese who, in the early 1980s, became
involved with PCs to such an extent that their knowledge of what comprised
PCs and how they worked surpassed that of cohorts in many other countries.
Their knowledge can be credited with helping to diffuse IT internally within
the country—largely in urban centers—alongside the initiatives taken by the
government. Honghong Tinn has documented the existence of a tech-savvy
community of young Taiwanese who went to electronic component shops and
bought all the parts required to build their own PCs, or to engage workers in
these small electronic supply stores to do this assembly for them. One observer
of the local PC scene of the early 1980s commented to Tinn that perhaps as
many as 90 percent of all PCs built in Taiwan were made this way in the early
1980s. As late as 1995, one survey he cited reported that as many as one third
of all PCs sold in Taiwan that year were still fabricated this way. These machines
worked, and they were less expensive to acquire this way than by buying
branded products from Apple and other vendors. In particular, these people—
Tinn referred to them as Tinkers and thus early adopters—put together Apple
II class machines, and copied the necessary software, treating these systems
not as copies of Apple II patented products, but more like a generic class of

machines. From Apple Computer's perspective, of course, these machines were illegal copies and since so many were made, it is no surprise that the American firm sued to address the problem and engaged the American government to use its diplomatic resources to convince Taiwanese officials to eliminate the practice of locally-made machines of this type. Tinn credited the Tinkerers with spreading knowledge about the technology through Taiwanese society, suggesting they—Taiwan's IT industry—had to go through a stage of imita-tion—copying—in order to learn hove to make more mature IT industries thrive.[127]

As development of small component manufacturing in computing took hold, and young Taiwanese experimented with PCs, large organizations were installing and using computers in growing numbers. Many began by doing their data processing work at the IBM data center until such time as they could economically justify their own systems and staffs. IBM, as a major sup-plier of such equipment, sold systems largely to public institutions, over a third just to the military and higher education in the 1970s. One internal report cataloged 35 installations as of the end of 1974 at universities, banks, ship building, telecommunications, civilian, and military agencies. In 1976, IBM had grown to 167 local employees, an increase of over 160 since 1960.[128]

But the takeoff in the number of computers installed in Taiwan began in the early 1980s, as their use spread further into many large government agencies and companies. IBM's lists of customers in the 1980s documented this expanded use of IT, which occurred at the same time as a similar spread in use of IT in other Asian economies. Table 8.11 lists some of IBM's installations of the 1980s. Observe that these systems were not small, nor inexpensive. All were the kind used to run concurrently a variety of applications designed for use by large organizations, which suggests that Taiwanese adoption of com-puting was already sophisticated and state-of-the art at the time. In short, knowledge transfer regarding the use of IT, a specific objective of the govern-ment, was yielding positive results rather quickly.

The more important story for global diffusion of IT is what happened on the supply side in the 1980s. The manufacture of components for other IT brands and the introduction of products by Taiwanese firms grew dramatically. In 1983, the combined total of these two classes of products amounted to $600 million, in the following year, over $1 billion. From 1984 through 1987, IT exports accounted for 60 percent of all Taiwanese sales outside the island, making IT Taiwan's third largest industry by 1988, after electronics (such as TVs, radios), and textiles. Small local firms had stepped in to make and sell these products as foreign vendors moved production to countries with lower labor costs.[129] This big shift involved more than just moving the assembling of parts and subassemblies to Taiwan by other manufacturing firms to designing and manufacturing their own components, largely for all facets of PCs, from motherboards to mice.[130] During this decade the manufacture and export of

Table 8.11 IBM COMPUTER INSTALLATIONS IN TAIWAN, 1980–1986

Customer	System Type	Year
Chang Kung Hospital	4331	1980
Far Eastern Textile Company	4341	1980
China Airlines	Unknown	1980
Hitachi	S/38	1981
China Steel Company	4341 (2)	1981
First Commercial Bank	4341	1982
Hua Nan Commercial Bank	4341	1982
Cooperative Bank of Taiwan	4341	1982
Taiwan Power Company	3033	1982
Veteran's General Hospital	3033	1982
Philips Taiwan	S/36	1983
IBM Computer Center	3083	1984
AKCC*	3090	1985
Ministry of Education	4341	1986
Department of Health	4341	1986

*AKCC = Asahi Kasei Corporation
Source: Ken W. Sayers, "A Summary History of IBM's International Operations, 1911–2006," October 20, 2006, pp. 439–441, IBM Corporate Archives, Somers, N.Y.

integrated circuits (ICs) also took off. Because semiconductors are a highly complex product Taiwanese success is all the more important to understand.[131]

By the early 1980s local and foreign-owned IC design, packaging, and manufacturing companies were beginning to operate in industrial parks. Taiwan also was home to two fabrication plants. ERSO, the public sector institute, operated one of these while the second was a spin-off of ERSO, the United Microelectronics Corporation, known better as UMC.[132] The semiconductor business was small in the early 1980s but by the mid-1990s had grown to rank fourth in the world. Semiconductor manufacturing had its origins when the Industrial Research Institute (ITRI) was established in 1973. ITRI drove forward the development of more complex design and manufacturing of ICs in the 1970s and 1980s.[133] When in 1974 ERSO was established, it was given the mission of creating IC capabilities in Taiwan, meaning also bringing to the island knowledge on how to fabricate ICs using Japanese and American know-how, which it did. It promoted the transfer of knowledge about the technology to Taiwan by finding U.S. trained Chinese engineers and persuading them to work in Taiwan. This strategy became the most important employed by ITRI in the 1970s and 1980s. ERSO sent engineers to live in the United States for months at a time to absorb knowledge and to connect with other Taiwanese and Americans, all the while improving the travelers' command of the English language and understanding of American business practices. Taiwanese firms licensed

Table 8.12 TAIWANESE SEMICONDUCTOR INDUSTRY'S TIES TO
FOREIGN CORPORATIONS, 1976–1995

Foreign Firm	Year of Initial Partnership	Technology
RCA	1976	IC fabrication
Philips	1986	VLSI IC fabrication
Texas Instruments	1989–91	DRAMs
Oki	1994	DRAM fabrication
MEMC	1994	Silicon wafers
HP	1994	RISC processors
MIPS (Sun)	1994	RISC processors
IBM	1994	PowerPC processors
Kanematsu	1995	Silicon wafers
Oki	1995	DRAMs
Mitsubishi	1995	DRAMs
Toshiba	1995	4M DRAMs

Source: John A. Mathews and Dong-Sung Cho, *Tiger Technology: The Creation of a Semiconductor Industry in East Asia* (Cambridge: Cambridge University Press, 2000): 188–189.

much technology (mainly in the 1970s). Many companies concentrated their manufacturing at the Hsinchu Science-based Industry Park (1980–88).

Over the next decade firms were established that specialized in specific aspects of IC manufacture, such as design, fabrication, or testing, honing their skills and efficiencies to such an extent that these enterprises were linked into the emerging global supply chain for IT products, and most importantly, for IT components used in multiple products, not just ICs or PCs. Table 8.12 presents a sample list of the connections made with foreign companies that enabled the Taiwanese to embed themselves in the global manufacture and diffusion of IT parts and products. It clearly demonstrates that Taiwan was able to participate in state-of-the-art manufacturing with important global players by the mid-to-late 1980s, and was able to sustain that capability right into the 1990s.[134]

The government continuously played a facilitative role in providing land for factories, tax credits to local and foreign firms, and friendly investment regulations for foreign companies. The two institutional channels of support, and focus were ITRI and ERSO. They facilitated development of collaborative R&D programs, establishment of the Hsinchu Park and, by the early 1990s, coordinated activities across the industry. Unlike Japan, where government agencies collaborated with large electronics firms, Taiwan's government worked instead more closely with small and mid-sized companies. Taiwanese enterprises never grew in size to the scale seen in South Korea or in Japan, but there were many, highly specialized in what they did. Spinoffs of ERSO launched the semiconductor industry, creating, in effect, what two students of the process called "public-sector enterprises in a high-technology field." This strategy ran contrary to the practice seen in the West in which it was widely believed by officials, company executives, and economists that only the private sector could be innovative and productive with such products.[135]

However, the Taiwanese really had no choice. For one thing, they did not have a large and sophisticated electronics industry upon which to build a local semiconductor industry, as did Japan, Korea, and the United States. Linking small firms to large foreign ones was a viable option for Taiwanese officials, using ERSO as either the umbrella organization to nurture such relationships or to create spinoff firms for the purpose. For another, with the creation of such local firms helped by the government, other indigenous businesses saw demonstrations (what technologists often called "proof points") that small enterprises could succeed in the world market, thereby encouraging them too to join in the opportunity. In short, government "seeded" and nurtured effectively the flowering of the industry and the rapid sharing of knowledge and expertise, particularly regarding assemblage, fabrication, and manufacture of IC components and sub-assemblies.[136] Finally, one should note that over the course of the last three decades of the twentieth century, Taiwan also improved the quality of all its telecommunications infrastructures, from land-line telephony to Internet and broadband, as part of its comprehensive strategy to facilitate manufacture and use of IT.[137]

One seasoned observer of the local scene concluded that since 1988 Taiwan had a mature IT industry, subject to global market conditions, economic realities, and, by the mid-1990s, to the negative forces of a global oversupply of DRAMs which led to declining costs, hence profits. This latter circumstance forced Taiwanese firms to broaden quickly their areas of capabilities into other components that were higher-profit steps in the production of semiconductors (such as design and testing). This was done to diversify their portfolio of products. Measured in revenues, the IT industry did well, growing from $2.3 billion 1986 to $5.7 billion in 1988, and to $6.8 billion in 1990. Its various sources of sales allowed local firms to grow in spite of global recession and oversupply of DRAMs in the 1990s, from $7.7 billion in sales in 1991 to $15.8 billion in 1995. Its volumes in 1995 made Taiwan the third largest manufacturer of IT products (with a healthy software component too) after the United States and Japan.[138] In the 1990s and early 2000s, diversity in the design and manufacture of IT products led to other OEM and branded product introductions with sustainable businesses at work in such market segments as flat screens and LCD panels, for instance. By then it had become normal for 90 percent or so of all hardware made in Taiwan to be exported.

These goods went to the United States, Europe, and to other Asian nations. For example, in 1998, Taiwan shipped 37.6 percent of its IT exports to the United States, and another 31.6 percent to Western Europe. Asian markets grew rapidly too in the 1990s, from absorbing 15 percent of its exports in 1995 to 19.4 percent in 1999.[139] Regarding semiconductors, Taiwan exported 50.3 percent of these in 1998 with roughly half to the United States and Japan in equal amounts. By then Hong Kong and mainland China were buying nearly 23 percent of its output. Only 8.1 percent of locally produced software was exported, the rest went into the domestic market. However, the kinds of software exported were not always dependent on users knowing Chinese. These included antivirus products, multimedia tools (for image processing and editing for example), and a variety of IC design and testing packages.[140]

The upgrading to world-class status by the Taiwanese IT industry that began in the 1980s and sustained itself to the present was the result of many forces at work. First, as the world-wide IT industry embraced open manufacturing processes and supply chains, the opportunity existed to step in with highly specialized capabilities. Taiwan chose in which ones to specialize, leveraging small firms that could focus on narrow products and services, such as testing ICs or putting motherboards into PCs as OEM manufacturers. That degree of specialization allowed firms in the United States and Japan to take advantage of lower labor costs in Taiwan, attracting businesses and investment funds to the island. In the process it became easier to transfer IT knowledge to Taiwanese individuals and firms. Second, government officials also implemented highly targeted industrial policies designed to make specific aspects of the IT industry core competences of the island. Over time Taiwanese manufacturers upgraded their skills and broadened their capabilities, hence creating for themselves new opportunities for greater profits. The continuous flow of OEM orders fueled local companies with revenues, profits, and know-how about ever-more sophisticated products. Over the same, local companies could design, develop, and sell directly more advanced components and products themselves.[141] As their knowledge of IT grew, their role in the global supply chain expanded in ever more in the international hardware and components markets, and of course with semiconductors, as evidenced by their ever-growing market share for specific products in the 1990s.

Taiwan's ability to maintain momentum in manufacturing and trade continued to evolve as the Asian IT eco-system did, with Taiwan right in the thick of things. For example, as China's prowess in manufacturing and export of IT products of all types expanded in the 2000s, Taiwan developed close ties to firms and agencies on the mainland. Taiwanese enterprises exported "know-how" on IT manufacturing to China, shifting production to China proper where labor costs were lower. As early as the 1990s, for example, Taiwanese firms began moving the production of notebooks to China, with the consequence for China that it soon became the leading producer of this class of computers.[142] Both the Taiwanese and Chinese governments encouraged intertwining of both their electronics industries and increased interdependencies, even in the face of growing competition in the creation of patents and selling abroad, largely to satisfy several conditions. First, China needed Taiwanese expertise in order to build up its own domestic capabilities and Taiwan was willing to collaborate. Second, China has the long term objective of folding Taiwan back administratively and politically into Mainland China, and so creating economic co-dependencies and habits of collaboration supported that political objective. Taiwan needed access to less expensive labor as the per capita GDP of its workers rose, putting Taiwanese cost competitiveness at risk in the global IT market. By 2010, Taiwan had become an important component of China's IT industry.[143]

However, there were limits then—and now—that constrained Taiwan. First was the size of the island; it is a small place with relatively small companies. Even local firms that had grown substantially in the 1980s and 1990s encountered limits in how much they could expand in their home base, such as Acer and Mitac.[144] A second constraint concerned the limited capacity of Taiwanese companies to absorb the massive and rapid growth in information about all manner of IT that began surfacing in the 1980s and which has shown no signs of slowing. There just were not enough human and financial resources to absorb all that information, other than for specialized IT products and services. As product life cycles shortened in the 1980s and 1990s, the amount of time local firms had to learn, catch-up, and innovate shrank and became more difficult.[145] This was the same problem faced by the Japanese, South Koreans, and so many European companies. Finally, since so many of Taiwan's customers were other IT firms, it had limited direct access to the end customers of Taiwan's customers (the firms that sold to individuals and users of IT in companies and government agencies). This lack of direct access to the ultimate customer became a growing problem since it was an ever more knowledgeable end user community around the world that pushed vendors to innovative, that is to say, Taiwan's customers, to continue transforming. That circumstance of being separated by one to two degrees from ultimate users of their products left the Taiwanese one step behind in the process of anticipating new requirements.[146]

But, Taiwan ultimately facilitated the global diffusion of IT because, in the words of one student of the process:

> Taiwanese manufacturers, as a collective entity, were able to lower production costs and bring down the price of existing or emerging products of leading companies in advanced industrialized countries, thereby making these products popular.[147]

It did this by developing and sustaining a "superb manufacturing capability."[148] In practical terms "superb" means efficient, reliable, quality products that were competitively priced. Most of this capability—IT firms—was clustered close to each other, as in Silicon Valley. By 2004, some 10,000 small local PC and integrated circuit firms existed within 50 miles of the Hsinchu Park.

Meanwhile, Taiwanese residents continued to become extensive users of IT in their own lives and work. The young Taiwanese who made their own PCs in the 1980s continued to use IT and in time their children did too. Using 2004 as an example, the number of installed machines makes this clear. That year the island had 12 million PCs being used, 8 million of which had access to the Internet. Taiwanese used 23 million cell phones. In proportion to its population, these were as many users as in other teledense nations, such as Sweden. One local Taiwanese economist concluded that "the most important driver behind e-commerce in Taiwan appears to be international competitive pressure," which he noted was "especially evident in the manufacturing sector," but also spilled out into all manner of digital products. Taiwanese had rapidly become extensive users of IT at the individual level too, not just their employers.[149] Clearly, Taiwan had entered the twenty-first century as a mature Wave One society, rushing quickly to this status in a matter of approximately three decades.

SINGAPORE A CITY STATE GONE DIGITAL

Tiny Singapore, a city-state made up of just over 5 million residents (2010) is important to the story of IT's diffusion for several reasons. First, a collection of companies manufactured IT components that made it into the global IT supply chain, as occurred in so many other Asian countries. Second, the city evolved into a proto-information society in which its citizens, government, and firms became extensive users of computing by the end of the century, thus, a model for the region. Third, and most important, building on its historic role as a regional trading center with its deep water port located off the southern tip of the Malay Peninsula in Southeast Asia, it became a conduit for the movement of IT and other goods into and out of the region to the entire world. Singapore's IT-dependent services for the region also became important by the end of the century, such as its banking industry.

The city had several unique characteristics that also affected its role in IT's diffusion. Having been an English colony during the nineteenth century, English took root as one of the official languages of the nation, and was clearly in wide use by local businesses in dealing with customers around the world. Singapore also has a large population that speaks Chinese and trades with China, thus is able to work in the second most important trading language of the early twenty-first century. It has long had a high literacy rate as well and its leaders have usually invested extensively in education. Singapore has almost no commercially viable natural resources, such as oil or minerals. It has a small land mass, so its agricultural base is diminutive; it even has limited supplies of fresh water. And, of course, it had such a tiny domestic market that to gain the necessary economies of scale to compete in world markets, it had to go global, in other words, implement effective export programs. Over many centuries, it really had few options but to rely on commerce, and in the second half of the twentieth century IT-centered services and trade to thrive. Much of the history of Singapore since 1965, when it became an independent state from Malaysia, is similar in many respects to the story of other Asian nations: focused government leadership in developing key industries, especially those that consumed less energy after the oil crises of the 1970s; extensive reliance on exports; evolution from cheap labor manufacturing to higher value-added production; movement from producing textiles to assembling technologies; reliance on IT for manufacturing, infrastructure, and services; and developing an ICT infrastructure in support of its trading role. By the early 2000s, manufacturing and financial services combined accounted for nearly half the total GDP of the nation.[150]

As in other societies around the world that became extensive manufacturers, exporters, and users of IT, Singapore's economy evolved and prospered enough that it could afford to participate in the IT arena. In 1980 its GDP was $25.12 billion; in 1990 it was $66.78 billion, in 2000 it reached $159.84 billion, and in 2010 climbed to $309.40 billion. At the individual level, in 1980 per capita income sat at 39.65 percent of that of an American citizen, then rose steadily through the next thirty years to 82.13 percent of an American's per capita income.[151]

The government for the majority of the second half of the century was dominated by one party, the People's Action Party (PAP) and by its dynamic leader, Lee Kuan Yew, so public administration and policies remained consist from one year to the next. A highly skilled public sector workforce and relatively sound management fostered an environment of commercial competition and advantageous business environment, making this city state one of the most business-friendly to work in the world, and most importantly in Southeast Asia. A Faustian bargain was struck by the state and its citizens: in exchange for strong economic growth and standards of living citizens made compromises in their access to civil and political liberties. The government allowed foreign companies to establish headquarters and welcomed foreign workers, creating a society in which they could live and work in peace. It welcomed foreign investments on the one hand, and on the other, forced its citizens to save more of their earnings than evident in most countries in the world, with the result that funding became available for local businesses while additional investments in companies outside of Singapore became possible too.[152]

Beginning in the 1970s, the government began emphasizing the importance of using computers and telecommunications in the activities of the nation—not simply their manufacture and export—as part of its campaign to build a modern infrastructure suitable for businesses and for management of its ports. Lee Kuan Yew waxed on this strategy in his memoirs:

We computerized the whole government administration to set the pace for the private sector. We gave income tax incentives by allowing rapid depreciation for computers. That decision has given us a lead over our neighbors. It seeded our plans for an "intelligent island," completely linked up with fiberoptics and directly connected with all the main centers of knowledge and information—in Tokyo, New York, London, Paris, and Frankfurt, and also our neighbors, Kuala Lumpur, Jakarta, Bangkok, and Manila.[153]

More important was the fact that these, and other initiatives, made Singapore a major global trading nation, indeed the eleventh largest by the end of the century. Historically, it traded in Asia, but in the second half of the century that commerce declined in part due to Singapore's actions but also to the increased capability of other local nations to manufacture goods of all kinds.[154] Scooped up in that pattern of changed proportions was IT. Overall, in 1965 some 45 percent of Singapore's exports went to other Asian nations; by the end of the 1990s these accounted for only 28 percent of its exports, albeit with larger overall volumes. Asian imports to Singapore amounted to 34 percent in 1965 and by the end of the century, 29 percent. In short, by the end of the century Singapore's total trade with Asia still remained high—nearly 60 percent of total—but was also now about 18 percent with the United States, followed by slightly less (roughly 15 percent) with Europe.[155]

Singapore attracted foreign companies eager to use less expensive labor than available in the United States and Europe to assemble semiconductors, beginning in the late 1960s. By the 1970s Singapore had become a principal site for manufacturing these products in Asia, along with other less sophisticated electronic parts and products of a non-IT nature.[156] Public policy to upgrade its manufacturing began in 1973 after full employment had been achieved and when it was becoming evident that the comparative advantage of cheap labor would decline soon and probably, rapidly. Like Japan—and during the same years of the 1970s and 1980s—it reacted to the high price of oil, restructuring its economy to encourage manufacturing of cleaner products requiring less energy to make: telecommunications equipment, semiconductors, and IT peripheral gear and components. Between the early 1970s and the early years of the new century, a series of economic development plans built on this fundamental focus of higher valued manufacturing and exports.[157]

Policies also differed from the neo-mercantilist ones in evidence in such Asian countries as Japan, South Korea, and China. As reported by two knowledgeable observers in 1989 when commenting on the "openness" of its economy:

"Despite rapid diversification away from its original entrepot trade status, the economy has retained its essential openness; total trade as a proportion of GDP remains one of the highest in the world," noting that "there are virtually no restrictions on international trade," with "only a few import tariffs imposed."[158] Its ability to attract foreign manufacturers of IT proved successful with the result that by the end of the century 62 percent of all its exports were concentrated in several high-value industries: office machines, electronic products, and telecommunications equipment—all goods that contained digital components.[159]

Many additional steps implemented elsewhere were taken in Singapore too. One of great importance involved the government establishing a science park next to its National University, which became home to a dozen small firms by the early 1980s. They worked on robotics, circuit design and assembly, relying on foreign experts and funding.[160] At the same time, companies could take tax credits for investing in IT. Tax incentives and normal business reasons for automating work began paying off. In 1979, some 400 local enterprises used computers. By the end of 1982, the number of firms with computing had risen to 1,800. Then in the 1983–1984 budget the government permitted the full depreciation Lee Kuan Yew mentioned in his memoirs.[161]

We can obtain yet another view—from the ground where the buying and selling of computers took place—by looking briefly at IBM's activities. The firm had an office in Singapore since 1952, populated in the beginning by three employees. In 1957 the company opened a service bureau to sell data processing, but not yet computers. In 1962 the first IBM computer was ordered by the Central Provident Board (an IBM 1401 system), installed in 1963. Sales picked up in 1964 when, for example, the Ministry of Finance acquired a 1401 system and the Lee Rubber Company an IBM 1440 system. Importantly, the Port of Singapore, which became one of the most digitized in the world by the end of the century, contracted with IBM to process its 7000-person payroll in 1965, beginning an important relationship. Shell Eastern Petroleum Ltd. became the first local user of an IBM System 360 in 1968, in the same company that installed one of the first IBM System 370s in Singapore in 1972. In 1976, IBM's revenues reached $10 million, which suggests that over half of its revenues came from adding peripheral equipment to existing systems and upgrading older computers to newer models. The rest came from leasing and selling computers to new, probably first time users of computers. By the early 1980s, Singaporeans were installing IBM's current products at the same time and rate as customers in other countries: 3031s, beginning in 1981, Personal Computers in 1984, and the first 3090 in 1987. By the end of 1986, annual revenues had reached $82 million, quite respectable for the size economy Singapore had at that time. The firm had 538 employees. The following year's revenues of $100.96 million helped to improve the labor productivity picture for the local IBM company.[162]

A major increase in the role of IT in both Singaporean government policies and commercial activities began after the recession of 1985–86. In addition to continuing its focus on select industries and using policies and taxation programs to encourage growth of IT manufacture and use, the core policy was now to concentrate on higher value-added IT, a principle central to the present. All industries were encouraged to use IT to upgrade their own international performance. Second, more complex, hence more profitable, sectors of IT were nurtured: IT in general, robotics, microelectronics, artificial intelligence, and communications. To strengthen national focus, in 1991 the government established the National Science and Technology Board (NSTB) to carry out these objectives. It soon published the first national IT plan, called the National Technology Plan for Singapore, in 1991. It called for investments, including a substantial $2 billion in R&D, to be spent over five years. A second plan in 1996 called essentially called for more of the same, but now intentionally to achieve "world class" status and scope.[163]

While Singapore overall met many of its objectives, some segments of its IT world performed erratically, which encouraged policy makers to support simultaneously multiple parts of the IT industry. Semiconductors, which had been the first major high-tech product area was subject to global economic and IT events. While manufacturing revenues expanded through the 1980s— as in most other countries making ICs—local manufacturers suffered a downturn in 1981–82 and yet another in 1985, resulting in declines of output ranging from 5.5 percent to 7.5 percent. These downturns translated into layoffs of between 10,000 and 13,000 workers. Furthermore, this was not the first time layoffs had occurred in the semiconductor business; in 1974–1975, local firms shrank by 15,000 employees.[164] All these layoffs were enormous for such a small country, which in these years had a total population closer to 2.5 million souls.[165]

Several lines of development and use emerged as essential elements in Singapore's diffusion of IT. The first in long-term importance was the arrival of the Internet to this island nation. The first users were faculty and students at the National University of Singapore in late 1980. In 1992 the National Computer Board, the government agency set up to promote use of computing and to monitor rate of progress, established Technet, expanding use of the Internet by secondary schools, beginning in 1993. In July 1994, SingNet opened its virtual digital doors as the nation's first Internet access service provider for the private sector. This same provider also made the Internet accessible to individuals. In July 1995, Technet was privatized and renamed Pacific Internet. In March 1996, Cyberway came into existence. Competition among providers in the 1990s drove down costs of access, thereby encouraging businesses and individuals to access the Internet. By early 1995 there were 52,000 users and barely two years later, 150,000.[166]

Singaporean officials have probably collected more data on diffusion of IT in their nation than those in any other country in the world, with the possible exception of the United States. Their surveys teach us a great deal about Singapore's patterns of adoption. Table 8.13 displays recent data on the percent of households with PCs—the primary means of accessing the Internet until

Table 8.13 PERCENT OF HOUSEHOLDS WITH PCS AND INTERNET
ACCESS IN SINGAPORE, 1988–2004, SELECT YEARS

Year	PCs	Internet
1988	11.0	
1990	19.1	
1992	20.2	
1993	26.6	
1996	35.8	8.6
1999	58.9	42.1
2000	61.0	50.0
2001	63.9	57.0
2002	68.4	65.0
2004	74.0	65.0

Source: *Annual Survey on Infocomm Usage in Households and by Individuals*, various years (Singapore: Information Development Authority of Singapore, and other agencies, 1988–2005).

the early years of the new century—and the percent of households with access
to the Internet. The data demonstrates that Internet usage by individuals
(households) is essentially an event that occurred during the second half of
the 1990s, but use of PCs had started in the 1980s. Never do all PC users also
access the Internet, a telling point because the orthodox view would hold that
by the late 1990s "everyone" who owned a PC was accessing the Internet all
over the world. The Singaporian data demonstrates the opposite, and the
more reasonable conclusion that not every user of a PC required, or wanted,
access to the Internet. 1996 is a useful year to look at because by then utilization of the Internet around the world had begun its great takeoff. A third of
the Singaporean households had a PC but less than 9 percent subscribed to
Internet services. Five years later (2002) they almost caught up (68.4 percent
of households with a PC and 59 percent subscribed to an Internet service). We
could argue that the catch up to the Internet occurred quickly—in less than a
half dozen years—but also one cannot assume that if the enabling infrastructure existed people would move to the next type of computing. The rate of
adoption of both technologies and the extent of diffusion paralleled patterns
evident among the most extensive users around the world, such as Sweden,
Denmark, and the United States.[167]

How and for what purposes did businesses adopt and react to the two technologies? The first observation one can make is that diffusion of PCs extended
through the 1980s, but deployment of the Internet was a 1990's event for
them. By 1994, nine out of ten establishments employing ten or more people
used computers, reflecting a twelve percent increase over 1992, suggesting
that takeoff was underway. In 1994, ranked in order of applications of these
systems from most to least the top five were finance and accounting (by 96

Table 8.14 PERCENT OF BUSINESSES USING COMPUTERS, LAPTOPS
OR WORKSTATIONS, INTERNET AND BROADBAND IN SINGAPORE,
1982–2002, SELECT YEARS

Year	IT Appliance	Internet Penetration	Broadband Access
1982	13.0		
1985	35.0		
1987	59.0		
1989	68.0		
1998	70.0	n/a*	n/a*
2000	79.9	72.1	14.9
2002	83.3	78.3	41.2

*Not available in table because data on Internet and broadband usage was collected in different ways, making like comparisons between some years difficult to make. Internet was not used by businesses prior to the 1990s.
Source: *Annual Survey on Infocomm Usage in Households and by Individuals*, various years, also under various other titles, such as *Singapore IT Usage Survey* (Singapore: Information Development Authority of Singapore, and other agencies, 1992–2003).

percent), general administrative (95 percent), human resource development (79 percent), procurement and purchasing (51 percent), and product development (41 percent)—all quite normal patterns of usage when compared to the most teledense societies at that time and none that required use of the Internet. These were support activities. Looking at primary business activities of the private sector, operations accounted for 59 percent usage, followed by sales and marketing (45 percent), inbound logistics (52 percent), distribution (47 percent) and after-sales service (21 percent)—all quite normal for an advanced economy—and that increasingly required use of the Internet or other forms of communications in order to be used.[168]

Table 8.14 summarizes similar data as the prior table. Companies in general were the early adopters, regardless of the number of employees. By 2004, 99 percent of all firms with over 200 employees used computers, the Internet, and broadband. Even small firms had become extensive users. Among those with less than 10 employees, 62 percent had a computer of some sort, 44 percent reported using the Internet, and 23 percent had broadband services.[169] For most practical purposes, nearly all firms had become some of the most vested in IT of any set of national enterprises in the world. The number of IT professionals in these firms and in the IT industry across the economy expanded slowly. In 2000, they comprised 4.8 percent of the total workforce, a percentage that grew only slightly over the next several years.[170] In earlier decades the largest firms and agencies were the first to adopt computing, with manufacturing and finance leading the way, making it possible to adopt the Internet more achievable, and by their employees and their families at home as well.

Before leaving the story of Internet usage, it should be noted that all these technologies—PCs, mainframe systems, networks, Internet—were used for the same purposes as around the world. One survey from the mid-1990s made this quite obvious: e-mail was the most widely used, followed in descending order the Internet and file transfers. The same survey noted that the top five benefits of using the Internet by enterprises, hence providing incentives to deploy its use, were in descending order easy access to worldwide information, worldwide virtual presence, extended global market reach, new business opportunities, and improved customer service. Reasons constraining use were the same as evidenced in the pan-Atlantic community: fear staff would waste time surfing the Internet, lack of expertise, and insufficient relevance of the Internet to the organization's business.[171] Over time, telecommunications traffic on the Internet outpaced that over private networks.

The port of Singapore was, and is, the most important natural economic asset of the nation. Central to the national strategy of becoming an integral hub for the region and the world in the globalizing logistics and supply chains from the 1970s to the present involves effective use of this port. Important activities in support of that strategy involved use of computers in the management of these facilities and the IT connections to global logistical systems and networks. By the early 2000s, Singapore had one of the most advanced supply chain infrastructures in the world and certainly the most efficient port operations, as measured by the speed with which ships came in and out of port, unloaded and reloaded cargoes, even in the wake of 9/11 when inspections and other security measures were adopted worldwide. Paralleling the examples of such ports as Long Beach, California, Tokyo, Japan, Kaohsiung, Taiwan, and Rotterdam, Netherlands, Singapore early and intensely applied IT to logistics and port management.[172]

The operating entity managing the harbor is the Port of Singapore Authority (PSA). PSA transformed the port into a facility that could accept standard cargo containers, dock ships and unload/load them, quickly get them on their way, while collecting information on each step. It could insert this data into the various ICT systems of the product transporters, ships, and customs authorities in various countries. Since the mid-1970s, the volume of traffic coming through this port has been some of the highest in the world, as occurred in other Asian ports that also automated much of their operations, including Hong Kong, Kaohsiung, and Pusan, for example. Singapore became the world's busiest container port in 1997, the result of concerted effort by the PSA to modernize and automate the majority of its operations, which it began to do in the mid-1970s.

Beginning in the mid-1980s, PSA began a campaign of technological upgrading, embedding IT into the routine functions of the port and to improve continuously performance of the large complex, an extension of the national IT infrastructure then beginning to be installed across the city-state. The collection and use of import and export documentation was automated, starting at the end of the 1980s. By the early 1990s some 8,500 companies and 20 public agencies had been linked through this system.[173] Automated container terminals were constructed in the 1980s and 1990s as well, making it increasingly possible to track goods and containers from ship to port and back to ship faster than most ports around the world.[174] In the 1990s three major systems were fully implemented. The first, TELEPORT, provided shippers with such information as birthing schedules and detailed cargo data before ships arrived in Singaporean waters. It linked to similar systems at other ports—a unique feature at the time of Singapore. Second, the Computer Operations Systems (CITOS) automated and directed container handling tasks in real-time for fast unloading and loading of cargos, making it possible to unload 40 ships with a total of 30,000 containers per day, while storing on average 100,000 containers each day. By the early 2000s, CITOS was handling daily 800,000 job instructions. In 1989, PSA had begun using a third system called PORTNET as its network to connect all its systems and to communicate with companies, ships and other ports. By 1993, it had 1,500 subscribing companies and agencies from around the world. In 1997 it became an interactive web site, making it possible for the PSA to boast that its port was a "global maritime information hub."[175]

By the early 2000s, the port was digitally linked to warehouses all over the nation that made it possible to manage more storage in a more integrated fashion to handle larger volumes of goods and to move these over land, water, and air over larger swaths in the region than previously possible. One student of the process assessed the results this way: "In its relentless pursuit to maintain its position amidst keen competition [as the best seaport in Asia] from neighboring ports, PSA has harnessed IT strategically to improve the efficiency and effectiveness of port operations not only for its own staff but also for other port users," adding "through a business-driven IT development, aligning business and IS [information systems] plans, maintaining a flexible and extensive IT infrastructure."[176]

Elsewhere in Singapore, diffusion of IT continued apace during the 1990s, following the rapid rate evident in Taiwan and South Korea, although arguably some would say even faster. Smaller businesses were slower to adopt, as in so many countries around the world for similar reasons.[177] In the early 2000s, schools that had started to embrace IT in the 1990s were almost universally "wired" and had become extensive users.[178] But, by the late 1990s, surveys were suggesting that just over half the businesses doing businesses with other firms (B2B) routinely used IT as their central conduit for such transactions with a large portion of such adopters in various services businesses. These types of enterprises by the nature of their work often lent themselves to the movement of digital data and the use of inexpensive computing in a thick teledense environment.[179] Perhaps the one application of IT in the early 2000s that received the most attention around the world was Singapore's use of real time systems to alter the flow of street traffic by cars and trucks through a real-time congestion management system and electronic road pricing scheme that caught the attention of hundreds of cities struggling with their own congested traffic.[180]

Finally, beginning in the late 1990s, but flowering in the early 2000s, existing wireless networks began supporting mobile commerce in the city. As in all societies that extensively used IT, Singapore had an excellent, modern telecommunications infrastructure.[181] Recall that by the end of 1997, it was fully privatized with the near immediate effect that competition led to improved services and sharp drops in fees. By 2003, there were 46.7 telephone lines for every 100 residents. At the same time 81 percent of the population had access to mobile telephony.[182] In 2000 the government began to improve wireless communications involving 30 wireless products, such as location-based services, wireless enterprise functions, and a mobile payment system. The government engaged the three largest mobile providers in the country to expand mobile functions: Singtel Mobile, MobileOne (MI), and StarHub.[183]

Like other Asian nations examined in this chapter, Singapore had the private-public sector discipline and focus required to transform its economy on the one hand, and on the other, the desire to embrace the same kinds of uses evident in American and European states. Singapore's process of adoption began substantively in the 1970s, took off in dramatic fashion during the 1980s, and became extensively diffused across all its society by the mid-1990s to such an extent that Singapore was widely recognized as one of the most teledense societies in the world. Singaporeans demonstrated that the size of their nation was not as much of a factor as other considerations such as per capita and national levels of GDP, ability to attract foreign capital, skilled workers, technology, and focused and expert collaboration between the public and private sectors. In particular, collaboration occurred far beyond what normally was evident in the Western economies of Europe and North America. The nation's smallness may have facilitated diffusion because there were fewer leaders and interest groups to mollify when compared to what had to be done in such larger economies as those of Japan, China, the United States, and Germany, for example. To be sure, the lack of natural resources forced its hand. The diminutive size of Singapore meant that large constituencies were not rivals for focused attention and use of such resources as funds, manpower, and managerial attention that otherwise would have had to be diverted from the promotion and deployment of IT, or as cause for delaying the start of Singapore's transformation into an extensively digitized society, because they hardly existed. Taiwan exhibited a bit of that same advantage of smallness too. In both cases, however, they fundamentally had to do an effective job in attracting capital and know-how, while also developing local skilled workforces and business-friendly environments.[184]

CONCLUSIONS

These three national case studies can shed a great deal of light on how IT spread in Asian societies, and more than with most other country examples, how governments can accelerate the diffusionary process. They all faced similar challenges, but individually overcame most of them in very local ways. These included acquiring the technical know-how on how to make and use IT, which they appropriated largely from Japan and the West. Attracting funding adequate enough to migrate to the new technologies was another, fixed by forced internal savings by citizens, directed investments by government agencies and local firms, and through attraction of foreign capital, normally via investments in local firms. Another issue was the development of a local industry that could support indigenous data processing needs and provide jobs, and that required scaling up IT firms to sell products outside their relatively small national markets. When these various issues were addressed two patterns of behavior became obvious. First, they succeeded. Nations created local IT industries that competed in the world market, particularly by the early 1980s. Second, local users of IT embraced the technology for the same purposes and reasons as institutional and individual customers did around the world.

But, unique to these cases was the role of national governments. Western scholars subscribe overwhelmingly to the two notions that in a capitalist economy firms thrive when competition is allowed to thrive and when customers can choose what to buy. Commercial enterprises were central characters in this play, as Alfred D. Chandler, Jr. and others have described.[185] To be sure, these companies were instrumental agents of change in Asia, once they reached a point of critical mass where they could thrive in a global market. Western scholars have not always recognized readily enough that when a nation wanted to develop a new industry in the second half of the twentieth century, it already faced rivals in other countries that had made the kinds of Chandlerian investments required to succeed, such as did the Americans earlier than the West Europeans and all Asian economies in product development, manufacture, distribution, and support services, such as maintenance, training, and sales. In recent years Western economists in particular and a generation of Western trained Asian scholars, have begun to reshape that earlier interpretation. It is their work which has so informed the interpretation of events presented in this book. As a result of these new perspectives it is increasingly understood that to overcome the existence of a rapidly integrating international IT market, East Asian governments chose to implement neo-mercantilist programs designed to shelter their fledgling industries until they could compete at a global level. This approach was evident in Japan, Korea, Taiwan, Hong Kong, and Singapore; later too with the next round of entrants into the world of computers, such as the Philippines, Malaysia, and Thailand.[186]

More than in the West, governments played a central nurturing role that protected industries that yet also facilitated competition within their national borders. Both protectionism and competition made possible firm-level prowess to develop needed to compete successfully in a global economy. The United States had an open economy where the Darwinian notion of the fittest thrive—the Chandlerian model—while the West Europeans opted for national champions, meaning governments supporting one or a very few companies within a nation. The American strategy worked, Europe's did not. Asians proved more successful than the latter and increasingly in recent years comparable to the former in select areas of IT, largely with hardware, less so with software. The Asian approach valued domestic competition within the contest of a protectionist firewall. That protectionism was only maintained until local firms could compete, or when diplomatic and economic pressure from the Americans, Europeans, and the IMF compelled a change in strategy. Even when that change was forced upon them, Asian IT industries had progressed fast enough to overcome influxes of Western competition and as stronger enterprises to thrive in the way Chandler argued corporations did.[187]

It would be quite easy to accept the idea that at some point the IT industry on the one hand, and the use of IT computing by institutions and people on the other, acquired some sort of unstoppable momentum. After all, the statistics on revenues and volumes presented in this and in the previous chapter could lull one into believing that there was momentum at work. However, upon closer examination of the events the opposite was more the case. These national industries were always fragile, were subject to international economic and market crises, and required intense involvement by government officials to thrive, year-after-year in so many East Asian countries. To be sure, all the players learned how to promote the production and use of IT, and so improved their performance over time, but problems never went away. When producers of DRAMs made too many of them in the 1990s, thereby flooding the global market, they watched the price of their ICs drop. But the semiconductor manufacturers moved quickly up the value chain to produce new, more complex products in shorter periods of time, a circumstance which in turn compelled public administrators to step in again with facilitative programs, because by now IT had become too important a part of their national economies to allow these to falter. The rapid decline in per capita GDP in South Korea in the 1990s made it clear to officials and to business executives in general that economic fortunes could change quickly in a highly integrated global economy that was also always in the spotlight and that was also subject to dramatic swings in high technology industries.

Customer firms and individual consumers did not make it easier for high-technology vendors, or governments either, because as they learned more about digital technologies they sought the most useful, best priced products they could find around the world. As the percent of a nation's GDP and exports that relied on IT increased, the greater the pressure on all concerned not to let the goose that laid golden eggs get sick. If anything, pressures mounted on all to perform well, indeed, continuously better. Every year brought new problems and opportunities, and every year governments had to play an enormous role; indeed, a far more proactive—and in hindsight a more effective one— than any other group of nations after the 1960s.

The timing of when these Asian countries embraced computing played a role too. While all three tinkered with computers in the 1960s—late adoption by Western and even Japanese standards—assimilation was slow in the beginning. Many reasons accounted for the slow start. The technology was complex, expensive, and not necessarily productive given the nature of what these economies were doing. The relatively low cost of labor was fine on the supply side, but an inhibitor on the demand side, since low wages meant less income to spend on digital goods and communications and less incentive to automate cheap human labor. Inadequate technical skills remained a chronic problem all through the 1960s and into the 1970s, then a lingering one to the present as the technologies underwent rapid and frequent changes, only taking off in the 1980s. But, as the technology began appearing in smaller, hence, less expensive and simpler (that is to say, easier to use) forms, beginning in the late 1970s, and massively so in the 1980s and 1990s, the situation changed. Personal computers and minicomputers were proportionately more appropriate for many individuals and customers. The hardware worked and there were many software products that proved easy enough to use by ever-growing numbers of less-technically trained workers. Even mainframes and their peripheral equipment continued to drop in cost to acquire and to operate. They also came in various sizes, which too made them economically and functionally more attractive. So, the early adopter problems with this technology experienced by the Americans and Europeans in the 1950s and early 1960s were less intense and often were avoided to a large extent by Asian users in the 1970s and 1980s. As software and hardware became easier and less expensive to use in the 1980s and 1990s, the faster one could install and use computers. Add in the capabilities of the Internet which increased all through the late 1990s and early 2000s, and one has to draw the conclusion that much was now available to speed-up the broad appropriation of IT. What is remarkable is how extensively that happened in the late 1990s and early 2000s. This pattern of adoption suggests that these societies had leapt ahead of the pan-Atlantic community and the Japanese in becoming telematic societies of some sort, if as yet not well defined, as evidenced by the Korean experiences generated by online gaming and high-speed broadband communications.[188]

Several trends stand out when looking at the role and effects of economists. First, while a number of industries were chosen by governments to be nurtured, the most prominent where those in computing and telecommunications.

In some nations the two lines of technology were often developed within the same firm, although government support and regulations normally spread across multiple agencies; nonetheless, collaboration was more often in evidence in Asia than in the pan-Atlantic community among the affected institutions. Second, the choice of IT was a sound one, because in these three and in other countries, the role proved so great that these technologies lifted a nation's GNP and GDP nationally and at the individual level. Textiles, agriculture, ship building, oil refining and other industries important in the 1950s and the 1960s did not have the same GDP lifting effects. In time these other industries declined in importance. Third, by focusing on exporting strategies, these nations were able to leverage their limited resources to gain economies of scale, comparative advantages with respect to local skills, capabilities, state-of-the-art factories, and ultimately the opportunity to market some of the most profitable IT products of their day. Fourth, the consumer electronics industry, which was largely based on products with built-in IT components by the end of the 1980s, was the largest, earliest, and fastest growing in the world, with the positive result of making these societies very teledense, and made affordable by the rising incomes of their citizens. By 2010, the economic role and effects of information technologies was profound, integral to how these societies now functioned, and often made these economies global pace-setters.

Culture and local style counted for much as well. The French wanted to preserve their culture and use IT to help do that, while the Americans spoke about information superhighways and increased employment, and the Soviets of a new technocratic socialist utopia. The Japanese wanted new sources of economic growth and jobs, and technological independence from the United States. They specialized, primarily in PCs and networking technologies. Japan also wanted to have a multimedia information society. The South Koreans thought in terms of participating in the computer revolution, worried about Japanese and American technological hegemonies, wanted to compete for economic superiority in Asia, and ultimately sought high status among industrialized nations of the world. The Singaporeans spoke about living in an "intelligent island" in which citizens prospered yet lived as a communitarian culture, some would say authoritarian state. The strategies for how to achieve all these visions were more closely similar but, as the histories of the experiences of each demonstrated, were executed uniquely. Government agencies dedicated to IT adoption proliferated: MITI in Japan, Ministry of Information and Communications in South Korea, and NCB and other agencies in Singapore, to mention a few; Taiwan and Hong Kong had these too. The Americans and Europeans also had such agencies, although they functioned with less authority and influence, and rarely as cabinet level departments as frequently was the case in Asia. Great Britain, the United States, and Japan liberalized their economies in the 1970s and 1980s to promote competition in IT, telecommunications, and a number of other industries, such as utilities, banking, and various modes of transportation. Japan did this far less than the Americans. The Koreans promoted market competition, as did the Singaporeans and Taiwanese, but with focused steps within measured phases.

Cold War politics, Middle Eastern oil embargos, and regional military tensions played such important roles between the 1950s and the end of the 1980s in the diffusion of IT that these activities can no longer be ignored by historians of IT, and indeed, beginning with the work of Paul N. Edwards, they have started to place the story of computing into this broader context.[189] Japan, South Korea, and Taiwan received considerable American foreign aid in the 1950s and 1960s. The flows of American technologies were subject to national security priorities in Asia as in Europe all through the period. As in Europe, the Korean War stimulated economic development among a number of Asian countries with the obvious exception of the two Koreas.

The oil price hikes of the 1970s profoundly affected Asians, causing widespread shift away from energy consuming economic development to more high-tech, less energy demanding industrial activity. While oil prices hurt Europe too—and less so the North Americans—Europeans did not transform their economies in response to this fuel shortage to the extent as did so many Asian states. In short, the Arab oil exporting nations inadvertently nudged the Asians toward their IT interests faster than might otherwise have happened. Without the hyper concerns over energy consumption, the Asians may have simply gone along putting together low value added components for high-technology equipment designed and finally assembled in the United States, for example, rather than to challenge the advanced industrialized nations to make IT core elements of their national economies.

Military tensions remained in the region, continuing to the present, and while that situation encouraged countries to build up modern military organizations with information systems, none of these countries spent as great a percent of their national GDPs on arms as did the United States and the Soviet Union, with two exceptions—the Koreas. The rising tide of Chinese military power in the first decade of the new century continues to keep the Taiwanese situation in stress, while Sino-Japanese economic rivalries hold out the possibility of potential future problems in the region among the most advanced economies. Even developing India has its military issues with Pakistan, about which more will be said later.

As Japan, South Korea, Taiwan, Singapore, and Hong Kong embraced computing at home, and as a strategy for export driven economic development, a second tier of nations in the region were cp-opted into the manufacturing eco-system. These included Malaysia, Thailand, the Philippines, and Vietnam among others. Just in the period 1985–1995, the Asian computer market expanded from 17 percent of the world's IT market to 32 percent, and later surpassed Europe. Today, the Asians continue to expand their use of IT across the region faster than any other part of the world, surpassing even the United

States and Western Europe which were the original early adopters.[190] While the top spenders on IT in Asia by the mid-1990s were Japan, Australia, South Korea, and China in descending order, many other Asian nations individually now spent annually a billion dollars or more on IT too, such as New Zealand, Taiwan, Hong Kong, Singapore, Malaysia, Indonesia, Thailand, China and India; only the Philippines spent less ($573 million in 1995).[191] IBM sold over a billion dollars of goods and services in the region in 1996, the first year it achieved that level of sales—and this was not the last year for such large sales. But the point to be drawn from this data is that such large volumes of business were visible signs that the region was no longer a backwater arena in the world of computing.[192] In less than forty years, Asia had become a full player in Wave One computing. It now even posited the question of whether some Asian societies were beginning to introduce a Wave Two era. The three nations studied here were instrumented in driving forward that Asian transformation.

The rise of the East has become a global conversation because in the last three decades of the twentieth century and during the first decade of the next, massive economic growth occurred in the region, while a substantial shift in the world's wealth started to shift to the Pacific Rim, first in the form of U.S. Treasury bonds but more pertinent to our focus, stock equities and capital investments in Asian firms. Their national histories make perfectly clear that a significant source of their prosperity came directly from their diffusion of computing and related technologies in telecommunications both on the supply and demand sides of the economic equation. Put in blunt terms, IT played a greater role than any other industry in the emergence of modern Asia as a major player in the global economy, a point we revisit in chapter eleven. That role was replete with very high national and personal GDPs, and either dominated or ranked as one of the top suppliers of numerous high technology value-added products and some IT-based services, such as call centers. That reality has now to be woven into our understanding of contemporary Asian affairs.

Before we can muse further on these and other issues related to the Asian experience as a whole, the roles of China and India must be accounted for, if for no other reason than they are homes to just over one third of the world's population. They too had vibrant IT experiences which, by the early 2000s, were profoundly influential on the world's continued appropriation of computing, even though the percent of their populations directly using computing was in the same category as such second tier IT users as Vietnam, Thailand, and Malaysia as of 2012. That circumstance is an interesting twist to our story of Asian ascendancy to the extensive usage of modern computing that must be accounted for. It is to their history that we turn to next to help round out our understanding of events in modern Asia.

CHAPTER 9

China

Embracing IT in Changing Times

It is essential to equip information institutions with modern facilities in the shortest possible time.

—Fang Yi, *Member Central Committee of the Chinese Communist Party, Vice-Premier of the State Council, 1978*[1]

The two most populous nations on earth are China and India, which combined accounted for over a third of the world's population of 7 billion in 2012 and for 15 percent of the globe's economy. Together they were poised to dominate the world's economy and it was widely believed they would become massive users of IT in the early decades of the twenty-first century. These two nations have long been major subjects of political, economic, military, and diplomatic interest around the world. Both had flirted with computing for many decades, each as early as the 1950s. But, not until nearly the end of the century did the two societies start using computing in substantial quantities. There are many issues to address specific to each of these countries while, at the same time, many others will appear similar to those faced by other nations as they individually embraced information technologies. The biggest endogenous variable is size, of course, with a complex system of diverse local political and cultural styles at work too. These two nations each already had over a billion residents during the period in which they began to deploy extensively computing and modern telecommunications. This is a far different circumstance than faced by such relatively tiny countries as Singapore or Sweden, for example. Both have some of the largest landmasses on earth to govern, yet even this factor is deceptive. In China's case, the vast majority of its population is crowded into the eastern half of the nation, because the western part is largely arid and desert. Geography is as important to our story as it is for Singapore and its harbor and for economic evolution in all countries in all periods.

Once an observer of either country gets over the issue of how big these two nations are by almost every conventional measure—land mass, population, size of their economies—one has to deal with the issue of speed of adoption of all manner of technologies, including ICTs. There are two aspects to the issue of how fast each adopted information technologies. The first is the most obvious and talked about: how quickly they began using the Internet, while emerging either as manufacturing centers for companies all over the world (largely China) or a fast growing center for software development, outsourcing, and call centers (India). These are seen as events that began in the 1990s and kept expanding rapidly in the early 2000s. To be sure that occurred. But there is a second element of speed that is routinely ignored, the slow adoption of computing in each country that began in the early 1950s, in short, over the course of some 40 years before the takeoff that became the subject of so much attention. Long before the Internet, export-oriented IT manufacturing, or outsourcing, Chinese and Indian scientists and engineers began working with computers, and slowly over the course of many years, a few, indeed limited, diffusion of IT took place. The contrast between the two periods—that before the mid-1990s and that of the subsequent era—will need to be explained, as well as how the experience of one period built upon the earlier era. Many of the insights gained from that story suggest these two nations went through many of the same phases of diffusion as did other countries and encountered similar problems, such as the inhibiting features of Soviet-styled technology adoptions in the earlier period, and indigenous political and cultural mores across the entire time.

By the early 2000s, it had become the practice in the West to discuss China and India together, largely because of their size, although they are quite different societies, each warranting separate, albeit linked, analysis.[2] Most recently, their rapidly imposing presence in the World's economy caused observers to lump them together. Increasingly, there is a growing realization in many countries that the combined effect of Sino-Indian economic behavior, that is to say, a more collaborative relationship, has enormous implications for all other national economies in the Pacific region and for such industries as manufacturing, finance, telecommunications, and IT suppliers. As one observer of the Asian scene insightfully noted, "What China is good at, India is not, and vice versa. The countries are inverted mirror images of each other."[3] Regional leaders understand the implications. In 2005, China's premier Wen Jiabao commented that, "it is true India has the advantage in software and China in hardware. If India and China cooperate in the information technology industry, we will be able to lead the world," an alliance that would "signify the coming of the Asian century of the information technology industry."[4]

For that reason alone it is useful to discuss China's experience with computers in considerable detail followed immediately by a similarly probing exploration of India's IT history to provide a fuller description of the role of information technologies in the greater Asian region.

A brief snapshot of the resources and proportions such a combination comprises provides more evidence of the premier's comments being important justification for examining developments of both nations in close tandem. Cataloged in table 9.1 are the proportions of people, GDP, and investments in R&D (much of which are going into IT and related fields) in Asia. Add the Japanese and the Asian presence and IT increases sharply, particularly when one takes into account Japan's GDP and expenditures on IT. The aggregate GDPs of the regional groups—Asia, Europe, North America (most significantly, the United States)—are almost equal in size. Combined, the three large Asian countries make up the second largest group of investors in R&D, with all of Europe collectively in third place. As this book demonstrates, a great deal of R&D is being devoted to information technologies in many nations, an investment expected to increase in Asia in the coming years. In the five years that followed the data presented in this table, China's economy became the second largest in the world after that of the United States, while India's experienced more rapid growth than Europe's or that of the United States. In contrast to its expansion, however, Japan's economy languished after several decades of its own heated growth. Trade between India and China increased during the 2000s, leading to the emergence of an economic ecosystem in Asia hinted at in the last chapter, but that in this one becomes an evident force influencing adoption of IT across Asia and diffusion of its products around the world.[5]

Table 9.1 COMPARISONS OF ECONOMIC PROFILES IN ASIA AND THE REST OF THE WORLD, CIRCA 2006 (PERCENT OF WORLD)

Country/Region	% of Population	% of GDP	% of R&D
China	20.5	10.3	13.5
India	17.4	4.3	3.8
Japan	2.0	6.0	12.6
These Three Combined	39.9	20.3	29.9
United States	4.7	21.7	32.5
European Union	7.7	22.8	22.1
Russia	2.2	3.6	2.2
EU (27) and Russia Combined	9.9	26.4	24.3

Source: Data drawn from statistics in Ernest H. Preeg, *India and China: An Advanced Technology Race and How the United States Should Respond* (Arlington, Va.: Manufacturers Alliance/MAPI, 2008): 267–268.

Because of the massive sizes of these nations, economic evolution was not uniform across either country, nor simultaneous. Parts of India and China entered the Second Industrial Revolution in the 1970s–1990s, while other regions seemed like tenements of the First Industrial Revolution. Both also had very large agricultural sectors, bigger than those of the United States, Western Europe, or Russia, for example, as measured either by number of farmers or GDP. Each had tiny services sectors as well. These nations had industries that were often both very behind and quite modern, when compared to those in the pan-Atlantic community or in Japan. Banking in both was—and are—still very paper-based; yet, banks are also large consumers of IT. In some corners of other industries the use of IT is not as evident as in the same industries in the United States or in Western Europe. In short, neither country was as monolithic in their economic development as any other nation we have examined. However, economists in Asia and in the West are in almost universal agreement that both were still undergoing economic development into what they referred to as "advanced economies," even in the early years of the twenty-first century. Specifically, both were in the process of completing their transitions into industrial states with uneven commitments of resources and varying levels of commitment to the efforts by the elites of both societies.[6]

The two have also been moving away from their socialist, quasi-Soviet models of public administration and economic development, each starting their journey at the end of the 1970s–early 1980s. Each set their economic strategies and what they perceived to be in their best national interests, applying their world views and political values and practices. Thus, a fundamental dual transition—political and economic—was still underway in the second decade of the new century, engaging information technology and its diffusion with it. To address the complexity of IT diffusion this and the next chapter will review the role of IT in these two countries, focusing one chapter on each nation.

The history of IT in China provides insights on how technology is woven into the fabric of societies undergoing fundamental change. When computers were first built, China was a war-torn communist authoritarian state. When the Internet came to China, it was a nation evolving rapidly into a competitive quasi-capitalist economy, in part because of the expanding role of IT. China's case demonstrates the extent and limits of the role that can be played by a cluster of technologies; also how more basic issues continue to sway events, such as political authority or premodern agricultural society, and are affected by the realities of education and economic structures. It is also a case study of differing levels of development within a nation that was never monolithic in culture, and certainly not in geography or technological advancement. The story ends with IT perched ready to be one of the core foundation pillars of future China. How that all happened within one generation is the central focus of this chapter.

ORIGINS OF INFORMATION TECHNOLOGY IN CHINA, 1950S–1970S

Like many Asian countries in the mid-twentieth century, China experienced a great deal of warfare and economic upheaval. The introduction and use of twentieth century technologies were slowed by wars—Japan's invasion in the 1930s, World War II (1939–1945), civil war from the 1920s to 1949, and the Korean War (1950–1953); then by political upheavals, some of which were violent and widespread in the 1960s, 1970s, and less so in the 1980s. Governmental practices, policies, political ideology, and administrative exigencies dictated the speed and rate of IT adoption. These influences applied to all manner of information and telecommunications technologies that continued to the present. As overtly, if not more, than in the Soviet Union, ideological and world views influenced events on China's experience with IT. Yet too, as in Cold War Communist Europe and in other communist nations, the uses to which Chinese deployed IT reflected many of the same patterns.

Beginning in the 1920s, civil unrest in China increased to the level of civil war, with the ultimate winners local communists under the leadership of Mao Zedong, coming to power in 1949. He established the People's Republic of China (PRC). As in other parts of Asia, portions of China had been occupied by Japan (1931–1945) with warfare between the two nations in the 1930s folded into the larger conflagration of World War II. Unlike the experience of South Korea and other parts of Asia, which had been under Japanese tutelage for a half century or more, Japan's occupation of parts of China was so disturbed by warfare that many of the managerial practices exported from Japan to other Asian nations essentially did not take firm root in China.

With the establishment of a communist regime, the new national Chinese leadership turned to the Soviet Union—most specifically, Russia—for considerable guidance on such matters as economic development, a bit on international affairs, and extensively on technological adoptions. Early Chinese computing was profoundly influenced by Soviet insights and support, much as the United States had served as a source of knowledge to Western Europe and to Japan. While the history of Communist China is currently the subject of a great deal of scholarly and journalistic research, suggesting much has yet to be learned, several guiding principles have been identified in play during the 1950s through the 1970s. The Chinese Communist Party set policies and governed through the national government in an autocratic manner with Chairman Mao exercising virtually unlimited authority. After his death in 1976, and an attempt by pro-Mao stalwarts to retain power, political reforms were introduced to protect the authority of the party while reforming the bureaucracy of the state and the economy. The state remained subservient to the Communist Party, while the latter operated within a political authority informed by its own local brand of Communist and nationalist ideology.

At the end of the 1970s fundamental changes in Chinese worldviews began to be slowly reflected in such new behaviors as a willingness to adopt more fully Western economic and technological practices, moving from class struggles to economic reforms. Following the collapse of Communist regimes in Eastern Europe, beginning at the end of the 1980s, Chinese officials sped up the process of introducing semicapitalist economic practices. They delegated some political, economic, social, and educational authority to the provinces, with the result that by the end of the century China was successfully well on its way to becoming a major economy in the World, one gingerly evolving politically away from the highly centralized top-down Soviet-styled practices of the past and toward an increasingly more open bottoms-up approach, albeit limited in scope. Russia's inability to transform its own economy smoothly in the 1990s and early 2000s from the Soviet model to a more market-driven one was clearly a lesson for Chinese officials to transform in a more programmatic, yet slower fashion. All of these events affected the rate of China's adoption of IT. Consistently from the founding of the modern state senior party officials wanted to make China's economy, hence its technological policies, independent of those of any other country, a strategy clearly also evident in other Asian national economic strategies. Their autarkic, nationalist aspiration was understandable, given China's recent experience with the Japanese, but these views also reflected the rapidly intensifying Cold War that posed the threat of occupation (or influence) over many nations by the Russians or the Americans. Like the Soviets, Chinese leaders wanted to build up rapidly an industrial state while at the same time shore up agriculture so their people could be fed and political stability maintained. Their long held perspective centered on the reality that famines and disgruntled farmers in China led to civil wars. Chairman Mao spoke frequently about maintaining national independence. That approach required the nation to rely on its own initiatives.

Tied to those notions of economics and local efforts was the equally important one of maintaining national security. The Korean War, in which it was heavily involved in the early 1950s against the United Nations' forces, reinforced the rationale for maintaining a strong military capability. U.S. efforts to "contain China" in the 1950s and 1960s, border disputes with the USSR in the 1960s and 1970s, and the added dangers emanating from the Vietnam War between the 1950s through the 1980s with the French, Americans, and finally countries contiguous to Vietnam, contributed further to the rationale for this military policy. China and the Soviet Union broke apart their close relations at the start of the 1960s, which slowed the subsequent flow of information about IT into China. Meanwhile, its old enemy, Japan, became an important, growing economic and military power in the region during the 1970s, and a major global economic and military force by the end of the 1980s. Such developments led to economic and technological policies and programs that might otherwise have been contrary to accepted practices in how an economically poor, technologically underdeveloped nation might be improved. But, these actions were tied to bolstering national security, to preservation of the new regime, and to feeding citizens. These three priorities required focus, discipline, and the submission of activities to these necessities.[7]

Following the Soviet pattern of articulating formal economic plans, a practice adopted by almost all Asian governments during the second half of the twentieth century, China introduced its first five year plan in 1952, modeled on Soviet styles. It called for rapid economic development through extensive allocation of GDP (20–25 percent to planned investments), despite its low national GDP. Officials emphasized heavy industry, but most important for the future of IT, adopted the concept of central economic planning and allocation of resources. The Soviets supported this Chinese plan with exports of machinery for industry to their Asian neighbor. However, the first plan did not lead to the increased levels of agricultural production so needed by the nation.

In the second plan, better known as the Great Leap Forward (1958–1961), officials expanded agricultural and small-scale industrial production. As one contemporary account of the process described it:

> They [officials] began to ask themselves, which was preferable, the latest technology that requires large-scale establishments or the traditional technology that preserves small-scale production.[8]

The answer: do both, including expanding labor-intensive components of the Chinese economy, which proved innovative (but not effective) in that, in theory, they addressed the problem that the nation had little capital but much labor. Backyard blast-furnaces became an icon of the failed approach, one intended to increase agricultural and industrial output. The approach was also intended to create a communist society at the same time built on the theory of collectivization. Hundreds of millions of Chinese were put to work in collectivized state-owned farms. Local industrial development was encouraged, leveraging the large number of available workers, rather than machinery. Education suffered as students, teachers, and professors went to work in the fields just at the time when in many parts of the world labor was moving to schools, factories, and cities.

This second plan proved to be an unmitigated disaster. Production of food fell so sharply that famine came harshly in 1959–1960. Estimates on the number of deaths due to this famine range from over 15 million to some 45 million people; there is yet no consensus on the total. But, either range made this five year plan a major national catastrophe.[9] Lack of food created excess industrial capacity due to inadequate economic inputs into industry, such as labor and food to feed industrial workers and their families, leaving some plants producing less than planned. China had to import food. Imports of whole manufacturing and oil refinery facilities and machinery slowed sharply so that foreign exchange could be used to buy food. Retrenchment in policies and implementation of programs to end the crisis eventually made it possible to again import machinery in the late 1960s. By the early 1970s, China was implementing industrialization programs initially crafted in the 1950s in support of both large and, now, also small-scale industrial development.[10] Imports of technologies in the 1960s and 1970s did not lead to innovations within Chinese industries as these economic inputs were quite low. Labor productivity remained at levels evident in the 1940s and 1950s. Growth in output came more from making additional products than by improving productivity through innovative practices and new goods and services.[11] Against this backdrop, Chinese officials concluded once again that substantive changes were needed, leading to a new strategy after 1978, which profoundly influenced the role of IT in China.

While 1978 is widely seen as the most important turning point in contemporary Chinese economic history, much also remained constant in subsequent years. Affecting all economic and technological developments were ingrained attitudes that influenced actions, including those affecting the adoption of telecommunications and computers. Public officials involved in economic development believed in general that institutional organizations were culturally neutral, that is to say, technologies and processes were "classless" and thus operated the same in all societies and nations. Not until the demise of the Soviet Union at the start of the 1990s did events lead them to realize otherwise, even though massive imports of Soviet and East European technologies in the 1950s had done little, if anything, to improve productivity. That experience should have prompted them to alter their attitudes and practices to reflect the reality that cultural, social, and political factors were at least as important as machines and other technologies. Economic growth had come from adding labor than by increasing the use of improved methods, resulting, in the words of one student of the process, "in the virtual stagnation of technical innovation in many industries," including all the ones that could have been obvious users of computers, such as manufacturing and banking.[12]

For many years after the failure of the First Five Year Plan, suspicions about the effectiveness of the Soviet approach toward centrally planned economics had crept into conversations about economic development, but clearly not yet into IT activities; that would come later. One byproduct of concerns about Soviet approaches was the attitude of being as Chinese in sufficiency as possible to tailor "institutions to existing technology," leading to the modification of organizations and practices to such an extent that technological innovation had to wait until a different time and attitude. Officials also shared an old Chinese nervousness about adopting Western technologies for fear of corrupting local values. All decisions and activities should be in service to the state, implementing political rather than economic or technological objectives and influences. Until new attitudes came after 1978, the need for IT proved less than the aspirations of some engineers and officials might otherwise have suggested.[13]

Before turning to events of post 1978, we need to appreciate China's earliest encounters with computing, because the history of Chinese IT is not just a post-1980 story, as many commentators suggest. As in so many other countries in the world, there were individuals and institutions interested in information technologies, which extant records documented as starting in the early 1950s.[14] China's early experience is an additional example of a nation in which information about IT transferred and was shared across borders, a transnational pattern of behavior that underpinned the diffusion of computing in the post-World War II period around the world to a far greater extent than historians and economists have yet recognized. Furthermore, as in many other countries, precomputer technologies had been used in China in the 1920s and 1930s by large government agencies and companies, such as IBM's products. This company had established its first permanent office in China in 1928, and installed equipment in the country by the early 1930s.[15]

Interest in computing surfaced early, in fact as early as 1951, when a mathematician in the Chinese Academy of Sciences (CAS), Hua Luogeng, proposed that a group of mathematicians be organized to look at the possibility of doing computational mathematics, quite in line with the Chinese Communist Party's decision of 1949 to seek Soviet advise on technology and economic development. During the 1950s, largely by the CAS, Chinese mathematicians, scientists, and engineers acquired knowledge about computers through a series of visits to Soviet computer engineers, and by sharing information gained with colleagues in China through formal training programs. These visits to the Soviet Union mimicked others occurring at the same time by Westerners visiting each other and the Soviet Union, and Russian experts doing the same in the West. The diffusion of IT knowledge into China thus paralleled both the timing and manner in which such know-how spread across dozens of countries. The Soviets were willing to share information as part of their broader effort to bring China into their political and military block. In practice, while sharing knowledge about IT proved effective from China's perspective, it was increasingly constrained by the end of the 1950s as the two governments diverged in their views on a broad range of issues, not the least of which were border disputes (among a series of issues) that accelerated further in the 1960s and 1970s. In the 1960s, the Chinese were on their own as ties to the Soviet Union were now severely restrained, and with Russian advise on computing virtually halted. However, in the critical period of the mid-1950s the Chinese had learned a great deal from the Soviets.

The Chinese had sent engineers to the Soviet Union to learn about computing as part of a much broader initiative to learn about many things, of which IT was only a small one. Over 15,000 technicians and other workers were trained in Russia, while some 11,000 Russian technicians came to China in the 1950s. Just as Soviet IT flows into China came to an end in the early 1960s, the same occurred for all manner of technologies, such as the acquisition of whole plants in the petroleum and manufacturing industries.[16] Put into the more parochial terms of IT, while embracing Soviet IT practices in the 1950s facilitated future adoption of computing in China, so too did the events of the Cultural Revolution slow profoundly adoptions in the 1960s and 1970s. Training in all kinds of non-agricultural technologies almost halted entirely, and education declined dramatically, while the millions of workers who went into the fields came out of industries that could have used computers. In the language of economics, the Cultural Revolution retarded "greatly Chinese efforts at technology acquisition," placing China's slow embrace of IT in the 1960s squarely at the doorstep of the economic problems of the day.[17]

The first delegation of experts from the CAS visited Russia in early 1953, and within a year Chinese mathematicians and engineers were studying Soviet computing practices and machine designs. In October, 1954, at a Soviet-sponsored industrial exhibit at Beijing (at the time also known as Peking) three Soviet computers were put on exhibit. Members of the CAS obviously showed interest in these, as did officials in various agencies of the Chinese government. In January, 1956, the Central Committee of the Chinese Communist Party had what was probably the highest level discussion about the role of computers in China held up to that time, resulting in computers being added to the list of priority technologies to be worked on during the first economic plan. Conversations with Soviet experts led to a series of recommendations and training programs in the mid-1950s in which Soviet economists, computer scientists, and diplomats advised the Chinese to build computers.

A second CAS delegation went to Moscow for training in September, 1956, and met with key Soviet computer engineers, including S. A. Lebedev, then the USSR's leading expert in the field. The delegation visited 21 Soviet organizations ranging from academies to educational institutions, and government agencies, where they saw machines, such as the M-20, the BESM, Strela, M-2 and M-3, and other systems, and just as important, toured the factory that manufactured the M-20. They attended 22 lectures on computing in Moscow given by various Soviet experts. Subsequently, CAS members held a series of public lectures in China about computing, relying on Soviet approaches to the design and use of such technologies. So far, there were no computers in China, but the small groups of experts were recommending that these be built, using Soviet designs.[18] No evidence has yet surfaced to suggest to what extent these individuals knew anything about Western designs, such as those from IBM in the 1950s, but we can presume that their base of knowledge rested overwhelmingly on that provided by the Soviets. They probably also had some access to a few Western technical publications, such as those of the Institute of Electrical and Electronics Engineers (IEEE) and the Association

for Computing Machinery (ACM), which could be acquired by Chinese em-
bassies, for example. Most important, as a consequence of this second visit,
the delegation now had enough confidence in their knowledge to recommend
that an Institute for Computing Technology be established, and that Soviet
versions of the BESM and M-20 computers be built.[19] The institute was established and quickly became the focal point for the earliest Chinese initiatives
in information technology.

A third delegation visited Moscow in the fall of 1957 to negotiate a formal
agreement for sharing of information on a variety of technological issues, not
just computers. On December 11 the Soviet and Chinese governments signed
an agreement defining the steps to be taken for sharing such knowledge. It
called for the exchange of experts by both countries, and most importantly for
China, Soviets resident in China. These experts brought with them technical
specifications for the BESM. The Chinese quickly sent 20 young engineers to
the Soviet Union to learn more, focusing on the BESM, BESM-II, and its relative, the 104. In China training of experts proceeded. Between 1956 and 1962,
the CAS trained some 800 people on Soviet-styled computing.[20] Construction
of the earliest Chinese computers began in March, 1958, at the 738th Factory
of Beijing, where the first machine was completed by September, a version of
the Soviet's M-3 system. The Chinese machine underwent several changes in
names that could confuse historians: *Youle* (means "we have a computer for
the first time"), *August 1st* (named after the founding date of the People's Liberation Army), and lastly, to the 103 in 1959. Small batches of these machines
were produced at the end of the 1950s and early 1960s. The 103—the most
important of the earliest devices—was essentially a copy of the Soviet M-3, a
small vacuum tube system. By September, 1959, the Chinese had finished construction of their first BESM system, which can be considered the country's
first large digital all-purpose mainframe system, named the 104. Experts
familiar with the technology likened it to an IBM 704 system, circa 1956.[21]

By relying on existing Soviet designs of computers, the Chinese were able to
build their own in a reasonably short period of time, harnessing Soviet expertise in design, manufacture, maintenance, and use of the M-3 and BESM-II
class systems. The earliest machines were turned over to the military for their
use, but even civilian applications were in evidence, such as weather forecasting, mechanical structural mathematics, and other calculations for meteorology, dam hydrology, architecture, design of electrical power grids. The
design of China's first atomic bomb was done with the aid of this class of computer systems. These were all applications of computing in evidence in the
Soviet Union and in the West at the same time, which leveraged the capabilities of existing technologies, such as calculating/computing power but limited
data memories. New computers appeared throughout the 1960s, including the
107 in April 1960, the 119 in April 1964, the 109 Model II in 1965 as China's
first transistor system, and, in 1967, the 109 Model III.[22] What is remarkable is

that all of this activity took place while the nation was going through a period of enormous change and churn, not the least, a great famine. These early systems roughly paralleled chronologically when other main-frame computers first began appearing in Israel, Poland, Denmark, Italy, and Japan, suggesting strongly that China's adoption of computers must be seen as part of a larger process of diffusion concurrently underway around the world.[23] The CAS had established a series of departments and institutes in the 1950s to facilitate acquisition and application of IT knowledge, with further diffusion of expertise occurring in the military and, after 1958, at two universities devoted to science and technology in Beijing and Shanghai. All these organizations collaborated in one fashion or another with each other during these early years. An early account of Chinese computing published in *Datamation* in 1968 stressed this feature of Chinese computing and other technological projects, such as development of their atomic bomb, functioning "as an integral part of the research and development process." Local successes with the development of such weapons were attributed to collaboration in part "to developments in the computer field."[24] But as Zhang Jiuchun and Zhang Baichun pointed out in their work on early Chinese computing, Soviet collaboration was even more important in the 1950s,[25] while in hindsight we can conclude with confidence that internal Chinese collaboration proved significant in the 1960s and 1970s, following the break with the Soviet Union.

The retirement of Soviet experts in 1960, followed by the failure of the Great Leap Forward, clearly slowed down development and deployment of Chinese computers. There are yet no definitive data as to how many were built and used in the 1960s, but Chinese and Western reports documented that new machines were developed, nonetheless, all through the 1960s. The *People's Daily*, a key source of publicly available information, continued to publish news stories and photographs of new systems into the 1970s. One Western expert, Russell Nyberg, concluded in the 1970s that Chinese computer technology lagged that of the West by as much as 15 years, and for those years we have no significant understanding of the extent of deployment, which was probably quite low since the earliest systems were mostly built one at a time.[26] That meant few systems were exactly the same; cost more to manufacture than a mass-produced one, and took longer to assemble. In short, there was no scaling up to produce many machines in a timely fashion. That reality proved to be an important brake on diffusion of IT in China, as it had for a while in Communist Europe in the 1950s and 1960s, and even earlier in the United States and Western Europe in the 1940s and 1950s. An internal report written at IBM in March, 1974, commented that "China is even less computerized than Russia . . . roughly one computer per one million of population" (versus one for every 35,000 residents of the USSR, one for every 2,000 in the United States). That calculation would have suggested there were roughly 820; but this calculus is at best a guess. The IBM report noted "no sales yet, but exploring possibilities," with problems similar to those the company faced in the USSR: "purchasing mechanism controlled by the state," "difficulty in dealing with end user as customary in the west," "shortage of foreign exchange," and "political uncertainties."[27]

American visitors to China in 1972 reported new machines, the models 111, C2, and 709—all third generation systems. In August, 1974, the DJS-18 was publicly displayed, a system that became a workhorse for Chinese computing. Other Western witnesses reported various related systems in the early to mid-1970s.[28] The first computer built with integrated circuits appeared at this time, the DJS-11 in early 1974. It provided multiprogramming and large storage (four 32k word memory units) and was used to replace analog systems that had done seismic data analysis. The new system was to be used for weather forecasting as well. In the 1960s and early 1970s, manufacturing of computers took place at two plants located in Beijing (Peking): the Peking Wire Communications Equipment Plant with 5,000 employees (circa 1975), where the DJS series were made, and at the Radio Plant Number 3 that employed about 1,000 individuals. Reports of other smaller production projects made clear that components were being manufactured at the Shanghai Radio Factory Number 3, at the Peking Institute of Computing Technology, at the Shanghai Computing Research Institute, and at an electronic manufacturing plant in Canton, where the DJS-17 process control computer was also being built. In all, this was a small production capacity for a nation as large as China, but clearly more than Western scholars normally thought existed.[29]

Given the turmoil of the 1960s and 1970s in which educated Chinese were sent to collectives to work in agriculture, the shuttering of universities (1966), and the persecution of scientists and technologists, how was it that China could design and build new computer systems? The answer lay in the fact that during those years, most of the work done on computer development was in service to the military, to a Chinese version of the "military–industrial" complex President Dwight D. Eisenhower so famously pointed out existed in his nation just a few years earlier. The connection of computer engineers to the People's Liberation Army (PLA) gave them protection and funding, sparing many from the ravages of the Cultural Revolution. These computer experts could continue to design and build systems. Not until 1979, after the worst of the political persecutions were over, did China begin to send students abroad to study all manner of technology and science, including computing, with results of that investment becoming evident in the mid-1980s, when IT experts were valued, their numbers expanded, and some managers were promoted into leadership positions who understood the potential value of computing. Thus, a small number (exact figures not available), of probably less than 2,000, trained by the Soviets in the 1950s, curated and exercised what little knowledge of computing existed in the country in the late 1960s and all through the 1970s while dozens of other countries were moving rapidly to develop and use knowledge about computing to apply IT across whole industries.

Problems with quality and insufficient production of enough reliable peripheral equipment plagued their early systems, slowing adoption of computing in China. These issues paralleled those clearly in evidence in the Communist economies of Eastern Europe at the same time. One reliable report (circa 1975) on Chinese computing, for instance, noted that "there are no discs in use except for a few units recently sold with a Honeywell Bull HIS 61/60 computer."[30] Punch cards were hardly in use even as late as the mid-1970s, with preference given to paper-tape input. However, magnetic tape technologies were now comparable to those in the West and in wide use.

Evidence of Western influences demonstrates that the Chinese were also learning from new sources after their breakup with the Soviets, beginning in the 1960s. For example, the European programming language, ALGOL-60, was one of the most widely used in China, also popular in Eastern and Soviet Europe too. In 1973, a Chinese delegation of 14 computer scientists visited several American companies and universities where they saw large mainframe systems and advanced applications. They visited IBM, DEC, Honeywell, CDC, Univac, Texas Instruments, Xerox, Hewlett-Packard, Xerox, Chase Manhattan Bank, Bolt Beranek and Newman, American Airlines, Bell Laboratories, 3M, Stanford Research Institute, and Fairchild Camera and Instruments—all important centers of contemporary IT developments and uses.[31]

By the mid-1970s, other circumstances were changing in China that facilitated the acquisition and use of IT. China had oil which it was selling, generating sufficient foreign reserves to make computers affordable, despite the IBM report suggesting otherwise. One byproduct of that available foreign currency was the acquisition of a few systems in the early 1970s, such as a complete data processing center sold by the French company Compagnie Generale de Geophysique (CGG) to do oil field computing, using two CDC CYBER 172 systems. Honeywell Bull reported making the sale of its exhibition system shown at a French trade fair in July, 1974, while minicomputer vendors in Europe, Japan, and the United States made similar claims in the mid-1970s.[32]

Because historians have accepted the notion that Chinese history underwent a significant shift after 1978, when the government departed from traditional Soviet-style management of its economy, it is worth pausing to note the large number of IT-related activities that had occurred in China prior to its substantive takeoff in computing. Table 9.2 catalogs some of the more important events from the mid-1950s to the mid-1980s, while table 9.3 lists some of the earliest Chinese computers and their estimated dates of introduction. The conclusion one can draw is obvious: there was much activity. However, we do not yet know how many copies of these machines existed; extant evidence suggests that even in the early 1970s, diffusion of IT had only reached levels

Table 9.2 MAJOR EVENTS IN CHINESE INFORMATION
TECHNOLOGY, 1956–1985

1956	Establishment of Institute of Computing Techniques at Chinese Academy of Sciences
1958	*August 1* computer unveiled in Beijing
1959	Computer research institutes established in Shanghai, Tsinan, Shenyang and Chengtu and computing projects at universities of Beijing, Shanghai, Nanching, Shonyang and Shantung
1962	Manufacturing begans of DJS-1 computers
1962	Soviet technical support ended
1963	Manufacturing begins of DJS-2
1964	First use of semiconductors and first transistorized computers built
1966	British ELLIOTT 803 installed for medical research
1967	British ICL 1903 and ICL 1905 installed in China
1968	DJS-7 exhibited for first time
1970	DJS-6 exhibited for first time
1970	First use of Chinese integrated circuits in Model 111 computer
1974	Honeywell Bull sold HIS 61/60 to Chinese government
1974	DJS-18 displayed for first time
1977	Peking Wire Factory began manufacturing computers
1978	Chinese Communist Party launches new era of reforms
1983	First Chinese super computer developed
1984	Development of the TQ-0671 microcomputer system
1985	Development of NCI-2780 super mini-computer

Source: Modified from data in Bohdan O. Szuprowicz, "China's Computer Industry," *Datamation* (June 1975): 84; Harvey L. Garner, "Computing in China, 1978," *Computer* (March 1979): 81–95; U.S. Congress, Office of Technology Assessment, *Technology Transfer to China*, OTA-ISC-340 (Washington, D.C.: U.S. Government Printing Office, July 1987): 96.

evident in Western Europe in the late 1950s and, perhaps, those in Eastern Europe in the early to mid-1960s.

The evolution of computing prior to 1978 has sometimes been described as sharing contradictory aspects: extensive investments in R&D resulting in substantive productive capacities in many types of electronics aimed largely at satisfying the needs of the military, yet the lack of adequate technological innovation to meet the overall needs of the nation. In this and in so many other industrial projects, Chinese operations mimicked Soviet models of behavior: verticalization, barriers between R&D and manufacturing, limited dialogue between consumers and developers of products, and disconnected operations among suppliers of peripheral equipment, mainframes, and software, and inadequate technical support. In short, cross collaboration did not exist to the extent evident in the West. Because of the military orientation of these early systems, officials paid less attention to

Table 9.3 INTRODUCTIONS OF KEY CHINESE COMPUTERS, 1958–1978

Year	System	Comments
1958	August 1	Similar to URAL-1
1962	DJS-1	Copy of Soviet M-3 (circa 1953)
1963	DJS-2	Similar to BESM 2 (vintage 1959)
1965	109C	32K memory
1965	DJS-7	Small ferrite core system
1966	DJS-6	Up to 32K memory
1966	DJS-21	Solid state, ferrite core
1968	C2	Solid state
1970	111	Integrated circuits
1971	709	Integrated circuits, expanded C2
1973	DJS-17	8K system built in Canton
1974	DJS-11	IC, semiconductor
1974	DJS-18	IC, built in Peking
1976	DJS-120	32K memory
1976	DJS-130	64K memory
1978	DJS-210	Similar to other DJS systems
1978	DJS-240	Built at various locations
1978	DJS-260	Builder unknown

Source: Modified from data in Bohdan O. Szuprowicz, "China's Computer Industry," *Datamation* (June 1975): 85; Harvey L. Garner, "Computing in China, 1978," *Computer* (March 1979): 81–95.

their costs or to improving their efficiencies than might otherwise had been the case if these had been built for the private sector. Production remained specialized and was done in small batches, which probably accounts for why there were so few systems evident in accounts of early Chinese computing. Applications privileged scientific "number crunching" systems, favoring systems that had high speed yet small memories, with less emphasis on advancing developments of peripheral equipment. By mid-to late 1970s, one estimate held that just over 200 different models of computers had been built, almost as one-of-a-kind; only 10 were manufactured in quantities of 50 like systems or more. The military orientation also limited access to foreign IT that might have been adopted (or imported) in the 1960s and 1970s due to Cold War issues that blocked Western and Soviet imports into the country.[33]

While the story just told about the rise of the supply side of computing might lead to the conclusion that China's record of diffusion was greater than might otherwise be assumed, the reality proved quite different. First, the number of projects just discussed was quite small. Second, the systems built in the 1960s and 1970s were few, no mass production here. Third, and most important, the structure of organizations, industries, and the economy at large were not yet ready to embrace computing. Manufacturing plants in many industries, supplied with some Soviet and even Western machines, were either small or highly unproductive. One student of the economy of those years commented that "Chinese factories are notoriously inefficient. Machines sit idle, finished products accumulate in large inventories, and production lines move slowly and stop frequently because of shortages of parts."[34] These mimicked problems evident at the same time in Comecon countries in Eastern Europe. After acknowledging that Chinese officials understood this problem, the same observer criticized them for "barely" doing anything about it: "Factories in China are automated only in the most rudimentary sense of the word"; additionally, "no facilities are computer-regulated from start to finish."[35] In fact, it was not until the National Science Conference held in March 1978, did officials so explicitly announce that computing had to be embraced and used. One observer dismissed computing with the quip that "computers are still little more than handcrafted toys in the P.R.C."[36] But, as our supply side of the story suggests, at least a start had been made with some projects and with growing emphasis on training people in modern industrial practices and technologies.[37]

Then came 1978. The Cultural Revolution, and its attendant famine, was followed by political protests that resulted in an estimated 10 million people dying and as many as 100 million denounced, retrained, or otherwise persecuted. Mao had died in 1976, followed by political churn. The Chinese Communist Party (CCP) was clearly in trouble due to its recent excesses and, so, in 1978, its leader, Deng Xiaoping, decided that a new course of economic and social action was needed. He initiated a large number of reforms aimed at creating a socialist market economy. Early elements of his reforms included making the state bureaucracy more professional, re-establishing the rule of law, and imposing regulations on what would turn out to be an emerging market economy.[38] One of the most important of the early reforms was to allow farmers to sell on the open market any food they raised over and above the state-mandated quotas. Productivity in food production grew dramatically very quickly, addressing the problem of insufficient supplies of food while reinforcing the notion that market-based economics might be the way to go in other industries, including manufacturing.

While it would be easy to jump to the conclusion that a new era in computing began that year with Deng's various declarations and actions, many of the seeds of China's enhanced use of computers in evidence in the 1980s had already been planted. For example, after the cutoff of Soviet aid and dialogue in the early 1960s, the Chinese made a concerted effort to acquire Western knowledge about computing. A delegation of members of the IEEE Computer Society who visited China in September–October 1977 for 21 days produced a very detailed report of their experiences in which one member of the American delegation reported that at Tsinghua University, its "library appeared to be well stocked. The technical periodical room—available only to faculty—contained about 400 journals. Some were original copies and were current, but most (including all of the IEEE publications) were copies produced by the Chinese government. These were less current, since the copying process takes about six months."[39] At Peking University, the Americans learned that its library had 3.1 million books of which 400,000 were in English, presumably some about computers. They reported that five universities were just starting computer science departments and were already training students. By the time the Americans had arrived, the Chinese had a clear understanding of much Western technology—a result of efforts undertaken in the 1970s—and as one visitor on the trip noted:

> The knowledge displayed by a number of the Chinese technologists was impressive. These individuals spoke fluent English and were conversant with current US computer research and development. Many technically competent individuals may have escaped our attention simply because they could not communicate directly in English.[40]

Access to published materials, understanding of foreign languages, and work in computer science departments were all essential elements for the diffusion of Chinese computers, as these were in the Soviet Union and more generally in the West.

This delegation concluded, however, that manufacturing and diffusion still lagged behind that of many countries, and that machines in use were comparable to Western systems circa "early-60's" (second generation U.S. systems). It estimated that as of 1976 there were some 1,000 in total in the country.[41] As in the Soviet Union, one agency had ostensibly the authority to set overall direction for the development and manufacture of computing, the Fourth Ministry of Machines Building, which predated the post-1978 reforms. Yet, there also existed the highly influential Peking Institute of Computer Technology of the Chinese Academy of Sciences, which had been functioning for some 20 years. In 1978, it had a thousand employees doing research, product development, manufacturing, and providing service-bureau services. Finally, it should be pointed out that "mass production" of 50 or more computers a year was already underway at the Peking Wire Factory, beginning in 1977.[42]

CHINA EMBRACES INFORMATION TECHNOLOGY, 1978–1980S

Asian approaches to technological developments are clearly evident in Chinese actions, particularly in regard to reliance on both a formal plan and public statements of intent, providing focus and guidelines for investments. At the Chinese National Conference on Science and Technology, hosted by the national government in March, 1978, in Beijing, a broad plan for all manner of technological and scientific plans was presented, best known as the "Outline National Plan for the Development of Science and Technology, Relevant Policies and Measures." One Western commentator noted two years later, "the Plan launched a momentous program—comparable in ambition to the post-Sputnik national science program in the U.S.—that within a few short months was already being implemented at the lowest organizational levels throughout the Chinese provinces."[43] Topics covered in the plan included agriculture, materials, space, lasers, high-energy physics, genetic engineering, and computers. There was a sense of urgency to get going in implementing this plan, borrowing technological know-how from abroad. In the area of computing, it would be in developing manufacturing capabilities in integrated circuits and microcomputers, and as the epigraph starting this chapter noted, quickly.[44] In the area of integrated circuits, China had been manufacturing devices since 1968, and had built up enough experience and appreciation for this IT building block to make it an important priority by 1978. The Chinese had switched to Western sources for the devices and necessary know-how following the break with the Soviet Union, saving, in the words of one American government official, "several years in bridging that technical gap."[45] These objectives were reaffirmed in new versions of the plan in the mid-1980s, with particular emphasis on the use of all manner of technologies by industry.[46] At the core of China's strategy, however, was the notion of "buying hens, not eggs," that is to say, acquiring knowledge about how to design, manufacture, and use IT, and not simply to buy Western or Japanese computer products.[47]

A report to the U.S. Congress several years later described uneven progress, but progress nonetheless. Components were still hand soldered on circuit boards twenty years after Western producers had moved to wave soldering techniques; on the other hand, the country had 10 factories producing computers in 1983, making a range of computers from PDP-11 class systems to IBM S/360 style computers. Agreements had been signed to acquire technologies from the Japanese company NEC and with the American firms Univac, Wang Laboratories, and Honeywell.[48] These transfers of technologies and goods were facilitated by changes in American foreign policy aimed at driving a wedge between China and the Soviet Union during the Cold War, a thaw that began as early as 1972, with an initial change from a total ban of exports of technologies to China to a partial one. Sales to China in the 1970s, however, were largely limited to agricultural goods, basic commodities, and some industrial goods. By the end of 1983, the American government had established broader guidelines permitting some export of computers, computerized instruments, microcircuits, and various other electronics to China. Sales to China across all these fronts (not just for IT-related goods) doubled between 1983 and 1985, while applications to export went from 4,300 in 1983 to 10,200 in 1985.[49]

Thanks to the local manufacture of systems and imports from other countries, the population of computers in China began to rise in the 1980s, albeit slowly. As late as 1993, one reliable assessment placed the total population of mainframes at 400 systems, although in hindsight that seemed a low estimate as it was not clear at the time how many were in use by the military, or how local manufacturing was developing, while the majority of these known 400 systems were imported.[50] IBM was an early provider, sending its first system to China in 1978 (a system 370 Model 138) to the Shenyang Turbine Factory, which was subsequently upgraded twice, first to an IBM 4331, then in 1986 to an IBM 4381—all state-of-the-art systems, used for computer-aided design (CAD) and to operate numerical control (N/C) equipment, both current applications worldwide of the time.[51] In the early 1980s, Japanese vendors entered the Chinese market, most notably Hitachi, selling several copies of its M series to such diverse organizations as Northern Jiaotong University in support of a railway system in Beijing, and to the Central Bureau of Meteorology for weather forecasting. One assessment held that mainframe sales to China in 1983 had reached $76 million, with IBM the primary supplier. Other participants included Fujitsu, Hitachi, Unisys, and CDC. U.S. exports of mainframe digital computers remained modest: $5.2 million in 1981, $11.3 million in 1982 and similarly in 1983, $25.3 million in 1984, climbing substantially beginning in 1985 to $80 million in products.[52] In short, the market was still small by Japanese and American standards of the time (6,000 Japanese installed systems, 18,000 in the United States, and some 400 in China).[53]

Banks, large national government agencies, some universities and the military were the primary users of IT in the early 1980s. However, there were problems with uses similar to those encountered in both the Soviet Union and Cold War Eastern Europe. Numerous systems were underutilized, installed to make an agency look modern and progressive for political purposes, many without having an underlying operational or economic justification. As these systems aged, thus became more difficult to maintain and operate, the less incentive there existed to use them even for a few hours per day. The same problems faced by potential users in Eastern Europe with software also plagued large systems users in China: lack of application software, good programmers, and technical support.[54]

The minicomputer market in the 1980s was larger, in existence since the early 1970s, consisting of some 10,000 installed systems by 1993. Foreign systems included DEC's VAX and PDP lines, H-P's various products, IBM's 32s, 36s, 38s, and AS/400s, and locally made systems as well. By the early 1990s, some 60 percent of all installed minicomputers had come from DEC, accounting partially for the rapid deployment of so many of these systems in China.[55]

Desktop workstations using the UNIX operating system became one of the most rapidly developing markets in China by the end of the 1980s, along with PCs. Yet as of 1990 there were only about 1,000 installed. Then that population of machines began growing at 30–40 percent each year during the early years of the new decade.[56] The largest population of computing devices comprised the PC, with some 1.4 million installed by 1993, jumping to over 2 million the following year.[57] Chinese demand for this technology mimicked what was happening elsewhere in Asia, most notably in South Korea and Taiwan, and in other developing economies, such as Brazil. As in those countries, installations of PCs were moderate in the 1980s as they were new, not yet well known, and expensive (see table 9.4). Take-off in PC sales began in 1991, when demand increased sharply, leading to annual growth rates hovering in the 40 percentile range in the early 1990s. In 1994 sales exceeded 718,000 units (US$1.3 billion), suggesting a total population of PCs in the country of around 2.4 million systems.[58]

These were not all imports, because there were four local manufacturers supplying 25 percent of the market: China Great Wall Computer Group Company, Beijing Legend Computer Group Company, Langschao Electronic Information Industry Group Company, and Changjiang Computer Group Company, the latter which accounted for 75 percent of all systems made in the country. The most visible foreign suppliers included IBM, Compaq, AST, HP, and Acer—all with reputations for high quality, yet expensive products.[59]

BIRTH OF CHINESE SOFTWARE, 1960S-EARLY 1990S

But it all takes software to work, and despite the difficulties Chinese users faced with software, their use of this essential class of IT increased over time. As in the rest of Asia, software products and markets emerged after local

Table 9.4 PC INSTALLATIONS IN CHINA, 1981–1994, SELECT YEARS

1981	1,400
1984	45,800
1988	80,000
1990	85,000
1991	180,000
1992	250,000
1993	450,000
1994	718,000

Source: Data in Jeff X. Zhang and Yan Wang, *The Emerging Market of China's Computer Industry* (Westport, Conn.: Quorum Books, 1995): 33–35.

hardware markets and production, a sector of the Chinese IT world still characterized as in its infancy by observers even in the early 2000s.[60] Chinese and foreign suppliers of computers included operating systems, compilers, programming languages, various software utilities, and application software as part of the sale of hardware. As a result, until the 1990s software was not considered separate products—an independent economic asset (commercial goods) apart from hardware—by either vendors or consumers of IT, just as in most other parts of Asia through the 1960s to the 1980s. Application software was written to order in-house, creating tailored made software perfect for one agency or company but that was not necessarily marketable to other firms, a pattern of development also in wide evidence, for example, in Japan and South Korea.

Then there was the notorious lack of appreciation of the importance of intellectual property protections on the part of both Chinese officials and users that inhibited development of software as products for sale. This lack of protection from pirating plagued all software and digital products right into the twenty-first century.[61] The first modern patent law did not go into effect until 1984, the first copyright law not until 1991. Until 1991, therefore, there was not even the hope of any protection, so understandably no local software industry in the 1970s or 1980s. The history of software in China as marketable commercial products that could be diffused widely is part of post-1990 events.

Systems in use during the 1970s and 1980s, nonetheless, required software to operate. Legally and illegally acquired IBM systems in these years were equipped with IBM's operating systems, such as MVS and the globally popular transaction management system, CICS. Many mid-range systems from IBM, DEC, and Unisys often had locally developed operating systems, or, copies originally written by the manufacturers of the hardware. UNIX had been mandated as the operating system of choice by the government throughout the 1980s and 1990s with the result that variations of these were in wide use; by 1993, over 40,000 copies.[62] Many applications written to work in UNIX were more portable than those developed for large mainframes. Many of these software packages were developed under the guidance and funding of government agencies. For PCs, MS-DOS rapidly became the most widely adopted operating system before 1995. Too many PCs were too small to handle the more advanced Windows operating system from Microsoft, thus the older, smaller package was used for more years than in such countries as South Korea and Japan. Additionally, development of so many application software packages under MS-DOS precluded users from converting to other systems.

As in South Korea, Taiwan, and Japan, there was also the issue of Asian languages to contend with in all systems. Much work was done at various institutes and universities to develop printing capabilities in the language and application software which, often, was highly dependent on what operating system they ran.[63] Vendors of hardware and software products also attempted to deal with the issue. In the world of PCs, Microsoft took the action of developing a Chinese version (traditional Chinese characters) of Windows 3.1 to encourage sale of its new package, introducing it in June 1993, although company officials were aware that the market for this version would likely be centered in Taiwan and Hong Kong, not in the PRC.[64] As PCs became more affordable, beginning in 1993 (40 percent less expensive in 1993 than in 1983), the market for more advanced products from Microsoft and others finally took off.

Unlike in some other Asian localities that could work in English, such as Singapore and Hong Kong, the Chinese could only use their own language; therefore, widely used applications had to be developed in Chinese, most notable word processors. In fact these packages were ubiquitous wherever there were any forms of computing in the 1980s and 1990s. Fangzheng Group Company of Beijing University produced word processors, and desktop publishing systems, with 70 percent of the nation's 2,000 newspapers using these and some 3,000 other institutions and companies by the early 1990s. Additional products in Chinese for spreadsheets and database management came from the same group, beginning in the mid-1980s. Suntendy Electronic Institute, which started up in 1992, produced Windows-based Chinese language products, claiming it had 300,000 customers by the end of 1994. By then several dozen software houses produced products for the Chinese market that can be described as office applications.[65]

Commercial applications' software was well received by mid- to large enterprises by the early 1990s, and so they sought out Chinese language editions where possible for such things as accounting and manufacturing tools. One survey from 1993 of 368 large state-owned enterprises suggested that 18 percent used full MRP-II systems, while another third used some modules of the MRP-II manufacturing system. On the other hand, the other 52 percent reported not using computers at all for any manufacturing processes, which by the standards of the early 1990s in Asia, was behind conventional practice.[66] Finally, with respect to database management software, Chinese products began appearing in the 1980s, largely for PCs, and a class of products that expanded more fully during the 1990s. The American company Oracle became the first foreign enterprise to establish a local software subsidiary, completing the transaction in 1986, its arrival date suggesting how late the local market was for such products. It became the dominant vendor of such software in the late 1980s and early 1990s.

Observers of the software environment in China commented in the mid-1990s on the situation as it stood at that time:

> Most Chinese users are busy installing the new computer systems and trying to get them to work properly in their changing business environment. At this stage, the value of the systems development and management tools has not been recognized by the ordinary users.[67]

It was a complaint that had been voiced many times since the 1970s. Even in the mid-1980s, when it had become clear for years that a problem existed with software development, there were only 30 software development centers in China, and an estimated 10,000 people who could write software, out of an estimated employment base of 1.4 million workers in the electronics industry. Prior to 1984, those centers were fewer in number and there was no coordination of activities even within the industry.[68]

UNDERLYING STRATEGIES FOR IT DIFFUSION IN THE 1980S AND EARLY1990S

Part of the reason why Chinese computing took off in the second half of the 1980s can be accounted for just through the sheer managerial and operational efforts it took to implement the plans initiated in 1978, such as allocating budgets, setting up laboratories and factories, and expanding training of experts on IT. But public officials also implemented a combination of state-ownership (hence government funded) efforts and processes that encouraged decentralized planning and implementation of the national plan, and as the data on installations suggested, the strategy worked to get things started. So too did the simultaneous application of an import strategy by which valuable foreign exchange was allocated to the acquisition of foreign technologies in these early years. The cumulative effect of these various seeding activities of the 1980s was that by the early years of the next decade individuals could take personal and institutional initiatives to embrace IT. As two observers at the time noted:

> Chinese customers are becoming more mature. Thanks to the open door policy
> and the rapid development of information and communication technology, they
> can obtain up-to-date information on technical and marketing developments in
> the rest of the world much faster than ever before.[69]

Coupled to an embryonic competitive market, China had the makings of an attractive buyers' market.

The successes that began to be felt in the early 1990s in the use of IT were long in coming, however. Local conditions constrained adoption of IT for many years. We need again to mention education. One consequence of the Cultural Revolution was that the shuttering of schools and universities had a devastating effect on education. Success in careers became more dependent on loyalty to the party in the 1960s and 1970s than to skills and meritocracy, with the result that as new IT became available in the 1980s many middle managers who would have been called upon to make decisions to acquire IT simply were not capable of appreciating adequately enough why they should do this, let alone how to use the technology. An inadequate telecommunications infrastructure also slowed adoption of IT in the 1980s, an issue discussed more fully below. The Asian focus on hardware and insufficiently on software added another cause for slow adoption of IT. These were all well-known problems which senior public officials began to address in the 1980s and continued to focus on right into the new century.

EMERGENCE OF THE CHINESE IT INDUSTRY

Before exploring more fully results in the 1990s, it is essential to understand early results of post-1978 reforms. We can tally some of the results achieved in the second half of the 1980s when IT finally took off in measurable terms in China. By using the same types of metrics as we applied to other countries, one can see that Chinese use of computing was more extensive than might otherwise be thought, even though smaller proportional to the size of the country's population and economy than in many other nations. On the demand side, as in all of Asia, there was never a computer industry, rather an electronics industry, which produced many things from radios and television sets to computers and semiconductors. Sales grew in all facets of the industry, beginning in the 1980s, although IT always remained a small part of the larger electronics industry.

One IT industry survey completed in 1991 began with a strong, yet accurate, declaration: "In the ten years since the beginning of China's post-Mao modernization efforts there is probably no industry that has been transformed as radically as the computer industry," citing "new organizational entities, both state and privately run," manufacturing and using these technologies. The same observers noted that in 1986, the total value of computer sales in China had reached nearly one billion dollars (US$913 million/RMB 3.4 million), representing a growth rate since 1981 of 45.5 percent compound, and with volumes essentially taking off in 1985 over prior years. Two-thirds of all systems installed in these early years of IT usage came from other countries with the exception of PCs in which the reverse circumstance held, with two-thirds of all systems manufactured in China.[70] IBM shipped between 20 and 25 large mainframes to China each year in the 1980s, while several thousand IBM PCs also made it into the country, often, however, through "gray markets," and not directly through the American firm.[71]

The military dominated the local industry and its products in the 1970s, but by the mid-1980s there existed an additional 460 organizations developing and making IT systems, with roughly half offering software and other IT services, and another 181 making equipment and over 30 serving as research institutes. A survey of 430 of the 460 enterprises reported that combined they employed 84,882 people, of which 75 percent were involved in manufacturing activities.[72] Prior to Mao's death in 1976 most manufacturing and use was either done or controlled by the Ministry of Electronics Industry (MEI). By the mid-1980s, many of these functions had also spread to other ministries as well, even to provincial and city governments.[73] Their expansion into IT projects was gated, of course, by the supply of properly trained experts in the field, which remained at levels far below what was needed.[74] However, observers of the Chinese situation in the 1980s were still commenting on how local supply chains operated inefficiently. For example, one American report noted that "the mentality of solving supply problems through vertical integration, a legacy of the Soviet-style economy, is deeply entrenched and biases solutions to the supply problems in the wrong direction," suggesting that products and needs were not being matched adequately, with the Chinese continuing "to resist the kind of specialization that would lead firms to seek product niches and search for the right technologies to achieve some sort of comparative advantage."[75] By the mid-1980s it was becoming clear in other Asian economies that specialization might be the way to go, and by the early 1990s had proven to be the correct approach in such places as Hong Kong and Taiwan, localities watched closely by mainland Chinese officials.

By the mid-1980s some non-state enterprises began to appear as well, approved by provincial and national government ministries, but not necessarily funded by government. Some were technical collective enterprises, which represented embryonic commercial enterprises that would become more important, but not until the 1990s.[76] These should not be confused with another cluster of institutions called computer group corporations dominated by one organization that planned what was to happen (usually a government agency) and with other entities attached that made, sold and used its products, implementing the core plan. They tended to have strong regional associations in their beginning, such as the Great Wall Computer Group Corporation, founded in 1986 in Beijing. These were not small enterprises; Great Wall had 65,000 employees by the end of the 1980s and 67 subentities, often called members.[77]

The economic and regulatory environment in which they functioned changed in the mid-1980s as officials shifted slowly away from neo-Soviet styles of management. The U.S. Office of Technology Assessment reported in 1987:

> The current direction of Chinese institutional experimentation involves greater use of decentralized market mechanisms to stimulate efficiency and innovation as well as more attention to central planning controls. The basic principles of socialist ownership are to be maintained, as are socialist fairness in distribution and socialist welfare and social security principles (although the mechanisms for providing the latter may change). The sanguine view of China's future assumes that these often contradictory elements in China's institutional quest can, in principle, be reconciled into a Chinese form of market socialism.[78]

What is quite clear from the evidence we have (circa mid-1980s) is that the Chinese government was already reforming the economy in ways that departed from the Soviet model many years before the downfall of the communist states in Eastern Europe.[79] That action stands in contrast to the widely held view that China began its journey in becoming a more market driven economy *after* the demise of European communism. The evidence presented here with respect to the Chinese experience with computing suggests that reforms had been underway toward liberalizing the economy as early as the first half of the 1980s. To be sure it was a painfully slow and uncertain process, one that took over two decades to unfold, but, nonetheless, facilitated the diffusion of IT as part of a larger process of economic and technological transformation.

Many groups participated in that effort, such as the lead ministry for much of this effort the Ministry of Electronics Industry (MEI), the central governmental management entity providing the most guidance and funding in these early years. It set policies for the entire electronics industry, not just for ICs or computers, implemented regulations, designed and manufactured products, and directed scientific and product research (e.g., through the Computer and Information Bureau).[80]

By the end of the 1980s, China also had user groups that interacted with vendors and MEI. These included the China Computer Industry Federation, the China Computer Users' Association, and the China Computer Software Industry Federation, among at least a dozen such institutions. An industrial computer fair held in June 1986 in Beijing, gave a hint of how these users were deploying computers. We know that applications in the 1970s and 1980s were overwhelmingly scientific and military, so the public display of 1986 is an early quantifiable window into civilian uses. Table 9.5 lists the top 15 places in the economy in which IT was being used and the number of cases (applications) exhibited. This list presents no surprises as the government's economic development focus had been on heavy industry and agriculture. What is more interesting is what appeared lower in the rankings. Textile was ranked 16th (37 applications), banking at 18th (38 uses), public security was listed as 20th, and economic information management 23rd, with 15 applications used to help manage the economy.[81]

This second tier of users suggests as much about the sophistication (or lack thereof) of some industries at that time, such as banking. As of 1987, there were some 10,600 computers of all sizes from PCs to mainframes in the country with 30.4 percent in government agencies, another 25.7 in industry and mining, another 31.5 percent in scientific and technical institutes, and the remaining 12.4 percent in higher education. This pattern of distribution follows a fairly typical pattern of deployment during those years in most countries in the world with the exception of the USA, where a larger percentage were installed in the private sector.[82] However, China's willingness to accept

Table 9.5 NUMBER OF APPLICATIONS OR PLACES IT WAS USED IN
THE CHINESE ECONOMY, CIRCA JUNE 1986

Frequency	Application or Industry	Number of Applications
1	National defense	195
2	Water conservation and power	141
3	Machinery	121
4	Oil and petroleum chemical industry	94
5	Shipbuilding	93
6	Electronics industry	90
7	Metallurgy industry	89
8	Railways and transportation	75
9	Scientific research	71
10	Education	64
11	Materials supply	56
12	Communications, broadcast and television	46
13	Geology and mining	42
14	Government administration	42
15	City construction	38

Source: International Data Corporation, "China Computer Industry Review and Forecast, 1986–1991"
(Framingham, Mass.: IDC, 1988): 7, CBI 55, Box 60, Folder 24, Charles Babbage Institute, University of
Minnesota, Minneapolis.

Western technologies may have been the most important influence in causing
adoption of IT to increase so much in the 1980s, particularly for microcomputers. Liu Jianfeng, Vice Minister of the Electronics Industry, acknowledged
in 1985 that a good third of all Chinese electronics manufacturing companies
had embraced Western technologies.[83]

Users did not work in the best environment, nor did the economy as a
whole appear optimized for IT expansion. American analysts reported to the
U.S. Congress a number of constraining factors: "Present problems in
the Chinese computer and electronics industry include lack of experience
in the field, technology not up to international standards, and too little use
of Chinese products."[84] While the government pushed hard for local production across the entire electronics industry in the 1980s, "they are not satisfied
with their efforts, since they have imported much technology at considerable
cost and their products, especially computers, are not up to international
standards," in particular for the most complex products, such as mini and
mainframe computers.[85] Yet, another China-watcher, Joseph Y. Battat, concluded at the same time that the Chinese government was "serious" about
promoting the use of IT, despite the challenges faced.[86] Officials made their
determination clear through repeated public policy pronouncements in the
1980s.[87]

Still there were many problems facing the manufacturer of IT goods. For example regarding semiconductors, production methods evident to Western observers visiting Chinese plants in the 1980s, resembled those the United States and Japan operated in the 1950s and 1960s. Inadequate environmental controls, lack of technical expertise, unreliable supplies of both raw materials and even of electricity hampered production.[88] A second problem was supplying adequate amounts of application software for those areas the government deemed critical to the nation, largely in management applications, but also in banking, railroad operations, weather forecasting, and the operations and management of the electrical grid among others. A third challenge that surfaced each decade was the issue of how to develop and deploy some sort of easy-to-use ways for entering and reading data written in Chinese characters, a problem not resolved effectively until the 1990s, after which popularization of computing expanded substantially. A fourth one, mentioned several times in this chapter, was the nearly non-existent support for end users by vendors of hardware and software. One student of the problem described it this way:

> Without technical support and marketing, computer purchasers were pretty much on their own to learn how to use their machines and to develop their own applications. Nevertheless, many managers rushed to buy or to import computers when government and party leaders proclaimed that computers would be an essential part of China's modernization.[89]

This rush to the computer helps to explain why so many reports circulated in the 1970s and 1980s about underutilization of installed IT. The same observer, writing for the Harvard Center for Information Policy Research, opined that there remained many inhibitors to the further diffusion of IT in the country:

> With many managers appointed for their political reliability rather than their managerial or technical expertise, with no profit incentive, with very little advertising allowed, with no marketing or sales networks, with institutional barriers to lateral information flows between similar enterprises, with no patent law in place until 1985 and no copyright law to protect computer software developers, there were few, if any, incentives built into the Chinese system to promote the spread of good ideas.[90]

It was a situation that only began to change slowly during the second half of the 1980s.

Table 9.6 provides an estimate of how various groups distributed these systems by type during the 1980s, suggesting the relative diffusion patterns that continued into the 1990s. These data reflect a low national GDP, a limited affordability for systems, and an underdeveloped economy from the perspective of IT use. In other words, it made more sense for more smaller systems to be installed earlier than larger ones, particularly since China's adoption of computing took off at a time when there were available several options in size of systems, from PCs to large mainframes. Table 9.7 presents important evidence on the rate of diffusion, in part made possible because the Chinese government reversed its strategy of trying to be independent of foreign technology to one of becoming temporarily more dependent on it. The large volume of imports contributed directly to the increased diffusion (think availability) of IT in these early years. While a busy table with numbers, it makes clear that over time the Chinese were able to increase their domestic production with the volumes achieved suggested in the previous table (9.6). That increase in supply made it possible for Chinese officials to maintain confidence that in time they would be able to break their dependence on technological developments on other countries. Finally, table 9.8 illustrates with the 15 largest manufacturers of computers that China's capacity to make its own systems had expanded rapidly in the 1980s. To be sure the data cannot be assumed to be absolutely accurate, but one can accept that they are generally correct in representing relative proportions across technologies and extent of relative adoption over time.

CHINA'S CONTINUING EVOLUTION TO AN OPEN ECONOMY, 1986–1990S

Despite the progress made in economic development in the early to mid-1980s, and certainly in starting to make and use computers, among many technologies, much work had yet to be done to reform China from a centrally planned socialist economy into some sort of mixed market one. So, reforms in the second half of the decade included another round of changes in national technology policies designed to speed up the "catch up" with the West, and continued into the 1990s. In combination with economic reforms, these actions led to enormous growth in the economy into the new century. The end of the Cold War made it easier to import technologies of all kinds, to encourage foreign capital investments, and to establish new alliances with non-Chinese vendors. Almost every observer of the Chinese scene of the second half the 1980s and the first half of the 1990s used the word *pragmatism* to describe government policies. Officials borrowed strategies from neighbors, specifically learning from Japan, South Korea, Taiwan and Singapore, and like those nations, embraced "informatization" of many industries, beginning as early as 1984, with the realization that it would take years to "catch up."[91]

Table 9.6 DISTRIBUTION OF COMPUTERS IN CHINA, 1981–1986 (NUMBER OF UNITS SHIPPED)

Year	PCs	Minis	Mainframes	Total Units
1981	1,400	240	32	1,672
1982	7,208	574	45	7,827
1983	20,792	657	69	22,518
1984	45,800	770	92	46,662
1985	54,200	930	101	55,231
1986	59,600	593	87	60,280

Source: Drawn from data in International Data Corporation, "China Computer Industry Review and Forecast, 1986–1991" (Framingham, Mass.: IDC, 1988): 9, CBI 55, Box 60, Folder 24, Charles Babbage Institute, University of Minnesota, Minneapolis.

Table 9.7 DOMESTIC VS. FOREIGN SHIPMENTS OF COMPUTERS IN CHINA, 1981–1986 (EXPRESSED AS PERCENTS OF TOTALS BY MACHINE TYPE)

System Type	1981	1982	1983	1984	1985	1986
PC domestic	36.0	20.6	26.1	26.2	56.5	65.7
PC foreign	64.0	79.4	73.9	73.8	43.5	34.3
Mini & mainframe domestic	68.8	38.9	49.6	44.2	27.7	39.9
Mini & mainframe foreign	31.2	61.1	50.4	55.8	72.3	60.1

Source: Drawn from data in International Data Corporation, "China Computer Industry Review and Forecast, 1986–1991" (Framingham, Mass.: IDC, 1988): 11, CBI 55, Box 60, Folder 24, Charles Babbage Institute, University of Minnesota, Minneapolis.

As previous tables demonstrated, diffusion of IT began in the 1980s to ramp up, a process that continued in the 1990s. Between 1989 and 1992 production of Chinese computing equipment grew at an annual rate of about 30 percent. Exports were few, however, with most consumption going to China's own companies and joint ventures. The central pattern of behavior, however, was the continued role of government interventions to direct the way forward for the economy, industry after industry, while simultaneously promoting the growth of a competitive market economy.

One should approach the numbers detailing the volume of machines made and installed with some caution, however, not so much because the data produced by government agencies were suspect—a problem, to be sure—but for more serious reasons evident in other countries, most notably in Communist Europe in the 1960s-1980s. Reports from visitors to China in the 1970s and 1980s made evident that many computers were not used. They sat idly in a

Table 9.8 TOP 10 CHINESE COMPUTER MANUFACTURERS, 1987, AND THEIR DATES OF INITIAL PRODUCTION (RANKED BY VALUE OF OUTPUT IN RMB M)

Rank	Name and Year of Founding	Value of Output
1	China Computer Development Corp. (1986)	636.19
2	Nanjung Wired Comm. Factory (1986)	194.67
3	Beijing No. 3 Computer Factory (1985)	176.43
4	Shandong Computer Service Corp. (1986)	134.00
5	Weifang Computer Corp. (1985)	85.02
6	Shanghai Electronic Computer Factory (1985)	80.94
7	Fujian Computer Factory (1985)	69.27
8	Beijing Wired Comm. Factory (1984)	63.93
9	Stone Group Corp. (1984)	63.00
10	Tianjin Computer Factory (1985?)	61.84

Source: Drawn from data in International Data Corporation, "China Computer Industry Review and Forecast, 1986–1991" (Framingham, Mass.: IDC, 1988): 20–23, CBI 55, Box 60, Folder 24, Charles Babbage Institute, University of Minnesota, Minneapolis.

data center or in a warehouse, because there were insufficiently skilled operators to use the equipment, for the lack of application software, peripheral equipment with which to work with computers, or even due to the lack of supplies of printer paper and repair personnel. One credible report from 1988 concluded that "large, mainframe computer centers in China do not operate 24 hours a day (as most do in the West), but often close down for the night."[92] And, of course, there was the problem of many middle rank and senior managers simply not understanding the potential uses of computers, dampening demand for the technology.

Although largely an ad hoc process, government actions resulted in economic growth in the early 1990s such that each year the economy grew by double digits, making it possible to afford more computing. A key infrastructure essential for diffusion of IT was a labor force educated in the manufacture and use of the technology.[93] Education, however, remained poor into the 1990s, producing a tiny percentage of qualified people for IT as a percentage of the total population. But, the absolute numbers were high, due to the large size of the population. At the start of the 1990s, for example, China had about 990,000 programmers. To put that figure in perspective, the United States had the most at 1.7 million, and Japan 850,000. South Korea, Taiwan, and Singapore combined employed 368,000. India had an estimated 1.3 million.[94]

Telecommunications was another infrastructure required to make IT attractive to use, and was in terrible shape even in the early 1990s. Only about

20 percent of the population had access to telephone services and even those mainly only along the coast and in southern China. That situation changed in the 1990s as China moved aggressively to fix that problem with upgrades in function and capacity, discussed further below.

Finally, China entered the new decade with a large electronics industry, although relatively backward in its ability to sustain modern IT industries, such as computer manufacturing. To address that last point, in 1992 the national government established China Electronics Corporation, made up of 100 manufacturing companies and 37 research institutes, all linked to seven universities and other academic institutions. Based on 1991 revenue figures from all these enterprises, the new conglomerate accounted for 30 percent of the nation's total production of electronics goods. Within this conglomerate— also sometimes Chinatron—were two key computer manufacturers, Great Wall Computer Group and the Computer Systems Engineering Company (CSEC). Several China watchers characterized the new organization as having "the look of a business conglomerate such as the Japanese *keirutsu* or the Korean *chaebol*," but without links those had to trading companies, financial institutions, and related industries that made it possible for these two other Asian nations to internationalize their IT ecosystem so quickly.[95] Also, it should be acknowledged that capital was not as readily accessible to this new conglomerate to the extent available to its counterparts in South Korea and Japan in the early 1990s.

China's increased evidence of commitment to ramping up its scientific and technological capabilities was in evidence by the early 1990s. It began with the "863" plan in 1986 that identified eight industries for attention, including computers, extended to another in 1988 called the "Torch Plan" to develop high tech products, and then into the organizational reforms such as Chinatron, among others. In addition to the changes made all through the 1980s in development of an indigenous IT industry was the opening of the national market in the early 1990s to foreign trade, because the government wanted to import both large computers (e.g., mainframes and minis) and to manufacture PCs and peripheral equipment in China—actions in response to a slow growth in the diffusion of IT in the nation during the late 1980s.

The leading agency responsible in the early 1990s to drive more rapid diffusion of IT continued to be MEI, which had a complex of departments responsible for this mission. It owned 216 computer factories, of which 36 made PCs. Five of the total number of facilities were so large that combined they manufactured 82 percent of the nation's entire output of computers in 1992.[96] It had taxing authority and used it to lower import duties in 1993 from the prior year, from 82 percent to 35 percent of a product's value, to attract foreign goods into China. MEI continued to support and to expand technology parks (which included many IT enterprises) begun in the 1980s with investments in infrastructures (e.g., telecommunications) and local companies, and through loans to startup businesses.

All of these efforts were intended to get local companies to become more competitive through increased IT-induced increases in productivity, a continuation of the intent established as early as 1978. By the early 1990s, the largest success had occurred in the diffusion of PCs, mentioned earlier regarding events in the 1980s, and that continued as a pattern of adoption during the early 1990s. Of the total IT market in 1993 at $1.5 billion, $560 million went into sales of PCs and only $76 million into mainframes; imports accounted for $1 billion in 1992.[97] China was still in a very early phase of Wave One computing.

The industry itself was also tiny at the start of the new decade. In 1991, there were approximately 200 local firms making hardware, one joint venture with DEC making PDP-11s and VAXs; 5 firms dominated 80–82 percent of PC manufacturing. There were 216 firms providing software and information services, with 20 percent at least of the latter owned by the government. Politically significant to note is that businesses owned by the government controlled 80 percent of all databases.[98]

To drive greater use of IT in the 1990s the challenge for national leaders was to find ways to coordinate public policies as MEI was not a monolithic organization. Like its counterparts in South Korea and Japan, it had interagency rivals and there always existed differing points of view on what to do. Lack of coherent policies and slowness in response to changes in the technology and market realities were known threats to success, based on the experiences of Australia, New Zealand, and South Korea.[99] Positive cases known to China, however, taught what needed to be done, as happened in Japan in the 1960s and 1970s, and in Taiwan and in Singapore in the 1980s, all of which demonstrated what was possible when policies and programs were coordinated and implemented in collaboration with the private sector.[100]

On the supply side of IT diffusion, the infrastructure to make, deploy, or export IT grew through the 1990s as part of China's broader initiative to modernize its economy.[101] This was true for all manner of technologies, not just ICT, so when looked at in their totality, one sees a fundamental modernization process underway. The signs were everywhere. For example, the number of high-tech incubator centers in China increased from a handful in the 1980s to 110 in 1999. In 2001 there were 324 and the following year 436. These were campuses for many high tech firms, with 2,670 enterprises in 1997 and over 23,000 in 2002, and a population of employees that reached 45,600 in 1997 but that expanded to 415,000 by 2003. Over 20 universities either were in these parks or tethered to them in some collaborative process, ensuring the geographic spread of these parks and their enterprises around the country.[102]

Many enterprises working in IT only did assembly and operations work, particularly with computers and flat panels, meaning low-profit, low-skilled work, not new product development. With the encouragement by China of foreign investors in various IT manufacturing, after 1995 product development slowly increased, bringing in fresh ideas.[103] Between 1995 and 2001, imports increased at annual rates of just over 12 percent, while exports hovered closer to 19 percent for all manner of high tech products; for IT the patterns were similar.[104]

Beijing remained the center of many IT activities in the 1990s. Products made and used in this zone included personal computers, minis, and work stations, manufactured by such enterprises as PCs by Legend and Taiji for minis and workstations. Boards and cards were also produced for such uses as office applications, television receivers, image and publishing systems. Industrial control machines and telecommunications equipment were also assembled and used as well. Similar production profiles existed in other parts of the country too.[105] An assessment of the IT landscape published in 1999 reported that computers were still largely foreign made and designed and that local manufacture was based on imported technologies. Chinese character-based printing software and machines were now locally designed and made, and dominated the Asian market for Chinese-specific publishing and typesetting systems. Computer parts manufacturing was now dominated by local firms. Computer-aided design and manufacturing systems were being implemented in modern plants for high-tech devices in the economic zones, often with assistance and investments by foreign manufacturers.[106] Telecommunications manufacturing was heavily reliant on foreign technologies, such as that from the United States.[107] While adoption of IT was spreading slowly through government and commercial organizations, the PC was making the most headway in these and in ever smaller enterprises. The result of these activities was the growth in the size of the IT market in China, which expanded ninefold in the decade to $168 billion in 2000, the fastest growth rate of any of the Chinese industrial sectors.[108]

China entered the new century with far more IT in use internally than it had at the start of the 1990s, although with overall adoptions still reflecting "emerging economy" features with pockets of intense use (e.g., military and electronics industries, and lower adoptions by retail and banking firms). Policies and programs launched as plans and early projects in the 1980s took off in the 1990s. The market for computers of all sizes grew from a small $744 million per year in 2000 to $2.8 billion by 2008. Exports of IT kept growing, from nearly $2 billion in 2001 to over $2.5 billion in 2007. Telecommunications, now highly digitized, thus computer-intensive, had grown to a market size of nearly $1 billion at the turn of the century, and grew to $2.1 billion by the end of 2007.[109] Its role was now crucial to the story of Chinese deployment of IT.

CHINESE TELECOMMUNICATIONS AND ARRIVAL OF THE INTERNET

By the early 1980s it had become quite clear that China's information infrastructure would be shaped to a large degree by the integration of telecommunications with computing. By the late 1990s, it was also becoming evident that China would not be immune to the effect of the Internet over the lives of at least a third of the nation. Yet, as one student of IT in developing economies cautioned his readers, the story of the Internet in China is "subtle and nuanced" and also "driven by competitive jockeying among powerful groups in agencies, ministries, government-owned companies, and private bodies."[110] Another observer thought Chinese public sector leadership in promoting use of the Internet in the beginning deserved credit, with access "granted to a quickly expanding number of the country's population," at least in eastern and southern China.[111] We can say the same was true for all manner of IT in the country. China had come a long way from owning one of the smallest, most fragile telecommunications infrastructures in the world in the 1950s to being one of the largest users of mobile phones and the Internet in the early 2000s, making Chinese, for example, the second most widely used language on "the Net," after English. As late as the mid-1980s China's telecommunications infrastructure was limited, technologically backward, and unable to serve the data processing needs of contemporary computing. In 1985, there were only 5–6 million telephones in the country, of which 1 million were in large urban centers; Beijing had only 250,000 and it was the nation's capital with 9 million residents. There were less than 2,000 international telephone connections; IBM's office had to use a low-baud data line to connect to its office in Hong Kong to communicate with the rest of the firm.[112]

Development of a modern telecommunications infrastructure facilitated the spread of computing, particularly online systems in companies and public institutions, and later, by individuals, in many ways following a pattern of adoption evident in dozens of countries, albeit later than the most industrialized economies. Already by the mid-1980s, however, modernizing the nation's telecommunications had been made one of the primary objectives of the Seventh Five-Year Plan with the Ministry of Posts and Telecommunications (MFT) in pursuit of digital switches, digital transmission media (such as satellites), technology to transmit data, and fiber optics among various modern technologies. Although indigenous manufacturing capabilities were coming online, such as fiber optics (since the 1970s), like production of semiconductors these were slow and of poor quality, hence the hunt for foreign technologies and partnerships with firms outside of China. The MFT was both the monopoly provider and regulator at the time. The MFT moved very slowly in the 1980s, spending only about 0.2 percent of GDP, upgrading to digital switches in some urban areas with the result that at the end of the decade there still was only one telephone per 100 residents.

Circumstances changed in the 1990s, however, when Chinese telephony finally became part of the IT diffusion story. By the early 1990s modern tele-communications had been recognized as both a vehicle for economic development and as a necessary component of the nation's modern business structure desired by the Communist Party. In 1993 the MFT created China Telecom to provide services and kept the regulatory function within the ministry. Another state enterprise to compete against China Telecom was established in 1994 as a joint-stock company called China United Telecom (better known as Unicom). Prices were lowered and competition increased, leading to better and faster service. Meanwhile the percentage of GDP invested in telecommunications rose, approaching 1.5 percent by 1993, a rate continued through the 1990s. To jump ahead briefly to see results, by the end of 2004, the nation had 312 million landline accounts (customers) and 335 million mobile phone subscribers, making both networks the largest national systems in the world. Penetration had moved from 1 per 100 residents in the late 1980s to 51 per 100 by the end of 2004, a remarkable achievement in its own right, made more impressive by the fact that these new users relied on modern technologies. This network included some 645,000 kilometers of fiber optics installed between urban centers and telecommunications satellites. In 2004, China had 94 million users of the Internet and some 30 million broadband subscribers.[113] The highest densities for each of these various infrastructures were in cities, hence in the eastern and southern parts of the nation, leaving many rural communities with more limited access to modern communications.

Observers of the Chinese telecommunications scene site this nation as a classic illustration of the dynamics of leapfrogging technologies the notion that implementing a technology that someone has already invented takes less time and is less expensive to do than for early adopters who have to invent it, experience the cost of displacing an earlier technology, and improving the new one to the point where it was practical and cost effective.[114] Yet, China had to create the necessary nontechnical environment to make this work too, in its case two telecommunications providers pitted against each other to compete in a rapidly developing market-driven economy. Officials also needed to retain a sufficient monopolistic presence to husband profits for re-investment in the infrastructure. The effort was funded through domestic savings, rather than relying on foreign investments in the early stages. The rising incomes of Chinese citizens in large urban centers helped generate both demand and the financial wherewithal to pay for the new services. One writer for the *Financial Times* noted as early as 1993, that demand for telephony was part of a broader appetite for information of all kinds, including paper-based publications, creating a dilemma for the government since, on the one hand, it needed to make information accessible, yet on the other hand, it wanted to control what information to make available. The dilemma led the writer to conclude that "no matter how hard the government might strive to contain the information virus, the authorities are fighting a losing battle."[115]

Table 9.9 provides a snapshot of the types of telecommunications diffused in China over the years, including PCs since they, like mobile phones, could transmit and receive data. Prior to China's entry into the WTO in 2001, at which time it agreed to open its telecommunications market to foreign firms, the local market had been served overwhelmingly by Chinese enterprises. Subsequently, foreign providers began entering the market. As the evidence from the table suggests, however, takeoff into Wave One computing was already well underway. Several observations can be gleaned from the data in this table important to our understanding of Chinese adoption of computational devices. Early adoptions of PCs were overwhelmingly by companies, the military, and civilian agencies, not individuals and households until almost the end of the century, because, as the per capita GDP suggests, they were largely unaffordable by families. More affordable were fixed line telephones, and to satisfy the demand of firms to transmit data, installations picked up in the 1980s, and became substantial by the mid-1990s. Mobile phone adoption is an example of leapfrogging at work, but not until after the turn of the new century. When both fixed and mobile telephone subscriptions are combined, one might think a great diffusion had taken place; in fact, that diffusion occurred largely in urban communities and across the eastern half of the country, leaving hundreds of millions of people with very limited access to telephony, hence little ability to connect to the Internet or to transmit data. The information provided in the table accounts for trends averaged across the entire population of China, yet as noted earlier, over 600 million people lived and worked in small agricultural communities while ICT concentrated in urban centers. One can assume that the average personal GDP for those Chinese who had PCs and telephones was actually higher than the data would indicate, making it possible for people to acquire these technologies. Those individuals lived in cities.

But diffusion data have another side to it, the population either late to the Internet or simply deciding not to use it. We have already mentioned the lack of local access and costs, so a digital divide of sorts existed. Surveys of nonusers parallel the reasons given in many countries, particularly in the early stages of the Internet's arrival. In the case of China surveys done in the early 2000s flagged the top three reasons for not using the Internet in descending order of importance: lack of knowledge of computers and the Internet, no facilities for accessing the Internet, and no need or desire to use it.[116] These are the kinds of responses one would expect early in the life cycle of a new technology. This survey was done in 2003, which other diffusion data in this chapter affirmed was still an initial stage of adoption in China.

What makes the adoption of telephony critical to our story is the ability to transmit data by companies and subsequently even more importantly, to access the Internet by both businesses and individuals. Before discussing how the Internet came to China, a quick review of the extent of deployment helps us understand the similar pattern of adoption evident with telephony. There existed a great number of users; but, as a percentage of the nation's total population, they still represented only a developing market with less diffusion as a percentage of the total population or size of the national GDP when compared to that evident in such neighboring countries such as Japan and South Korea. As the data in table 9.10 indicate, access to the Internet by institutional and individual users remained low, with approximately a third of the population gaining access. Since so much telephony is concentrated in urban centers, diffusion of the Internet in eastern China, for example, reached levels comparable to Japan and South Korea. Given the uncertain quality of the evidence on China, these statistics may understate access.[117] The primary form of access to the Internet was by mobile phones, and the number of mobile phone users was probably over twice that for formal subscribers to the Internet.[118]

Adoption of online computing by China mimicked patterns evident in Western and Asian countries, with the national government taking the earliest and most important steps. Eric Harwit, who has studied China's experience with telecommunications, has used the word *revolution* in describing this nation's adoption, because it happened so fast and so radically changed how hundreds of millions of people communicated and were now using computers and the Internet. As the statistics in tables 9.9 and 9.10 suggest, such rapid adoption would not have occurred without effective use of the administrative

Table 9.9 DIFFUSION OF KEY TELCO/IT DIFFUSION IN CHINA, 1985–2012, SELECT YEARS

Year	GDP Per Capita US$	PCs/100 Households (units)	Fixed Phone Subscribers (millions)	Mobile Subscribers (millions)	Mobile/100 Households (units)
1985	$502		3.12		
1989	749		5.68	0.01	
1991	888		8.45	0.05	
1993	1,182		17.33	0.64	
1995	1,350		40.71	3.63	
1997	1,848	2.6	70.31	13.23	1.7
1999	2,162	5.9	108.72	43.30	7.1
2000	2,376	9.7	144.83	84.53	19.5
2002	2,878	20.6	214.22	206.01	62.9
2005	4,102	41.5	350.45	393.41	137.0
2007	5,553		365.45	547.29	
2012				1,000*	

When data was not available, spot on the chart left blank. GDP Per capita based on purchasing power parity in US dollars. *Estimated by Chinese Government.
Sources: Data drawn from Chinese sources presented in Denis Fred Simon and Cong Cao, *China's Emerging Technological Edge: Assessing the Role of High-End Talent* (Cambridge: Cambridge University Press, 2009): 261; GDP data from http://www.earth-policy.org/datacenter (last accessed 6/18/2010). For another listing that shows per capita GDP at roughly half the size as those listed here, but which grew over time at the same rates, see Simon and Cao, *China's Emerging Technological Edge*, p. 257.

Table 9.10 INTERNET ADOPTIONS IN CHINA, 1997–2012,
SELECT YEARS

Year	Internet Users (millions)
1997	0.6
1998	2.1
1999	8.9
2000	22.5
2002	59.1
2004	94.0
2006	132.0
2008	298.0
2010	450.0
2012	500.0

Data drawn from Chinese sources presented in Denis Fred Simon and Cong Cao, *China's Emerging Technological Edge: Assessing the Role of High-End Talent* (Cambridge: Cambridge University Press, 2009): 261; Sherman So and J. Christopher Westland, *Redwired: China's Internet Revolution* (London: Marshall Cavendish, 2010): 8.

and political powers of the state, leading him to conclude rightly that "to a great extent, the Chinese government deserved praise for rapidly building the data network and seeing that access was being granted to a quickly expanding number of the nation's population."[119] While the results appeared in our tables only late in the century, there was a long run up to the results, as also was the case in every other country that became extensive users of telecommunications and, specifically, the Internet. As in other countries, traditional fixed telecommunications and early networks in China were the product of MPT's efforts, while mobile communications and Internet services the result of relatively private sector initiatives.[120]

Beginning in the 1980s, the earliest data networks were centered in academic institutions and in the military, as happened in the United States, for example. The China Academic Network (CAnet) and the High Energy Physics (IHEP) were the first computer networks established in China in 1987. In the following year China had its first link to an external data link when CANet connected to Karlsruhe University, sending e-mail through a German gateway; China also adopted "cn" as the nation's Internet domain name. Other academic and government data networks were established in the early 1990s, such as in 1992 when the leading universities of Peking and Tsinghua set up the PUnet and TUnet campus networks for researchers. In 1994 IHEP conformed to the rapidly standardizing international Transmission Control Protocol/Internet Protocol (TCP/IP) standard, making it possible to link to the growing patchwork of many other international networks. Then in February, 1996, all these various Chinese networks were brought together under the control of the CAS into one network, China Science and Technology Network (CSTnet). Outside the academic environment, MPT was also assembling a packet-data network, called CHINAPAC, beginning in 1993, which it launched in 1995 as ChinaNET, to provide public commercial services to transmit data from one computer to another. State corporations, private firms, and a few wealthy individuals became early users of this network. By the end of 1993, however, MPT was not the only agency building a data network. The Ministry of Electronics Industry (MEI), competed with a state corporation it established called Jitong. It used satellite communications linked to MPT's land-based wired telephony network. Then in 2000 China Network became a third provider to provide data transmission services. Before the end of the year, however, four other networks appeared, ranging from China Mobile's CMNet for mobile communications to the military-controlled China Great Wall's CGWNet. The Ministry of Foreign Trade and Economic Cooperation also entered the market with CIETNet and the Ministry of Railway with CRNET (not to be confused with CRnet, China's early research network). Each pursued specific sets of customers, such as CMNet wireless access to the Internet, CIETNet international trade and foreign e-commerce, and the military network for defense. By the middle of 2007, there were at least 10 interconnecting data transmission networks providing the core backbone of China's computer-linked telecommunications.[121]

Elements of China's pattern of diffusion proved similar to that of other nations: initially academic networks were followed by other government-sponsored networks for public and private traffic, subsequently private firms competing for commercial and consumer business. In the early 2000s direct sales of services began moving quickly to private firms—Internet Service Providers (ISPs)—making such organizations as ChinaNET more of a wholesaler of services. As the demand for Internet services grew in the late 1990s and early 2000s, the market for ISPs did too, beginning with the Beijing PTA in May, 1995. Many began to appear, often too many for the yet slowly expanding market in the late 1990s; but, overall this increase in providers led to declining costs for services which made these more affordable to organizations, firms, and, of course, individuals. At first the MPT used its monopolistic power to generate revenues from the ISPs, which resulted in some customers demanding lower fees, pleading their cases to the government, such as academic users. Premier Zhu Rongji addressed the problem in 1999, starting a process of liberalizing the market and lowering prices for these services, with some initial price cuts of over 25 percent for international services, which also benefitted individual consumers in the early 2000s.[122]

On the commercial side, applications too mimicked those evident in other countries: business data transmissions, e-mail, and e-commerce. The number of.cn registered domains on the Internet grew from some 122,000 in 2000 to 1.8 million in 2007.[123] Foreign providers of information were given access to China's network, such as Yahoo and Microsoft, because the government felt foreign-run ISPs could be controlled and they were providing e-commerce services, deemed crucial to the economic modernization of China. Early individual users (1990s–early 2000s) were largely young men living in eastern cities, often students who later went on to become professionals in various industries.[124]

Table 9.11 suggests where the majority of activity occurred, reinforcing the point that most computing concentrated in urban eastern China. The five largest urban centers accounted for the greatest concentration of users in the nation, resulting in over 11 percent of the population in eastern China accessing the Internet by the end of 2006, over half the citizens in central China between 7 and 11 percent, and most of western China below 7 percent. Eric Harwit's research demonstrated that higher income regions and people were more extensive users of ICT than lower income areas. In the cities listed in table 9.11, for example, in 2006 per capita incomes ranged from a low of $161 in Tianjin to a high of $238 in Shanghai, while in many cities not listed, and where utilization was low, incomes ranged from $98 to $110. Because the highest incomes existed in cities, one would expect that the smallest number of users were in rural China, which is exactly what happened. In 2003, the Chinese Ministry of Science and Technology reported that roughly 600,000 users lived in rural areas out of a population of some 60 million users nationwide, in other words, only 1 percent.[125] The problems of spreading Internet usage to rural areas often mimicked similar challenges in other parts of the world: lack of reliable sources of electricity, affordability, and cost of building out appropriate telecommunication infrastructures across vast land masses. In some communities, inadequate literacy levels and poor educational levels also slowed diffusion of ICT.[126]

Table 9.11 LOCATIONS OF INTERNET USERS BY PERCENT OF POPULATION IN CHINA, 2000–2006, SELECT YEARS (IN PERCENT OF TOTAL POPULATION)

City	2000	2002	2004	2006
Beijing	21	29	27.6	30.4
Shanghai	12	26	25.8	28.7
Tianjin	5.8	14	19.1	24.9
Guangdong	2.6	6.6	14.9	19.9
Zhejiang	3.3	7.2	11.4	19.9
National average	1.8	4.8	7.2	10.4
Median	1.4	3.7	6.0	9.2

Source: Drawn from Chinese data and modified from table in Eric Harwit, *China's Telecommunications Revolution* (New York: Oxford University Press, 2008): 170.

Because I believe Wave One begins, in part, to evolve into Wave Two as consumer uses of IT rises extensively, appreciating how Chinese access the Internet is as important as what they use it for. During the 1980s and most of the 1990s, people accessed the Internet through terminals connected to mainframes, which in turn had access to networks, or via personal computers with modems. How many PCs in China in the 1980s and 1990s had modems (to connect to networks) remains unclear, but probably, at least, very few during the 1980s. (The earliest PCs in all countries could not connect to any network.) In the late 1990s and then in the early 2000s, handheld mobile telephones became a widely used instrument for accessing the Internet by individuals either for personal or professional life. Personal use has been the subject of considerable study in recent years because of concerns about the free flow of information in China, Internet-based criticisms of government policies, and as part of the evolution of the lifestyle of the emerging middle and upper classes in China.[127]

Clearly, however, much news, weather, and other sought-after information (e.g., on health and prices for goods) is now accessed via mobile phones, and not just through such traditional technologies as television or radio, or even PCs. Mobile phones are essentially an open-data outlet that allow people to comment and communicate about topics, from where is the best nearby restaurant to how one could get the government to change a law or policy. Increasingly, individuals are consuming digital products as well, such as games (as also occurring all over Asia), and digitally delivered news, stock information, and weather reports. Chinese have embraced more aggressively than people in many countries in using cell phones for short communications, using SMS messaging, suggesting that quite soon mobile phones could become a business delivery platform. As two observers noted, "the major bottleneck impeding the takeoff of mobile media as a business platform in China lies in the acceptance of mobile payments. Chinese consumers are still not inclined to pay large amounts using their thumbs."[128] Censorship challenges exist for the government trying to control flow of information, which proved easier to do with print media, and most difficult (so far) with digital media.[129]

There was growing recognition that Chinese citizens were using the Internet in ways that paralleled somewhat those in evidence in Japan and South Korea, although as of the early 2000s, not quite as intensively. The most obvious was social production of content and discussions, almost from the earliest days of Chinese Internet activity. Chinese users were proving to be avid bloggers, for example. One estimate suggested there were some 230,000 active blogs in China in 2002, and over 7.6 million four years later.[130] One can safely assume there are substantially more today, since the number of mobile phone users continued to increase and the concept of discussing issues in communities organized by subject is very compatible with Chinese interpersonal patterns of behavior. Chinese government agencies have also set up websites that at first were intended to push out information, but increasingly became interactive sites too, moving from 982 in 1999 (first year of such sites open to the general public) to over 8,000 gov.cn sites in 2003. However, in the early years of their existence, they were not influential, many were outdated, or inactive, as people sought their information elsewhere, or simply did not trust official sources.[131]

Chinese civic associations were slow to establish a presence on the Net, although many had PCs by the early 2000s to handle normal office work; one survey showed that less than 2 percent did not have access to some form of computing (although that number seems high given the other diffusion data presented earlier).[132] One surveyor concluded that these associations were comparable in their volume of use and percentage of installations of IT to other developing countries, but less than in the economically most developed nations, characterizing their Internet capacity as "minimal."[133] So the data on diffusion remains murky. Most went online for the first time in 1999–2004, paralleling when people and other organizations first established their presence on the Internet. Civic organizations also paralleled applications in government and business: e-mail (the most), search functions, and access to other Web sites. They became extensive users of electronic newsletters and BBBs (bulletin-board systems).

But, the key trend was the diversification of uses of networks. Surveys done of these suggest that between 1999 and 2004, the top applications as percent of total uses were e-mail, followed in descending order newsgroups, BBBs, and online shopping. Between 2003 and mid-2007, e-mail went from nearly 92 percent of the total to 55.4 percent while BBS applications rose from nearly 23 percent of the total to 70 percent. Blogging ranged from 20 to 25 percent in this second group of years; only 25 percent did online shopping.[134]

In addition to inexpensive mobile phones, one should not overlook an earlier channel of IT diffusion, especially into the Internet, observed in many other countries around the world: Internet cafés. When the Internet became available publicly in China during 1995, one of the most widely accessible ramps into it was the Internet cafés largely for e-mail, with both Chinese and foreigners active users, such as university students. By 2008 these were now cybercafés that gave people access to online gaming, streaming video, online telephony, and social venues filled with young people. The earliest facilities used personal computers, but a decade later the devices were often specialty items designed for collective entertainment, complete with waterproof keyboards, and online games. As one veteran user described the evolution in the mid-1990s, "it was called *wangluo kafeiwu*, meaning literally 'network coffee house,' a place of enlightenment, culture and taste for the brainy and foreign minded. Ten years later, it is known simply as *wangba* or 'Net bar'."[135]

Users evolved from an elite group in urban centers to a more working-class clientele, more like bars, as people acquired other paths into the Internet.

Users, much like those in South Korea and Taiwan, were (and still are) young, better-than-average educated males who live in urban centers, communities most able to provide access to the Internet, community, and entertainment. Some unknown percentage had their own Internet access codes but could not afford to acquire a PC. Increasingly, access was also through cell phones. Sociologist Jack Linchuan Qiu identified many of these people as a large class of "have-less" rather than "have-not" users, thus reducing dependency on such bars for e-mail and other non-entertaining uses of ICT.[136] But these facilities were popular points of access to computing, with surveys suggesting that in 1999 there may have been 60,000 users, but by 2008 there were over 71.2 million, a population that initially made up about 3 percent of all Internet users in 1999 but nearly 40 percent by 2007. Home users of the Internet made up 44 percent of all Internet users in 1999 and just over 67 percent by 2008, suggesting a growth curve driven by government incentives for use and marketing by ISPs. Interesting, access from work declined in the same period from 50 percent to 24 percent, indicative of the alternative avenues of access to the Internet. The total number of cafés is not fully known, although estimates place the number at between 110,000 and 200,000 as of early 2003, and 1.8 million a year later, so the accurate number is not really known; but, that there were many such facilities can be accepted. In the case of one survey of 112,000 such facilities in 2006, this group had more than 6 million computers and employed 786,000 people, further evidence of the unique form of IT diffusion underway in China.[137] By 2010, government surveys were reporting that in the nation's 60 largest cities many Chinese spent as much as 70 percent of their leisure time online, 50 percent did so in rural communities, and with 3G services now rapidly increasing, replacing TV as the most popular technology in use for digital leisure time activities.[138]

CHINESE FOREIGN TRADE PATTERNS

China's foreign trade in ICT and services is another facet of the global computer diffusion story as are also the cases of Taiwan, Singapore, South Korea, and the United States. As with those other nations, exports of IT products raised utilization internally while making products available to users around the world. International trade in high-tech products is a post-1980s chapter in Chinese economic history. The overall international Chinese trade experience is significant if for no other reason than by 2005 it had become the third largest trading nation on earth (following the United States and Germany), an important milestone not lost on both political leaders and IT industry executives and users. Trade liberalization began in earnest, however, in 2001 with China's entry into the World Trade Organization (WTO). Space does not permit a detailed history of this aspect of Chinese IT history, but several features of that story should be acknowledged.[139] Following trade liberalization and currency devaluations in the 1980s, reform of tariffs, and creation of incentives for foreign investments and exports, foreign trade rose sharply, beginning in the second half of the 1980s and expanding through the 1990s and beyond, leading economists to start using the term "open economy" to characterize the turn-of-the-century economy of China.[140]

One feature of this trade was the growing exports of IT products, such as laptops,[141] but the bulk of China's ICT exports in the 1990s and early 2000s concentrated on goods that were finally assembled in China with parts imported from other countries (e.g., Taiwan), leveraging cheap labor, not technology-intensive knowledge or facilities. This kind of trade represented medium-skilled work, not high-tech work, so far.[142] Leading initially with a strategy that leveraged low wages and medium-skilled workforces was working. By 2005, for example, about 37 percent of the World's supply of mobile phones was made in China, and a similar percent of all personal computers. Other consumer-oriented digital devices were increasingly manufactured there too, such as cameras.[143] Table 9.12 suggests the extent of China's IT exports as of 2004. While the percentages were low, given that China's trade was then one of the largest in the world, the actual volume of goods shipped around the world were high. Products became more technologically intensive over time, beginning in the early 1990s. In 1992, office and computer equipment accounted for only 12 percent of Chinese exports, but by the early 2000s, these approached a third.[144] If we look at import–export traffic, between 1996 and 2002, for example, one can see that imports of ICT goods into China grew annually on average by 26.5 percent, while exports did by an almost similar amount (25.1 percent), both growth rates increasing faster in these years than in any other country in the world.[145] One can conclude from these data that China's economic development was evolving in a similar trajectory evident in those parts of Asia advancing into high-technology product assembly 10–15 years after launching market-based reforms, followed by knowledge-based products and services.[146]

In an advanced IT sector in a country with relatively low cost labor outsourcing becomes an attractive extension of local IT diffusion, as was occurring in India in the early 2000s. The question came up with respect to China which had a growing export business and was expanding its use of IT substantially, along with building a modern telecommunications infrastructure—all prerequisites for the high-value outsourcing market. One OECD study completed in 2007 concluded, however, that, "China is not a major supplier of these offshored services," with demand for such services remaining a "potential" future source of jobs and revenues. China had not entered this market yet because "it has not yet developed the specialized firms and human resources, including foreign language resources, or the stock of inward specialized services investment to supply these services globally," necessary to be a competitive player.[147]

Table 9.12 CHINA'S IT EXPORTS AS PERCENT OF TOTAL CHINESE
EXPORTS, 2004

Type	Chare of Exports (%)
Parts for data processing equipment	4.0
Digital data-processing equipment	4.0
IT input/output equipment	4.2
Integrated circuits	1.9
Computer data storage units	1.5
Automatic data processing machines and units	0.9
Digital computers and some peripherals	0.8
Total	17.3

Source: Data in Shahid Yusuf and Kaoru Nabeshima, "Strengthening China's Technological Capability,"
Policy Research Working Paper 4309 (Washington, D.C.: The World Bank, August 2007): 5.

CONCLUSIONS

How far had China gone in its absorption of information technologies by the
end of the first decade of the twenty-first century? An analyst at the International Telecommunications Union declared that the country had "the world's
largest online community, more than the entire population of the United
States."[148] The investments in broadband infrastructures in China were the
largest in the world as well, posing a competitive threat to the United States
and to other IT-advanced economies as new startup businesses became possible in an ever larger swath of the country.[149] These uses of technology—
Internet and broadband—were significant contributions to the economic
development of China.[150] Much of the investments in IT in the early 2000s
were also going into rural China, the nation's last ICT frontier.[151] This suggests
that over time, inland provinces would be able to integrate more fully into international supply chains, a process underway, but had much farther to go,
implying that early urban entrants into manufacturing and use of IT would
continue their lead position for years to come.

The remarkable elements of China's experience can be summarized briefly.
The planned economy approach postulated by the Soviets did not work, while
the political turmoil of the 1960s and 1970s reversed hopes for economic development. Political will to change proved sufficient at the end of the 1970s to
make it possible to slowly evolve China's economy into a market-driven variant in the 1980s and 1990s, and into a competitive capitalist version in the
2000s. IT's diffusion in China reflected this mega-pattern of economic transformation over the course of the past half century. While the political fallouts
of more information and IT available in the country have not been discussed
in this chapter, these, nonetheless, are progressing along lines evident in other

countries as citizens gained access to information not otherwise available or allowed to them. The diffusion of IT in China, and in particular in the eastern more densely populated half of the country, was sufficient-enough to lead two highly knowledgeable China-watchers to conclude that IT has "been a major force in recent years for dramatic social, economic, and political changes in China."[152]

Leaving aside the social and political impact for others to debate, and finally the role of the Internet already discussed above, we should recognize that commercial enterprises are both intensive users of IT and still monuments to pre-data processing processes and paper-dependent practices. Family run enterprises and small businesses still outnumber massive organizations, although today some of the largest enterprises in the world are Chinese. Much has yet to be done to support further development and use of IT, most notably protection of digital products, such as software and publications. Therefore, China is a nation that has only travelled partly through Wave One adoption and use of IT, and still remains behind Western users, Japan, South Korea, and even Taiwan, and certainly Singapore, New Zealand, and Australia. But the Chinese have demonstrated the skill and discipline to move quickly, catching up most of the way in less than three decades. It demonstrated that being more open, and with more players involved from around the world, did have a positive impact on the adoption of modern technologies, while not threatening the demise of long-standing Chinese cultural and political practices, at least so far.

Finally, we confront size for it still mattered. China's leaders worried about its effects on national and international affairs; IT vendors around the world salivated at the potential markets possible in China, while local leaders knew that size meant Chinese technologists could someday establish technical standards and set the pace for much IT product distribution. Officials recognized their limited ability to run the economy and acknowledged that the future of IT supply, sale, and uses would come from the growing entrepreneurial capabilities cropping up in many parts of China. China has proven relatively effective in picking portions of economic and political practices from other nations useful for adopting internally, and willingly continued to change and adopt new ones as they became useful. As two observers of the Chinese situation characterized the nation's way to evolve its economy and use of technologies of all kinds, "in the past . . . 'muddling through' has been a fairly effective approach for" coping "with China's diversity and dynamism."[153] It is also a reasonable characterization of China's experience with information technology.

And what about India, the other country always compared to China? That is a different story, worthy of its own chapter, because for many contemporary observers of the global diffusion of IT India represents the next wave of adoption, one with the potential of becoming a tsunami of Wave One computing.

CHAPTER 10

India and the Limits of Digital Diffusion

The automation equipment that are being imported into this country on a very wide scale include those frighteningly monstrous machines called electronic computers.
—K. Anirudhan, *Communist leader, Indian Parliament, 1967.*[1]

India represents the latest actor to step onto the world stage in the use of information technologies, entering early Wave One diffusion within the lifetime of middle-aged Indian adults. The takeoff in IT use dates to the period following significant reforms that liberalized India's economy enough to make that possible and, in turn, that permitted the development of an internationally important IT software and services industry. The emerging environment made India more attractive to foreign companies to work in, led to international commerce, and to a market-driven capitalism. These changes took off largely in the 1990s, much as economic reforms in China after 1978 stimulated demand for computing in that country. As in China, diffusion of IT was less a function of the nature of the technology itself than a consequence of fundamental transformations in the economic and public policies and the subsequent actions that made use of IT more attractive than before. Despite these changes, internal diffusion of IT remained lower than one might have expected, setting up a paradox where India became an important player on the international IT stage but remained a developing economy with diminutive IT utilization at home.

As in every other country examined in this book, Indians in small groups learned about computers very early in the life of the technology; in India's case, at the start of the 1950s. As with so many other nations in Eastern Europe, Asia, and Latin America, a long multiyear gap existed between when a few engineers and academics built, or acquired, their first computers and when this technology went into commercial use. In the case of India, the time gap was quite wide—in effect some four decades in duration—constituting one of the most extensive lapses between the acquisition of initial knowledge and wide deployment. Even relying on the word *wide* to describe diffusion of the technology after 1991 is questionable and will be qualified, because the use of IT concentrated in urban centers and in a few states. The notion of a gap is important to keep in mind as historians, economists, and apologists for national innovations speak of "firsts," because gaps in time always existed in all countries between when people learned about IT and when they used it. The lengths of these gaps were more often influenced by nontechnical issues, such as those suggested throughout this study: levels of national and personal GDP, size of enterprises, what industries there were, levels of literacy and education, cost of the technology, funding for its operations, and effect of government attitudes and policies, among others. Functions and features of the technology (that is to say of the devices and software), while obviously important factors, were normally less influential in the deployment of information technologies. India provided another proof point that just focusing on the evolution of computers into more complex, larger, or less expensive forms is insufficient for explaining how these diffused through societies, both rich and poor.

This chapter reviews the Indian experience as if it was a well-established long-understood history. However, it is not, because the activities described are more current events than historical experiences, more unfolding patterns of deployment than documented realities. Rather, India's experience is informed by similar practices in evidence in other countries with longer-term experiences with IT. Much of what is reported below is tentative, but also at the heart of where much IT diffusion is occurring during the second decade of the twenty-first century.

The Indian experience with information technology has much to teach us, particularly about the limits of the role of government in the diffusion of technologies. In each country studied governments played important facilitative roles and, while their results varied from highly successful (such as in the United States and Taiwan) to quite limited (as in the Soviet Union and in the German Democratic Republic), programs were purposefully developed and executed (as in China), which were based on local ideological, political, and economic realities. In the Indian case we have an example that can be bluntly described as often ignorant of the value and effects of technologies for at least a quarter of a century, and the result of the mismanagement of public administration, involving multiple rivaling agencies regarding digital matters over a period of several decades.[2] This is not an issue of corruption or divergent perspectives—to be sure those circumstances existed—but more overtly ineffective, often ignorant or ineffective public administration, particularly before the mid-1980s.

Apologists for Indian public administration are numerous, yet harsh realities need to be acknowledged and addressed by government agencies and other Indian communities if Indian IT is to progress in the decades to come. The majority of the discussion across many academic disciplines about Indian economics and IT center on the supply side of the story, that is to say, on the export of software and IT and business process outsourcing services, operations of industrial research and development centers, and especially of Indian call centers, all of which are important parts of the story of how these goods and services helped to diffuse IT around the world, most notably in the United States. Yet, the demand side of the story—internal use of computing in India—has largely been overlooked for good reason: many parts of India remained a digital desert with remarkably few computers installed, let alone used, during the first seven decades' existence of these technologies.

What statistics are available on the number of computers installed in India show that only a small number of systems ever existed in this country, with the data appearing even tinier when compared to the number of humans living in this very large nation. One's impulse is to question the validity of the data. How is it possible for a nation as large as India to have so few computers as the data on installed systems suggested was the case? One might logically conclude that we got the facts wrong. For example, India really had less than 500 computers in the 1980s, when one might reasonably think the number should at least be 5,000 or more credibly, 50,000, if we included large mainframes and minicomputers. But, 500 is not a typographical error; in fact, it is too large a number. It is that paucity of systems in a nation as large as India at a time when well over half the countries in the world were installing and using computers that must be addressed. Because governments were often the most enabled by legislative authority and endowed with economic and human assets to encourage national diffusion of IT, the comments just made about state and national governments not being effective in India's use of computers might seem harsh. In the gentle and elegant language of scholarly parlance, this circumstance speaks to the "limits" of public administration in the diffusion of any information technology.

To be fair to Indian officials, and as suggested in the introduction to the prior chapter, India lacked many of the environmental prerequisites crucial for the local flowering of computing. The paucity of such essentials as electricity, necessary levels of personal GDP, sufficiently large-enough modern agencies and businesses to justify use of computers, and the development of those industries most inclined to use the technology, were comparable to the sorts of deficiencies evident in other areas of the world that too were limited users of IT, such as those in Central Africa, North Korea, and the Philippines. To Western observers changes occurred more slowly in India than in many other parts of the world, which may help to account for what seems to be an incredibly slow start in India, contributing a plethora of cultural and legal nuances to the story.

The evidence clearly demonstrates several realities. First, there was much going on within government and in the tiny supply side of IT, beginning in the 1950s and running through the 1980s, even though the results can almost seem trivial. These activities created the necessary knowledge base and facilities needed for the dramatic changes that came after 1990 and that are still unfolding. Without these previous developments, it is difficult to imagine those that came after 1990 occurring at all. Second, the story of IT in India is one of enormous contrasts, paradoxes and a range of counter currents at work. IBM's or Microsoft's facilities in Bangalore are beautiful and as modern as anything one can see on the West coast of the United States, for example. Yet not far away are traffic jams on poorly maintained roads.[3] In these and other IT campuses some businesses have to generate their own reliable and sufficient supplies of electricity, while brownouts and blackouts occur regularly in the city and around the country. Indian technology centers are filled with hundreds of thousands of skilled IT workers respected around the world for their capabilities, but there are thousands of Indian villages that still do not have electricity to recharge cell phones. Seen as the latest global player in the IT market, India has the biggest agricultural sector amongst the 80 largest economies of the world, the sector that historically worldwide had the least number of uses for computing when compared to such others as manufacturing, retailing, finance, and public administration. So we must keep in mind that where diffusion occurred in India was more often than not in urban pockets, on "islands of automation," to resurrect an old term often used in the West in the 1960s and 1970s to describe departments within companies that became early adopters of computing.

Because every country had unique experiences while adopting IT, yet also shared common ones too, understanding India's is vitally important today. There is an urgency to the task because many global IT companies—suppliers of so much computing to nearly 200 other countries—are investing enormous sums, human capital, and managerial effort in developing technological and service-based capacities there. By 2012 IBM, for example, had reportedly over 100,000 employees in India.[4] Other Western companies also have made massive investments in Indian IT. Meanwhile, Indian firms moved from low-value-added activities, for example, such as programming and call centers, to such value-added tasks as design and delivery of sophisticated software systems. Infosys, Wipro, and Tata are currently three Indian firms that are recognizable brands in the world economy. More will undoubtedly join that short list. Numerous American and European IT, pharmaceutical, and other technology firms are, to use an old American slang phrase, putting so many of their "eggs in one basket" that understanding the implications of that strategy, and, concurrently, the risks associated with such an approach, requires a far clearer understanding of India's IT history more than ever before.

The story told below will seem more similar to that of Communist East Europe, and far more different than those of Japan, Taiwan, South Korea, Singapore, and China. The reason is simply that India's experience with computing was sufficiently unique from that of other Asian, American, and European societies that making comparisons to them can possibly be misleading, as occurs in books and articles that deal with China and India in some simultaneous manner.[5] I begin to demonstrate this point by discussing the infertile ground in which one had to attempt to plant information technologies.

ECONOMIC, POLITICAL, AND SOCIAL PRECONDITIONS FOR IT ADOPTION

During the more than seven decades during which computers have been available, India's economic, business, and social circumstances have overwhelmingly not been the most optimal for organic growth of IT use. Put in other words, this nation proved a hostile environment for the diffusion of IT. India was saddled with circumstances often overlooked, or inadequately taken into consideration by students of failed public policies that want to blame officials for being either critical toward the use of computers or for implementing policies that retarded their spread. Those kinds of analyses, while relevant, nonetheless need to take into account endemic realities that we know are harmful to the spread of IT based on studies of the experiences of other countries and that serve as a brake on diffusion, regardless of governmental actions.

Beginning our analysis by looking at the sectors making up India's economy suggests some of the problems India faced. Agriculture and its allied industries were not users of IT for the same reasons farmers did not need computers in other countries, until at least the arrival of the twenty-first century in India. These devices were far too expensive for Indian farmers until at least the start of the second decade of the new century, and even then only small devices were affordable, such as cell phones. Often farmers were illiterate; most usually did not have electricity with which to run the machines. Using 1950 as convenient starting point, 57 percent of India's GDP came from agriculture. Over the next few decades that percent dropped to 21 percent of the economy by 2004–2005; but even in the late 1980s it still hovered at 33 percent. It was not uncommon in many decades for some 65 percent of the workforce to be in agriculture. Manufacturing industries, the most natural early candidates for computing, occupied a paltry 9 percent of the economy in 1950 and only reached 17 percent by 2004, thus not a fortuitous circumstance for IT. The services sector, which included banking, insurance, and government among other industries, began the 1950s at 28 percent of GDP; reached 41 percent by the late 1980s, when some computing was beginning to take off; but barely broke above 50 percent in 2004–2005, when it contributed 52 percent of GDP. All other industries made up the remaining 15 percent (1950) to 27 percent (2005) of GDP. These data speak only to organized industries, not to the economic activities undocumented by official sources.[6] While the temptation is great to write a brief economic history of India, since those grander currents affected directly IT's diffusion, the discussion on what the government did will be more focused. It is enough now to call out the fact that the economy began the 1950s, and continued for many decades, to contain more sectors least amenable to the use of information technology than in many other countries.

In Wave One personal incomes had to be high enough to make it possible for people to afford computers, and for companies to have customers who could afford to buy products that in turn could be manufactured, leveraging uses of IT in their production and delivery. Indian per capita GDP remained low all through the period with poverty the central economic and social problem faced by India throughout the second half of the twentieth century and extending into the new millennium. Its reduction remained the single most consistent intention of public officials. Accurate data on levels of poverty remain difficult to acquire but even those that exist tell of a harsh reality. Using 1980 for a snapshot—when mainframes were in wide use in many countries (but not in India), along with minicomputers (in limited use in India), and only a handful of PCs, the average annual per capita income in India was under $100, while some 45 percent of the populace had an income of $50 or less per capita. Some 90 percent of the population had incomes of under $150 per annum.[7] These terribly depressed levels of income did not change to any substantive degree until at least the end of the 1970s. Leaving aside definitions of poverty, official Indian government statistics indicated that roughly half the country was poor in 1980, and still so in 1993 with a third of the population still in poverty, when computing significantly began to spread. At the dawn of the new century poverty hovered at 26 percent. Urban and rural poor remained about equally balanced throughout the previous six decades. The population grew all through these decades so the absolute numbers remained high: about 330 million poor in 1980, and 260 million at the start of the new century, when the larger majority of the indigent lived in rural communities.[8]

Rates of illiteracy of the poor also remained high; while within the continuously expanding middle class, and the financially well-off and social elite, illiteracy declined throughout the period. To reinforce the point that illiteracy remained widespread, in 2000 the literacy rate had reached 65 percent—an historical high and an important achievement for India—and by 2011 had expanded to 74 percent; yet, that still left 26 percent of the population illiterate.[9] Indian economist Amiya Kumar Bagchi began arguing toward the end of the century that illiteracy would slow Indian adoption of computers and, as late as 2005, was still calling out this reality, labeling it "another major obstacle against the development of an information society." He observed that "computer literacy has spread fast only in those states or regions where the people possess a high degree of literacy, such as Karnataka, Tamil Nadu, Pune, Delhi, Kolkata and Hyderabad." The problem of illiteracy was also compounded "when people are literate in their mother tongue, they may not be able to use computer facilities because of their lack of command over the English language," constraining diffusion of the Internet, for example, in rural parts of the nation.[10] Literacy did not mean everyone had a Ph.D. in computer science, were proficient users of IT, or were engineers, however; that more skilled demographic remained under 3 million in the new century, and only in the hundreds of thousands in the 1970s–1980s.

India also has long had a relatively immobile workforce, which constrained optimizing allocation of labor to growing industries. The nation has 350 languages and dialects, each with their own radio and print media, while just less than 10 percent of all Indians speak English. Roughly half of all internal migrants were women who moved for reasons of marriage, and they always represented a minority in the organized workforce of India. Even at the dawn of the twenty-first century only 10 percent of women migrated for job reasons. Over 60 percent of all internal migrations occurred within a women's home district, and 20 percent within their state. The rest moved to other parts of the nation.[11] The majority lived in rural areas and as late as 2009, only 30 percent in urban centers, the location of almost all of India's computers and telecommunications infrastructures.[12]

One of the lessons about Wave One diffusion of IT is that it occurs slower and later if manufacturing and services businesses are small, that is to say, employ less than 100 workers, because these companies do not have enough labor costs that could be offset by the productivity gains promised by computing. This was especially the case when the cost of IT remained so high between the 1950s and mid-1980s (or later in some countries, such as India) relative to the expense of labor and capital.

A quick test about the level of possible receptivity of computing is to measure the size of firms, especially in manufacturing during the early decades of Wave One. In the case of India, extant evidence, and only on those firms registered with the government, thus known about, suggests India had a large number of enterprises not likely to be candidates for computing until long after the arrival of microcomputers. One OECD study indicated that some 87 percent of all employees in manufacturing worked in firms that had less than 10 employees as late as 2004. Hardly any country tracked by the OECD had such a high concentration of small firms in manufacturing. The report noted that "the small scale of Indian industry arose in part by design" of the government which favored having one "national champion" in an industry, along with many other small firms. Although the policy was being dismantled at the turn of the century, it was not completely gone. For decades the government put caps on how large a firm could be, requiring permission to change those limits. Economists at the OECD observed that as a result of the large percentage of small firms, "India is not . . . reaping gains from scale economies," which include, "often the use [of] newer technologies."[13] Combined, these firms only produced a third of the nation's manufacturing output. Outsourcing tasks internally within India by larger enterprises to smaller ones continued the fragmentation of work, hence, opportunities to improve productivity. Policies that depressed growth of firms made it possible to employ inexpensive labor but at the price of keeping work tied to low value-add activities with little possibility to take advantage of economies of scale, which were ways of working that favored use of computers. The same report noted that whatever productivity gains were achieved, and there were some in the 1980s and 1990s, were due less to the use of IT and modern equipment to optimize labor productivity than to economic reforms initiated by the government, discussed further below.[14] It was always more important to policy makers to encourage employment of people than to improve productivity. The latter effort would have required larger enterprises and use of computers and other automation equipment, possibly displacing workers.

Finally, there is the famously frustrating general issue of all manner of infrastructure. No student of modern India can long resist complaining about the notoriously poor supplies of clean water, bad sewer systems, crowded roads (too many in need of paving or repair), unreliable rail systems, crowded airports, and erratic supplies of electricity. One has only to visit the country once, even in the twenty-first century, to marvel at the extent of such problems. But for our purposes, one infrastructural problem rises to the fore often overlooked as a prerequisite for the diffusion of all manner of IT: electricity, because in countries that use computers more than India this form of energy routinely exists in adequate amounts; it is ubiquitous. Not so in India. Computers can be dragged from shipside (or airport) to a user on a bullock cart going down a dusty road. It seems every non-Indian computer vendor operating in this country has photographs of such delivery methods in their archives!

But, the simple truth is that without electricity one cannot have smart phones, PCs, mainframes or any other modern technological marvel. With the exception of North Korea, India remains one of Asia's weakest suppliers of electricity and is routinely ranked one of the lowest in the world in having reliably sufficient amounts. For computers to work properly they must have a continuous supply of electricity without brownouts or blackouts; either condition, when not planned for, can cause a computer to shut off or to be damaged. A mainframe can suffer hundreds of thousands of dollars in problems and take hours to restart all its software after such an incident. Lack of electricity for a few hours a day makes use of expensive computers highly impractical, as occurs quite frequently in the poorest countries in the world. On a smaller scale the same problems occur with PCs, laptops, and other digital equipment. "Spikes" (surges) of electrical power down a wire into a machine can also cause damage and often is a bigger threat to the use of IT than complete outages. Inadequate supplies of electricity discourages potential users to install computers, or, as occurs frequently in Indian technology parks, requires them to generate their own supply for mainframes and users' terminals and laptops. This is not their core competence and can only be cost justified as long as Indian labor and other local expenses remain low.

Beginning in 1981, as India's economy in the non-agricultural sectors grew, and especially in the 1990s and later, demand for electricity grew annually by an average of 3.6 percent. Demand always outpaced supply, actually with a growing gap between demand and supply in the early 2000s annually at over 7 percent each year, approaching 8.5 percent by 2010. In short, this was a serious, indeed chronic, problem, because lack of adequate supplies slowed economic development. By 2010, some 400 million residents in India were experiencing blackouts on a regular basis. A related problem is that supplies of electricity are concentrated in urban centers, which is fine for today's users of computers (firms and their employees). This is not so for many villages where, as late as 2008, some 7 percent did not have at least one electrical line, which meant their residents could not even have cell phones, let alone use PCs and other digital products.[15] All these data suggest a condition that predated Wave One circumstances in most countries, yet lingers in India to such an extent that the issue of the availability of electricity must be included in the mix of inhibitors to the diffusion of IT in the country.[16] The related electrical issue—telephony, also a major infrastructure problem throughout the period in India—is discussed below in the context of computing in the 1980s and beyond, when insufficient availability of telecommunications became another brake slowing diffusion of IT.

To suggest in the briefest of terms how far India had gone in creating the kind of environment one would expect to see in Wave One diffusion of IT, and the extent of deployment of some components of that information infrastructure, by the early years of the new century one could turn to various global rankings that had become available. The World Economic Forum ranked India 54th out of 75 countries in its use of networks, such as the Internet and telephony; 65th in overall information infrastructures. It reported that there were 3.20 telephones installed per 100 Indians and less than 5 personal computers per 1,000 people. Of those PCs, about 75 percent were connected to the Internet and the number of users of the Internet was roughly comparable to the number of PC owners. Use of cell phones was lower (0.35 per 100 people). Most PCs—the type of computing in greatest use at the time—were largely owned by businesses located in urban centers, leaving the countryside digitally barren. The causes cited are the ones discussed throughout this chapter for the entire previous half century: weak telecommunications regulation, legacies of failed statist economic policies of the past, slow functioning and complex bureaucracy, political strife, and "monumental social challenges."[17]

According to the World Bank, in 2004 India had achieved an uptick in telephone usage (43 lines per 1,000 people) and 48 users per 1,000 of cell phones. Yet, users of the Internet remained low (23 per 1,000), while only 11 people per 1000 had a PC, most owned by businesses. It also reported that the nation as a whole was spending about 3.7 percent of its GDP on all manner of ICTs, which was below the level at that time expended by the South Asian region (4.1 percent) and at least 50 percent less than most countries in the advanced economies.[18]

In 2010, yet a third annual survey reported essentially the same findings: India's use of ICTs remained below regional and global levels, with extremely limited use of Internet connectivity, for example, and equally low levels of IT adoption by businesses and the public at large. In adoption India ranked in the bottom quartile of 70 nations after China and most of the Middle East and Latin America, which historically had low rates of adoption too.[19]

In short, part of the answer as to why India embraced computing at slower and lower levels was caused by fundamental environmental issues. These endogenous circumstances existed as pervasive realities for years, demonstrating that their influence on Indian IT behavior remained in force to the present, despite an enormous, indeed impressive, growth in the indigenous IT industry in the 1990s and early 2000s. These data serve collectively as a cautionary reminder that the story of a nation's engagement with a technology involves both demand and supply side experiences, and broader environmental realities. One cannot rely simply on the supply side of the economic equation to dominate views of any nation's experience; there were too many other influential factors in play too.

This was a point made specifically in regard to India in 1992 when an economist, Robert Schware, discussed the strategic reasons why a government would concentrate first on the development of a local industry and then encourage its diffusion within the wider local economy before expanding into an export strategy, what was the pattern of IT evolution across all of Asia, except in India.[20] In India, the reverse was the case. Normally, local markets provide feedback into new product development and increase economies of scale to prepare for international competition, usually with domestic revenues. Schware used the phrase "walking on two legs" to describe this strategy. Arvind Panagariya claimed to have first discussed the notion, and certainly had by the early 2000s, when it was a topic of discussion regarding India's economic development. Panagariya viewed the "two legs" as about manufacturing and information technology, rather than an issue of domestic and foreign markets for a particular industry.[21]

Jason Dedrick and Kenneth L. Kraemer, however, get to the nub of the issue affecting India encountered throughout this chapter: when there is an

IT industry growing based on cheap labor, it remains vulnerable to even less expensive labor from another nation or, simply, from automation. As a result, such an industry fails "to develop project management capabilities, or to develop applications which can be packaged and sold to a large number of users." The direct and practical consequence is that it becomes "difficult to institutionalize the knowledge and experience gained by programmers working abroad, so that knowledge is wasted if programmers leave the company" (particularly while on projects outside of the country) and shuts off the opportunity for such skills to seep into other parts of the local economy.[22] That was the problem faced by India. This situation remains relevant today. It is against that backdrop that we can better understand India's introduction to computers, appreciate why it could enjoy such a positive "high tech" reputation in the early 2000s, but still remain a slow and limited adopter of IT.

EARLY ENCOUNTERS WITH COMPUTERS

As in other countries, India had used data-processing equipment before the arrival of the computer, albeit in very limited amounts. The Indian Statistical Institute (ISI) in Calcutta was one of the first users of such technology, when it installed an IBM Electric Accounting Machine in 1951, while in the following year the New India Assurance Company did the same. It was the first user of any kind of punch-card equipment in India. Yet, this insurance firm had used mechanical desk top calculators since the 1930s.

In the early 1950s, ISI constructed up to 20 data-processing card-handling devices, largely for the Defense Ministry. Professor P.C. Mahalanobis at ISI became aware of computers right after World War II and in 1950 met John Von Neumann and Howard H. Aiken—two important computer pioneers of the 1940s and 1950s—while on a visit to the United States. He took the knowledge gained to build a small analog electronic computer designed by S.K. Mitra at the institute, which they constructed out of surplus World War II electronic parts. They built the system to solve simultaneous linear equations in economic development. Their initial effort was followed by construction of a digital computer. India had its first digital computer running in 1954 or 1956 (there is some confusion about exactly when), called the HEC-2M (Hollerith Electronic Digital Computer 2M) at the ISI.[23] By the end of 1959, a Soviet URAL-1 had also been installed. As one Indian computer pioneer recalled, "by 1959–1960 ISI's Electronic Computer Division was the de facto National Computer Centre of the country," dating "the manpower development programme on electronic digital computers" to the mid-1950s.[24] Following a common pattern evident in other countries in the 1950s, ISI brought two Americans to India for six weeks to help design a computer-based system to use with the existing computer and to transfer other knowledge about IT to the local staff. Then in 1961, ISI ran a course on computers, considered the first on the subject taught to civilians in the country, with a class of 10 students, all of whom went on to careers in computing.

In 1953 IBM opened a data-processing service bureau in Bombay, a clear indication there was growing demand for some use of IT, largely for administrative applications. Important to the future of IBM's relations with the Indian government, the first IBM 1401 installed in India went into the Standard Vacuum Oil Company in Bombay during 1961, followed soon after with a few other installations during the 1960s.[25] Other suppliers of pre-computer equipment included Powers' Samas (British), Hollerith (British) and possibly from Burroughs, although the latter's archives do not mention its presence; years later it would play an important role with computers in India.

Several other projects began in the early years as well. In 1961 engineers at Jadavpur University (JU) and a team from ISI began construction of a second generation transistor computer they named the ISIJU-1. This project demonstrated the further diffusion of American knowledge about computers into India because Nicholas C. Metropolis, professor of computer science at the University of Chicago, spent four weeks in India working on this system. Then in 1962 the first national public meeting on computers in India was held, the Conference on Automation and Computation Scientists of India (CACSI), with some 30 attendees. Other events were subsequently held over the next several years, some hosted by computer experts from Jadavpur University.[26] In 1965, the Computer Society of India came into existence and, at the time of its silver jubilee in 1990, had grown to 12,000 members, making it one of India's earliest and most important gathering points for computer scientists and engineers. Other early IT projects were launched at the Tata Institute of Fundamental Research (TIFRAC), located in Bombay, initially on advanced components.[27] In the late 1950s and 1960s the Indian Institute of Science (IIS) at Bangalore offered training programs. Others were held at the Indian Institute of Technology (IIT) at Kanpur, at the IIT in Bombay, at the IIT at Madras in the 1970s, and at the military academies and other training centers.[28]

Who were these early users, those from the 1950s through to the mid-1960s? They were a handful of engineers and professors, and later scientists working on India's nuclear arms program. By the early 1960s, it appears there were also businesses and government agencies using these systems and the few data processing vendors in India (such as IBM) to do limited processing, largely clerical work, some accounting, and linear programming. The military services also were users, although we know almost nothing about their IT activities.[29] Almost from the beginning, however, fear spread in some organizations that these machines would cause people to lose their jobs, threatening to push them into the vast pool of India's poor. So, from the very start, there existed hostility to the technology, slowing IT diffusion for decades. One user recalled of the 1960s that a "very strong anti-automation agitation of white-collar workers" sprang up, "which was particularly successful in Calcutta in preventing installation of electronic computers anywhere for any purpose." These were government employees and, despite assurances that no jobs would be lost, they prevailed for years.[30]

Another early problem constraining use of IT was the lack of components and peripheral equipment, particularly in the 1960s as some deployment began of mainframes. It remained a chronic condition through the 1970s, reflecting largely the same kinds of issues faced in Communist European countries where state-operated suppliers failed to respond adequately to the needs of their customers with timely and sufficient quantities of equipment and support.[31]

As normally was the case in the very earliest stages of IT diffusion in any country, the birth of an indigenous computer industry occurred after the successful experiences with pilot projects, such as those located in universities and national laboratories, and as experiments in such high-technology Indian firms as telephone companies and in the pre-existing electronics industry served as proof-of-concepts. India reflected that pattern, which is why when one speaks about the Indian computer industry, the conversation begins with events of the early 1970s. However, in the 1960s IBM sold a few refurbished 1401 computer systems and ran its small service bureau, while the British firm ICL was also beginning to make some inroads. Best estimates hold that between 1960 and 1966 IBM installed 31 computers in India, accounting for 74 percent of all systems. ICT (later renamed ICL) claimed 2; another 2 came from the Soviet Union, and 7 others from Indian and other sources, for a total of 42. In the period 1967–1972, another 145 systems were installed of which 106 came from IBM, 17 from ICL; 12 from Honeywell, another Soviet system, and 4 other miscellaneous non-Indian systems. As in the earlier period IBM "owned" 73 percent of the market, ICL nearly 12 percent, Honeywell just over 8 percent—all in what can only be described at best as a tiny proto-industry.[32]

The Tata Institute of Fundamental Research, Jadavpur University, and the national government's Atomic Energy Establishment (AEE) were involved too, but in the 1960s in developing one-of-a-kind systems, so not really in commercial products. That was essentially the extent of the computer industry in India in the 1960s–very early 1970s. In many countries the protocomputer industry often originated in the local electronics industry, which, in India, was weak, largely focused on radios, television sets, and other small electronics in the 1950s and 1960s. All data about the electronics industry's growth in revenue during these two decades can confidently be considered not to have included the sale of computers, peripheral IT equipment, and software.[33] The story of an indigenous commercial computer industry began for all practical purposes in the 1970s.[34]

PUBLIC POLICIES, GOVERNMENT ACTIONS, AND THE INDIAN IT INDUSTRY, 1960S–MID-1980S

If the root cause of India's slow adoption of IT was structural, that is to say, tied to the shape of India's economy, infrastructure, and to its educational limitations, then the lack of effective public policies and programs that could have facilitated the diffusion of this new technology acerbated the influence of inhibiting features that circumscribed the use of computers in the subcontinent. There have been many critics of public policy toward computing and, in general, about India's inadequate economic development; but before recognizing these, a brief overview of what occurred during the first quarter century of India's independence helps to explain the slow embrace of modern information technologies.

After 1947, the government's initial economic programs were directed toward developing internal autarkic heavy industry, following a combined quasi-Marxist/nationalist approach to economic development, mimicking strategies implemented by China and other Asian countries. Often characterized as policies of "economic repression" during the 1950s and 1960s, they led to the "quiet crisis" in which Hindu economic growth hovered at 3.5 percent while the population grew at 22 percent into the 1980s, which resulted in growth in personal GDP of just over 1.3 percent per annum. In other words, this was an anemic result for a government committed to attacking the problem of poverty, potential famine, and modern industrialization.[35] Shifting from heavy industry to expanding agricultural output—the Green Revolution—in the 1970s, with accompanying expenditures of foreign currencies for such purposes, meant that computing, while not ignored, was clearly less significant to officials than to governments at that time in Japan, Taiwan, and Western and Eastern Europe. That neglect of IT began to change during the 1970s, however, when officials sought links to foreign sources of technologies, including IT. But, officials sought terms that would allow India to control these, such as insisting on substantial percentages of Indian ownership of local branches of foreign companies and the establishment of complex rules to ensure that foreign exchange came into India that could be used by the government for its priorities. India's defeat by China in the border war of 1962 also called attention to the need to modernize and expand the electronics industry at large, not just computing, for weapons, and to create military command-and-control processes.

Economists who have studied India of the 1950s–1980s often minimized greatly the nation's structural problems, focusing more on the fundamental policy regimes of these years. There is rough consensus among them and other observers that India's commitment to socialism, and its extreme implementation of autarkic economic policies did much to constrain modernization and growth of the economy. One can acknowledge the shocks of wars, droughts,

and rising oil prices, but still be critical of government's approaches. Economist Arvind Panagariya argued "that the performance during 1965–1981 was
worse than during any other period of the five and half decades" since independence, despite the policy intention of trying "to achieve an equitable
distribution of income and wealth."[36] In the process government policies
"effectively killed the incentive to create wealth at various levels."[37] Large
firms were constrained from competing in the world economy. Small enterprises could not participate either, since they often needed the channels available to larger firms through which to access national and international
markets. These conditions, when added to structural problems of the Indian
economy and society, contributed to slowing the rate and extent of adoption
of IT.

The fundamental strategy applied to both electronics and computing was
to leapfrog generations of earlier technologies, avoiding the requirement of
having India go through the phases of evolution in adoption of new tools
as occurred in other countries, such as moving from low value-add labor-
intensive manufacturing to more complex product development. The Bhabha
Committee, commissioned to develop a national strategy in the mid-1960s,
accused foreign firms of transferring obsolete products, technologies, and fabrication to India, arguing that this resulted in the nation being both behind
other countries and dependent on these suppliers. That included IBM and its
1401s. The committee advocated leapfrogging as an approach "to become
self-sufficient in the manufacture of computers of a wide variety," in addition
to other electronics.[38] It concluded that India's demand for computers could
be satisfied by local manufacturers for about $4.2 million and that this fund-
ing should immediately be allocated to the effort. These recommendations
were naive, and would be laughable except for the fact that by 1967, they had
become the official goals of Indian policy. The committee and many enamored
with the proposed approach failed to take into account that one could not just
leapfrog over one generation of hardware and software to another. Involved
were other critical issues not factored realistically into their thinking, such as
the managerial expertise needed to run different systems, new programming
skills to acquire, others for the design of more sophisticated applications using
far more complex software tools, such as database management and telecommunications, moving from batch to online systems, and as mentioned later,
having the right kinds of uses and volumes of transactions to pump through
more advanced systems. India was not ready for the leapfrogging that seemed
such a short cut to IT modernity.

Predictably, Indians faced problems with implementation. In the mid-
1960s, the Indian government approached IBM about the firm sharing partial
local ownership of IBM India. IBM rebutted that its internationally integrated
product development and manufacturing processes made this impossible
to do, as its work required centralized management of its operations. IBM

executives threatened to retire from India in 1968 if pressed further on the matter. Government authorities backed down.[39] In 1968, ICL, however, agreed to an Indian-owned manufacturing unit, but not for the local sales operation; that latter organization became the sole distributor of its products, which in effect ensured British control. In short, as of the late 1960s, the Indian government was not able to wrestle control over the foreign dominated tiny computer industry. Between then and 1972, the government was also unable to create an Indian computer industry. Perhaps only two locally built systems were sold out of a population of some 187 systems.[40] One student of the situation, Joseph M. Grieco, has argued convincingly that "prior to 1968 India imported almost all the computers it required, and in doing so it could search for the system on the market that was best for India." But after 1965, when the government had begun to pressure foreign vendors to manufacture locally (which they were reluctant to do), these firms were willing and able to bring in older systems, refurbish, and sell or lease these locally. That approach meant the latest technology would not be an attractive set of products to bring into the nation. Pressuring local users to embrace aging technology remained an important risk to government policy. Grieco concluded that the Indian government failed to implement its three objectives of controlling foreign currency exchange for such products, developing an indigenous industry, and promoting local use.[41]

In 1973–1974, the government again came to IBM to discuss the wholly owned subsidiary issue. The negotiations that extended to 1978 were long and protracted, but IBM executives did not budge, although they offered to expand manufacturing of some current products in India. These IBMers believed Indian computing had to go through various stages of growth and sophistication as had other nations. Leapfrogging did not seem practical to them, nor did surrendering part ownership of local operations. Grieco interviewed IBM executives in the 1980s on the issue to arrive at this conclusion.[42] IBM's corporate records on the matter reinforced his observations. While the temptation to divert at this point to a discussion of economic and technology development theory is great, suffice it to say that theory and history were on IBM's side. Leapfrogging required at least a modicum of indigenous capabilities which India did not have.[43]

Nonetheless, the matter with IBM came to a crisis in 1977, when the government pushed hard, while IBM made various proposals to ward off the inevitable. One report to senior IBM executives at their corporate headquarters referred to the prime minister, Moraji Desai, who IBMers met with, as "pro-west," and that he had a "reasonable attitude to foreign investments." After a June, 1977, meeting with him, senior IBM executives heard a presentation that suggested the situation was "difficult but not impossible" to resolve.[44] Meanwhile, business for IBM came to a virtual halt as everyone waited for the outcome. Tata and Burroughs negotiated an arrangement allowing Burroughs to operate in India, since Tata would participate as the local equity stake-holder. Ignored by all commentators was the extent of IBM's business in India; it was actually quite small. Its last previous order for data-processing equipment came in December, 1974; rental and other revenues were less than $6 million per year; its manufacturing staff hovered at 219 employees, with a total from all divisions at 803 people as of June, 1977. More meetings were scheduled for that fall, but the IBM archival record shows that the company also began to plan its retirement from India.[45]

By late 1977, it was clear to IBM that senior Indian officials had the "strong belief that IBM presence will impede development of local [DP] industry," while the IBM senior executive team concluded that "ownership is the key issue." If IBM gave in to the Indians, governments around the world would demand that IBM do the same in their countries, making it impossible for the firm to manage centrally a world-wide integrated company.[46] IBM was willing to give up its existing customer base of 362 data-entry accounts, 479 unit record customers, and 139 1400 installations, involving a total of 7,881 machines with a net book value of $5.6 million. Executives decided that, if India did not compromise, the firm would sell off its local installed base of equipment, stop supporting these products, and leave. The cost of separation expenses for employees, and other expenses would hardly exceed the sales value of installed equipment ($5.9 million vs. $5.6 million value). IBM developed a plan for evacuating the Indian market. In November, 1977, negotiations broke down, IBM announced it would leave, and in June, 1978, did so.[47] The government believed IBM's retirement would help in the development of a local industry, especially since the Tata–Burroughs arrangement had been completed in 1967 and had begun consolidating a position in the local industry during the early 1970s.

One important reason for India's stance with IBM resulted from its establishment of a local national champion, the Electronics Corporation of India Limited (ECIL) in 1967 as a government wholly owned firm, which it counted on to fill the gap left by IBM. By the end of the 1970s, four other Indian firms came into existence to assemble and sell systems: Hindustan Computers Ltd (HCL), DCM Dataproducts (DCM), Operations Research Group (ORG), and International Data Machines (IDM), established by former IBM employees. By 1980 ECIL had about 1,000 employees, Burroughs and Tata combined another 600; all the other firms some 1,900, for a total of 3,500 people comprising India's computer industry.[48]

The government required all customers and suppliers in the 1960s and 1970s to request permission to order systems, or to build and import them. Both suppliers and customers complained that the rules were onerous, and confusing, since they involved multiple agencies (often in rivalry with each other for control), all through the 1950s, 1960s, 1970s, and deep into the 1980s. Agencies and their regulations proved always slow to work with, often resulting in delays of up to 18 months from the time a request for a system was submitted to when it was approved. No doubt other potential users of computers may have concluded that the effort required to get a system was just not worth the aggravation of dealing with the government agencies in some long drawn out process. In the 1970s, the process of government intervention in acquisitions increased with the establishment of the Central Evaluation and Procurement Agency for Computers within the Department of Electronics, which was created to manage the process in a nation already populated with some 800 million residents. Even the Soviets at the height of their command-and-control management of their economy did not create such a tiny keyhole through which an entire nation's computing market had to pass through as in India's case, "with the sole responsibility of coordinating all activities relating to the import of computers," an authority it used to promote local vendors.[49]

There were consequences for these government policies and actions, on the one hand, and for vendors on the other hand. The heavy-handed control of whom and when computers could be made, imported, or installed clearly slowed adoptions, despite attempts of potential users to circumvent regulations. Second, the hostile policies of the government toward foreign computer vendors ensured that the inflow of modern large technologies would slow initially, especially for mainframes while newer minicomputers were manufactured or imported. One study of the age of systems in India in the 1960s and 1970s showed that the lag between older and newer technologies went from 4.4 years between 1960 and 1966, to 8.3 years between 1967 and 1972, then came down to 3.7 between 1973 and1977. The lag of 1967–1972 was largely due to IBM and ICL placing into the market older equipment, such as IBM's 1401, and the decline in the lag that followed was caused by the reduction of IBM's sale of older systems as it began to retreat from the market, replaced by newer systems, largely minicomputers.[50] Third, the creation of new firms provided the basis for Indian protocomputer and software industries that flowered in the 1980s, the most significant consequence of IBM's departure to the supply side of India's IT experience in the 1970s-1980s.[51]

Table 10.1 documents the number of systems in India from various sources up to 1980. A local, albeit small industry, was slowly gaining a presence, and that in turn gave officials confidence in dealing with IBM and in pursuing a strategy of local computing.[52] Officials thought ECIL could fill the gap left by IBM; rather the gap was filled by increased imports of equipment. Fourth, what is not clear is how many of the later installations replaced earlier ones as those of the 1960s became out-of-date and so were no longer as useful as newer models. But, as the data on lag times suggest, there were replacements so the total cited in the table should not be read as the total number in use; the actual number in operation was probably closer to 1,000 or less. Local Indian machines did not start trickling into the market until the end of the 1960s, actually more realistically in the early 1970s, and it took the departure of IBM in 1978 and the manufacture of local mini systems to open that local market. As one Wipro executive put it, "When IBM left, it created a vacuum so we decided to zero in on info tech."[53]

Pushing IBM and Coca Cola out of India, and in the process discouraging other foreign enterprises from attempting to enter the Indian market, resulted in more restrictions on foreign investments (the ownership issue for example, but also capital inflows and outflows regulations) that slowed the infusion of investments and technology into India. The effects were so immediate that public officials began to question the wisdom of their policies just as IBM was leaving the country.[54] Finally, and discussed more fully below, IBM's departure led to demand for local programmers who could write software for Indian-designed computers, which could no longer rely on IBM for their programs. Ultimately, that growth in demand for programmers was the most important consequence, indeed an unintended yet positive result, of IBM's departure.[55]

Grieco's assessment was correct that the government's policies proved relatively successful in encouraging development of a local industry in the late 1970s and early 1980s.[56] However, he was also critical as "the government clearly was unnecessarily inflexible with regard to IBM in mid- to late 1977, and as a result India's computer industry and user community suffered unnecessary costs." The government underestimated the value of having IBM in the local market. Officials also underestimated their ability to restrain imports, hence the cost in foreign currency. Officials were too concerned about appearing as appeasing foreign companies, leading Grieco to conclude that "the Indian government's actions toward IBM clearly deserve criticism."[57]

Overlooked by observers of Indian computing in the 1970s and 1980s was a fundamental effect on users—hence directly on the diffusion of IT in India—the role played by IBM's (and other foreign firms') sales forces in explaining to potential customers the value of using computers, helping them to rationale their operations, and teaching them how to use these systems. In every country

Table 10.1 COMPUTER SYSTEMS INSTALLED BY YEAR IN INDIA, 1960–1980

	All Systems	Indian Systems Only
1960–1966	42	0
1967–1972	153	2
1973–1977	317	77
1978–1980	962	840
Total	1,474	919

Source: Data accumulated by Joseph M. Grieco, *Between Dependency and Autonomy: India's Experience with the International Computer Industry* (Berkeley, Cal.: University of California Press, 1984): 41.

that had extensively embraced computers, the evidence of the importance of such a role existed. In countries where sales operations were limited, or non-existent, diffusion remained low, as occurred across all of Eastern Europe during the Cold War, and in China in the years prior to about 1980. Given the experiences of so many other countries, one can confidently conclude that having effective local sales forces permitted to expand their business as fast as local market conditions justified, was an important knowledge transfer process in evidence in Wave One nations. Just as academics and small groups of engineers facilitated knowledge and use of IT in the 1950s and early 1960s, so too did sales and service organizations in subsequent years, initially, for institutional customers who installed complex, large and mid-sized computer systems and, later, networks of terminals attached to these computers. These activities were all hallmarks of Wave One computing.

The inability of computing to expand its footprint in the 1960s and 1970s involved vendors and users in other ways too. By the late 1960s, Indian policies had led companies like IBM to offer the local market refurbished 1401s, since there were limits on how much foreign technology was allowed in the country, while constraints were placed on foreign firms that had more modern technologies on how much of their earnings from these they could take out of India. Costs to end users proved significant. By 1972, for example, at the height of IBM's installations of these systems, bigger, more reliable machines were available on the world market for a third of the cost; additionally, IBM was selling its S/360s, which had proven extremely successful around the world, but barely available in India. The big price/performance revolution that began in the late 1960s and that extended across the 1970s was the emergence of minicomputers, which were so much less expensive than mainframes and existed in smaller, hence, more affordable units. Very few of these came into India in the 1970s, leaving most Indian customers to sit out that significant evolution in the technology. Lack of adequate supplies of imported manufacturing capabilities insured while other Asian nations were learning how to build newer machines, India did not. Japan, Taiwan, Singapore, and other Asian societies did. The lack of manufacturing maturity was a major reason why in the early 2000s India did not have a competitive hardware IT manufacturing industry.

Many users were reluctant to acquire ECIL's TDC-12 computers between 1971 and 1975 as they were unproven, while at least IBM's old 1401s had a record of reliability. As a result, they only took delivery on some 35 of the locally made systems, with the vast majority shipped to government agencies, which had been pressured to support the Indian national champion.[58] One contemporary analyst of the Indian computer scene reported that by mid-1977, in total India had only 391 computers installed in the entire country, with the overwhelming majority second generation systems. Of these, 154 came from IBM and another 42 from ICL. ECIL installed more systems after

1977, bringing its total population in the country to 87, while almost all others were American minicomputers. The same report noted that in 1980 the population of installed systems had only increased to 600 country-wide.[59]

Why did the government fail to be more effective in facilitating diffusion of computers in these years? Publicly available information about what other governments were doing existed that could have informed Indian officials. IBM and other foreign firms wanted to find ways to operate in India. Poor performance in the 1960s can be attributed to many causes. Indian officials blamed IBM for pushing its 1401s onto users,[60] while other critics complained that there were inadequate supplies of technical expertise in agencies that had to approve acquisitions of computers by users to explain options for other systems.[61] The first reason seems a weak argument, but the second not so, because during the 1960s and early 1970s, agencies and publicly owned enterprises (both comprising the primary market for computers at the time) were regulated by the Department of Statistics within the Ministry of Home Affairs (later reporting to the Planning Commission). This ministry relied on the Computer Center within the Planning Commission for guidance and advice, inspecting potential uses, assuming technical requirements of systems, determining benefits, and establishing if the proposed system for a potential user would do what was required of it. It never looked for optimal alternatives. Even if it could offer alternatives, no agency had the authority to do that let alone implement such conclusions. Finally, inadequate knowledge of alternative systems and methods of assessing their viability stymied regulators. Multiple agencies trying to influence deployment of computers also slowed diffusion to a trickle, confusing and delaying users and installation of systems.[62]

Non-Indian economists looking at the situation of the 1960s and 1970s later argued that potential users could not always acquire low-cost computers, because prices remained 2.5 times higher than world prices for the same products; thereby limiting access to the economic and functional benefits of computing. Government agencies protected domestic manufacturers of computers, even constraining the emergence of a software industry in these years, for the same reason: inadequate access to hardware software developers needed. Higher prices were caused by import protection practices, limiting local demand for software too. These problems lingered into the 1980s, when, for example, the highly fragmented tiny computer industry led to some 200 small producers of PCs, with no economies of scale, and so remained uncompetitive in the world market. Manufacturing was restricted by fragmentation to assembly of imported parts, essentially for PCs; indeed, a classic example of low value-added work.[63]

In defense of the officials controlling the acquisition of computing, in the 1970s, they wanted to husband their foreign reserves and develop a local computer industry, placing high hopes that ECIL could do just that while it also built up local capacity. To buy time for the local national champion, they constrained the growth of IBM, ICL, and others by setting limits on the volumes of systems they could bring into the country. As occurred in other countries that supported local firms, government agencies funded projects at the ECIL and pressured users to acquire its systems. Yet in the late 1970s, local enterprises began to appear that challenged the primacy of ECIL. Officials tolerated that development as its new IT policy in 1978 after these upstarts had already become a reality. Officials restricted the new entrants to manufacturing PCs and minicomputers, not mainframes, although they also protected ECIL's own microcomputer product, the Micro-78. Users were directed to use Indian-built minicomputers instead of imported mainframes, or to use locally manufactured mainframes, which users perceived as less reliable than foreign products. The new entrants were able to grow and compete despite government policies to constrain (or not support) them, such as DCM and HCL. By the end of the 1970s, however, government support for ECIL began to erode.[64] As table 10.2 indicates, ECIL had not been aggressive in developing and selling products that were competitive both in price and quality to those available in the world market, largely explaining why government officials began allowing the new upstarts room to thrive. In short, the government had interfered too extensively in those normal market forces that routinely created demand and computer products. This situation remained relatively unchanged in India until the mid-1980s.

Tables 10.1 and 10.2 also indicate that not only was the Indian computer scene nescient, indeed positively slowly growing by world-wide standards, the circumstance clearly also reflected what can reasonably be called the very start of its Wave One computing, even as late as 1978–1980. As in other countries at the start of their Wave One, the small user institutions were heavily dominated by academic institutions and government agencies. Again as in so many other nations, India essentially had one national champion, and although that number grew during the late 1970s and early 1980s, the group remained tiny. The Indian results reflected a pattern evident in all other countries at the start of their Wave One experience, suggesting the Indian government had a large responsibility to play in both the level of diffusion achieved and in providing what it would take to progress through this wave.

If the Indian government had done more to encourage use of computers, might it had made a significant difference in the 1970s and early 1980s? The evidence suggests that users would have acquired more systems, largely foreign, essentially killing off ECIl and possibly limiting the results of other local firms. We know more foreign systems would have come into the country because government officials had applications for a couple dozen requests for imports that they either delayed approving or denied in any given year in this period. We have very little data, however, on another important component of the story: the potential users of these systems. Several dozen universities

Table 10.2 ECIL SYSTEMS USERS, 1971–1978

Type of User	Number Installed
Higher Education	28
Government-owned Companies	18
Central Government	14
Atomic Energy agencies	12
ECIL itself	12
Private companies	4
State Government	3
Others	7
Total	98

Source: Data accumulated by Joseph M. Grieco, *Between Dependency and Autonomy: India's Experience with the International Computer Industry* (Berkeley, Cal.: University of California Press, 1984): 127.

were possible candidates for such systems, certainly banks, large state and national government agencies, and such quasiprivate institutions as transportation systems, like the railroads. Yet the number of potential users remained small and their desire to acquire computers grew "slowly," to use the phrase observers often utilized to explain the demand. National government agencies were—and might have continued to be—early adopters, since they were extensive collectors and users of data and for such diverse applications as missile flight control to treasury accounting, inventory management, and to model flood control practices. The Indian Railways was perhaps India's largest user of computers in the 1970s and early 1980s for such traditional railroad applications as rolling stock inventory and accounting. The Indian military was also an early adopter for training applications, in planning, accounting, and intelligence. Some crime detection using computers also began.[65] But all summed up, the market would have remained small, just not as small as it turned out.

CHANGING CIRCUMSTANCES IN THE SUPPLY OF IT, EARLY 1980S–1993

As the 1980s unfolded, diffusion sped up. Government policy makers became more active in reversing some of their restrictive regulations of the 1970s, creating what one economist called economic "liberalism by stealth" in the 1980s with circumstances changing in small increments.[66] An embryonic IT industry began to pick up some momentum, personal computers came onto the market, and users of IT embraced computing more enthusiastically than in prior years. By the early to mid-1990s, both software and IT services industries had come into existence, while problems with telecommunications in India presented both opportunities and issues more serious than in prior years. By 1990, the Indian computing scene had the feel of what had existed in the Asian Tigers at least a decade earlier and in Japan in the 1970s.

During the 1980s the national government implemented a series of policy and regulatory reforms that favored a more liberal, free-market economy and, in the process, began opening India to foreign investments and inflows of technologies. Products imported into India began to be freed from some licensing requirements, beginning in 1978, and essentially more liberalized throughout the 1980s. Yet, import tariffs remained some of the highest in the world all through the 1980s, especially after 1984–1985 to generate revenues for government agencies. To begin facilitating some exports, India established export processing zones (known as EPZs), having experimented with different forms of these since the 1960s. By the 1980s the earliest ones were operating reasonably efficiently. They soon proved useful in the export of software in the 1980s and continued to do so in subsequent decades. The tone of these and other reforms were described by one Indian economist: "None of these policy changes was profound. Yet, the implementation of the existing policies seemed to have been more relaxed substantially by officials than a change in the policy," the notion of "liberalism by stealth," mentioned earlier.[67] Tactical steps helped too as in 1985, when the government ended its monopoly on the manufacture of telecommunications equipment, which had supplied products to one of the world's worst national telephone infrastructures. One of the first technology parks opened in Bangalore, and imports of electronic equipment were now encouraged.

Specific to IT, officials began to pay far more attention to modernizing this IT industry in the 1980s than they had in prior years. Experts at the time estimated the industry was a good 15 years behind the global one both in its research and production of equipment and software.[68] In particular, officials wanted to encourage the export of small peripheral IT equipment and software. Toward these ends, the New Computer Policy of 1984 and the 1986 Policy on Computer Software Export, Software Development and Training were introduced. The 1984 policy came from the Department of Electronics to promote the manufacture of modern technologies at internationally competitive prices, such as micro- and minicomputers that now could be produced by any company. To help them, tariffs on foreign machines were kept high, at approximately 200 percent of a product's value, but with the understanding that these charges would come down over time. Existing regulations that limited the amount of machines one could manufacture were eliminated, which made it possible for local suppliers to produce enough machines to achieve more optimal economies of scale. One could see that while liberalization was underway, officials still wanted to protect the local industry and not fully open India's market to the world.[69]

The actions taken in 1986 were in hindsight quite important. These were done to promote the development of a local software industry and to encourage the use of computers for economic development. On the software side, products from outside could come into India, which officials expected would result in Indians producing software products that they could export. This thinking was based on the limited success software developers had already enjoyed in the 1970s and 1980s in sharp contrast to their nearly complete failure in local hardware development. Software could become an engine of economic growth, while the inadequate availability of hardware could not be allowed to constrain this growth of the software business. No national champions in software were favored, unlike the prior and failed experience with hardware.[70]

On the backs of these two new policies came a series of actions by government. Financial and regulatory support for both IT R&D and training of computing experts expanded in the 1980s and the 1990s. In 1988, the National Informatics Center established NICNET, a communications network using satellite-based technologies that connected 439 urban centers in support of an export software services business for India, and to facilitate the diffusion of computing at national and local levels of government. A series of demonstration projects were also launched, such as the use of CAD/CAM with computer networks, installation of computers in such industries as cement, coal, steel, electrical power, education, health, and other public administration.[71] An effort was launched to make computers publically visible, such as the computerized Railway Reservation system, another for airlines, electrical billing, and accounting for pension programs. As two American economists noted, however, despite these efforts, there remained "a notable lack of incentives, such as tax breaks or accelerated depreciation rules, to encourage private sector use."[72] These were effective policy tools used by governments in many countries to promote the diffusion of IT. One result was that some 60 percent of adoptions of IT in the late 1980s and 1990s were government agencies, not private firms. Agencies were required to use local products when available before reaching out to global providers, mimicking a pattern evident in the 1960s and 1970s in France and in all of Eastern Europe in the 1960s through the 1980s.[73]

In the 1980s the government recognized the need to expand the pool of skilled IT workers and as early as 1983 had launched its Programme on Development of Manpower for Computers, stimulating a 10-fold increase in the number of organizations and universities that trained people. But it started from a small base, resulting in the development of 10,000 experts, up from 1,000 per year at the start of the decade. The increased availability of skilled IT workers satisfied about 50 percent of the demand. Supplies of skilled IT workers remained constrained by inadequate facilities, insufficient budget for the effort, and too few instructors. Had India satisfied the demand, the total number of workers would still have represented a tiny population so small that we cannot consider them as agents of IT diffusion with much effect in such a large country, unable to have the influence as did their cohorts in Europe, for example.[74]

Two economists who assessed India's efforts concluded that activities in the 1980s just loosened up existing regulations, "with minimal attention given to improving the IT infrastructure or directly promoting IT production or use." This resulted in low demand for computing, but, "by maintaining high barriers to computer imports, the government has created a situation where it is most profitable for hardware makers to simply assemble imported components for resale." On the software side the lack of adequate amounts of cost effective modern equipment led local firms to export their IT workers to such places as the United States and Western Europe to do contract programming. Many of these workers chose not to return to India.[75]

Nonetheless, there were appreciable results. Local firms could import components to build PCs to sell locally. Total sales of computers in India in 1979–1980 reached $12 million; but, in the following two years volumes more than doubled, then doubled again in the next two years. In 1985–1986 sales reached $180 million and in 1989–1990 climbed to $930 million.[76] Local firms now existed, and were beginning to become substantial enterprises, such as HCL, CMC, Wipro, ECIL, Pertech, and Tata Consultancy Services among some 250 manufacturers and service providers across India.[77] Regulations still constrained the consolidation of many of these into firms large enough that they could compete in the global economy or provide local customers with price competitive systems. What few exports existed went to the Soviet Union.

The software industry enjoyed export-based growth. In 1980 these amounted to only $3 million in revenues, but grew each year during the 1980s such that in 1990 these reached $128 million.[78] These data are, however, misleading too since they included revenues generated from Indian programmers working in other countries, leveraging low Indian wages in comparison to those prevailing in the United States and Great Britain, for example. The number of actual exports of software—not people—remained low in the 1980s, dominated most notably by Tata Consultancy Services, Wipro, and Infosys Consultants. In the language of the economist, "India's software industry has competed mainly on the basis of low-cost skilled professionals," a strategy continued during the 1990s.[79] Sales of software in India in the 1980s and 1990s were low for three primary reasons. First, the number of customers remained small; second, software piracy proved rampant with inadequate legal protections for producers of software products. High tariffs on foreign software encouraged piracy, of course. In the years 1988–1989, one local estimate placed the amount of pirated software at $30 million.[80] A third constraint was the lack of venture capital, which two observers described as "virtually non-existent in India," a concern of sufficient importance that officials had to address it, which they did with the economic reforms of the early 1990s. These reforms made it possible for foreign investors to begin to operate in India, such as IBM, and venture capitalist firms as well.[81] Combined, the three problems made it difficult for software firms to flourish, let alone procure adequate capital for investment in new products.

Despite these circumstances restraining expanded use of computing in India, the number of users began to increase during the second half of the 1980s. Total local expenditures on IT hardware, software, and services were an estimated $373 million in 1985 and $959 million in 1990 (see table 10.3). As measured by IT expenditures as a percent of national GDP, Indians were spending 0.40 percent on these technologies by 1990–1991. As table 10.4 demonstrates, spending on IT continued on the low side when compared to other Asian countries at the time.[82] In aggregate, all the data suggested that the adoption of PCs—the most popular class of systems in use in India then and now—remained relatively low in the 1980s and 1990s, even among the middle and upper classes that could afford these (1 PC per 750 of these people). That translated into 1 machine for every 4,000 Indians. In contrast, in Taiwan the ratio was closer to 1 per every 35 people. The rate of adoption increased toward the end of the 1980s, reaching 20.3 percent by the end of the decade, faster than any other Asian nation with the exception of South Korea which hovered at 25 percent, explained by India's late arrival to a process well underway elsewhere.[83]

In the 1980s and early to mid-1990s other barriers slowed diffusion. The two highly knowledgeable economists of Asian technologies, Jason Dedrick and Kenneth L. Kraemer, described many of the problems with the government's regulations, the nation's poor infrastructures, labor unions opposing automation of work, and the widespread belief that IT offered little or no value. They pointed out the lack of competition, which meant that local PCs, for example, cost between two and 2.5 times the world price for this technology, import barriers, and the widespread concern that adoption of computing would widen gaps between social classes. Nonetheless, they observed

Table 10.3 TOTAL IT EXPENDITURES IN INDIA, 1985–1990, SELECT YEARS (US $MILLIONS)

Year	Hardware	Software	Services	Total
1985	299	32	42	373
1987	450	49	67	566
1989	637	77	101	815
1990	743	91	125	959

Source: Jason Dedrick and Kenneth L. Kraemer, "India's Quest for Self-Reliance in Information Technology: Costs and Benefits of Government Intervention," Asian Survey 33, no. 5 (1993): 489–490.

Table 10.4 ASIAN IT EXPENDITURES AS PERCENT OF NATIONAL
GROSS DOMESTIC PRODUCT, 1990

Australia	2.44
New Zealand	2.25
Singapore	2.04
Hong Kong	1.51
South Korea	1.06
Taiwan	0.97
Malaysia	0.83
Indonesia	0.27
Philippines	0.24
India	0.40

Source: Jason Dedrick and Kenneth L. Kraemer, "India's Quest for Self-Reliance in Information Technology: Costs and Benefits of Government Intervention," *Asian Survey* 33, no. 5 (1993): 488–490.

that progress was finally made in the 1980s. As a result of local conditions and regulations, however, by the mid-1990s Indian suppliers of IT were more interested in exporting products and services than in serving their domestic economy. The world-wide shortage of programmers in the 1980s and 1990s encouraged this bias as Indian firms rushed to fill global demand, a major focus of Indian computing in the second half of the 1990s and the 2000s.[84] Those were resources diverted away from potentially promoting use of IT.

Before turning to the new phase in Indian computing that emerged after the early 1990s, the role of local telecommunications has to be accounted for because around the world PCs and many computers were rapidly being connected together through networks during the 1980s. India could not avoid the consequences of this convergence of communications and computing. Yet Indian access to good telephony was essentially nonexistent for several decades. The supply of national telephone service came through a state monopoly under the control of the Ministry of Post and Telegraph (P&T). it also owned three subsidiaries to ensure only Indian-made telephony was deployed: Indian Telephone Industries (ITI), Hindustan Teleprinters (HTL), and Telecommunications Consultants India (TCIL), the latter a consulting firm. P&T was famous for always being unable to meet India's demand for telephones, while its management defended their operations by arguing P&T never had enough budget, and adequate foreign exchange to import equipment, and that these requirements were of low priority in the broad scheme of India's economic development. Telephones, famously known as a luxury, a toy of the rich, in some Indian political circles for decades, sufficient appreciation of the significance of this kind of infrastructure hardly existed. Not until the 1980s were Indians able to enjoy a sharp increase in the supply of telephones. What few phones there were largely existed in urban centers.[85]

Table 10.5 provides information on the demand, installed number, and waiting time for delivery of a telephone. Leaving aside the inconvenience of not having a telephone for private uses, it becomes quite evident that businesses did not have them either, dramatically slowing the rate of communications and, of course, transactions, not to mention precluding the possibility of using online computing by hundreds of millions of Indians. However, the data also show that the total acknowledged demand was quite low given the size of India's population. This circumstance reflected many of the structural problems inherent in India's society and economy discussed at the start of this chapter. Existing demand grew at single digit levels until the second half of the 1980s, when it then expanded. Finally, the Indian telephone industry was not able to begin reducing waiting times until the 1990s and even then, marginally.

The Indian telephone system began using information technology in the 1970s, initially in some of its automated switches, expanding extensively use of that technology in the 1990s. Sam Pitroda, an American-trained telecommunications expert, is credited with personally pushing forward new initiatives to improve India's telecommunications infrastructure. He served as advisor to the Indian government in the 1980s and in 1984 established the Centre for Development of Telematics (C-DOT) to develop new local switching technologies, which it did.[86] A change in national government soon after meant he was out of power, and so his influence came to an end. By 2000, India had 14.5 million lines, and enjoyed some exports of a few telecommunications products to other emerging economies. Senior public support for expanding telecommunications depended on who was in political power, with changes in administration often leading to more or less support all through the 1980s and 1990s. After India's major financial crisis of 1990, which resulted in the IMF demanding greater liberalization of the local economy in exchange for help, telecommunications began reluctantly, often with visible hostility of some officials, opening up to competitors for landlines, and for cell

Table 10.5 INDIAN TELEPHONE INSTALLATIONS, WAITING LIST, AND WAITING TIMES, 1982–1992, SELECT YEARS (IN MILLIONS OF INDIANS AND TELEPHONES, YEARS ENDING MARCH 31ST)

Year	Telephones	On Waiting List	Waiting Times (Months)
1982	2.30	0.59	47.2
1985	2.90	0.84	43.8
1988	3.80	1.29	49.9
1992	5.81	2.29	37.1

Source: Data drawn from annual reports of the Department of Telecommunications and reported in Ashok V. Desai, *India's Telecommunications Industry: History, Analysis, Diagnosis* (New Delhi: SAGE Publications, 2006): 42.

phone services by the mid-1990s. The history of cell phones in this period is complicated and not yet fully clear; however, services were initially poor and expensive, due to government regulations and pricing rules. Ashok V. Desai, who has studied India's telecommunications, blamed government regulators for causing costs of cell phone service to be too high for a provider to offer and still make a profit. Wired service was also poor. Desai blamed the government for this as well.[87] Not until new regulations were issued in 1999 did liberalization expand, giving private suppliers more flexibility to offer their services. Soon after, the Delhi High Court overturned many of its covenants, restoring numerous prior governmental controls over the industry, "bringing the new entrants to the threshold of bankruptcy."[88]

The number of landlines continued to increase during the 1990s, albeit slowly, to nearly 20 million in 1998 and to over 45 million around 2005. Despite enormous regulatory headwinds, the number of cell phone subscriptions grew too, such that by the late 1990s there were 889,000 installed in 1998; 3.5 million in 2001, and in 2005, 53 million.[89]

Desai was harsh on the government's telecommunications regulators. For example, he charged:

> Its excessive workforce, and the impossibility of reducing it, reduced the incumbent's ability to cope with changes emanating from the market and made it averse to competition; its ownership structure gave it privileged access to policy makers, but also deprived it of access to risk capital, and consequently slowed down its response to competition.[90]

Others too spoke critically of the system. As recently as in 2000, Gurcharan Das was characteristically blunt:

> The government has made a mess of bringing competition in both basic and long-distance services, as is being done around the world. With all its improvements, India's telecom is still unresponsive and below par. Because of their monopoly, the employees of the Department of Telecommunications are still arrogant and corrupt. They remain the ugliest face of the License Raj.[91]

One Indian economist presented a more positive account. Arvind Panagariya characterized developments in the 1980s as "a modest beginning," although [he] had to confess that events in the 1990s made for "a messy road to private entry."[92] Panagariya concluded that the government "has achieved success in telecommunications, raising the teledensity from less than 3 percent in 1999–2000 to more than 16 percent by the end of 2006."[93] But, his own statistics affirmed the opposite result after a half century of activity, with only 16 percent penetration.

The root cause of the slow deployment of telephone services in India can be found in the government owning the incumbent telephone system, not with the private sector or potential users. The regulatory regime was fragmented and unresponsive to market needs, causing new entrants to rely on political influence over senior leaders to improve their situation. Beginning in the mid-1990s, these two circumstances—fragmented regulatory practices and lobbying influences—also led to rapid concentration, resulting in four cell phone operators dominating 66.6 percent of the market by 2002.[94] The government had never been a neutral regulator, as its agencies were swayed by the changing and highly parochial political interests of prime ministers, local parliamentarians, and the industry. Despite these problems, teledensity in India went from 1 telephone per 100 Indians in the early 1980s to 1 per 34 by 2000. Yet, this technology still remained out of reach of most villages and market towns, leading to a unique Indian solution: "call offices." Some 650,000 opened all over the country, much like a community public telephone office (not an unmanned telephone booth as in the West).[95]

GROWTH OF INDIAN IT INDUSTRIES IN THE 1990S AND EARLY 2000S

The slow diffusion of computers and telephony in the 1980s and early 1990s began to pick up the pace when India's adoption of computing and use of the Internet increased. Its software and services businesses also became major contributors to the country's foreign trade. The greatest successes in Indian IT involved the development of an export business for software and services. Government initiatives had increasingly focused attention on promoting the local software industry, beginning in the 1980s. This attention became the dominant priority for public officials in the 1990s. Before analyzing the burgeoning software industry, taking stock of India's IT situation as of 1993–1995 helps to set the stage on which India's diffusion of computing played over the next two decades. This helps us to understand the circumstances that continued to limit internal, but not stop, deployment of ICT.

In 1994 the World Bank completed a careful analysis of the IT situation in India, describing a harsh reality: "India faces pervasive forms of information poverty, its infrastructure and financial services are in need of substantial modernization," although it recognized the potential for the export of "labor-intensive software" and the leveraging of inexpensive workers for information services.[96] Nagy Hanna, an economist at the World Bank who went on to become an important expert on the use of IT in developing countries, argued in 1994 that India still lacked domestic demand for IT and an infrastructure that could facilitate its diffusion. One significant result was that 60 percent of university graduates with degrees in computer science left India for jobs in other countries. He criticized the government for having stiff tariffs on personal computers and software, which he characterized as "extremely high by international standards." The brain drain would constrain diffusion if demand picked up, while he called the telecommunications infrastructure "rudimentary."

In most countries, the national and many state governments used IT for their own operations, not only to improve their own productivity and efficiency, but also as an effective method to encourage use of computing by other sectors of society as a byproduct of dealing with public agencies. Hanna observed little of that occurring in India, or within the government-dominated banking industry, describing both as "limited and ineffective" adopters.[97] But he also made similar comments about the use of IT by small and medium-sized enterprises, because they too were hardly aware of the potential benefits of using this class of technology. That lack of knowledge was due to various causes: not enough associations to promote the use of IT, lack of adequate sales and support coverage by vendors, and management reluctance to spend money on consultants to advise them on what to do. With respect to software—upon which the Indian IT world was beginning to pin its hopes—Hanna criticized software firms for being "deficient in marketing, productivity, and quality, mainly because of the predominance of small software houses with limited capacity for marketing and in-house training," charges reiterated by other observers of India's software world of the 1980s and early 1990s.[98]

India still had one of the most complex regulatory environments affecting all industries, not just IT. For example, in the financial sector—banks and stock brokerage firms—hundreds of laws and additional enabling regulations were enacted *each year* that affected the activities of these enterprises, creating confusion, and delays in adopting IT, collectively negatively affecting their performance. The Reserve Bank of India documented these concerns in an attempt to reduce their complexity, working with various national and state government agencies and legislatures. Both in government and in these industries, management and labor resisted changes that would have facilitated adoption of computing. Similar underdevelopment in the use of IT existed in manufacturing industries.[99] The results are displayed in table 10.6, with comparative data from several neighboring countries. These data are ranked in order of percent consumed, which in India's case involved the largest occurring in government and manufacturing, with a sharp drop in the acquisition of IT by the financial sector when compared to patterns of adoption evident in the region. Finally, note the total size of the domestic market as compared to that of other nations we have studied; it was small, tiny to use the word I have employed elsewhere in this chapter.

Exploring the evolution of the software industry against this challenging background allows us to understand three circumstances: the relative position of domestic and export roles of IT within the national economy; effect of IT public policies and actions in the 1990s and early 2000s; and diffusion of

Table 10.6 INDIA'S DOMESTIC IT MARKET COMPARED TO OTHER
ASIAN MARKETS, 1992 (PERCENTAGE OF TOTAL ACQUISITIONS OF IT
BY SELECT SUBMARKETS)

User Community	India	Japan	Singapore	South Korea	Taiwan
Government	15.2	7.9	20.0	18.6	29.0
Manufacturing	23.5	25.7	16.0	44.1	12.0
Transportation	11.5	—	9.0	—	2.0
Distribution	10.1	9.4	8.0	—	8.0
Education & Research	9.1	—	2.0	1.8	5.0
Service Industry	8.6	5.5	—	1.8	6.0
Finance	8.1	26.8	27.0	25.3	25.0
Home & Individual	2.1	—	4.0	—	2.0
IT Market (US$M)	$880	$56,278	$1,152	$3,633	$665

Source: data extracted from statistics in Nagy Hanna, *Exploiting Information Technology for Development: A Case Study of India* (Washington, D.C.: World Bank, 1994): 15–16.

emerging technologies of the period, most notably use of the Internet in India. India's experience with software represented a positive development, both for the nation's economic performance as a whole, and as an enabling force that facilitated the diffusion of IT in other countries. It is this latter shining development within this nation's IT experience that helped to justify devoting a chapter to India in our analysis of how IT diffused around the world. While much of what went on within the industry remained isolated from many mainstream economic activities in India—the contradictions between a high-tech Bangalore and bullock carts on dusty roads, even in major cities—it is an important part of modern Indian economic history.

A few statistics make this circumstance more obvious. Since the late 1980s, for over two decades this industry grew each year by over 30 percent. Most of its sales and services were exported to some 60 countries, with about two-thirds to the United States. In the late 1980s annual exports hovered at $50 million, grew to $200 million by 1993, and to some $6 billion in 2001. By 2008 these had approached $60 billion. In 1995, software exports accounted for only 2 percent of all Indian exports, but by 2008 approached 25 percent. Measured as a percent of Indian GDP, by 2004–2005 it had reached 4.1 percent, and it continued to gain share of total GDP over the next several years. By the early 2000s Indian outsourcing firms enjoyed 3.3 percent of the global outsourcing market.[100] To put things in perspective, however, table 10.7 compares domestic to export volumes; note that the domestic market remained relatively small and slow-growing in recent years when the Indian IT industry was considered to be the most vibrant in modern Indian history and dynamic internationally. As one would expect under these circumstances, the percent of total sales of the software business going to domestic users should have declined as the export business grew faster than local volumes. Foreign sales grew at double digit rates from a small base each year. In 1987/88—the first year for which there are reliable data—exports totaled $2 million; a decade later (1997/98) these were nearly $1.8 billion.[101]

Students of India's software's history date the birth of this IT sector to 1970 when Tata Consulting Services (TCS) began to provide programming outsourcing services. With IBM's departure from India in 1978, the notion of selling—and buying—outsourcing services, such as programming, installation, and operation of applications took hold in India, and later as an exportable offering. The Central Bank of India and Bombay Telephones became early adopters of such services. In the early 1980s TCS sent engineers to Burroughs Corporation in the United States for training and then placed them in "bodyshop" programming assignments in the United States and in Germany, through a joint partnership called Tata–Burroughs.[102] In the same decade other firms began to offer such services as well. These events all became possible because Indian-trained programmers were in larger supply than could be assimilated within the local Indian IT economy, actually leading to workers being successfully exported as early as 1968. These Indians demonstrated that they were competent and were also far less expensive than European and American programmers, if the later were available as there always seemed to be a shortage of these kinds of workers in the 1970s through the 1990s in the West. Many of these individuals remained in the United States and elsewhere, creating an Indian diaspora that resulted in a tight network between Indian and American IT experts.[103] "Bodyshopping" (as it was called) eventually evolved into higher value offerings with Indian software firms developing programs for American clients, beginning in the 1980s. This signaled that India's software capabilities were maturing and were being recognized as reliable and competitive in other countries.[104] Indian firms began wisely to adhere

Table 10.7 INDIAN DOMESTIC AND EXPORT SOFTWARE SALES, 1996–2007, SELECT YEARS ($BILLIONS)

Year	Domestic	Export	Percent Domestic of Total Sales
1996/97	0.8	1.1	42
1999/00	1.9	3.4	36
2002/03	2.8	7.1	28
2004/05	4.3	12.2	26
2006/07	8.2	31.2	21

Source: NASSCOM's data in Sabhash Bhatnagar, "India's Software Industry," in Vandana Chandra (ed.), *Technology, Adaptation, and Exports: How Some Developing Countries Got It Right* (Washington, D.C.: World Bank, 2006): 51; NASSCOM, *Indian Software Directory* (various years).

to international quality standards in IT, which helped to sell work. In 1988 the National Association of Service and Software Companies (NASSCOM) came into being to facilitate that process and to promote India's software industry.[105] The growing availability of the Internet encouraged both the export of talent and also the import into India of software work.[106]

Already by the mid-1980s, national government officials began to note the success software firms were enjoying, in sharp contrast to the disappointing performance of hardware vendors (the group officials most wanted to promote at the time). Briefly put, prior to 1984 officials and economists did not recognize the existence of a local software industry, so bodyshopping developed essentially on its own. As mentioned earlier, the impetus for local growth in the late 1970s and extending into the 1980s was IBM's departure from India and its successful replacement with new local firms. In the 1980s, national officials began to liberalize its import/export rules and other regulations to encourage the growth of this embryonic business, largely with two major initiatives in the 1980s: the Computer Policy of November 1984 and the Computer Software Export, Development and Training Policy of December 1986, both of which encouraged further bodyshopping.

In 1990 the government began the process of establishing Software Technology Parks (STPs), sites devoted to the software and outsourcing businesses, complete with modern telecommunications facilities in order to move work in and out of India from other countries. Meanwhile, education of programmers and other computer experts continued apace at Indian institutes and universities. The STPs proved instrumental in facilitating the expansion of Indian software firms, particularly in bringing software work into India to these locations. In 1990—the last year before the availability of STPs—only 5 percent of India's software export revenues came from offshored services. By the end of 1994, offshored worked done in India accounted for 30 percent of India's software export revenues; a percentage of the total that grew in the 1990s and to just over 50 percent in 2001/2002. That business kept growing in subsequent years.[107] In the 1990s the government began encouraging foreign enterprises to establish centers in India and to collaborate with local firms. IBM came back to India in 1991, initially through a joint venture with Tata Industries, with a facility established in Bangalore the following year, and five years later with another in New Delhi.[108]

A result of the various factors, not the least of which was the changed regulatory practices of the government, was the spread of industrial software parks to various parts of India, most notable at Bangalore, what one Indian economist called "India's Silicon Valley," the nation's first STP.[109] This city already had higher education facilities, a cadre of skilled workers, and had long enjoyed a reputation as having a good climate and life style. By 2000, similar pockets of firms and skilled workforces existed in other parts of India. As measured by percent of software exports Bangalore controlled 27 percent of the

total with 160 firms headquartered there; the New Delhi/Gurgaon/Noida cluster accounted for just over 15 percent with 106 companies; Chennai just over
10 percent with 72 firms; Hyderabad enjoyed 7 percent (that share grew in
subsequent years to 61 enterprises), and the Mumbai/Navi Mumbai areas
nearly 6 percent and with 148 companies.[110] The rest of the exports came from
other cities. Two points to make is that the industry now had footprints in
various parts of India and that these firms were highly concentrated in STPs
with the majority of their work directed out of India.[111]

Indian officials were pleased, because these STPs contributed significantly
to exports, hence to the accumulation of foreign currency reserves that could
be used for other projects, while giving India the patina of a new Asian High-
Tech Tiger at work. Yet, most observers of the Indian scene continued to note
the contrast of high-tech versus bad infrastructure and poverty in the nation,
and potentially not having enough IT workers, raising questions about the
sustainability of India's IT supply side prosperity. Two Indian industry analysts concluded in 2007 "while ICT Superpower status is a realistic aspiration,
government intransigence on the two primary constraints on India's progress—
infrastructure and an expanding well-educated workforce—is expected to
continue."[112] With the worldwide recessions of 2001–2002 and 2007–2009,
companies in many countries sought to outsource work to India where software support was now available, reliable, and far less expensive than at home.

The U.S. recession of 2001–2002, however, also had the reverse effect of
causing some 35,000 Indians to return to India, often spurred by government
programs and perceptions of growing opportunities in India. Their return
home brought back to the nation skills in software, the entrepreneurial experiences gained in Silicon Valley in establishing and running firms, and many
personal connections through which to sell back to their prior locations work
now that could be done in India.[113] These returnees played an important role
in developing India's software industry, while providing positive images of
Indian programming and software design, and they were able to hire Indians.
In 2000, for example, Indians ran or were senior executives in 973 firms in
Silicon Valley, which accounted for $50 billion in sales and some 26,000 jobs
in the United States. They had been at it for some time; between 1980 and
1985, they led 3 percent of all new technology firms and a decade later some
10 percent of Silicon Valley startups.[114] Many were not actively interacting
with Indian IT enterprises. So, the potential to expand the connection was
substantial.

The government was not able to cause high levels of concentrated pockets
of IT skills to spill out into the rest of the economy. Market demands for these
resources were foreign, far in excess of local opportunities, having the effect of
denying the Indian economy access to these resources. There was also too
small of a group of people who could create that demand, such as salesmen.
Foreign vendors, who were good at this, like IBM, were not yet adequately

engaged in a domestic market, and the few that were, did so in insufficient numbers to have any appreciable effect on local demand. This is a circumstance that stands out when compared to patterns of behavior in countries that did enjoy high IT deployment. In addition to macro economic conditions and supportive government efforts, a stronger vendor presence was there and in those countries where governments encouraged use of IT. Any emphasis on the role (or lack of) by the Indian government is only part of the story; the absence of a vibrant local industry selling to local businesses is another crucial element.

The will of government officials and Indian diffusion did not mean there was no activity. The World Bank cataloged some of the data on diffusion, circa 2006, exhibited in table 10.8, highlighting issues that encouraged the Indian government to implement various policy initiatives since the 1990s to address. A National Task Force on IT and Software Development, established in 1998, made 108 recommendations regarding improvements in broadband, other telecommunications infrastructures, and financial incentives, among others, while The New Telecom Policy (1999) was intended to bring forth state-of-the-art telecommunications for the software industry. Then in 2004 a Broadband Policy was introduced with enabling legislation, building on the Information Technology Act of 2000, for example, which legally recognized and allowed electronic commerce and filing of public documents via the Internet. Next came the National e-Governance Plan in 2006, intended to facilitate implementation of online access to public services. Students of the development of India's software industry generally gave the government good marks for its efforts of the 1990s and early 2000s.[115]

Parthasarathy contended that the government's positive role could be attributed more to these different approaches than those which proved such a failure in the 1970s and 1980s. Until the mid-1980s, the government operated on its own, discounting any possible role for the private sector; it used its own

Table 10.8 ICT CONSUMPTION IN INDIA AND THE WORLD, 2006

	India	World
ICT expenditures as percent of GDP	5.15	6.56
Personal computer per 100 people	2.79	15.32
Internet users per 100 people	6.85	18.64
Broadband subscribers per 100 people	0.21	4.90
Fixed-line and mobile telephone subscribers per 100 people	18.64	62.13

Source: Data from World Bank, cited in Balaji Parthasarathy, "The Computer Software Industry as a Vehicle of Later Industrialization: Lessons from the Indian Case," *Journal of the Asia Pacific Economy* 15, no.3 (August 2010): 261.

ECIL, while allowing IBM to leave the country. By the late 1980s sector-specific policies and initiatives began to appear, along with new voices providing qualified input, such as the IT industry association NASSCOM. The software industry itself also embraced effectively software standards and tools, such as Unix, which allowed firms to work flexibly with companies and various technologies all over the world, and to improve their economies of scale by moving up the value chain from simple "bodyshopping" to more profitable software design and delivery, all buttressed with public policies in support of these changes in the industry.[116]

Good statistics on the number of IT workers in India in the 1960s through the 1980s are hard to come by, but, nonetheless, important information as a gauge of knowledge about IT in India. We know that the various Indian technology institutes and universities were training thousands of students per year.[117] By the 1990s, the volume was impressive. Using 1998 as an example, there were just over 660 institutions of higher learning, enrolling over 150,000 students in engineering (although other forecasts range as high as 300,000), most of whom learned various amounts about programming and computer science.[118] At the turn of the century the number of Indian workers in the IT and IT-enabled services sectors was estimated by NASSCOM at about 284,000; that number rose to just over 1 million by 2005, with about 175,000 new entrants into the IT workforce by that year.[119]

Few domestic diffusion success stories have yet emerged, although, implementation of embedded IT in various manufacturing operations was one of them. Parthasarathy's own view was that "as the informational needs of the export markets of the Indian industry had little social relevance for a vast segment of the Indian population, the requirements of the latter went unaddressed."[120] Nor was the IT sector big enough to unilaterally uplift India's national economy.[121]

IT DIFFUSION WITHIN INDIA, 1990S–2010

Given the growth of IT capabilities in the nation that began in the 1980s, as the government slowly eased its neo-Soviet styled controls over the economy, it became possible for the supply side of the IT economic equation to flower. What occurred on the demand side, that is, among users, once the newly formed IT industry began to expand in the 1990s? I have already argued that the demand side lagged severely in its use of IT from the 1950s through at least the mid-1980s, but what about afterward? In fact, deployment of IT picked up, albeit still slower than in many other nations. Already by the mid-1980s, major institutions were often installing their first or second computer, such as Indian Airlines, Air India, and various large banks. The World Bank had insisted that Indian Railways improve its operations by using computers in exchange for loans from the international body, such as in reservation and rolling stock inventory processes. The Computer Maintenance Corporation (CMC), formed to support IBM's customers upon the departure from India of the American firm, became an important agent of change in the late 1980s, because it provided software programming services and project management to local companies, such as to the railroad industry. The railroad projects went well, while Indian Airlines, and its Sperry computer and software, poorly executed its installation of new IT systems with too few terminals. That circumstance created problems for passengers trying to get on flights, and led to negative press coverage of computers in general in the late 1980s, not just at Indian Airlines.[122]

Banks in many countries were some of the earliest adopters of computers, and to a limited extent also in India. As early as 1961 the Reserve Bank of India (RBI) and the State Bank of India began using computing, soon after IBM 1401s and Honeywell Bull 400 systems. By the end of the 1960s, however, banks still faced severe problems using computers, because their labor unions resisted automation for fear of losing jobs. During the 1970s the RBI took the lead in importing a few computers and making them available to its industry, easing the trenchant stance against the use of such technologies that even some management at banks supported in the 1960s and 1970s, an attitude that did not begin to dissipate until the early 1980s. In an historic agreement a group of Indian bank unions agreed to allow computerization to begin in commercial banks, documenting which processes could be computerized and which were off limits, such as no computing in branch offices—a unique restriction not in evidence in any other banking industry studied for this book. Not until the 1990s did such restrictions begin to disappear.[123] Finally, some insurance companies, the postal system, and a few large hotels also began to automate in the 1980s.

Yet, to quote a World Bank report from the mid-1990s, "India has an enormous pool of potential users. Well-established capital-goods, consumer electronics and pharmaceutical industries, large transport and distribution networks, and a growing financial system constitute a substantial source of potential demand for software," a demand not yet satisfied.[124] The same report noted that potential users still did not have many of the basic resources required to make investments in IT: teaching management about the benefits of using IT, availability in the domestic market of IT experts (most were drawn to the export-oriented IT sector), loans from banks for investments in enterprises, and access to consultancy services. To be more precise, in India these did not exist to the extent required:

> Inadequate policies and management skills slow down the adoption and lower the benefits of modern information and communications systems. A substantial lack of awareness exists of the potential benefits and associated costs of IT,

particularly among small and medium-sized enterprises and public institutions. The physical and institutional infrastructures for IT diffusion are at an early stage of development.[125]

A World bank official blamed the national Indian government for its efforts to-date, characterizing them as "fragmented," advising that "the activities of various ministries and industry associations need to be coordinated and strengthened through a coherent national strategy—to broaden the domestic market," for example.[126] Missing from this criticism of the government was any mention of the personal responsibility of Indian business executives to learn about emerging technologies and managerial practices from around the world, a duty routinely carried out by senior managers in industries in most countries. There also continued to exist widespread reluctance on the part of local managers to use IT and management consultants for intangible services, such as educating employees on best practices in IT adoption and about managerial practices. Indian management's insularity was part of the problem, leading to misjudgments about the role of IT, and after the reforms of 1991, to many ill-conceived business ventures. Both were circumstances that diminished slowly over the next 15 years, however.[127]

Other problems extended into the 1990s of a more immediate nature. For example, public officials favored local hardware manufacturers, driving up the cost of Indian and foreign-made computers. Yet Indian hardware and software suppliers invested less in the education of their customers than evident in other countries. One economist summarized a fundamental problem visible in both government agencies and in the private sector: "Informatics professionals lack business orientation, while business organizations invest little in understanding the technology." To compound problems, "computer and information literacy is scant and business managers are poorly prepared to manage the skill and institutional changes associated with the introduction of IT."[128]

The public sector came in for special criticism beyond policy and regulatory practices. From the perspective of government as a user of computers, "Indian public administration is notorious for unnecessary paperwork, excessively hierarchical management, inflexible personnel policies, and limited attention to services," made worse by "resistance to the use of information systems," often out of fear of job losses, a concern shared with workers in the government owned banks.[129] In 2003, a RAND report leveled similar charges of inadequate use of IT, this time with respect to e-government applications, declaring that the national government was failing to make it "a priority," and where examples of deployment did occur, they were spotty and in the majority implemented at the state level.[130] An Indian government report in 2006, after presenting an elegant study on the use of IT in nine ministries, concluded that the use of IT still remained low, "for example, transaction of official documents or use of digital signatures is not common in spite of intranet availability." Furthermore, the only language on these websites was Hindi, only one of a large number of languages spoken in the country. Final observation: "The institutional systems do not appear to be fully conducive towards effective promotion of e-Readiness."[131]

These were all early Wave One issues, and included additional manifestations such as widespread skepticism about the value of IT, fear of the unknown, need for pilot projects, proof-of-concept demonstrations, and inadequate management of installations. As a reminder to the reader, the situations described in the last two paragraphs were of circumstances of the mid-1990s, long after many of these had been put behind them in other countries. The ultimate problem, already mentioned, but that continued through the 1990s, and beyond, was the lack of a porous interaction between the vibrant export-oriented software and outsourcing industries in India and a domestic market. The World Bank's economist observing the Indian scene in the mid-1990s explained, "until recently, government policies and promotional efforts in the software industry have focused almost exclusively on exports of software manpower," with the result that knowledge of IT did not diffuse back into the domestic economy, let alone result in local installations," a situation that changed only slowly beginning in the late1990s.[132] As in other societies in early stages of Wave One diffusion, knowledge transfer and skills development remained important conditions that always took time to either conduct or develop.

Finally, what happened with the Internet? If PCs were too expensive for the average Indian to acquire, or for small and mid-sized Indian firms to use, and larger systems were slow to be implemented because of insufficient understanding of IT or government impediments, what about the Internet, the one new technology that, if adequate telecommunications infrastructure was in place, individuals and companies could flock to either through their PCs or cell phones? The question should be asked because the Internet spread around the world (both early at the start of the 1990s and more widely worldwide in the late 1990s and early 2000s). Often it did so as much by businesses as individuals, so the story of India's slow embrace of the Internet seems initially to be an anomaly. Its experience with the Internet remains a complex story, but one that aligns with many of the patterns evident in the use of other forms of ICT. The diffusion of the Internet followed many of the same trajectories evidenced in other countries too, only slower in comparison. Adoption evolved over roughly four short phases, from its introduction in the mid-1980s to the present.[133]

Prior to the arrival of the Internet, there had existed telenetworks in India, in spite of the local underdeveloped telecommunications infrastructure. In 1986 the CMC established what, perhaps, became the first network in the country, INDONET. It used initially telephone leased lines that connected to eight cities by 1998. In 1987 the National Informatics Centre (NIC) began implementing a satellite-based network, linking state capitals with the nation's capital. Over the next several years—what we can conveniently refer to as the first phase (1986–1991)—a few more networks were established. The first Internet connection came in 1986 and over the next few years universities began to exchange e-mail, with the first e-mail connection to the United States occurring in 1988. The most important of the early links was ERNET (Education and Research Network), established in 1986 that connected the Department of Education and seven other users, all of them institutes of science and technology in various cities.

In 1989 India obtained the domain name.*in.* Toward the end of 1991 and early 1992, a second period began that extended to 1995 in which e-mail and private networks spread as a result of regulatory changes allowing these, largely for use by some firms and other institutes for a total of 13 closed user groups by 1995. Future historians may well come to agree with Gurcharan Das that reforms initiated in 1991 were "golden," to use his characterization of this time, in stimulating the arrival of a new era in Indian ICT.[134] Yet, for all practical purposes use remained contained within a tight circle.

A third phase began in 1995 that extended to about 2001. In this period Internet and network access was provided by commercial organizations. Internet service providers (known as ISPs) provided international gateways. Access spread to additional cities and other communities. By the end of 1998, there was an excess of 200,000 users of the Internet in India.[135] The change in telecommunications regulations in 1999 made it possible for private suppliers to compete directly with the Department of Communications, followed by other legislative and regulatory acts that began to encourage more actively the use of data transmission via telecommunications. This was largely the case for the private sector, and, of course, for use with the Internet. The number of communities with some access to the Internet had climbed to roughly 125 by 2001; but, by the end of the following year had jumped to over 450, scattered widely around the country.[136] Because of their importance for the software industry and business process outsourcing, it should be noted that by 2002 international connectivity through commercial ISPs occurred through six international gateways, located where many technology firms were headquartered: Bangalore, Calcutta, Chennai, Delhi, Mumbai, and Pune. These reached out to Japan, the United States, Italy, and Singapore, later joined by other networks. Demand always exceeded capacity, reaching crisis status in 2000 when the lack of sufficient bandwidth became a national cabinet level topic of discussion.[137] Additional capacity was added, much more fiber-optic cable laid, and new ISPs licensed to operate in India, ushering in the next—and fourth—phase in the evolution of the Internet in India. Yet, in these early years, most subscribers were businesses and other organizations. Table 10.9 provides an estimate of the number of subscribers during the key period of early adoption.

Table 10.9 ESTIMATED NUMBER OF INTERNET SUBSCRIBERS AND
USERS IN INDIA, 1995–2006, SELECT YEARS

	Subscribers	Users
1995	2,000	10,000
1997	50,000	450,000
1999	350,000	1.4 million
2002	3.8 million	10.6 million
2006		86.0 million

Source: Data on subscribers tabulated by Peter Wolcottt and Seymour Goodman, "Global Diffusion of the Internet I: India: Is the Elephant Learning to Dance?" *Communications of the Association for Information Systems* 11 (2003): 607–608; data on users from multiple sources collected by R.C. Mascarenhas, *India's Silicon Plateau: Development of Information and Communication Technology in Bangalore* (New Delhi: Orient Blackswan, 2010):123.

Among citizens, users came from the middle and upper classes. To access the Internet they needed to be literate, to afford a personal computer and a telephone line, and to speak English, or at least read it, as very little content became available in Indian languages until the turn of the century. By about 2000, India had a middle class of between 150 and 200 million people (criteria for membership varies by estimate), living largely in some 150 towns and cities across India. As two observers of the scene commented, "In 1999, the Internet moved from an exotic technical accessory to a relatively common-place component of middle class day-to-day life," although as the deployment data in table 10.9 demonstrates, while perhaps commonplace for some, it was not for most others in the middle class.[138]

Formidable barriers to access existed. At the turn of the century, first poverty continued to be widespread and severe. That circumstance alone would have made it impossible for millions of people to even afford a telephone. But, there were other issues as well. A second one was the existence of very few telephone connections, indeed only 26 million in 2000 in a nation rapidly reaching a billion people in population. Existing lines proved too often unreliable or too low in capacity to transmit data, largely because of the lack of a national data backbone network. Third, the cost of Internet access remained expensive. Fourth, other necessary infrastructures were also in poor condition, such as those that would not allow one to take advantage of the higher speed transaction of business over the Internet, such as bad roads and unreliable train service, which cancelled out any of the advantages obtained by faster transaction processes to integrate effectively into supply chains.[139] While the number of commercial websites is not accurately known, one creditable estimate held that as of 2000 there were at least a few hundred, and probably none were yet profitable, again a tiny number.[140]

Business-to-consumer websites (B2C) could not function well in the late 1990s and early 2000s because the number of potential customers proved too low. There were less than 4 million PCs in India in 2000 and only 20 percent of those were used by households, logical candidates for B2C business. That left the other 80 percent, however, for business-to-business website (B2B), yet even that group of potential users was miniscule too. The third potential set of users, business-to-government websites (B2G) was a possibility since the public sector was spending over 28 percent of all IT expenditures in the nation at the time; but as already discussed, government agencies were also slow adopters of IT. The biggest markets, explored in the next section of this chapter, were providers of Internet-enabled services to companies in other countries, such as software development and maintenance, and call centers. In India those services alone provided up 100,000 local jobs using the Internet, generating some $500 million in export revenues.[141] Yet overall, demand for Internet services in India grew as the nation entered the new century. Educated members of the middle class wanted access to information; middle class government did too for use at work.

Even people who could not afford a PC and a telephone line wanted access to the Internet and, as happened in so many other Asian countries, cyber cafes appeared quickly to satisfy part of this demand. One early survey suggested that by April, 2001, there were some 8,000 of these facilities in India and by June of the following year, 12,200. To put this information in perspective, another study suggested that 60 percent of all users of the Internet accessed it through cyber cafes, implying that the data presented in table 10.9 understated the number of users. We just do not yet know.[142] Another group for which we do not have good data on diffusion and use involved corporations, because most international firms (and some local ones as well) hosted their websites in other countries; indeed one NASSCOM survey suggested 91 percent.[143] They did this for many reasons: Indian networks were highly unreliable, inefficient and slow; the United States had greater bandwidth; and hosting a website in the U.S. was less expensive than in India.[144]

In 2002 the government legalized Internet telephony.[145] Mobile phones represented the least expensive form of digital technology introduced into India in the early 2000s and cell phones in general became quickly both a status symbol and a useful tool. India's largest mobile service provider, Bharti Airtel, also was unique in that the firm outsourced almost all its infrastructure to American and European firms, such as IBM (the latter maintained its IT infrastructure), making this company the only telco in the world at that time to outsource its core functions.[146] The number of subscribers increased dramatically during the fourth phase of Internet diffusion in India, from 0.4 subscribers per 100 people to just over 45 by 100 in 2009, rapidly making this technology the platform of choice for accessing the Internet by 2010. Mobile networks still only covered about 60 percent of the population, so there was much room for further diffusion of the technology.[147] Online gaming had been one of the first nonconversational applications, followed by accessing the

Internet for information, and most recently, the kinds of applications first made accessible by Apple's iPhone technology.

We can conclude that the middle class was moving faster through Wave One than the private or public sectors with respect to telecommunications, although just as slow when it came to the general use of computing. Its members responded to the evolution and availability of the technology much like their peers in other countries, buying what made sense for them to use, and what they could afford. To a large extent the same pattern of acquisition applied to the private sector as well, with businesses adopting IT when it made sense to them to do so and when these technologies became affordable. The fact that businesses were slower to adopt IT than in many other countries was due largely to these twin issues—what would contribute to the productivity of their businesses and what was cost-justifiable—all influenced by government regulations that affected the price of technology (such as causing PCs to cost more due to tariffs), and the relative costs of labor. At the same time, the very poor had not entered Wave One, nor had a large number of small to mid-sized businesses by the early 2000s.[148] However, both slow adopting government agencies and large enterprises were well into Wave one. Then there was a small segment of India's economy that possibly was moving into Wave Two, providers to the world of software services and business process support, such as call centers.

INDIAN SUPPLY SIDE GOES GLOBAL—EFFECTS ON THE WORLD'S ADOPTION OF IT

The jewel of Indian economic performance in the new century is its software IT exports and its business process outsourcing (BPO). Combined, the two are the face of a services-based economic development process underway in India. Liberalization of regulations in the 1990s and even more so in the early 2000s, coupled with effective government initiatives in support of this new business made it possible for multinational corporations to establish call centers, software development laboratories, and back office processing facilities in India, such as Intel, Microsoft, IBM, and a raft of American banks and insurance companies.[149] Good telecommunications infrastructures in the economic and science parks made it possible to use ICT to provide these various services and activities by Indian companies as well and the two sets of employers to hire hundreds of thousands of IT-trained Indians. The growth of these activities has been so dramatic that one can understand why much commentary about IT in India focuses on the supply of these services to the rest of the world, and most extensively to American, British, and European clients, in that order of activity.[150] While the veracity that comes with hard statistics are subject to some debate, essentially by 1997 Indian firms had captured 62 percent of the

World's IT outsourcing business and ended 2007 controlling 65 percent. Its BPO share climbed from 39 percent in 1997 to 45 percent a decade later.[151] Both lines of business continued to grow despite the global economic slow-down caused by the recession of 2007–2009. By 2003, this market was already generating over $1.3 billion in revenue, and by 2010 approached $8 billion, a result of good quality, properly skilled Indian resources, and low local salaries in comparison with those prevailing in the pan-Atlantic world.[152] The number of Indians engaged in all manner of IT-related outsourcing grew sharply as well, from 284,000 by 2000 to 1,287,000 in 2006; in other words, adding a million jobs in 6 years. If one included the additional indirect jobs created as well, such as drivers, household servants, retail workers, suppliers of other goods and services, then there is an additional estimated three million positions.[153]

Where did all these IT workers come from? In societies where extensive use of IT had been the norm for many years, workers came from universities, the military and private sector, which trained their own experts. If demand was sufficient, emigrants joined the pool too. All of these circumstances were most in evidence in the United States, and to varying, but far lesser extent, in other economies that had extensively used computers. In India, IT was so new that the majority had initially to come from universities. The number of engineering colleges in India grew from 158 in 1980 to 1,346 by 2005, while over 2,000 IT training institutes and other educational providers were pro-ducing over a half million IT-skilled workers of varying qualities by the early 2000s.[154] A second source were Indians who returned to India. They did this either because of job losses and shrunken business that resulted from the American recession and Dot.com bust at the start of the new century, or from perceived opportunities in the home country, although surveys indicated that young Indian IT professionals in Bangalore and elsewhere in India still aspired to spend some time working outside India after the recession was over.[155]

Users of such outsourcing services were motivated by several factors in the 1990s and 2000s which made sending work to India attractive. Simply put, the recessions of the early 1990s, early 2000s, and 2007–2009 put enormous pressure on firms to reduce costs in their operations, especially Japanese and American companies. Process reengineering they did in the 1990s made it pos-sible to detach portions of their work and either automate or outsource these. These developments—pressure on profits and results of process reengineering—led many to move work to India, where salaries were often two-thirds (or more) lower and where the existing technological infrastructure made it pos-sible to communicate work and support back and forth.[156] A third factor was the lack of adequate supplies of programmers and other technical staff in the West during the 1990s to implement major systems, such as ERP and to do the remediation work required to respond to Y2K. The importance of Y2K cannot be minimized because, in the words of one Indian reflecting on the experience, "Y2K allows us to expand our target client list," and without this impending problem "many medium-sized firms that would not otherwise have considered Indian software firms were forced to get to know them as a result of the shortage of U.S.-based programmers in the run up to Y2K. Sending work to India made sense, again because of the competence of existing IT workforces, their low cost, and just as important, their availability," if in the long term only providing a bump upwards in revenues as that kind of work ended with the arrival of 2000.[157]

As these activities occurred, word of their success spread and now one could read of hundreds of such experiences. By 2010, U.S. clients were reportedly saving $5 billion by sending work to India. GE announced that it had outsourced some 900 processes, saving $350 million in costs. Reportedly, the American financial sector enjoyed between 7 and 10 percent lower operating costs than their European rivals, thanks to outsourcing back office work to India. Indian observers were arguing (not necessarily with hard evidence) that "offshoring has resulted in quality and productivity gains of 15%–20% and customer satisfaction of almost 85%," in the United States.[158] The OECD reported that demand was growing from its members at nearly 25 percent compounded growth rate between 1995 and 2004.[159] Such reports suggested that India was playing an important role on the global stage in supporting the further diffusion and updating of IT in other countries. This performance makes it possible to add Indian activities to others as evidence of how IT was diffusing globally late in Wave One, possibly in early Wave Two.[160] These activities also demonstrate that broadening the scope of offerings also caused demand to increase, as India went from just doing low-level inexpensive programming to higher-value software development, research, and operations too. India's activities provided further evidence that the diffusion of IT during Wave One continued to be a highly integrated global process, rather than a series of isolated national experiences. The Indian experience during this wave also demonstrated that officials knew of actions of other governments but could not manage their nation's IT experiences in isolation of larger global trends, such as the aspirations of some citizens to use IT and the effects of changes in technologies and their costs in the world market.

By 2005, India had become the favorite destination for Japanese, American, and European firms to outsource software and business processing. It surpassed other nations that also had low cost labor, good ICT infrastructures, and improving business climates, such as China (often ranked second in popularity) and in descending order Malaysia, Philippines, Singapore, and Thailand in Asia, and Russia, and Eastern Europe in the West.[161] The kinds of activities outsourced to India as the most popular destination of choice by the 2006 included, in order of most popular to least in IT and BPO services, call center help desks, business research analytics, finance and accounting, human resources, and industry-specific R&D. All required extensive use of ICTs.[162] By then the world was spending over $8 billion in Indian services.[163]

Of particular relevance to our study is software outsourcing services. By early 2006, the largest users of such services were Japanese firms, accounting for 59 percent of the market. The Americans and Europeans accounted for nearly another quarter of the demand, services that grew in popularity as a percent of the total in subsequent years.[164] In addition to the relevant skills in IT and the availability of English-speaking workers for the pan-Atlantic world was the relative cost savings. Demand for Indian technical labor increased during the first decade of the new century at increasingly faster rates than Indian universities could supply, resulting in rising salaries (in some years by 15 percent) but the differences between Indian and U.S. or Japanese costs remained significant. Table 10.10 provides a comparison of some salaries as of 2002, an important year because a great takeoff in demand had begun as companies like IBM, Intel, and Microsoft started to hire rapidly thousands of local Indians and as many non-Indian firms expanded their reliance on Indian call centers. The data are self-explanatory; operating from very low salaries and even accounting for inflation of Indian wages, India's workforce proved financially very attractive. It lists the favored destinations to send work to at the time for this kind of assignments, although not all of them. India, Israel, and, to an extent, Russia had English-speaking programmers; while other emerging outsourcing destinations at the time less so, such as China, Malaysia, Poland, and Hungary.

Users increased their dependence on India for the preparation of software and support services after 2000 for new reasons as well. In 2000 the Indian government began implementing a new law, the Information Technology Act, which facilitated the use and security of electronic information. Users of Indian outsourcing services were concerned about the protection and use of

Table 10.10 SALARIES OF SOFTWARE PROGRAMMERS, CIRCA 2002, SELECT COUNTRIES (U.S. DOLLARS)

India	4,800–8,000
USA	60,000–80,000
China	8,952
Russia	5,000–7,500
Ireland	23,000–34,000
Israel	15,000–38,000

Source: Organization for Economic Co-operation and Development, *Is China the New Centre for Offshoring of IT and ICT-Enabled Services?* (Paris: Organization for Economic Co-operation and Development, April 5, 2007): 25.

their information as in prior decades elsewhere. The law established guidelines and practices for securing electronic records and detailing how digital signatures and authentication would be done.[165] Then in 2006, a new law called the Special Economic Zone Act detailed a variety of initiatives designed to encourage companies around the world to outsource ICT-enabled services to various parts of India housed in these economic zones. In short, the Indian government was continuing to liberalize its regulatory practices to encourage private sector ICT services to grow in the 2000s, which in turn stimulated expansion of services that helped suppliers and institutional users of IT in the West, and in Japan, to use more IT.

Observers of the Indian situation have commented frequently on the efforts made by local firms to adhere to international best practices and certifications to credential their capabilities. In addition to the government's regulatory changes of the 1990s and 2000s, which helped to create the necessary business climate and sense of operational continuities that Western IT users sought, Indians conformed to such credentialing practices as those of the International Standards Organization (ISO), especially its ISO 9001 and ISO 14001, the Carnegie Mellon software certification classes, and use of "six sigma" practices, among others. The use of such technical standards as the SELCMM level 5 were very important for programming; just over half of all the companies certified at level 5 were Indian by 2006.[166] All of these efforts were designed to remove fears of riskiness, incompetence, and low quality about the Indians and that strategy worked.

Yet, the environment for this and other kinds of work with software remained a dangerous one. As late as 2003, two observers of the Indian IT scene reported that the "environment simply does not seem ripe yet" for this sort of business, arguing that "software created there [India] has negligible copyright protection, services are hard to scale, the domestic market is tiny, and manufacturing has not taken root. Political instability and a discouraging venture formation environment are added hurdles."[167]

The move to India was a relatively late event for the rest of the world. Firms that began to send work there were in the mature stages of Wave One, possibly on the doorstep of Wave Two. After the series of regulatory reforms were implemented in India between 1991 and 1993 that encouraged foreign investments in the country far more than in the previous decade, Western firms began to come to India to have work done, to establish local facilities, and to hire Indian IT people. Firms that did this included GE, Citicorp, HSBC, Texas Instruments, IBM, Motorola, Hewlett-Packard (H-P), Dell Computer, and American Express, among others. American CEOs endorsed publicly the offshoring strategy, such as GE's highly respected chairman, Jack Welch, Michael Dell from Dell Computing, and Carly Fiorina, while she was CEO at H-P. Initial offshoring involved the most routine services, such as call centers, programming Y2K changes to COBOL programs, and mundane software repairs and customer support. Risks to Western companies, should that not work, proved minimal as these activities could be moved quickly and electronically to other countries if problems developed. That portability of service work remains a potential risk for India today should a war, economic disaster, or some other natural calamity afflict the nation. Meanwhile, the story presented here continues to unfold based on the path dependencies of prior events and circumstances still prevailing. For instance, having a near 12 hour difference in time zones with the United States, in particular, also made it possible to have Indians work on many software maintenance issues during India's daytime while American and West European users were already at night, with results ready for them the next day.[168]

As these activities succeeded, more complex work was sent to India. CAD and accounting activities were followed by product development, web development, IT administration simultaneously with BPO by about 2002–2003, followed by ERP and financial applications by 2004–2005.[169] Some firms became quite dependent on Indian IT services in order to diffuse their own IT around the world. For example, IBM—still the world's most significant supplier of IT services and software—now sourced a great deal of its product development, support, and services out of India, with over 25 percent of its workforce now resident there. India is yet another example of the process of up-scaling its capabilities in IT from mundane technical activities to more sophisticated profitable work, and now to the development and marketing of new products and services to the world.[170] In short, India now plays a very serious role in the global diffusion of IT outside of its borders.

The core activity that seems to permeate its IT and other services was handily described by two observers as "the conversion of written information into a digital format," followed by extensive IT-enabled services. This is an important observation because it leads to other IT diffusion as individuals become reliant on digitized records with which to do their work. Further on their observation, "Organizations around the world want to digitize blueprints, maps of water, sewage, and power infrastructure systems, aerial photographs, newspaper and magazine archives, and many other items within information encoded on paper."[171] These are labor intensive activities, so sending them to India to get done played to that nation's relatively inexpensive labor for ongoing work, and also if a one time job, avoided a Western firm having to hire additional staff which might not be needed after a project was completed. The same applied to the fast-spreading use of IT for medical transcription. All of these activities await their historians; all are still unfolding as relatively new activities in Indian IT. To put these services to the world in some perspective, by 2010 one could sense the proportions of IT activities involving the rest of the globe from the data presented in table 10.11[172]; although, the recounting of events that led to these accomplishments is continuing.[173]

Table 10.11 PROPORTION AND VALUE OF INDIAN IT OFFSHORING
SERVICES AND BPO, 2010 (U.S. DOLLARS BILLION)

IT Service	Revenue	Percent of total services revenue
Application development and Maintenance	13.7	39
Traditional IT	9.0	26
R&D services	6.6	19
Systems integration	4.8	14
Consulting	0.7	2
Totals	$34.8	100
BPO	Revenue	Percent of total BPO revenue
Banking	7.6	31
Insurance	4.8	19
Horizontals	4.5	18
Manufacturing	2.0	8
Telecom	1.7	7
Travel & transportation	1.3	5
Pharmaceuticals	0.9	3
Others	2.5	9
Totals	$24.3	100

Source: Data extrapolated from tables in Jayashankar M. Swaminathan, "Outsourcing," in Jayashankar M. Swaminathan (ed.), *Indian Economic Superpower: Fiction or Future?* (Singapore: World Scientific, 2009): 26.

CONCLUSIONS

India's experience in adopting information technologies was unique in many ways from that of other Asian and pan-Atlantic nations. It has become quite convenient to speak about an Asian way of computing; we saw many similarities among the experiences of other Asian countries, from technologically advanced Japan to the city-state of Singapore. However, it is difficult to drop India neatly into that same mold. Asian nations generally adopted an export strategy, relying on hardware sales evolving from low value-added fabrication to higher value-added design and manufacturing, particularly of semiconductors and other advanced machines. When it did implement export-oriented IT strategies, India did so for services and software, far less so for hardware.[174] While the others were successful with hardware, India was not; the others struggled with software while India finally figured out how to compete worldwide with it. Other Asian countries opened up their economies to capitalism and competition, welcomed foreign investments and firms, and moved to build up non-IT infrastructures needed to develop both a local IT presence and to promote its diffusion across various sectors of their economies. They did all these tasks more effectively and quicker (once begun) than did the Europeans. India did not; it was late, slow, and inefficient with the process. Because in all countries national governments played a central role in diffusion of IT, including in such iconic capitalist economies as the United States and South Korea, both big and small, any inadequate performance by a national government would cause deployment of IT to wane or in some other way falter, as occurred across Communist Eastern Europe. India's national government has not done as well as its neighbors; critics comparing its performance were especially harsh of officials, often with good reason.

As easy as it would be to blame the government for its very late interest that it could have developed earlier about IT, waiting for nearly four decades after independence before taking it on in a serious manner, and even then in highly fragmented and politically charged ways, it had significant headwinds with which to contend. It had a poor population that was massively large in extreme poverty. Its illiteracy rate was high, compounded by the fact that it extended to a population that spoke so many languages and had numerous highly diverse cultures that even thinking of India as a country, rather than a subcontinent of countries, can mislead economists, political scientists, public officials, historians, non-Indian business leaders, and other observers of the IT scene. To be sure, Indian adherence to a socialist economic approach, coupled with a badly executed strategy for self-sufficiency for as long as it did (late 1940s–early 1980s), was a big failure for the economy, but focusing on poverty and feeding people was a reasonable priority, given the realities faced by two generations of leaders. Critics of Indian policy sometimes overplayed their hand, just as apologists for Indian economic policies often did too. Ultimately, problems of execution, even of faulty economics and public policy, were India's next set of problems after poverty and social conditions, issues future economic and political historians will need to explore.

Mitigating the harshness of the slow adoption of IT were the countervailing influences of the drop in the world-wide, and even Indian, cost of hardware that took place, with their largest decline beginning in the 1970s and early 1980s. This first came with a new generation of mini computers, next with the arrival of ever-less expensive PCs. Most recently smart phones became available that could do some of the work of earlier IT, such as e-mail, Internet queries, and other "on-demand" functions. Arvind Singh and Everett M. Rogers were some of the earliest commentators to acknowledge this influence, despite their frequent criticisms of Indian government policies. Writing in 1989:

> Why is the information revolution beginning to occur now in India? One reason
> is technological, centering on the applications of microelectronics innovations
> in computers and telecommunications. India's information revolution is en-
> couraged by the declining cost of microelectronics technology worldwide, and

by the increased availability of high-quality microelectronics products in the country.[175]

This influence of costs on adoption parallels similar effects of pricing and affordability sweeping across Asia, Latin America, and parts of the Middle East and Africa, facilitating the initial introduction of IT to millions of users worldwide. This influence is also further evidence that around the world and in India too, knowledge about developments in the technology and its economics were quickly shared across a society, not just among IT experts.

But, what if the Indian government had done what advocates of IT diffusion wanted: aggressively establish and support financially an indigenous computer industry, beginning in the 1950s, feeding it resources as its military encounters with China and Pakistan in the 1960s and later turned to war, reinforced by the global Cold War and its tense relations with the United States government in the 1970s and 1980s, and allowed foreign vendors to sell whatever they wanted in the country, would things have turned out differently? If one accepts a fundamental premise of this book that an environment of preconditions is required more urgently than a particular information technology to stimulate use of IT, then we must conclude that only some things would have changed. There were too many small businesses that could not afford computers; the telephone network was so bad across such a vast expanse of the earth's surface for so many decades that it would have taken a long time to bring it up to acceptable levels; increasing substantially levels of literacy and knowledge of English would have taken several generations to accomplish, just as it took Europe, North America, and other parts of Asia. Where was the money to come from to pay for all of this? It could certainly not come from a nation that entered modern era of post-World War II with one of the lowest per capita GDP rates in the world.

Then there was the culture of how businesses and government agencies were run—the bureaucracies—that were so roundly and routinely criticized by one and all.[176] Many of those practices had been inherited from the British, but in fairness to the colonial rulers, also from the various Indian states through which Great Britain ruled the sub-continent. These practices were often antithetical to the use of modern IT which threatened to cost jobs, destroy careers, and disrupt political and institutional advantages. This large and very complex collection of problems was not going to be resolved by a democratic India; if anything, they were exacerbated by that pluralistic form of government, making them seem intransigent far longer than needed.

Summed up, then, progress would have been made at the margins. More PCs that were less expensive would have been sold, both locally and internationally made products. If the banks had become privately held institutions, they would have embraced computers earlier and more extensively to improve productivity, largely by displacing workers, maybe. I say "maybe" because salaries were so low in so many industries that a fundamental reason for using computers over the past six decades—to drive down the high cost of labor— was not there. It was often cheaper to hire a room full of human "calculators" than to automate their work on an IBM or Japanese computer during the mainframe, minicomputer, and PC eras.[177] Those agencies of government that should have been logical early adopters indeed were, such as parts of the military, universities, the science and engineering institutes, and later, data intensive agencies of government, all of which acquired computers, so one might have expected that they would have used more of them and earlier. All of the marginal activities indeed took place after India finally opened its doors widely enough to welcome IT after 1991. Based on available diffusion data we can conclude that the nation seemed to be acquiring IT in ways and at speeds similar to what had happened in the United States and Great Britain in the 1950s–1960s, Japan and Western Europe in the 1960s–1970s, South Korea and Taiwan in the 1970s–1980s, and Eastern China in the 1990s, only later. Our summing up, therefore, brings us to the most basic conclusion, that India reflected many of the limits of diffusion that could be imposed on a nation during Wave One.

On the other hand, India is not a cut-and-dry case of limits, slowness, or failures. Its software and services export business is remarkable and grew very quickly once the Indian government began to liberalize its economy and, in effect, get out of the way of market capitalism. So, we need to set aside frivo- lous debates about whether Indians could be entrepreneurial or not; even Sil- icon Valley in the United States was populated with many successful Indians. Entrepreneurs are to be found in any economy that presents opportunities for people to succeed in business, even if there was insufficient capital and myriad bad infrastructures and social problems with which to contend.

The services and software experiences suggest two more important insights. First, that it is possible to have nation-sized levels of economic development driven by innovations in services, rather than by the more tra- ditional form of manufacturing serving as the engine of growth. Making ser- vices a particularly attractive alternative to manufacturing is that it requires less capital investments, hence less time to start, made even more attractive by the fact that India really never could afford to fundamentally transform its massive agricultural workforce into factory workers. Even richer China is learning that lesson. Second, India's experience raises an interesting ques- tion: Is IT services-driven economic development a feature of Wave Two IT adoption? It is more than an academic question, because we have all of Africa and sizeable portions of Latin America and the Middle East yet to enjoy per-capita GDPs comparable to the top fifty percent of the world's most pros- perous nations. These areas were burdened in many instances with turbulent states and chronic levels of high unemployment among young educated, lit- erate, cell-phone and social-media savvy citizens, impatient for opportunities to enjoy the comforts and pleasures of the middle class. Can the Indian experience teach them how to evolve without having to build factories in every town?[178]

Already the hunt is on for answers. Studies coming out now on how the work of Indian enterprises and managers were increasing are examples of the process, just as such studies appeared when the Japanese seemed to know more than others, and, of course, as have also been done about China as it rapidly evolves into a market-driven economy.[179] The debate about the effectiveness of public policy in nurturing, or hobbling, India's economic development is a popular subject of debate.[180] But as suggested here, while public policies were profoundly influential in the evolution of the economy and in the diffusion of IT, the scope of the discussion about technological adoption must be opened farther to account for a larger set of issues. The role of state governments has been understudied by economists and historians; so too the histories of individual Indian companies, such as those in banking, some services, and manufacturing; and the use of pre-computer information handling methods and tools, such as desktop calculators and adding machines, and, of course, paper-based work processes. When those studies are conducted, we will understand better how distant Indian society was from the entrance to Wave One computing when this new technology became available to them.

India is not the only nation in Asia to enter Wave One very late in the twentieth century. Other nearby nations did too, while some participated almost as early as those in the pan-Atlantic community. It is to the broader Asian experience that we now turn in order to complete our composite view of how computing diffused across Asia.

CHAPTER 11
How Asia Embraced
Information Technologies

East Asia succeeded, in part, because it provided strategic, selective and consistent support to industries which were in the best position to facilitate their economies' transition
from a labor-intensive to a knowledge-intensive industrial structure.

—World Bank, 1996[1]

Asian nations began adopting information technologies later than countries on both sides of the Atlantic Ocean, yet once started some moved
very quickly, caught up, and in some instances, went beyond the West, such
as South Korea in its use of broadband-delivered IT services to consumers.
As a general statement about diffusion of IT, however, the start dates,
extent, and nature of the diffusion proved far less even than in Western
Europe and North America. The variety of experiences proved so great that a
long standing discussion about an "Asian way" of adopting IT will have to
give way partially as our detailed understanding of the experiences of individual countries makes clearer that each had unique experiences on their
journey toward the use of IT. It is true that they also shared many common
experiences, for example, their wide use of exporting manufactured IT goods
as an economic development strategy that, as a byproduct, resulted in use of
IT in many Asian societies, but not all. The breadth of their speed of adoption often appeared far greater than what occurred in the pan-Atlantic community once started, especially when compared to all of Europe. That
difference with the West needs to be understood, because diffusion of IT has
yet to be substantially done to a similar extent across Latin America, Africa,
Middle East, and pockets of the poorest nations of Central Asia. For all these
regions the East Asian experience is influential. We can now conclude that

the move to adopt information technologies across all of Asia is massive and shows no signs of slowing. Technological advances and retreats in Asia have a long history, going back thousands of years, but in our time, and with ICTs in general, we are in a period of unquestioned adoption that is fundamentally changing the face of over a dozen nations, involving nearly 3 billion people.[2]

As its counterpart on Western European diffusion of IT, chapter 5, this chapter explores broad patterns and consequences of adoption in Asia, largely by nations along the Pacific Rim. The chapters on individual countries documented many of the specifics; in this one we step back to review the shaping forces at work, key patterns, influences, and sources of innovation and adoption. The experience of Asia reaffirms that adoption of IT has been—and is—essentially a two-step process so far, what I have called Wave One and the start of Wave Two. Europe and the United States took a long time to enter Wave Two, while many parts of Asia remain in Wave One. Other parts are, I believe, well into Wave Two, perhaps even more so than a goodly portion of the pan-Atlantic community. We need to understand how that became possible, and in part the answer involved a combination of timing and availability of smaller, more compact technologies to the Asians. The Americans and Europeans had to start with developing all of the original and subsequent technologies, then using mainframes and minis, because there were no PCs and smart phones when they began; Asians embraced IT later when they had those newer options as well as earlier forms of computing. So, their experiences with Wave One and Wave Two would expectedly be different. The pan-Atlantic world could more afford IT in every decade than the less affluent Asian nations, although with convergence in per capita GDP in some Asian economies with those of the West now an impending inevitability, future trajectories of IT adoption may also converge, bringing us back to a technological balance-of-power which so frequently characterized human existence both East and West over thousands of years.[3] While all West European and North American societies have passed through Wave One and are just entering Wave Two, doing this in a half century, only a few Asian nations have accomplished this, most specifically Singapore (really a city state), Japan, clearly South Korea, Hong Kong, and a growing portion of Taiwan. The rest are largely either in early stages of Wave One (such as many parts of China and India), or, so far, have experienced insufficient domestic diffusion to even suggest they are in Wave One, most notably North Korea. Between those who we should not characterize as being at the gates to Wave One, or are just passing into this wave, are nations that are adopting many of the diffusionary practices of the Asian Tigers as their way to national economic growth, such as the Philippines and Indonesia.[4] We will have more to say about who is in what wave and why as we proceed through this chapter.

TAKING STOCK OF ASIAN IT DIFFUSION BY THE NUMBERS

Asia consists of over two dozen countries, from massively large India and China to sparsely populated Tibet and Mongolia. One can argue over definitions of what nations are Asian, particularly those on the periphery of East Asia in what is routinely called Central Asia, bordering Muslim states to the south and Christian states to the west, to use a long standing way of delineating societies. The point is, Asia is as varied in culture, economic prosperity, and political structures as are the four dozen states that make up Africa, so generalizations about some Asian style of IT that we can credibly accept are hard to accept with a great deal of confidence and are often overstated. The first step is to see if we can categorize all these nations into groups where cohorts share more common experiences with each other than with other clusters of countries, using such metrics as per capita GDP, extent of IT deployment, and the amount nations spend on all manner of ICTs. Since conversations about diffusion always end up in tabulating how much of some technology was deployed, hence how the rationale for why the topic is discussed, we begin with some score keeping. Using 2004–2005 as our baseline, one can ask to what extent IT had spread around Asia. That is a good period, one that came after the various Asian financial crises of the 1990s, Y2K in the West, and the global Dot.com bubbles when exogenous jolts to more normal patterns of adoption had minimized. The West was also recovering from its millennium recessions. Additionally, since we want to know if they are in Wave One or Wave Two, a few statistics of enough types of diffusion data suggest how far they might have gone, who got there, and if any are in Wave Two-like circumstances.

We can conveniently divide Asian societies into three groups based on the extent of their diffusion. The first is made up of the earliest and most extensive adopters of IT, such as Japan and the Asian Tigers studied in previous chapters. They shared in general several characteristics:

- They began using computers earlier than other Asian nations
- They diffused these across many industries and sectors of society
- They used a large variety of technologies, from mainframes to the Internet and cell phones
- They enjoyed high per capita incomes sufficient to pay for their adoptions of IT
- They had many large institutions, agencies, and companies that could afford IT, such as large manufacturing and data-driven government agencies and could use it for sound business reasons
- They were major users and exporters of IT
- They also had strong commitments to the diffusion of IT from their national governments that remained constant from one regime to another over decades.

In every previous chapter we discussed how important (or not) physical characteristics of a technology proved to be in a society, the "materiality" effects so often the focus of historians of technologies of all kinds. In the Asian experience "bells and whistles" were never as important conditions for diffusion as were so many other non-technical elements, such as levels of national GDP and public policy in support of diffusion. A technology's improvement and price performance was almost a necessary condition but not emphatically or always so, which is why the Soviets and Indians could function with older technologies. It is partly why some IBM executives were concerned about Indian public policy and local needs for types of data processing than about technological currency. Experiences in all these countries suggested that prerequisite conditions had to be in place. We see evidence of that reality in table 11.1, which shows tools of Wave Two were extensively deployed by nations spending considerable amounts on IT. Internet and PC usage usually could not take place unless other ICT preconditions were already met, such as digital transmission over modern telephone networks, which required use of computers to function, a cadre of IT experts who ran systems on mainframes, minicomputers, and PCs at work because with such high, if varying levels of GDP devoted to ICTs, it would have been nearly impossible to spend all of that money just on PCs and cell phones. We can also safely conclude that there was a small group of nations that embraced IT more extensively than neighboring countries who I call Early Adopters. That the speed of adoptions and time it took varied is made clear by the detailed country studies presented earlier in this book.

These same insights apply to these two nations listed here, but not discussed earlier, Australia and New Zealand. Had we devoted attention to their experiences, the narrative would have read much like that about the European examples, with Australians working with the British during World War II, then visiting early computer pioneers in Great Britain and in the United

Table 11.1 ICT DIFFUSION BY EARLY ADOPTERS IN ASIA, CIRCA 2004 (EXTENT OF DEPLOYMENT IS NUMBER PER 1,000 PEOPLE)

Country	Mobile subscribers	Internet Users	PC Users	ICT (% of GDP)
Japan	669	606	425	7.4
South Korea	760	656	558	6.6
Singapore	891	559	565	10.4
Hong Kong	1,192	508	453	8.4
Australia	887	497	616	5.9
New Zealand	811	526	417	10.0

Source: World Bank, *Global Trends and Policies* (Washington, D.C.: World Bank, 2006).

States, followed by building experimental systems in universities, vendors familiar to Europeans selling systems in the 1950s and beyond to the same kinds of institutions as in the pan-Atlantic world, and for the same uses at the same time, using the same equipment.[5] A similar tale would have been told about New Zealand, with the exception that it began to appropriate IT later and, like the Indian software industry of the 1990s and early 2000s, acquired a software specialty, in the case of the former, software development for movies, as used in *Lord of the Rings*.[6] But, unlike India, many industries in New Zealand had already become users of computers before the emergence of a local software industry tied to film making.

A second group, labeled Recent Adopters, in many instances heard about computers soon after World War II, such as India, but for the reasons discussed in earlier chapters, began to adopt computers later than the Early Adopters (see table 11.2). They shared several characteristics as well:

- They began using computers later than the Early Adopter Asian nations
- They diffused these across some industries and sectors of society, largely in manufacturing and government first
- They used a larger proportion of newer technologies, particularly PCs, and later, and to far lesser extent, the Internet and cell phones
- They experienced far lower per capita incomes which often were inadequate to pay for extensive adoption of IT across multiple industries and economic groups
- They had a large proportion of small enterprises, institutions, and government agencies which could not as readily justify the use of such technologies as in wealthier economies
- They tended to be major exporters rather than extensive domestic users too of IT

Table 11.2 ICT DIFFUSION BY RECENT ADOPTERS IN ASIA, CIRCA 2004 (EXTENT OF DEPLOYMENT IS NUMBER PER 1,000 PEOPLE)

Country	Mobile subscribers	Internet Users	PC Users	ICT (% of GDP)
China	258	73	40	5.3
India	48	23	11	3.7
Malaysia	573	392	170	6.9
Thailand	420	112	74	3.5
Indonesia	141	52	19	3.0
Philippines	387	58	29	5.9
Vietnam	53	65	11	—

Source: World Bank, *Global Trends and Policies* (Washington, D.C.: World Bank, 2006).

- They had strong commitments to diffusion of IT from their national governments, but this support came later and was not always as steadfast as with Early Adopters.

Their prerequisite requirements for adoption also varied. They tended to have inadequate training or supplies of IT experts, although these societies valued education along the lines espoused by Confucian and Buddhist world views. Literacy was a particular problem in the modern period too, as contradictory as that might sound when describing societies that valued education and knowledge. Those sectors which world-wide were the least able to adopt IT were also in large presence in this second tier of nations, the most important of which were agriculture, followed by local and state public administration still densely paper-based and underfunded and, third, elementary and secondary education. One could argue that Recent Adopters also had fewer of the complex industries that the earliest required for the use of automation of all types, such as automotive or aircraft manufacturing, when computers were the most expensive in their history and, one might argue, the most challenging to use.

But, the biggest difference between this group and the first concerned labor costs; the earliest adopters entered the era of the computer with higher wages, or these wages subsequently increased rapidly, than the second group of adopters. Both Asian groups leveraged their relatively low costs of labor to compete on the global stage in manufacturing for export IT and other forms of telecommunications and electronics. This resulted in export-centered industries using IT themselves. Their adoption spilled over into other segments of society, albeit at various speeds and in differing amounts. As wages rose in Japan, beginning by the 1970s, and in China in the early 2000s, manufacturing of IT, or that used IT in their processes of producing exported goods, were shipped to other Asian countries where the cost of labor remained lower. That process sometimes involved "pushing" IT into a country before its local citizens or governments wanted to "pull" in IT for economic development. The results proved similar in that later adopters embraced the economic benefits of IT and consequently supported its diffusion, often using preexisting approaches, most notably Japan's.[7] It is a process still underway, most recently with Chinese and Japanese work moving to Vietnam and the Philippines, and even some of India's software development to Latin America and Africa.

What about the rest of Asia, the Laggard Adopters, how behind where these countries? While not discussed in detail in this book, it is a fair question to ask and much as there was difficulty in documenting Chinese and Indian IT diffusion during their very early stages, the same challenge exists with these states. Table 11.3 lists most of these to demonstrate the variety of levels of diffusion in Asia, reinforcing our contention that too much generalizing about Asian IT practices can be misleading. This third cohort of nations shared several characteristics:

Table 11.3 ICT DIFFUSION BY LAGGARD ADOPTERS IN ASIA, CIRCA 2004 (EXTENT OF DEPLOYMENT IS NUMBER PER 1,000 PEOPLE)

Country	Mobile sub-scribers	Internet Users	PC Users	ICT (% of GDP)
Bangladesh	27	2	4	2.7
Cambodia	63	3	2	—
Kazakhstan	174	20	—	—
Lao	35	3	3	—
Mongolia	129	58	28	—
Myanmar	1	1	5	—
Nepal	10	9	4	—
Pakistan	52	13	5	7.3
Sri Lanka	114	14	13	5.7
Tajikistan	7	1	—	—
Turkmenistan	2	2	—	—
Uzbekistan	13	19	—	—

Source: World Bank, *Global Trends and Policies* (Washington, D.C.: World Bank, 2006).

- They had virtually no extensive deployment of IT as of 2004; although many are today aggressively pursing ICT
- They use cell phones more frequently than any other IT platform, but these were not smart phones in 2004
- They had so few PCs installed that one can assume they were located in government agencies and companies
- They spent so little on IT that the World Bank was not able to calculate how much their expenditures were on such technologies
- They have some of the lowest per capita GDPs in the world
- They began the age of the computer with little or no national telecommunications infrastructures.

Their statistics demonstrate the enormous variety of diffusion rates across all of Central and Eastern Asia. The more one approached the two ancient silk routes of Central Asia, the poorer these societies were, and, thus, the later one would expect IT to begin. However, even in these regions there is evidence that knowledge of IT had seeped in, reinforcing our point that just because someone in a nation knew, indeed even built a computer, did not mean diffusion had realistically begun.[8] Rather, one can conclude that citizens and their institutions did not yet see the relevance of using IT, at least not until the late 1990s.

Extant evidence shows, however, that everywhere there was activity, with new technologies absorbed into all these nations by the early 2000s. Table 11.4, based on data collected by the World Bank, shows that the rate of change in

Table 11.4 RATE OF DIFFUSION OF ICT IN ASIA, CIRCA 2004–2005 (EXPRESSED AS PERCENT CHANGE FROM 1999 TO 2004–2005)

Region	Internet Users	PCs	Digital Cellular Subscribers
East Asia and Pacific	48	26	54
South Asia	66	29	88

Source: World Bank, *Global Economic Prospects: Technology Diffusion in the Developing World, 2008* (Washington, D.C.: World Bank, 2008): 72.

the early years of the new century in adoptions was extensive, even off a low base. It collated data in some odd fashions, which can be misleading too. For example, it has a category of users for the kind of data displayed in table 11.4 called "Europe and Central Asia," which makes little sense, but even there we see the Internet spreading with a 48 percent growth rate. Knowing what we do about Western Europe's adoption of the Internet one can conclude that Central Asia was adopting the Internet at a pace similar to that of East Asia. Adoption of PCs for that combined area was 20 percent, which makes sense since Europeans experienced their great growth of adoption in the 1990s, suggesting that Central Asians were not acquiring many PCs. All regions of the world were adopting digital cellular telephones at very high rates, ranging from a low of 42–43 percent in Europe and Latin America (early adopters of these in the 1990s) to the Middle East and North Africa (70 percent) and South Asia (88 percent). In short, PCs diffused quite slowly, while use of the Internet and cell phones grew at very impressive rates.[9]

There is a fundamental differentiator that existed among the wealthiest and also later adopters suggested in several earlier chapters. The earliest and later most aggressive adopters were major users of IT, not simply producers for export, most specifically Japan, South Korea, Taiwan, and Singapore. In some instances, extensive adopters focused more attention on leveraging this general purpose technology to improver the overall performance of their economies than just on manufacturing for export. For example, Hong Kong, and Singapore wanted to leverage their ports, trading capabilities and banking industries to generate trade, but did some manufacturing too. Australia and New Zealand did very little manufacturing of IT, but became extensive users of the technology much in the style evident in Western Europe. Some of the most recent entries into the digital arena were more manufacturers than users of IT, such as Malaysia, Thailand, Philippines, and Vietnam. By the early 2000s, it was quite evident that both India and China, which had approached IT as an export opportunity, beginning largely in the 1980s, had started to use IT for their own internal activities by the early 2000s, when public programs encouraged diffusion of this technology into domestic markets.[10] The nations listed in table 11.3 were not significant players on the world stage in the manufacture of IT hardware or in providing software services, explaining why they were normally left out of discussions about how IT diffused in Asia.

THE "JAPAN MODEL" VERSUS AN EAST ASIAN WAY?

There has been a debate underway since the 1970s regarding whether or not there was some "Japanese Model" or an "Asian Model" for economic development in the region. Western observers began noticing in the late 1960s how quickly Japan's economy had progressed, with dramatic rates of growth, increased national GDP per capita, evolution to modern export manufacturing, all affecting many domestic industries, regions of the country and all classes. It became common to label what happened there as the "Japan Model," and it was observed by officials and business leaders in other Asian nations. Then in the 1970s and 1980s, the Asian Tigers took off economically, also with double digit economic growth, rising per capita GDPs, and manufacturing for export. So now observers of the Asian scene began to see the "Japan Model" replicated in places like South Korea, Singapore, Hong Kong, and Taiwan, but not in Australia or New Zealand.[11] By the 1990s, as nuances in economic development in these various countries became visible, the conversation extended to the possibility of some sort of mega-"Asian Model" that subsumed the Japanese experience within it, and that differentiated East Asia from that of the pan-Atlantic community, especially as Japan's star status declined in the 1990s and that of the North Americans and West Europeans did too in the early 2000s.[12]

What really happened? We have three possible answers: that the Asians all copied Japan, that they developed their own pan-Asian hybrid model, or, that each one followed its own path. As the previous several chapters illustrated with respect to IT, all three styles were in evidence simultaneously. Furthermore, in the cases of China and India, Soviet economic development models were highly influential during the Cold War but had been essentially discarded by the time the Soviet Union collapsed. Often the discussions about economic and public policy models centered on information technologies, but also extended to other industries, such as automobile manufacturing (e.g. Japan and South Korea), inexpensive consumer goods (e.g., Taiwan and China), and textiles across Asia. These discussions invariably came back to examinations of the role of complex technologies, such as computers and telecommunications. The reasons are obvious: fast double-digit economic growth, which for decades exceeded results in the United States or anywhere in Europe; transformation of technologically backward nations, or societies heavily dependent on agriculture into modern societies; and the speedy recovery of economies, societies, rates of literacy, and health of nations devastated by the civil and world wars of the 1930s and 1940s, and regional wars in Korea and Vietnam.

The most enduring model among economists is Japan's, which has been described in such detail that one could almost present it as a "plug-and-play" process. As described in chapter 7, however, Japan's experience was a complex and iterative journey from losing World War II to threatening the destruction of whole industries in the pan-Atlantic community in the 1960s–1980s. Two economists writing for the RAND Corporation as late as 2003 observed that "Asian IT producers have generally followed the 'Japan Model' of progressively sophisticated production technology, beginning with labor-intensive, low-value manufacturing," arguing that it represented "a compromise of sorts between European top-down regulation and U.S. bottom-up entrepreneurialism."[13] But then they literally presented a flow chart, with description of the process, detailing their thesis, making the Japanese experience seem as if it were a static cook-book recipe when in reality it was not.[14] Representations of a Japanese model included features in evidence in those Asian economies where development of export manufacturing took place. The most important of these was government support for the nurturing, protection, and promotion of industries that could export manufactured goods to the West, including IT equipment. That support proved successful in its objective. It was also disciplined and sustained over several decades.

As early as the first half of the 1980s, it had become evident that the national governments in Japan and Asian Tigers had clearly targeted IT as crucial to their national economic development strategies. In most instances, they also encouraged their various industries to use the technology in their continuously upgrading of manufacturing capabilities—the diffusion of IT to improve productivity and sophistication of production in their export industries. Governments also sought ways to import the "know-how" to do all of these things so as to avoid the time and expense of new R&D.[15] The results were increases in Japanese exports of machinery, computers, steel, and electronic products, while in other countries, the variant of this model included textiles, consumer electronics, plastic products, and smaller proportions of the heavy industrial exports evident in Japan.[16]

That Japanese ways would be influential should not come as a surprise. While most observers of the Asian scene point to the impressive growth rates of Japan's economy in the 1960s–1980s as the reason, there were others besides recent results. Japan knew more about industrialization than other Asian states, having committed to transforming its economy into a modern industrial one as early as the 1860s, with results evident by the end of the 1870s, while some Asian countries did not make similar commitments to industrialize until the 1950s, or later. So Japan had expertise and experience. Japanese ways of doing things, including management practices, were familiar to many Asians since Japan had made a number of these countries its colonies, dating back to the late nineteenth century. In some nations, almost all of the most senior business executives had been Japanese nationals for many decades, as in Manchuria and South Korea. In the post World War II period, when Japan recovered economically it quickly resumed investments in other national economies, ensuring it could export both manufacturing know-how and managerial practices as well. By the early 1980s, for example, Japanese investments in Indonesia represented 35 percent of all foreign investments in that country, similarly in Singapore (30 percent), and in Malaysia it dominated. Trade among these nations with Japan was similarly impressive with the Japanese involved in roughly 20 percent of the market in Southeast Asia.[17] In short, Japan had a large presence, hence influence, in Asia.

Arguments in support of an "Asian Model," are largely influenced by the results of supply side ICT and consumer electronics in exports. Some of the most articulate advocates of this approach have included Jason Dedrick and Kenneth L. Kraemer. As discussed earlier, they were not alone in presenting this point of view, although they presented it with clarity and confidence. They argued that this "Asian Model" had four features common to those Asian economies that had modern economies, were extensive exporters of ICT products, and enjoyed relatively fast GDP growth. First, these countries generally—but not always—continuously attempted to upgrade their capabilities to produce more complex value-added products for export. They specialized in that process too, such as South Korea in semiconductors and Taiwan in PCs, with neither attempting seriously to compete against each other's core competence. Second, these economies embraced an export mentality, although implemented their objectives in various ways. For example, Japan and South Korea protected their local industries from foreign investments and competitors, while Taiwan, Singapore, and Hong Kong welcomed foreign trade and investments to a far greater degree. Third, there existed government policies and coordination among agencies, and most effectively, among organizations within the private sector to implement national strategies. This was an explicit, indeed overt process, about which we will say more below. Fourth, the "Asian Model" emphasized production and export over local consumption, which is why in previous chapters we have had to distinguish discussions about the existence of local export IT industries from local diffusion. These were two tracks of IT activities, and not always tightly integrated with one another. This last, fourth, feature of the Asian approach stands in sharp contrast to pan-Atlantic practices where domestic diffusion was far more evident than in Asia.[18]

Either model proved generally successful when it came to producing and selling IT products largely for export. Between 1985 and 1995, the years when Asia as a whole took off in IT and consumer electronics' exports did too, Hong Kong's percent of GDP devoted to these products always ranged between nearly 2–2.5 percent of its GDP; Singapore's was more spectacular with 10–30 percent devoted to IT exports; while South Korea, Taiwan, and Japan were more similar in the 1980s with 1–2 percent of GDP devoted to IT and in the 1990s with Taiwan taking off with 3.5 to just over 6 percent devoted to these products. In comparison during these years, that of the United States hovered closer to one percent.[19]

All these discussions about Asian models largely focused on the supply side of IT. But what can be said of the demand side, that is to say, about the appropriation of IT within a country? Is there an Asian model? The evidence presented in previous chapters demonstrate that the answer is no, that diffusion internally followed patterns evident in Europe and North America as well: Tax incentives for acquiring modern equipment, funding of PCs for secondary education, sponsorship and budgeting for university programs, establishment of technical institutes, and use of IT by governments themselves. The one very important exception was the role of the military as both developer and consumer of IT. The various military establishments played a major role in North America and across all of Europe, far less so across Asia in the first several decades of the computer's existence. Not until the 1990s did the activities of these communities become more visible. By the early 2000s, as one military observer in the West wrote, "none has sought to import U.S. methods wholesale. Rather, each has begun to develop its own, unique approach to information-age warfare."[20]

In general, diffusion internally proved far slower than in the pan-Atlantic community. There were many reasons for this, not the least of which were the presence of small enterprises that could not justify the expense of IT, industries that were not natural candidates for the technology (e.g., agriculture), lack of adequate sophistication of processes that lent themselves to its use, insufficient supply of expertise and, in some instances, the technology. But most important across a good half century was the relative low cost of labor across Asia. One of the most important reasons why IT diffused so quickly in North America and Western Europe was due to their high cost of labor. They needed to reduce labor content in work to drive down costs, improve profits, and to be competitive, at least in the West. All of these objectives led to rapid automation and computerization of tasks and processes.[21] In Asia, it was often cheaper to use people than machines; indeed it was inexpensive labor that made export manufacturing so attractive as the fundamental economic advantage—hence strategy—for economic development in the region from the 1950s to the present.

Yet, domestic consumption of IT occurred as its use in export-oriented activities spilled over into internal economic activities, albeit slowly unless governments promoted its use, which many did. Diffusion of IT internally became critical, for example, in attracting foreign companies in many industries to set up call centers and software centers in India, and to use Singapore's port and warehousing capabilities in their global supply chains.[22] Table 11.5 displays in a very simple fashion evidence of the results of domestic expenditures on IT during the decade when this technology's domestic diffusion appreciated most significantly over all others over the past half century or so. Three observations

Table 11.5 ASIAN IT EXPENDITURES AS PERCENT OF GDP, 1985–1995, SELECT YEARS AND COUNTRIES

	1985	1987	1989	1991	1993	1995
Japan	1.15	1.48	1.69	2.06	1.75	1.89
Singapore	2.15	2.19	3.05	2.23	2.07	2.25
South Korea	1.64	1.34	1.35	1.33	1.60	1.97
Hong Kong	1.11	1.02	0.96	0.79	1.25	1.31
Taiwan	1.02	0.82	0.83	0.88	0.76	0.84

Source: Data collected in Jason Dedrick and Kenneth L. Kraemer, *Asia's Computer Challenge: Threat or Opportunity for the United States and the World?* (New York: Oxford University Press, 1998): 323.

are in order, however. First, the percentage of a nation's GDP spent on IT increased over time, albeit quite slowly. Second, when compared to what happened in the United States during the same period, this occurred at roughly half the rate as in the USA, which routinely devoted between 2.5 to just over 3 percent to IT.[23] Third, when IT was used in Asian countries, the economic benefits were similar in type as in the West both for economic development and productivity.[24]

For domestic diffusion of IT, a number of universal conditions had to prevail and to the extent they did, IT diffused faster or slower than in other countries. By the early 2000s, this insight had become so evident that one could compare Asian performance to that of other nations, in such ways as the Economist Intelligence Unit's E-Readiness rankings reports from those years. These rankings are attractive because they consolidate 100 attributes, ranging from GDP, to levels of education, patent and copyright protections, extent of available communications networks including the Internet, stability of the legal environment, government support, and consumer business adoptions, all of which affected use of IT worldwide. Using data from 2010 allows us to see the accumulation of results over a long period of time. When compared to 70 countries, and ranked against all of them from most extensive users of IT to least, table 11.6 demonstrates that more than the technology or one or two variables were essential to the process of diffusion. The top twenty ranked nations included West European, both North American countries, and Asian countries. All shared high levels of pre-existing infrastructures, such as networks (connectivity), and healthy business, social, legal, and cultural environments, leading to both strong consumer and business adoptions of IT. One could argue that this group was busily going about the process of entering Wave Two.

The group from Malaysia down to China had very strong business environments, educational infrastructures, but lesser quality connectivity. Both business and consumer adoptions were far less than the top listed, often ranked at half the rates of the leaders in adoptions. They were actively going about the

Table 11.6 E-READINESS RANKING OF SELECT ASIAN
NATIONS OUT OF 70 COUNTRIES, 2010

Hong Kong (7th)	Thailand (49th)	Kazakhstan (67th)
Singapore (8th)	Philippines (54th)	Azerbaijan (70th)
Australia (9th)	China (56th)	
New Zealand (10th)	India (58th)	
Taiwan (12th)	Vietnam (62nd)	
South Korea (13th)	Sri Lanka (63rd)	
Japan (16th)	Indonesia (65th)	
Malaysia (36th)	Pakistan (66th)	

Source: Economist Intelligence Unit, *Digital Economy Rankings 2010: Beyond e-readiness* (London: Econo-mist Intelligence Unit, 2010): 23–24.

business of living in Wave One. The lowest ranked were lacking in each cate-gory of the required ecosystem by a wide margin, ranked with African and some Latin American states.[25] Most were either just entering Wave One or were about to step through the threshold into it.

In summary, major parts of Asia were progressing as the most extensive adopters were doing so too, overall creating the environment in which IT could flourish, while those that lagged did so for reasons very similar to those evident in Eastern Europe, the Middle East, and in parts of Latin America. If we just ranked them by consumer and business adoption and leave out all the other elements, the rankings remain the same, additional evidence that it is the composite of all these elements that count, reaffirming what many econo-mists have argued for decades were prerequisites for modern economic devel-opment.[26] The experience with IT reaffirmed their views and evidence.

ROLE OF GOVERNMENTS

One of the distinguishing features of Asian diffusion of IT was the overt, indeed commanding role and presence of national governments in directing their nations' adoption of computers. Most observers of the Asian IT scene have acknowledged this singular feature of Asian practices.[27] The national gov-ernment of every nation discussed in detail in this book played proactive, important, usually pivotal roles, in the diffusion of IT. Asian officials did more of that than across the pan-Atlantic Community and were certainly far more effective than Communist officials in Eastern Europe during the Cold War. Ob-serving the sweep of public administration across more than a half century in Asia, one is struck by the discipline imposed on the adoption of IT when com-pared to other corners of the world. We can largely charge this off to the au-thoritarian nature of many of the regimes; but not all, because authoritarianism came in many stripes. In Japan it was a dominant party that held power for decades, providing a form of strong control over public policy and administration for years, in effect a very mild form of disciplined or authoritarian rule. To be sure it was also sloppy, as when multiple agencies competed for control over the regulation and support of computing and telecommunications.

At the far other extreme was the government of North Korea that essentially blocked the flow of information, and their underpinning technologies of computers and telecommunications, from the majority of its citizens for decades. Nearly as authoritarian, and perhaps the most disciplined in its implementation of public policies was China, contributing to the reason why this nation could make profound changes over the half century from agriculture to manufacturing and from a Communist command-and-control economy to what is rapidly emerging as a market-driven one. In the muddy middle of our spectrum of disciplined government administration were the Asian Tigers—South Korea, Taiwan, Singapore, and Hong Kong—that all had strong public administrators and competent administrations, and were often rather consistent in their commitment to shared values and approaches to economic development. The exceptions to this model of government participation in Wave One diffusion were New Zealand and Australia, which followed strictly the West European approaches to public administration and to the role of government in promoting the use of IT.

Social and cultural features of various Asian societies facilitated the role of government in many aspects of life, not just in economic development or in the diffusion of IT. South Korea's military heritage helped both military and civilian authorities to marshal the support and actions of this society needed to implement quickly economic initiatives. The Japanese had a heritage of trusting their national government to do what was best for the nation and so people and organizations also more often than not fell in line. Singapore—a city state—although culturally pluralistic, was small enough to build consensus around economic policies. Taiwanese shared a common culture, and shared political values, and with the military threat of mainland China always looming over them, came together more quickly than perhaps they might otherwise have done. The antithesis of these various disciplined behaviors, shared values, and obedience to public authorities could be seen in other Asian societies that came to IT later, such as in the Philippines, Thailand, Indonesia, and Vietnam; but this is a statement that historians will want to qualify and test when they eventually look back at the experience these latter nations had in adopting IT. The key point to make, however, is that the adoption of IT was an important part of a larger process of economic modernization and transformation underway across East Asia, in particular.

The entire lot of Asian economies entered the second half of the twentieth century with profoundly severe problems. All had experienced wars, civil wars, other forms of strife, some essentially for a century (such as China). The West had won World War II, was wealthier, recovered economically in short order, and embraced a consumerism to an extent never before seen in human history. The massive spread of the Cold War around the globe imposed a dangerous set of conditions for Asia that could not be ignored. Asians realized quickly that they had a lot of economic catching up to do, new opportunities to seize upon that leveraged their low costs of labor, and national security to protect in the swirling events of the Cold War. It thus should be no surprise that national political and military leaders would assert themselves, taking sides in the Cold War, seeking the transfer of knowledge and funding from the West, where it made sense, and in general implementing economic development with an export strategy. All the myriad tasks required to implement that grand strategy have been discussed in considerable detail in earlier chapters, and thus need not be rehashed here; suffice it to acknowledge that linking to the development, use, manufacture, and export of information technologies and products that used these ICTs became an important activity of the region since roughly the 1970s. The process was not always efficiently executed, as we saw with agencies competing for control over the effort in Japan and South Korea, for example. But, ultimately the result was the emergence of a style of administration that sociologist Frederic C. Deyo characterized as "a strong, developmentalist state."[28] Looking back from the early 2000s at these efforts, most Asian states demonstrated the "capacity to implement well-chosen development strategies," to use Deyo's words again, to leverage whatever local advantages and resources existed.[29]

Styles varied from one nation to another among the first adopters of IT, but are also evident in the work of subsequent emerging adopters. Nagy Hanna and his colleagues neatly summarized these differences in 1996 worth borrowing for our discussion:

> It [government] played several roles in developing the electronics industry: coach and coordinator for the private sector (Japan), creator of private conglomerates to compete abroad (Korea), incubator and supporter of SMEs (Taiwan), integrator and strategist (Singapore), and infrastructure provider (Hong Kong).[30]

One can continue their thumbnail characterization of public policy by adding that governments selected industries and regions to nurture and support (China), and industries and companies to constrain then support (India) during the early phases of Wave One adoptions in Asia. They offered more recent arrivals to the world of computing models of behavior and public administration that generally worked from which to draw upon for their own economic development, such as to Indonesia, Thailand, Malaysia, Philippines, and Vietnam. For all, the senior model on what to do and how was Japan's, since it was the first to embrace computing, and also had been the source of much managerial and governmental leadership in the region for nearly a century before the arrival of the computer.

The Asian experience reinforces important lessons about economic development and technological diffusion. First, rapid economic transformation requires clear vision of what a state wants and that is sustained for several decades. It is implemented with a constancy of purpose that deviates very little from one generation of politicos and administrations to another, because it takes time to do things. Even the highly disciplined transformation of South Korean and Chinese economies has taken over a generation to implement, and neither would argue that they see some end in sight. Structural transformations are difficult to do, involve tens or hundreds of millions of people who, in the beginning may be ill prepared by training, experience, education, or inclination to assume new roles. Second, such transformations require massive economic investments by the largest institutions in society, which in Asia essentially meant national governments. That required them to prioritize where to spend their domestic funds and scarce supplies of foreign currencies required to import technologies and know-how. In turn, that meant saying no to some needs and desires of their societies so as to favor others. In hindsight some decisions were wise, others perhaps not. The Indians would have done themselves a service to avoid the fashionable investment in heavy industries that they made in the 1950s, although a similar simultaneous commitment by the Japanese proved effective and positive.

Third, nurturing domestic industries, until they could compete internally and in the world market, called for myriad programs, ranging from protectionist policies to equipping industrial parks with electricity, telecommunications, endowing firms with tax incentives, inexpensive rental fees for land, and supplying sufficient quantities of properly skilled labor. As a nation's capabilities to manufacture ever-more sophisticated technologies increased, or similarly in the enhancement of IT services as occurred in India, new skills are required, calling for governments to improve their nations' capabilities to train engineers, scientists, and various IT and managerial experts. Developing knowledgeable workforces continued to be an evolving and challenging activity for public administrators, and regardless of whether in Wave One or Wave Two, remains a moving target.

Because governments frequently embraced export-oriented economic development across multiple industries, not just in information technologies, they voted, in effect, to participate in the global economy. The economy became increasingly more integrated as technologies in information, communications, and transportation reduced the inhibiting forces of distance during the second half of the twentieth century. These Asian governments were thus introduced to what Peter Evans has conveniently called the "new internationalization," and what journalist Thomas L. Friedman cleverly labeled the "flat" world.[31] Asian governments learned quickly that they could not operate independently of a rapidly integrating, delicate, global supply chain. An economic crisis on one continent affected people negatively on another. If global prices for semi-conductors dropped, their manufacture moved rapidly from one nation to another as vendors sought lower production costs. The same process is now apparently at work with IT-related services, with India currently the benefi-ciary of that process. If one Asian country experienced a financial crisis, even-tually they all did, as happened in the 1990s. If one's customers slowed their purchase of consumer electronics, workers in far-away places felt it too, as American consumers pulled back their spending in 2009–2011, lowering demand for products made in various parts of China, South Korea, Taiwan, and Vietnam. With IT having become such a large part of an Asian nation's exports, they became increasingly subject to the behavior of that technology's economics and to the actions of its vendors and users. It was an old story, of course. The British devalued their currency in the late 1940s to promote exports of their products with some temporary successful results; the Chinese manipulated the value of theirs in the new century for similar reasons.

While policies developed by governments for the promotion of IT industries and deployment led to uneven results from one country to another. One pat-tern jumps out regardless of what country we look at, and that is the concentra-tion of both suppliers and users of IT in large urban centers in all these countries. The geographic footprints made by IT were in places that had large institutions, such as state and national capital cities, communities with technical institutes and higher education, normally port cities and manufacturing centers. The same applied to many localities in North America and across Europe and Latin America as well. IT's history is thus very much a story about cities. Where IT sprouted successfully, it had spillover effects in upgrading the economic condi-tion of a region, whether in Silicon Valley in California, or in Bangalore, India. These were the places that had the best preconditions required for IT, such as skilled labor, electricity, telecommunications, government officials, other tech-nologies, and funding. In many Asian economies the urban feature of IT's diffu-sion was even more intensely concentrated with various industrial parks, which explains, for example, why in India, one could see modern corporate IT cam-puses and a paucity of IT just a few blocks away. Concentration within cities, or economic development zones, facilitated development of both suppliers and users of IT. Finally, while IT was diffusing around the world during those same decades, hundreds of humans were moving off the farms to urban centers with the result that by 2010, just over half the world's population lived in towns or cities. It was inevitable, therefore, that major technological innovations would probably have to start where all those people lived.[32]

Before discussing the role of other sectors of Asian society in diffusing IT, we should ask why did so many governments decide to focus on developing local IT industries? Several reasons present themselves rather quickly. By the late 1960s, it had become quite clear to knowledgeable officials, and to many in business, that computers represented new economically attractive products to make, sell, and use. Demand in the West made it manifestly obvious that inexpensive Asian labor and the ability of American and European companies to export jobs, knowledge, and capital to Asia made this class of technologies attractive to pursue. The variety of components and products that could be manufactured also was nothing less than vast, certainly by the late 1970s.[33] Nations that had an industrial base lent themselves more to development of IT suppliers, thanks to local skills, need, and funding, making it easier for officials to opt to promote IT as the next economic force, as happened very early in Japan, followed by the Asian Tigers, and most recently by other Asian societies. Decisions to grow local economies through manufacturing export strategies provided yet another impetus. To be sure, the export strategy involved many industries, not just IT, such as steel, commercial ships, textiles, and inexpensive consumer products. So IT was part of a larger tapestry of economic development, reinforced by successes achieved with other exports that gave officials and whole industries both the confidence and know-how to engage successfully in export driven economic development. Organizations such as the World Bank, International Monetary Fund, American universities, and Western technology companies complimented the growing knowledge base about both export driven economic development and the role that ICTs could play in such grand strategies. To be sure, there were limits to what governments could do, particularly in stimulating the demand side of diffusion; nonetheless, as a whole we can conclude that Asian governments proved quite successful in creating IT industries nearly from no base to a point where by the early 2000s over a third of the world-wide bulk of these technologies were being manufactured by a dozen Asian nations.[34]

ROLE OF THE PRIVATE SECTOR

The private sector played an important role too in the supply side of IT diffusion in Asia, but to a lesser degree in the demand side. It made computers and provided services. Often to a lesser extent it was either slower to use this technology in their own operations, or came to it later than in the West for reasons discussed in earlier chapters. Unlike the West, where the private sector quickly became the most prominent developer and user of IT by the end of the 1970s, in the East it was the public sector that played the most prominent role. The private sector frequently took direction from officials, depended on them for investment wherewithal, and for trade and fiscal protections against competitors from other countries. Developers of IT—the supply side of diffusion— were newly created with less heritage of inherited technical and marketing skills. If a company like IBM (founded 1911) operated in a country, with permission of the local government, then skills were quickly transferred; but, in many cases, local companies were created from scratch. In the West, office appliance and electronic products firms were the earliest suppliers of IT to their local and later international markets, while in Asia the origins of IT surfaced first from telecommunications firms and close behind these, local consumer electronics enterprises. The former were normally single national monopolies or extensions of the public sector until the 1990s or later, while the latter were usually smaller entities than in the West, making less sophisticated products, such as radios and television sets. All of the telecommunications firms of the 1950s and 1960s were technologically and managerially weak in comparison to their Western counterparts, but quickly improved their capabilities and modernized, beginning by the 1980s to such a degree that in some countries they surpassed the West, as with broadband in South Korea and Japan's cell phone infrastructure. Asia embraced the Internet with speed and passion by the early 2000s.

Particularly unique to the Asian experience was the existence of industrial conglomerates, such as the *keiretsu* in Japan. That style of organizing companies made it possible to shift resources (people and budget) from one firm to another within such a group to facilitate the development of some capability, such as the manufacture of IT goods. These organizations were big enough to make the large investments required to enter the computer market. In turn, they became extensive users of IT as they discovered the benefits of using such devices. Where these organizations functioned, it was common for public officials to work directly with these institutions to launch and expand projects involving IT export initiatives. Individual regulators and other public officials often served as employees of these firms in retirement (or before entering government service) or sat on their boards of directors. So the flow of information and influence back and forth facilitated the engagement of these organizations in the process of diffusing IT, in some instances rather quickly, as occurred in South Korea and in some industries within the Asian Tiger economies. Conversely, they were also able to advise officials on what policy and regulatory actions could facilitate the manufacture, and diffusion, of IT, such as protections against foreign competitors and creation of effective industrial parks, as occurred in China and among the Asian Tigers. Associations were also productive agents of diffusion, such as the Taiwan Computer Association which represented small manufacturers of IT components and products, in working with public officials and larger enterprises in the promotion of manufacture and use of computing. In Japan the Key Technology Research Center played a similar role, facilitating knowledge transfer and funding for IT diffusionary projects in the private sector.[35]

The private sector embraced computers for the same reasons as firms in other countries: to make products requiring use of IT in the production process, to link to global supply chains, to improve quality and productivity, and to reduce labor content in work. The largest enterprises were routinely the first adopters, but unlike in the West, these organizations always had to weigh the benefits of using IT against the option of deploying very low cost labor. While low wages was clearly one of Asia's competitive advantages in entering the world of computers during both Wave One and Two, it also served as a drag on adoption until the cost of computing devices dropped enough to reduce or cancel out the advantages of low wages, as happened with the arrival of PCs and later IT-laden automation equipment in many industries, all of which happened late in the twentieth century. As new and less expensive and easier-to-use IT tools became available at the same time that the cost of labor began to rise, these trends reinforced the attraction to use more IT, certainly a pattern evident by the dawn of the new century. The issue of labor costs also led the private sector to move work around Asia as quickly as did the West into Asia, such that by the early 2000s Japanese companies were manufacturing IT and other consumer goods in China, for example. A new tier of nations followed the Asian Tigers into the manufacture of IT, such as Malaysia and Vietnam, to mention two. That hunt for less expensive labor led to the diffusion of IT manufacturing into new regions and in turn that action built up local IT knowledge, which in turn led to some seepage of these capabilities into other industries. As affordability of IT improved, as measured by per capita GDP, IT began to diffuse into the larger economy, a process underway for many nations.

It was not until the end of the 1980s in the United States that government officials essentially lost the power to influence the nature of IT development and the rate and nature of its use in the American economy. In earlier chapters I argued that European governments in the West had less influence on who used IT and for what since the 1960s. But in Asia, the retirement of public sector influence on how the private sector uses IT is not so clear. The Tigers still get much direction from their governments, while in Japan public officials still wield profound influence. These influences enjoyed a rebirth with the arrival of modern telecommunications, mobile phones, and the Internet, all of which required regulations if for no other reason than to facilitate the creation of a networking environment where all these technologies could interact. The newest arrivals to the diffusion of IT still need the hand holding that governments provided to early adopters of the 1960s and 1970s. So, the dependence of the private sector on public institutions remains high, even though since the 1980s it has become manifestly evident that Asian economies have rapidly transformed into increasingly market-driven forms, as we are witnessing in China, for example.

ROLE OF ENGINEERS, ACADEMICS, AND IT PROFESSIONALS

One had to know about information technology in order to build computers, manufacture information technologies beyond simply assembling semiconductors and PCs, and to use these in the daily work of companies and agencies.

Early and recent adopters recognized the need to develop locally these kinds of skills. Every country either beefed up the teaching capabilities of existing universities or established new ones, along with technical institutes to do the job. Public officials encouraged citizens who had these skills living in the West to return, as occurred with the efforts of both the Chinese and Indian governments. Almost from the dawn of the computer, officials were told, and quickly agreed, that local capabilities needed to be developed. By the 1980s, economists and officials were tracking the development of new skills in engineering, science, and IT specifically. Dedrick and Kraemer, in their analysis of the development of such Asian skills, which were needed for any nation to be competitive in the world economy, concluded that those skills were directly linked to the number of engineers trained and deployed. Such know-how allowed whole industries to move from just assembling low-tech components to designing and building complex IT devices, and later to write software. Japan eventually produced as many scientists and engineers as the Americans had as a ratio of total population, followed closely by the Asian Tigers, such as South Korea and Taiwan.[36] To be sure, quality, volume, and nature of the knowledge varied over time but the development of these skilled workers was recognized as an essential component of every Asian IT strategy.

Linked to that strategy, but to a far lesser extent, were investments in R&D, a topic so vast that it would require its own book. However, what can be said with confidence is that throughout Wave One, the vast majority of technology innovation and know-how about manufacturing was borrowed from the West. One saw it with the technology and managerial know-how transfer initiatives in Japan in the 1950s and 1960s, and we could see the process still at work in the 1990s and 2000s with Dell Computers, Apple, IBM, Microsoft, and Google transferring their capabilities to various Asian suppliers and users.

The role of institutions, both public and private, and foreign organizations were important in the absorption of IT into Asian societies. Economist Richard R. Nelson, who has studied technology transfer for decades, drew the same conclusion, "in my judgment . . . institutions matter, and the ones that are appropriate for economic growth tend to change as the key driving technologies change," calling this process "right-headed," even though he cautioned that economists did not yet have good theory "to enable us to judge just what are the institutions that are appropriate for an era."[37] To a large extent, I would add, that is why using historical analyses of specific national experiences is essential in enhancing our understanding, the fundamental subtext of the book you are reading. Economists were, however, on the hunt for insights. At the same time that Nelson drew his conclusion about the lack of sufficient economic theoretical insights, others were at work on the problem. Evidence was mounting that governments could pick technological winners (Asians did this better than Europeans), that investments in training were enormously effective, and that a combination of general practices and policies in conjunction with specific programs to "push" or promote a class of technologies works. The result is the emergence of best practices for both public and private sectors are now being documented, in part based on Asian experiences with information technologies.[38] It is a delicate issue because Western economic orthodoxy holds that governments do a poor job in selecting technology winners, when in fact the Asians did fine. That success calls into question the widely accepted notion that entrepreneurship would identify more accurately winners and losers in a market-driven economy, an idea so accepted by economists going back to at least Joseph Schumpeter in the early twentieth century.[39] Raising the issue of the efficiency of market-driven economies should not be seen as an assault on capitalism; rather, the possibility that governments have a greater role to play in complex, highly integrated, global economies than possibly recognized, hinting instead of a modernized slice of Keynesian economics.

WHAT WERE THE RESULTS?

The results so far are difficult to address with the same confidence as we can articulate for the pan-Atlantic community. This is in large part due to Asia embracing information technologies in an expansive manner beginning in the 1980s, some three decades after the Americans and the West Europeans. Asia's overall initial experience with computers is still unfolding, with the majority of East Asia working its way through Wave One and a few entering Wave Two in bits and pieces, such as consumers in South Korea and Taiwan, but not necessarily all other sectors of their societies. Japan was at one end as the longest involved with computers and large swaths of South Asia and all of Central Asia early in their journeys. With those caveats in mind, what were the results?

In those countries where the manufacture for export of IT goods, and later services, was done, national GDP growth occurred. These nations proved able to attract foreign investments, hence precious foreign currencies to use for other national priorities, and manufacturing know-how that made it possible to employ workers coming out of agriculture and universities. For another, these nations were able to upgrade their exports from simple fabrication of IT products to more profitable, sophisticated production of high-technology components, and even later, to the design of new technologies and products, such as more advanced semiconductors and later, flat screens. This upgrading of the types of value-added exports contributed to the rapid economic growth of many nations, in tandem with other similar developments in such industries as pharmaceuticals, automotive, all manner of consumer electronics, aircraft, and CAD/CAM tools. In some fields, Asia dominated the world market, such as for most consumer electronics, flat screens, laptops, and digital photography, to name a few.

On a more negative note, so much attention was paid to the development of export oriented manufacturing that the seepage of knowledge about IT into other sectors of their societies remained low. This circumstance contradicts the widely accepted notion among economists and historians about how domestic that diffusion of a technology occurs and among experts in public administration and economic development who have long argued that the creation of a high technology industry causes diffusion into broad sectors of society. Theirs is a fundamental canon of most of the economists cited in this book, especially from the World Bank. How are we to explain away this problem? First, IT industries managed to consume a nation's total available IT workforce year-after-year as this export oriented part of their economies grew so fast that even local universities could not keep up with the demand for programmers, scientists, and engineers. There were few exceptions to this imbalance of supply of people, especially after a nation entered Wave One, with the result that outside this IT export-oriented world, the amount of knowledge about computers remained quite low for a very long time. A second problem was the cost of computers—they remained very expensive relative to local affordability in most countries—until the 1990s. Even then, it was often cheaper not to use them since labor was inexpensive, and management in many industries remained less informed about these technologies than their cohorts in the pan-Atlantic community, including in many (but not all) government agencies.

Execution of plans and desires is always an issue in any conversation about the diffusion of something new in society. We saw experts on China commenting on how disciplined the government was; others criticized India's early efforts; still others were amazed at the lack of IT initiatives in North Korea and the lateness of interest expressed by the poorest countries in Asia. However, such generalizations need to be qualified because in most societies, there were pockets of knowledge and effective deployment, as well as less in others. Some Japanese firms in a *keiretsu* were backward in their use of IT, such as banks for some time, while member companies in the same *keiretsu* in manufacturing were as sophisticated (or more) than rivals in the West. The same applied to government agencies. Services industries lagged across Asia in their use of IT, but that can be largely explained by the small sizes of these enterprises and their undercapitalization if they wanted to compete in national or global markets, both conditions which slowed, but did not stop, implementation of IT, particularly after the arrival of very inexpensive PCs, digital cash registers, and, of course, cell phones.

What the history of deployment of IT across Asian societies tell us, therefore, is that just because someone has a great export business in laptops or flat screens, for example, and its young citizens use cell phones to text each other and play online games, does not mean that their society has fundamentally embraced IT and appropriated it evenly across the nation. It does suggest, however, where it did occur was often in large urban settings. The history of Asia's experience with all manner of telephony and computing indicates that generalizations are dangerous, that these nations all had different experiences, therefore, requiring in depth investigation, and that so many are new to this technology that insufficient time has passed for some of them to have a history with IT.

CHAPTER 12

Diffusion of Information Technologies

Results and Implications

Technology changes. Economic laws do not.
  —Carl Shapiro and Hal R. Varian, *1999*[1]

The journey taken from the invention and early uses of office appliances in the late nineteenth and early twentieth centuries, followed by the development and deployment of computers, telecommunications, to today's social media software, has been a long one involving billions of people and costing trillions of dollars. The historical evidence and available information about current and anticipated developments, suggests there is more use to come for information technologies. This is remarkable since so much has been done to essentially make IT ubiquitous in so many nations. With inexpensive smart phones we are on the verge of having the entire planet "wired up," regardless of a country's level of economic and social development. Pausing to look at what the historical record had to say about humankind's experiences with IT to inform the activities of current and future adopters provided the essential justification for this book. By looking at what happened over the past seven decades in the three regions of the world which earliest adopted IT, there is a sufficient base of knowledge to discuss the relevance of the technology to our time, when IT was seen as the iconic symbol of our informationalizing age.

This chapter summarizes key findings about IT's diffusion and suggests implications of past experience for key actors in this long history. Participants included vendors, IT users, public officials, business managers, and scholars of technology and economic development. I also want to come back to the question posed in the first chapter about the volume of information (and its underpinning digital technologies) in the world, because the evidence is mounting that IT may be starting to have an affect on how humankind's minds and bodies are evolving. If indeed that is happening, answering in this book the blunt question "So what?" transcends in importance everything discussed so far. In short, it is essential to learn what we can about human engagement with these evolving classes of information technology.

FROM OFFICE APPLIANCES THROUGH WAVE ONE TO WAVE TWO

Asians and Europeans spent several thousand years organizing information into clay tablets, scrolls, books, 3 x 5 cards, file cabinets, journals, and whole libraries. By the mid-nineteenth century, a process had begun to mechanize rapidly some of the activities required to organize ever-growing mountains of information and publications. In business and governments in the West, large enterprises were also collecting and managing ever greater volumes of information in order to do their work. Historian Alfred D. Chandler, Jr., for example, argued that one could not have railroads crisscrossing the United States without managing the flow of information required to operate such complex enterprises, hence the rise of a new Mandarin class to do this—largely middle managers—people who quickly embraced the use of adding machines, calculators, and tabulating equipment by the end of the 1800s.[2] This process of adoption spread during the first half of the new century across many industries and countries, leading to ever-larger organizations using those technologies to help run these to increase their size and to expand roles. They faced problems with managing so much information and so had turned to automation of data to gain, or maintain, control over both information and their daily work. James Beniger referred to this adoption of early office appliances as a "Control Revolution."[3] Work by subsequent scholars filled in details that reinforced the desire to manage large organizations with information, initially in the West, but increasingly also in Asia.[4]

In these seven decades, governments were also consumers of the technology—users—to track soldiers, to conduct censuses, and by World War II, to monitor and document economic events in the West, but not yet in East Asia. The latter needed to get through several decades of civil and world wars to have the resources to develop a focus on their own "Control Revolution." Higher levels of economic wherewithal, larger supplies of engineers and mathematicians, and the exigencies of World War II code breaking, development of firing tables, and invention of atomic bombs joined with many scientific research efforts to make it possible to invent digital computers in the pan-Atlantic community.[5]

The approach taken here to understand how information technologies spread around the world could have centered on the evolution of specific technologies, or by describing the history of key suppliers of computers, or by examining how consumers of this technology came to embrace it. I have applied each of these approaches before. The story could have been overwhelmingly economic in form too, or a collection of wonderful tales of heroic and brilliant efforts by individuals in engineering, computer science, business, and government. These are all reasonable approaches to the subject; however, I chose as our approach the experiences of whole nations and regions, and secondarily the role of institutions, such as governments, industries, and firms. Under that large national-regional umbrella one could neatly include many other slices of the story, from the role of engineers and users of computing to that of large and small vendors, and other individuals. By using a nation, and secondarily a region, continents, and major institutions, as our fundamental units of measure of who acquired and used IT, we could showcase the more complicated set of variables that affected the arrival and diffusion of computing. The role of national governments became clearer, global players their efforts, the flow of knowledge within whole industries and professions across borders highlighted, and account for local cultural, social, historical, legal, and economic realities. The technique of exploring in detail the experiences of over a dozen individual nations reinforced the value of this approach. Each country had unique experiences yet also shared common ones too. By taking a national perspective we could begin to delineate these, identifying what was unique, yet also cataloging broad patterns of shared behaviors. With that strategy in mind, we can generalize with greater confidence than before about how this complex general purpose technology spread so far so fast in barely a half century.

Motives for inventing, building, and using computers evolved over time. In the beginning (1930s–1950s) it was largely an issue of wartime needs, in anticipation of war, the occurrence of the Second World War, and of the unfolding Cold War. The risk of failure in the early development of equipment and the funding needed for such very expensive, hence, financially risky work for the private sector, and technologically complex projects, called for government involvement. Government leadership began in the 1940s in support of R&D for information technologies; then extended to the development of such civilian IT-dependent applications as space travel. By the 1960s, a second motivation for using computers harkened back to the office appliances of the late nineteenth and early twentieth centuries, to provide control, reduce the cost of labor, and by the end of the 1980s, to do work that could not be done solely by humans, such as precision welding of automobiles, processing of credit card transactions, and facilitating rapid communications now called e-mail. Computing in the decades of the 1950s through the 1970s evolved in the shadow of the Cold War while commercial enterprises selling computers formed into IT industries across North America, most of Latin America, all of Western Europe, and in parts of East Asia, most notably in Japan. As the technology became smaller and less expensive, individuals embraced computing by using personal computers that they could afford to purchase, beginning in the late 1970s. That form of adoption expanded first in the West then in the East from the 1980s to the early 2000s. Now reliance on PCs is waning as less expensive, more portable devices can do much of their work, such as iPads and smart telephones.

The arc of diffusion of IT in East Asia, and later in Latin America, Africa, and currently in parts of Central Asia, was different enough to call out a unique motivation for using computers. To quote a World Bank economist writing in 2006, reflecting much of the motivations for whole nations finding computers attractive, "In most developing countries, technological adaptation is indispensable for rapid economic growth, especially export-led growth."[6] But this fundamental belief in the economic benefits of using and exporting IT appeared first in the United States in the 1950s, spread to Western Europe by the mid-1960s when the region was shaping its economy to compete with the Americans. By the 1970s the same mantra had expanded into Asia. By the end of that decade if a national government had an multiyear economic plan, invariably it included the role of IT and often articulated specific initiatives to create an indigenous computer or IT industry.

In country after country a pattern of diffusion emerged that proved similar to that of one's neighboring societies. Governments and interested technical experts developed the original and subsequent technologies in the beginning and played an important, if not primary, role in introducing these into their economies. They nurtured local IT industries. As these became strong enough to function in local and international markets, local firms increasingly took the lead in pushing the technology into various corners of an economy, leading the charge in introducing IT to users. There were exceptions for a while to this pattern, such as during the Cold War in Eastern Europe and in China, until even these regions began transforming their economies into market-based ones, most dramatically China. In those exceptions, there were few computer sales people walking the hallways of potential users trying to persuade them to embrace the technology. The evidence demonstrates clearly that without this commercial push, rates of adoptions were measurably lower. In market driven economies users needed to improve continuously their productivity, and to drive labor costs out of the work of manufacturers and other labor intensive firms across many industries where salaries were higher than in other countries, and where IT could be used to reach new customers. By the late 1980s, American government officials watched as private sector IT firms now drove the innovation agenda for information technologies, not large public agencies; European governments had already lost that influence over their own companies, if they ever really had much, while the Asians were quick to borrow innovations from the West. Asian economies proved able to adopt the products themselves, including methods required to manufacture ever-more complex goods that were more profitable and generated higher salaries for their own workers.

Across the world one could see a rolling momentum building in the use of
IT over the decades from West to East and most recently to Latin America,
Africa, and the Middle East. Firms, agencies, and academic and technical
institutes acquired one or a few systems; built, bought, or leased equipment
and software to run one or a few applications. Over time they increased the
volume of jobs processed through their systems, adding new applications as
they came along. World-wide users deployed an incremental approach to embracing computers. As they learned what they could do with IT, they added
applications. Over the past seven decades, computers became less expensive,
acquired more functions, often became easier to use, and came with more
application software relevant to a user's work activities. As the cost and functions of computing improved, access to computers spread from the wealthiest
nations able to afford these technologies (1950s–1960s) to mid-level economies (1970s–1980s), and then to others with lower levels of GDP, especially
after the wide diffusion of relatively inexpensive personal computers, telephony, and the Internet, the latter beginning in the 1990s. As costs changed
relative to affordability over time, one could see unfold a global pattern of
adoption. At first only the largest institutions acquired computers, then mid-
sized private and public sector organizations. Next, wealthy and middle class
individuals embraced IT, and finally "everyone" it seemed could afford to gain
access to mobile phones that by the end of the 1990s could do some of the
work previously performed using laptops, personal computers, and even
earlier, terminals connected to mainframes and minicomputers. Users also
incrementally and continuously replaced older hardware, software, and piecemeal work practices over the past seven decades. Today, even young children
have smart phones with "apps" and they too can be expected to upgrade to
newer devices as they come along, adding new uses as well, such as just-in-time
education.

To borrow the term crafted by Arnulf Grüber to describe the diffusion of
transportation technologies over hundreds of years, whole nations were
involved in a "social learning process," one that made it possible to identify
waves of IT adoption and to forecast the role of this class of technology.[7] While
the notion of multiple infrastructures proved essential in understanding the
process of how a general technology diffused through a society like IT, it seems
a bit early to attempt to forecast too aggressively where this class of technologies is going as it remains very much in a developmental stage. But, like
transport systems, whole nations embraced IT when it had relative advantages over other technologies, although it, like transport, never displaced fully
earlier ones. People drive cars but some still have horses for pleasure riding,
individuals write letters by hand yet send e-mails too. In more economic
terms, like other technologies, IT was overwhelmingly adopted when its use
improved productivity over older options. The exceptions were the forced
marches, as occurred in Eastern Europe during the Cold War when some users

were ordered to take delivery and to use computers. The effort failed because many could not comply; some simply left the machines in their original packing cases. The muted adoption in Communist Europe highlighted the importance of having many infrastructures supporting computing in order for this technology to spread through an economy.

In Wave One several long-term patterns of activity were in evidence. First, it flourished in an age when the technology was initially created and continued to evolve rapidly, never reaching some stasis or unchanging stable maturity. Incremental improvements became the norm, as happened with adding machines and calculators and with punch-card tabulating equipment by the 1930s. The world went from a German making a computer in his parent's living room and radio hams doing the same at American universities in the 1930s to massive projects involving the development of Whirlwind at the Massachusetts Institute of Technology (MIT) at the dawn of the Cold War, to IBM's Watson super computer in the early 2000s, and to Apple's voice-activated personal assistant software, Siri, introduced in 2011. If one thought mainframes represented the apex of evolution, they were sadly mistaken, because minicomputers came, then the more radical innovation, personal computers, and most recently the highly mobile, small, powerful cell phones and Apple products of the early 2000s. As of this writing (2012) there was no slowdown in evidence in the evolution of the technology.

Buried in that broad pattern of technological evolution was the world-wide upgrades and innovations that occurred in all manner of telecommunications from the humble twisted copper-wired telephone to development of satellites, data transmission, and, of course, the Internet. Tied to that evolution in telecommunications was the convergence of IT and communications, beginning in the 1960s that became so intertwined that today we more frequently think of ICT as never having started out on different technological trajectories.

Second, governments played a far greater role in promoting the development and use of all manner of ICTs than previously understood. For most nations during the entire time of Wave One governments were *the* primary institutions in societies driving the diffusion of IT. This was so much the case that we may need to conclude that a feature of Wave Two adoption is the relatively diminishing role played by governments when compared to that during the first wave. In every country studied public officials nurtured, promoted, and protected local IT industries—the supply side of the story—and more often than not, stimulated demand for adoption, although far less so, particularly after the 1960s in the pan-Atlantic world and after the mid-1990s in East Asia. While historians and economists knew that governments had been extensive users of IT, as mentioned frequently throughout this book, officials looked at events in one country or one geographic region almost with blinders on in the early decades of IT diffusion. They acted normally on their own experiences and local circumstances for their individual and national betterment.

But, when we look at many nations at the same time, one can see clearly similar activities carried out by governments across the world. Furthermore, we witnessed some officials sharing information and insights with each other across governments and regions, although scientists and engineers did most of that. By the 1980s international organizations were facilitating the transfer of information to officials across countries on how best to leverage IT, such as the World Bank, the Organisation for Economic Co-operation and Development (OECD) in Europe, NATO, of course, with the various national military establishments to foster coordination of its work, and such quasi-academic and industry associations as the Institute of Electrical and Electronics Engineers (IEEE) and the American Federation of Information Processing Societies (AFIPS).

The third grand pattern in evidence involved the ever-expanding role of individual users. They had computers thrust upon them at work, the result of IT advocates and later line management embracing the technology. In time, users became managers who, in turn, acquired IT systems for their organizations. As technologies became less expensive and accessible, they and their employees also acquired computers for their personal use at home. Over the decades, access spread to families, most notably to children, while computing began to appear in other non-work environments, such as in entertainment, churches, and, to a limited extent schools. As the number of consumer products which used IT appeared, beginning in the 1980s, and exploding in variety during the 1990s, consumers influenced the courses of innovation and adoption around the world. In some instances their actions caused structural shifts in both an industry and affected an economy. For example, the giant American camera company Kodak nearly went out of business because it ignored the rapidly emerging attraction of digital cameras, which moved from being manufactured by the photography and camera industries to consumer electronics industries, most notably those in Japan. That kind of shift, both rapid and massive, occurred essentially within one decade. That type of shift encouraged various economies to continue and to enhance their export-oriented development, particularly to satisfy the appetites of increasingly IT-savvy consumers in wealthy countries. It is a story of South Korean and Taiwanese economic successes, and earlier that of the United States and Japan.

During the early decades of Wave One, expenditures on IT were small and concentrated on the supply side and a few large users. By the end of Wave One domestic consumption far outweighed investments made on the supply side in the most advanced economies. But, as one studies the diffusion of this, or any other technology, there were also other sets of activities underway: the development of the supply side—which had to happen first in every instance—and then the demand side. At the risk of offending those who have studied diffusion of IT, often the distinction was not made emphatically enough nor was the relationship and effects of one on the other properly accounted for. It was most noticeable with India, which currently is considered a hot bed of IT supply side activities, but, when we peer more closely had less use of IT inside the nation for non-export work than we might have thought otherwise. Going forward, we will need to treat both supply and demand sides of IT activities as contiguous well-defined lines of investigation, while simultaneously recognizing that over time they become increasingly intertwined.

WHAT HUMAN ADOPTION OF IT CONTRIBUTES TO THEORIES OF DIFFUSION

All studies and theories about the diffusion of technologies have centered on how humans embraced them. There have been far fewer studies about the use of tools by primates, such as using sticks and rocks. It is so obvious that scholars have not qualified their studies by saying theirs' are limited to human consumption. Even this book is about people using computers. But, we may find by the end of Wave Two that humans are not the only consumers of IT, as was largely the case in Wave One. Already there are more devices using the Internet than people, such as computerized traffic lights, security video cameras, and meters and sensors tracking and reporting movements, temperatures, and speeds of liquids, gases, and physical objects. Most complex equipment also have embedded within them computers as well. As discussed further in this chapter, computers are becoming so intelligent and so endowed with the authority and capability to make decisions independently of people that it is reasonable to assume that they may soon start making decisions about when and how to embrace future IT, to create their own innovations, and to use them. These are points being increasingly made by responsible observers of technological evolutions.[8]

An immediate implication is the effect on jobs. Computers were used to replace and augment human workers in manufacturing and office environments during the past half century. In the most advanced economies, that is to say, those with the most expensive workers, the use of IT caused some jobs to move to where there was less expensive labor, most notably to Asia, or caused them to be eliminated as automation displaced human labor, particularly in manufacturing and process industries. That process of work trading off between humans and machines is continuing in Asia and will in other parts of the world, as new products and work are created for humans, and other activities are automated. IT-intensive devices involved in this process have included robotics, numerically controlled machines in manufacturing, computerized inventory control (one of the first IT applications to surface in the early 1950s), recently voice recognition, and, of course, online commerce. There is some controversy among economists as to whether machines are winning some sort of race for jobs against humans, displacing more workers faster than they are creating new jobs.[9] But that the shifting is continuing is not in question. But now with the emergence of ever more sophisticated software that understand human speech, can be used to translate languages, and recognize patterns, the technology holds out the possibility of taking over the work of call centers, sales; even some marketing activities. In short, automation is moving into the service sector of both mature and emerging economies, first in the pan-Atlantic world, but also in Asia. In some countries, where services jobs account for as much as 70 percent of GDP, that could result in significant reordering of work, as might happen in the United States, for example.[10]

Taking initially human-centered adoption as our starting point for building perspectives on IT diffusion is a reasonable step, but it is not an assumption to continue maintaining as the only approach after the twenty-first century has passed. Changes come fast today in ICTs, and if one compares the profound evolutions that occurred with metal-based computers over the past seven decades, we can only imagine feebly what could happen over the next seven decades. Some of those changes could quite probably involve technologies or new forms of human-made biological entities also diffusing technologies.

Even the definition of technologies could change. In the West the use of the word technology normally refers to machinery and their components. This routinely means complex devices. Historians and social scientists are more inclusive, adding into the mix earlier innovations, such as axes and wooden plows and the processes dependent on them. In the past several decades, ICTs have crowded out many earlier tools and innovations when people in the pan-Atlantic community thought of technologies. However, much of the economic and public policy studies and pronouncements about technological diffusion in East Asia written by Asians in the same period involved non-IT or ICT technologies. So, a warning to students of Asian IT, beware when reading about some Asian technological theme not to assume that the author meant just computers or telecommunications, as often it could mean the expansion of textile production in the 1960s, the Green Revolution in food production in the 1970s, and most recently, pharmaceuticals. Biological issues are increasingly being mixed into the discussion with the term technology—often now called biotechnology—also deployed to account for what happens with pharmaceuticals, for example. The experience with IT suggests that the meaning of technology, and of the diffusion of a technology, will continue to evolve, requiring that we constantly challenge the meaning of the term and how its transformation affects our understanding of the theories of diffusion.

A second lesson experience with IT can teach us about diffusions in general that we may begin to see technologies doing the same things. Multiple differing types of devices might be able to perform the same tasks as others, as occurs today with music playing on laptops, iPads, iPods, and more traditionally on radio and CD players. Learning technologies may also change as educational software begins to challenge fundamental precepts of pre-digital pedagogy. Economists, biologists, and brain experts, and others knowledgeable about how learning occurs, will need to spend more time collaborating with each other than in the past. Given what is known today, one should resist speculating too much about this fascinating aspect of diffusion. Instead, we turn to a more conventional discussion of how IT fits into accepted theories of diffusion, because for the immediate future that is what will be most practical and our reality. But, one should also anticipate the possible next big step in the diffusion of IT, watch developments, listen to commentators speculate, and, in effect, prepare for such a possibility. Already computer scientists, futurists, and biologists are doing this.[11]

The starting point for most scholars interested in the diffusion of technology is the set of perspectives introduced by Everett M. Rogers in the early 1960s that he continued to enrich over the next four decades, already discussed in chapter one and again with respect to India in chapter 11. Part of the attraction of his thinking was his categorization of adopters as early, late, and laggard. In the last edition of his seminal work, *The Diffusion of Innovations*, published in 2003, he integrated the early stages of the Internet's and other telecommunications' deployment into his ideas.[12] I embraced his concept of early to late adopters, even some of his language, because his approach provided convenient scaffolding for organizing much of our discussion. That also makes it possible for those interested in other technologies who use his worldview to compare IT's experience to those. His definition of diffusion— essentially that it was a process by which a technology was "communicated" through various institutions into a society—while essentially my use of the term (diffusion), could be made more precise about IT.

The experience of American, European, and Asian societies demonstrates that diffusion of IT was a process for creating and improving technologies over time as a result of technological, scientific, political, and economic dynamics. Next—and simultaneously—it involved pushing these technologies into an economy and across a society through such major institutions as governments and private sector suppliers of the technology. This was followed by the transformation of that process into one where organizations and individuals demanded supplies of a technology and took charge of the efforts of creating more innovations and supplies of a technology to appear as they acquired the capacity to absorb these. The shift to demand dominance of adoption emerged from the growing knowledge about the use and benefits of a technology, and from increased dependence on it over time as central to how one worked and played. Underlying that process of adoption were several behaviors from innovation to acceptance.

First, there is a growing body of knowledge about a technology that expands first from tiny sources (for example, a few computer scientists) to governments and IT vendors; then, moves more broadly into a larger community of

users (firms, agencies, subsequently individuals). The technology is sufficiently complex that casual or minimal methods of knowledge transfer prove
insufficient to cause adoption. A lesser form of complexity of adoption might
be a father teaching his son the skilled use of a chain saw or carpentry. IT's
experience teaches us that the knowledge transfer requires a great deal more
explicit knowledge, such as engineering and mathematics, and more precisely,
programming along with many other bodies of knowledge in addition to how
to use tools and other complex infrastructures. However, tacit knowledge is
absolutely required as well, such as managerial skills, an understanding of the
work processes to be modified by the use of IT, appreciation of the potential
effects of computing on institutional behaviors, firms, agencies, and societies.[13] Those countries that had the earliest and greatest amount of explicit
and tacit knowledge about a technology and its underpinning from other
technologies and scientific principles could be expected to provide IT innovations the earliest and be the most sustained over time, as occurred in the
United States. This country had more engineers, university graduates, and
practitioners than any other nation during most of Wave One. So, IT's diffusion teaches us that the role of knowledge transfer is far greater and more
structured than for many earlier technologies, hence quite important to the
overall process of deployment. Inherent in that process of knowledge transfer is both explicit and tacit forms, revealing a type of information path
dependency.

A second underlying component of IT's diffusion, often only discussed by
economists concerned with economic development in countries that are poor
or only moderately productive in comparison to the "developed" economies, is
the economic wherewithal to pay for IT. For the truth is, IT has always been
expensive. In absolute dollars, a million dollar machine is expensive even in
the wealthiest economies, causing an executive at a Fortune 500 company to
think carefully why to spend so much money. It has always been so, indeed
even more so when IT was far more expensive in earlier decades. But everything is relative. A digital smart phone in Sweden may seem inexpensive to a
local user, but in Malaysia the same device remains very expensive, and in
Central Africa might require a village to invest in just one. The cost for the use
of information technologies has often exceeded that of the purchase price of a
device or software. We can conclude on the basis of IT's diffusion that the
more complex a technology is the more expensive it will be, at least until the
characteristics of the so-called Moore's Law take effect to drive down relative
costs. That means it will be adopted the earliest by societies that can most
afford it and last when its cost has declined sufficiently to make it broadly
affordable. The diffusion of a technology will be impeded or propelled as either
the cost of the technology declines (as happened with computing) or as an
economy's prosperity improves its ability to acquire the technology (as happened in Europe, Japan, and across East Asia with IT and in recent years with

medical technologies). In other words, the features of the technology are not always as important a consideration as is its affordability. We saw that in many instances: Russians using 15 year old technology which worked just fine for them for a while, China relying longer on labor instead of computers because people were less expensive to use.[14] In short, affordability must be accounted for in the diffusion of a technology.

Third, there is the functionality of the technology itself as an undercurrent of diffusion, because affordability aside, one still needed IT to do desirable tasks reasonably well. Functionality ("feeds and speeds") has been the one aspect of IT that has attracted the greatest amount of attention on the part of historians of IT, engineers, computer scientists, and many industry watchers. Back to an idea first introduced in chapter one, information technologies are general purpose. That means they can be put together and be used in a variety, indeed nearly infinite, number of ways, giving it the flexibility necessary to fit into so many activities engaged in by people and machines. That malleability, enhanced through so many innovations in IT over the past seven decades, contributed to the ability of people, and their organizations, to "apply," "adapt to," and "deploy" IT.[15] These are phrases used widely to explain the acts of IT adoption. IT represents the most flexible, adaptive body of technologies to emerge since the harnessing of electricity and for those closest to the continuing evolution of it, are convinced that there is much yet to happen with computing.[16] Across all societies studied in this book malleability of IT proved quite important. IT could do things in ways people liked, and the ability to do this improved continuously enough to draw in more users. Those who embraced IT did so for the same reasons all over the world, normally to reduce operating costs, to collect information needed to make decisions, to perform tasks that humans were less capable of, to have fun, to communicate, and in response to market pressures from customers and competitors. In short, more than any other recent class of technology, IT facilitated control and increased productivity.

But, and in a partial departure from conventional thinking about why IT diffused so rapidly, I presented considerable evidence from across the world to suggest that the capabilities of IT was not the single most important reason why people embraced computing. Too often, students of diffusion begin with the qualities of the artifact; even Rogers did that. While something also has to have a function useful to people, it has to be relevant, affordable, reasonably safe, legal, and convenient to use. Those attributes extend beyond the device itself to other requirements such as funding, skills, laws, and time to use it. On the other hand, if the technology does not work, or is impractical, people will be less inclined to use it unless forced to by their management, regulators, or by a lack of attractive alternatives. This insight was uncovered while examining computing in Eastern Europe during the Cold War, for example.

The fundamental lesson from the story of IT's diffusion is that it spread because of a combination of factors at work, of many moving parts in society. That means unilateral explanations are inadequate to the task of explaining how modern complex technologies have, and will continue to diffuse across whole nations and through various industries and organizations. In fairness to earlier students of a technology's acceptance, who probed deeply into one or more facets of a technology's acceptance, they contributed to our list of variables to consider. When K.J. Arrow argued in 1962 that new technologies were embedded in new capital investments and that the technologies could be applied by using them, it was a start.[17] When nearly a half century later other economists noted the ability of East Asians to find and learn about new technologies, our understanding expanded.[18] That insight led to other studies of how knowledge was upgraded to more complex levels as industries and whole nations moved up value chains.[19] Other scholars of diffusion looking at the steps and actions taken to encourage foreign direct investments, licensing agreements, imports of capital goods and other inputs, local industrial development, contractual terms and conditions, impact of taxation and protectionist policies, local and foreign R&D, use of technology parks, and leveraging diasporas, identified actions in evidence in all the countries we have examined. These played different roles with varying degrees of influence that shifted over time, demonstrating that as technologies became more varied and intricate, the greater the importance of a more complex ecosystem, clusters of influences, that were required to explain how a technology spread around the world.

That observation—the growing complexity of simultaneously dynamic elements at work—represents perhaps one of the most important contributions to theories of technology diffusion that can be gleaned from examining the experience with information technologies. As tempting as it is to compare quite directly the evolution of IT to stone tools, telecommunications to the Roman short sword, to the myriad technologies developed in China, as much conventional thinking might suggest,[20] we should recognize that IT's history indicates that our theories of how many recently developed technologies of the past several centuries emerged will have to be partially reframed—not dismissed—to incorporate a broader consideration of multiple influences.

To make matters even more complicated, we may not be able to separate cleanly external from internal influences, that is to say, exogenous from endogenous circumstances because those two notions imply that there is a separation. But look at the Indian experience, when the technology parks are near replications of communities of practice in Silicon Valley, are those centers exogenous in India's economy or are they part of India's society? A visit to Bangalore should cause one to scratch their head as they contemplate how to answer the question. The inability of West European computer vendors to collaborate, or at least their reluctance to obey their local national governments to compete effectively in European-wide markets, raises the same question: local or foreign influences, which affected these firms the most? In Europe local influences dominated, while in Asia less so, but not to the exclusion of domestic realities by any means. We know that local and foreign influences always affected suppliers, policy makers, and users too. This means economists need to understand what political scientists and historians have to think about the subject, not just what other economists have said. Public officials have to understand economic, political, social, and historical perspectives or suffer unanticipated consequences, as occurred to so many public officials who did not understand the power of social media in 2011 with the Arab Spring, and more recently riots in London and sit-ins on Wall Street in New York City. It also means that vendors cannot just sell the functionality of their products, but must also become students of regional cultures and economics.

For these reasons the rest of this chapter focuses on the implications of the greater complexity posed by diffusion of a sophisticated general purpose technology (such as IT) to various constituencies interested in such a technology. This is less history, although a few tentative insights are posited, based on historical perspectives.[21]

INSIGHTS FOR ECONOMISTS AND HISTORIANS

The most obvious insight one can glean from IT's global diffusion is that it was tied to many economic activities, with the result that as both economists and historians examine the affairs of nations that occurred since the 1950s or 1960s, the role of computing will need to be taken into account. Conversely, with so much IT installed in so many nations, how that happened—and is happening—has to studied within the context of so many other factors, such as economic events, politics, social and cultural practices, and how businesses and governments are managed. What may appear on the surface to be simple and obvious observations become possible, but they have complex underpinnings. For example, Amar Bhidé, a business school professor, argued that "rich countries tend to make greater and more effective use of IT and other advanced technologies than do poor countries."[22] It is when an economist reveals the data beneath such a statement that we begin to see evidence of how the dynamics affecting IT diffusion work. In his case, he revealed that "IT expenditure—to-GDP ratio in Western Europe is 15–20 percent lower than in the United States, and in Japan it ranges from 10 to 30 percent lower than in the United States," leading him to conclude that "U.S. businesses are relatively stingy in their conventional investments but are exceptionally venturesome when it comes to IT."[23] Such an observation, in turn, encourages more IT goods and services to be developed for an advanced economy, leaving developing economies to catch up, rather than for the latter to contribute new technologies. There is a great deal of work for economists and historians to do to validate, or contradict, and certain qualify, such statements and to document their implications.

Issues of productivity and competitiveness of companies were very importantly tied up in the diffusion of IT in the most market-driven economies. That class of economies was also the most extensive and earliest users of IT. A key argument in this book is that affordability—both national and personal GDP—affected the amount and speed of adoption of IT. The wealthiest nations on earth were capitalist, making it possible for them to be the most extensive users of computing. IT enabled gains in productivity, enhanced access to new customers, and generated new products and services, all of which translated into revenues and profits in such economies. For both historians and economists, that dynamic suggests new lines of research and insight. This is important because across North America, Western Europe, and East Asia, how large private sector organizations reacted to IT was—and is—crucial to the story of diffusion of these technologies. This is especially the case across the pan-Atlantic community where the hunt for profits and productivity in its modern forms originated and the practices to pursue these were developed.[24] In East Asia a similar process unfolded in the Asian Tigers and earlier in Japan, but with more collaboration and coordination amongst such enterprises, often with the influential hand of public administrators who determined who the players would be, the extent of their participation, the rules by which they would operate, and the benefits they would obtain.

That comparison raises as many problems for students of the process as it resolves. We saw the ebb and flow of public administration's role in diffusion in many countries. Selecting IT national champions in Western Europe essentially failed to make Europe an IT powerhouse in the global economy or to free it from dependence on American technologies and products. The process worked better (but not perfectly) in Asia, where the strategy was to co-opt and align with American innovations to provide competitive products. The Americans generally avoided the game of picking national champions; they overwhelmingly let market forces do the job.

A challenge for economists and political scientists is to understand that dynamic of public-private collaboration, and how it can be done better, because we now live in a time when officials are more actively involved in how national, and indeed, the global economy functions. Just as there was a convergence of technologies with IT and telecommunications that affected how work and play was done by the end of the twentieth century, there occurred simultaneously convergence of economies, compelling increased involvement by governments concerned about national security, economic welfare of its citizens (particularly in nations that elected their leaders), and economic growth. As economists have been observing for some time, economic convergence is now obvious and its features more definable.[25] That line of investigation requires additional inclusion of the role of IT into the mix to enrich our understanding of how that convergence was occurring, and that is still underway. Enough time and activities have passed for historians to participate in that work.

If IT has now become so integral to the lives of over half the world's population—as I contend—then there are at least three issues warranting further understanding. The first involves the effects of inadequate education and training of citizens and, more specifically, workers to function in a world filled with IT, because future technologies will emerge that will invariably have extensive information content. What are the economic effects of inadequate knowledge both on how to use the technology and other bodies of skills and of expertise required to support the diffusion and use of a new technology? IT's experience tells us that it is more than reliable supplies of electricity or literacy. Second, therefore, what is the full list of requirements, and what are the implications of having, or not having these? I worry about nations that do not invest enough in the education of their children, about places that have poor telecommunications infrastructures (although that is changing rapidly), and about the lag in economic development in the poorest nations on earth. How can IT and other emerging technologies affect those conditions? Third, and an issue often on the minds of economists at the World Bank, whose studies played an important role in informing this book, how is economic development facilitated by IT? Today few conversations are held about a nation's economic development without discussing the previous and future role of IT. Every country that has a documented economic development plan—and most do—discuss the role of IT, right alongside with what to do about natural resources, education, industrial development, and expansion of their service economies. I am not aware of any exceptions.[26]

Research by economists and historians can be supported by the ever-growing amounts of data becoming available to them; it is one of the pleasant discoveries made in writing this book. While one can grouse about the lack of good data, and I did from time to time in this one, nonetheless, international economic bodies are collecting vast amounts of useful information, such as the OECD and the World Bank, but others too on education levels and infrastructures, such as by the United Nations. Access to enough archival records that deal with the activities of individual firms, government agencies, and individual players remains a problem to fill out our account of how IT diffused so rapidly. Western sources are the most voluminous, Asian ones available, but far less so, while the Middle East, Africa, and Latin America represent nearly blank spaces in our hunt for evidence and insights about the diffusion of IT in their regions. Part of the problem is, of course, that they came to IT much later so they have had less interaction with the technology. That lateness in itself is fodder for the historian as it can provide additional evidence that it was not the technology's features and costs necessarily that affected adoption (although surely important factors), rather other issues, such as levels of GDP, types of industries in one's economy, level of political stability, and skills of its work forces. The implication for both historians and economists is that these regions may provide additional insights on why IT spread so fast in the West and the East, and how that might occur with emerging new sciences and products, such as biologics and nano technologies.

One final issue needs addressing: How our findings about Wave One adoption fit into the larger discussion long held by economists and economic historians about long waves. We need to briefly situate the IT story into that broader debate. For nearly a century economists discussed the possibility that there existed waves of economic activity, such as prosperity and depression, or inflation and deflation, with those advocating for such an inherent process arguing that these tended to run in roughly 50 year cycles. One could imagine very easily that the half dozen decades following the end of World War II represented one of these cycles. Since the 1960s, economists have examined the possibility with renewed interest in debating its contours, usually measuring a wave by what happens to prices and more recently to rising GDPs and economic growth.[27] While space does not permit for an extended discussion of waves, suffice it to point out two widely accepted features of these. First, a wave can be initiated and be sustained by the expansion and evolution of one or more sectors in an economy, such as manufacturing did for the First and Second Industrial Revolutions.[28] Second, these waves often had a strong technological innovative process in play, such as steam in the 1700s, railroads in the 1800s, along with electricity straddling the nineteenth and twentieth centuries. When technologies emerged, spread, and declined in importance has been the subject of much discussion since Joseph Schumpeter attempted to describe that process in the 1930s.[29] IT can be seen as playing a similar role during the second half of the twentieth century while straddling the next one. The growing consensus is that technologies as a whole are important to the process of waves. To use one economist's blunt statement, "the technology diffusion process" quickly "becomes quite central to any complete theory of long waves."[30] In recent years, this point has variously been made by a distinguished list of economists, such as Richard Nelson, Sydney G. Winter, Davendra Sahal, Giovanni Dosi, and Carlotta Perez.[31] Technologies make it possible for new industries to be born and flourish, as we saw with the office appliance and computer industries, to mention just two rooted in information technologies. Clearly, IT joins a long list of earlier technologies feeding evidence into studies on waves.

In addition to the economic and business environmental features that help to shape a wave's form, and thus provide fodder for much debate, a second controversy is the discussion around the length of a wave. Some students of the process have documented 25 year waves, others 40–50 year waves, still others 55 year waves, and so forth. There is no firm consensus, although if polled, probably most wave historians would favor a 50–60 year cycle. The vast majority of the evidence for any wave is drawn from European and American experiences; in other words, from that of the pan-Atlantic world, with far too little from Asia for the modern period.[32] The experience with IT does not yet offer important evidence in favor of one length of a wave or another because it is still unfolding with no serious computer scientist or IT vendor prepared to declare that the technology has stabilized and that it has diffused as far as it can. Our comparative study has shown that there is plenty of room for further adoption of IT in eastern Europe, all of Central Asia, large parts of India and China, North Korea, a large number of late adopters on the Pacific Rim, broad swaths of Latin America, and the vast majority of Africa and the Middle East. If anything, it is quite possible that IT will upend current thinking, suggesting that waves might be far longer than 50–60 years if we accept the notion that technology is a primary component of any economic wave.

But the obvious is true for any wave. First, there is some sort of technological breakthrough to a new technology, such as what happened after the development of the transistor and subsequently, integrated circuits. Thousands of incremental changes and improvements clustered subsequently around those breakthroughs. Then comes a period of new uses (applications), improvements, spin-offs, and so forth that in time become identified as a "technological trajectory," to use a term Nelson and Winter introduced in 1977.[33] Third, a period emerges characterized by widespread adoption of some technology and its diffusion to all logical corners of the world. Clearly in Wave One IT we have gone through the first event, the second is continuing, as is the third (diffusion) with much yet to be done before what an economic historian might call an economic wave involving IT having concluded. That is why the current experience with IT, which is massively being adopted around the world and is also taking longer than 50 years to diffuse to its practical limits, may require wave economists and historians to revisit earlier assumptions and evidence. IT's "technological life cycle," to use a very convenient term, is still unfolding. This suggests more work has to be done, for example, in understanding the lag between diffusion of information about computing and the buildup of all the various infrastructures that facilitate the adoption of a technology, in our case IT.[34] While business writers like to shout to their readers that IT is a disruptive technology requiring them to revamp massively their business models, so too our history of IT's diffusion requires one to possibly recalibrate to some extent their conceptions of wave theories in economics and in economic history. When examined in detail every technology seems unique, and the exception to some generalization. IT fits this pattern, but most, if not all technologies, also share common attributes, which means that dignifying IT with a mantel of exceptionalism, unreasonably challenges the findings of historians of human tools and toys.

IMPLICATIONS FOR BUSINESS MANAGERS

Large corporations are overwhelmingly participating in the global economy because of their ability to use ICT. That is no longer a novel observation, merely a statement of fact. The global supply chains are populated by mid-sized firms all over the world that provide specialized goods and services, and they too are able to do this because of ICTs. That observation is increasingly becoming the case. But often it is the largest members of a supply chain that dictate who participates the most and how, despite various business school case studies of small start up firms making their fortunes by selling over the Internet. But, a question for managers: Will their supply chains become so ubiquitous that no single company will dominate it, as usually occurred in the past when, for example, General Motors could pick its suppliers and countries to draw them from, but decreasingly so as new providers of automobiles came online, tapping into the same automotive supply chains, as is occurring with Asian automobile makers, for example? These supply chains transcend borders but are also affected by local conditions, such as war, poverty, European Union regulations, North American tax systems, labor strikes and supplies of workers (and their costs) in Asia, among others.

IT's diffusion around the world forced participants in the globalizing economy to take into account local cultures and conditions on how they accepted and used the technology. American companies simply could not force everyone to learn English or to adopt North American economic and political values. Nor was everyone interested in adopting Chinese or Indian ways. In fact, there are risks in doing that. China may not develop a middle class consumer society fast enough to absorb its manufacturing output, leading to potential economic problems in the years to come, while India does have serious infrastructure problems that it has not been able to resolve, along with long standing problems with Pakistan which could result in more wars. Brazil needs to do more to educate its citizens, while Sub-Saharan Africa is coming into the global economy slowly with massive social and security problems of its own. The world is flat and global yet local conditions still count.[35] One of the lessons for management is that IT diffuses at differing rates in different ways in most countries, even though the reasons why individual organizations, firms, and people might acquire a particular form of computing is more similar than not around the world. As time passes, both institutions and individuals are becoming increasingly savvy about how to embrace new forms of IT and their uses, than in prior times.

That is an important insight for managers because they will need to continue tailoring their offerings and how they manage their supply chains to suppliers and customers as they step through one country to another. IT's history demonstrates that variety is possible, indeed even more so than ever, particularly with services that are information-laden, and that citizens will want things their way. Africans have innovated in their development of terms and conditions for the use of mobile phones, for example, to make them more affordable and available to ever-larger groups of users.[36] Americans have found more ways to use the Internet for business transactions than Europeans or Asians; while Asians have pushed further the boundaries of social media and broadband than any nation in North America or on the European continent. Increasingly too, everyone knows what others are doing, as you are by reading this book, for instance.

This discussion leads to another implication for management. It used to be that innovations were largely a dominant activity of the pan-Atlantic world, as measured by the number of patents issued to Western firms. That is rapidly becoming a fiction as corporations disperse their R&D and product innovation activities to centers all over the world. This is occurring now with centers in India set up by IBM, Microsoft, Google, to cite a few obvious examples. The patents resulting from these efforts are usually held by American firms and thus are counted as U.S. achievements, but are they if they were developed in Bangalore, Zurich, or Sao Palo by local employers or by teams of workers working in different countries, as happens frequently at IBM?

Additionally, there is the issue of the use of innovations. Economists and many technology executives learned a long time ago that inventing something new is not always as important as adopting an innovation someone else has created. That is why, as this book was being written early in the second decade of the twenty-first century, there was a global footrace underway among ICT firms trying to buy and control patents; Google in telecommunications and IBM in all manner of software dealing with analytics, security, and the movement of digitized data, for example. The footrace started several decades ago, largely within the pharmaceutical industry among its largest firms, and spilled over into other industries. IT continues to facilitate the footrace, indeed to characterize much of it. Nations are more concerned about innovations today because officials know that it is a way of nurturing local high-tech, high-paying jobs, protecting their economies from declining, and from being unable to afford providing its citizens the services residents demand of them. So, while the pressure is on governments to localize activities, participants in global supply chains roam around the world, much like a global R&D and innovation bazaar, squeezing the fruit and picking up the best deals along the way. As the world becomes more networked, better educated, and increasingly endowed with the facilities and other assets necessary to produce new products and services, this globalized hunt and diffusion of innovations in most industries should speed up and, at a minimum, become more diverse. The implication for management becomes quite obvious: they need to deal with that rapidly emerging reality. In turn, as they respond, their actions will thrust IT into new nations that initially did not have as extensive prior commitments to it as did East Asia and the pan-Atlantic community and now by Malaysia, Philippines, Rumania, Turkey, Central Asia, most of Latin America, and parts of Africa.

IMPLICATIONS FOR PUBLIC OFFICIALS

The banking crisis that began in 2007 made it painfully obvious that regulators and political leaders had to collaborate together more fully on economic matters than they had in recent decades. Platitudes simply were not going to be enough; they had to start a difficult process of reducing sovereign debts in synchronized ways and, to use a widely deployed way of describing similar regulatory regimes, to "harmonize" international rules governing economic, banking, and business practices. This all had to occur while some of their citizens rioted over the lack of jobs, increased taxes, inadequate social benefits, and perceived unfair economic circumstances across Europe, the United States, and sporadically in Asia. It was an ugly public display of the heightened mediating role officials had to play in stabilizing the world's rapidly integrating, yet seriously troubled, economy. It was made more visible and dangerous by the use of ICT to display events for all to see, through the use of social media IT to organize flash mobs, protests, and even the revolutions of the Arab Spring (2011).[37] Officials had always wanted to control their national economies by setting rules for firms and to promote their own domestic welfare even at the expense of others. During the Cold War Russia and the United States did an excellent job in essentially forcing large portions of the world to align in one camp or another, with the "unaligned" nations dubbed the "Third World." Often these unaligned countries were the poorest and, therefore, of lesser interest to the two super powers unless geographically located closer to more strategic locations, like Cuba for the Soviets and Vietnam for the Americans, or if they had natural resources, such as oil and minerals. Asian nations entered the second half of the twentieth century willing to take a more overt command-and-control approach to their economies than the pan-Atlantic community as the latter struggled with the Cold War on the one hand but on other with how much governmental involvement there should be in a market-driven economy. It is a relevant issue for historians and economists to continue debating.

For officials, the outcome was clear; they had become increasingly involved in more issues related to the diffusion and use of ICTs, and in the workings of their economies than they had with earlier technologies prior to World War II. Citizens frequently will not reelect a president or support a prime minister if unemployment is deemed too high; financial crises destroy the careers of senior Japanese officials, while economic failures cause the Chinese Communist Party to push public officials into retirement, or worse. In short, by the early 2000s, public officials were playing a major—if not dominate—role in the evolution of IT and its diffusion, and in the operations of their economies. To be sure, it was an uncomfortable role for many officials, but the implication was also obvious for them: their central participation would continue for the foreseeable future.

Declarations that international corporations would dominate the world's economy have been made since at least the 1950s; yet, this has not happened, despite efforts by some firms to dominate their markets. IBM's dominance of the pan-Atlantic markets for mainframes in the 1950s–1980s, for instance, was not fully duplicated across Asia, and by the early 1990s, new technologies and entrants into an even larger IT market than existed earlier meant that IBM could only serve extensively one portion of that global market, even though it remained one of the largest enterprises in the world. Google has often been cited as the next giant to make the attempt.[38] But none of these could for two fundamental reasons: they were not big enough, and they did not control all of the infrastructures discussed in this book that proved so essential to the diffusion and use of IT. Governments are bigger and control large geographic footprints with laws, traditions, police, armies, cash, ownership of local businesses, and public support.

All of these circumstances draw us to the conclusion that public officials will need to become better students of how complex general-purpose technologies emerge, are used, and spread through their nations, spilling out into others as well. This is no longer an economic development conversation; it involves national security, the physical, economic, and spiritual welfare of citizens; and it concerns the environmental health of the planet as a whole. This is not about who is going to build the next atomic bomb in a politically unstable part of the world, but rather how will public administration function and collaborate in an ICT-laden world in the decades to come. These circumstances are already forcing officials to do what seems strange to many: encouraging market-driven economic behavior in Communist China; implementing "open government" practices in the West, even in Central Europe and in Africa; tolerating freer movement of information about all manner of topics in authoritarian regimes; using PCs and the American dollar in Communist Cuba; deploying online voting in Brazil; dealing with worker migrations into the United States and across all of Europe with or without the consent of local officials; and monitoring the movement of funds across borders with inadequate abilities of officials to track, let alone, block undesirable transactions. The list is rapidly becoming quite extensive.

All of this draws us to the conclusion that national governments have played a greater role in industrial policy and economic development in the second half of the twentieth century than in the prior half century, resulting in a large body of knowledge about how this is done. Our book has documented many examples of this kind of work done by officials promoting local supply and demand IT capabilities. Economists and political figures have learned that complex technologies require mastering a variety of capabilities, not just IT to make a go of it. Initially, investments in new technologies and new economic sectors do not have rigorously defined returns on investments and, thus, must be done as a strategic initiative on the part of national and state governments. Asian governments also demonstrated another necessary practice: a consistent, intense and coordinated commitment to the complex diffusion of IT to an extent not evident in other parts of the world. In short, it is complex work promoting development of IT in one's country. That was not so well understood until at least the 1980s. Dani Rodrik has studied extensively economic growth and industrial policies and has drawn several conclusions relevant to the challenges faced by public officials dealing with the kinds of economic activities discussed in this book. One of these is that picking winners is hard work but not as crucial to get right than the ability to cut one's losses when a bad choice has been made. The Europeans had difficulty mastering both skills, the Asians did a better job (except India with its heavy industry choices), while the Americans avoided overtly picking winners and losers. Rodrick pointed out the delicate problem many developing economies face, namely, that in those economies no matter what officials do, "entrepreneurship in new activities has high social returns but low private returns."[39]

IBM's conversations with several thousand public officials between 2005 and 2012 flushed out several related issues. First, senior officials recognized increasingly over time that even the fundamental ways of public administration were being challenged and would have to evolve, often into forms they were not quite sure they understood. Second, they reported being inadequately prepared to face these changes, let alone to deal with them. Third, their citizens were increasingly becoming impatient in demanding services that paralleled in convenience many that IT had already made available to them in the private sector, such as 24 hour-a-day access to services, opportunities for education, access to health, and jobs.[40] Officials overwhelmingly recognized that their responses would involve IT, economic development, greater transparency in their work, and in information they shared with citizens.

Failure to respond was not an option, as illustrated by events in Egypt, Libya, and Tunisia. They also learned that there was an upside to the diffusion of IT as well. Just as the technology drove down the cost of doing work in businesses, so too, IT could do the same for governments. The technology allowed services and information to flow to citizens less expensively when individuals wanted these, improving everyone's productivity and levels of satisfaction with results. The same tools used by businesses to understand what their customers wanted and liked were the same ones being tentatively used by officials to appreciate what their citizens desired and thought of their performance. Social media tools could spark a revolution in the Middle East or rally support for the election of an official as occurred in the successful effort by Barack Obama to become president in 2008. In short, officials were becoming familiar with the nuances of IT diffusion and use as had business executives at almost the same time. Officials long understood the power of IT and developed deep pockets of knowledge, whether at MITI in Japan in the 1970s and 1980s, at the cabinet level in South Korea in the 1990s, or in the military forces of the pan-Atlantic and East European countries. All data intensive departments and agencies long had such insights, such as a census bureau or a labor tracking agency going back to before World War II. This kind of insight was now spreading to departments and agencies that traditionally had been minor users of IT when compared to other organizations. Discussions about IT as national issues rose to cabinet levels as a routine topic of conversation in the countries explored in this book. These discussions went from being exceptions, as occurred in Japan and France in the 1960s and in China in the late 1970s–early 1980s, to permeating many conversations about all manner of public issues. Gone were the days when tiny groups of engineers, professors, or data processing professionals buried in the bowels of an agency advocated using IT, as occurred in each of the regions in the late 1940s–early 1950s.

Finally, we might ask what kinds of governments facilitate or constrain the diffusion of complex technologies like ICTs? From democracies to authoritarian regimes, governments have demonstrated the ability to promote effectively the development of supply and demand for ICTs, and in particular computing; we also have examples of failures by either type. Authoritarian governments can cause greater marshalling of resources faster by edict than can democracies, which must persuade various constituencies to collaborate. However, the experiences described in this study suggest strongly that while officials played an enormous role all over the world, and in some societies more than in others, the more critical success factor was not the form of government but the form of economy. Market-driven economies always outperformed centrally-controlled economies in diffusing IT. That was so much the case that numerous authoritarian regimes converted to market-driven practices not only to facilitate use of IT but for the overall improvement of their economic performance, such as all of Eastern Europe and India, beginning in the 1990s, and spectacularly in recent years in China. That conversion required the state to be supportive of the private sector, creating the necessary legal, technological, economic, financial, and social environments in which IT could thrive either as a source of exports, jobs, and national prosperity or as consumers of these technologies. Democracies often have expensive lumbering states saying grace over economic development, but so too do authoritarian regimes and those governments that are partially one or the other type. Knowledge of sound industrial policies can exist in any form of government, and in our detailed case studies, there existed such pockets of insight and willingness to promote the use of IT. In every case there were clusters of officials who were willing to promote innovations in various forms, although ideological perspectives affected fundamentally what actions they were prepared to take. Yet for all, promoting the diffusion of a complex and expensive technology deemed essential to the welfare of a nation proved difficult to do, was expensive, and took a long time to accomplish. These circumstances still exist today.

IMPLICATIONS FOR CONSUMERS AND THE HUMAN CONDITION

In addition to IT vendors, commercial users, and governments playing pivotal roles in the future of IT's diffusion, there is that of citizens, those individuals who have led us most to debate when Wave One diffusion ended and Wave Two began. Suggested across many countries, once individuals started to acquire and use computers for their personal uses, mainly PCs, and then other digital consumer electronics, diffusion of IT moved forward and deeper into society than before. When many of these devices used by individuals were interconnected through communications, mainly the Internet, what they were used for and how frequently also took a significant leap forward in the amount of IT used. When the computing and communicating done by individuals were integrated into uses of computing by businesses, government agencies, and other institutions in society, yet further intensification of use occurred. Human users shaped the next wave of diffusion, suggesting that with more historical perspective, their extensive movement into IT diffusion may define the boundaries of our two waves. People will have to vie for attention of future historians when it becomes more obvious how machines and other non-human devices and entities also will have used computers, suggesting new applications, more uses, and possibly different consequences. But, that is all in the future, even if it is beginning to arrive today. More central to the discussion of Wave One and Wave Two is what people are doing today.

There are three activities underway that have profound implications for people. First, how they are acquiring digital technologies is changing. Second, there is the matter of how IT is being used to teach and train children and adults. Third, and now increasingly being taken seriously in scientific circles, are possible effects of IT on how the brain works and human bodies evolve. We have seen evidence of the first two trends at work throughout this book, while the third one reaches beyond the normal purview of the historian and so will only be touched upon to remind readers to pay attention to the issue in the years to come.

As the personal GDP of individuals, and by default, that of their local economies rose during the second half of the twentieth century people could afford to buy more expensive, indeed more complex products. As the world became wealthier after World War II, literacy and educational levels expanded too, making it possible for people to use computers and other digitally-based goods. The list is long and obvious, and includes such iconic devices as microwave ovens, digital watches, hand calculators, personal computers, digital cameras, laptops, mobile phones of all generations, portable music equipment, and iPads. Applications began as standalone, meaning without access to networks, but by the end of the 1980s, cell phones and PCs were linking massively to the globally improved telecommunications infrastructure implemented in so many countries, followed by the fundamentally important Internet. With that last technological innovation people connected all manner of devices to do e-mail, play games, gather information, publish, share photographs and music, view films and video, and to create and profit from businesses *sans* "bricks and mortar." Broadband hurried up these processes and even made it possible to have digital avatars, digital personas on "the grid," as occurred so extensively in South Korea. But earlier they had learned to heat up their food quicker and to let a calculator do much of their math, and later PCs to facilitate much of their writing, for instance by correcting spelling and grammatical errors.

With individuals now having acquired various digital devices since the early 1970s one could see a pseudo-Moore's Law form of behavior in which individuals sped up their acceptance of every new technology faster than previous ones across all nations and continents, an idea suggested in chapter one. As individuals became familiar with digital technologies, and had relatively positive experiences with earlier ones, they learned how to buy better the next digital product (or upgrade) that came along and they more willingly accepted the idea that digital devices deserved their attention. If they followed such an observable pattern, then we would expect them to conclude that in the first year a new digitally-based product appeared it would be more expensive than if they waited a year or two to acquire it, understanding (or sensing) that its price would drop and that the performance or volume of activity it could assume would increase, such as greater clarity in a digital camera or a musical device that could hold more music and video for the same purchase price as an older product. Very few consumers or other users of digital products know about Moore's Law, so it must be inferred that they acquired this behavior through experience. The first digital watch was less expensive than a Swiss mechanical one; little hand held TI calculators dropped in price rapidly in the 1970s too; PCs went from $5,000 to under $500 in a decade; digital cameras doubled their pixel capacity and storage in less than five years, while nearly halving their cost. Flat screens in the late 1990s went from $5,000 for the largest consumer products to less than $600 for the same size in roughly five years. Cell phones are frequently swapped out for newer models and styles in some countries almost every six months, especially along the Pacific Rim.

The implication of this behavior is that people are increasingly becoming comfortable acquiring and using various forms of digital technologies. On the basis of that consumer behavior one can assume they will embrace others for use both in their private and professional lives. Indeed, even differentiating private from public lives is already a staid Wave One paradigmatic way of looking at utilization as the two blur together with people using the same devices in both. Also, as they have with other information technologies, people will continue to add and use newer digitally-based ones alongside earlier ones already in service. Books and laptops, snail-mail and e-mail, and bricks-and-mortar shopping alongside e-commerce are examples of uses of IT which continue to spread in societies that are more recent adopters of the technology, while the most experienced ones deepen their reliance on the news but also uses the older ways. Convergence continues to be an increasingly pervasive feature of how individuals are using the technology too. For example, using a digital camera to take a photograph, which is telephoned to a laptop, which is connected to a flat screen used for television in the home where it can be displayed, and later embedded into a document is already a common practice. That concept of further integration suggests new activities dependent on the existence of IT already in a person's possession and that can be upgraded, replaced with newer forms, and be supplemented with different devices. The introduction, or replacement, of various Apple products over the past 15 years, and that included replacing older devices (such as personal computers), adding new ones (iPods, iPads), novel applications (iTunes) and software to interconnect all of these devices together is symptomatic of the observable emerging behavior.

A second issue to face is the effects ICTs are having on education and training. It is an important topic for several reasons. Literacy, education, and relevant skills are essential preconditions for diffusing IT in any society, so far. There is a direct correlation between the existence of these social attributes across large swaths of the population and extensive use of computing. Furthermore, while training methods for adults have relied extensively on IT since the 1980s, and are now widely used, this is clearly not the case with formal education of children and, to a lesser degree, of university students. Yet children in many countries routinely use computers to play games, for other uses, and to interact with each other, just less so in the classroom.[41] In countries that are already mature Wave One adopters, educators have done far less with computing than other industries. But pedagogy is finally beginning to feel the effects of IT, as tech-savvy educators and experts on how children learn begin to develop tools and techniques that depend on the use of IT.[42]

One can expect two trends to intensify that are already in progress. The use of digital games by children is now so widespread in those nations where personal GDP levels are high enough to make these affordable to families that researchers examining the process by which children play and learn are turning to this technology for new ways to teach.[43] This has already been a path taken by adult trainers, most notably in the U.S. Department of Defense to train several million young soldiers over the past two decades, adapting training methods to individuals who grew up using gaming software. One can reasonably expect this use of IT to diffuse more extensively to students around the world. There is, however, a growing debate about how the technology is affecting the way young people think, learn, and most interesting, use their brains. The latter point, a subject of controversy (even concern) is only now becoming evident.[44] Although, since the heart of current uses of IT in education is to teach students by having them do things as the way to learn, this mode of using computing reinforces a way of teaching that had long been in use before the invention of schools and modern pedagogy.

The last implication for individuals concerning IT, for both children and adults, involves the effects of this technology on the human race itself. It appears that IT may be affecting the species itself, although IT has been around for such a short period of time that such a statement seems more hypothetical, but bears watching over time. Ian Morris, an archeologist, in his history of humankind covering hundreds of thousands of years, and the diffusion of many technologies from stone implements to computers, recently pointed out how quickly humans evolved physically. Improved diets and medical treatments in the twentieth century made it possible for people to double their life spans over those of their ancestors of the eighteenth century, and in one century to increase their height by an average of six inches.[45] Now we are on the verge of genetic reengineering, holding out the possibility of correcting problems or changing the nature of a person's bodily functions, replacing organs, and altering their appearances, leading to "partial immortalization."[46] The brain is not immune from such a transformation. I have already mentioned the notion of singularity in which brains and computers either become as smart as each other, or converge (carbon–silicon), the event predicted by Ray Kurzwell, even making it possible for technology to evolve faster and smarter than people.[47] Whether he is correct or not will not be resolved for some time, but people have long augmented their memory and thinking with earlier versions of IT, such as writing and paper so one did not have to memorize and remember everything, later used digital calculators to do simple mathematical calculations, and now computers to make decisions about myriad issues. Today experimental molecular computing exists, constructed out of enzymes and DNA.[48] This is all heady stuff.

Will people improve their response times and ability to sift through large bodies of data quickly as they need to in order to play computer games? Will they diminish their abilities to focus on one issue for an extended period of time, as some argue is already occuring?[49] Some merger of man and machine would represent the ultimate diffusion of IT, from laboratory to market, across societies, and into every individual. That is the ultimate implication and potential of IT diffusion. There is now enough knowledge about computing and installed ICTs to consider that a real possibility in the future. Perhaps that is Wave Three, or an advanced stage of Wave Two.

WHY DID INFORMATION TECHNOLOGIES SPREAD
SO MUCH SO FAST?

This is the central question posed by this book. Information technologies spread around the world because the technology was developed, societies could afford it, and it proved profoundly versatile in facilitating the work and play of people. The world had accumulated enough scientific and technological know-how to make computing possible by the 1940s and it experienced such a growth in wealth that it could fund the development and diffusion of this general purpose technology at the same time as it created massive quantities and varieties of new products and services that lent themselves to the adoption of computers. The fundamental supporting causes for the technology spreading so fast can be found in its rapidly declining costs, variety of possible applications, and to its improved ease of use. ITs came in ever more practical sized packages and dropped in cost for the function received faster than any other modern invention. Even the decline in the cost of communications and transportation can importantly be attributed to developments in the field of information technologies, especially in the case of telecommunications.

But, we must keep in mind that no single answer exists to our question. A combination of factors facilitated and constrained the extent and speed of diffusion. The most critical ones in both the pan-Atlantic world and in Asia included affordability, proactive government support and promotion, the right kinds of education, types of industries and companies functioning in an economy, knowledge transfer, and necessary legal environments. Diffusion was never just about economics, public policy, business practices, scientific path dependencies, or culture. It was a function of many features of modern life, and that is why information technologies have become so central to any appreciation of what humans were doing in modern times. But the result is quite clear: the world experienced its first digital flood.

Preparing a Global Diffusion History: Lessons Learned, Paths Not Taken

Increasingly, historians are conducting research on world-wide issues, often within the context of "world history," but most frequently about more singular topics as played out on a global stage. They are facing a number of challenges, all of which were evident in the current examination of the diffusion of digital technologies around the world. So, formal, effective approaches for conducting such historical investigations remain in flux, yet evolving. The challenges they face include:

- Quantity, quality, and unevenness of information they would want to examine
- Understanding and explaining accurately the various contexts in which events occurred in different parts of the world
- Working with a few (or even) dozens of languages
- Overcoming the limits of what one or two historians can do within a reasonable period of time
- Reducing their findings to one (or few) volumes without giving up the rich narratives we expect historians to provide.

These are not trivial issues, and others could be added, such as the exponential increases in time and budget required to travel around the world to various libraries, archives, and other institutions, ability to present findings in different countries, often in various languages, and the lack of the right kinds of materials in one's own country to examine. Understanding each of these challenges helps to explain how this study on the diffusion of information technologies was conducted, suggesting as well that the approach proved serviceable.

QUANTITY, QUALITY, AND UNEVENNESS
OF INFORMATION

In the case of this book, which required examining information covering events and trends from the 1930s to approximately 2012, several problems presented themselves early on. First, the amount of data available of a statistical (numerical) nature that helped to provide answers to such questions as "How many by when?" proved uneven, sometimes incomplete, inaccurate, or simply unavailable. The best data begin with the late 1940s and become richer in quantity as the decades passed and as more government agencies, vendors, and observers collected information on either the diffusion of this technology or on its economic impact. Governments in the most economically advanced countries collected more and better statistical data earlier than others; thus, it is possible to understand in considerable detail how many computers were installed in the United States in the 1940s–1960s, but we must estimate for almost all of Latin America and China what happened even as late as the 1980s. In our particular case existing evidence could be found in several places.

For each nation—and that is the unit of scope chosen for looking at global diffusion—one had to hunt around to see what government agencies collected data on computing or office appliances. It varied from one country to another. With the data of the 1980s and beyond, however, one could generalize and say that the United Nations to some extent, the World Bank to a greater extent for poor countries, and the OECD were routinely gathering data and presenting it in comparative tables and reports. European communist countries represented one collective exception, particularly for the years of the 1960s through the 1980s, for which most international agencies could not collect information. Prior to the 1980s for most countries in the OECD, comparative data proved hard to come by, although American government agencies and a few writers attempted to collect such information. In our case a wonderful source were IBM's internal marketing reports for the period 1950s through the 1980s, which for many countries described the number of machines installed, who where its competitors, and the economic and business environment affecting the rate of adoption of this technology. The data are made even more useful by two conditions: they were meticulously gathered on the ground by IBM sales and marketing employees intimately familiar with the individual adoptions by enterprises and governments, and second, by the fact that in many countries IBM enjoyed market shares of between 50 and 80 percent, which meant its records accounted for 50 to 80 percent of the market, and usually for the entire market, since the firm meticulously tracked activities of its competitors. Archives of other vendors had similar information, for example, Sperry Univac in the United States for the 1950s and 1960s, and in Europe, Machines Bull for even later years. For some countries, these were the only accessible sources of information on the number of installations.

A second strategy for obtaining information on diffusion trends and rates was to assume that some government agency collected such data and, therefore, to query local historians and librarians about these. For example, the Swedish government published annual detailed reports on how many computers it had installed and their costs; but these reports are nearly impossible to find outside of Sweden and thus arrangements had to be made to obtain copies of these without which it would have been impossible to write about Sweden's use of computers in government in any confident way. The same situation applied to Singapore for early years, while for recent ones such information was available on the national government's website. One quickly learns that there are sources one should expect to tap into about the topic at hand. In our case national banks, staffed with statisticians and economists, began routinely to collect, analyze, and publish reports on information technology in the 1960s. By the 1990s this source resulted in the publication of several thousand white papers and articles, often with trend analysis of local and multinational comparisons covering the period from 1980 to the end of the century. For every country studied, and for at least four dozen others not reported on in this book, such materials existed, proved reasonably accurate, and useful. This was particularly the situation for the years after 1980, when more countries than before were installing computers, and studied by more economists on the ground, who collected data and wrote about various issues associated with information technology, and not simply its adoption. But clearly there were two lessons:

- A great deal of data is organized by country, less by region or globally, particularly for the earlier decades of the 1940s through the 1970s/early 1980s (pre-1985). That suggests global studies of many types should probably begin by examining portions of a history in national units, country-by-country, then aggregate findings to regions then globally, but always starting at the national level because of how initial sources of information are collected, organized, and made accessible to scholars.
- Information of a contextual nature, such as political and economic history, so essential for setting our story into its proper place was also largely national in scope, but not always. Regional economic histories abound, but may not be rich enough in detail with which to understand events in a country. Thus, a general economic history of Europe, while essential to our research, did not eliminate the requirement to read more detailed economic histories of each country examined. Those local studies are not always available or as current as one would like in English, let alone in another language.

Related to the issue of primary source material—in our case more in the form of numerical data than classic archival ephemera—are publications on a particular region and time. While some obvious sources that cover many countries and periods are available, in our case such journals as the *IEEE Annals of the*

History of Computing and *Culture and Technology*, others were not and they did not necessarily surface when using such bibliographic tools as Google Scholar or more traditional published bibliographies. Yet, from a researcher's perspective, one has to assume that something was published and that they simply do not know about it yet. For example, not all books on a topic were published with ISBN numbers and even when some were, they often proved nearly impossible to obtain, even in their country of origin, because of either initial small print runs, which made them scarce, or because they were not available for sale or wide distribution in the first place. I found, for example, over a dozen country histories of IBM, none of which were for sale, yet they contained valuable information; universities, other firms, and government agencies celebrating anniversaries also published such material. That occurred in scores of countries in various languages, most of which could only be learned about if cited in an obscure endnote, or through word of mouth. This pattern applied even in the countries most known for their use of computers, such as the United States, Great Britain, Germany, France, all of the Nordics, and Japan.

Just because one does not have a great deal of information about events in one country does not necessarily suggest little occurred in these countries. In the case of all of Eastern Europe—that is to say, those European nations that had been Communist during the Cold War—the amount of information available in the West and additionally in English or German proved frustratingly too minimal, yet upon close examination of extant regional publications and conversations with local experts, it was amazing how much had been going on behind the Iron Curtain in such places as Bulgaria, Romania, and the more obvious locales to explore, such as East Germany, Poland, and Russia. The same applies to assumptions about the existence of, or access to, archival materials. That someone was saving something all the time in every country proved to be a good operating assumption in this project.

UNDERSTANDING AND EXPLAINING VARIOUS NATIONAL CONTEXTS

Perhaps the greatest contribution historians make before all other disciplines in the humanities is the attention they pay to explaining the context of events and through their ability to link these with the narratives they are writing. I could not simply state how many computers went into West and East Germany in the 1960s or 1970s, for instance, without also describing what was happening with the local economy, national politics, relations with other European nations, reflecting the influence of the Cold War, the specific local role of IBM or Siemens (often Europe's largest computer vendor in this period), activities in the universities, and so forth. Yet how is one to know the national histories of scores of nations to write a global history? Remember, historians often have to study events in neighboring countries not chosen to be included in a book so as to understand why a country should be left out of the narrative, as I did for such countries as Canada, Vietnam, Norway, Malaysia, and Spain, for instance. Very early on in such a project, one has to develop a method for understanding local context at a level of detail sufficient enough to explain what was happening with respect to their individual subject—in this case information technologies—without spending years mired in one or two countries, when there are potentially 200 to understand, each at a comparable level of detail, with emphasis on the notion of striving for comparable striving for understanding of events and circumstances.

To accomplish this, one has largely to rely on synthetic histories to provide the context. In our case this meant reading political, economic and business histories of a country and region, discussing the historic period of the project (1930s–1990s), and a bit more on before and after these years in order to sustain the thesis that the epoch studied was in fact a logical era, bounded at either chronological end by relevant macro-events. That is how one can arrive at such contrasts as Wave One and Wave Two, for instance. In our case it was usually World War II at the beginning and the combination of the rapid diffusion of the Internet, beginning largely for most in the 1990s, both coupled to a broad-based round of economic crises or transformative political events in the 1990s. Often one has to rely on just one, two or three economic histories of a country, or two or three business histories. Usually there are "standard" works in the field one can consult; however, since many of these were written by scholars who were not native to the country about which they wrote, it behooves one to also read one or more books and many articles as possible written by natives of the country in question as they usually have different interpretations of causality and even unique data sets. Thus, while contemporary local literature and primary materials are essential to examine for a specific topic, for contextual understanding, secondary materials often proved most adequate. They also took almost as much time to study (not just read) as primary or more specific monographic materials. But, it proved to be an essential exercise if one was to write global history at a general level yet with enough specificity to set a topic in a credible context.[1]

WORKING WITH A FEW OR EVEN DOZENS OF LANGUAGES

There are several strategies that can be employed to address this very serious problem. First, the use of multiple historians on a project, which combined have a working knowledge of many languages, is an obvious solution and already deployed in Europe on several projects. Historians will increasingly have to rely on this strategy and it may have to become a more common practice with respect to global history. A second approach, where detailed local research has to take place, one used frequently in this book, is to locate bilingual individuals in each country who could describe, translate, and otherwise facilitate overcoming a language barrier. By the same token, one should encourage such individuals to write their articles and books in English, or to have them published in translation, to reach larger audiences. This is especially necessary for all the Nordics, which in the case of computer history, are grossly underrepresented in the global story yet once coxed out of people who can speak English or German, turn out to be extraordinarily important narratives, hitherto not understood well outside the region. The problem is multiplied in Eastern Europe, China, Japan, and the most recent Asian nations to appreciate IT where even working knowledge of English or other West European languages is limited. Chinese academic institutions and publishers have an appetite to fund translations of Western publications into Chinese, but not theirs into Western languages; the same applies to Japan. Help can come on occasion, however, from interested user groups or historical associates willing to assist scholars in other countries identify materials and even to translate small portions of these. Choosing wisely what to translate is essential since so little can be done.

The way many historians overcome the problem of languages is to rely on whatever has been written in their language, which often means journalistic content weak on assessments, insufficient content and evidence, and too contemporaneous to events for a scholar's taste. The bigger problem that emerges from that kind of dependence is, therefore, that the level of reliable detail and analysis is sacrificed with which to write with confidence the more summarized accounts required in a pan-world narrative. The other way to get around the problem, and to a rare extent reluctantly used here, is to select country case studies for which accessible materials exist. Hopefully, poor choices were avoided. In our case we had very little on Croatia, Vietnam, and about all of the Muslim republics that comprised part of the old Soviet Union. On the other hand, surprises abounded. Lithuanians and Estonians speak more English than one might otherwise have thought and, therefore, served as conduits for information and reviews of written narratives about Eastern Europe. The farther one went back in time, the less material could be found in a familiar language. Therefore, an historian must come to the global project able to work in at least 4 to 6 languages, even if only in some faltering manner.

OVERCOMING THE LIMITS OF A SINGLE HISTORIAN'S CAPACITY TO WORK ON A PROJECT

Big books take a great deal of time to research and write, most so to conduct the quantity of research required to support the broader "tip of the iceberg" narrative that can fit into a single volume and that relies on the use of many languages. To write enough detail that is convincing and authoritative requires examining events in sufficient quantity to allow for the generalized account. To do that requires collecting far more information than can be presented in one book. Economic historians and political scientists rely on extensive use of numerical tables and statistics, but most historians still have to write a narrative, to tell a story, to argue a point of view, and to assess consequences. To do that with confidence requires looking at a vast array of data in normal circumstances. But, what about a global history involving dozens of countries, many decades, and very different national cultures and practices, what is to be done? It is understandable why so many historians of subjects, such as of information technology, prefer to focus on far narrower research projects. But the latter also need broad synthetic works based on empirical research in order to set their narrower projects into the larger context of their subfield. With interest continuously growing in writing world histories on various topics, the historical community faces some severe methodological problems not yet resolved adequately by their profession.

Can one individual do all the necessary research for a credible global history within a reasonable period of time? There is traditionally one way to do this in order to arrive at a positive answer and that is to take a very narrow sliver of an issue and trace it through various nations largely through published materials, or selectively in a few sets of archival materials so that the volume of research approaches the 2,000–5,000 hours more commonly invested in a monograph. The second way is to borrow from the physical scientists their technique of organizing a virtual team of historians and divide the work up; in other words, the Hollywood Model as movies are made. The only problem with this approach is that most historians are evaluated for career advancement and tenure based only on the work they do as solo performers. The scientific community fixed that problem, but so far not the historians. So, we are left with the heroic warrior model where one individual must conquer the world.

However, there is a third path, the one chosen for this project honed in an earlier one, the three-volume *The Digital Hand* (New York: Oxford University Press, 2004–2008). That third way borrowed successfully a page from large business and engineering project management. In this way, the historian acts as much as project manager as researcher/scholar/writer, directing, in effect, much like an orchestra leader. Once the historian has identified the topic to study, its scope, issues to explore, and so forth, the novel next step is to map out who to involve and how the work is to be done. The bigger the scope of a project, the more people who will have to be persuaded to assist in conducting bits of research, supplying ephemera (documents, books, and so forth), then insights drawn from their tacit knowledge, all as if cameo appearances in a big movie. Today, historians routinely have colleagues critique chapters and whole manuscripts but largely do all the research themselves, or with the assistance

of a few graduate students. I am suggesting that an ecosystem be created around a large project with individuals coming in and out, helping in small but vital ways, some paid, and others largely through *pro bono* activities. In the case of a global study rooted in multiple country-level studies, a micro-circle of assistance is required in each nation. For this information technology study, typical participants included the following for most countries to one degree or another:
Experts on the topic who could:

Advise on what archives and records to examine
Advise on the existing local bibliography to consult
Provide access to archivists, librarians and so forth through personal connections and with contact information
Provide copies of key documents in their personal files
Critique manuscript texts
Advise on best bookstores for acquiring materials, old and new

Archivists and librarians willing to work electronically or via mail in:

Identifying key materials to consult
Willing and able to make copies to send to the historians

Public officials and participants in the story:

Willing to be interviewed by the historian
Willing to share copies of documents and photographs
Able to connect the historian to other participants
Able to contribute published materials to the historian

Book dealers:

Able to identify and acquire on one's behalf older publications, not just new ones
Able to identify and sell quickly in-print publications, especially those produced by government agencies and privately published (e.g., by institutes).

Each has to be nurtured and communicated with on a regular basis and be given updates on the progress of the research initiative so that their enthusiasm and support for it remains fervent for the life of the project, which can last many years. Newsletters do not seem to work as well as personal e-mails and telephone conversations. If one is able to visit a country being studied, then a carefully structured agenda involving face-to-face meetings with as many members of the local eco-system is essential and would need to be planned a couple of months in advance.

All of these activities seem natural and obvious, but are not. For example in our case, while working on the chapters dealing with Western Europe it was tempting to ignore Eastern Europe and Asia. Nonetheless, while working on Western Europe (involving the early chapters in the book) I had to start building a support group across Eastern Europe, and begin the process of acquiring materials on Asia. While working on Western Europe, Eastern Europe was a year away and Asia 18 months away. Experience demonstrated that it was not too early to start constructing the eco-system two years in advance because finding people, then engaging them in dialogue, and acquiring materials electronically or by mail simply took a long time.

A practice taken from *The Digital Hand* that proved quite useful was to seize the moment when it presented itself to acquire materials and to write down flashes of insight and information that would not be needed until much later for future chapters. For example, when examining a source, say, a specific journal, review prior issues of that journal at that one time so that whatever it has published over the years on any country of interest to your entire project you acquire copies at that moment and file these away in relevant country folders so that they are already available to you when the time comes to study that nation's experience. This step saves an enormous amount of time because it avoids retracing steps literally through a library, going back to a website, or returning to an archive. As many follow in business, one's operating mode should be "touch something only once," in our case a complete run of a journal or international multi-national dataset. Besides improving one's efficiency, a crucial byproduct is that along the way a historian learns about other areas and topics while focusing on earlier ones, acquiring comparative perspectives and an emerging world view of one's global project.

What were the costs of project management for the lead historian? In the case of that particular book, about 20 percent of the time invested in the project went toward project management and interacting with the network. All other tasks were the traditional ones historians engage in, research, writing, presenting ideas at conferences, finding a publisher, and so forth. But that 20 percent made it possible to expand the scope and depth of detail required to build a base for the story that sits on with a further level of detail than had hitherto been done.

Applying these basic project management techniques (among others) made it possible for one historian to "scale up" his work on a global project. The alternative would have been to have multiple historians, but as long as the historical profession favors individuals conducting research and publishing, the proposed project management approaches, when combined with the creation of a support group, provides a way for individuals to engage in global research at a level of detail that their discipline expects. That work has to be based on considerable empirical evidence presented in rich narrative detail, and it must balance specific events with the inevitable generalizations that global history compels one to produce.

One might ask how is that different from prior methods of doing world history? The more traditional approach involves writing a world history of an issue relying overwhelmingly on monographic work of others. That is a useful approach if there is a large body of monographic materials available, such as exists for diplomatic history. But, it is a flawed approach if you are attempting to write a global history for which there is not an even amount of monographic materials on each area of the world, which is normally the case with the most recent decades, and with new topics, such as the history of computers and their uses. We still face the problem of filling in where there is a paucity of monographic literature, or where it is weak or inadequately informed, as was often the case with the history of the diffusion of computing in most countries. In such circumstances, the strictures of good historical research did not absolve the historian from consulting primary and contemporary materials, including archival records. In short, historians of global history are required to adhere to the same standards of research and narration as those who write on far narrower topics. That is what makes the approach used in writing this book relevant and, perhaps, portends of things to come in the historian's craft.

PATHS NOT TAKEN IN WRITING A GLOBAL HISTORY OF INFORMATION TECHNOLOGY

Writing a global history on any topic presents its own unique challenges. The most obvious—indeed central—is how to deal with such a vast quantity of material, so many countries, times, people, and topics? What does one include, or leave out, and at what level is a subject covered? These are questions one faces even before the daunting task of comparing and contrasting activities in one part of the world to those in another, let alone mastering a world-view of events. Histories of the world tend to be long and complicated; we have not mastered the best way to do these yet. Nobody is ever quite satisfied, the writer who had to leave things out, or, the reader who thinks the author generalized too much, or simply did not accurately characterize a situation. The strategy normally selected, therefore, is to pick an approach, or a theme, and pursue it around the world, as I did in this book by looking at the role of information technology from economic, managerial, and business perspectives, and consciously only touching lightly on the rich history of the technologies themselves. But paths not taken are worthy of discussion because they too represent part of the process of creating world histories, and that of IT is no exception. Let's discuss several not taken to illustrate the process.

There is a rich tradition of writing histories of the technologies themselves, particularly about the development and evolution of semiconductors and individual lines of computers, including their component sub-systems. The literature on this topic is vast if we include all the reports written by the engineers and scientists who worked on these. Writing a history of the global evolution of this technology is a viable strategy and could follow the approach taken by Paul Ceruzzi in his study, but that concentrated largely on the American experience.[2] Such a study would demonstrate that in dozens of countries people built machines that borrowed heavily from one another in the 1940s and 1950s, clear indications of simultaneous interests in IT and that everyone was reading each other's technical literature, a free flow of information around the world. It would show that these activities became more channeled, indeed formalized in the 1950a and 1960s as universities and corporations essentially took over the task of continuing development of the technology. In recent years, attention has been growing around the development of software, yet historical research on the topic remains primitive at best, in part because scholars have not yet figured out what the history of software should look like, an issue that bedeviled outstanding scholars such as Michael S. Mahoney and Martin Campbell-Kelly for several decades.[3]

Another path not taken here, but important, with its own cadre of supporters is sociological and anthropological in forms. Looking at advocates promoting the use of computing allows historians to project themselves into the decision-making process that led to the adoption and use of IT, as we saw in the case studies of Dutch institutions.[4] Some historians in the United States have also embraced this approach by looking at the roles of groups of individuals, typically technical communities.[5] These are a hybrid of classic institutional case studies and very important because they allow one, in effect, to acquire a core sample of the way technology is accepted or resisted. The problem, of course, with this approach is that one would need hundreds and probably thousands of these from which to extract generalizations that would make up the soup of a global history. It takes a minimum of an article or chapter to describe the activities of such a group and we do not have enough of these yet to say that a foundation exists upon which to build a global history. To generalize on the backs of extant studies of these types poses the risk business school professors experience when they conduct one or few case studies about an issue then generalize that "everyone does it this way." Historians find generalizing on a few case studies anathema, and since global history requires generalizations, those must sit on top of a massive iceberg of information and cases to be creditable.

A variant of the anthropological approach involves studying firms and industries. Examining what a company, and typically several within an industry have done with respect to IT is quite manageable, but normally is restricted to the events in one country. The only IT-related industry that has been studied in considerable detail at the global level is semiconductors, providing approaches that could be quite useful to historians of global IT because they combine discussion of the technology's evolution with the managerial and institutional evolutions required to bring forth new products. It was truly a global industry and thus lent itself quickly to world-wide historical treatment by historians and economists.[6] Similar histories of computer manufacturing cannot really be said to exist today, although there are anthologies of articles with chapters devoted to events in individual countries. Integrated histories of the industry, however, are only just starting to appear.

There has been a blossoming of studies on how individual industries have embraced IT, most notably in the United States. These could become the basis of similar studies that look at industries on a global basis. JoAnne Yates, for example, has looked at how the life insurance industry in the United States embraced all manner of IT from tabulating equipment up to the PC; it is a wonderful model for doing such studies.[7] I too have used a variant of that approach to look at nearly three dozen American industries from the 1940s to the end of the century.[8] However, to generalize based on American industry-centered experiences around the world runs into the same problem as relying on too few case studies. Furthermore, we already know that how industries use IT varies from one country to another, much the way they vary from one nation to another. At the moment, we have very few studies, and those that exist are often weak on historical research and are more oriented toward reporting on contemporary events, such as the software industry in India. But, what has been done suggests that looking at how an industry embraced IT around the world makes a great deal of sense, and should be exercised.

Historians have become enamored with theory, to be sure not as much as colleagues in other social sciences and the humanities. They engage in discussions and use theoretical constructs and models. Yates demonstrated in her study of the insurance industry how that approach to history can be applied. The strength of such an approach for a global history of any topic is that it allows one to create a structure upon which to hang their story, and so it has much to commend itself. Its limitation as an approach is that it imposes too much of a predetermined perspective on a topic that screams for variety as one moves around the world across languages, historical differences, and enormously varied cultures and world views, as we saw with how capitalist societies approached computing versus those in Eastern Europe, and China's versus India's experiences. This approach proved quite effective for Paul N. Edwards to explain Cold War computing in the United States which, by implication, suggested an approach that could be used in dealing with Soviet military computing or perhaps a whole national approach, such as that in Japan.[9]

So arriving at a theoretical construct that works requires that a great deal of earlier work has been done by numerous scholars to validate the theory employed, as done by the French scholars of the *Annales* school in the 1950s–1970s; otherwise, it becomes an artificial construct that may or may not help an historian explain what happened. It is a problem economists run into all the time, leaving their analysis often anemic and open to criticism that they are not providing important insights, let alone a reflection of a reader's reality. Theory-based world views run right up against core beliefs of historians, most notably the one that holds all history is story-telling and that no two stories are alike. But, in defense of theory-centered scholarship, when historians look at a variety of events, such as what one does in writing global history, patterns of activities begin to surface that must be described. Those patterns can become the grist of other historians who want to use those as an indicator of trends possibly in play elsewhere and in different times, hence the birth of a theory. Discussions about that topic have engaged historians for almost two centuries. Suffice it to point out that I personally believe that global histories need to rely more on story telling than structuring a narrative that is measured by the extent to which it fits a norm or that is told within the context of a theory. Yet anarchy must not prevail, so some construct or strategy is required; in our case it is the two waves and the similar issues discussed about all nations.

Another path, and one taken often by business management professors doing quasi-historical research, and more notoriously in IT by journalists, is writing corporate histories. Hardly a month goes by without a history appearing on some company, and these are often global in scope because their subjects operate on an international plane. All the obvious candidates are covered: IBM, AT&T, ICL, Apple, DEC, Philips, Google, and Fujitsu—the list runs into the hundreds of firms. But one, or a few firms, does not a global history make, since even an industry-centered study would indicate that the cast of characters is numerous and most of them not yet having been properly studied. In short, the gaps are too big. Historians have written very few corporate histories of firms providing IT products, and even fewer about institutions that used those products. In fact, one can reasonably conclude that all such work lies ahead for historians of information technology working in the twenty-first century. Examples of where that has been done are just that, examples, exceptions to an otherwise barren landscape. Two useful models, however, of how to do this can be found in the history of ICL by Martin Campbell-Kelly and Pierre E. Mounier-Khun on French institutions developing IT.[10]

All possible approaches considered for possible use in our history of IT shared the common limitation that they have been insufficiently applied on a global basis so far, and thus not enough upon which to rest a world history. For the truth is, the history of IT is embryonic and too fragmented to lend itself fully to a particular approach, let alone to a richly fact-based debate about one view or another. It is a debate historians desperately would love to engage in, but cannot because they lack sufficient information about such basic issues as who had computers, where and when did they get these. That is why the history I wrote is simple in its approach, almost analogous to a pencil sketch that will have to be colored in and later be replaced with magnificent oil canvases. But ultimately, there is no right way to approach world history. A diversity of approaches actually is the best way.

Wave One and Wave Two Compared

The purpose of this appendix is to summarize key similarities and differences between Wave One and Wave Two of information technologies (IT) that guided the discussion in this book. It displays the features in a simple typological format so that others looking at the diffusion of IT, or some other modern complex technology, can use it as an initial check list of issues to examine. In this book features of both waves have been described, or alluded too, with the focus primarily on deployment of computing. This appendix does not provide a formal theory of waves for Wave One and Wave Two. The reason for not doing it has been mentioned in the book numerous times—we are too close to the events being described to know precisely the boundaries of one to the other and the elements that make up each wave. The discussion here, therefore, has two purposes: to explain further what was meant by the two paradigms and to offer up a potentially useful framework for others examining the same issues.

Table B-1 catalogs the most fundamental features of both waves as we can see them now. Wave One encompasses the initial introduction of the technology into a nation and its diffusion to a fairly extensive point, characterized by nearly ubiquitous use of IT by all large and mid-sized organizations, and most small ones, between 60 and 75 percent of workers in mid- to large organizations on a regular basis, and by over half the population of a country in various forms: cell phones, PCs, digital cameras, and most important, with over a third accessing the Internet. In each country, Wave One takes on its own personality, and flowers at its own speed. Nations began entering the earliest phases of Wave One as early as World War II in the United States and Great Britain and as late as the 1990s in some African societies. There are still some nations that have not started Wave One as well, mostly in Africa and in some remote corners of Asia. Wave One is also the time when deployment of IT can either be concentrated in large urban centers or diffused to wide areas of a nation. Some Wave Two countries also have large expanses of rural areas with limited or no cell phone or broadband diffusion, such as the United States in the northern plains and even in parts of Central Virginia, just 80 miles from Washington, D.C. These attributes should be seen as general guides to the waves, not firm characteristics as those will probably evolve over time as we learn more about the diffusion of IT and others add issues and attributes relevant to them.

Table B-1 CHARACTERISTICS OF WAVE ONE AND WAVE TWO DIFFUSION OF INFORMATION TECHNOLOGIES

Wave One

Earliest Computers Developed

Extensive initial and sustained investments by government in IT

Establishment and Expansion of IT supplier firms, agencies and industries

Massive deployment of computing to mid-sized and large organizations in most industries

Extensive development of IT-literate workforce

Extensive acquisition of PCs by individuals (e.g., minimum one-third of households)

Society becomes extensive user of networks, such as Internet

Some use of small digital electronics (e.g., cell phones, cameras, laptops)

Majority of workers and students (above the age of 7) have some working knowledge of computing and other digital technologies (e.g., video games)

Wireless communications begin

Academics, officials, and commentators see their society as now in Information Age, Post-Industrial, or some form of Post-Fordist economy, beginning mid-way through the wave

R&D evident in most forms of information and communications technologies by late Wave One

Growing levels of per capita income and prosperity

Began 1940s, earliest adopters exited in mid-to late 1990s; most nations still in mid- to late stages of Wave One

Wave Two

All Wave One features are present

Over 75 percent of citizens use the Internet; over 95 percent have wireless telephones

Broadband available to minimum 75 percent of the population

Vast majority of workers use some form of IT in their work

Service Sector of national economy at least as large as manufacturing, and is very IT intensive

Digital technologies are embedded extensively in national infrastructures, such as traffic management, law enforcement, education, and training at all levels and ages

Majority of installed technologies are current generation IT

Consumers display Moore's Law-type behavior

Use most advanced forms of IT available in weapons, transport, medical practices

Growing levels of per capital income and prosperity

Began in 1980s—Wave Two not over yet

Wave Two always includes most of the features of Wave One; in fact, they overlap. That overlap—transition from one to the other—takes a decade or longer, with 15 years apparently more the norm so far. Wave Two fits the model more closely to what Manuel Castells has been pre-announcing for over a decade in that it is teledense and that the digital technologies are shaping the character of a nation's political, economic, cultural and social behavior.[1] The emblem of Wave Two is the Internet; however, the Internet also plays a profoundly important role late in Wave One. Most important, Wave Two is more than just about using the Internet, it speaks to a profoundly indigenous use of information by people, machines of all kinds, and organizations. It is not clear if all the features of Wave Two are identifiable for any inventory like table B-1, because it is a phase in human history barely 15 years old, so still in an embryonic stage.

Major facets of these waves include appropriation of information technologies, and institutions, users, applications of IT, and initial perceptions societies interacting with the technology. In our book the social structure has been kept simple on purpose, consisting of a traditional hierarchy of nations, industries, enterprises (including public sector institutions), and individuals. This book represents an early attempt to suggest a new supra-hierarchy—global. No attempt has been made to dissect this construct into more finite forms, although that could be done for more in-depth studies, as I did in *The Digital Hand* about the experience of the United States with IT. However, it is important to note that each wave has non-technical characteristics that cannot be ignored. The ones most obvious, perhaps important, for this study, are catalogued in table B-2 with their relative importance weighted. A simple High, Medium, and Low ranking were applied to describe in general terms global features of a wave; no attempt was made to segment those evaluations by country. I normally ranked something High if it was widespread or very influential and visible in a society. At the other extreme Low implied an embryonic status, as one might see for many of these features in parts of Latin America and Central Africa. I did not rank something as absent, because it is quite difficult to find any society today without some form of IT in use or effect, even in remote villages in Asia and Africa, where cell phones are increasingly widespread.

Table B-2 applied a variety of characteristics which are widely used by economists and public officials in this, or variant forms. They all share in common the belief that such characteristics either facilitate or retard diffusion. Democracies tend to facilitate diffusion of IT quicker than authoritarian regimes, because they permit a greater free flow of information and normally have market-driven economies that promote innovation and individual and entrepreneurial initiatives. Because governments play such an important role in diffusion of IT, its characteristics, policies, and programs have long been the subject of study by economists and political scientists, less so by historians so

Table B-2 NON-TECHNICAL CHARACTERISTICS OF SOCIETIES IN
WAVE ONE AND WAVE TWO (HIGH, MEDIUM, AND LOW)

Wave One

Feature	Ranking by Presence
Business Environment	Low to High
Social & Cultural Environment	Medium to High
IT Localized	High
IT Geographically Widespread	Low to Medium
Legal Environment Stable	High
Government Policy & Vision	Medium to High
Representative Government	Medium to High
Authoritarian Government	Medium to Low
Government Effectiveness	Low to Medium
Business Adoption	Medium to High
Consumer Adoption	Medium to High
Education Levels of Citizens	Medium to High
Post-Industrial Economy	Medium
Per Capita Gross Domestic Product	Medium to High

Wave Two

Feature	Ranking by Presence
Business Environment	Medium to High
Social & Cultural Environment	Medium to High
IT Localized	Low
IT Geographically Widespread	High
Legal Environment Stable	High
Government Policy & Vision	High
Representative Government	Medium to High
Authoritarian Government	Low
Government Effectiveness	Medium to High
Business Adoption	High
Consumer Adoption	High
Education Levels of Citizens	High
Post-Industrial Economy	Medium
Per Capital Gross Domestic Product	High

far, so all rankings and assessments take into account their role. The same applies to those economic features deemed crucial to the successful functioning of a modern economy, specifically, rule of law, enforceability of contracts, transparency in negotiations, minimal or no corruption, ease of establishing a new business, physical safety of goods and employees, quality transportation, access to capital and skilled workers, and a reasonable tax structure.

Economists also place great store in a nation investing in R&D—often also referred to as innovation—while historians are more concerned with existing technologies and their effects on people and their societies.

The variables used in table B-2 provided input into the ranking of nations presented in table B-3 by wave. These rankings are the result of 10 years' of analysis done by the Economist Intelligence Unit in collaboration with the IBM Institute for Business Value and your author, using an econometric model developed by the EIU, along with analysts' assessments. Rankings as of about 2009–2010 were chosen to reflect nations clearly in Wave Two and many in a wide range of positions in Wave One. An examination of the annual reports, called *E-Readiness Ranking*, provide statistical scores on the weighting of nearly 100 variables on a country-by-country basis. While this is not the only such global ranking, it is one of the longest running one, begun in 2000. Over time nations moved up and down in the rankings. Wave One nations typically ranked statistically with differences in annual scoring quite broad with big jumps over the years, while Wave Two nations saw the variability in their scores diminish quickly, as was the case with Wave One with the highest scores (See table B-4). This statistical pattern suggests that there is a global convergence rapidly underway in the teledensity and use of IT. Even the minimal data presented in this table, which is ranked by numeric score (data not shown) also demonstrates clearly that clusters of Wave One and Wave Two nations exist by region.[2]

Table B-3 SAMPLE NATIONS IN WAVE ONE, CIRCA 2010

Wave One

Portugal	South Africa	Philippines
Slovenia	Brazil	Venezuela
Chile	Turkey	China
Czech	Jamaica	Egypt
Lithuania	Argentina	India
Greece	Trinidad & Tobago	Russia
United Arab Emirates	Bulgaria	Ecuador
Hungary	Romania	Nigeria
Slovakia	Thailand	Ukraine
Latvia	Jordan	Sri Lanka
Malaysia	Saudi Arabia	Vietnam
Poland	Colombia	Indonesia
Mexico	Peru	Pakistan

Source: Adapted from model of Economist Intelligence Unit and IBM Institute for Business Value annual E-Readiness Rankings; for description, Economist Intelligence Unit, *Digital Economy Rankings 2010: Beyond e-Readiness* (London: Economist Intelligence Unit, 2010).

Table B-4 SAMPLE NATIONS IN WAVE TWO, CIRCA 2010

Wave Two

Denmark	United Kingdom	Malta
Sweden	Austria	Estonia
Netherlands	Switzerland	Spain
Norway	France	Italy
United States	Taiwan	Israel
Australia	Germany	
Singapore	Ireland	
Hong Kong	South Korea	
Canada	Belgium	
Finland	Bermuda	
New Zealand	Japan	

Source: Adapted from model of Economist Intelligence Unit and IBM Institute for Business Value annual E-Readiness Rankings; for description, Economist Intelligence Unit, *Digital Economy Rankings 2010: Beyond e-Readiness* (London: Economist Intelligence Unit, 2010).

The simple models provided here hopefully will lead others to decompose them further into phases within each of the waves. That requires specific experiences be added of more countries not reviewed in this book to our historical corpus, requiring a tripling of the number of the nations that should be studied. That richness of detail will also make it possible to dig deeper into crucial elements of the story barely possible here, most notably the roles of different professions and interest groups in promoting the awareness of and use of information technologies, a story whose players change over time. The biggest uncertainty is the definition of the borders between Wave One and Wave Two. Figure B-1 proposes a graphical way of approaching the issue. Note that Wave One begins with almost no IT, while a large portion of a nation moves into Wave Two at its beginning. What that phrase "a large portion" means has yet to be defined; I have usually thought in terms of more than half of a population, nation, firm, or industry. Furthermore, we know from extant, if limited research on the pre-computer Office Appliance Industry and users of such devices as tabulators and adding machines that these were normally most in use in those nations early in Wave One and Wave Two, so it is quite possible than some historian may want to begin his or her narrative with three waves instead of my two. That would be consistent with one fundamental finding historians of IT have been discovering, namely, that the use of devices and techniques for managing the collection, organization, and use of data has been going on at least since the Renaissance, and most intensely beginning in the early eighteenth century with no gaps in innovations and no sign of slowing, let alone in evolving and expanding.

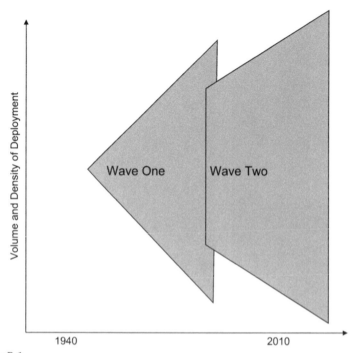

Figure B-1
Transition From Wave One to Wave Two, 1940–2010

NOTES

PREFACE

1. Koji Kobayashi, *The Rise of NEC: How the World's Greatest C&C Company Is Managed* (Cambridge, Mass.: Blackwell Publishers, 1991): xx.

CHAPTER 1

1. David E. Nye, *Technology Matters: Questions to Live With* (Cambridge, Mass.: MIT Press, 2006): 226.
2. Montgomery Phister, Jr., *Data Processing Technology and Economics* (Santa Monica, Cal.: Santa Monica Publishing, 1976) and *Data Processing Technology and Economics, 1975–1979 Supplement* (Santa Monica, Cal.: Santa Monica Publishing Co., 1979); Martin H. Weik, *A Survey of Domestic Electronic Digital Computing Systems* (Aberdeen Proving Ground, Md.: Ballistics Research Laboratory, 1955); Frederick G. Withington, *The Computer Industry, 1969–1974* (Cambridge, Mass.: Arthur D. Little, 1969).
3. World Bank, IBM Institute for Business Value, and World Information Technology and Services Alliance (WITSA), the latter at http://www.witsa.org/v2/media_center/pdf/DigitalPlanet2008_ExecutiveSummary.pdf (last accessed 2/17/2012); International Telecommunications Union (ITU); Royal Pingdom, "Internet 2010 in Numbers," http://royal.pingdom.com/2011/01/12/internet-2010-in-numbers/ (last accessed 1/17/2011); on the USA, http://pewinternet,org/Static-Pages/Trend-Data/Daily-Internet-Activities-20002009.aspx (last accessed 1/19/2011); David Goldman, "Smartphones Have Conquered PCs," *CNNMoney*, February 9, 2011, http://money.conn.com/2011/02/09/technology/smartphones_eclipse_pcs/index.htm (last accessed 2/17/2012).
4. Ryan Block (September 5, 2007). "Steve Jobs live—Apple's "The beat goes on" special event." *Engadget* (Last accessed 3/10/2008). Statistics in what might be the early stages of a Third Wave are enormously larger than any we can cite for First and Second Wave computing. For example, just using Apple: In one three day period in 2009, the company sold one million iPhone 3Gs, 3 million in one month; over 50,000 applications available for use on these telephones; over 70 percent of all new cars in 2007 offered iPod functions; while in one quarter in 2007, Apple sold 1.6 million Macintosh computers; in 2009 Apple's iTune Store accounted for 85 percent of all songs downloaded in the United States. Sources for these data: http://www.iphone footprint.com/2009/06/one-million-iphone-3gs-sold-in-three-days/; http://www.mlearninghub.com/ipod_sales.html (last accessed 9/21/2009).

5. A key finding reported in James W. Cortada, *The Digital Hand*, 3 vols (New York: Oxford University Press, 2004–2008).
6. Now routinely reported on by national economic agencies, such as the U.S. Bureau of the Census and U.S. Department of Commerce, by international bodies such as the United Nations and the Organization for Economic Co-Operation and Development (OECD), and private research enterprises, such as the Economist Intelligence Unit.
7. This guidance also applies to many other technologies and their use too, such as telephones, electricity, and transportation of all types. In recent years some of the most useful research on this issue has been done by Diego Comin and with various colleagues, "Cross-Country Technology Adoption: Making the Theories Face the Facts," *Journal of Monetary Economics* 51, no. 1 (2004): 39–83, "The Intensive Margin of Technology Adoption," HBS Working Paper 11-026 (2010). Both have extensive bibliographies of related studies by them and others.
8. James E. Short, Roger E. Bohn, and Chaitanya Baru, *How Much Information? 2010 Report on Enterprise Server Information* (San Diego, Cal.: Global Information Industry Center, April, 2011): 7.
9. I have commented more extensively on this issue and will not discuss it in this book, James W. Cortada, *Making the Information Society: Experience, Consequences, and Possibilities* (Upper Saddle River, N.J.: Financial Times, 2002) and *Information in the Modern Corporation* (Cambridge, Mass.: MIT Press, 2012).
10. Well described by Paul E. Ceruzzi, *A History of Modern Computing*, 2nd edition (Cambridge, Mass.: MIT Press, 2003).
11. Based on author's continuing conversations with IBM computer scientists in 2008. On nanotechnology, see, for example, Stanley Schmidt, *The Coming Convergence* (Amherst, N.Y.: Prometheus Books, 2008): 147–183, but also David M. Berube, *Nano-Hype: The Truth Behind the Nanotechnology Buzz* (Amherst, N.Y.: Prometheus Books, 2006).
12. A process well documented by Martin Campbell-Kelly, *From Airline Reservations to Sonic the Hedgehog: A History of the Software Industry* (Cambridge, Mass.: MIT Press, 2003) and its excellent bibliography.
13. Some of the most thoughtful studies on the long-term prospects for computing prepared by computer scientists often working on these evolutions can be found in Peter J. Denning and Robert M. Metcalfe (eds.), *Beyond Calculation: The Next Fifty Years of Computing* (New York: Copernicus/Springer-Verlag, 1997) and Peter J. Denning (ed.), *The Invisible Future: The Seamless Integration of Technology into Everyday Life* (New York: McGraw-Hill, 2002).
14. In recent years one commentator, not well versed in IT, made sensational claims that the technology has so stabilized that one should not expect future opportunities to create new businesses of a radically different type because of this circumstance, Nicholas G. Carr, *Does IT Matter: Information Technology and the Corrosion of Competitive Advantage* (Boston, Mass.: Harvard Business School Press, 2004).
15. Between 1993 and the end of 2007, IBM was awarded over 38,000 U.S. patents, more than any other corporation in the world in any of those years; in 2007 alone it received 3,125 U.S. patents, http://www.ibm.com/ibm/licensing/patents/portfolio.shtml (Last accessed 8/24/2008).
16. For information by those promoting use of this technology, see http://www.laptop.org/ (Last accessed 8/24/2008).
17. Paul E, Ceruzzi, *A History of Modern Computing*, discusses these and has extensive citations.

18. Dale W. Jorgenson, Mun S. Ho, and Kevin J. Stiroh, *Productivity*, vol. 3, *Information Technology and the American Growth Resurgence* (Cambridge, Mass.: MIT Press, 2005): 2–3.

19. Daniel E. Sichel, *The Computer Revolution: An Economic Perspective* (Washington, D.C.: Brookings Institution Press, 1997): 16.

20. A subject I explored more fully in James W. Cortada, *How Societies Embrace Information Technology* (New York: John Wiley & Sons, 2010): 27–70.

21. Moore's Law has been the subject of increased study and analysis. See, for example, Robert R. Schaller, "Moore's Law: Past, Present, and Future," *IEEE Spectrum* 34, no. 6 (June 1997): 52–59; Ethan Mollick, "Establishing Moore's Law," *IEEE Annals of the History of Computing* 28, no. 3 (July-September 2006): 62–75; Jorgenson, Ho and Stiroh, *Productivity*, 1–3.

22. A central issue in the debate over the Productivity Paradox of the 1990s and early 2000s, mostly focused on the US experience, Graham Tanaka, *Digital Deflation: The Productivity Revolution and How It Will Ignite the Economy* (New York: McGraw-Hill, 2004): 318–326. Productivity of labor has always been measured narrowly as the number of hours a human took to do something or to make a product. In the automotive industry, the number of man hours required to make each model was always measured and compared to the performance of similar workers making the same model even, in other plants within the enterprise.

23. Paul E. Ceruzzi, *A History of Modern Computing* (Cambridge, Mass.: MIT Press, 1998): 13–24.

24. EDP began as electronic data processing machines (EDPM), a term promulgated by IBM to describe one of its earliest computers, the Defense Calculator, introduced on April 7, 1953, also known as the 701. Often a manager or department would be named after a machine, such as the Tabulating Department, the IBM Manager, or IBM Room. With the introduction of the 701 and its sequels, EDP became the term widely used to describe the function. For the origins of EDPM as a term, see Charles J. Bashe, Lyle R. Johnson, John H. Palmer, and Emerson W. Pugh, *IBM's Early Computers* (Cambridge, Mass.: MIT Press, 1986): 162. It quickly became an industry-wide phrase. One author from the period (not working at IBM) defined it as "the use of electronic computers and data processing machines to aid in the following business operations: lower-level management decision-making operations; issuing the necessary paperwork to instruct the organization in accordance with those decisions; measuring the actual progress and feeding it back for management control," all functions still done with computers in the early 2000s, Richard G. Canning, *Electronic Data Processing for Business and Industry* (New York: John Wiley & Sons, 1956): 4.

25. Paul E. Ceruzzi, *A History of Modern Computing* (Cambridge, Mass.: MIT Press, 1998):, 48.

26. The specific origins of the term are unclear; however, it was already in use in the 1960s within the data processing community to describe the main data center where a company or agency had its large computer. Often, to demonstrate the modernity of the company or agency, the wall of the room facing a hallway or outer wall of the building would be made of glass so that people walking or driving by would see whirling tape drives and blinking lights, sights of the modern enterprise at work. At its corporate headquarters at 590 Madison Avenue in New York City, IBM maintained a glass-walled data center in the 1960s and 1970s to show off its newest hardware. Your author used to take customers there to see the "new iron" at work. Of course, by the early 1980s, with security of a data center so

important, glass walls gave way to concrete ones, and often data centers went underground, with doors protected by guards and security badges.

27. Also an attempt to report at much higher levels in the organization—in the C-suite—hence Chief in the title, Marianne Broadbent and Ellen S. Kitzis, *The New CIO Leader: Setting the Agenda and Delivering Results* (Boston, Mass.: Harvard Business School Press, 2005): 5–6.

28. Advertising copy in the possession of the author, James W. Cortada.

29. Data for tables and this paragraph located at www.swivel.com/data_columns/ spreadsheet/382294 (last accessed 4/19/2009). Similar data is routinely created by ITU and made available at its website, *www.itu.int/ITU-D/ict/statistics* (last accessed 2/17/2012).

30. There is no modern full history of the Internet written by a professional historian, although there are several million white papers and articles, dozens of short book-length histories, memoirs, and other documents about aspects of the story. For population counts by country, one normally consults the United Nations, World Bank, and the U.S. Department of Commerce, all which post these kinds of statistics on their various websites. Additionally, the data varies in reliability from one country to another.

31. Increasingly specialized devices are being connected wirelessly to the Internet, such as systems in automobiles to signal when the vehicle has been in an accident and RFID-based systems to track inventory levels.

32. Organisation for Economic Co-Operation and Development (OECD), *The Future of the Internet Economy: A Statistical Profile* (Paris: OECD, June 2008), available at http://www.oecd.org/document/49/0,3343,en_2649_37441_9217329_1_ (last accessed 4/20/2009).

33. Table 5.11, *2007 World Development Indicators*, March 25, 2007, http://sitesources. worldban...STICS/Resources/table5_11.pdf (last accessed 4/20/2009).

34. While many organizations track the growth in cell phone populations, the most widely quoted source is the International Telecommunications Union, better known as the ITU.

35. Little historical research on cell phones and their manufacturers has been done. However, see Dan Steinbock, *The Nokia Revolution: The Story of an Extraordinary Company That Transformed an Industry* (New York: AMACOM, 2001); less historical but informative, Alex Lightman and William Rojas, *Brave New Unwired World: The Digital Big Bang and the Infinite Internet* (New York: John Wiley & Sons, 2002); Louis Galambos and Eric Ambrahamson, *Anytime, Anywhere: Entrepreneurship and the Creation of a Wireless World* (Cambridge: Cambridge University Press, 2002); James B. Murray, *Wireless Nation: The Frenzied Launch of the Cellular Revolution in America* (Cambridge, Mass.: Perseus, 2001); Paul Levinson, *Cellphone: The Story of the World's Most Mobile Medium and How It Has Transformed Everything* (New York: Palgrave, 2004).

36. Colin Blackman, "The Public Interest and the Global, Future Telecommunications Landscape," *Info* 9, nos. 2–3 (2007): 10.

37. "Special Eurobarometer: Roaming Special Eurobarometer," 269 /*Wave 66.1-TNS Opinion & Social* (Brussels: European Commission, March 2007).

38. ITU, *Telecommunication/ICT Markets and Trends in Africa 2007* (Geneva: ITU, 2007): 1–2.

39. Figure 1.18, *OECD Information Technology Outlook 2008* (OECD, 2008), unpaginated.

40. Everett M. Rogers, *Diffusion of Innovations*, 5th Ed. (New York: Free Press, 2003): 5.

41. Ibid., 15–16.
42. A.S. Alesina, G. Nicoletti Ardagna, and F. Schiantarelli, "Regulation and Investment," *Journal of the European Economic Association* 3, no. 4 (2005): 791–825; Paul Conway, Donato de Rosa, Nicoletti Giuseppe, and Faye Steiner, *Regulation, Competition and Productivity Convergence* (Paris: OECD, September 4, 2006); the series of essays in Erik Brynjolfsson and Brian Kahin (eds.), *Understanding the Digital Economy: Data, Tools, and Research* (Cambridge, Mass.: MIT Press, 2000); James W. Cortada, "Economic Preconditions that Made Possible Application of Commercial Computing in the United States," *IEEE Annals of the History of Computing* 19, no. 3 (1997): 27–40.
43. The literature is substantial in quantity, but for some good examples, see F.M. Scherer, *New Perspectives on Economic Growth and Technological Innovation* (Washington, D.C.: Brookings Institution Press, 1999); OECD, *ICT and Economic Growth: Evidence from OECD Countries, Industries and Firms* (Paris: OECD, 2003); Maximo Torero and Joachim von Braun (eds.), *Information and Communication Technologies for Development and Poverty Reduction: The Potential of Telecommunications* (Baltimore, Md.: Johns Hopkins University Press, 2006); Vandana Chandra (ed.), *Technology, Adaptation, and Exports: How Some Developing Countries Got It Right* (Washington, D.C.: World Bank, 2006); Henry S. Rowen, Marguerite Gong Hancock, and William F. Miller (eds.), *Making IT: The Rise of Asia in High Tech* (Stanford, Cal.: Stanford University Press, 2007).
44. For examples, Nina Hachigian and Lily Wu, *The Information Revolution in Asia* (Santa Monica, Cal.: RAND National Defense Research Institute, 2003); D.J. Peterson, *Russia and the Information Revolution* (Santa Monica, Cal.: RAND National Security Research Division, 2005); National Research Council, *Bridge Builders: African Experiences with Information and Communication Technology* (Washington, D.C.: National Academy Press, 1996); Shahid Yusuf, M. Anjum Altaf, and Kaoru Nabeshima (eds.), *Global Production: Networking and Technological Change in East Asia* (Washington, D.C.: World Bank, 2004); Andrea L. Kavanaugh, *The Social Control of Technology in North Africa: Information in the Global Economy* (Westport, Conn.: Praeger, 1998); OECD, *Reviews of National Science and Technology Policy: Italy* (Paris: OECD, 1992).
45. Superbly explained by Vernon W. Ruttan, *Technology, Growth, and Development: An Induced Innovation Perspective* (New York: Oxford University Press, 2001).
46. Espen Moe, *Governance, Growth and Global Leadership: The Role of the State in Technological Progress, 1750–2000* (Aldershot, Hampshire, UK: Ashgate Publishing, 2007): 199–248; all the essays in Miles Kahler and David A. Lake (eds.), *Governance In a Global Economy: Political Authority in Transition* (Princeton, N.J.: Princeton University Press, 2003).
47. A central concern was to use IT to create more or better jobs, an appreciation of what computing and other digital technologies could do, which was recognized as far back as the 1970s. For one of the earliest thorough discussions of the issue, see Christopher Freeman, John Clark, and Luc Soete, *Unemployment and Technical Innovation: A Study of Long Waves and Economic Development* (Westport, Conn.: Greenwood Press, 1982).
48. Nicely summarized for many of the areas mentioned in World Bank, *Global Economic Prospects: Technology Diffusion in the Developing World: 2008* (Washington, D.C.: World Bank, 2008).
49. James W. Cortada, "Public Policies and the Development of National Computer Industries in Britain, France, and the Soviet Union, 1940-80," *Journal of Contemporary*

History 44, no. 3 (2009): 493-512; Pierre E. Mounier-Kuhn, *L'informatique de la seconde guerre mondiale au Plan Calcul en France: L'émergence d'une science* (Paris: Presses de L'université Paris-Sorbonne, 2010).

50. The U.S. Government is the best case study of this process at work, described in a series of case studies, Mark A. Abramson and Roland S. Harris III (eds.), *The Procurement Revolution* (Lanham, Md.: Roman & Littlefield, 2003) and Jacques S. Gansler and Robert E. Luby, Jr. (eds.), *Transforming Government Supply Chain Management* (Lanham, Md.: Roman & Littlefield, 2004).

51. Erja Mustonen-Ollila and Kalle Lyytinen, "Why Organizations Adopt Information System Process Innovations: A Longitudinal Study Using Diffusion of Innovation Theory," *Information Systems Journal* 13 (2003): 275–297.

52. Recently summarized by Richard R. Nelson, *Technology, Institutions and Economic Growth* (Cambridge, Mass.: Harvard University Press, 2005): 111. My personal work at IBM helping governments do economic development using IT policies confirms his conclusions.

53. The first major economic studies of the issue were initiated by Fritz Machlup, and reported in a shelf-full of books, beginning with his three volume study co-authored with Kenneth Leeson, *Information Through the Printed Word*, 3 vols. (New York: Praeger, 1978); see also James W. Cortada (ed.), *Rise of the Knowledge Worker* (Boston, Mass.: Butterworth-Heinemann, 1998). For introductions to modern economics of knowledge, see Dominique Foray, *The Economics of Knowledge* (Cambridge, Mass.: MIT Press, 2000) and the more useful collection of essays, Brian Kahin and Dominique Foray (eds.), *Advancing Knowledge and the Knowledge Economy* (Cambridge, Mass.: MIT Press, 2006).

54. Joel Mokyr, *The Gifts of Athena: Historical Origins of the Knowledge Economy* (Princeton, N.J.: Princeton University Press, 2002): 1.

55. Nicely summarized in three collections of essays (among many such anthologies), Uwe M. Borghoff and Remo Pareschi (eds.), *Information Technology for Knowledge Management* (New York: Springer-Verlag, 1998); Rob Cross, Andrew Parker, and Lisa Sasson (eds.), *Networks in the Knowledge Economy* (New York: Oxford University Press, 2003); and Marleen Huysman and Volker Wulf (eds.), *Social Capital and Information Technology* (Cambridge, Mass.: MIT Press, 2004).

56. One of the key driving forces behind IT innovations in the United States from the 1860s to the present that I uncovered in my early research on computing, James W. Cortada, *Before the Computer: IBM, NCR, Burroughs, and Remington Rand and the Industry They Created, 1865–1956* (Princeton, N.J.: Princeton University Press, 1993), especially 128–136, and *The Computer in the United States: From Laboratory to Market, 1930–1960* (Armonk, N.Y.: M.E. Sharpe, 1993): 102–140.

57. For yet another way of looking at the typology of diffusionary activities in which global, nations, and firms are the central players, see a study that examines U.S. and British experiences, Alan Booth, *The Management of Technical Change: Automation in the UK and USA Since 1950* (New York: Palgrave, 2007): 171–178. My view of a typology is similar with the addition of the role of industries; see my Appendix B for more details.

58. Peter Hall and Paschal Preston, *The Carrier Wave: New Information Technology and the Geography of Innovation, 1846–2003* (London: Unwin Hyman, 1988).

59. This may sound like a ridiculously obvious statement, but it has been the subject of extensive debate as historians have fought language such as "computers did. . . ." For a balanced, clear discussion of the issue, which is very pertinent to the themes of our book, see David Nye, *Technology Matters*, 17–31. Nye states the

position overwhelmingly held by historians: "Technological determinism lacks a coherent philosophical tradition, although it remains popular." Essentially he is referring to people other than historians, and continues, "A variety of thinkers on both the right and left have put forward theories of technological determinism, but the majority of historians of technology have not found them useful," calling such views, "misguided," Ibid., 31.

60. Ibid., 25.

61. Carlota Perez, *Technological Revolutions and Financial Capital: The Dynamics of Bubbles and Golden Ages* (Cheltenham: Edward Elgar, 2002): 8–21.

62. Ibid., 18.

63. I have explored in some detail the issue of what constitutes an historical era, and about our own time, in "Do We Live in the Information Age? Insights from Historiographical Methods," *Historical Methods* 40, no. 3 (Summer 2007): 107–116.

64. For a wave-centric discussion, Chris Freeman and Francisco Louçã, *As Time Goes By: From the Industrial Revolutions to the Information Revolution* (Oxford: Oxford University Press, 2001): 301–335; for a manager's perspective on how management makes decisions to acquire IT, see James W. Cortada, *The Digital Hand: How Computers Changed the Work of American Manufacturing, Transportation, and Retail Industries* (New York: Oxford University Press, 2004): 382.

65. Lawrence J. Lau, "The Sources of Long-Term Economic Growth: Observations from the Experience of Developed and Developing Countries," in Ralph Landau, Timothy Taylor, and Gavin Wright (eds.), *The Mosaic of Economic Growth* (Stanford, Cal.: Stanford University Press, 1996): 65–66, 89–90 suggests that results might not yet be reflected in significant increases in a nation's productivity, while providing a brief overview of the concept; William J. Baumol, Sue Anne Batey Blackman, and Edward N. Wolff, *Productivity and American Leadership: The Long View* (Cambridge, Mass.: MIT Press, 1989): 89–102. The word leveraging was not chosen lightly. The concept that a technology, or innovation in work processes, albeit small, can allow a society to leverage it to its benefit was classically described and illustrated by Joel Mokyr, *The Lever of Riches: Technological Creativity and Economic Progress* (New York: Oxford University Press, 1992).

66. International Telecommunication Union, *Telecommunication/ICT Markets and Trends in Africa 2007* (Geneva: ITU, 2007): 10–12.

67. Everett M. Rogers, *Diffusion of Innovations* (New York: Free Press, 2003): 474.

68. For an introduction to the concepts from a vast literature, see D.W. Jorgenson, M.S. Ho, and K.J. Stiroh, *Productivity*, 3 vols (Cambridge, Mass.: MIT Press, 1995–2005).

69. Rudi Volti, *Technology Transfer and East Asian Economic Transformation* (Washington, D.C.: Society for the History of Technology and American Historical Association, 2002): 2.

70. A conclusion reached while writing *The Digital Hand*, 3 vols (New York: Oxford University Press, 2004–08).

71. The case for new technologies having to compete with existing technologies was recently made by David Edgerton, *The Shock of the Old: Technology and Global History Since 1900* (Oxford: Oxford University Press, 2007), in which demonstrates that many technologies remain in use for decades, even centuries, long after they were seen to have been displaced by something newer.

72. The famous case of the Picturephone was discussed in the now classic article by A. Michael Noll, "Anatomy of a Failure: Picturephone Revisited," *Telecommunications Policy* 16, no. 4 (May–June 1992): 307–316.

73. The process of a revolution informed by the findings of Thomas S. Kuhn, *The Structure of Scientific Revolutions* (Chicago, Ill.: University of Chicago Press, 1962).

74. James W. Cortada, "Patterns and Practices in How Information Technology Spread around the World," *IEEE Annals of the History of Computing* 30, no. 4 (October–December 2008): 4–25. Briefly listed the eight were: government sponsored/ private sector driven, national champion driven, Asian private sector driven, planned economy driven, industry driven, corporate driven, application driven, and technology driven.

75. Most vociferously described recently by Thomas L. Friedman, *The World Is Flat: A Brief History of the Twenty-First Century* (2nd ed., New York: Farrar, Straus and Giroux, 2006) and for a perspective that argues he overstated the case for globalization, Harm De Blij, *The Power of Place: Geography, Destiny, and Globalization's Rough Landscape* (Oxford: Oxford University Press, 2009).

76. Diego Comin and Bart Hobijn, "Cross-Country Technology Adoption: Making the Theories Face the Facts," *Journal of Monetary Economics* 51, no. 1 (2004): 39–83; findings summarized and discussed as well in World Bank, *Global Economic Prospects: Technology Diffusion in the Developing World 2008* (Washington, D.C.: World Bank, 2008): 87–90.

77. Most emphatically articulated by Ray Kurzwell, *The Singularity Is Near: When Humans Transcend Biology* (New York: Penguin, 2006).

78. Largely by James W. Cortada, *Before the Computer*; Yates, *Structuring the Information Age* Lars Heide, *Punched-Card Systems and the Early Information Explosion, 1880–1945* (Baltmore, Md.: Johns Hopkins University Press, 2009); and even earlier by James Beniger, *The Control Revolution: Technological and Economic Origins of the Information Society* (Cambridge, Mass.: Harvard University Press, 1989); and William Aspray (ed.), *Computing Before Computers* (Ames, Iowa: Iowa State University Press, 1999).

79. I explain more fully the problems of naming an era, and also the benefits of doing so with respect to computing in "Do We Live in the Information Age? Insights from Historiographical Methods," *Historical Methods* 40, no. 3 (Summer 2007): 107–116.

80. I am grateful to Dr. Mary Keeling, an economist at IBM, for collecting and validating these data through her sources at the World Bank, IMF, and the World Information Technology and Services Alliance (WITSA). This latter source represents almost all IT associations in the IT industry, and is a good source for information on the topic at http://www.witsa.org. Available at that site, for example, is its *Digital Planet 2006 Executive Summary* (May 2008) that contains extensive data by global region and is updated from time-to-time.

CHAPTER 2

1. Richard R. Nelson, *The Sources of Economic Growth* (Cambridge, Mass.: Harvard University Press, 1996): 271.

2. The Economist Intelligence Unit has tracked data on dozens of countries, which they have published each spring since 2000, working in collaboration with IBM. See, for example, The Economist Intelligence Unit, *E-Readiness Rankings 2009: The Usage Imperative* (London: Economist Intelligence Unit, 2009). The United Nations has also tracked this kind of data, but more useful than the U.N.'s is that from the World Bank, for example, *2006 Information and Communications for Development Global Trends and Policies* (Washington, D.C.: World Bank, 2006), one

of hundreds of such reports published by this organization about diffusion around the world.

3. The literature on the history of the Internet's development is extensive; however, more interesting is the work currently starting on how people used the Internet and to what extent. For that see, William Aspray and Paul E. Ceruzzi (eds.), *The Internet and American Business* (Cambridge, Mass.: MIT Press, 2008) and William Aspray and Barbara M. Hayes (eds.), *Everyday Information: The Evolution of Information Seeking in America* (Cambridge, Mass.: MIT Press, 2011).

4. On the Polish role, Christopher Andrew and David Dilks (eds.), *The Missing Dimension: Governments and Intelligence Communities in the Twentieth Century* (London: Macmillan, 1984) and Marian Rejewski, "How Polish Mathematicians Deciphered the Enigma," *Annals of the History of Computing* 3, no. 3 (July 1981): 213–234; on the United States, begin with Martin Campbell-Kelly and William Aspray, *Computer: A History of the Information Machine* (New York: Basic Books, 1998): 79–95, then see Ladislas Farago, *The Broken Seal: "Operation Magic" and the Secret Road to Pearl Harbor* (New York: Random House, 1967); Ronald Lewin, *The American Magic: Codes, Ciphers, and the Defeat of Japan* (London: Farrar, Straus Giroux, 1982), and Paul Ceruzzi, *Reckoners: The Prehistory of the Digital Computer from Relays to the Stored Program Concept 1935–1945* (Westport, Conn.: Greenwood Press, 1983); on the British, John M. Bennett, "World War II Electronics and the Early History of Computers," *Annals of the History of Computing* 18, no. 3 (1996): 66–69; Jennifer Wilcox, *Solving the Enigma: History of the Cryptoanalytic Bombe* (For George G. Meade, Md.: Center for Cryptologic History, 2004), and two essential books from Jack Copeland, *Colossus: The First Electronic Computer* (Oxford: Oxford University Press, 2003) and his edited volume, *Colossus: The Secrets of Bletchley Park's Code Breaking Computers* (Oxford: Oxford University Press, 2006).

5. Best sources on these early years are still the writings of some of those technicians, for example, John Atanasoff, "Advent of Electronic Digital Computing," *Annals of the History of Computing* 6, no. 3 (July 1984): 229–282; Alice R. Burke and Arthur W. Burke, *The First Electronic Computer: The Atanasoff Story* (Ann Arbor, Mich.: University of Michigan Press, 1988); reports and memoirs in N. Metropolis et al. (eds), *A History of Computing in the Twentieth Century: A Collection of Essays* (New York: Academic Press, 1980). *Annals of the History of Computing* published a considerable number of memoir articles all through the 1980s and early 1990s as well.

6. Stan Augarten, *Bit by Bit: An Illustrated History of Computers* (New York: Ticknor and Fields, 1984): 225–251; Ernest Braun and Stuart Macdonald, *Revolution in Miniature: The History and Impact of Semiconductor Electronics* (Cambridge: Cambridge University Press, 1982): 33–44; Paul Ceruzzi, *A History of Modern Computing* (Cambridge, Mass.: MIT Press, 1998): 64–65.

7. A point made recently by Arthur L. Norberg, *Computers and Commerce: A Study of Technology and Management at Eckert-Mauchly Computer Company, Engineering Research Associates, and Remington Rand, 1946–1957* (Cambridge, Mass.: MIT Press, 2005): 6–7.

8. The subject of locality and computing in the United States has recently become the subject of growing interest; see in particular several excellent introductions to the issues, Annalee Saxenian, *Regional Advantage: Culture and Competition in Silicon Valley and Route 128* (Cambridge, Mass.: Harvard University Press, 1994); Chong-Moon Lee et al. (eds), *The Silicon Valley Edge: A Habitat for Innovation and*

Entrepreneurship (Stanford, Cal.: Stanford University Press, 2000); Christopher Lécuyer, *Making Silicon Valley: Innovation and the Growth of High Tech, 1930–1970* (Cambridge, Mass.: MIT Press, 2004); Margaret Pugh O'Mara, *Cities of Knowledge: Cold War Science and the Search for the Next Silicon Valley* (Princeton, N.J.: Princeton University Press, 2005); Paul E. Ceruzzi, *Internet Alley: High Technology in Tysons Corner, 1945–2005* (Cambridge, Mass.: MIT Press, 2008).

9. I. Bernard Cohen, *Howard Aiken—Portrait of a Computer Pioneer* (Cambridge, Mass.: MIT Press, 1998); G. Pascal Zachary, *Endless Frontier: Vannevar Bush, Engineer of the American Century* (Cambridge, Mass.: MIT Press, 1998): 61–88; Ernest A. Guillemin, "Network Analysis and Synthesis," in Karl Wildes and Nilo Lindgren (eds.), *A Century of Electrical Engineering and Computer Science at MIT, 1882–1982* (Cambridge, Mass.: MOIT Press, 1985): 154–159; William Aspray, "Was Early Entry a Competitive Advantage? U.S. Universities That Entered Computing in the 1940s," *Annals of the History of Computing* 22, no. 3 (July–September 2000): 42–87. An Wang wrote about these years with Eugene Linden, *Lessons: An Autobiography* (Reading, Mass.: Addison-Wesley Publishing Co., 1986).

10. Nancy B. Stern, *From ENIAC to UNIVAC: An Appraisal of the Eckert-Mauchly Computers* (Bedford, Mass.: Digital Press, 1981) remains the standard reference; but see also James W. Cortada, "The ENIAC's Influence on Business Computing, 1940s–1950s," *IEEE Annals of the History of Computing* 28, no. 2 (April–June 2006): 26–28. Historians and others are fixated on "firsts" when it comes to computing technologies and patent lawsuits. For decades the ENIAC was considered the first publicly known operating digital computer. However, as evidence of the workings of the British Colossus systems, in operation during World War II, become known, recognition grew that ENIAC might not be as first as once thought.

11. Two biographies discuss von Neumann's work, yet we are still missing a full history of early computing activities in the Princeton, N.J. area; William Aspray, *John von Neumann and the Origins of Modern Computing* (Cambridge, Mass.: MIT Press, 1990) and Norman Macre, *John von Neumann* (New York: Pantheon Books, 1992).

12. William Aspray, "Introduction," *Proceedings of a Symposium on Large-Scale Digital Calculating Machinery, Harvard University, January 7–10, 1947* (Cambridge, Mass.: MIT Press and Tomash Publishers, 1985): ix.

13. N.C. Davis and S.E. Goodman, "The Soviet Bloc's Unified System of Computers," *Computing Surveys* 10, no. 2 (June 1978): 100–105.

14. Emerson W. Pugh, *Building IBM: Shaping an Industry and Its Technology* (Cambridge, Mass.: MIT Press, 1995): 150–161, 167–182.

15. Paul E. Ceruzzi, "The Early Computers of Konrad Zuse, 1935–1945," *Annals of the History of Computing* 3, no. 3 (July 1981): 241–262.

16. Kenneth Flamm has studied the role of the US Government in the early years of computing most extensively. See, his *Targeting the Computer: Government Support and International Competition* (Washington, D.C.: Brookings Institution, 1987): 174.

17. Ibid., 6.

18. Ibid., 42.

19. Ibid., but also consult his other book on the subject, Kenneth Flamm, *Creating the Computer: Government, Industry, and High Technology* (Washington, D.C.: Brookings Institution, 1988); I have relied extensively on both of these to describe the role of the U.S. Government in the early post-war decades. Others have looked at the role of government as well, in particular Arthur L. Norberg, "The Shifting

Interests of the U.S. Government in the Development and Diffusion of Information Technology Since 1943," in Richard Coopey (ed.), *Information Technology, Policy: An International Perspective* (Oxford: Oxford University Press, 2004): 24–53; Computer Science and Telecommunications Board, National Research Council, *Funding a Revolution: Government Support for Computing Research* (Washington, D.C.: National Academy Press, 1999).

20. Flamm, *Targeting the Computer*, 86.
21. Ibid., 96.
22. Paul N. Edwards, *The Closed World: Computers in the Politics of Discourse in Cold War America* (Cambridge, Mass.: MIT Press, 1996): 1.
23. Ibid.
24. US Census Bureau, *Statistical Abstract of the United States*, annual Editions.
25. A major finding of James W. Cortada, *The Digital Hand: How Computers Changed the Work of American Manufacturing, Transportation, and Retail Industries* (New York: Oxford University Press, 2004): 17–19, 76–86.
26. Calculated off 1982 constant dollars, US Department of Commerce, *National Income and Product Accounts of the United States, 1920–1982* (Washington, D.C.: Government Printing Office, 1986): 254–255.
27. Alfred D. Chandler, Jr., *The Visible Hand: The Management Revolution in American Business* (Cambridge, Mass.: Harvard University Press, 1977): 482–483; James W. Cortada, "Economic Preconditions That Made Possible Application of Commercial Computing in the United States," *IEEE Annals of the History of Computing* 19, no. 3 (1997): 27–28.
28. Discussed in considerable detail by Fritz Machlup, *The Production and Distribution of Knowledge in the United States* (Princeton, N.J.: Princeton University Press, 1962).
29. A key finding of my *The Digital Hand*, 3 vols (New York: Oxford University Press, 2004–2008).
30. James W. Cortada, *Before the Computer: IBM, NCR, Burroughs, and Remington Rand and the Industry They Created, 1865–1956* (Princeton, N.J.: Princeton University Press, 1993): 187–221.
31. The most aggressive and successful of these firms was IBM, Nancy Foy, *The Sun Never Sets on IBM: The Culture and Folklore of IBM World Trade* (New York: William Morrow, 1975): 15–61; Robert Sobel, *IBM vs. Japan: The Struggle for the Future* (New York: Stein and Day, 1986): 113–156; David Mercer, *The Global IBM: Leadership in Multinational Management* (New York: Dodd, Mead and Company, 1988): 251–295; Cortada, *Before the Computer*, 228–230.
32. K. Flamm, *Targeting the Computer*, 10–12; K. Flamm, *Creating the Computer*, 80–202.
33. Amar Bhide, *Venturesome Economy: How Innovation Sustains Prosperity in a More Connected World* (Princeton, N.J.: Princeton University Press, 2008).
34. For a substantive introduction to the issue and literature, written by an economist, however, see Vernon W. Ruttan, *Technology, Growth, and Development: An Induced Innovation Perspective* (New York: Oxford University Press, 2001).
35. IDC, *EDP Industry Report* (Waltham, Mass.: International Data Corporation, August 1974): 2.
36. M.R. Rubin and M.T. Huber, *The Knowledge Industry in the United States, 1960–1980* (Princeton, N.J.: Princeton University Press, 1986): 163.
37. Martin Campbell-Kelly, *From Airline Reservations to Sonic The Hedgehog: A History of the Software Industry* (Cambridge, Mass.: MIT Press, 2003).

38. Extracted from data in Montgomery Phister, Jr., "Computer Industry," in A. Ralston and E.D. Reilly, Jr. (eds.), *Encyclopedia of Computer Science and Engineering* (London: International Thomson Computer Press, 1983): 343.

39. US Department of State, *Bureau of Intelligence and Research Report* (Washington, D.C.: Government Printing Office, 1974).

40. Almost no serious historical research has been conducted on how consumers embraced computers. One notable early exception is William Aspray and Donald deB. Beaver, "Marketing the Monster: Advertising Computer Technology," *Annals of the History of Computing* 8, no. 2 (April 1986): 127–143, although these advertisements were aimed at corporate and public sector purchasers. The notion that consumers might behave with a Moore's Law approach has not been studied at all.

41. For a substantial bibliography of this kind of literature, although very incomplete, see James W. Cortada, *Second Bibliographic Guide to the History of Computing, Computers, and the Information Processing Industry* (Westport, Conn.: Greenwood Press, 1996): 285–377; Nathan Ensmenger, *The Computer Boys Take Over: Computers, Programmers and the Politics of Technical Expertise* (Cambridge, Mass.: MIT Press, 2010).

42. For a view of that population in the years after the first wave of computer works, see Peter Freeman and William Aspray, *The Supply of Information Technology Workers in the United States* (Washington, D.C.: Computing Research Associates, 1999); on the ongoing dispersal of people over the country, Joel Kotkin, *The New Geography: How the Digital Revolution Is Reshaping the American Landscape* (New York: Random House, 2000); for a look from the 1920s to the 1980s, Frederik Nebeker, "Computers in Use," in Atsushi Akera and Frederik Nebeker (eds.), *From 0 to 1: An Authoritative History of Modern Computing* (New York: Oxford University Press, 2002): 148–160.

43. John Backus, "Programming in America in the 1950s. Some Personal Impressions," in N. Metropolis, J. Howlett, and Gian-Carlo Rota (eds.), *A History of Computing in the Twentieth Century* (New York: Academic Press, 1980): 127.

44. The exchanges of this early period, which extended to other forums all through the 1950s, and 1960s are described by William Aspray (ed.), *Proceedings of a Symposium on Large-Scale Digital Calculating Machinery* (Cambridge, Mass.: MIT Press, 1985). At the fall 1945 Harvard/MIT computing conference, approximately 70 people attended and they represented many of the luminaries of the early computer community, "Conference on Advanced Computation Techniques," *Mathematical Tables and Other Aids to Computation* 2, no. 14 (April 1946): 65–68.

45. There is a growing body of literature, including excellent memoirs and academic studies. Among the memoirs see Martin Goetz, "Memoirs of a Software Pioneer: Part 1," *Annals of the History of Computing* 24, no. 1 (January–March 2002): 43–56, and "Memoirs of a Software Pioneer: Part 2," Ibid., 24, no. 2 (October–December 2002): 14–31; Eldon C. Hall, "From the Farm to Pioneering with Digital Control Computers: An Autobiography," Ibid., 22, no. 2 (April–June 2000): 22–31; Joe F. Moore, "Creating Profit with Computers: My Life as CEO of Bonner & Moore Associates," Ibid., 25, no. 3 (July–September 2003): 30–47; Sherman N. Mullin, "Into Digital Computing through the Back Door," Ibid., 25, no. 3 (July–September 2003): 20–28; and among the academic discussions, see Martin Campbell-Kelly and Daniel D. Garcia-Swartz, "Economic Perspectives on the History of the Computer Time-Sharing Industry, 1965–1985," Ibid., 30, no. 1 (January–March 2008): 16–37 and Jeffrey R. Yost, "Maximization and Marginalization: An Examination

of the History and Historiography of the U.S. Computer Services Industry," *Enterprises et Histoire* no. 40 (November 2005): 87–101.

46. William Aspray, Andrew Goldstein, and Bernard Williams, "The Social and Intellectual Shaping of a New Mathematical Discipline: The Role of the National Science Foundation in the Rise of Theoretical Computer Science and Engineering," in Ronald Calinger (ed.), *Vita Mathematica: Historical Research and Integration With Teaching* (Washington, D.C.: Mathematical Association of America, 1996); William Aspray, "Was Early Entry a Competitive Edge? US Universities That Entered Computing in the 1940s," *Annals of the History of Computing* 22, no. 3 (July–September 2000): 42–87; Gopal K. Gupta, "Computer Science Curriculum Developments in the 1960s," Ibid., 29, no. 2 (April–June 2007): 40–54.

47. Arthur L. Norberg, "Oral History Interview with Gene M. Amdahl," Charles Babbage Institute, 1986–87, www.cbi.umn.edu/oh/pdf.phtm?d=4 (last Accessed 2/17/2012).

48. Unpublished survey results conducted in 1987, now housed at the MIT Archives, provided by Deborah G. Douglas, Curator of Science and Technology, MIT Museum, February 4, 2010.

49. Your author has had the practice of checking academic library catalogs, online and manual, for over twenty years to see what they had in the way of computing literature, including journals. I have not documented this exercise nor published a formal report, and so my comments can only be offered as anecdotal evidence. Careful studies of the pertinent literature of the field have yet to be done by historians.

50. On the early collaboration across business and academic sectors, which extends beyond national borders, see, D. Mowery, *International Collaborative Ventures in U.S. Manufacturing* (Cambridge, Mass.: Ballinger, 1988); on the more recent past, Martin Kenny and John Seely-Brown (eds.), *Understanding Silicon Valley: The Anatomy of an Entrepreneurial Region* (Stanford, Cal.: Stanford University Press, 2000); on the Dot.com bubble, the journalistic account by John Cassidy, *dot.com: The Greatest Story Ever Sold* (New York: HarperCollins, 2002).

51. Because the semiconductor industry has been the subject of intense study, I have chosen not to cover that ground again in this chapter. However, for useful introductions that informed my comments, see Richard N. Langlois et al., *Micro-Electronics: An Industry in Transition* (Boston, Mass.: UNWIN HYMAN, 1988) and a sequel, Richard N. Langlois and Paul L. Robertson, *Firms, Markets, and Economic Change: A Dynamic Theory of Business Institutions* (London: Routledge, 1995); Andrew Goldstein and William Aspray (eds.), *Facets: New Perspectives on the History of Semiconductors* (New Brunswick, N.J.: IEEE Center for the History of Technology, 1997); Daniel Holbrook et al., "The Nature, Sources, and Consequences of Firm Differences in the Early History of the Semiconductor Industry," *Strategic Management Journal* 21 (2000): 1017–1041.

52. Norberg, *Computers and Commerce*, 267–280.

53. Ibid., 69.

54. Ibid., for the fullest and most balanced account of the role Remington Rand played in these acquisitions, 209–266.

55. "The Computing Industry in Minnesota," http://www.cbi.umn.edu/resources/mnbib.html (last accessed 2/17/2012).

56. P. E. Ceruzzi, *A History of Modern Computing*, 13–16, 22–23, 25–27; M. Campbell-Kelly and W. Aspray, *Computer*, 110–112; J. R. Yost, *The Computer Industry*, 31–32; J. W. Cortada, *The Computer in the United States*, 91–92.

57. A. L. Norberg, *Computers and Commerce*.

58. J. R. Yost, *The Computer Industry*, 30.

59. A. D. Chandler, Jr., *Inventing the Electronic Century*, 84–86, 96–99, 113–114, 117.

60. K. Flamm, *Creating the Computer*, 80–133. ·

61. Arthur L. Norberg and Jeffrey R. Yost, *IBM Rochester: A Half Century of Innovation* (Rochester, Minn.: IBM Corporation, 2006): 11–15.

62. P. E. Ceruzzi, *A History of Modern Computing*, 179, 182–187.

63. Don Hoefler, "Semiconductor Family Tree," *Electronic News*, July 8, 1971. Hoefler, a journalist, is best remembered for coining the term, "Silicon Valley" to describe the area near San Francisco, California, where semiconductor firms clustered together. For the roots of Fairchild, see Richard S. Tedlow, *Andy Grove: The Life and Times of an American* (New York: Portfolio, 2006): 85–86.

64. For discussion of the role of physical proximity, see Manuel Castells, *The Informational City* (Oxford: Blackwell, 1989): 33–125; M. Campbell-Kelly, *From Airline Reservations to Sonic the Hedgehog*, 29–120.

65. Historians have recently started to look at computing on an industry-by-industry basis. Two recent examples include JoAnne Yates, *Structuring the Information Age: Life Insurance and Technology in the Twentieth Century* (Baltimore, Md.: Johns Hopkins University Press, 2005) and J. W. Cortada, *The Digital Hand*, 3 vols.

66. Ian Inkster, *Science and Technology in History: An Approach to Industrial Development* (New Brunswick, N.J.: Rutgers University Press, 1991): 1–31.

67. WWW recounted by its lead developer, T. Berners-Lee, *Weaving the Web* (New York: Harper-Collins, 1999); on CDs, Kees A. Schouhamer Immink, "The CD Story," *Journal of the AES 46 (1998): 458–465.*

68. M. Castells, *The Informational City*, 33–125.

69. P. E. Ceruzzi, *A History of Modern Computing*, 51.

70. Described in considerable detail by Charles J. Bashe, Lyle R. Johnson, John H. Palmer, and Emerson W. Pugh, *IBM's Early Computers* (Cambridge, Mass.: MIT Press, 1986). The Whirlwind project was clearly one of the most important ever done in the history of computing. Details of its evolution are emerging in considerable detail; see, for example, Kent C. Redmond and Thomas M. Smith, *From Whirlwind to MITRE: The R&D Story of the SAGE Air Defense Computer* (Cambridge, Mass.: MIT Press, 2000).

71. Loren R. Graham, *What Have We Learned About Science and Technology from the Russian Experience?* (Stanford, Cal.: Stanford University Press, 1998) includes many details of large Soviet projects; Sergei P. Prokhorov, "Computers in Russia: Science, Education, and Industry," *Annals of the History of Computing* 21, no. 3 (1999): 4–15.

72. P. E. Cerruzi, *A History of Modern Computing*, 53.

73. Ibid., 57–65.

74. A. D. Chandler, *Inventing the Electronic Century*, 2–4.

75. George T. Gray and Ronald Q. Smith, *Unisys Computers: An Introductory History* (Self published, 2008): 42 and their article in which they presented the original census of this machine, "Sperry Rand's Transistor Computers," *Annals of the History of Computing* 20, no. 3 (July–September 1998): 16–26.

76. Franklin M. Fisher, James W. McKie, and Richard B. Mancke, *IBM and the U.S. Data Processing Industry: An Economic History* (New York: Praeger, 1983): 16–18.

77. Flamm, *Creating the Computer*, 79.

78. P. E. Ceruzzi, *A History of Modern Computing*, 200.
79. C. Bashe et al., *IBM's Early Computers*, 185, 297, 299–300, 303, 308; P. E. Ceruzzi, *A History of Modern Computing*, 70, 200.
80. The near-definitive history of S/360's development is Emerson W. Pugh, Lyle R. Johnson, and John H. Palmer, *IBM's 360 and Early 370 Systems* (Cambridge, Mass.: MIT Press, 1991), and on early issues and efforts, Ibid., pp. 39–40, 113–174.
81. Ibid., 113–114.
82. For example, K. Flamm, *Creating the Computer*, 102–103.
83. F. M. Fisher, J. W. McKie, and R. B. Mancke, *IBM and the U.S. Data Processing Industry*, 141, 153.
84. To give one a sense of the momentum in play, five years later, in 1975, that number had climbed to 225,000 (included minisystems), based on research done by James W. Cortada at the time and documented in his, *EDP Costs and Charges: Finance, Budgets, and Cost Control in Data Processing* (Englewood Cliffs, N.J.: Prentice-Hall, 1980): 37.
85. E. W. Pugh, *Building IBM*, 324.
86. Ibid.
87. A. D. Chandler, *Inventing the Electronic Century*, 8.
88. K. Flamm, *Creating the Computer*, 99.
89. A point increasingly being made by a few historians, Atsushi Akera, "Engineers or Managers? The Systems Analysis of Electronic Data Processing in the Federal Bureaucracy," in Agatha C. Hughes and Thomas P. Hughes (eds.), *Systems, Experts, and Computers* (Cambridge, Mass.: MIT Press, 2000): 191–220; Nathan L. Ensmenger, "The 'Question of Professionalism' in the Computer Fields," *Annals of the History of Computing* 23, no. 4 (October–December 2001): 56–74.
90. Even the names hark back to an earlier time. Here is text from an IBM advertisement of May, 2009, "Previewing z/S Version 1 Release 11: This latest release not only builds on time-tested technologies, but introduces new-to-the-industry dynamic capabilities. Functions such as predictive failure analysis, identity propagation, and responsive networking can help deliver improved service, reduced risk, and enhanced flexibility, all of which can ultimately help you reduce costs and drive profit," http://www-03.ibm.com/systems/z/os/ (last accessed 5/21/2009). To be sure, in the nearly five decades since S/360 was announced much changed with hardware, its architecture, and myriad software, but there was never a full abandonment of the past, because of the enormous requirement for compatibility. Thus, changes in the past five decades came incrementally, with users transforming their applications software also in pieces, with customers and vendors changing roughly at the same speed.
91. Possibly being repeated in the early 2000s by Apple with its family of increasingly compatible "i" products: iTunes, iPhone, and iPad, for example.
92. Thomas J. Watson Jr. and Peter Petre, *Father Son & Co.: My Life At IBM And Beyond* (New York: Bantam, 1990): 242.
93. A. L. Norberg, *Computers and Commerce*, 167–208.
94. The Soviets serve as an example of the consequences of not having vendors at working selling computers, discussed in chapter 6.
95. F. M. Fisher, J. W. McKie, and R. B. Mancke, *IBM and the U.S. Data Processing Industry*, 47–50.
96. J. R. Yost, *The Computer Industry*, 61–62; but largely by the American trade press in the 1960s.

97. Buck Rodgers, *The IBM Way: Insights into the World's Most Successful Marketing Organization* (New York: Harper & Row, 1986); David Mercer, *The Global IBM: Leadership in Multinational Management* (New York: Dodd, Mead & Company, 1987): 180–184.

98. For example, N. L. Ensmenger, "The 'Question of Professionalism' in the Computer Fields," 56–74.

99. In the spirit of open disclosure, your author was an IBM salesman and later a sales manager, selling large mainframes and "managing" accounts all through the 1970s and 1980s. On IBM's role as vendor, James W. Cortada, *The Computer in the United States: From Laboratory to Market, 1930 to 1960* (Armonk, N.Y.: M.E. Sharpe, 1993): 77–89.

100. Specifically my three volume *The Digital Hand*; as did Yates, *Structuring the Information Age.*

101. I have recently explored this issue in some depth, James W. Cortada, *How Societies Embrace Information Technology: Lessons for Management and the Rest of Us* (Hoboken, N.J.: John Wiley and Sons and IEEE Computer Society, 2009): 129–161.

102. K. Flamm, *Creating the Computer*, 238.

103. Ibid.

104. Daniel E. Sichel, *The Computer Revolution: An Economic Perspective* (Washington, D.C.: Brookings Institution Press, 1997): 44–45.

105. Montgomery Phister, Jr., *Data Processing Technology and Economics* (Santa Monica, Cal.: Santa Monica Publishing Company, 1979): 24–27.

106. Which I explore in more detail in James W. Cortada, "Economic Preconditions That Made Possible Application of Commercial Computing in the United States," *IEEE Annals of the History of Computing* 19, no. 3 (1997): 34.

107. I explain the strategy in some detail, J. W. Cortada, *The Digital Hand*, vol 1, 389–394.

108. I started the process of analyzing this literature in, "Using Textual Demographics to Understand Computer Use: 1950–1990," *Annals of the History of Computing* 23, no. 1 (January–March 2001): 34–56.

109. *Books in Print* is the source for these comments about volumes; however, your author personally collected several thousand books published during this period and observed that many came from series that I did not yet have, concluding that there were many more volumes than he had collected. This collection of publications has been donated to the Charles Babbage Institute at the University of Minnesota for use by future historians of computing.

110. A personal note: when I sold a computer in the 1970s and 1980s, along with the mainframe there arrived about 12 linear feet of publications from IBM about the machine and how to operate it; if I sold a full system, which would have included printers, DASD, tape drives, control units, and software, and so forth, we had to be ready for the arrival of a truck loaded with publications. Data centers employed "librarians" to manage this material.

111. The Charles Babbage Institute collects bibliographies and has posted some to its Web site, http://www.cbi.umn.edu/hostedpublications/index.html (last accessed 5/21/2009). In 2009, it also acquired three large collections of these kinds of books, most of which were published in the United States from roughly 1950 to 2006.

112. Edmund C. Berkeley, *Giant Brains or Machines That Think* (New York: John Wiley & Sons, 1949); JoAnne Yates, "Early Interactions between the Life Insurance and Computer Industries: The Prudential's Edmund C. Berkeley," *Annals of the History of Computing* 19, no. 3 (1997): 60–73.

113. William Aspray, "The Mathematical Reception of the Modern Computer: John von Neumann and the Institute for Advanced Study Computer," *MAA Studies in Mathematics* 26 (1987): 166–194; Julian Bigelow, "Computer Development at the Institute for Advanced Study," in N. Metropolis et al. (eds.), *A History of Computing in the Twentieth Century: A Collection of Essays* (New York: Academic Press, 1980): 291–310.

114. Jennifer Bayot, "John Diebold, 79, a Visionary of the Computer Dies," *New York Times*, December 27, 2005.

115. Your author and most of his friends in high school in the 1960s were voracious readers of this literature and it was through these novels that he was first introduced to computers and what they could do, leaving a very favorable impression subsequently reinforced when he learned to play various games on an IBM System 360 in college.

116. J. W. Cortada, *The Computer in the United States*, 122.

117. A similar phenomenon occurred when IBM's highly advanced super computer, named Watson, won a television quiz contest (*Jeopardy*) in February 2011, followed very quickly with a book-length history of the machine, Stephen Baker, *Final Jeopardy: Man vs. Machine and The Quest to Know Everything* (Boston, Mass.: Houghton Mifflin, 2011).

118. A story that should be put into the context of many uses of computing in flight, Paul E. Ceruzzi, *Beyond the Limits: Flight Enters the Computer Age* (Cambridge, Mass.: MIT Press, 1989): 94, 97–98.

119. J. W. Cortada, *The Digital Hand*, vol. 1, 52–59.

120. Richard L. Nolan, "Information Technology Management Since 1960," in Alfred D. Chandler, Jr. and James W. Cortada (eds.), *A Nation Transformed by Information: How Information Has Shaped the United States From Colonial Times to the Present* (New York: Oxford University Press, 2000): 224–226.

121. A key finding across two dozen American industries in J. W. Cortada, *The Digital Hand*, vols. 1 and 2.

122. Catherine L. Mann and Jacob Funk Kirkegaard, *Accelerating the Globalization of America: The Role for Information Technology* (Washington, D.C.: Institute for International Economics, 2006): 27–37, 74–88.

123. For an example, see J. R. Yost, *The Computer Industry*, 178–180; every history that I have read of PCs of the period make this similar comment about the legitimization of the technology that came as a result of IBM's entry into this market. My comment is not intended as a criticism, because it was my personal experience at the time as an IBM salesman selling PCs that IBM's introduction of a microcomputer made it possible for many anti-glass house computing in companies to say, in effect, "see even IBM says its an OK technology," such as engineers and those individuals who could not seem to get enough support for their computing needs from the centralized data processing organizations which were often seen as beholding to the accounting and financial executives to whom they reported at the expense of end users in other parts of an organization who, at the time, were massively increasing their demands for computing.

124. Yost, in particularly does an effective job in explaining this development, Ibid., 176–178.

125. Lee S. Sproull, "Computers in U.S. Households Since 1977," in A. D. Chandler and J. W. Cortada, *A Nation Transformed by Information*, 260–262; J. W. Cortada, *The Digital Hand*, vol. 1, 33–34 for statistics.

126. Although we do not yet as of this writing have solid empirical histories of the microcomputer, I relied largely on the quasi-encyclopedic work, Paul Frieberger and Michael Swaine, *Fire in the Valley: The Making of the Personal Computer*, 2nd ed. (New York: McGraw-Hill, 1999).

127. L. S. Sproull, "Computers in U.S. Households Since 1977," 262.

128. Ibid., 257–280.

129. Described and compared to that of other products, notably the automobile, by Joseph J. Corn, *User Unfriendly: Consumer Struggles with Personal Technologies, From Clocks and Sewing Machines to Cars and Computers* (Baltimore, Md.: Johns Hopkins University Press, 2011): 88–119, 177–202.

130. Gunther Rudenberg, *World Semiconductor Industry in Transition: 1978–1983* (Cambridge, Mass.: Arthur D. Little, 1980): 11, 31.

131. *Dataquest*, various reports (1982–1983).

132. For a series of articles rich in content and bibliography, see the entire issue of *IEEE Annals of the History of Computing* 31, no. 3 (July–September 2009), also available at http://www.computer.org/annals.

133. For recent examples, see Don Tapscott, *Growing Up Digital: The Rise of the Net Generation* (New York: McGraw-Hill, 1999); John Palfrey and Urs Gasser, *Born Digital: Understanding the First Generation of Digital Natives* (New York: Basic Books, 2008); Kathryn C. Montgomery, *Generation Digital: Politics, Commerce, and Childhood in the Age of the Internet* (Cambridge, Mass.: MIT Press, 2009).

134. For a current explanation of today's measurement issues and metrics, see Erik Brynjolfsson and Brian Kahin (eds.), *Understanding the Digital Economy: Data, Tools, and Research* (Cambridge, Mass.: MIT Press, 2000); for a demonstration in current economic analysis, see Robert E. Litan and Alice M. Rivlin (eds.), *The Economic Payoff from the Internet Revolution* (Washington, D.C.: Brookings Institution Press, 2001).

135. Examples include, World Bank, *2006 Information and Communications for Development, Global Trends and Policies* (Washington, D.C.: World Bank, 2006); Yutaka Kurihara, Sadayoshi Takaya, Hisashi Harui, and Hiroshi Kamae, *Information Technology and Economic Development* (Hershey, N.Y.: Information Science Reference, 2008); James W. Cortada, Ashish M. Gupta, and Marc Le Noir, *How Nations Thrive in the Information Age: Leveraging Information and Communications Technologies for National Economic Development* (Somers, N.Y.: IBM Corporation, 2007) and their, *How Rapidly Advancing Nations Can Prosper in the Information Age: Leveraging Information and Communications Technologies for National Economic Development* (Somers, N.Y.: IBM Corporation, 2007).

136. J. R. Yost, *The Computer Industry*; in addition to the many citations in this chapter, see also Timothy Breanahan and Shane Greenstein, "Technological Competition and the Structure of the Computer Industry," *Journal of Industrial Economics* 47, no. 1 (1999): 1–40; Gerald W. Brock, *The Second Information Revolution* (Cambridge, Mass.: Harvard University Press, 2003); Computer Science and Telecommunications Board, *Keeping the U.S. Computer Industry Competitive: Systems Integration* (Washington, D.C.: National Academy Press, 1992); Stephen P. Bradley, Jerry A. Hausman, and Richard L. Nolan (eds.), *Technology and Competition: The Fusion of Computers and Telecommunications in the 1990s* (Boston, Mass.: Harvard Business School Press, 1993); John Zysman and Abraham Newman (eds.), *How Revolutionary Was the Digital Revolution? National Responses, Market Transitions, and Global Technology* (Stanford, Cal.: Stanford University Press, 2006); Shane Greenstein, *Diamonds Are Forever Computers Are Not: Economic and Strategic Management in Computing Markets* (London: Imperial College Press, 2004).

137. Catherine L. Mann and Jacob Funk Kirkegaard, *Accelerating the Globalization of America: The Role for Information Technology* (Washington, D.C.: Institute for International Economics, 2006): 70–95.

138. The logic and rationale for this line of thinking has been best explained by Graham Tanaka, *Digital Deflation: The Productivity Revolution and How It Will Ignite the Economy* (New York: McGraw-Hill, 2004): 31–52.

139. D. E. Sichel, *The Computer Revolution*, 2.

140. Classic sources on measures in the American economy used by economists of the period and subsequently by historians include John W, Kendrick, *Productivity Trends in the United States* Princeton, N.J.: National Bureau of Economic Research, 1961) and his sequel, *Postwar Productivity Trends in the United States, 1948–1969* (New York: National Bureau of Economic Research, 1973); for discussion of measures and metrics, see National Bureau of Economic Research, *The Rate and Direction of Inventive Activity: Economic and Social Factors* (Princeton, N.J.: National Bureau of Economic Research, 1962). How economists measure the American economy evolved, and continues in part as a direct result of the enormous effects IT has had on it since at least the 1990s. For an overview, circa late 1990s, see Norman Frumkin, *Tracking America's Economy*, (Armonk, N.Y.: M.E. Sharpe, 1998, 2004) and his more recent, *Guide to Economic Indicators* (Armonk, N.Y.: M.E. Sharpe, 2005).

141. Summarized in James W. Cortada, "*The Digital Hand*: How Information Technology Changed the Way Industries Worked in the United States," *Business History Review* 80 (Winter 2006): 755–766.

142. Victor R. Fuchs, *The Service Economy* (New York: National Bureau of Economic Research, 1968).

143. Jack R. Triplett and Barry P. Bosworth, *Productivity in the U.S. Services Sector: New Sources of Economic Growth* (Washington, D.C.: Brookings Institution Press, 2004): 4–5. By then these economists could declare with confidence, based on data on US economic activities beginning in the late 1980s that "IT has been a major contributor to recent U.S. productivity growth. We confirm this at the industry level. . . . IT in services industries accounted for 80 percent of the total IT contribution to U.S. labor productivity growth since 1995 and 2001," (p. 2).

144. Vernon W. Ruttan, *Technology, Growth, and Development: An Induced Innovation Perspective* (New York: Oxford University Press, 2001): 115.

145. Ibid., quote p. 316, for discussion 316–317.

146. Ibid., 317. This was the same behavior I observed in 36 industries while writing, *The Digital Hand*.

147. Ibid., 329.

148. K. Flamm, in particular, did repeated counts presented in his two books, *Creating the Computer* and *Targeting the Computer*.

149. Richard N. Langlois and W. Edward Steinmueller, "The Evolution of Competitive Advantage in the Worldwide Semiconductor Industry, 1947–1996," in David C. Mowery and Richard R. Nelson (eds.), *Sources of Industrial Leadership: Studies of Seven Industries* (Cambridge: Cambridge University Press, 1999): 19–78; Nico Hazewindus and John Tooker, *The U.S. Microelectronics Industry: Technical Change, Industry Growth and Social Impact* (New York: Pergamon Press, 1982); P.R. Morris, *A History of the World Semiconductor Industry* (Stevenage, U.K.: Peter Peregrinus, 1990).

150. Richard C. Levin, "The Semiconductor Industry," Richard R. Nelson (ed.), *Government and Technical Progress: A Cross-Industry Analysis* (New York: Pergamen Press, 1987): 19.

151. W.E. Steinmueller, "The U.S. Software Industry: An Analysis and Interpretive History," in D.C. Mowery (ed.), *The International Computer Software Industry: A Comparative Study of Industry Evolution and Structure* (New York: Oxford University Press, 1996): 28.
152. Dale W. Jorgenson and Kevin J. Stiroh, "Information Technology and Growth," *American Economic Review* 89 (1999): 109–115; see also their book written with Mun S. Ho, *Productivity*, vol. 3, *Information Technology and the American Growth Resurgence* (Cambridge, Mass.: MIT Press, 2005): 1–48.
153. V. W. Ruttan, *Technology, Growth, and Development*, 359.
154. D. W. Jorgenson, M. S.Ho, and K. J. Stiroh, *Productivity*, 42–47.
155. K. Flamm, *Targeting the Computer*, 36–39.
156. U.S. Bureau of Labor Statistics, http://web.its.doc.gov/ITI/iti-Home.ns (last accessed 11/18/2010).
157. For a rigorous economic study supporting all the statements made in this chapter, see David Autor, *The Polarization of Job Opportunities in the U.S. Labor Market: Implications for Employment and Earnings* (Washington, D.C.: Center for American Progress and the Hamilton Project, April, 2010).
158. Gerard Alberts, "Appropriating America: Americanization in the History of European Computing," *IEEE Annals of the History of Computing* 32, no. 2 (April–June 2010): 4–7. The issue of *Annals* in which his article appeared is devoted to this issue in one way or another. It is a path breaking issue of this journal.

CHAPTER 3

1. Simon Nora and Alain Minc, *The Computerization of Society: A Report to the President of France* (Cambridge, Mass.: MIT Press, 1980): 3.
2. While writing this book I read hundreds of economic reports about European IT and almost every one of those contained a short discourse on the unreliability or insufficiency of data on both IT and their impact on a nation's economy.
3. Norman Davis, *Europe: A History* (Oxford: Oxford University Press, 1996): 7–46.
4. This follows a long tradition of historians, most notably in business history with Alfred D. Chandler, Jr., *Scale and Scope: The Dynamics of Industrial Capitalism* (Cambridge, Mass.: Harvard University Press, 1990).
5. http://europa.eu/bulletin/en/200012/p103005.htm (last accessed 7/5/2009).
6. http://www.photius.com/rankings/gdp_2050_projection.html (last accessed 7/5/2009).
7. Excellent sources include the World Bank, United Nations, and the OECD, the two latter which maintain extensive statistical information on their websites. Their data are both segregated by country and comparative to regions and globally. The most accurate and comprehensive data for any country's diffusion of ICT, capital investments, and extent of uses of specific machines, exist for the post-1980. For prior periods, the data are quite sketchy and the best are largely limited to the United States and Great Britain, and to a lesser extent France.
8. Lars Heide, *Punched-Card Systems and the Early Information Explosion, 1880–1945* (Baltimore, Md.: Johns Hopkins University Press, 2009). Until his work appeared in print, we were largely left with minimal data which could lead one to the false conclusion that only the United States had this kind of extensive use, possibly created largely by my own work, James W. Cortada, *Before the Computer: IBM, NCR, Burroughs, and Remington Rand and the Industry They Created, 1865–1956* (Princeton, N.J.: Princeton University Press, 1993) and most recently on Great Britain by Jon Agar, *The Government Machine: A Revolutionary History of the*

Computer (Cambridge, Mass.: MIT Press, 2003). Neither of us included significant amounts of comparative information about activities in other countries.

9. Lars Heide, *Hulkort og EDB I Danmark, 1911–1970* (Århus, Denmark: Systime, 1996) and for an English summary of his key points, "Punched-Card and Computer Applications in Denmark, 1911–1970," *History and Technology* 11, no. 1 (1994): 77–99. His work demonstrates one of the great challenges for historians of pan-European history. His book is far richer in detail than his English articles, yet its accessibility is very limited because of the language in which it is written, while his next book, published in English, enjoyed a far wider readership but proved a challenge to write since English is not his native language.

10. Peter J. Bird, *LEO: The First Business Computer* (Workingham: Hasler Publishing, 1994) provides a detailed history of the firm and its machine. All other accounts focus just on the technology.

11. For a recent survey that includes some of the key bibliography see Richard Coopey, "Empire and Technology: Information Technology Policy in Postwar Britain and France," in Richard Coopey (ed.), *Information Technology Policy: An International History* (Oxford: Oxford University Press, 2004): 144–168.

12. Jack Copeland, *Colossus: The First Electronic Computer* (Oxford: Oxford University Press, 2003) and *Colossus: The Secrets of Bletchley Park's Code Breaking Computers* (Oxford: Oxford University Press, 2006).

13. John Hendry, *Innovating for Failure: Government Policy and the Early British Computer Industry* (Cambridge, Mass.: MIT Press, 1989): 33; Martin Campbell-Kelly, *ICL: A Business and Technical History* (Oxford: Oxford University Press, 1989): 115–123.

14. M. Campbell-Kelly, *ICL*, 123–125.

15. Ibid., 126–143.

16. Ibid., 136.

17. Quoted in Ibid., 139.

18. James W. Cortada, *The Computer in the United States: From Laboratory to Market, 1930–1960* (Armonk, N.Y.: M.E. Sharpe, 1993): 64–101.

19. Recently the subject of a major review by B. Jack Copeland, "The Manchester Computer: A Revised History Part 1: The Memory," *IEEE Annals of the History of Computing* 33, no. 1 (January–March 2011): 4–21 and "The Manchester Computer: A Revised History Part 2: The Babt Computer," Ibid., 22–37.

20. Martin Campbell-Kelly and William Aspray, *Computer: A History of the Information Machine* (New York: Basic Books, 1996): 99–104. Much has yet to be learned about development of British computers in the 1940s and early 1950s, in part because some projects were shrouded in secrecy; yet information on events from these years continue to seep out, even in the early 2000s. For example, in early 2010, the BBC reported that as late as 1954, the government had sponsored a secret project team that built a computer system called the Oedipus, which at the time was considered perhaps the most advanced in the world in terms of processing speed and in its online data storage. One historian who was recently allowed to interview members of the project, Simon Lavington, was quoted by the BBC: "For particular tasks it was about 10,000 times faster than the fastest commercially-available computer at that time—an IBM 704," additional evidence that the British still were leaders in the rapidly expanding evolution of computer technology, Mark Ward, "Secrets, Spies and Supercomputers," BBC News, February 1, 2010, http://news.bbc.co.uk/2/hi/technology/8490464.stm (last accessed 2/3/2010); Martyn Clark, "State Support for the Expansion of UK University Computing in the

1950s," IEEE Annals of the History of Computing 32, no. 1 (January–March 2010): 23–33.
21. M. Campbell-Kelly, *ICL*, 168.
22. J. Hendry, *Innovating for Failure*, 1.
23. Ibid., 21–22.
24. R. Coopey, "Empire and Technology," 145.
25. B. Eichengreen, *The European Economy Since 1945*, 262–263.
26. J. Hendry, *Innovating for Failure*, 74, but see the entire book for a description of the early market.
27. Isaac L. Auerbach to Nelson Blackman, February 3, 1961, and with attached report, "European Information Technology: A Report on the Industry and the State of the Art," Technical Report 1048-TR-1, January 15, 1961, Blachman Papers, Box 217, Smithsonian Institution Archives.
28. Bernardo Bátiz-Lazo and J. Carles Maixé-Altés, "Organizational Changes and the Computerization of British and Spanish Savings Banks, circa 1950–1985," in Bernardo Bátiz-Lazo, J. Carles Maixé-Altés, and Paul Thomes, *Technological Innovation in Retail Finance: International Historical Perspectives* (New York: Routledge, 2011): 144–145.
29. For CAD, in particular, Erik Arnold and Peter Senker, "Computer-aided Design: Europe's Role and American Technology," Margaret Sharp (ed.), *Europe and the New Technologies: Six Case Studies in Innovation and Adjustment* (Ithaca, N.Y.: Cornell University Press, 1986): 27–32. CAD applications are clear indicators of a second generation of computing applications going into a manufacturing firm since manufacturers always installed accounting and financial systems first, before providing end-users applications relevant to their roles as producers.
30. David J. Jeremy, *A Business History of Britain, 1900–1990s* (Oxford: Oxford University Press, 1998): 215; B. Bátiz-Lazo, J. Maixé-Altés, and P. Thomes, *Technological Innovation in Retail Finance*.
31. M. Campbell-Kelly, *ICL*, 206–207.
32. Ibid., 226.
33. For explanation of "technology gap," Organization for Economic Cooperation and Development (OECD), *Gaps in Technology: Electronic Computers* (Paris: OECD, 1969).
34. It is difficult to arrive at an exact figure since R&D on the system began in the early 1960s and extended to the end of the decade. However, historians know that the technology in this system generated over $100 billion in revenue over the next quarter century, Emerson W. Pugh, Lyle R. Johnson, and John H. Palmer, *IBM's 360 and Early 370 Systems* (Cambridge, Mass.: MIT Press, 1991): 641–642.
35. M. Campbell-Kelly, *ICL*, 248.
36. Quoted in Ibid., 264.
37. *The Prospects for the United Kingdom Computer Industry in the 1970s* (London: HMSO, 1971).
38. M. Campbell-Kelly, *ICL*, 288.
39. Ibid., 300.
40. Martin Campbell-Kelly and Ross Hamilton, "From National Champions to Little Ventures: The NEB and the Second Wave of Information Technology in Britain, 1975–1985," R. Coopey, *Information Technology Policy*, 169–186.
41. Paul Jowett and Margaret Rothwell, *The Economics of Information Technology* (New York: St. Martin's Press, 1986): 47–69.

42. Brian Oakley and Kenneth Owen, *Alvey: Britain's Strategic Computing Initiative* (Cambridge, Mass.: MIT Press, 1990).
43. Nagy Hanna, Ken Guy, and Erik Arnold, *The Diffusion of Information Technology; Experience of Industrial Countries and Lessons for Developing Countries* (Washington, D.C.: World Bank, 1995): quote p. 46; this paragraph is based on Ibid., 43–47.
44. Brian Oakley, "An Overview of Research and Co-operation in Advanced Information Technology," in Arthur Cotterell (ed.), *Advanced Information Technology in the New Industrial Society: The Kingston Seminars* (Oxford: Oxford University Press, 1988): 22–23.
45. Martin Campbell-Kelly, "Information Technology and Organisation Change in the British Census, 1801–1911," in JoAnne Yates and John Van Maanen (eds.), *Information Technology and Organizational Transformation: History, Rhetoric and Practice* (Thousand Oaks, Cal.: Sage, 2000): 35–58; Marie Hicks, "Compiling Inequalities: Computerization in the British Civil Service and Nationalized Industries, 1940–1979" (Unpublished Ph.D. dissertation, Duke University, 2009): 1–32, 289–294.
46. Agar, *The Government Machine*, 201–261.
47. See Ibid for an excellent account of the role of the British military, particularly air force and navy in these years, 263–292.
48. Ibid., 291.
49. The role of British universities has only been studied sporadically, often in the form of memoirs. See for one of many early examples, Andrew D. Booth, Computers in the University of London, 1945–1962," in N. Metropolis et al (eds.), *A History of Computing in the Twentieth Century* (New York: Academic Press, 1980): 551–561. On the other hand, more formal studies are beginning to appear, see for example, Frank Verdon and Mike Wells, "Computing in British Universities: The Computer Board 1966–1991," *The Computer Journal* 38, no. 10 (1995): 822–830. An early study that deserves attention is P. Drath, "The Relationship Between Science and Technology: University Research and the Computer Industry, 1945–1962" (Unpublished Ph.D. dissertation, Manchester University, 1973). More recent studies are of individual university activities, for example, Richard Giordano, "Institutional Change and Regeneration: A Biography of the Computer Science Department at the University of Manchester," *Annals of the History of Computing* 15, no. 3 (1993): 55–62.
50. J. Agar, *The Government Machine*, 295.
51. For a list of many of these early systems, see Ibid., 316–317.
52. M. Hicks, "Compiling Inequalities," 306, 308.
53. Ibid., 372–380.
54. Alan Booth, *The Management of Technical Change: Automation in the UK and USA Since 1950* (New York: Palgrave, 2007): 88.
55. Ibid., 89–91.
56. Ibid., 93–94.
57. Ibid., 128–129.
58. Ibid., 174–176.
59. I accept Agar's numbers, Ibid., 331, although once a more thorough inventory is someday completed of systems from various vendors, the actual number may prove to be slightly higher.
60. A situation repeated in various papers in B. Bátiz-Lazo, J. Maixé-Altés, and P. Thomes, *Technological Innovation in Retail Finance*.

61. A.K. Watson to T.J. Watson, Jr., December 13, 1967, Folder 1, RG 11: Employees. The J.J. Watson, Jr. Papers/Administrative Rescords/Personnel/Watson, A.K., Box 229, IBM Corporate Archives, Somers, New York.

62. Data from Bernardo Batiz-Lazo, "Emergence and Evolution of Proprietary ATM Networks in the UK, 1967–2000," June 2007, http://mpra.ub.uni-muenchen. de/3689 (last accessed 8/5/2009). This paper and correspondence with the author was the basis of this paragraph. I am grateful to him for sharing this good work. See also the volume he edited with more material, *Technological Innovation in Retail Finance*.

63. Ibid., 341.

64. Kenneth Flamm, *Creating the Computer* (Washington D.C.: Brookings Institution, 1988): 148–150.

65. Oakley, "An Overview of Research and Co-operation in Advanced Information Technology," 16.

66. First quote, Ibid., p. 13, other quotes, p. 27. It should be noted that the British government was no different than many others in examining the IT situation at home and recommending new policies and programs. After Oakley's effort, there was, for example, the Bide Committee's report, *Information Technology: A Plan for Concerted Action* (London: Her Majesty's Stationery Office, 1986). Private sector engaged in similar activities, see G. Sashe, P. Jowett, and J. McGee, *The Software Industry in the U.K.*, Project Report, Centre for Business Strategy (London: London Business School, 1986).

67. All data drawn from a unpublished report prepared by Ken W. Sayers, "A Summary History of IBM's International Operations, 1911–2006," October 20, 2006, 135–152, IBM Corporate Archives, Somers, N.Y.

68. M. Campbell-Kelly, *ICL*, 349.

69. J. Agar, *The Government Machine*, 367–389; D.C. Pitt and B.C. Smith, *The Computer Revolution in Public Administration: The Impact of Information Technology on Government* (Brighton, Eng.: Wheatsheaf Books, 1984); Christine Bellamy and John A. Taylor, *Governing in the Information Age* (Buckingham: Open University Press, 1998), and see their extensive bibliography, 176–190; Norio Kambayashi, *Cultural Influences on IT Use: A UK-Japanese Comparison* (New York: Palgrave Macmillan, 2001).

70. C. Bellamy and J. Taylor, *Governing in the Information Age*, 7; see page 8 for how those expenditures broke out by department.

71. OECD, "The Information Economy," *PIIC* 8 (1986). For more details, see Baudouin Durieux, *Online Information in Europe* (Calne, Eng.: European Association of Information Services, 1989).

72. For a detailed discussion of the problem, see Ian Miles and contributors, *Mapping and Measuring the Information Economy* (London: The British Library Board, 1990). For an example of what was available on diffusion, see *National Computing Centre, Information Technology Trends* (Manchester: National Computing Centre, 1986).

73. Robert Taylor, *Britain's World of Work—Myths and Realities* (Swindon, Eng.: Economic and Social Research Council, undated, 2000).

74. U.S. Department of Commerce, *Showcase Europe: Guide to E-Commerce Markets in Europe* (Washington, D.C.: U.S. Government Printing Office, 2001): 44–46.

75. R. Coopey, *Information Technology Policy*, 161–165; Jacques Jublin and Jean-Michel, *Quatrepoint, French Ordinateurs, de l'affaire Bull à l'assassinat du Plan Calcul* (Paris: Editions Alain Moreau, 1976):17–27; Charles Le Bolloch, "L'intervention de l'Etat dans l'Industrie Eléctronique en France de 1974 à 1981" (Ph.D. diss., Université de rennes I, VER de Sciences Economiques, 1983).

76. AUERBACH, "European Information Technology," 51.
77. R. Coopey, *Information Technology Policy*, 157.
78. Frank Cain, "Computers and the Cold War: United States Restrictions on the Export of Computers to the Soviet Union and Communist China," *Journal of Contemporary History* 40, no. 1 (2005): 131–147.
79. Best documented in considerable detail in Jublin and Quatrepoint, *French Ordinateurs*; Vassiliki N. Koutrakou, *Technological Collaboration for Europe's Survival* (Aldershot: Avebury, 1995): 228–230; Jean-Pierre Brulé, *L'Informatique Malade de L'etat Du Plan calcul à BULL Nationalisée: Un fiasco de 40 milliards* (Paris: Les Belles Lettres, 1993): 72–78.
80. Pierre E. Mounier-Kuhn, *L'informatique de la seconde guerre mondiale au Plan Calcul en France: L'emergence d'une science* (Paris: Presses de l'université Paris-Sorbonne, 2010): 553.
81. Ibid., 553–557.
82. Ibid., 556.
83. Ibid., 594.
84. R. Coopey, *Information Technology Policy*, 158.
85. Pierre E. Mounier-Kuhn, "Le Plan Calcul, Bull et l'Industrie des Composants: Les Contradictions d'une Strategie," *Revue Historique*, CCXC, no. 1 (1995): 125–126; but see also his, "Un Exporteur Dynamique Mais Vulnerable: La Compagnie Des Machines Bull (1948–1964)," *Histoire, Economie et Societe* 4 (1995). France had a long tradition of using advanced IT equipment, see Lars Heide, "Monitoring People: Dynamics and Hazards of Record Management in France, 1935–1944," *Technology and Culture* 45, no. 1 (January 2004): 80–101.
86. For the text of the actual first *Plan Calcul*, J.-P. Brulé, *L'Informatique malada de L'etat*, 327–344.
87. R. Coopey, *Information Technology Policy*, 159–160.
88. For an excellent early account of the French response, Jean-Michel Treille, *L'Économie mondiale de L'Ordinateur* (Paris: Éditions du Seuil, 1973): 139–155.
89. Kenneth Flamm, *Targeting the Computer* (Washington, D.C. Brookings Institution, 1987): 155.
90. Patrick A. Messerlin, "France," in Benn Steil, David G. Victor, and Richard R. Nelson (eds.), *Technological Innovation and Economic Performance* (Princeton, N.J.: Princeton University Press, 2002): 148–177.
91. Ibid., 161.
92. The only major defense of Bull came from Jean-Pierre Brulé in his account of the events, *L'Informatique malada de L'etat*.
93. Ibid., 306.
94. Ibid., 305–319.
95. J. Jublin and J.-M. Quatrepoint, *French Ordinateurs*, 17–27, 62–68.
96. R. Coopey, *Information Technology Policy*, 163.
97. Data drawn from Frédérique Sachwald, "Colbertism in ICT: Lessons from the French Experience," 1997, p. 5, http://www.tcf.or.jp/data/19971011_Frederique_Sachwald.pdf (last acccessed 7/19/2009).
98. Andrew Odlyzko, "The History of Communications and Its Implications for the Internet," June 16, 2000, AT&T Labs—Research, p. 117, http://www.research.1tt.com/~amo (last accessed 8/15/2004).
99. For details, see Jacques P. Chamoux, *Télécoms, la fin des privilèges* (Paris: Presses Universitaires de France, 1993); E. Cohen, *Le colbertisme "high tech."* (Paris: Hachette, 1992); William L. Cats-Baril and Tawfik Jelassi, "The French Videotex

System Minitel: A Successful Implementation of a National Information Technology Infrastructure," *MIS Quarterly* 18, no. 1 (March 1994): 1–20; Odile Challe, "Le Minitel: la télématique à la française," *The French Review* 62, no. 5 (April 1989): 843–856; Marie-Christine Monnoyer and Jean Philippe, "Using Minitel to Enrich the Service," in Ewan Sutherland and Yves Morieux (eds.), *Business Strategy and Information Technology* (London: Routledge, 1991): 175–185; OECD, *France's experience with the Minitel: lessons for electronic commerce over the Internet* /Working Party on the Information Economy (Paris, OECD, 1998), and useful bibliography, pp. 35–37. For data on numbers installed along with discussion of the Minitel's success see, L. J. Libois, *Les telecommunications, technologies, réseaux, services* (Paris: Eyrolles, 1994); M. Marchang, *La grande aventure du Minitel* (Paris: Librairie Larousse, 1987); A. Gonzalez and E. Jouve, "Minitel: histoire du réseau télématique française," *Flux 2002* 1, no. 47 (2002): 84–89.

100. R. Moreau, *The Computer Comes of Age: The People, The Hardware, and the Software* (Cambridge, Mass.: MIT Press, 1984): 63–64. Lest one think that all important books on the history of IT are published in English, Moreau published his study first in France, *Ainsi naquit l'informatique: Les homes, les matériels à l'origine des concepts de l'informatique d'aujourd'hui* (Paris: BORDAS, 1981). At the time, Moreau was the Director of Scientific Development in IBM France.

101. Ken W. Sayers, "A Summary History of IBM's International Operations," 182–195. This report was prepared for internal use within IBM's archive in preparation for future projects related to the centenary of the firm in 2011.

102. Compagnie IBM France, *Réalisations 1979 Compagnie IBM France* (Paris: Compagnie IBM France, 1980): 8; but see also, Compagnie IBM France, *Bilan social d'enterprise Année 1978 Année 1979* (Paris: Compagnie IBM France, undated [circa 1979]).

103. K. W. Sayers, "History of IBM's International Operations," 191. On activities late in Wave one, see "IBM France: Le changement," *EHQ News*, no. 5 (1986), unpaginated, IBM Archives, Somers, New York.

104. Richard Coopey (ed.), *Information Technology Policy: An International History* (Oxford: Oxford University Press, 2004).

105. This topic requires more research; however, an important step in support of that effort was taken by Abdelkerim Ousman, "The Transatlantic Discursive Regime and French Policy in Information and Computer Technology: 1945–1981" (Unpublished Ph.D. dissertation, Carleton University (Canada), 1996).

106. S. Nora and A. Minc, *The Computerization of Society*. The original French version was, *L'Informatisation de la Société* (Paris: La Documentation Française, 1978). For two learned examples of the French discussions which extended over a decade, see Anne Mayère, *Pour Une Economie de L'Information* (Paris: Editions du Centre National de la Recherche Scientifique, 1990), and Jean Lojkine, *La Révolution Informationnelle* (Paris: Presses Universitaires de France, 1992); each has extensive bibliographic citations of the period.

107. Jean Jacques Servan-Schreiber, *American Challenge* (New York: Atheneum, 1968), original edition published in France as *Le Défi Américain* (Paris: Denoël, 1967).

108. S. Nora and A. Minc, *The Computerization of Society*, xv.

109. Ibid., 6.

110. Ibid., 8.

111. These discussions launched a debate about telematic society that continued right through Wave One France. For an example, see Alain Bron and Laurent Maruani,

La démocratie de la solitude: De l'économie politique de l'information (Paris: Desclée de Brouwer, 1996): 23–26, 139–224.

112. G. Brémond, *La Révolution Informatique: Dictionnaire Thématique* (Paris: Hatier, 1983): 97–99.

113. Alfonso Molina, "The Nature of Failure in a Technological Initiative: The Case of the Europrocessor," *Technological Analysis and Strategic Management* 10, no. 1 (March 1998): 23–40. For French perspectives, however, see G. Thèry, *Les auto-routes de l'Information* (Paris: La Documentation Française, 1994); T. Miléo, *Les réseaux de la société de l'Information* (Paris: Commissariat Général du Plan, 1996); P. Lafitte, *La France et la société de L'Information*, Rapport de l'Office Parlementaire d'Evaluation des Choix Scientifiques et Technologiques, Paris, Sénat, 7 février [1997?]; P. Martin-Lalande, *L'INTERNET: un vrai défi pour la France* (Paris, Report to the Government, 1997). For more global perspectives, begin with OECD, *Politiques en matière de technoligues de l'Information: structure organisationnelle dans les pays membres*, Documents de travail, no. 43 (Paris: OECD, 1995); M. Sharp, "The Single Market and European Technology Policies," in C. Freeman, M. Sharp, and W. Walker (eds.), *Technology and the Future of Europe* (London: Pinter, 1991); L. Guzzetti, *A Brief History of European Union Research Policy* (Luxembourg: Office for Official Publications of the European Communities, 1995).

114. For a useful description of these various programs, Coopey, *Information Technology Policy: An International History*. Prior experiences with telephony continued to influence French responses to new innovations, such as the arrival of the Internet, Pierre-Jean Benghozi and Christian Licoppe, "Technological National Learning in France: From Minitel to Internet," in Bruce Kogut (ed.), *The Global Internet Economy* (Cambridge, Mass.: MIT Press, 2003): 153–190.

115. Bart van Ark, Johanna Melka, Nanno Mulder, Marcel Timmer, and Gerard Ypma, *ICT Investments and Growth Accounts for the European Union*, Research Memorandum GB-56 (Brussels: University of Groningen and Paris: Centre d'études prospectives et d'informations internalizes (CEPII), March 2003): 32.

116. Simon Nora and Alain Minc, *L'informatisation de la société*, vol. 2, *Industrie et services informatiques* (Paris: La Documentation Française, 1978): 23; but see also, Agence de l'Informatique, *L'Etate d'informatisation de la France* (Paris: Economica, 1986): 135–152, and passim.

117. Alain Beltran, "Arrivée de L'informatique et Organisation des Entreprises Français (fin des années 1960-début des années 1980)," *Entreprises et histoire* no. 60 (2010): 122–137; Pierre-Eric Mounier-Kuhn, "Les Clubs d'Utilisateurs: entre syndicats de clients, autils marketing et 'logical libre' avant la lettre," Ibid., 158–169.

118. GFCF is not a well-known term outside of economics; it stands for gross fixed capital formation, which is a measure of net new investments within a national economy and is a popular indicator of economic trends, particularly in forecasting business activities, hence patterns of economic growth. This measure has been in wide use since the 1930s and is calculated in time series that are quarterly and annual by government statistical agencies. Using CAD as an indicator of second generation applications, we see that the French embraced the relevant technologies at the same time as most Western European markets, all supported by facilitative government policies and programs, E. Arnold and P. Senker, "Computer-aided Design: Europe's Role and American Technology," 19–22.

119. Armand Mattelart, *Technology, Culture, and Communication: A Report to the French Minister of Research and Industry* (Paris: Ministère de la recherché et de l'industrie, 1985, English ed., Amsterdam, N.Y.: North-Holland, 1985).
120. Ark, et al., *ICT Investment and Growth Accounts for the European Union*, 66.
121. Gilbert Cette and Jimmy Lopez, "What Explains the ICT Diffusion Gap Between the Major Advanced Countries? An Empirical Analysis," *International Productivity Monitor* 17 (Fall 2008): 28–39; Gilbert Cette, Jacques Mairesse, and Yusuf Kocoglu, "Dissemination of Information and Communications Technologies and Economic Growth: The Case of France Over the Long Term Period (1980–2000)," *Croissance* 6 (2002): 1–20; Yusuf Kocoglu and Jacques Mairessee, "An Exercise in the Measurement of R&D Capital and Its Contribution to Growth: Comparison Between France and United States and with ICT," unpublished paper, August 18, 2004, http://www.iariw.org/papers/2004/mairesse.pdf (last accessed 8/2/2009). These papers also have extensive bibliographies of contemporary discussions.
122. A suggestion made by G. Cette and his colleagues in, "Dissemination of Information and Communications Technologies and Economic Growth: The Case of France Over the Long Term Period (1980–2000)," 1–20. Besides the concern about over-investing leading to the charge that IT did not help productivity at the end of the century, others have argued that French and German companies did not invest in as much IT as the Americans, causing a depressed rate of labor productivity when compared to the US, along with ineffective or too much government regulations that stifled innovation and competitive prowess, Diana Farrell, *The Productivity Imperative: Wealth and Poverty in the Global Economy* (Boston, Mass.: Harvard Business School Press, 2006): 62–65.
123. Eric Brousseau, "E-Commerce in France: Did Early Adoption Prevent Its Development?," *The Information Society* 19 (2003): 45–57.
124. Ibid., 48.
125. Ibid., 49.
126. Ibid., 49-50.
127. Ibid., 51.
128. Ibid., 53.
129. U.S. Department of Commerce, *Showcase Europe*, 18.
130. On French telecommunications developments in the 1990s and beyond, Martin Fransman, *Telecoms in the Internet Age: From Boom to Bust to . . .?* (Oxford: Oxford University Press, 2002): 158–184.
131. Ibid., 18–20.
132. On the occasion of the merger of the two Germanies, the new Germany renounced claims to Polish territories that had large German populations, formally acknowledging the Oder-Neisse border as permanent, thereby reducing potential tensions in the region in future years for the new German state.
133. John Gimbel, *Science, Technology, and Reparations: Exploitation and Plunder in Postwar Germany* (Stanford, Cal.: Stanford University Press, 1990).
134. B. Eichengreen, *The European Economy*, 56–58.
135. For example, in June 2006 during the World Cup finals, held in Germany, resulted in the national German team making it almost to the finals, winning a number of games over a two-week period. As that happened citizens all over Germany celebrated, waved German flags, and sang patriotic music—all normal behavior in such a circumstance. As that happened, discussions in the media, and by academics, for instance, turned on whether or not this burst of national fervor was a

good thing to allow, given the country's Nazi past. Your author was in Germany at the time and heard these concerns on television and in conversation with academics in various parts of the country.

136. Ibid., 67.

137. Ibid., 93.

138. A recent study of the banking industry demonstrates these patterns at work, Paul Thomes, "Is There an ICT Path in the German Savings Banking Industry?," in B. Bátiz-Lazo, Carles Maixé-Altés, and Paul Thomes, *Technological Innovation in Retail Finance*, 119–136.

139. Harmut Petzold, *Moderne Rechenkünstler: Die Industrialisierung der Rechentechnik in Deutschland* (Munich: Verlag C.H. Beck, 1992): 63–116.

140. L. Heide, *Punched-Card Systems and the Early Information Explosion*, 145–152, 180–192.

141. Edwin Black, *IBM and the Holocaust: The Strategic Alliance Between Nazi Germany and America's Most Powerful Corporation* (New York: Crown Publishers, 2004); Petzold, *Modern Rechenkünster*, 125–128; Sayers, "History of IBM's International Operations," 203–204; T. Driessen, *Von Hollerith zu IBM, Zur Fruhgeschichte der Datenverarbeitungs-technik von 1880 bis 1970 aud wirtschaftswissenschaftlicher Sicht* (Cologne: Muller Botermann, 1987).

142. Wilfried de Beaclair, *Rechnen mit Maschinen: Eine Bildgeschichte der Rechentechnik* (Berlin: Springer, 2005): 73–92; Paul E. Cerruzi, "Die fruhen Arbeiten von Konrad Zuse im Kontext der Erfindung des digitalen Computers, 1955–1950," *Deutsche Museum, Wissenschaftliches Jahrbuch 1992/93*, but see the entire issue for other articles on Zuse as well, celebrating the 50th anniversary of the first operation of the Z3; Paul E. Ceruzzi, "The Early Computers of Konrad Zuse, 1935 to 1945," *Annals of the History of Computing* 3, no. 3 (July 1981): 241–262; Karl-Heinz, Czauderna, *Konrad Zuse, der Weg zu seinem Computer Z 3* (Munich: R. Oldenbourg, 1979); A.P. Speiser, "The Relay Calculator Z4," *Annals of the History of Computing* 2, No. 3 (July 1980): 242–245.

143. Paul E. Ceruzzi, *A History of Modern Computing* (Cambridge, Mass.: MIT Press, 1998): 83–84.

144. Konrad Zuse, *The Computer—My Life* (Berlin: Springer-Verlag, 1993). The original edition appeared in German, *Der Computer—Mein Lebenswerk* (Berlin: Springer-Verlag, 1984).

145. The story told by Petzold, for example, in *Moderne Rechenkünstler*.

146. Zuse, *The Computer*, 34–35.

147. Ibid., 42.

148. Ibid., 64.

149. Ibid., 106.

150. Ibid. Dr. Gerhard Dirks worked on the development of a drum memory in Germany in the 1930s and 1940s. His work is discussed by Zuse, Ibid., 111, 142, who he did not know of it at the time, but later met Dirks.

151. Ibid., 107.

152. Ibid., 123.

153. Ibid., first quote, p. 123, second quote, p. 124.

154. Many of these conferences published proceedings, for example, H. Cremer (ed.), *Probleme der Entwicklung programmgesteuerter Rechengeraete und Integrieranlagen* (Aachen: Mathematisches Institut, Lehrstuhl C., 1953); *Vortraege ueber Rechenanlagen, gehalten in Gottingen, 19. bis 21. Maerz 1953* (Gottingen: Max-Planck-Institut fuer Physik, 1053).

155. Wallace D. Hayes, "A Second Progress Report on German Computer Work, June 5, 1953, U.S. Office of Naval Research, Box 216, Mathematical Branch, Smithsonian Institution.

156. Nelson M. Blachman, "Some Automatic Digital Computers in Western Europe," *IRE Transactions on Electronic Computers* EC-5, no. 3 (September 1956): 162–163.

157. For example, N. Winer's seminal book was published in Germany as *Mensch und Menschmaschine*, translated by G. Walther (Frankfurt/Main: Metzner Verlag, 1952), W. J. Eckert R. Jones, *Faster, Faster: A Simple Description of a Giant Electronic Calculator and the Problems it Solves* (New York: IBM, 1955) was published in German, translated by E. Aikele, under the title, *Schneller, schneller. Eine allgemeinverstaendliche Beschreibung einer electronischen Grossrechenanlage und der von ihr loesbaren Probleme* (Sindelfingen: IBM, 1956). The earliest and most comprehensive local publication was by H. Rutishauser, A. Speiser, and E. Stiefel, *Programmgesteuerte digitale Rechengeraete (elektronische Rechenmaschinen)* (Basel: Verlag Birkhauser, 1951), sponsored by the Institute of Applied Mathematics of ETH Zurich, along with another from this institute written by A. Speiser, *Entwurf eines electronischen Rechengeraetes unter besonderer Beruecksichtigung der Erfordernis eines minimalen Materialaufwandes bei gegebener mathematischer Leistungsfauehikeitt* (Basel: Verlag Birkhauser, 1950, 1954). And for users of the technology, books began to appear in German on data processing as well, for example, E. Aikele et al., *Handbuch der Lochkarten-Organisation* (Frankfort-on-Main: Agenor Druck-und Verlagsgesellschaft, 1956) and F. Martin, H. Becker, L. Heinlein, and N. Kinner, *Electronische Automation im Buro* (Cologne: Mnemoton Verlag, 1956). All through the 1950s dozens of articles also appeared and by the early 1960s, routinely dozens of books each year. Updates on projects and bibliography also appeared on a regular basis; for examples see, *Stand des elektronischen Rechnens und der eletronischen Datenverarbeitung in Deutschland* (Darmstadt: Deutschen Arbeitsgemeinschaft für Rechenanlagen (DARA), 1961 and and from the same source, *Stand des elektronischen Rechens und der elektronischen Datenverarbeitung in Deutschland—2. Folge*, published 1962, both in Box 213, Mathematical Branch, Smithsonian Institution.

158. Ibid., 151–152.

159. Ibid., 152.

160. AUERBACH, "European Information Technology," 71.

161. Ibid., entire report.

162. Beauclair, *Rechnen mit Maschinen*, 9. The other machines were distributed as follows: Benelux 120, Switzerland 110, all the Scandinavian countries combined 85, and everyone else 80, Ibid., 9.

163. M. Breitenacher et. al., *Elekstrotechnische Industrie* (Munich: IFO-Institut für Wirtschaftsforschung, 1974), focusing on semiconductor and computer vendors of the 1960s and 1970s; on the products available in the early 1960s, see Rolf Haske, *Einführung in die Informations- und Dokumentationstechnik unter besonderer Berücksichtigung der Lochkarten* (Leipzig: Bibliographisches Institut, 1965); on the structure and size of the computer industry, N. Kloten et al., *Der EDC-Market in der Bundes Republik Deutschland* (Tübingen: Mohr, 1976) but see also A. Rösner, *Die Wettbewerbverhältnisse auf der Markit fur electronische Datenverarbeitungsanlagen in der BRD* (Berlin: Duncker-Humblot, 1978); "West German Data Processing Industry," *Data Processing* no. 3 (May–June 1963): 166–183; Flamm, *Targeting the Computer*, 158.

164. Eike Jessen, Dieter Michel, Hans-Juergen Siegert, and Heinz Voigt, "The AEG-Telefunken TR 440 Computer and Large-Scale Computer Strategy," *IEEE Annals of*

the History of Computing 32, no. 3 (July–September 2010): 20–29 and their "Structure, Technology, and Development of the AEG-Telefunken TR 440 Computer," Ibid., 30–38.

165. Market share data drawn from Norbert Klotten, Alfred E. Otto, Willi Gösele and Rolf Pfeiffer, *Der EDV-Markt in der Bundesrepublik Deutschland* (Tübingen: J.C.B. Mohr, 1976): 100, 114–115, 167.

166. K. Flamm, *Targeting the Computer*, 158.

167. Ibid., 159.

168. A formal history of the firm does not yet exist; however, there is a lengthy biography of its founder with considerable information on his company, Volker Werb, *Heinz Nixdorf* (Paderborn: Schoeningh Ferdinand, 2007); and a shorter one by Klaus Kemper, *Heinz Nixdorf: Eine Deutsche Karriere* (Landsberg: Verlasg Moderne Industrie, 1987).

169. Nagy Hanna, Ken Guy, and Erik Arnold, *The Diffusion of Information Technology: Experience of Industrial Countries and Lessons for Developing Countries*, World Bank Discussion Papers, 281 (Washington, D.C.: World Bank 1995): 146.

170. Baudouin Durieux, *Online Information in Europe* (Calne, Eng.: European Association of Information Services, 1989): 45–47.

171. Karl E. Ganzhorn, *The IBM Laboratories Beoblingen: Foundation and Build-up: A Personal Review* (Sindelfingen: Privately printed, 2000): 13–24.

172. Sayers, "History of IBM's International Operations," 204–210.

173. We lack a formal history of post-World War II IBM in Germany. However, a series of books focusing on the technical achievements of the company in Germany were privately published by retired IBM employees that are highly informative, but nearly impossible to obtain. I cite them here mainly to preserve a reference to them, although I consulted my own personal set in preparation for writing about IT in Germany. In addition to Ganzhorn, *The IBM Laboratories Boeblingen*, see Albert Endres, *Die IBM Laboratorien Böblingen: System-Software-Entwicklung* (Sindelfingen: Privately printed, 2001); Albrecht Blaser, *The Heidelberg Science Center: User Oriented Informatics and Computers in Science: An Overview* (Sindelfingen: Privately printed, 2001); Helmut Painke, *Die IBM Laboraterien Böblingen: System-Entwicklung: Ein Persönlicher Rückblick* (Sindelfingen: Privately printed, 2003); Horst E. Barsuhn and Karl E. Ganzhorn, *The IBM Laboratories Boeblingen: Semiconductor and Chip Development* (Sindelfingen: Privately printed, 2005); R. Beyer et al., *IBM Informsationstechnik für Banken und Sparkassen im 20. Jahhundert* (Sindelfingen: Privately printed, 2006).

174. Flamm, *Targeting the Computer*, 167.

175. W.R. Smyser, *The Economy of United Germany: Colossus at the Crossroads* (New York: St. Martin's Press, 1992): 177–178, 194–197, 212–214; Claire Annesley, *Postindustrial Germany: Services, Technological Transformation and Knowledge in Unified Germany* (Manchester: Manchester University Press, 2004): 6–8.

176. Arik Arnold and Peter Senker, "Computer Aided Design: Europe's Role and American Technology," in Margaret Sharp (ed.), *Europe and the New Technologies: Six Case Studies in Innovation and Adjustment* (Ithaca, N.Y.: Cornell University Press, 1986): 23–27.

177. Ernst-Jürgen Horn, Henning Klodt and Christopher Saunders, "Advanced Machine Tools: Production, Diffusion and Trade," in Ibid., 50.

178. Ibid., 50–54; OECD examined the topic closely as well, OECD, *Trade in High Technology Products. An Examination of Trade Related Issues in the Machine Tool Industry*, DSTI/SPR/83. 102, OSTI/IND/83. 40 (Paris: OECD, March 22, 1984).

179. Horn, Klodt and Saunders, "Advanced Machine Tools: Production, Diffusion and Trade," 57; see also Arndt Sorge, Gert Hartman, Malcolm Warner, and Ian Nicholas, *Microelectronics and Manpower in Manufacturing: Applications of Computer Numerical Control in Great Britain and West Germany* (Aldershot, Eng.: Gower Publishing Co., 1983): 24–25.

180. E.-J. Horn, H. Klodt, and C. Saunders, "Advanced Machine Tools: Production, Diffusion and Trade," 59.

181. OECD, *ICT and Economic Growth: Evidence From OECD Countries, Industries and Firms* (Paris: OECD, 2003): 21.

182. Alessandra Colecchia and Paul Schreyer, *ICT Investment and Economic Growth in the 1990s: Is the United States a Unique Story? A Comparative Study of Nine OECD Countries* (Paris: OECD, 2001): 10.

183. E. Arnold and P. Senker, "Computer-aided Design: Europe's Role and American Technology," for CAD, 22–27, and on robots, 61–63.

184. For a review of German economic events of the 1990s, Eichengreen, *The European Economy Since 1945*, 318–328.

185. Bart van Ark, Johanna Melka, Nanno Mulder, Marcel Timmer, and Gerard Ypma, *ICT Investments and Growth Accounts for the European Union*, Research memorandum GD-56 (Groningen: Groningen Growth and Development Centre, March 2003): 59, http://www/niesr.ac.uk/research/gd56-2.pdf (last accessed 6/19/2009); Nadim Ahmad, Paul Schreyer and Anita Wölf, "ICT Investment in OECD Countries and Its Economic Impacts," in OECD, *The Economic Impact of ICT: Measurement, Evidence and Implications* (Paris: OECD, 2004): 67.

186. Susanne Schnorr-Bäcker, "The New Economy And Official Statistics," *Wirtschaft und Statististik* 3 (2001): 5.

187. U.S. Department of Commerce, *Showcase Europe*, 21–22. Trying to go global with such items as telecommunications occurred at the same time but with enormous difficulty due to global competition, especially for Deutsche Telecom, Fransman, *Telecoms in the Internet Age*, 124–157.

188. Timo Leimbach, "The SAP Story: Evolution of SAP Within the German Software Industry," *IEEE Annals of the History of Computing* 30, no. 4 (October–December 2008): 60–76; Hasso Plattner, *Anticipating Change: Secrets Behind the SAP Empire* (Rocklin, Cal.: Prima Publishing, 2000).

189. The literature on semiconductor economics is massive, but for a wonderful, yet a bit dated introduction covering the years when Europe lost its opportunity to excel in this field (1940s–70s) see, Franco Malerba, *The Semiconductor Business: The Economics of Rapid Growth and Decline* (Madison, Wis.: University of Wisconsin Press, 1985), especially pp. 6–7. See also, Richard N. Langlois and W. Edward Steinmueller, "The Evolution of Competitive Advantage in the Worldwide Semiconductor Industry, 1947–1996," in David C. Mowery and Richard R. Nelson (eds.), *Sources of Industrial Leadership: Studies of Seven Industries* (Cambridge: Cambridge University Press, 1999): 19–78.

190. Im Blickpunkt, *Informations-Gesellschaft* (Stuttgart: Metzler-Poeschel, 2002).

191. Gilbert Cette, Jacques Mairesse, and Yusuf Krocoglu, "Diffusion of ICTs and Growth of the French Economy over the Long-Term, 1980–2000," *International Productivity Monitor* 4 (Spring 2002): 28.

192. Ibid., 33, 35.

193. Gilbert Cette and Jimmy Lopez, "What Explains the ICT Diffusion Gap Between the Major Advanced countries? An Empirical Analysis," *International Productivity Monitor* 17 (Fall 2008): 28–39.

194. N. Ahmad, P. Schreyer, and A. Wölfl, "ICR Investment in OECD Countries and Its Economic Impact," 78.

CHAPTER 4

1. In particular, see Angus Maddison, *Monitoring the World Economy, 1820–1992* (Paris: OECD, 1995): 23.
2. "Population Movements in the European Union in 2000," http://europa.eu/bulletin/en/200012/p103005.htm (last accessed July 5, 2009).
3. A. Maddison, *Monitoring the World Economy*.
4. K.J. Allen and A.A. Stevenson, *An Introduction to the Italian Economy* (London: Martin Robertson, 1974): 30.
5. Ibid., 31–32.
6. Barry Eichengreen, *The European Economy Since 1945: Coordinated Capitalism and Beyond* (Princeton, N.J.: Princeton University Press, 2007): 115.
7. Between 1955 and 1971, over 9 million Italians migrated within Italy alone, providing largely inexpensive, if frequently illiterate, laborers for Italy's evolving economy, Paul Ginsberg, *A History of Contemporary Italy, 1943–1988* (London: Penguin, 1990): 2–3; see also the discussion of the social changes in James N. Cortada and James W. Cortada, *Can Democracy Survive in Western Europe?* (Westport, Conn.: Praeger, 1996): 90–101.
8. K. J. Allen and A. A. Stevenson, *An Introduction to the Italian Economy*, 8.
9. B. Eichengreen, *The European Economy Since 1945*, 129.
10. K. J. Allen and A. A. Stevenson, *An Introduction to the Italian Economy*, 11.
11. Ibid., 13–14.
12. The ability of so many European nations to control inflation in wages in the 1940s–early 1960s is considered by economic historians as a major factor that contributed to the recovery from the war and the rapid development of the continent's recovery. For an explanation of this feature of post-war European economic activities, see B. Eichengreen, *The European Economy Since 1945*, 32–35, 42–43, 269–270.
13. Note the lack of attention to the subject in economic development plans published when many other nations had aggressive strategies documented and in the process of being implemented, Ministerio del Bilancio a della Programmazione Economica, *Programma Economico Nazionale 1971–75, Allegato Secondo Programma 1966–71: Obiettivi e Risultatti* (Rome: Ministerio del Bilancio, 1972).
14. Sabrina Pastrrelli, *Quaderni dell'Ufficio Ricerche Storiche* (Rome: Banca D'Italia, December 2006): 20. This report claims, however, that a few investments were made, first in 1958, then again on a regular basis, beginning in 1966, although still small by the Istituto per la Rcostruzione Industriale (IRI), Ibid., 38. IRI was the publicly owned holding company that invested in targeted industries by supporting or owning companies, a model for economic development unique in Western Europe and a derivative from the Fascist government of pre-war Italy. The author of this study concluded that IRI was not a notable success since publicly owned companies often had to make decisions and take actions that accounted for political and social considerations that would not constrain the effectiveness of a "for profit" firm.
15. Data on IBM drawn from various unpublished records at the IBM Archives, Somers, New York.
16. Luigi Dadda, "Il Centro di Calcoli Numerici e l'introduzione della discipline informarische al Politecnico di Milano," in *Atti de Convegno Internazionale sulle Sotria e*

Preistoria del Calcolo Automatico e dell'informatica (Milan: AICA, August 1991): 7–23 and his memoirs of computing in Italy, *Ricordi di un informatico in La cultura informatica in Italia: Riflessioni e testimonianze sulle origini—1950–1970* (Fondazione Adriano Olivetti, Bollati Boringhiere, April 1993): 61–106; Giuseppe DeMarco, Giovanni Mainetto, Serena Pisani, and Pasquele Savino, "The Early Computers of Italy," *IEEE Annals of the History of Computing* 21, no. 4 (October–December 1999): 29–30; Lorenzo Soria, *Informatica un'occasione perduta. La direzione elettronica dell'Olivetti nei primi anni del centrosinistra* (Torino: G. Einaudi, 1979); S. Torrisi, "Discontinuità delle strategie di ingresso nel settore informatico," in C. Bussolati, F. Malerba, and S. Torrisi (eds.), *L'evoluzione delle industrie ad alta tecnologia in Italia: Etrata temperstina, declino e opportunità di recupero* (Bologna: Società editrice il Molino,1996): 97–133.

17. Richard R. Weber, "Plans for a Computing Machine at the Istituto Nazionale Per Le Applicazioni del Calolo," Office of Naval Research, London, Technical Report ONRL-89-53, June 29, 1953, Box 217, Mathematical Division, National Museum of American History, Smithsonian Institution.

18. G. DeMarco et al., "The Early Computers in Italy," 30–32; C. Bonfanti, "L'affaire FINAC tra Manchester e Roma (1935–1955) ed alcuni documenti inediti ad esso relativi," in *Atati del Congresso annuale dell'AICA* (Palermo: AICA, September 1994): 35–64; P. Ercoli, "From FINAC to CINAC," in *Atti del Convegno Internazionale sulla Storia e Preistoria del Calcolo Automatico e dell'informatica* (Milan: AICA, August 1991): 57–68; E.L. Aparo, "Mauro Picone e l'Istituto Nazional per le Applicazioni del Calcolo," in *Attiti del Convegno Internazionale sulla Storia e Preistoria del Calcolo Automatico e dell'informatica* (Milan: AICA, August 1991): 47–55.

19. Quoted in G. DeMarco et al, "The Early Computers in Italy," 32. This same article was published, also in English, in a slightly modified form by the same authors as "First Computers in Italy," a white paper by the Consiglio Nazionale dell Richerche (Pisa: CNUCE, 1998); Center for the Study of Electronic Computers (CSCE), University of Pisa, "Description of Electronic Digital Computer," Translated from the Italian by Dr. John Pasta, U.S. Atomic Energy Commission, and Arthur Borsei, University of California, Los Angeles, undated, circa 1953–56, Box 216, Smithsonian Institution; University of Pisa, "General Information Concerning the Center of Research on Electronic Computers (CSCE)," March 2, 1960, Box 216, Smithsonian Institution.

20. G. DeMarco et al, "The Early Computers in Italy," 32–33; Piero Maestrini, "La Calcolatrice Electtronica Pisano (CEP): Una storia che sembra una leggenda," undated paper released by the Dipartimento di Informatica, Università di Pisa, in author's posession; J.I.F. D Ker, "A Survey of New West-European Digital Computers," *Computer and Automation*, 12 (1963): 27–28, Archives of Charles Babbage Institute, University of Minnesota at Minneapolis.

21. G. DeMarco et al, "The Early Computers in Italy," 33–34; F. Filapazzi and G. Sacerdoti, "Progetto ELEA: Il Primo Computer Made in Italy," in *Atti del Convegno Internazionale sulla Storia e Preistoria del Calcolo Automatico e dell'informatica* (Milan: AICA, August 1991): 187–203.

22. G. DeMarco et al, "The Early Computers in Italy," 34. Olivetti's role in computing has not been fully studied yet; however, for some details on this new machine and other activities, see Camillo Bussolati, Franco Malerba, and Salvatore Torrisi, *L'Evoluzione Delle Industrie Ad Alta Technologia In Italia: Entrata tempestiva, decline e opportunità di recupero* (Bologna: Società Editrice Il Mulino, 1996): 20–28; Salvatore

Torrisi, "Discontinuità e Credibilità delle Stragtegie di Ingresso nel Settore Informatico," in Ibid., 102–120. On its new small computer, P.G. Perotto, "Olivetti della P101 in Avanti," in *Atti del Convegno Internazionale sulla Storia e Preistoria del Calcolo Automatico e dell'informatica* (Milan: AICA, August 1991): 218–221 and also his *Programma 101. L'invenzione del personal computer. Una storia appassionante mai raccontata* (Milan: Sperling & Kupfer, 1991). The one major work in English on the company in general is by S. Kicherer, *Olivetti: A Study of the Corporate Management of Design* (London: Trefoil Publications, 1990), but clearly this is not a full corporate biography.

23. Piero Maestrini, "La Calcolatrice Electtronica Pisano (CEP): Una storia che sembra una leggenda," undated paper released by the Dipartimento di Informatica, Università di Pisa, p. 10.
24. AUERBACH Electronics Corporation, "European Information Technology," January 15, 1961, p. 61.
25. "Summary of Latest Economic Developments in Key WTC Countries, 1971" Record Group 11, Learson Papers, "Marketing and Finance: World Trade Corporation," Box 10, Folder 10, IBM Corporate Archives, Somers, New York.
26. Ibid.
27. Drawn from a unpublished report prepared by Ken W. Sayers, "A Summary History of IBM's International Operations, 1911–2006," October 20, 2006, 260–266, IBM Corporate Archives, Somers, N.Y.
28. L. Soria, *Informatica: un'occassione perduta* (Turin: Einaudi, 1979), in which the author also expresses concern about the state of the supply side of the industry in Italy, including the lack of adequate government support, particularly for the semiconductor industry.
29. Margaret Sharp, *Europe and the New Technologies* (Ithaca, N.Y.: Cornell University Press, 1986): 63–64.
30. Graziella Fornengo, "Manufacturing Networks: Telematics in the Automotive Industry," in Cristiano Antonelli, *New Information Technology and Industrial Change: The Italian Case* (Dordrecht: Kluwer Academic Publishers, 1988): 33–56; Enzo Rullani and Antonello Zanfei, "Networks Between Manufacturing and Demand: Cases From Textile and Clothing Industries," in Ibid., 57–96, but see the entire book for other insights relevant to the discussion.
31. Stephen E. Siwek and Harold W. Furchtgott-Roth, *International Trade in Computer Software* (Westport, Conn.: QUORUM, 1993): 34–38.
32. Alessandra Coiecchia and Paul Schreyer, *ICT Investment and Economic Growth in the 1990s: Is the United States A Unique Case? A Comparative Study of Nine OECD Countries*, DSTI/DOC (2001)7 (Paris: OECD, October 25, 2001): 10.
33. Marcel P. Timmer, Gerard Ypma, and Bart van Ark, *IT in the European Union: Driving Productivity Divergence?*, Research Memorandum GD-67 (Groningen: Groningen Growth and Development Centre, University of Groningen, October 2003): 47, 51; Paul Conway, Donato de Rosa, Giuseppe Nicoletti, and Faye Steiner, *Regulation, Competition and Productivity Convergence*, Economics Department Working Papers No. 509 (Paris: OECD, September 24, 2006): 43.
34. Di A. Bassanetti, M. Iommi, C. Jona-Lasinio, and F. Zollino, *La crescita dell'economia italiana negli anni novanta tra ritardo tecnologico e rallentamento della produttività*, Temi di discussione 539 (Rome: Banca D'Italia, December 2004): 13–14.
35. Organization for Economic Co-Operation and Development, *ICT and Economic Growth: Evidence From OECD Countries, Industries and Firms* (Paris: OECD, 2003): 22–23.

36. Andrea Brandolini, "A Note on the Slowdown of the Italian Economy," p. 125 [undated, circa 2005–06], http://www.bancodemexico.com.mx (last accessed December 10, 2010).
37. Bart van Ark, "The Renewal of the Old Economy: An International Comparative Perspective," July 2001, http://www.stat.go.jp/english/info/meetings/iaos/pdf/ark.pdf (last accessed September 24, 2009): 2–3.
38. Andrea Bassanini and Stefano Scarpetta, "Growth, Technological Change, and ICT Diffusion: Recent Evidence from OECD Countries," *Oxford Review of Economic Policy* 18, no. 3 (2002): 329–330.
39. Ibid., 330.
40. For discussion of the Italian innovation experience, see Daniele Archibugi, Sergio Cesaratto, and Giorgio Sirilli, "Sources of Innovative Activities and Industrial Organization in Italy," *Research Policy* 20 (1991): 299–313; and for a more theoretical discussion, Giulio Sapelli, *Economia, tecnologia e direzione d'impresa in Italia* (Torino: Piccola Biblioteca Einaudi, 1994).
41. Cinzia Colapinto, *L'Innovazione nel settore informatico in Italia: L'attività di corporate venture capital del Gruppo Olivetti negli anni ottanta*, Working Paper No. 2006–36 (Milan: Università degli Studi de Milano, Dipartimento di Scienze Economiche, Aziendali e Statistiche, November 2006): 23.
42. Salvatore Torrisi, "Discontinuità e credibilità della strategie di ingresso nel settore informatico," in Camillo Bussolatti, Franco Malerba, and Salvatore Torrisi, *L'Evolucione delle industrie ad alta technologia in Italia: Entrata tempestiva, declino e opportunità di recupero* (Bologna: Società Editrice Il Mulino): 126–127.
43. Renato Giannetti, *Tecnologia e sviluppo economico italiano, 1870–1990* (Milan: Società editrice il Mulino, 1998): 192; Emilio Gerelli, *La Politica per L'Innovazione industriale: Problemi e Proposte* (Milan: Franco Angeli Editore, 1982): 35–37, 55, 195–201.
44. E. Gerelli, *La Politica per L'Innovazione industriale*,195–196.
45. Alessandro Capocchi, *The Role of Information Communication Technologies in the Reform Process of Italian Local Authorities* (Milan: University of Milano—Bicocca, 2008).
46. C. Bussolati, F. Malerba, and S. Torrisi, "L'evoluzione delle industrie ad alta tecnologia in Italia," 52–53.
47. Franco Malerba and Laura Pellegrini, "Entrata iniziale, declino competitivo e creazione di competenze avanzate della SGS nella microelecttronica," in C. Bussolati, F. Malerba, and S. Torrisi, *L'Evoluzione delle industrie ad alta tecnologia in Italia*, 63–95.
48. Simon Nora and Alain Minc, *L'informatisation de la société* (Paris: La Documentation Française, 1978).
49. See, for examples, Camillo Bussolati, Vittorio Chiesa, and Luca Mari, *Economia del cambiamento tecnologico* (Milan: Guerini, 1996); Camillo Bussolati, Franco Malerba and Salvsatore Torrisi, *L'evoluzione delle industrie ad alta tecnologia in Italia: Entrata tempestiva, declino e opportunità di recupero* (Bologna: il Mulino, 1996); Franco D'Egidio, *L'Economia digitale el il culture change: Come prosperare nella Nuova Economia* (Milan: Franco Angeli, 2001). For a good, early exception that focused on Italy see Assinform, *Rapporto Assinform sulla situazione dell'informatica in Italia* (Milan: Assinform, undated, circa 2000).
50. Francesco Garibaldo and Mario Bolognani, *La Società dell'informazione: Le nuove frontiere dell'informatica e delle telecommunicazioni* (Rome: Donzelli Editore, 1996): 143.

51. Ibid., 143–161.
52. U.S. Department of Commerce, *Showcase Europe: Guide to E-Commerce Markets in Europe* (Washington, D.C.: U.S. Government Printing Office, 2001): 28–30.
53. Maddison, *Monitoring the World Economy*, 23.
54. Jan L. van Zanden, *The Economic History of the Netherlands, 1914–1995: A Small Open Economy in the "Long" Twentieth Century* (London: Routledge, 1998): 3–26.
55. Ibid., 36.
56. Based on Dutch government statistics reported in Ibid., 49.
57. Jan van den Ende, *Knopen, kaarten en chips* (Voorburg/Heeerlen: Utigave Centraal Bureau voor de Statistiek, 1991): 15–60, and for early introductions of digital computing, 61–74.
58. AUERBACH Electronics Corporation, "European Information Technology," in which an analyst wrote that "contributions in the information processing field from the Netherlands have been high for the total computer activity in the country," 83; The Centrum's staff consisted of some 65 people, p. 84.
59. Richard R. Weber, "Electronic Computer Development in the Netherlands," June 12, 1953, pp. 1–3, Blackman Papers, Box 216, Smithsonian Institution, Washington, D.C.
60. Edna Kranakis, "Early Computers in the Netherlands," *CWI Quarterly* 1 *Special Issue* (1988): 61–67; Dirk de Wit, *The Shaping of Automation: A Historical Analysis of the Interaction Between Technology and Organization, 1950–1985* (Rotterdam: Erasmus Universiteit, 1994): 33; Jan van Den Ende, *The Turn of the Tide: Computerization in Dutch Society, 1900–1965* (Delft: Delft University Press, 1994): 92–93.
61. E. Kranakis, "Early Computers in the Netherlands," 67; Edsger W. Dijkstra, "A Programmer's Early Memoirs," in N. Metropolis, J. Howlett, and Gian-Carlo Rota, *A History of Computing in the Twentieth Century* (New York: Academic Press, 1980): 563–573.
62. AUERBACH Electronics Corporation, "European Information Technology," 85.
63. Quote in E. Kranakis, "Early Computers in the Netherlands," 69.
64. Ibid., 70–77; Ende, *The Turn of the Tide*, 93; Wit, *The Shaping of Automation*, 91–123.
65. E. Kranakis, "Early Computers in the Netherlands," 77–80; Ende, *The Turn of the Tide*, 93–95; D. Wit, *The Shaping of Automation*, 187.
66. A point made emphatically by E. Kranakis, "Early Computers in the Netherlands," 82.
67. Alan J. Hoffman, "New Computers in France and the Netherlands," July 25, 1957, p. 1, Box 216, Backman Papers, National Museum of American History, Mathematical Branch, Smithsonian Institution, Washington, D.C.
68. Ende, *The Turn of the Tide*, 237, and for uses of these systems, 133–217; Ende, *Knopen, kaarten en chip*, 61–92.
69. D. Wit, *The Shaping of Automation*, 372.
70. D. Kuin, "De computermarkt: strijdperk voor de grootste concerns," *Het Financieele Dagblad*, speciale editie, December 7, 1967.
71. IBM, *IBM World Trade Corporation Annual Report 1969* (Paris: IBM World Trade Corporation, 1970): 25, IBM Corporate Archives, Somers, N.Y. For a description of the attractiveness and problems of S/360 in the Netherlands, see Wit, *The Shaping of Automation*, 215–219.
72. Sayers, "History of IBM's International Operations," 321–326.
73. OECD, *Gaps in Technology: Electronic Computers* (Paris: OECD, 1969): 11.
74. Ibid., 20, also for the growth and market share statistics.

75. Ibid., 41.
76. Ibid., 128. The same report measured the value of computers installed in Europe in 1967 by vendor and concluded that IBM accounted for 68%, next Siemens with 9%, followed by Univac and Bull/GE each with 6%, and CDC and Honeywell each with 3%, all others accounted for only 5%. By the same unit of measure, IBM had 61% of the Dutch market and Bull/GE 12%, followed by Univac at 8%, and the local Philips-Electrologica firm with only 5%, the same share as CDC at the time, Ibid., 164.
77. Banks became important users of IT and their role has recently begun to receive attention of historians, Joke Mooij, "Rabobank: An Innovative Dutch Bank, 1945–2000," in Bernardo Bátiz-Lazo, J. Carles Maixé-Altés and Paul Thomes, *Technological Innovation in Retail Finance: International Historical Perspectives* (London: Routledge, 2011): 173–193.
78. D. Wit, *The Shaping of Automation*, 247–248.
79. Ibid., 260. Defining those social and institutional restraints on the adoption of computing is the great contribution of this important book.
80. Ende, *Knopen, kaarten en chip*, 93–112.
81. U.S. *International Trade Administration, Computers and Peripheral Equipment, The Netherlands* (Washington, D.C.: U.S. Department of Commerce, 1982).
82. On how the Dutch government invested in R&D, Dirk de Vos, *Governments and Microelectronics: The European Experience* (Ottawa, Canada: Science Council of Canada, March 1983): 92–107.
83. Brian M. Murphy, *The International Politics of New Information Technology* (New York: St. Martin's, 1986): 105–108.
84. For details, see Alfonso H. Molina, "Sociotechnical Constituencies as Processes of Alignment: The Rise of a Large-Scale European Information Technology Initiative," *Technology in Society* 17, no. 4 (1995): 385–412.
85. This story is very authoritatively told by Jan van den Ende, Nachoem Wijnberg, and Albert Meijer, "The Influence of Dutch and EU Government Policies on Philips' Information Technology Product Strategy," in Richard Coopey (ed.), *Information Technology Policy: An International History* (Oxford: Oxford University Press, 2004): 187–208.
86. Hanna, Guy and Arnold, *The Diffusion of Information Technology*, 152.
87. Paul Conway, Donato de Rosa, Giuseppe Nicolli, and Faye Steiner, *Regulation, Competition and Productivity Convergence*, OECD Economics Department Working Papers No. 509, ECO/WKP(2006).37 (Paris: OECD, September 4, 2006): 43.
88. Sayers, "History of IBM's International Operations," 326–327.
89. Bart van Ark, *The Productivity Problem of the Dutch Economy: Implications for Economic and Social Policies and Business Strategy*, Research Memorandum GD-66 (Groningen: Groningen Growth and Development Centre, September 2003): 2, http://www.e-biblioteka.It/resursai/ES/memorandumi/gd66.pdf (last accessed September 6, 2009). His bibliography cites much of the economic literature on Dutch productivity and ICT.
90. Ibid., 1–21.
91. Frank C. Veraart, "Vormgevers van Personalijk Computergebruick De ontwikkeling van computers voor kleingebruikers in Nederland, 1970–1990" (Unpublished Ph.D. dissertation, Technische Universiteit Eindhoven, 2008).
92. All material in this paragraph drawn from Eli Skogerbø and Tanja Storsul, "Prospects for Expanded Universal Service in Europe: The Cases of Denmark, the Netherlands, and Norway," *The Information Society* 16 (2000): 135–146.

93. Henry van der Wiel, "ICT Important for Growth," 2002, http://www.Cpb.nl/nl/pub/cpbreeksen/cpbreport/2000_2/s2_.pdf (last accessed September 6, 2009).

94. Three Dutch observers were more blunt, "after the early 1980s, EU and Dutch policy did not affect the computer industry very much," Ende, Wijnberg, and Meijer, "The Influence of Dutch and EU Government Policies on Philips' Information Technology Product Strategy," 204.

95. U.S. Department of Commerce, *Showcase Europe: Guide to E-Commerce Markets in Europe* (Washington, D.C.: U.S. Government Printing Office, 2001): 32–34.

96. Each nation publishes this data on a regular basis through numerous EU, OECD, UN publications and through their national government websites and portals. Europeans tend to be more precise in their definitions of the area, in which Nordics refers to a formal organization of Denmark Norway, Sweden, Finland and Iceland and a few other smaller regions, while Scandinavia is viewed as a subset of the Nordics, the large countries.

97. See for example, Lars Heide, "Punched-Card and Computer Applications in Denmark, 1911–1970," *History and Technology* 11 (1994): 77–99.

98. Matti Virén and Markku Malkamäki, "The Nordic Countries," in Benn Steil, David G. Victor and Richsard R. Nelson (eds.), *Technology Innovation and Economic Performance* (Princeton, N.J.: Princeton University Press, 2002): 200–226.

99. A. Maddison, *Monitoring the World Economy*, 23.

100. Subhash Thakur, Michael Keen, Balázs Horváth Cerra, *Sweden's Welfare State: Can the Bumblebee Keep Flying?* (Washington, D.C.: International Monetary Fund, 2003): 1.

101. Various data drawn from OECD Online Statistics.

102. Ibid.

103. Thomas Kaiserfeld, "Computerizing the Swedish Welfare State: The Middle Way of Technological Success and Failure," *Technology and Culture* 37, no. 2 (April 1996): 252.

104. Thakur et al., *Sweden's Welfare State*, 8–20.

105. Explained in Ibid., 26–28.

106. Lars Magnusson, *An Economic History of Sweden* (New York: Routledge, 2000): 200.

107. Ibid., 203–205.

108. Ibid., 214–215.

109. Ibid., 215–216.

110. Nordic countries are unique in Europe in having had high levels of literacy dating back to the eighteenth century. In the case of Sweden, about 90 percent of its citizens were literate to some degree by 1850. One byproduct of that literacy was a commitment to education which stood the nation well in the twentieth century when its economy required a literate workforce with skills different from those of prior centuries. For more details on Sweden's literacy, Henri-Jean Martin, *The History and Power of Writing* (Chicago, Ill.: University of Chicago Press, 1988): 341–343.

111. Ibid., 260–266.

112. Computer Branch, Mathematical Sciences Division, *A Survey of Large-Scale Digital Computers and Computer Projects* (Washington, D.C.: Office of Naval Research, Department of the Navy, 1949): 22, Box 2, National American History Museum, Mathematical Branch, Smithsonian Institution, Washington, D.C.

113. Magnus Johansson, "Early Analog Computers in Sweden—With Examples From Chalmers University of Technology and the Swedish Aerospace Industry," *IEEE*

Annals of the History of Computing 18, no. 4 (1996): 27–33; see also, Hans de Geer, *På väg till datasamhället. Datatekniken I politiken, 1946–1963* (Stockholm: Royal School of Technology, 1992); Richard R. Weber, "A Survey of the Swedish Computing Machine Development," October 31, 1952, Blachman Papers, Box 217, National American History Museum, Mathematical Branch, Smithsonian Institution, Washington, D.C.

114. Per Lundin, Niklas Stenlås, and Johan Gribbe (eds.), *Science for Welfare and Warfare: Technology and State Initiative in Cold War Sweden* (Sagamore Beach, Mass.: Watson Publishing International, 2010): 31 and Tom Petersson, "Private and Public Interests in the Development of the Early Swedish Computer Industry: Facit, Saab and the Struggle for National Dominance," Ibid., 109–129.

115. Matematikmaskinnämnden (Stokholm: MMN, Undated).

116. M. Johannson, "Early Analog Computers," 28.

117. BARK Binary Arithmetic Relay Calculator) became operational in April 1950 and remained in use until September 1954, by which time BESK was online. For details on this machine, see Göran Kjellberg and Gösta Neovius, "The BARK: A Swedish General Purpose Relay Computer," *Mathematical Tables and Other Aids to Computation* (January 1951): 29–34; Carlsson, "On the Politics of Failure," in Bubenko, Jr. et al., *History of Nordic Computing*, footnote 5, p. 96.

118. Paul M. Marcus, "Swedish Automatic Relay Computer," July 20, 1950, Blachman Papers, Box 217, National American History Museum, Mathematical Branch, Smithsonian Institution, Washington, D.C.

119. Peter Klüver, "From Research Institute to Computer Company: Regnecentralen, 1946–1964," *IEEE Annals of the History of Computing* 21, no. 2 (1999): 31–43; Tom Petersson, "Facit and the 'BESK Boys': Sweden's Computer Industry, 1956–1962," Ibid., 27, no. 4 (2005): 23–30; Hans E. Andersin, "The Role of IBM in Starting Up Computing in the Nordic Countries," in Janis Bubenko, Jr., John Impagliazzo, and Arne Sølvberg (eds.), *History of Nordic Computing: IFIP WG9.7 Working Conference on the History of Nordic Computing (HiNC1), June 16–18, 2003, Trondheim, Norway* (New York: Springer, 2005): 42; Erik Bruhn, "Nordic Cooperation Within the Field of Computing," Ibid., 159–177.

120. First quote, William E. Hubert, "The Work of the Swedish Board for Computing Machinery," November 4, 1953, Blachman Papers, Box 217, National American History Museum, Mathematical Branch, Smithsonian Institution, Washington, D.C.; second quote, Nelson M. Blachman, "Some Automatic Digital Computers in Western Europe," *IRE Transactions on Electronic Computers* EC-5, no. 3 (September 1956): 160.

121. The centers of this interest since the early 1960s, and that continues to the present, were the Royal Institute of Technology and the Department for Computer and Systems Sciences (DSV) at the Stockholm University. Their combined history has been documented in Janis Bubenko, Jr., Carl Gustaf Jansson, Anita Kollerbaur, Tomas Ohlin, and Louise Yngström, *ICT For People: 40 Years of Academic Development in Stockholm* (Stockholm: DSV, 2006).

122. Tom Petersson, "Private and Public Interests in the Development of the Early Swedish Computer Industry," Paper No. 122, CESIS Electronic Working Papers Series (Uppsala: Uppsala University, March 2008): 9–10, http://cesis.abe.kth.se/documents/WP122.pdf (last accessed August 8, 2009); BESK was used for 1,200 computing problems between 1953 and 1956, a substantial number of projects for its day, AUERBACH Electronics Corporation, "European Information Technology," 87–91.

123. There are no formal histories of Saab's computer operations in English; however, a series of publications were published in Swedish, discussed in English in footnote 13 page 10, by Per Lundin, "Documenting the Use of Computers in Swedish Society between 1950 and 1980."

124. Andersin, "The Role of IBM in Starting Up Computing in the Nordic Countries," 42; Sayers, "History of IBM's International Operations," 414–417.

125. Petersson, "Private and Public Interests in the Development of the Early Swedish Computer Industry"; Tom Petersson, Facit and the BESK Boys: Sweden's Computer Industry (1956–1962)," *IEEE Annals of the History of Computing* 27, no. 4 (October–December 2005): 23–30; Private correspondence from Lars Arsenius with author, January 11, 2010; Tord Jöran Hallberg, *IT-Gryning: Svenk datahistoria frän 1840-till 1960-talet* (Pozcal, Poland: Studentlitteratur, 2007): 151–178.

126. Petersson, "Private and Public Interests in the Development of the Early Swedish Computer Industry," 18. On public policy, the early Swedish industry, and the rise of IBM in Sweden, Jan Annerstedt, Lars Forssberg, Sten Henriksson, and Kenneth Nilsson, *Datorer och politik: Studier I en ny tekniks politiska effekter på det svenska samhället.* (Lund: University of Lund Press, 1970); Hans De Geer, *På väg till datasamhället* (Stockholm: University of Stockholm, 1992); Kent Lindkvist, *Dataeknik och politik: Datapoliten I Sverige 1945–1982* (Lund: Lund University Press, 1984); Per Lundin, *Designing Democracy: The UTOPIA Project and the Role of Labour Movement in Technological Change, 1981–1986* (Stockholm: University of Stockholm, 2005).

127. Harold (Bud) Lawson, "The Datasaab Flexible Central Processing Unit," in Bubenko, Jr. et al, *History of Nordic Computing*, 190–195 and includes references to the specific technology; Pet-Arne Persson, "Transformation of the Analog: The Case of the Saab BT 33 Artillery Fire Control Simulator and the Introduction of the Digital Computer as Control Technology," *IEEE Annals of the History of Computing* 21, no. 2 (1999): 52–64.

128. Magnus Johansson, "Big Blue Gets Beaten: The Technological and Political Controversy of the First Large Swedish Computerization Project in a Rhetoric of Technology Perspective," *IEEE Annals of the History of Computing* 21, no. 2 (April–June 1999): 14–30.

129. Per Vingaard Klüver, "Technology Transfer, Modernization, and the Welfare State," in *History of Nordic Computing*, 61–66; Magnus Johansson, *Smart, Fast and Beautiful: On Rhetoric of Technology and Computing Discourse in Sweden, 1955–1995* (Linköping: Linköping University, 1997): 63–136.

130. Based on official statistics, reproduced in a table in Johansson, "Big Blue Gets Beaten," 15.

131. For an important discussion of some of the uses to which Swedish government put computers, see L. Ingelstam and I. Palmlund, "Computers and People in the Welfare State: Information Technology and the Social Security in Sweden," *Informatization and the Public Sector* 1, no. 1 (1991): 5–20.

132. Statskontoret, *Statliga Dataorer 1981: State-Managed Computers* (Stockholm: Statskontoret, 1981), 7.

133. Janus Bubenko, Jr., "Annotations of First Generation Systems Development in Sweden," in John Impagliazzo, Timo Järva, and Petri Paju (eds.), *History of Nordic Computing 2: Second IFIP WG 9.7 Conference, HiNC2 Turku, Finland, August 21–23, 2007 Revised Selected Papers* (Berlin: Springer, 2009): 101.

134. See, for example, Johansson, "Big Blue Gets Beaten," 14–30. In fairness to this author, he concentrated the majority of his article on IBM-government affairs in

the 1960s. In subsequent decades, the Swedish government acquired IBM products.

135. Sayers, "History of IBM's International Operations," 414–416.

136. "Annual Report," IBM Svenska Aktiebolog, February 24, 1965, RG 6: Legal Records/Country Files/Sweden, "Stockholders Meetings Minutes, 1960–1975," File 6, Box 88, IBM Corporate Archives, Somers, N.Y.

137. IBM EAHQ Market Planning and Research, "Industry Sales Report (Area Consolidations) Situation as of End of November 1964," RG 6: WTC Marketing and Advertising/Marketing/Market Planning and Research, File 16, Box 19, IBM Corporate Archives, Somers, N.Y.

138. "Annual Report," IBM Svenska Aktiebolog, February 24, 1965, RG 6: Legal Records/Country Files/Sweden, "Stockholders Meetings Minutes, 1960–1975," File 6, Box 88, IBM Corporate Archives, Somers, N.Y.

139. Various market analysis reports conducted by IBM, all located in RG 6: WTC Marketing and Advertising/Marketing/Market Planning and Research, File 16, Box 19, IBM Corporate Archives, Somers, N.Y. The 75 percent figure was also mentioned in the minutes of the Swedish IBM company's minutes of a board meeting held on June 15, 1964, entitled "Minutes No. 46 (1954–1964), RG 6, "Legal Records/Country Files/Sweden—Directors Meetings Minutes, 1955-64," File 4, Box 87.

140. E-mail from Lars Arosenius to author, January 10, 2010.

141. Ibid.

142. Ibid.

143. Ingemar Dahlstrand, "The Development of University Computing in Sweden, 1965–1985," in Impagliazzo et al., *History of Nordic Computing 2*, 130–137; Bubenko, Jr. et al., *ICT for People*, 405–413.

144. Janis Bubenko, Jr., "Information Processing—Administrative Data Processing: The First Courses at KTH and SU, 1966–67," in Impagliazzo et al., *History of Nordic Computing 2*, 138–148.

145. T. Kaiserfeld, "Computering the Swedish Welfare State, 249–279.

146. Franklin D. Kramer and John C. Cittadino, *Sweden's Use of Commercial Information Technology for Military Applications* (Washington, D.C.: National Defense University, 2005), a short overview of 8 pages.

147. Sten Henriksson, "When Computers Became of Interest in Politics," in Bubenko, Jr. et al., *History of Nordic Computing*, 416.

148. Erik Arnold and Peter Senker, "Computer Aided Design," in Margaret Sharp (ed.), *Europe and the New Technologies: Six Case Studies in Innovation and Adjustment* (Ithaca, N.Y.: Cornell University Press, 1986): 32–35.

149. The narrative for CAD relies entirely on Erik Arnold and Peter Senker, "Computer-Aided Design: Europe's Role and American Technology," in Margaret Sharp (ed.), *Europe and the New Technologies: Six Case Studies in Innovation and Adjustment* (Ithaca, N.Y.: Cornell University Press, 1986): 32–35. These authors include coverage of developments in all the Nordic countries, 1960s-early 1980s.

150. The combined telecommunications and computer industry began to occupy a position of occupational prominence by the 1970s when about 4.2 percent of the nation's work forces were in these industries. They remained a stable component of the employment scene, growing to 4.7 percent of he total only by 1999. For details see Dan Johansson, *The Dynamics of Firm and Industry Growth: The Swedish Computing sand Communications Industry* (Stockholm: Department of the Organization and Management, KTH, 2001), and for context, Gunnar Eliasson,

The Macroeconomic Effects of Computer and Communications Technology: Entering a New and Immediate Economy Stockholm: The Royal Institute of Technology, 2002): 44–61.

151. The story of ICT diffusion in the 1980s and 1990s draws heavily from Thakur et al., *Sweden's Welfare State*, 29–34.

152. Nagy Hanna, Ken Guy, and Erik Arnold, *The Diffusion of Information Technology: Experience of Industrial Countries and Lessons for Developing Countries* (Washington, D.C.: World Bank, 1995): 153.

153. Ibid.

154. Dirk de Vos, *Governments and Microelectronics: The European Experience* (Ottawa, Canada: Science Council of Canada, March 1983), for quote, 81, for export statistic 82.

155. This diffusion of decision-making is described in H.J. van Houten, *The Competitive Strength of the Information and Communication Industry in Europe* (The Hague: Martinus Nijhoff Publishers, 1983): 101–109, which argues that so many agencies slowed the ability of the government to better coordinate its policies. This was the same problem evident with so many agencies doing the same in the United States.

156. Ibid., 85.

157. For example, Swedish National Board for Technical Development, *Technology for the Future: STU Perspectives 1979—Ideas and Groundwork for the Emphasis and Planning of Technological Research in Sweden* (Stockholm: Swedish National Board for Technical Development, 1979).

158. Data are from Sweden's national bank, Sveriges Riksbank, Villy Bergström, "Can We Be Best Again? The Role of Capital Formation in Long-Term Growth," *Sveriges Riksbank Economic Review* 2 (2004): 17–18.

159. W.C. Boeschoten and G.E. Hobbink, "Electronic Money, Currency Demand and Seignoragie Loss in the G10 Countries," DNB Staff Reports, No. 1 (Amsterdam: De Nederlandsche Bank, 1996): 9.

160. The data are drawn from Table 12.3 in Eichengreen, *The European Economy Since 1945*, 404.

161. Ibid., 404–405; on the 20 percent greater cost of IT, see Martin Neil Baily and Jacob Funk Kirkegaard, *Transforming the European Economy* (Washington, D.C.: Institute for International Economics, 2004).

162. Björn Andersson and Martin Ådahl, "The 'New Economy' and Productivity in Sweden in the 2000s," *Economic Review* 1 (2005): 48–72, especially pp. 58–59; Magnusson, *An Economic History of Sweden*, 168.

163. For many examples of the arguments, see various papers in *History of Nordic Computing, History of Nordic Computing 2*.

164. Sten Henriksson, "When Computers Became of Interest in Politics," in Bubenko, Jr. et al., *History of Nordic Computing*, 419.

165. The project to collect these oral histories has been documented by Per Lundin, "Documenting the Use of Computers in Swedish Society between 1950 and 1980," supported by the National Museum of Science and Technology in Stockholm. The Website for the project is http://www.tekniskamuseet.se/1/192.html (in Swedish) (last accessed January 26, 1010). Lundin's report (in English) can be found at http://tekniskamuseet.se/download/18.4d755928124885167d48000869/Documenting_the_Use_of_Computers_in_Swedish_Society.pdf (last accessed January 26, 2010

166. Scholars and a large group of veterans of the IT community have begun to shift their attention to what users of computers did, which is how the 100+ interviews

came about. It remains for the historians now to catch up to provide a more rounded view of Swedish IT activities.

167. U.S. Department of Commerce, *Showcase Europe*, 41–43.
168. A primary finding of my work on the United States, James W. Cortada, *The Digital Hand*, 3 vols (New York: Oxford University Press, 2004–2008).
169. M. Hilson, *The Nordic Model*, 84–86.
170. Matteo Bugamelli and Patrizio Pagano, "Barriers to Investment in ICT," *Applied Economics* 36 (2004): 2275–2286.
171. It appears from extant, yet extensive evidence, that these lessons were first learned in the United States, but were quickly learned by international companies and their smaller business partners. For an introduction to the organizational lesson, see two books edited by Jerry N. Luftman (ed.), *Competing in the Information Age: Strategic Alignment in Practice* (New York: Oxford University Press, 1996) and *Competing in the Information Age: Align in the Sand* (New York: Oxford University Press, 2003, 2nd ed.). For the lesson about redesigning work processes to optimize the use of IT, while many publications appeared describing how to do that, the one which most influenced managers around the world was written by Thomas H. Davenport, *Process Innovation: Reengineering Work through Information Technology* (Boston, Mass.: Harvard Business School Press, 1993).
172. If one wanted to pick a time or event that opened the new era of process reengineering in general, it was when executives all over the world read and acted upon the ideas about process redesign with extensive use of ICT in a book by James Champy and Michael Hammer, *Reengineering the Corporation: A Manifesto for Business Revolution* (New York: HarperBusiness, 1993), while Davenport's book, *Process Innovation*, less radical, gave more specific examples of how best to implement the kinds of ideas introduced in the Champy-Hammer volume.
173. Olof Zaring and C. Magnus Eriksson, "The Dynamics of Rapid Industrial Growth: Evidence from Sweden's Information Technology Industry, 1990–2004," *Industrial and Corporate Change* 18, no. 3 (2009): 507–528.

CHAPTER 5

1. The size machine he describes had less computing power and space to store data than a credit card-sized calculator does today, indeed vastly less power even by the standards of the time. The author was an early computer technologist who worked at the Computer Center at the Norwegian Technical University. If we leave aside the physical size of the machine, and just measure its computing and memory capacities, he may have described the smallest computer ever built! For the quote, Norman Sanders, "Making Computing Available," in Janis Bubenko, Jr., John Impagliazzo, and Arne Sølvberg (eds.), *History of Nordic Computing* (New York: Springer, 2005): 318.
2. The role of mathematics in the conceptualization of the power and form of computing (hardware and software) in European history has yet to be fully acknowledged or studied, but received a large boost in that direction with the publication of a recent study on the origins of French computing which emphasized the important influence of mathematics on the subject, Pierre E. Mounier-Kuhn, *L'informatique en France de la seconde guerre mondiale au Plan Calcul: L'emergence d'une science* (Paris: Presses de l'université Paris-Sorbonne, 2010). Physics is important as it provided basic knowledge underpinning research on semiconductors and electronics in the five decades preceding the development of computers in the 1940s.

3. By the 1970s, the conversation had become a fine art, see, for example, Rodney L. Roenfeldt and Robert A. Fleck, Jr., "How Much Does a Computer Really Cost?," *Computer Decisions* (November 1976): 77–78; first book-length examination of the issue, and that sold around the world, was James W. Cortada, *EDP Costs and Charges: Finance, Budgets, and Cost Control in Data Processing* (Englewood Cliffs, N.J.: Prentice-Hall, 1980).

4. This is actually a serious problem. In preparing this book, I read hundreds of economic reports, which almost universally lamented the lack of adequate data sets recording the role of information technology's effects on national economies and industries around the world. For introductions to some of the current issues facing economists, see Erik Brynjolfsson and Brian Kahin (eds.), *Understanding the Digital Economy: Data, Tools, and Research* (Cambridge, Mass.: MIT Press, 2000); Jack E. Triplett and Barry P. Bosworth, *Productivity in the U.S. Services Sector: New Sources of Economic Growth* (Washington, D.C.: Brookings Institution Press, 2004); Graham Tanaka, *Digital Deflation: The Productivity Revolution and How It Will Ignite the Economy* (New York: McGraw-Hill, 2004): 1–30; Carol Corrado, John Haltiwanger, and Daniel Sichel (eds.), *Measuring Capital in the New Economy* (Chicago, Ill.: University of Chicago Press, 2005); also the extensive bibliography in Dale W. Jorgenson (ed.), *Econometrics*, vol. 3: *Economic Growth in the Information Age* (Cambridge, Mass.: MIT Press, 2002): 421–454. The issue concerning IT was also entangled in knowledge work, a discussion dating back to the 1950s, and summarized with sample texts of the debate in James W. Cortada (ed.), *Rise of the Knowledge Worker* (Boston, Mass.: Butterworth-Heinemann, 1998). For a discussion of what is needed in the way of data and metrics written when the world was deep into Wave Two IT, see Erik Brynjolfsson and Adam Saunders, *Wired for Innovation: How Information Technology Is Reshaping the Economy* (Cambridge, Mass.: MIT Press, 2010): 117–128.

5. The standard work on diffusions of all kinds remains Everett M. Rogers, *Diffusion of Innovations*, 5th ed. (New York: Free Press, 2003); see, in particular, his discussion about decision making and rates of diffusion measures, 219–239, economic considerations are almost completely absent from his book.

6. To be sure when a technology and its artifacts (products) are young and not stabilized, rapid churn in their forms are to be expected. Churn also represents good marketing strategies in many instances allowing a vendor to be the "first" with a new product. If a company can afford to be ahead of its rivals in bringing out new products, which of course displace earnings that might have come from continued sale of older products, there are economic market-gaining benefits, as IBM enjoyed repeatedly between the 1950s and the mid-1980s through its very aggressive replacement of older offerings with newer ones. It was a well understood phenomenon at IBM, one still followed, albeit without the same success as came with the S/360, for example.

7. Organization for Economic Co-Operation and Development, see *Gaps in Technology: Electronic Computers* (Paris: OECD, 1969): 74.

8. For a study set in a later period that is one of the very few on behavior, see G. Battisti and P. Stoneman, "Inter- and Intra-firm Effects in the Diffusion of New Process Technology," *Research Policy* 32 (2003): 1641–1655, but also, Chris Forman and Avi Goldfarb, "Diffusion of Information Technology and Communication Technologies to Business," in Terrence Hendershott (ed.), *Handbooks in Information Systems*, vol. 1, *Economics and Information Systems* (Amsterdam: Elsevier, 2006): 1–52, which also includes a superb bibliography on the subject.

9. AUERBACH Electronics Corporation, *European Information Technology: A Report On The Industry And The State Of The Art* (Philadelphia: AUERBACH Electronics Corporation, 1961).
10. In many of the unpublished American embassy reports of the 1950s cited in chapters three and four, there is attached a distribution list of who was to receive copies. They all included academics in important American universities, the military of course, security agencies, and scientists and engineers in other national government agencies, such as at national laboratories, often a circulation to over 100 individuals, and rarely to less than a couple of dozen, Blackman Papers, Boxes 216 and 217, National American History Museum, Mathematical Branch, Smithsonian Institution, Washington, D.C.
11. Herman H. Goldstine, *The Computer from Pascal to von Neumann* (Princeton, N.J.: Princeton University Press, 1972): 349–362.
12. Gisela Hürlimann, *Die Eisenbahn der Zukunft: Automatisierung, Schnelverkehr und Modernisierung bei den SBB 1955 bis 2005* (Zurich: CHRONOS, 2007). She also comments briefly on pan-European developments in transportation as well.
13. Composite population figures for all Europe are hard to come by, although both the OECD and national economic agencies began tracking such professions in the late 1960s and early 1970s.
14. OECD, *Information Activities, Electronics and Telecommunications Technologies: Impact on Employment, Growth and Trade*, vol. 1 (Paris: OECD, 1981): 13.
15. Ibid.
16. Martti Tienari (ed.), *Tietotekniik. An alkuvuodet Suomessa* (Helsinki: Suomen Atk-Kustannus Oy, 1993) explores the early years of information technology in Finland and on the role of IBM and its customers, see Pentti Anttila, *Big Blue Suomessa.o.y. International Business Machines A.b. 1936–1996* (Saolo: The Author, 1997).
17. OECD, *Trends in the Information Economy* (Paris: OECD, 1986): 9.
18. These became a core source for my three volume history of the use of IT in the United States, James W. Cortada, *The Digital Hand*, 3 vols (New York: Oxford University Press, 2004–2008).
19. E-mail from Ulf Hashagen to author, June 9, 2009. Hashagen was in the process of writing a book on scientific computing in Germany and thus had become familiar with the materials. I examined copies of the bibliography at the library at the IBM laboratory in Zurich later that month.
20. H.J. van der Aa (ed.s), *International Computer Bibliography: A Guide to Books on the Use, Application and Effect of Computers in Scientific, Commercial, Industrial and Social Environments*, 2 vols (Amsterdam: Studiecentrum voor Informatica, 1968–1971).
21. FUNDP NAMUR, *Belgian Computing Literature* (Brussels: FUNDP NAMUR, circa 2008).
22. H. Stachowiak, *Denken und Erkennen im kybernetischen Modell* (Wien: Springer-Verlag, 1965, 1969): 245–261.
23. I have made a serious attempt to acquire some of these on a country-by-country basis since most are not available in the United States.
24. Various statistics are culled from several charts in Alfonso H. Molina, "Technology Collaboration: 1992 and Beyond," in Gareth Locksley (ed.), *The Single European Market and the Information and Communication Technologies* (London: Belhaven Press, 1990): 230–232.
25. J. Lesourne, R. Lebon, K. Oshima, and T. Takushita, *Europe and Japan Facing High Technology. From Conflict to Cooperation?* (Paris: Euro-Japan Project on High Technology, 1988).

26. Alfred D. Chandler, Jr., *Inventing the Electronic Century: The Epic Story of the Consumer Electronics and Computer Industries* (Cambridge, Mass.: Harvard University Press, 2005): 180.

27. Vassiliko N. Koutrakou, *Technological Collaboration for Europe's Survival: The Information Technology Research Programmes of the 1980s* (Aldershot: Avebury, 1995): 5–23; William R. Nester, *European Power and the Japanese Challenge* (New York: New York University Press, 1993): 191–254.

28. The value of prior experience with a technology, mentioned in chapters one and two, has not received as much attention from historians of IT, as from economists. For an economic perspective, see L.G. Tornatsky and M. Fleischer, *The Processes of Technological Innovation* (Lexington, Mass.: Lexington Books, 1990); on the costs of switching technologies, P. Chen and C. Forman, "Can Vendors Influence Switching Costs and Compatibility in an Environment with Open Standards?," *MIS Quarterly* 30 (2006): 541–562; S. Greenstein, "Did Installed Base Give an Incumbent Any (Measurable) Advantages in Federal Computer Procurement?," *RAND Journal of Economics* 24 (1993): 19–39.

29. How dominant IBM was in the American market has been the subject of an enormous debate that began in the 1960s, extended through the 1970s as one of the central issues in the U.S. Government's 12 year antitrust suit against the company, and that continued for years afterwards in the academic literature. A group of economists hired by IBM in the 1970s to help it define share and then argue that it was smaller than claimed by the government wrote two books which are excellent discussions of the economic issues involved in defining market share. In the course of those deliberations, sympathetic to IBM, they presented an enormous amount of data about the construct of the U.S. computer industry, Franklin M. Fisher, James W. McKie, and Richard B. Mancke, *IBM and the U.S. Data Processing Industry: An Economic History* (New York: Praeger, 1983) and Franklin M. Fisher, John J. McGowan and Joen E. Greenwood, *Folded, Spindled, and Mutilated: Economic Analysis and U.S. v. IBM* (Cambridge, Mass.: MIT Press, 1983). A journalist wrote a very different history extremely hostile to IBM, that also discussed market share issues, Paul Carroll, *Big Blues: The Unmaking of IBM* (New York: Crown Publishers, 1993).

30. In addition to making these points, these two retired IBM executives described how sales forces worked with clients all over the world, useful accounts of how knowledge transfer occurred between IBM and its many customers, Buck Rodgers, *The IBM Way: Insights into the World's Most Successful Marketing Organization* (New York: Harper & Row, 1986); Jacques Maisonrouge, *Inside IBM: A Personal Story* (New York: McGraw-Hill, 1985). For a French perspective that provides a more detailed and balanced account, focusing on IBM in France, see Peter Halbherr, *IBM: Mythe et Réalité: La vie quotidienne chez IBM France* (Lausanne, Switzerland: FAVRE, 1987).

31. "We consistently outsold people who had better technology because we knew how to put the story before the customer, how to install the machines successfully, and how to hang on to customers once we had them," Thomas J. Watson, Jr. and Peter Petre, *Father Son & Co.: My Life at IBM and Beyond* (New York: Bantam, 1990): 242.

32. A.K. Watson to T.J. Watson, Jr., December 13, 1967, Folder 1, RG 11: Employees. The T.J. Watson, Jr. Papers/Administrative Records/personnel/Watson, A.K., Box 229, IBM Corporate Archives, Somers, New York.

33. A. D. Chandler, *Inventing the Electronic Century*, 82–131, 177–189; Emerson W. Pugh, *Building IBM: Shaping an Industry and Its Technology* (Cambridge, Mass.: MIT Press, 1995): 263–300.

34. The bibliography on IBM is extensive; however, key, useful—current—studies to consult include Pugh, *Building IBM*, and Kevin Maney, *The Maverick and His Machine* (New York: Wiley, 2003).

35. For a detailed and well informed description from the 1970s see, Rex Malik, *And Tomorrow . . . The World? Inside IBM* (London: Milligan, 1975): 159–473; the only book on World Trade is the short work by Nancy Foy, *The Sun Never Sets on IBM: The Culture and Folklore of IBM World Trade* (New York: William Morrow, 1975).

36. Watson and Petre, *Father Son & Co.*, 176.

37. My view of how corporations worked in the IT industry is influenced profoundly in particular by two books written by Alfred D. Chandler, *The Visible Hand: The Managerial Revolution in American Business* (Cambridge, Mass.: Harvard University Press, 1977) and *Scale and Scope: The Dynamics of Industrial Capitalism* (Casmbridge, Mass.: Harvard University Press, 1990).

38. Pugh, *Building IBM*, endnote 32, p. 380.

39. Ibid., 258; on the complexity of the accounting, finance, and trade aspects, Malik, *And Tomorrow . . . The World?*, 299–332.

40. Maney, *The Maverick and His Machine*, first quote, p. 378–379, second quote, p. 379.

41. These findings are not new; in fact, they were being made by European observers at the time, but obviously with little influence on public policies with respect to supporting national champions. Jean-Michel Treille, for example documented the rapid penetration of IBM into the European market, *L'Économie mondial de L'Ordinateur* (Paris: Éditions du Seuil, 1973): 54–55.

42. I recently described the global IT public policy phenomenon, James W. Cortada, *How Societies Embrace Information Technology* (New York: Wiley, 2010): 27–127; for a largely European centered discussion by an economist, Vernon W. Ruttan, *Technology, Growth, and Development: An Induced Innovation Perspective* (New York: Oxford University Press, 2001); on Asia, Masahiko Aoki, Hyung-Ki Kim, and Masahiro Okuno-Fujiwara (eds.), *The Role of Government in East Asian Economic Development: Comparative Institutional Analysis* (Oxford: Clarendon Press, 1996).

43. Nagy Hanna, Ken Guy, and Erik Arnold, *The Diffusion of Information Technology: Experience of Industrial Countries and Lessons for Developing Countries* (Washington, D.C.: World Bank, 1995): 39–41.

44. Rebecca Marschan-Piekkart, Stuart Macdonald, and Dimitris Assimakopoulos, "In Bed With a Stranger; Finding Partners for Collaboration in the European Information Technology Programme," *Science and Public Policy* 28, no. 1 (February 2001): 68–78.

45. Paul Jowett and Margaret Rothwell, *The Economics of Information Technology* (New York: St. Martin's Press, 1986): 5.

46. Manuel Castells, *The Informational City* (Oxford: Blackwell, 1989) and *The Information Age*, 3 vols (Oxford: Blackwell, 1996–1998).

47. Jowett and Rothwell, *The Economics of Information Technology*, 5–8.

48. For quote and discussion of the larger issue of Europe's economic security, Klaus W. Grewlich, "Technology: The Basis of European Security," *German Foreign Affairs Review* 32, no. 3 (1981).

49. Quoted in Dirk de Vos, *Governments and Microelectronics: The European Experience* (Ottawa, Can: Science Council of Canada, 1983): 11, but extracted from a report highly influential on European policy makers, Commission of the European Communities, *European Society Faced with the Challenge of New Information Technologies: A Community Response* (Brussels: Commission of the European Communities, 1979).

50. Which so informed many commentators on national economic development, for example, many of the contributors to Hendershott, *Economics and Information Systems*, written in the early 2000s.

51. Jowett and Rothwell, *The Economics of Information Technology*, 71–86 is particularly useful in explaining why European firms formed alliances; for samples of the debate, see E. Arnold and K. Guy, *Parallel Convergence: National Strategies in Information Technologies* (London: Frances Printer, 1986); M. Hobday, *Trends in the Diffusion of Application Specific Integrated Circuits* (Sussex: University of Sussex, 1988) which has considerable material on a Philips-Siemens joint project of the 1980s; European Economic Community, *L'espace scientifique et technologique European dans le contexte international. Resources et conditions de la competitivité de la communauté* (Brussels: EEC, 1988) which includes in its policy debates a great deal of information on the structure, status and characteristics of the West European information processing industry of the 1980s when it was under competitive duress; Commission of the European Communities, *Esprit '88: Putting Technology to Use*, Part 2 (Amsterdam: North-Holland, 1988) describes the most extensive of the pan-European projects just launching at the time; European Economic Community, *European Community Policies for Semiconductors* (Brussels: EEC, 1989) which offers an analysis of the multiple R&D initiatives around Europe and the results achieved so far; OECD, *Government Policies and the Diffusion of Microelectronics* (Paris: OECD, 1989) describes the changing nature of European government support for R&D in the late 1980s; OECD, *The Changing Role of Government Research Laboratories* (Paris: OECD, 1989) describes the rethinking these institutions were undertaking in the late 1980s, with special attention to semiconductors and robotics; M. Sharp and P. Holmes (eds.), *Strategies for New Technologies* (New York: Phillip Allan, 1989) is an example of the debate underway; but more importantly D. Foray, M. Gibbons, and G. Ferné, *Major R&D Programmes for Information Technologies* (Paris: OECD, 1989) which reviews government policies and makes recommendations regarding European R&D strategies, reflecting the extensive churn governments experienced with their IT strategies in the 1980s; but see also the collection of essays on the period in Michael S. Steinberg, *The Technical Challenges and Opportunities of a United Europe* (Savage, Md.: Barnes & Noble Books, 1990); European Economic Community, *L'industrie Européene de l'electronique et de l'informatique* (Brussels: EEC, 1991) provides a data-rich snapshot of the IT industry within the EEC and the context in which it functioned. For historical perspectives, Franco Malebra's analysis of how product innovation was done in Western Europe for IT during the 1980s, at the height of Wave One, "The Organization of the Innovative Process," in Nathan Rosenberg, Ralph Landou, and David C. Mowery (eds.), *Technology and the Wealth of Nations* (Stanford, Cal.: Stanford University Press, 1992): 247–278; see also the extremely useful and well researched chapter by Dimitris Assimakopoulos, Rebecca Marschan-Piekkari, and Stuart MacDonald, "ESPRIT: Europe's Response to U.S. and Japanese Domination of Information Technology," in Richard Coopey (ed.), *Information Technology Policy: An International Perspective* (New York: Oxford University Press, 2004): 247–263.

52. Koutrakou, *Technological Collaboration for Europe's Survival*, 5–22.

53. For an excellent account of these collaborative initiatives, see Jowett and Rothwell, *The Economics of Information Technology*, 71–86.

54. Arthe Van Laer, "Developing an EC Computer Policy, 1965–1974," *IEEE Annals of the History of Computing* 32, no. 1 (January–March 2010): 44–59.

55. Christopher Layton, *European Advanced Technology: A Programme for Integration* (London: Allen & Unwin, 1969) is an early example of the debate beginning in the 1960s; see also Commission of the European Communities, *Industrial Policy in the Community: Memorandum from the Commission to the Council* (Brussels: CEC, 1970).

56. Eda Kranakis, "Politics, Business, and European Information Technology Policy: From the Treaty of Rome to Unidata, 1958–1975," in Coopey, *Information Technology Policy*, 209–246.

57. Americans held the same mistaken view of Japanese automotive manufacturers in the United States in the 1980s–2000s, because well over 95 percent of all employees making and selling Japanese cars in the USA were American citizens. Additionally, the vast majority of the revenue these companies generated in the USA remained in the United States.

58. Ibid., 239.

59. Most observers subsequently concluded that this was the fundamental strategic error; that the markets for these two items had been conquered permanently by the Americans and Japanese and instead, Europeans should have turned their attention to other possibilities. For an example of this logic, see Steinberg, *The Technical Challenges and Opportunities of a United Europe*, 133–136.

60. Assimakopoulos, Marschan-Piekkara, and MacDonald, "ESPRIT: Europe's Response to U.S. and Japanese Domination in Information Technology," in Coopey, *Information Technology Policy*, 247–263.

61. Quoted in Andrew J. Pierre (ed.), *A High Technology Gap? Europe, America and Japan* (New York: Council on Foreign Relations, 1987): 6.

62. Assimakopoulos, Marschan-Piekkara, and MacDonald, "ESPRIT: Europe's Response to U.S. and Japanese Domination in Information Technology," 247–263.

63. Cost of a particular machine or software was—and is—less of a factor in the expense of IT because in every decade the expense of staffs to install and operate the technology always expanded year-over-year both in number required and for their salaries and benefits. When looking at the total cost of equipment, software, and of a project, users normally took all these personnel expenses into account in arriving at what they thought something would cost.

64. Sharp, *Europe and the New Technologies*, 264.

65. Ibid., 281.

66. Gannon, *Trojan Horses and National Champions*, 333.

67. Scholars have started to explore the role of rhetorical features of the conversation. For example, there is the case study of Sweden by Magnus Johansson, *Smart, Fast and Beasutiful: On Rhetoric of Technology and Computing Discourse in Sweden, 1955–1995* (Linköping: Linköping University, 1997).

68. Atlantic Institute for International Affairs, *The Impact of Technological Change in the Industrial Democracies: Public Attitudes Toward Information Technology* (Paris: Atlantic Institute for International Affairs, 1985): 4–6, 10–15.

69. Ibid., 1–24.

70. As recently as 2011, it took an American journal to bring together a discussion of European computing and even then, the contributors largely focused their individual discussions on their own countries. See, for example, two issues of the *IEEE Annals of the History of Computing*, 33, no. 1 (January–March 2011), no. 3 (July–September 2011).

71. James W. Cortada, *How Societies Embrace Information Technology* (New York: Wiley & Sons, 2010): 129–157.

72. Khuong Vu, "Measuring the Impact of ICT Investments on Economic Growth," Program on Technology and Economic Policy, Harvard Kennedy School of Government, undated, circa 2000, quote p. 16; all data in this paragraph is from this paper.
73. Explained in William J. Baumol, Sue Anne Batey Blackman, and Edward N. Wolff, *Productivity and American Leadership: The Long View* (Cambridge, Mass.: MIT Press, 1989): 85–102.
74. Dirk Pilat and Anita Wolfl, "ICT Production and ICT Use: What Role in Aggregate Productivity Growth?," in *The Economic Impact of ICT—Measurement, Evidence, and Implications* (Paris: OECD, 2004): 85–104.
75. Francesco Caselli and W. John Coleman II, "Cross-Country Technology Diffusion: The Case of Computers," *American Economic Review* 91, no. 2 (May 2001): 328–335.
76. Cortada, *How Societies Embrace Information Technology*, 129–157.
77. Vu, "Measuring the Impact of ICT Investments on Economic Growth," Program on Technology and Economic Policy, 27.
78. James W. Cortada, Ashisha M. Gupta, and Marc Le Noir, *How Nations Thrive in the Information Age: Leveraging Information and Communications Technologies for National Economic Development* (Somers, N.Y.: IBM Corporation, 2007) and their, *How Rapidly Advancing Nations Can Prosper in the Information Age: Leveraging Information and Communications Technologies for National Economic Development* (Somers, N.Y.: IBM Corporation, 2007); Kenneth L. Kraemer, Jason Dedrick, and Nigel P. Melville, "Globalization and National Diversity: e-Commerce Diffusion and Impacts Across Nations," in Kenneth L. Kraemer, Jason Dedrick, Nigel P. Melville, and Kevin Zhu (eds.), *Global E-Commerce: Impacts of National Environment and Policy* (Cambridge: Cambridge University Press, 2006): 13–61.
79. Jonas A. Eriksson and Martin Ådahl, "Is There a 'New Economy' and Is It Coming to Europe?," *Economic Review* 1 (2000): 22–67.
80. Ibid., 61.
81. Barry Eichengreen, *The European Economy Since 1945: Coordinated Capitalism and Beyond* (Princeton, N.J.: Princeton University Press, 2007): 8.
82. Spain is a well documented case of the economic impact of Cold War politics, Samuel Chavkin, Jack Sangster and William Susman, *Spain: Implications for United States Foreign Policy* (Stamford, Conn.: Greylock Publishers, 1976): 31–52; R. Richard Rubottom and J. Carter Murphy, *Spain and the United States Since World War II* (New York: Praeger, 1984): 34–76.
83. Explored extensively by James W. Cortada, *Before the Computer: IBM, NCR, Burroughs, and Remington Rand and The Industry They Created, 1865–1956* (Princeton, N.J.: Princeton University Press, 1993) and Lars Heide, *Punched-Card Systems and the Early Information Explosion, 1880–1945* (Baltimore, Md.: Johns Hopkins University Press, 2009).
84. This has been the subject of a rapidly growing body of literature because the use of computers by the Allies during World War II is one of the biggest—and most important—revelations about events of that war to appear since the late 1970s. American and European exchange of information about codebreaking predated World War II, Stephen Budiansky, *Battle of Wits: The Complete Story of Codebreaking in World War II* (New York: Free Press, 2000).

CHAPTER 6

1. Seymour E. Goodman, "Computing and the Development of the Soviet Economy," in Joint Economic Committee, Congress of the United States, *Soviet Economy in a Time of Change* (Washington, D.C.: U.S. Government Printing Office, 1979), vol. 1, p. 524.

2. Slava Gerovitch, *From Newspeak to Cyberspeak: A History of Soviet Cybernetics* (Cambridge, Mass.: MIT Press, 2002): xi. For an earlier review of the discussion that focuses less on cybernetics and more on the views of various Soviet elite communities, see Marcia Anne Weigle, "Technology and Society: Ideological Implications of Information and Computer Technologies in the Soviet Union" (Unpublished Ph.D. dissertation, University of Notre Dame, 1988).

3. Holland Hunter, "Soviet Economic Problems and Alternative Policy Responses," *Soviet Economy in a Time of Change*, vol. 1 p. 37.

4. For a typical example, Vladimir V. Bitunov, "Economic Planning and Management," in *Soviet Economy Today* (Westport, Conn.: Greenwood Press, 1981): 63–76.

5. Vladislav M. Zubok, *A Failed Empire: The Soviet Union in the Cold War From Stalin to Gorbachev* (Chapel Hill, N.C.: University of North Carolina Press, 2009): xxiii.

6. I picked ten because in the West, a normal mainframe computer system would include at a minimum about nine peripheral pieces of equipment, such as a printer, at least one or two punch-card devices, one to two tape drives, often with a stand-alone control unit for these, one to two disk drives, and if connected to a network, at least one standalone telecommunications control unit. At a large data center, there might be several systems printers and dozens of tape and disk drives. Typically, when a data center manager complained of not having enough peripheral equipment he or she usually was complaining about the lack of data storage devices, most notably tape drives in the early years of computing, and later disk drives. For a non-technical discussion of what constituted a "configuration" of a system in the West, circa early 1980s, see James W. Cortada, *Managing DP Hardware: Capacity Planning, Cost Justification, Availability, and Energy Management* (Englewood Cliffs, N.J.: Prentice-Hall, 1983): 4–15.

7. *Status of Digital Computer and Data Processing Development in the Soviet Union*, ONR Symposium Report, ACR-37, and also reported under the same title in *Communications of the ACM*, 2, no. 6 (1958): 8.

8. On early Russian publications, see Eugene Gros, *Russian Books on Automation and Computers* (London: Scientific Information Consultants, 1967).

9. W.H. Ware (ed.), "Soviet Computer Technology—1959," Unpublished paper from the Rand Corporation, Santa Monica, March 1, 1960, quote, p. 47, and on number of machines built, p. 48; but see rest of report for details the trip, Box 217, Computer Documentation Collection, Computers and Mathematics Collection, Museum of American History, Smithsonian Institute; hereafter materials from this site will be listed by box number, Smithsonian.

10. Frank Cain, "Computers and the Cold War: United States Restrictions on the Export of Computers to the Soviet Union and Communist China," *Journal of Contemporary History* 40, no. 1 (2005): 131–147. Note, that the Americans even restricted the export of advanced technology to Western Europe as well, as we saw in the French case in the 1960s, James W. Cortada, "Public Policies and the Development of National Computer Industries in Britain, France, and the Soviet Union, 1940–1980," Ibid., 44, no. 3 ((2009): 493–512; on the 1980s, Daniel L. Burghart, "Technology Transfer, Export Control, and Economic Restructuring in the Soviet Union: The Case of Soviet Computers" (Unpublished Ph.D. dissertation, University of Surrey, 1992).

11. The number changed over time. For example, Yugoslavia broke its alliance with the Soviet Union in 1948, and Albania did the same in 1960. Other parts of Eastern Europe came into the Soviet Union in the 1920s: Byelorussia, Ukraine, Transcaucasian state; and into its sphere of influence in 1939, eastern Poland, Latvia, Estonia, Finland, and parts of Romania and Lithuania.

12. The last Soviet census before the collapse of the USSR was conducted in 1989, John Dunlop, Marc Rubin, Lee Schwartz, and David Zaslow, "Profiles of the Newly Independent States: Economic, Social, and Demographic Conditions," in Richard F. Kaufman and John P. Hardt (eds.), *The Former Soviet Union in Transition* (Armonk, N.Y.: M.E. Sharpe, 1993): 1025, 1030; David Lane, *Soviet Economy and Society* (Oxford: Basil Blackwell, 1985): 51. One should be careful not to confuse this data with the population of Russia, which was only one state within the larger USSR, and that had a population of about 146 million at that time, and that declined during the 1990s.

13. Barry Eichengreen, The *European Economy Since 1945: Corporate Capitalism and Beyond* (Princeton, N.J.: Princeton University Press, 2007): 131–133.

14. Soviet Union comprised of Armenia, Azerbaijan, Belarus, Estonia, Georgia, Kazakhstan, Kyrgyzstan, Latvia, Lithuania, Moldova, Russia, Tajikistan, Turkmenistan, Ukraine, and Uzbekistan.

15. See, for example, "History of Computer Developments in Romania," *IEEE Annals of the History of Computing* 21, no. 3 (1999): 58–60; Friedrich W. Kistermann, "Leo Wenzel Pollak (1888–1964): Czechoslovakian Pioneer in Scientific Data Processing," Ibid., 21, no. 4 (1999): 62–68; G.K. Stolyarov, "Computers in Belarus: Chronology of the Main Events," Ibid., 21, no. 3 (1999): 61–65; Zsuzsa Szentgyörgyi, "A Short History of Computing in Hungary," Ibid., 21, no. 3 (1999): 49–57; Laimutis Telksnys and Antanas Zilinskas, "Computers in Lithuania," Ibid., 21, no. 3 (1999): 31–37; G. Banse, C.J. Langenbach, and P. Machleidt (eds.), *Towards the Information Society: The Case of Control and Eastern European Countries* (New York: Springer, 2000); David H. Kraus, Pranas Zunde, and Vladimir Siamecka, *National Science Information System: A Guide to Science Information Systems in Bulgaria, Czechoslovakia, Hungary, Poland, Romania and Yugoslavia* (Cambridge, Mass.: MIT Press, 1972); Erich Sobeslavsky and Nikolaus Joachim Lehmann, *Zur Geschichte von Rechentechnik und Datenverarbeitung in der DDR 1946–1968* (Dresden: Tillichbau der Technischen Universität Dresden, 1996).

16. Others have commented on Russian (not necessarily Soviet) ways of working and that probably influenced how Westerners saw the USSR. For an example from a U.S. military participant, see David A. Wellman, *A Chip in the Curtain: Computer Technology in the Soviet Union* (Washington, D.C.: National Defense University Press, 1989): 21–70.

17. Eichengreen, *The European Economy Since 1945*, 133–145; David A. Dyker, *Restructuring the Soviet Economy* (London: Routledge, 1992): 6–8, 19; Lane, *Soviet Economy and Society*, 3–49.

18. A comment your author has heard repeatedly over the past 35 years about doing business in Russia from those attempting to do so was the lack of clarity about such things as who had to approve projects and contracts, who would pay for goods and services, and about the lack of any ability to enforce a signed contract. Since the arrival of the Putin era these problems of opacity have been compounded by the fact that the Russian government often uses trade and commercial activities in support of foreign policy objectives, such as denying the supply of oil and natural gas to customers in the West and to its ex-Soviet neighbors despite contractual obligations to do so.

19. The technology was also insufficiently advanced to do the job either, but technical considerations were less influential on the issue than political ones.

20. Lane, *Soviet Economy and Society*, 8.

21. Ibid., 9.

22. Eichnengreen, *The European Economy Since 1945*, 141.
23. Ibid., 144–146.
24. For examples, see Dyker, *Restructuring the Soviet Economy*, 55–78; Henry S. Rowen, "The Soviet Economy," *Proceedings of the Academy of Political Science* 35, no. 3 (1984): 32–48; R. W. Davies, "Economic Planning in the USSR," in Morris Bornstein (ed.), *The Soviet Economy: Continuity and Change* (Boulder, Col.: Westview Press, 1981): 7–35; Holland Hunter, "Soviet Economic Problems and Alternative Policy Responses," in Michael Ellman and Vladimir Kontorovich (eds.), *The Disintegration of the Soviet Economic System* (London: Routledge, 1992): 345–355; and not to be overlooked are the many anthologies of expert witness papers and testimony made at the request of the Joint Economic Committee of the Congress of the United States all during the 1950s through the 1980s of which one of the most useful was one very large 2 volume collection, *Soviet Economy in a Time of Change*.
25. Joseph S. Berliner, "The Prospects for Technological Progress," in Bornstein, *The Soviet Economy*, 299–300.
26. Ashok Deo Bardhan and Cynthia A. Kroll, "Competitiveness and an Emerging Sector: The Russian Software Industry and Its Global Linkages," *Industry and Innovation* 13, no. 1 (March 2006): 78.
27. Anne Fitzpatrick and Boris N. Mallinovsky, "The MESM and the Monastery," *IEEE Annals of the History of Computing* 24, no. 2 (April–June 2002): 91–93.
28. When one refers to a computer as a "production" system, this means it is a device, or collection of equipment, that can be built (manufactured) in a standard format in some quantity. Systems that are experimental, also known as "pone-of-a-kind" precede models which can be manufactured. All systems of the 1940s and many from the very early 1950s, fit this characterization. One occasion, these independently built early systems might be made in multiple copies, but typically they were not fully duplicates of each other and were made in a craft-like manner, rather than following a more formal mass production style of fabrication. Only with production systems could a firm, agency, or economy build enough systems fast enough for one to start talking about deployment of IT in a society.
29. For biographical information on Lebedev in Russian, V.M. Glushkov et al., *Sergei Alekseevich Lebedev* (Kiev: Naukova Durmka, 1978). On BESM, see G.G. Ryalov (ed.), *Ot BESM do superEVM* [From BESM to Supercomputer] (Moscow: Institute of Precision Mechanics and Computer Technology, 1988); but see also for BESM and other early Soviet computers, technical descriptions, draft report "Digital Computers in Eastern Europe," Box 217, Smithsonian, later published in a revised fashion as Alston Hourscholder, "Digital Computers in Eastern Europe," Computers and Automation 4, no. 12 (December 1955): 8, which also reported that some 500 individuals worked on the design and construction of this system.
30. Segei P. Prokhorov, "Computers in Russia: Science, Education, and Industry," *IEEE Annals of the History of Computing* 21, no. 3 (1999): 4–15.
31. For a detailed, early bibliography of publications available in the USSR, see John W. Carr III, Alan J. Perlis, James E. Robertson, and Norman R. Scott, "A Visit to Computation Centers in the Soviet Union," *Communications of the ACM* 2, no. 3 (1959): 8–20.
32. Seymour E. Goodman, "Computing and the Development of the Soviet Economy," in U.S. Congress, Joint Economic Committee, *A Compendium of Papers Submitted to the Joint Economic Committee Congress of the United States* (Washington, D.C.: U.S. Government Printing Office, 1979), vol. 1 p. 526; a variation of this paper

was published as "Soviet Computing and Technology Transfer: An Overview," *World Politics* 31, no. 4 (July 1979): 539–570.

33. The subject of a number of excellent studies by Slava Gerovitch, *From Newspeak to Cyberspeak*, 8–21, 105–197, 279–292; "'Mathematical Machines' of the Cold War: Soviet Computing, American Cybernetics and Ideological Disputes in the Early 1950s," *Social Studies of Science* 31, no. 2 (April 2001): 253–287; "'Russian Scandals': Soviet Readings of American Cybernetics in the Early Years of the Cold War," *The Russian Review* 60 (October 2001): 545–568.

34. Gerovitch, "'Mathematical Machines'," 272.

35. Prokhorov, "Computers in Russia," 7–10.

36. A number of Soviet memoirs about the period are shedding light on technological and managerial activities across several decades. On the earliest period (1953–63), there is Simon Berkovich, "Reminiscences of Superconductive Associative Memory Research in the Former Soviet Union," *IEEE Annals of the History of Computing* 25, no. 1 (January–March, 2003): 72–75; on the years 1965–95, see the important memoirs of Stanislav V. Klimenko, "Computer Science in Russia: A Personal View," Ibid., 21, no. 3 (1991): 16–30; and for a analysis to put their memoirs in context, see Boris Malinovokoy and Lev Malinovsky, "Information Technology Policy in the USSR and Ukraine: Achievements and Failures," in Richard Coopey (ed.), *Information Technology Policy: An International Perspective* (New York: Oxford University Press, 2004): 304–319, but see also Peter Wolcott, "Soviet Advanced Technology: The Case of High-Performance Computing" (Unpublished Ph.D. dissertation, University of Arizona, 1993) which takes the discussion from the BESM-class systems through projects of the late 1980s.

37. Ibid.; Gerovitch, *From Newspeak to Cyberspeak*; but also see his "InterNyet: Why the Soviet Union Did Not Build a Nationwide Computer Network," *History and Technology* 24, no. 4 (December 2008): 335–350.

38. Gerovitch, "InterNyet," 341–347.

39. Ibid., 342–344.

40. Ibid., 344 based on Soviet sources; Kathryn M. Bartol, "Soviet Computer Centres: Network or Tangle?" *Soviet Studies* 23, no. 4 (April 1972): 608–618.

41. Constantin A. Krylov, *The Soviet Economy* (Lexington, Mass.: Lexington Books, 1979): 74.

42. Prokhorov, "Computers in Russia," 10.

43. A.G. Dale, "Database Management Systems Development in the USSR," *Computing Surveys* 11, no. 3 (September 1979): 213–226, and includes a good bibliography of Russian sources; Thomas J. Richards, "Examination of Issues Concerning ADP Technology Transfer to the Soviet Union Using the Control Data Corporation Cyber 76 Situation As A Case Study" (Unpublished Ph.D. Dissertation, George Washington University, 1981) is a reminder that imports were not limited to IBM's products.

44. N.C. Davis and Seymour E. Goodman, "The Soviet Bloc's Unified System of Computers," *Computing Surveys* 10. no. 2 (June 1978): 93–122, still the best source on this topic.

45. Wade B. Holland, "Comments on an Article by M.E. Rakovskiy," *Soviet Cybernetics Review* (November 1971): 33.

46. McHenry, "The Absorption of Computerized Management Information Systems in Soviet Enterprises (Hardware, Software, Social Impact)," 225–227.

47. Davis and Goodman, "The Soviet Bloc's Unified System of Computers," 114.

48. Ibid., 115.

49. For an excellent description of these Soviet computers, how they compared to IBM's and the consequences of copying them, see Richard W. Judy, "The Riad Computers of the Soviet Union and Eastern Europe, 1970–1985: A Survey and Comparative Analysis" (Indianapolis, Ind.: Hudson Institute, March 1986), copy in author's possession. I am grateful to Richard W. Judy for giving me a copy of this report in 2010.
50. Ibid., 52.
51. Ibid., 99.
52. Ibid., 140.
53. Ibid., 116.
54. To see an example of how one of these sub-industries worked in the USSR, see James Grant, "Soviet Machine Tools: Lagging Technology and Rising Imports," U.S. Congress, *Soviet Economy in a Time of Change*, vol. 1, pp. 554–580.
55. On computing in education from a Soviet source, see Bryce F. Zinder (ed. and translator), *Computers and Education in the Soviet Union* (Englewood Cliffs, N.J.: Educational Technology Publications, 1975).
56. Goodman, "Computing and the Development of the Soviet Economy," vol. 1, pp. 531–535. The number of computer centers in universities were quite small, between 17 and 19 (1969–72), Louvan E. Nolting and Murray Feshback, "R&D Employment in the USSR—Definitions, Statistics, and Comparisons," U.S. Congress, *Soviet Economy in a Time of Change*, vol. 1, p.720.
57. Ibid., 537.
58. Information available to the author who worked at that IBM plant site in 1981–83.
59. "World Trade Story: Slide Update, March 19, 1974," World Trade Corporation, Annual Reports, Slide Presentations, Box 4, Folder 4–9 R-6., "WRC/Slide Presentations," IBM Archives, Somers, N.Y.
60. Kenneth Tasky, "Soviet Technology Gap and Dependence on the West: The Case of Computers," in *Soviet Economy in a Time of Change*, vol. 1, p. 512, quote, 514.
61. Goodman, "Computing and the Development of the Soviet Economy," 537.
62. I explained this process in considerable detail in James W. Cortada, *The Digital Hand*, 3 vols (New York: Oxford University Press, 2004–2008); see, in particular, the first volume which describes how IT was used in manufacturing, the sector that most attracted Soviet applications in the 1960s and 1970s for their production industries.
63. Martin Cave, *Computers and Economic Planning* (Cambridge: Cambridge University Press, 1980): 6–12; William Keith McHenry, "The Absorption of Computerized Management Information Systems in Soviet Enterprises," (Unpublished Ph.D. dissertation, University of Arizona, 1985): 16–23.
64. William K. McHenry and Seymour E. Goodman, "MIS in Soviet Industrial Enterprises: The Limits of Reform From Above," *Communications of the ACM* 29, no. 11 (November 1986): 1036.
65. Ibid., 1034–1043; Cave, *Computers and Economic Planning*, 10–12; William K. McHenry, "The Absorption of Computerized Management Information Systems in Soviet Enterprises" (Unpublished Ph.D. dissertation, University of Arizona, 1985): 31–72, 529–537.
66. On surveys, Cave, *Computers and Economic Planning*, 155–157.
67. McHenry, "The Absorption of Computerized Management Information Systems in Soviet Enterprises," 218.
68. Ibid., 215.

69. "Automated Production Management Systems," *Problems of Economics* (October 1971): 61–70.
70. Cave, *Computers and Economic Planning*, 11.
71. Ibid., 179.
72. McHenry and Goodman, "MIS in Soviet Industrial Enterprises: The Limits of Reform From Above," 1035.
73. McHenry, "The Integration of Management Information Systems in Soviet Enterprises," in U.S. Congress, *Gorbachev's Economic Plans*, vol. 2, *Study Papers Submitted to the Joint Economic Committee, Congress of the United States* (Washington, D.C.: U.S. Government Printing Office, 1987): 186.
74. Ibid.
75. McHenry, "The Absorption of Computerized Management Information Systems in Soviet Enterprises." However, for the most thorough discussions available in English on Soviet IT of the 1980s, the key works are two lengthy articles by Richard W. Judy and Robert W. Clough, "Soviet Computers in the 1980s: A Review of the Hardware," *Advances in Computers* 29 (1989): 251–323 and "Soviet Computing in the 1980s: A Survey of the Software and Its Applications," Ibid., 30 (1990): 223–306.
76. Seymour E. Goodman, "Computing and the Development of the Soviet Economy," in *Soviet Economy in a Time of Change*, vol. 1, 540.
77. Ibid., for an example of this view, 543.
78. Cave, *Computers and Economic Planning*, 119, 175–178.
79. McHenry and Goodman, "MIS in Soviet Industrial Enterprises: The Limits of Reform From Above," 1035.
80. Ibid.
81. McHenry, "The Integration of Management Information Systems in Soviet Enterprises," 199.
82. One of the most thorough of these studies, remarkably well informed given that it was done in the early 1970s, nonetheless, reflected ongoing realities right into the 1980s, Gertrude E. Schroeder, "The 'Reform' of the Supply Chain System in Soviet Industry," *Soviet Studies* 24, no. 1 (July 1972): 97–119.
83. Zubok, *A Failed Empire*, 298.
84. Paul R. Josephson, "Science and Technology as Panacea in Gorbachev's Russia," in Loren Graham et al. (eds.), *Technology, Culture, and Development: The Experience of the Soviet Model* (Armonk, N.Y.: M.E. Sharpe, 1992): 25–62.
85. Anders Åslund, *Russia's Capitalist Revolution: Why Market Reform Succeeded and Democracy Failed* (Washington, D.C.: Peterson Institute for International Economics, 2007): 31, and for an excellent analysis of economic conditions in the second half of the 1980s, 11–42, 67–73.
86. Ibid., 11–84.
87. Zubok, *A Failed Empire*, 322.
88. Ibid., 322.
89. George Holliday, "Technology and Science Policy," in *Gorbachev's Economic Plans*, vol. 2, 141.
90. Richard W. Judy, "The Soviet Information Revolution: Some Prospects and Comparisons," Ibid., 161–175; Peter B. Nyren, "The Computer Literacy Program: Problems and Prospects," Ibid., 200–210.
91. Holliday, "Technology and Science Policy," Ibid., 142.
92. Perhaps the most informed and articulate Western observers of Soviet technology in the 1980s was Seymour B. Goodman. Seminal articles by him include, "The

Prospective Impacts of Computing: Selected Economic-Industrial-Strategic Issues," Ibid., 176–184; "Technology Transfer and the Development of the Soviet Computer Industry," in Bruce Parrott (ed.), *Trade, Technology, and Soviet-American Relations* (Bloomington, In: Indiana University Press, 1985): 117–140; "The Partial Integration of the CEMA Computer Industries: An Overview," in John P. Hardt and Richard F. Kaufman (eds.), *East European Economies: Slow Growth in the 1980s*, vol. 2, *Foreign Trade and International Finance*, Joint Economic Committee, Congress of the U.S. (Washington, D.C.: U.S. Government Printing Office, 1986): 329–354.

93. Goodman, "The Prospective Impacts of Computing: Selected Economic-Industrial-Strategic Issues," 180.

94. For a bibliography of earlier commentators, see Cave, *Computers and Economic Planning*, 204–220.

95. Paul Cocks, "Soviet Science and Technology Strategy: Borrowing From the Defense Sector," in *Gorbachev's Economic Plans*, vol. 2, pp. 152–153.

96. Åslund, *Russia's Capitalist Revolution*, 52–82.

97. Nyren, "The Computer Literacy Program: Problems and Prospects," 200.

98. Ibid.

99. Loren R. Graham, "Science and Computers in Soviet Society," *Proceedings of the Academy of Political Science* 35, no. 3 (1984), *The Soviet Union in the 1980s*, 129–130.

100. Ibid., 130.

101. Åslund, *Russia's Capitalist Revolution*, 17.

102. Ibid., 131.

103. Hans Heymann, Jr., "Commentary," in *Gorbachev's Economic Plans*, vol. 2, p. 211.

104. Ibid; Morris H. Crawford, *Communications Networks for Finance and Trade in the USSR and Eastern Europe* (Cambridge, Mass.: Program on Information Resources Policy, Harvard University, 1991).

105. D. J. Peterson, *Russia and the Information Revolution* (Santa Monica, Cal.: RAND Security Research Division, 2005): xi–xvi, 8–9.

106. David A. Dyker, *Restructuring the Soviet Economy* (London: Routledge, 1992): 171–217.

107. Åslund, *Russia's Capitalist Revolution*, 284.

108. Raj M. Desai and Itzhak Goldberg (eds.), *Can Russia Compete?* (Washington, D.C.: Brookings Institution Press, 2008): 1–12.

109. William C. Boesman, "Science and Technology in the Former Soviet Uion: Capabilities and Needs," in Kaufman and Hardt, *The Former Soviet Union in Transition*, 610–628; Graham, "Big Science in the Last Years of the Soviet Union," 49–71.

110. Gertrude E. Schroeder, "Post-Soviet Economic Reforms in Perspective," in Kaufman and Hardt, *The Former Soviet Union in Transition*, 57–80; Nicholas Forte and Shelley Deutch, "Defense Conversion in the Former USSR: The Challenge Facing Plant Managers," Ibid., 730–735.

111. Seymour.E. Goodman and William K. McHenry, "The Soviet Computer Industry: A Tale of Two Sectors," *Communications of the ACM* 34, no. 6 (June 1991): 25–29.

112. Ken W. Sayer, "A Summary History of IBM's International Operations, 1911–2006" (Somers, N.Y., 2006): 368–369, IBM Archives, Somers, N.Y.

113. A modem in the 1980s was a physical component in a computer that made it possible for it to access a telephone or other communication line and transmit and receive data from other computers in other locations. When PCs were first produced in the late 1970s, they did not have this capability, but by mid-decade they had built-in modems.

114. Rafal Rohozinski, "Mapping Russian Cyberspace: Perspectives on Democracy and the Net," UNRISD Discussion Paper No. 115, (New York: United Nations Research Institute for Social Development, October 1999): 5–6.

115. Ibid., 6.

116. Ibid., 12–14.

117. Ibid., 22.

118. William McHenry and Leonid Malkov, "The Russian Federation's Y2K Policy: Too Little, Too Late?" *Communications of the Association for Information Systems* 2, Article 10 (August 1999): 7, 11–15; also includes a rich bibliography on Soviet computing and Soviet Y2K published in the 1990s, 31–38.

119. Andrey A. Terekhov, "The Russian Software Industry," *IEEE Software* (November/December 2001): 98–101.

120. Ibid., 99.

121. James W. Cortada, Ashish M. Gupta, and Marc Le Noir, *How Nations Thrive in the Information Age: Leveraging Information and Communications Technologies for National Economic Development* (Somers, N.Y.: IBM Corporation, 2007); Desai and Goldberg, *Can Russia Compete?*, 44; Bart van Ark and Marcin Piatkovski, "Productivity, Innovation, and Investment Climate Surveys in Old and New Europe," *International Economics and Economics Policy* 1, nos. 2–3 (2004): 215–246; World Bank, *Russian Federation: Country Partnership Strategy* (Washington, D.C.: World Bank, 2006).

122. One could argue that parts of East Asia did it in three, demonstrating that emerging economies can leapfrog more established technological regimes in deployment, a topic discussed more fully in subsequent chapters. Periodically, the World Bank publishes useful comparative data on diffusion and investments in IT, including for Russia. See, for example, World Bank, *2006 Information and Communications for Development: Global Trends and Policies* (Washington, D.C.: World Bank, 2006): 257 for Russia in 2000 and 2004, which indicated that expenditures on ICT as a percent of GDP was 3.5 percent in 2000 and 3.7 percent for 2004.

123. Even down to the debates about Marxism and cybernetics, Steffen Werner, *Kybernetik Statt Marx?* (Stuttgart: Verlag Bonn Aktuell GMBH, 1977): 20–43.

124. Friedrich Naumann and Gabriele Schade (eds.), *Informatik in der DDR—eine Bilanz* (Bonn: Gesellschaft für Informatik, 2006) and in particular within this volume, see Gerhard Merkel, "Computerentwicklungen in der DDR—Rahmenbedingungen und Ergebnisse," pp. 40–70; Birgit Demuth (ed.), *Informatik in der DDR—Grundlagen und Anwendungen* (Bonn: Gesellschaft für Informatik, 2008), which focuses largely on the many technological developments underway in East Germany; Wolfgang Coy and Peter Schirmbacher (eds.), *Informatik in der DDR: Tagung Berlin 2010* (Berlin: Tagungsband, 2010), which includes numerous case studies of use. Specifically on universities and related institutes and that also provides a new account of the supply side of events as well there is now the excellent survey by Christine Pieper, *Hochschulinformatik in der Bundesrepublik und der DDR bis 1989/1990* (Stuttgart: Franz Steiner Verlag, 2009).

125. Even down to the debates about Marxism and cybernetics, Steffen Werner, *Kybernetik statt Marx?* (Stuttgart: Verlag Bonn Aktuell, 1977): 20–43.

126. Friedrich Naumann, "Zur Entwicklung der Computerindustrie in den beiden deutschen Staaten," in *Der Wandel von Industrie, Wissenschaft und Technik in Deutschland und Frankreich*, 20 (Jahrhundert 2002): 143–164.

127. John P. Hardt, "East European Economies in Crisis," in *East European Economic Assessment*, 1–3; Laszlo Czirjak, "Industrial Structure, Growth, and Productivity

in Eastern Europe," in U.S. Congress, Joint Economic Committee, *Economic Developments in Countries of Eastern Europe* (Washington, D.C.: U.S. Government Printing Office, 1970): 434–447; J. Wilczynski, *Technology in Comecon: Acceleration of Technological Progress through Economic Planning and the Market* (London: Macmillan, 1974): 4–22.

128. Erich Sobeslavsky and Nikolaus Joachim Lehmann, *Zur Geschchte von Rechentechnik und Datenverarbeitung in der DDR 1946–1968* (Dresden: Hannah-Arendt-Institut, 1996): 13; this is the only history available on East German IT; Edwin M. Snell and Marilyn Harper, "Postwar Economic Growth in East Germany: A Comparison with West Germany," in *Economic Developments in Countries of Easter Europe*, 558–575.

129. Raymond G. Stokes, *Constructing Socialism: Technology and Change in East Germany, 1945–1990* (Baltimore: Johns Hopkins University Press, 2000): 1–18.

130. Ibid., 19–20; Rainer Karlsch, *Allein bezahlt?* (Berlin: Ch. Links, 1993): 282.

131. Stokes, *Constructing Socialism*, 21–30.

132. Ibid., 26–27.

133. Ibid., but also consult the earlier book by Vladimir Slamecka, *Science in East Germany* (New York: Columbia University Press, 1963), which provides a detailed snapshot of the research infrastructure in the GDR as of 1961–1962.

134. Sobeslavsky and Lehmann, *Zur Geschchte von Rechentechnik und Datenverarbeitung in der DDR 1946–1968*, 27–29; Nelson M. Blachman, "The State of Digital Computer Technology in Europe," *Communications of the ACM* 4, no. 6 (June 1961): 258.

135. Ibid., 29–33.

136. For a description of these machines, Nelson M. Blachman, "Central European Computers," *Communications of the ACM* 2, no. 6 (September 1959): 14–15; Sobeslavsky and Lehmann, *Zur Geschchte von Rechentechnik und Datenverarbeitung in der DDR 1946–1968*, 33–38, 160.

137. Stokes, *Constructing Socialism*, 61, 66–67.

138. Ibid., 95.

139. Doris Cornelsen, "The GDR in a Period of Foreign Trade Difficulties: Development and Prospects for the 1980s," in Joint Economic Committee, Congress of the United States, *East European Economic Assessment, Part 1—Country Studies, 1980* (Washington, D.C.: U.S. Government Printing Office, 1981): 299–324.

140. Coordinating Committee for Multilateral Export Controls was established by the Western powers to block the sale weapons and other advanced technologies to Communist nations. It was a major player in slowing and stopping the flow of various IT products from the West.

141. Raymond Bentley, *Research and Technology in the Former German Democratic Republic* (Boulder, Col.: Westview Press, 1992): 24, 101, 103–106.

142. Ibid., 93–108.

143. Slamecka, *Science in East Germany*, 14–16. For a description of the combine system, see Thomas A. Baylis, "East Germany's Economic Model," *Current History* (November 1987): 377–381, 393, 394; Irwin L. Collier, Jr., "GDR Economic Policy During the Honecker Era," *Eastern European Economics* (Fall 1990): 5–29.

144. Stokes, *Constructing Socialism*, 131.

145. Ibid., 148; Jerome Segal, "Cybernetics in the German Democratic Republic: From Bourgeois Science to Panacea," in Dieter Hoffman, Benoît Severyns, and Raymond G. Stokes (eds.), *Science, Technology, and Political Change* (Liège: Proceedings of the XXth International Congress of History and Science, 1999).

146. Wilczynski, *Technology in Comecon*, 109.

147. Ibid., 111.

148. Goodman, "Soviet Computing and Technology Transfer," 552; Stokes, *Constructing Socialism*, 181.

149. Simon Doing, "Appropriating American Technology in the 1960s: Cold War Politics and the GDR Computer Industry," *IEEE Annals of the History of Computing* 32, no. 2 (April–June 2010): 42, but see also the entire article, 32–45.

150. Goodman, "Soviet Computing and Technology Transfer," 551–553.

151. Ibid., 115–116. His statistics on the World include USA 344, Switzerland 145, West Germany 109, Canada 107, Netherlands 98, Great Britain, 91, Benelux 90, Sweden 86, Denmark 71, Australia 69, Norway 64, Japan 56, Ireland 52, Italy 48, Austria 47, Israel 43, New Zeland, 39, Finland 32, Ibid., 116.

152. Quote and data, Ibid., 133.

153. Complaints and lack of demand were even discussed in print during this period, Gerhard Merkel, "Mikroelektronik und wissenschaftlichtechnischer Fortschritt," *Einheit* 32 (December 1977): 1354–1360; Otfried Steger, "Mikroelektronik—ein wesentlichess Element der Wirtschaftsstrategie," *Neus Deutschland*, March 12, 1981, p. 3.

154. Geipel, "Politics and Technology in the German Democratic Republic," 168; on inability to support entrepreneurship in GDR, Irwin L. Collier, Jr., "Intensification in the GDR: A Postscript," *Studies in Comparative Communism* 20, no. 1 (Spring 1987): 72.

155. Sobeslavsky and Lehmann, *Zur Geschchte von Rechentechnik und Datenverarbeitung in der DDR 1946–1968*, 98–100.

156. André Steiner has looked at these issues for some time, see for example, "Zwischen Konsumversprechen und Innovationszwang. Zum wirtschaftlichen Niedergang der DDR," in Konrad H. Jarausch and Martin Sabrow (eds.), *Weg in den Untergang. Der innere Zerfall der DDR* (Göttingen: Vandenhoeck & Ruprecht, 1999): 153–192. The focus of all the essays in this volume is on the economic decline of the GDR

157. Geipel, "Politics and Technology in the German Democratic Republic," 19.

158. Ibid., 35.

159. Klaus Fuchs, "Warum Mikroelektronik?" *Spectrum* 8 (June 1977): 6; Helmut Koziolek, "The Economic Strategy of the Eleventh Party Congress of the SED and the New Stage in Science-Production Relations," *Eastern European Economics* 26, no. 2 (Winter 1987–88): 67.

160. Quoted in Geipel, "Politics and Technology in the German Democratic Republic, 1977–1990," 40–41.

161. A. Steiner. *Von Plan zu Plan: eine Wirtschaftsgeschichte der DDR* (Munich: Deutsche Verlagsanstalt, 2004).

162. For statistics on oil and manufactured goods' trade between the Soviets and GDR, see Geipel, "Politics and Technology in the German Democratic Republic,"69–70.

163. Stokes, *Constructing Socialism*, 171–173.

164. IDC data, 1990, in Geipel, "Politics and Technology in the German Democratic Republic," 155.

165. Based on East German data, Ibid., 156.

166. Ibid., 159.

167. Gary L. Geipel, Jarmoszko, A Tomasz, and Goodman, Seymour E., "The Information Technologies and East European Societies," *East European Politics and Societies* 5, no. 3 (Fall 1991): 425–429.

168. *Economics* 26, no. 2 (Winter 1987–88): 67. Geipel, "Politics and Technology in the German Democratic Republic," 309–310.

169. Ibid., 160.

170. Discussed in Ibid., 105–107.

171. Geipel estimated that possibly 100,000 East Germans had PCs at the start of the 1990s., Ibid., 123.

172. Paul Gannon, *Trojan Horses and National Champions: The Crisis in Europe's Computing and Telecommunications Industry* (London: Apt-Amatic Books, 1997): 17; Steven W. Popper, *East European Reliance on Technology Imports from the West* (Santa Monica, Cal.: RAND Corporation, 1988): 20–21.

173. Chalmers Johnson (ed.), *Change in Communist Systems* (Stanford, Cal.: Stanford University Press, 1970); Siegfried G. Schoppe, "Die intrasystemaren und die intersystemaren Technologietransfers der DDR," in Gernot Gutman (ed.), *Das Wirtschaftssystem der DDR* (Stuttgart: Gustave Fischer Verlag, 1983): 345–362, see other chapters in same book; Rüdiger Pieper, "The GDR System for Managing Change: The Managerial Instruments," in Margy Gerber (ed.), *Studies in GDR Culture and Society 9* (Lanham, Md.: University Press of America, 1988): 35–55; Bentley, *Technological Change in the German Democratic Republic*, 69–71.

174. Bentley, *Research and Technology in the Former German Democratic Republic*, 51–52, 176.

175. Geipel, "Politics and Technology in the German Democratic Republic," 190–200.

176. Quoted in Ibid., 249.

177. Ibid., 261.

178. Stokes, *Constructing Socialism*, 173.

179. Pieper, *Hochschulinformatik in der Bundesrepublik und der DDR bis 1989/1990*, who also demonstrates that these institutions collaborated with various local companies and other agencies.

180. Pieper, *Hochschulinformatik in der Bundesrepublik und der DDR bis 1989/1990*.

181. Yugoslavia, while an authoritarian state, went its own way during the 1960s and beyond, not functioning as part of the Comecon world described in this chapter.

182. A shared pattern noted by observers of East European economics as far back as the 1950s that remained in place for the entire period of Comecon economics, John P. Hardt, "East European Economic Development: Two Decades of Interrelationships and Interactions With the Soviet Union," in U.S. Congress, Joint Economic Committee, *Economic Developments in Countries of Eastern Europe* (Washington, D.C.: U.S. Government Printing Office, 1970): 5–40, and observed again by the same author a decade later, "East European Economies in Crisis," 1–3; Gary R. Teske, "Poland's Trade with the Industrialized West: Performance, Problems, and Prospects," U.S. Congress, Joint Economic Committee, *East European Economic Assessment: Part 1—Country Studies, 1980* (Washington, D.C.: U.S. Government Printing Office, 1981): 72–91.

183. Turnock, *The East European Economy in Context*, 15.

184. Leon Lukaszewicz, "On the Beginnings of Computer Development in Poland," *Annals of the History of Computing* 12, no. 2 (1990): 103–107 and "Outline of the Logical Design of the ZAM 41 Computer," *IEEE Transactions of Computers, The Computer Systems Issue, EC* 12 (1963): 609–612; R.W. Marczynski, "The First Seven Years of Polish Digital Computers," *Annals of the History of Computing* 2, no. 1 (January 1980): 37–48. In the 1930s, Poland played a major role in the development of cryptoanalytical machines, copies of which they made available to the British on the eve of World War II and that proved crucial in the ability of the

Allies to read Nazi encrypted military messages, now universally recognized by historians as one the major developments of the war and the ability of the Allies to defeat the Germans in Europe. The literature is vast on this topic, but for introductions to Poland's role and mathematical capabilities, see David Kahn, *Kahn on Codes: Secrets of the New Cryptology* (New York: Macmillan, 1984); Wladyslaw Kozaczuk, *Enigma: How the German Cipher Was Broken and How It Was Read by the Allies in World War II*, ed. and trans. By Christopher Kasparek (Frederick, Md.: University of Publishers of America, 1984); C.A. Deavours, "The Black Chamber," *Cryptologia* 4, no. 3 (July 1980): 129–132; Christopher Kasparek and Richard Woytak, "In Memoriam: Marian Rejewski," Ibid., 6, no. 1 ((January 1982): 19–25; for a fuller bibliography, see James W. Cortada, *A Bibliographic Guide to the History of Computing, Computers, and the Information Processing Industry* (Westport, Conn.: Greenwood Press, 1990): 303–306.

185. For descriptions of both the projects and technical features of these systems, see a series of reports by Nelson M. Blachman, "Central-European Computers," *Communications of the ACM* 2, no. 9 (September 1959): 14–17, "The State of Digital Computer Technology in Europe," Ibid., 4, no. 6 (June 1961): 256–265, "The State of Digital-Computer Technology in Europe," unpublished paper, February 14, 1962, Smithsonian Institution, Box 213; Alston S. Householder, "Digital Computers in Eastern Europe," *Computers and Automation* 4, no. 12 (December 1955): 8; Jan Oblonsky, "Some Features of . . . The Czechoslovak Relay Computer SAPO," *Datamation* (January–February 1958): 34–36, which, after describing this one machine being built in the entire country, ran on the last page of the report a news item announcing that the American firm, BENDIX had just completed manufacturing its 100th G-15 computer (page 36), a mid-sized general purpose system—what a contrast to the Comecon results.

186. Wilczynski, *Technology in Comecon*, 109–139.

187. Householder, "Digital Computers in Eastern Europe," 8. We now have a short history of SAPO and Czech computing critical to our understanding of East European computing, Helena Durnová, "Sovietization of Czechoslovakian Computing: The Rise and Fall of the SAPO Project," *IEEE Annals of the History of Computing* 32, no. 2 (April–June 2010): 21–31.

188. Wilczynski, *Technology in Comecon*, 109.

189. "Chairman's Overview," *East European Economic Assessment*, viii.

190. Zsuzsa Szentgyörgyi, "A Short History of Computing in Hungary," *IEEE Annals of the History of Computing* 21, no. 3 (July–September 1999): 49–57. The entire issue of the *Annals* is devoted to computing in Central and Eastern Europe.

191. Ibid.

192. Ibid., 54.

193. Ibid.

194. David Turnock, *The East European Economy in Context: Communism and Transition* (London: Routledge, 1997): 204.

195. Jane Hardy, *Poland's New Capitalism* (New York: Pluto Press, 2009): 12–27.

196. Andrzej Karpinsky, "Intellectual Background and the Technological Lag in Poland," in János Kovács (ed.), *Technological Lag and Intellectual Background: Problems of Transition in East Central Europe* (Aldershot, G.B.: Dartmouth, 1995): 131–141.

197. Ibid., 141.

198. Geipel, "Politics and Technology in the German Democratic Republic," 142; Marian Malecki, "Technology Transfer to Poland through Foreign Polonian Enterprises," in Jan Monkiewicz and Roland Scharff (eds.), *Reform, Innovational Performance*

and Technical Progress: The Polish Case (Erlangen: Institut für Gesellschaft und Wissenschaft, 1989).

199. Work on this subject has already begun, beginning with a useful dissertation by Andrzej Tomasz Jarmoszko, "Transformation of the Telecommunication Environment in Poland, 1989–1991" (Unpublished Ph.D. dissertation, University of Arizona, 1992), in which the author demonstrates that in that short period of time, the nation's telecommunication infrastructure was fundamentally transformed.

200. Steven W. Popper, *East European Reliance on Technology Imports from the West* (Santa Monica, Cal.: RAND Corporation, 1988): 18–20. For the discussion focusing on the 1990s and beyond, see Martin Srholec, "Global Production Systems and Technological Catching-Up: Thinking Twice About High-Tech Industries in Emerging Countries," in Piech and Radosevic, *The Knowledge-based Economy in Central and Eastern Europe*, 57–78.

201. Sayers, "History of IBM's International Operations, 1911–2006," 67, 115, 118–119, 237–238, 357–358, 367.

202. Andrea Szalavetz, "From Industrial Capitalism to Intellectual Capitalism: The Bumpy Road to a Knowledge-based Economy in Hungary," in Piech and Radosevic, *The Knowledge-based Economy in Central and Eastern Europe*, 189–191.

203. Ibid., 190.

204. Ibid., 191, using data from the Hungarian Central Statistical Office (CSO), *The Information and Communications Technology Sector in Hungary, 1998–2001* (Budapest: CSO, 2003).

205. Szalavetz, "From Industrial Capitalism to Intellectual Capitalism: The Bumpy Road to a Knowledge-based Economy in Hungary," 191.

206. Ibid., 191–193.

207. Marcin Piatkowski, "The Output and Productivity Growth Effects of the Use of Information and Communication Technologies: The Case of Poland," in Piech and Radosevic, *The Knowledge-based Economy in Central and Eastern Europe*, 82.

208. A major finding when I examined the U.S. Experience, James W. Cortada, *The Digital Hand*, 3 vols. (New York: Oxford University Press, 2004–2008).

209. Piatkowski, "The Output and Productivity Growth Effects of the Use of Information and Communication Technologies: The Case of Poland," 84; B. Van Ark, J. Melka, N. Mulder, M. Timmer and G. Yypma, "ICT Investments and Growth Accounts for the European Union, 1980–2000," *Research Memorandum* GF-56, Groningen Growth and Development Centre, September 2002, httip://www.ggdc.net/index-publc.html#top (last accessed 4/17/2010); U.S. Department of Commerce, *Showcase Europe*, 54–56.

210. Piatkowski, "The Output and Productivity Growth Effects of the Use of Information and Communication Technologies: The Case of Poland," 90.

211. Michal Jaworski, "ICT Use and Trends in Small and Medium Businesses in Poland—Growth Factors and Inhibitors," in *Visions on the Future of Information Society*, no date or publisher, circa 2006, pp. 113–128.

212. "Poland: Playing Catch-up," Economist Intelligence Unit, May 7, 2007, http://www.ebusinessforum.com/index.asp?layout=rich_story&;chanelid=4&;title=Pol and%3A+ . . . (last accessed 4/1/2010); "IT Star—Fistera Workshop: ICT and the Eastern European Dimension, October 22, 2004—Prague, the Czech Republic Final Report"; Maja Bučar, Metka Stare and Andreja Jaklič, "eStrategy and ICT Investment in Slovenia," *19th Bled eConference eValues, Bled Slovenia, June 5–7, 2006*, pp. 1–13.

213. Srholec, "Global Production Systems and Technological Catching-Up: Thinking Twice About High-Tech Industries in Emerging Countries," 57–78.

214. Ibid., 92.

215. Quoted in Wilson P. Dizard and S. Blake Swensrud, *Gorbachev's Information Revolution: Controlling Glasnot in a New Electronic Era* (Washington, D.C.: Center for Strategic and International Studies, 1987): 80–81.

216. Quoted in Simon Shuster, "Russia Plans a Silicon Valley," *Time*, April 19, 2010, p. Global 2.

217. Loren R. Graham, *What Have We Learned About Science and Technology from the Russian Experience?* (Stanford, Cal.: Stanford University Press, 1998): 36.

218. Ibid.

219. Ibid., 39.

220. OECD, *OECD Information Technology Outlook, 2008* (Paris: OECD, 2008), www.oecd.org/sti/ito (last accessed 4/18/2010).

221. Deniz Eylem Yoruk and Nick von Tunzelmann, "Knowledge Accumulation, Networks and Information and Communication Technologies: Evidence from Traditional Industries in Central and Eastern Europe," in Piech and Radosevic, *The Knowledge-based Economy in Central and Eastern Europe*, 124.

222. Hardy, *Poland's New Capitalism*, 93–94.

CHAPTER 7

1. Shohei Kurita, "Computer Use in Japan," paper presented at 1973 National Computer Conference, New York, N.Y., June 4–8, 1973, Record Group 62, Box 17, folder 16, Charles Babbage Institute, University of Minnesota, Minneapolis.

2. G.C. Allen, *A Short Economic History of Modern Japan* (London: Allen & Unwin, 1981): 187–230; Masataka Kosaka, *A History of Postwar Japan* (Tokyo: Kodansha International, 1972): 200–224.

3. Tom Forester, *Silicon Samurai: How Japan Conquered the World's I.T. Industry* (Cambridge, Mass.: Blackwell, 1993): ix.

4. Ikujiro Nonaka and Hirotaka Takeuchi, *The Knowledge-Creating Company: How Japanese Companies Create the Dynamics of Innovation* (New York: Oxford University Press, 1995), first quote, p. 16; second quote, p. 246.

5. Key themes well articulated and documented by Michael E. Porter, Hirotaka Takeuchi and Mariko Sakakibara, *Can Japan Compete?* (Cambridge, Mass.: Perseus, 2000) and Marie Anchordguy, *Reprogramming Japan: The High Tech Crisis under Communitarian Capitalism* (Ithaca, N.Y.: Cornell University Press, 2005).

6. William K. Tabb, *The Postwar Japanese System: Cultural Economy and Economic Transformation* (New York: Oxford University Press, 1995): 225–254.

7. Japan was finally knocked out of its long standing second position by China for the first time during 2010, stirring a flurry of commentary about the decline of the nation, the rise of China and the "New Asia"; "China's March Towards World's No. 2 Economy," *CNNMoney.Com*, August 16, 2010, http://www/CNNMoney.com (last accessed 9/2/2010).

8. The scope of the literature on these issues has yet to be defined, but for a start, there is Dawn E. Talbot, *Japan's High Technology: An Annotated Guide to English-Language Information Sources* (Phoenix, AZ: Oryx Press, 1991) and the non-annotated, yet detailed bibliography, Fransman, *Japan's Computer and Communications Industry*, 511–525; for bibliography that includes Japanese titles, Kazunori Minetaki and Kiyohiko G. Nishimura, *Information Technology Innovation and the Japanese Economy* (Stanford, Cal.: Stanford University Press, 2010): 218–223.

9. Gene Gregory, *Japanese Electronics Technology: Enterprise and Innovation* (Tokyo: The Japan Times, 1985): 7.

10. T.F.M. Adams and N. Kobayashi, *The World of Japanese Business: An Authoritative Analysis* (Tokyo: Kodansha International, 1969); on European views about Japanese practices, see William R. Nester, *European Power and the Japanese Challenge* (New York: New York University Press, 1993): 195–198, and for Japanese perspectives, 198–201. There were also more balanced reviews, such as by John Nathan, *Japan Unbounded: A Volatile Nation's Quest for Pride and Purpose* (Boston, Mass.: Houghton Mifflin, 2004); and even earlier during the near hysteria evident in the West, Rodney Clark, *The Japanese Company* (New Haven, Conn.: Yale University Press, 1979); Richard E. Caves and Masu Uekusa, *Industrial Organization in Japan* (Washington, D.C.: Brookings Institution, 1976).

11. I used the word partial because every personal experience is unique. Comparison of memoirs of startup firms illustrates the point. In addition to the Japanese memoirs cited elsewhere in this chapter, see Akio Morita with Edwin M. Reingold and Mitsuko Shimomura, *Made in Japan: Akio Morita and SONY* (New York: E.P. Dutton, 1986) and Takeo Miyauchi, *The Flame from Japan: A Story of Success in the Microcomputer Industry* (Tokyo: Sord Computer Corporation, 1982), then compare their sense of uniqueness to those in the West, such as those written by Konrad Zuse, *The Computer—My Life* (New York: Springer-Verlag, 1993) and Thomas J. Watson Jr. and Peter Petre, *Father Son & Co.: My Life at IBM and Beyond* (New York: Batam, 1990).

12. Ezra F. Vogel, *Japan As Number One: Lessons for America* (Cambridge, Mass.: Harvard University Press, 1979): ix.

13. On Japan's educational system, Brian M. Murphy, *The International Politics of New Information Technology* (New York: St. Martin's Press, 1986): 226–229. Most Japanese programmers learned their trade on the job; formal education on computer science at universities was quite limited in the twentieth century. Entrepreneurial initiatives by individuals skilled in IT remained low, in part because of social disapproval of risk and potential failure, and software was not treated as important as hardware; but also due to attractiveness of life-time employment practices by the largest firms, Anchordoguy, *Reprogramming Japan*, 140–143.

14. William Chapman, *Inventing Japan: The Making of a Postwar Civilization* (New York: Prentice-Hall Press, 1991): 94.

15. Patents are traditionally used as a surrogate measure of innovation in an industry or economy, and if we leave aside the fact that what constitutes a patent in the USA versus Japan, for the moment, the evidence suggests in general that by the late 1980s, of the top ten corporate recipients of patents in the IT communication, four were Japanese (1987): Canon, Hitachi (normally with the largest number), Toshiba, and Mitsubishi Electric and if one looks at the number per unit of sales, again the Japanese did well in the late 1980s-early 1990s, particularly Canon and Toshiba, Fransman, *Japan's Computer and Communications Industry*, 431–436.

16. Koji Kobayashi, *The Rise of NEC: How the World's Greatest C&C Company Is Managed* (Oxford: Blackwell, 1991): 62–75.

17. Makoto Ohtsu and Tomio Imanari, *Inside Japanese Business: A Narrative History, 1960–2000* (Armonk, N.Y.: M.E. Sharpe, 2002): 17–43; but see also for origins of consumer electronics, Gene Gregory, *Japanese Electronics Technology*, 141–191; Alfred D. Chandler, Jr., *Inventing the Electronic Century: The Epic Story of the*

Consumer Electronics and Computer Industry (Cambridge, Mass.: Harvard University Press, 2005): 217–220.

18. These various models emerged from such global influences as political alignments of nations with Communism, hence Communist economies, and Third World or "non-aligned" models which embraced socialism, or quasi-managed economic models; in short, not all advanced economies using IT were neo-classical cases of practicing capitalism. I argue that these varying models of economic organization influenced profoundly the adoption and use of IT all through the second half of the century.

19. Angus Maddison, *Monitoring the World Economy, 1820–1992* (Paris: OECD, 1995): 107.

20. Japan Government, www.stat.go.jp (last accessed 9/10/2010).

21. Mario Polèse, *The Wealth and Poverty of Regions: Why Cities Matter* (Chicago, Ill.: University of Chicago Press, 2009): 110; and impact cities have that continues, Richard Florida, *The Great Reset: How New Ways of Living and Working Drive Post-Crash Prosperity* (New York: Harper, 2010): 117–120.

22. T.F.M. Adams and Iwao Hoshii, *A Financial History of the New Japan* (Tokyo: Kodansha International, 1972): 64.

23. Thomas F. Cargill and Takayuki Sakamoto, *Japan Since 1980* (Cambridge: Cambridge University Press, 2008): 1–8.

24. MITI's name changed its name to Ministry of Economy, Trade and Industry (METI) in 2001.

25. The most detailed study about IBM in Japan was conducted by Robert Sobel, *IBM vs. Japan: The Struggle for the Future* (New York: Stein and Day, 1986): 143–190; but see also Martin Fransman, *Japan's Computer and Communications Industry: The Evolution of Industrial Giants and Global Competitiveness* (Oxford: Oxford University Press, 1995): 147–156, 165–166; Anchordguy, *Reprogramming Japan*, 126–136, 160–163, 187–188.

26. Leon Hollerman (ed.), *Japan and the United States: Economic and Political Adversaries* (Boulder, Col.: Westview Press, 1979); United States-Japan Trade Council, *Yearbook of U.S.-Japan Economic Relations* (Washington, D.C.: U.S.-Japan Trade Council, 1979-annual); Nester, *European Power and the Japanese Challenge*, 191–258.

27. Cargill and Sakamoto, *Japan Since 1980*, 27–100; Hiroshi Yoshikawa, *Japan's Lost Decade* (Tokyo: LTCB International, 2002): 4–5, 29–32, 135–139; Shahid Yusuf and Kaoru Nabeshima, "Japan's Changing Industrial Landscape," World Bank WPS3758, 2004, *www-wds.worldbank.org/servlet/. . ./WDSP/IB/. . ./wps3758.txt-Cached* (ast accessed 3/8/2012).

28. Cargill and Sakamoto, *Japan Since 1980*, 3.

29. Considerable attention was given to the issue by influential organizations, especially in the United States, National Research Council, Computer Technology Resources Panel, *The Computer Industry in Japan and Its Meaning for the United States* (Washington, D.C.: National Research Council, 1973); U.S. Congress, Committee on Science and Technology, *Background Reading on Science, Technology, and Energy R&D in Japan and China* (Washington, D.C.: U.S. Government Printing Office, 1981); U.S. Congress, Joint Economic Committee, *International Competition in Advanced Industrial Sectors: Trade and Development in the Semiconductor Industry* (Washington, D.C.: U.S. Government Printing Office, 1982); U.S. Congress, Office of Technology Assessment, *U.S. Industrial Competitiveness* (Washington, D.C.: U.S. Government Printing Office, July 1981), but most damning is U.S. International

Trade Commission, *Foreign Industrial Targeting and Its Effects on U.S. Industries, Phase I: Japan*, Publication 1437 (Washington, D.C.:U.S. Government Printing Office, October 1983).

30. Nester, *European Power and the Japanese Challenge*, 128–161.
31. Because the body of literature is so large, it is described more thoroughly in the bibliographic essay at the back of this book.
32. Fransman, *Visions of Innovation*, 167–197.
33. Exports represented a major portion of the business of key industries. For example, in 1950 exports comprised only 2.3 percent of Japan's total GDP, but jumped to 7.9 percent in 1973, and by the end of 1992, to 12.4 percent, actually less than for other Asian countries, Maddison, *Monitoring the World Economy*, 38. However, just a few industries routinely dominated Japan's export business. Using early 1990s, motor vehicles and consumer electronics led Japan's exports, over 48 percent of Japan's export business, while integrated circuits amounted to 2.63 percent in 1991–92, Fransman, *Japan's Computer and Communication Industry*, 8–9.
34. Anchordguy, *Reprogramming Japan*, 6.
35. Ibid.
36. Although increasingly over time, firms hired fewer permanent workers, made possible by automating working, outsourcing it to less expensive labor markets in Asia, and by hiring contract (temporary) workers in Japan. Almost all Western writers like to point out, however, that this policy institutionalized retention of employees who were not so productive, adding a drag on the cost of doing business, particularly in large enterprises. I used the word *increasingly* in the narrative to point out that in aggregate management sought for its total workforce to be more productive, an objective hardly met when measured at the national level, and discussed later within the context of economic effects of the use of IT.
37. Ibid., 13.
38. Ibid.
39. Anchordoguy, *Reprogramming Japan*, 50–54 and passim; William K. Tabb, *The Postwar Japanese System: Cultural Economy and Economic Transformation* (New York: Oxford University Press, 1995): 37, 41–43, Porter, Takeuchi and Sakakibara, *Can Japan Compete?*, 74, 151–152, 165.
40. Chalmers Johnson, *MITI and the Japanese Miracle: The Growth of Industrial Policy, 1925–1975* (Stanford, Cal.: Stanford University Press, 1982); Scott Callon, *Divided Sun: MITI and the Breakdown of Japanese High-Tech Industrial Policy, 1975–1993* (Stanford, Cal.: Stanford University Press, 1995): 182–207; Martin Fransman, *Japan's Computer and Communications Industry: The Evolution of Industrial Giants and Global Competitiveness* (New York: Oxford University Press, 1995); Porter, Takeuchi and Sakakibara, *Can Japan Compete?*, 1–68.
41. Anchordguy, *Reprogramming Japan*, 10.
42. Sobel, *IBM vs. Japan*, 150–154; Marie Anchordoguy, *Computers Inc: Japan's Challenge to IBM* (Cambridge, Mass.: Harvard Council on East Asian Studies, 1989): 22–24.
43. Anchordguy, *Reprogramming Japan*, 96–146; Tabb, *The Postwar Japanese System*, 43–44.
44. The issue of time-based competition in industries undergoing rapid change is a complicated story but crucial to ours, W. Mark Fruin, "Competing in the Old-Fashioned Way: Localizing and Integrating Knowledge Resources in Fast-to-Market Competition," in Jeffrey K. Liker, John E. Ettlie and John C. Campbell (eds.),

Engineered in Japan: Japanese Technology-Management Practices (New York: Oxford University Press, 1995): 217–233.

45. Nester, *European Power and the Japanese Challenge*, 110–112, 121–127.
46. Laura D'Andrea Tyson and David B. Yoffie, "Semiconductors: From Manipulated to Managed Trade," in David B. Yoffie (ed.), *Beyond Free Trade: Firms, Governments, and Global Competition* (Boston, Mass.: Harvard Business School Press, 1993): 37.
47. Chandler, *Inventing the Electronic Century*, 6–7, 64–65, 79–80; Anchordoguy, *Reprogramming Japan*, 144–146, 164–166, 219–222; Martin Fransman, *Visions of Innovation: The Firm and Japan* (Oxford: Oxford University Press, 1999): 93–96. A founder of Sony, Akio Morito, wrote a highly informed, well written memoir/ history of the firm, *Made in Japan* (New York: Dutton, 1986).
48. Cargill and Sakamoto, *Japan Since 1980*, 83–98.
49. Cargill and Sakamoto, *Japan Since 1980*, 13–15.
50. Fransman, *Visions of Innovation*, 8.
51. Ibid., 74–75.
52. Tabb, *The Postwar Japanese System*, 257.
53. Ryota Suekane, "Early History of Computing in Japan," in N. Metropolis et al. (eds), *A History of Computing in the Twentieth Century* (New York: Academic Press, 1980): 575–578; Sigeru Takashashi, "Early Transistor Computers in Japan," *Annals of the History of Computing* 8, no. 2 (April 1986): 144–154; Hidstosi Takahasi, "Some Important Computers of Japanese Design," Ibid., 2, no. 4 (October 1980): 330–337; Sigeru Takahashi, "A Brief History of the Japanese Computer Industry Before 1985," Ibid., 18, no. 1 (1996): 76–79.
54. Fuido Kricks, "A Brief Overview of the History of the Japanese Computer Industry," November 3, 1982, pp. 2–7, Charles Babbage Institute, University of Minnesota, Minneapolis; Kenneth Flamm, *Creating the Computer: Government, Industry, and High Technology* (Washington, D.C.: Brookings Institution, 1988): 172–179; Martin Fransman, *The Market and Beyond: Information Technology in Japan* (Cambridge: Cambridge University Press, 1990): 13–56.
55. Japan Electronic Development Association, *J.E.I.D.A. and Its Computer Center* (Tokyo: JEIDA, undated [circa 1959], CBI 32, Box 714, folder 5, Charles Babbage Institute, University of Minnesota, Minneapolis.
56. Everett S. Calhoun, "New Electronic Computer Developments in Japan and Europe," February 21, 1957, Blachman Papers, Box 217, Smithsonian Institution, Washington, D.C. but see also Nelson M. Blachman, March 9, 1953, Box 216.
57. Kent E. Calder, *Strategic Capitalism: Private Business and Public Purpose in Japanese Finance* (Princeton, N.J.: Princeton University Press, 1993): 116–117.
58. Fransman, *Japan's Computer and Communications Industry*, 134–136.
59. Described well by NEC's chairman, Koji Kobayashi, *Computers and Communications: A Vision of C&C* (Cambridge, Mass.: MIT Press, 1986): 61–76 and *The Rise of NEC: How the World's Greatest C&C Company Is Managed* (Cambridge, Mass.: Blackwell, 1991): 112–169; NEC Corporation, *NEC Corporation, 1899–1999* (Tokyo: NEC Corporation, 2002): 194–200, 237–279.
60. Fransman, *Japan's Computer and Communications Industry*, 136–138; Chandler, *Inventing the Electronic Century*, 194–211.
61. Kricks, "A Brief History of the History of the Japanese Computer Industry," 5–6.
62. Flamm, *Creating the Computer*, 178.
63. Quoted in Fransman, *Japan's Computer and Communications Industry*, 138.
64. Ken. W. Sayers, "History of IBM's International Operations, 1911–2006," (2006), pp. 273–275, IBM Corporate Archives, Somers, New York.

65. James W. Birkenstock, "Pioneering: On the Frontier of Electronic Data Processing, A Personal Memoir," *IEEE Annals of the History of Computing* 22, no. 1 (January–March 2000): 36; for further details, see Fransman, *Japan's Computer and Communications Industry*, 137–139.
66. Report by Ulric Weil to A.L. Harmon, December 19, 1967, World Trade Corporation Legal Records, Country Files, Japan, Box 61, folder 4, IBM Corporate Archives. The same records document IBM's frustrations with Japanese officials and activities of government agencies toward the firm in the 1960s with the most useful a detailed description of government policies by Richard W. Rubinowitz to William T. Ketcham, December 24, 1965, World Trade Corporation Legal Records, Country Files, Japan, Box 63, folder 7, IBM Corporate Archives. It documents Japanese policy in these years with insights about the difficulty of describing specifics: "the policy announced by the MITI official with whom we have been dealing is nowhere embodied in any provision of law or administrative regulation of which we are aware. Nor has there even been, to the best of our knowledge, any public disclosure of this policy."
67. Flamm, *Creating the Computer*, 184–185.
68. Ibid., 186–188; Fransman, *Japan's Computer and Communications Industry*, 140–141; Chandler, *Inventing the Electronic Century*, 191–192; Anchordoguy, *Reprogramming Japan*, 125–140.
69. Chandler, *Inventing the Electronic Century*, 192.
70. Fransman, *Japan's Computer and Communications Industry*, 139–141; Sobel, *IBM vs. Japan*, 157–173.
71. "JCM & IPS Status in Japan & U.S.," November 20, 1979, unpaginated, Record Group 5, "Business Planning CMC/Meeting Material, 1979-11/20," Box 46, folder 3/4, IBM Corporate Archives, Somers, N.Y.
72. Ibid.
73. Ibid.
74. Chandler, *Inventing the Electronic Century*, 190–210; reviewed in greatest detail by Nester, *European Power and the Japanese Challenge*.
75. Scott Callon, *Divided Sun: MITI and the Breakdown of Japanese High-Tech Industrial Policy, 1975–1993* (Stanford, Cal.: Stanford University Press, 1995): 166–181; Forester, *Silicon Samurai*, 194–201.
76. Anchordoguy, *Reprogramming Japan*, 146.
77. Taiyu Kobayashi, *Fortune Favors the Brave: Fujitsu: Thirty Years in Computers* (Tokyo: Keizai Shinposha, 1983): 77–110. The author was chairman of the board at this company during the crucial years when it committed itself to computers. He should not be confused with another Japanese chairman with the same name, Koji Kobayashi, who led NEC, both of whom overlapped in time their leadership of their respective firms.
78. For examples of contemporary discussions, see Hugh Patrick and Larry Meissner (eds.), *Japan's High Technology Industries: Lessons and Limitations of Industrial Policy* (Seattle, Wash.: University of Washington Press and Tokyo: University of Tokyo Press, 1986) and for more contemporary presentations, William H. Davidson, *The Amazing Race: Winning the Technorivalry with Japan* (New York: Wiley & Sons, 1984); Christopher Wood, *The End of Japan Inc. And How the New Japan Will Look* (New York: Simon & Schuster, 1994). For Japanese perspectives, see Naoto Sasaki, *Management and Industrial Structure in Japan* (Oxford: Pergamon, 1981); Michio Morishima, *Why Has Japan "Succeeded"?* (Cambridge: Cambridge University Press, 1981). For a more current retrospective, see Makoto Ohtsu, *Inside*

Japanese Business: A Narrative History, 1960–2000 (Armonk, N.Y.: M.E. Sharpe, 2002): 17–41.

79. Sheridan Tatsuno, *The Technolopis Strategy: Japan, High Technology, and the Control of the Twenty-first Century* (New York: Prentice Hall, 1986): 23–31; Y.W. Liu and I.R. Marchant, *The Japanese-American Struggle For Supremacy in the Computer Industry*, Occasional Paper No. 9 (Singapore: EPS Publishers for University of Western Australia, Centre for East Asian Studies undated [circa 1982–83]).

80. For the clearest and most compelling discussion of the issue, see Porter, Takeuchi and Sakakibara, *Can Japan Compete?*, 1–68.

81. Chandler, *Inventing the Electronic Century*, 209–210.

82. A central point made by Martin Fransman in all his books on Japan, see for example, *The Market and Beyond, Japan's Computer and Communications Industry* and more recently, *Visions of Innovation*.

83. Anchordguy, *Reprogramming Japan*, 127–129.

84. Paul Fannon, *Trojan Horses and National Champions: The Crisis in Europe's Computing and Telecommunications Industry* (London: Apt-Amatic Books, 1997): 187–195; Vogel, *Japan As Number One*; Fransman, *The Market and Beyond*, 9; Nester, *European Power and the Japanese Challenge*, 97–258.

85. Anchordoguy, *Reprogramming Japan*, 177–205.

86. Ibid., 187–188.

87. Gannon, *Trojan Horses and National Champions*, 195.

88. Richard Florida and Martin Kenney, "High-Technology Restructuring in the USA and Japan," *Environment and Planning* 22 (1990): 233–252, but see also their *Beyond Mass Production*, 50–55, which takes the story of Japanese dominance story to the end of the period of prosperity before the economy began declining in 1989–1990.

89. Yui Kimura, *The Japanese Semiconductor Industry: Structure, Competitive Strategies and Performance* (Greenwich, Conn.: JAI Press, 1988): 51.

90. Ibid. 52.

91. Ibid., 54–55.

92. Ibid., 56.

93. Ibid., 57.

94. Fransman, *Japan's Computer and Communications Industry*, 273–274.

95. Ibid., 276.

96. Chandler, *Inventing the Electronic Century*, 210–212; Jason Dedrick and Kenneth L. Kraemer, *Asia's Computer Challenge: Threat or Opportunity for the United States and the World?* (New York: Oxford University Press, 1998): 78–84. On the language problem see Kobayashi, *Computers and Communications*, 80–82.

97. For examples of these various perspectives, see Yasunori Baba, Shinji Takai, and Yuji Mizuta, "The User Driven Evolution of the Japanese Software Industry: The Case of Customized Software for Mainframes," in David Mowery (ed.), *The International Computer Software Industry: A Comparative Study of Industry Evolution and Structure* (New York: Oxford University Press, 1996): 104–130; Fransman, *Japan's Computers and Communications Industry*; and the perspective I most adhere to, Achordoguy, *Reprogramming Japan*, 147–176.

98. OECD, *Information Technology Outlook* (Paris: OECD, 2008): 118.

99. Achordoguy presents much of the evidence from multiple sources in a useful graph, *Reprogramming Japan*, 149.

100. Ibid., 150.

101. Fransman, *Japan's Computer and Communications Industry*, 186–187, 191.

102. Subject of much discussion in the West, Michael Cusumano, *Japan's Software Factories: A Challenge to U.S. Management* (New York: Oxford University Press, 1991); Yoshihiro Matsumoto and Yutaka Ohno, *Japanese Perspectives in Software Engineering* (Singapore: Addison-Wesley, 1989): 3–18; Kenney and Florida, *Beyond Mass Production*, 81–85.

103. Achordoguy, *Reprogramming Japan*, 155–157; Fransman, *Visions of Innovation*, 160–165.

104. Achordoguy, *Reprogramming Japan*, 159; Marie Anchordoguy, *Computers Inc.: Japan's Challenge to IBM* (Cambridge, Mass.: Harvard Council on East Asian Studies, 1989): 149–150.

105. Ibid., 162.

106. Ibid., 163.

107. The standard work consulted in the 1990s on FMS was by B. Joseph Pine II, *Mass Customization: The New Frontier in Business Competition* (Boston, Mass.: Harvard Business School Press, 1993).

108. Gregory, *Japanese Electronics Technology*, 300–301.

109. Ibid., 287–309.

110. Jon Sigurdson, *Industry and State Partnership in Japan: The Very Large Scale Integrated Circuits (VLSI)* (Lund: Research Policy Institute, 1986); Edward A. Feigenbaum and Pamela McCorduck, *The Fifth Generation: Artificial Intelligence and Japan's Computer Challenge to the World* (Reading, Mass.: Addison-Wesley, 1983): 99–148; Fransman, *The Market and Beyond*, 193–242.

111. Anchordoguy, *Reprogramming Japan*, 163–166.

112. Martin Campbell-Kelly, *From Airline Reservations to Sonic the Hedgehog: A History of the Software Industry* (Cambridge, Mass.: MIT Press, 2003): 284–285.

113. Ibid., 288.

114. Anchordoguy, *Reprogramming Japan*, 164.

115. Chandler, *Inventing the Electronic Century*, 70–71; Gregory, *Japanese Electronics Technology*, 96–99. Digital watches had a similar effect as video games, beginning in the 1970s.

116. For a definition of robotics from the perspective of manufacturers and users: "a reprogrammable, multifunctional manipulator designed to move material, parts, tools, or specialized devices through variable programmed motions for the performance of a variety of tasks," *RIA Robotics Glossary* (Dearborn, Mich.: Robots Institute of America, 1984): 28. See also a Japanese catalog of various types of robots, "Japan Industrial Standards (JIS) Classification of Robots," reprinted in Frederick L. Schodt, *Inside the Robot Kingdom: Japan, Mechatronics, and the Coming Robotopia* (Tokyo: Kodansha International, 1988, 1990): 45.

117. Schodt, *Inside the Robot Kingdom*, 39. On uses, Shimony Nof (ed.), *Handbook of Industrial Robotics* (New Yorlk: John Wiley & Sons, 1985).

118. Kuni Sadamoto (ed.), *Robots in the Japanese Economy: Facts About Robots and Their Significance* (Toko: Survey Japan, 1981): 2.

119. Ibid., 53, 94, 114–115.

120. Ibid., xv–xvi.

121. Scholdt, *Inside the Robot Kingdom*, 15.

122. Ibid., 111–130; *The Japan Industrial Robot Association, The Robotics Industry of Japan: Today and Tomorrow* (Tokyo: Fuji Corporation, 1982): 58–108, 133, 251; on uses in manufacturing, Gregory, *Japanese Electronics Technology*, 299–309.

123. Scholdt, *Inside the Robot Kingdom*, for description of costs and problems with performance, pp. 17–28, 115–116.

124. Porter, Takeuchi, and Sakakibara, *Can Japan Compete?*, 133.

125. Rogers W. Johnson, "Computer Progress in Japan," *Industrial Research* (December–January 1960–61): 66.

126. That is beginning to change, however. See, for example, Kiyoshi Murata, "Lessons From the History of Information System Development and Use in Japan," *Entreprises et Histoire* 60 (September 2010): 50–61.

127. Japan Productivity Center, *Computer Utilization in Management in Japan, July 1969* (Tokyo: Japan Productivity Center, 1969): 5–14, CB 32, Box 471, folder 3, Charles Babbage Institute.

128. Ibid., 15.

129. Ibid., 26.

130. Ibid.

131. Ibid., 27.

132. Japan Computer Usage Development Institute, *Computer White Paper: 1969* (Tokyo: Japan Computer Usage Development Institute, 1969): 42, CBI 32, Box 551, folder 4. For additional affirming evidence, Shohei Kurita, "Computer Use in Japan," June, 1973, CBI 62, Box 17, folder 16, Charles Babbage Institute, University of Minnesota, Minneapolis.

133. Franklin F. Kuo, "Computers in Japan," January 1971, pp. 4–5, CBI 32, Box 117, folder 6, Charles Babbage Institute, University of Minnesota, Minnapolis.

134. Kuo, "Computers in Japan," 10.

135. Fransman, *The Market and Beyond*, 26; Anchordoguy discusses JECC throughout her book, *Computers Inc.*, but a detailed description of its operations, see especially pp. 59–89.

136. Anchordoguy, *Reprogramming Japan*, 131.

137. Japan Electronic Computer Co., Ltd., *Progress of Computer Industry in Japan* (Tokyo: JECC, 1974): 19, CBI 32, Box 631, folder 22, Charles Babbage Institute, University of Minnesota, Minneapolis.

138. Ibid., 21.

139. Ibid., 24.

140. Japan Computer Usage Development Institute, *Computer White Paper* (Tokyo: Japan Computer Usage Development Institute, 1971): 57, CBI 32, Box 551, folder 2, Charles Babbage Institute, University of Minnesota, Minneapolis.

141. Ibid.

142. JIPDEC, "Outline of Computer Usage by Local Governments," Report No. 20 (Tokyo: JIPDEC, 1973): for quote, p. 15, for paragraph, 12–15, CBI 32, Box 613, folder 9, Charles Babbage Institute, University of Minnesota, Minneapolis.

143. Ibid.

144. Computer Technology/Resources Panel of Computer Science and Engineering Board, National Research Council, "The Computer Industry in Japan and Its Meaning for the United States," 13 (1973), CBI 32, Box 598, folder 21, Charles Babbage Institute, University of Minnesota, Minneapolis.

145. Arthur D. Little, Inc., "The World Computer Industry, 1973–1978," February 1974, p. 25, CBI 55, Box 9, folder 33, Charles Babbage Institute, University of Minnesota, Minneapolis. For a useful summary of types of systems adopted by type of technology (batch, then online) across industries and several decades, see Murata, "Lessons From the History of Information System Development and Use in Japan," 50–61.

146. H.J. Welke, *Data Processing in Japan* (Amsterdam: North-Holland Publishing, 1982): 122.

147. Ibid., 131.
148. Ibid., 135.
149. Ministry of International Trade and Industry, "Report of the Information Industry Committee," September 9, 1981, reprinted in Ibid., 161–182.
150. Ibid., 166.
151. International Data Corporation, "Japanese User Buying Intentions, 1987–1995," November 1987, CBI 55, Box 60, folder 27, Charles Babbage Institute, University of Minnesota, Minneapolis.
152. Jiro Kokuryo, "Information Technologies and the Transformation of Japanese Industry," unpublished paper (circa 1997), p. 4, http://www.jkokuryo.com/papers/1997003/pacis97.htm (last accessed 7/28/2010).
153. Ibid., 10.
154. Various MITI surveys, 1994.
155. Norris Parker Smith, "Computing in Japan: From Cocoon to Competition," *Computer* (March 1997): 26–33.
156. Discussions about IT in Japan in the 1990s and 2000s centered on the supply side, as had occurred in all other decades. For examples, see "Will Technology Leave Japan Behind?" *BusinessWeek*, August 31, 1998, http://www.businesswekk.com/1998/35/b3593052.htm (last accessed 6/2/2010); Minetaki and Nishimura, *Information Technology Innovation and the Japanese Economy*.
157. For an excellent and thorough discussion of these issues, see Robert E. Cole, "Telecommunications Competition in World Markets: Understanding Japan's Decline," in John Zysman and Abraham Newman (eds.), *How Revolutionary Was the Digital Revolution?* (Stanford, Cal.: Stanford Business Books, 2006): 101–124.
158. Takehiko Musashi, *Japanese Telecommunications Policy* (Cambridge, Mass.: Program on Information Resources Policy, Harvard University, 1985).
159. Kenji Kushida, "Japan's Telecommunications Regime Shift: Understanding Japan's Potential Resurgence," Ibid., 125.
160. Since these events lie outside the scope of this book, for details see Ibid., 125–147.
161. Tessa Morris-Suzuki, *Beyond Computopia: Information, Automation and Democracy in Japan* (London: Kegan Paul, 1988): 128.
162. Smith, "Computing in Japan: From Cocoon to Competition," 30.
163. An important change that had to be made before Wave Two computing could occur, with a major structural reorganization of NTT in 1999 to help drive down the costs of communications and make the firm more responsive to technological changes. High rates for phone calls were suppressing more extensive use of the Internet and until costs per minute were lowered (beginning in 2001) and new terms and conditions that allowed for more competition for using telephony were introduced in the early 2000s, progress would be held back. Further, as long as the communications proved reluctant to embrace non-Japanese technical standards and new technologies, yet another break on Internet diffusion would remain, a problem that remained during the first decade of the new century, Anchordoguy, *Reprogramming Japan*, 115–123.
164. Kushida, "Japan's Telecommunications Regime Shift: Understanding Japan's Potential Resurgence," 125–170; Mito Akiyoshi and Hiroshi Ono, "The Diffusion of Mobile Internet in Japan," *The Information Society* 24 (2008): 292–303.
165. John Beck and Mitchell Wade, *DoCoMo: Japan's Wireless Tsunami* (New York: AMACOM, 2003): 6–12.
166. Ministry of Public Management, Home Affairs, Posts and Telecommunications, Japan, *Stirring of the IT-Prevent Society* (Tokyo: General Policy Division, Information

and Communications Policy Bureau, Ministry of Public Management, Home Affairs, Posts and Telecommunications, Japan, 2002): 5–6.

167. "Outline of 2007 Information and Communications in Japan White Paper," *MIC Communications News*, 18, no. 13 (October 12, 2007): 2.

168. Mark B. Fuller and John C. Beck, *Japan's Business Renaissance: How the World's Greatest Economy Revived, Renewed, and Reinvented Itself* (New York: McGraw-Hill, 2006), 59.

169. Takanori Ida, "Broadband, Information Society, and the National System in Japan," in Martin Fransman (ed.), *Global Broadband Battles: Why the U.S. and Europe Lag While Asia Leads* (Stanford, Call, Stanford Business Books, 2006): 65–86.

170. Press release by Ministry of Internal Affairs and Communications, September 12, 2006, "Results of Fact-finding Survey of the Telecommunications Industry as of July 2006 (final), another two press releases, March 13, 2007, "Broadband Service Subscriber Trends (December 31, 2006), February 16, 2009, "State of Numbers of Subscribers to Telecommunications Services," which show, however, that the number of subscribers for telephones and ISDNs had actually leveled off and shrunk slightly, as users continued switching to Internet-based telephony, with 110.4 million users by the start of 2009.

171. Japan Computer Usage Development Institution, *Computer White Paper* (Tokyo: Japan Computer Usage Development Institution, 1971); this report was published annually in the 1970s and 1980s, CBI 32, Box 551, folder 2, Charles Babbage Institute, University of Minnesota, Minneapolis; James K. Imai, "Computers in Japan—1969," *Datamation* (January 1970): 147–152; see also annual reports by Arthur D. Little, Inc., *The World Computer Industry*, published in the 1960s into the 1990s, CBI 55, Box 8, folder 27 and others, Charles Babbage Institute.

172. Kuo, "Computers in Japan," unpaginated, Table 1.

173. Arthur D. Little, Inc., *The World Computer Industry: 1978–1983* (No city: Arthur D. Little, Inc., 1978): 21, CBI 55, Box 8, folder 10, Charles Babbage Institute, University of Minnesota, Minneapolis.

174. Arthur D. Little, Inc., *World Markets for Information Products to 1991* (No city: Arthur D. Little, Inc., April 1982): 30, CBI 55, Box 8, folder 27, Charles Babbage Institute, University of Minnesota, Minneapolis.

175. Abigail Christopher, "Apple in Japan," *Pacific Planning Service* (Framingham, Mass.: International Data Corporation, March 1990): 2, CBI 55, Box 104, folder 2, Charles Babbage Institute, University of Minnesota, Minneapolis.

176. OECD, *OECD Information Technology Outlook 2000: ICTs, E-Commerce and the Information Economy* (Paris: OECD, 2000): 64.

177. Anchordoguy, *Reprogramming Japan*, 141; Ministry of Internal Affairs and Communications, Japan, "2010 White Paper Information and Communications in Japan," July 2010; Kenji E. Kushida, "Leading Without Followers: The Political Economy of Japan's ICT Sector," BRIE Working Paper 184, December 14, 2008.

178. OECD, *OECD Information Technology Outlook 2000*, 232–233.

179. Ibid., 234.

180. Dennis Tachiki, Satoshi Hamaya, and Koh Yukawa, *Diffusion and Impacts of the Internet and e-Commerce in Japan* (Irvine, Cal.: Center for Research and Information Technology and Organization, University of California, February 2004); Chaojung Chen and Chiho Watanabe, "Diffusion, Substitution and Competition Dynamism Inside the ICT Market: The Case of Japan," *Technological Forecasting*

and Social Change 73 (2006): 731–759; Mito Akiyoshi and Hiroshi Ono, "The Diffusion of Mobile Internet in Japan," *The Information Society* 24 (2008): 292–303. For a perspective on broadband use, see Kenji Kushida and Seung-Youn OH, "Understanding South Korea and Japan's Spectacular Broadband Development: Strategic Liberalization of the Telecommunications Sectors," BRIE Working Paper 175, June 29, 2006, http://www.crito.uci.edu/papers (last accessed 8/10/2010).

181. Data extracted from graphs and tables in Kyoji Fukao, Tsatomu Miyagawa, Hak K. Pyo, and Keun Hee Rhee, "Estimates of Multifactor Productivity, ICT Contributions and Resource Reallocation Effects in Japan and Korea," RIETI Discussion Paper Series 09-E-021, April 2009, p.14.

182. Ibid., 19.

183. Paul Schreyer, *The Contribution of Information and Communication Technology to Output Growth: A Study of the G7 Countries*, OECD Science, Technology and Industry Working Papers, 2000/2 (Paris: OECD, 2002).

184. Dale W. Jorgenson, "Information Technology and the G7 Economies," *Rivista di Politica Economica* (January–February 2005): 25–56, quote, p. 43.

185. Ibid., 53.

186. Discussed in considerable detail by Dale W. Jorgenson and Kazuyuki Motohashi, "Information Technology and the Japanese Economy," *Journal of Japanese International Economics* 19 (2005): 460–481.

187. Jiro Kokuryo, "Information Technologies and the Transformation of Japanese Industry," 1997, http://www.jkokuryo.com/papers/19977003/pacis97.htm (last accessed 6/25/2010).

188. For both quotes, Adam S. Posen, "Japan," in Benn Steil, David G. Victor and Richard R. Nelson (eds.), *Technological Innovation and Economic Performance* (Princeton, N.J.: Princeton University Press, 2002): 103, but the entire chapter is devoted to Japanese economics and is excellent, 74–111.

189. Minetaki and Nishimura, *Information Technology Innovation and the Japanese Economy*, quote, p.75, but consult more broadly, pp. 53–76.

190. Masauki Morikawa, "Information Technology and the Performance of Japanese SMEs," *Small Business Economics* 23 (2004): 171–177.

191. Quoted in Takuji Fueki and Takuji Kawamoto, "Does Information Technology Raise Japan's Productivity?" Bank of Japan Working Paper Series No. 08-E-8 (Tokyo: Bank of Japan, September 2008), 3.

192. Ibid., 4, quote, 20. Data and conclusions are similar to those of other studies, see for example, Takahito Kanamori and Kazuyuki Motohashi, "Information Technology and Economic Growth: Comparison between Japan and Korea," RIETI Discussion Paper Series 07-E-009 [undated, 2007–08].

193. Porter, Takeuchi, and Sakakibara, *Can Japan Compete?*, 101.

194. Kimura, *The Japanese Semiconductor Industry*, 165–167.

195. Tabb, *The Postwar Japanese System*, 257.

196. Ibid., 258.

197. Gregory, *Japanese Electronics Technology*, 408–419.

198. Ibid., both quotes, 418.

199. For a description of how one Japanese electronics firm implemented the strategy, see NEC Corporation, *NEC Corporation, 1899–1999* (Tokyo: NEC Corporation, 2002): 174–175.

200. Dedrick and Kraemer, *Asia's Computer Challenge*, 211–215, 252–255.

201. Ibid., for a list of these firms, 217.

202. Ibid., 228–229.

203. OECD, *OECD Information Technology Outlook 2000: ICTs, E-commerce and the Infor-mation Economy* (Paris: OECD, 2000): 232.
204. "Leaving Home," *The Economist*, November 20, 2010, p. 74.
205. For a perceptive analysis of Japanese-European innovation strategies in the 1990s and beyond, see Fransman, *Telecoms in the Internet Age*, 235–259.
206. OECD, *ICT and Economic Growth: Evidence from OECD Countries, Industries and Firms* (Paris: OECD, 2003): 20–21.

CHAPTER 8

1. Quoted in International Telecommunications Union, *Broadband Korea: Internet Case Study* (ITU, March 2003): 53.
2. Nina Hachigian and Lily Wu, *The Information Revolution in Asia* (Santa Monica, Cal.: National Defense Research Institute/RAND, 2003): xi–xiv, xii–xiii.
3. Most scholars prefer to differentiate these nations' activities by discussing the Electronics Industry as a whole, of which IT is a part, while I focus more narrowly on the latter. Nonetheless, the broader discussion is an essential one as it provides rich context not to be ignored. An excellent example that influenced the narrative in this chapter was written by Dilip K. Das, "The Dynamic Growth of the Electronics Industry in Asia," *Journal of Asian Business* 14, no. 4 (1998): 67–99. Key to the thinking of many observers of the IT scene in comparative perspective, see Nina Hachigian and Lily Wu, *The Information Revolution in Asia* (Santa Monica, Cal.: National Defense Research Institute, 2003): 1–54; on rationale for the three nations studied in this chapter focusing largely on the 1990s, see Poh-Kam Wong, "National Innovation Systems for Rapid Technological Catch-up: An Analytical Framework and a Comparative Analysis of Korea, Taiwan and Singapore," Paper presented at DRUID Summer Conference on National Innovation Systems, June 9–12, 1999
4. Lawrence J. Lau, "The Role of Government in Economic Development: Some Observations from the Experience of China, Hong Kong, and Taiwan," in Nasahiko Aoki, Hyung-Ki Kim, and Masahiro Okuno-Fujiwara (eds.), *The Role of Government in East Asian Economic Development: Comparative Institutional Analysis* (Oxford: Oxford University Press, 1996): 41–73; all the essays in Frederic C. Deyo (ed.), *The Political Economy of the New Asian Industrialism* (Ithaca, N.Y.: Cornell University Press, 1987).
5. There is an excellent discussion and summary of these various perspectives regarding developmental economics as they relate to IT and Asia buried in a study of Asian semiconductors written by a Korean professor of public administration, Sung Gul Hong, *The Political Economy of Industrial Policy in East Asia: The Semiconductor Industry in Taiwan and South Korea* (Cheltenham, U.K.: Edward Elgar, 1997): 14–42.
6. This line of reasoning has also been widely adopted by others to explain Korea's economic performance, with emphasis on the role of large institutions in society, such as government and telecommunications. These arguments are often used as the reasons and sources for Korea's economic success. See, for example, Larson, *The Telecommunications Revolution in Korea*, 302–304; Gary G. Hamilton, "Patterns of Asian Network Capitalism," in W. Mark Fruin (ed.), *Networks, Markets, and the Pacific Rim: Studies in Strategy* (New York: Oxford University Press, 1998): 181–199.
7. He has discussed these ideas in many books, but the fullest argument applied to is in Alfred D. Chandler, Jr., *Scale and Scope: The Dynamics of Industrial Capitalism*

(Cambridge, Mass.: Harvard University Press, 1990): 594–605 for a short summary of his ideas.

8. Joonghae Suh, "Overview of Korea's Development Strategies," in Joonhae Suh and Derek H.C. Chen (eds.), *Korea as a Knowledge Economy: Evolutionary Process and Lessons Learned* (Washington, D.C.: World Bank, 2007): 24–27.

9. Ibid.

10. Every Asian nation's economic development effort has been compared to those of Japan. One useful—and short—example is both useful and illustrative: "Korea's income per capita in the mid-1970s was about the same as Japan's in the early 1950s and, if current growth rates continue, would match Japan's early 1960s income by the mid-1980s. However, Korea has both a much greater export share of GDP and a lower savings rate than Japan had in the early 1950s. Nevertheless, Korean growth has been substantial, in part because it has imported a relatively larger amount of foreign capital, and in part because it has used its lower volume of savings more efficiently than Japan did," Edward S. Mason et al., *The Economic and Social Modernization of the Republic of Korea* (Cambridge, Mass.: Harvard University Press, 1980): 122–123.

11. Hyung-Koo Lee, *The Korean Economy: Perspectives for the Twenty-First Century* (Albany, N.Y.: State University of New York Press, 1996): 91–94.

12. Linsu Kim, "Korea's National Innovation System in Transition," in Linsu Kim and Richard R. Nelson (eds.), *Technology, Learning, and Innovation: Experiences of Newly Industrializing Economies* (Cambridge: Cambridge University Press, 2000): 338–358.

13. Gary G. Hamilton, "Patterns of Asian Network Capitalism: The Cases of Taiwan and South Korea," in W. Mark Fruin (ed.), *Networks, Markets, and the Pacific Rim* (New York: Oxford University Press, 1998): 183–195.

14. Kim and Roemer, *Growth and Structural Transformation*, 153–154.

15. Hamilton, "Patterns of Asian Network Capitalism," 186.

16. Ibid., 188–189.

17. Chung-in Moon, "Democratization and Globalization as Ideological and Political Foundations of Economic Policy," in Jongryn Mo and Chung-in Moon (eds.), *Democracy and the Korean Economy* (Stanford, Cal.: Hoover Institution Press, 1999): 4.

18. Michael E. Porter, *The Competitive Advantage of Nations* (New York: Free Press, 1990): 465.

19. Ibid., 473.

20. Linsu Kim, Jangwoo Lee, and Jinjoo Lee, "Korea's Entry into the Computer Industry and Its Acquisition of Technological Capability," *Technovision* 6 (1987): 277–293, data drawn from p. 279.

21. Young-Iob Chung, *South Korea in the Fast Lane: Economic Development and Capital Formation* (Oxford: Oxford University Press, 2007): 37–40.

22. John A. Mathews and Dong-Sung Cho, *Tiger Technology: The Creation of a Semiconductor Industry in East Asia* (Cambridge: Cambridge University Press, 2000): 105–155.

23. Nicole Kim to author, December 22, 2010.

24. "Request for Special Review and Comment," August 27, 1976, WCT Legal Country Files, Box 65, file 5, IBM Archives, Somers, New York.

25. Hong, *The Political Economy of Industrial Policy in East Asia*, 6–7, 85–86.

26. Larson, *The Telecommunications Revolution in Korea*, 72–73.

27. Ibid., 85.

28. Hamilton, "Patterns of Asian Network Capitalism," 181–199; Jason Dedrick and Kenneth L. Kraemer, *Asia's Computer Challenge: Thresat of Opportunity for the United States and the World?* (New York: Oxford University Press, 1998): 116–122.

29. International Data Corporation, *Korea Multiuser Market Review and Forecast, 1989–1994* (Framingham, Mass.: International Data Corporation, 1990): 12, 15.

30. Jason Dedrick, Kenneth L. Kraemer, and Dae-Won Choi, "Korean Technology Policy at a Crossroads: The Case of Computers," *Journal of Asian Business* 11, no. 4 (1995): 9–10.

31. Ibid., 10.

32. Ibid., 12.

33. Mason et al., *The Economic and Social Modernization of The Republic of Korea*, 263–275, 484–489.

34. This narrative relies largely on Kim's version, *Imitation to Innovation*, 149–170.

35. One of the most useful is by Hong, *The Political Economy of Industrial Policy in East Asia*, 73–126.

36. Ibid., 157–158. Other versions of the semiconductor's Korean history include Dedrick and Kraemer, *Asia's Computer Challenge*, 122–126; Parvez Hasan, *Korea: Problems and Issues in a Rapidly Growing Economy* (Baltimore, Md.: Johns Hopkins University Press, 1976): 180, 182; James F. Larson, *The Telecommunications Revolution in Korea* (Hong Kong: Oxford University Press, 1995; *Technology Policies and Planning Republic of Korea*, 81–83.

37. Kim, *Imitation to Innovation*, 163–165.

38. How this industry worked has been explained well by Sung Gul Hong, *The Political Economy of Industrial Policy in East Asia: The Semiconductor Industry in Taiwan and South Korea* (Cheltenham: Edward Elgar, 1997): 75–162.

39. Methodical collection and sharing of information is an important feature of the technology diffusion story for all Asian countries, so much so that no history of technology in the region can ignore the topic. For a description of how this happened in South Korea in the 1960s-early 1980s see Asian and Pacific Centre for Transfer of Technology, *Technology Policies and Planning Republic of Korea* (Bangalore, India: Asian and Pacific Centre for Transfer of Technology, 1986): 77–91.

40. These five perspectives drawn from Ibid., 168–170.

41. Ibid., 81–87.

42. Dedrick and Kraemer, *Asia's Computer Challenge*, 131–132.

43. Even in the early 2000s developing economies invested more in upgrading their telecommunications infrastructure than in acquiring IT. I explore this issue in more detail in James W. Cortada, *How Societies Embrace Information Technology: Lessons for Management and the Rest of Us* (New York: John Wiley & sons, 2009): 71–100.

44. Keuk Je Sung, "South Korea: Telecommunications Policies into the 1990s," in Eli Noam, Seisuke Komatsuzaki, and Douglas A. Conn (eds.), *Telecommunications in the Pacific Basin: An Evolutionary Approach* (New York: Oxford University Press, 1994): 300–305; Chung, *South Korea in the Fast Lane*, 115–174; San-Chul Lee, "Korea," in Georgette Wang (ed.), *Treading Different Paths: Informatization in Asian Nations* (Norwood, N.J.: ABLEX Publishing, 1994): 107–108.

45. Dedrick, Kraemer, and Choi, "Korean Technology Policy at a Crossroads: The Case of Computers," 13.

46. Ibid., 12.

47. Described more closely in EIAK, *Electronics Industry Yearbook* (English version) (Seoul: EIAK, 1993).

48. Dedrick, Kraemer, and Choi, "Korean Technology Policy at a Crossroads: The Case of Computers," 36.
49. Dedrick and Kraemer, *Asia's Computer Challenge*, 131–132.
50. Lee, *The Korean Economy: Perspectives for the Twenty-First Century*, 31.
51. For example with Korea's largest electronics conglomerate, see Lee Dongyoup, *Samsung Electronics—The Global Inc.* (Seoul: YSM, 2006): 21–25; Tony Michell, *Samsung Electronics And the Struggle for Leadership of the Electronics Industry* (Singapore: John Wiley & Sons, 2010): 22–41, 170–175; Sea-Jin Chang, *Sony vs. Samsung: The Inside Story of the Electronics Giant's Battle for Global Supremacy* (Singapore: John Wiley & Sons, 2008): 15–16, 33–40.
52. Alice H. Amsden, *Asia's Next Giant: South Korea and Late Industrialization* (New York: Oxford University Press, 1989): 286.
53. Ibid., 305–306, 310.
54. Asian and Pactic Centre for Transfer of Technology, *Technology Policies and Planning Republic of Korea* (Bangalore, India: Asian and Pacific Centre for Transfer of Technology, 1986): 81.
55. Ibid.
56. Lee, *The Korean Economy*, quote on 128, 130.
57. Ibid., 130.
58. For a more detailed description of this firm's use of computers, see Linsu Kim, *Imitation to Innovation: The Dynamics of Korea's Technological Learning* (Boston, Mass.: Harvard University Press, 1997): 118–119, 121–127.
59. Morris H. Crawford, *Programming the Invisible Hand: The Computerization of Korea and Taiwan* (Cambridge, Mass.: Program on Information Resources Policy, Harvard University, 1986): 21–60.
60. Ibid., 138–142.
61. Ibid., 141–146.
62. Ken W. Sayers, "A Summary History of IBM's International Operations, 1911–2006" (2006), p. 398, IBM Archives, Somers, N.Y.
63. Young-Iob Chung, *South Korea In the Fast Lane: Economic Development and Capital Formation* (New York: Oxford University Press, 2007): 37–40, 74–76, 147–153, 172–185; Suh and Chen, *Korea as a Knowledge Economy*, 17–46; Myung Oak Kim and Sam Jaffe, *The New Korea: An Inside Look at South Korea's Economic Rise* (New York: AMACOM, 2010): 2–6; and a useful overview of the crisis, Jongryn Mo and Chung-in Moon, "Epilogue: Democracy and the Origins of the 1997 Korean Economic Crisis," in Jongryn Mo and Chung-in Moon (eds.), *Democracy and the Korean Economy* (Stanford, Cal.: Hoover Institution Press, 1999): 174–197.
64. Jonathan Clemens, "The Making of a Cyborg Society: South Korea's Information Revolution, 1997–2007" (Unpublished Masters thesis, University of Hawaii, 2008) for both the quote and the poll, p. 37.
65. Ibid., 41–42.
66. Ibid., 75.
67. Ibid., 44–46.
68. Ibid., 46.
69. Ibid., 47. IBM's own rankings of global performance of nations reached similar conclusions, labeling South Korea as an "established leader" by the early 2000s, James W. Cortada, Ashish M. Gupta, and Marc Le Noir, *How Rapidly Advancing Nations Can Prosper in the Information Age* (Somers, N.Y.: IBM Corporation, 2007): 16.
70. Hachigian and Wu, *The Information Revolution in Asia*, 85.

71. Clemens, "The Making of a Cyborg Society: South Korea's Information Revolution, 1997–2007," 47.
72. Duk Hee Lee and Gyoung-Gyu Choi, *The IT Industry in Korea: A Perspective of the Experience of Rapid Growth* (Seoul: Korea Institute for Industrial Economics and Trade, July 2001): 1.
73. Ibid., 17 for the quote, and for the statistics, 3–15.
74. Ibid., 32.
75. "Country Review: Korea," Gerstner Papers, Trip Files, Asia Pacific 1998, Box 248, file 3, IBM Archives, Somers, N.Y.
76. Ibid.
77. Ibid.
78. Y2K stood for Year 2000 and referred to the initiatives taken by companies and agencies all over the world in the late 1990s to modify their software to recognize the calendar dates 2000 and beyond since many programs had been written years earlier without any consideration of post 1999 dates.
79. Ibid.
80. Clemens, "The Making of a Cyborg Society," 48.
81. Ibid., 50 for data, 50–51 for quote.
82. Ibid., 54.
83. Dal Yong Jin, *Korea's Online Gaming Empire* (Cambridge, Mass.: MIT Press, 2010): 17.
84. Heejin Lee, Robert M. O'Keefe, and Kyounglim Yun, "The Growth of Broadband and Electronic Commerce in South Korea: Contributing Factors," *The Information Society* 19 (2003): 81–93.
85. Ibid., 3.
86. Ibid., 4.
87. Clemens, "The Making of a Cyborg Society," 56.
88. Jin, *Korea's Online Gaming Empire*, 23; but see also, Jun Suk Huhh, "The Bang Where Korean Gaming Began: The Culture and Business of the PC Bang in Korea," in Larissa Hjorth and Dean Chan (eds.), *Gaming Cultures and Places in Asia-Pacific* (London: Routledge, 2009): 102–116.
89. Jin, *Korea's Online Gaming Culture*, 23.
90. Ibid., 24.
91. Clemens, "The Making of a Cyborg Society," 58.
92. Ibid., 60.
93. Ibid., 63.
94. Ibid., 64.
95. Jin, *Korea's Online Gaming Culture*, 27.
96. Ibid., 44.
97. Kim and Jaffe, *The New Korea*, 159–162.
98. Ibid., 81–100.
99. Clemens, "The Making of a Cyborg Society," 72.
100. Ibid., for summary of these studies, 81.
101. Hachigian and Wu, *The Information Revolution in Asia*, 68–69.
102. International Telecommunications Union, *Broadband Korea: Internet Case Study* (ITU, 2003): 2–4.
103. Ibid., 7.
104. ITU World Communications Indicators database, the standard global source for such information, http://www.itu.int (last accessed 1/2/2011).
105. ITU, Broadband Korea, 15.

106. Ibid., 32.
107. Kim and Jaffe, *The New Korea*, 267.
108. Ibid., 46.
109. This is especially the case in the pan-Atlantic community except in three circumstances: when privacy of one's information in digital form is discussed, when security of such data is of concern, or when copyright and patent rights in general are debated.
110. Nicholas Eberstadt, *The North Korean Economy: Between Crisis and Catastrophe* (New Brunswick, N.J.: Transaction Publishers, 2007): 4.
111. It is your author's personal opinion that computers in use in North Korea by the military is limited to a few groups who are well steeped in advanced weapons systems, such as rockets and missiles, and the local air force, two small communities within the military eco-system. Record keeping is a paper intensive, often an abacus-supported process, observations based on his nearly forty years' experience working in IBM on global IT issues. More knowledgeable military observers at the U.S. Department of Defense are reluctant to disclose what they know, simply to wink in acknowledgment to my guesses, questions, and statements.
112. Ibid., 6–7.
113. Larson, *The Telecommunications Revolution in Korea*, 267.
114. ITU, *Broadband Korea*, 69.
115. Larson, *The Telecommunications Revolution in Korea*, 284–288.
116. Angus Maddison, *Monitoring the World Economy, 1820–1992* (Paris: Development Centre of the Organization for Economic Co-Operation and Development, 1995): 115.
117. Other global supply chains included automotive and other transport vehicles, pharmaceuticals, and women's clothing,
118. Dedrick and Kraemer, *Asia's Computer Challenge*, 241–249; Henry Wai-Chung Yeung, "From Followers to Market Leaders: Asian Electronics Firms in the Global Economy," *Asia Pacific Viewpoint* 48, no. 1 (April 2007): 1–25.
119. Po-Lung Yu and Chieh-Yow ChiangLin, "Five Life Experiences That Shape Taiwan's Character," in Chun-Yen Chang and Po-Lung Yu (eds.), *Made in Taiwan: Booming in the Information Technology Era* (Singapore: World Scientific Publishing, 2001): 302.
120. Gwo-Hshiung Tzeng and Meng-Yu Lee, "Intellectual Capital in the Information Industry," Ibid., 298–344.
121. Ibid.
122. Honghong Tinn, "Cold War Politics: Taiwanese Computing in the 1950s and 1960s," *IEEE Annals of the History of Computing* 32, no. 1(January–March 2010): 90–92.
123. Sayers, "History of IBM's International Operations, 1911–2006," 436–438. The previous paragraph was largely drawn on this IBM manuscript as well.
124. Teresa Shuk-Ching Poon, *Competition and Cooperation in Taiwan's Information Technology Industry: Inter-Firm Networks and Industrial Upgrading* (Westport, Conn.: Quorum Books, 2002): 52.
125. Ibid., 53; Dedrick and Kraemer, *Asia's Computer Challenge*, 87.
126. Poon, *Competition and Cooperation in Taiwan's Information Technology Industry*, 52–53. On Acer, there is a book-length history, Robert N. Chen, *Made in Taiwan: The Story of Acer Computers* (Taipei: McGraw-Hill, 1997); there is no comparable treatment on Mitac.

127. Honghong Tinn, "From DIY Computers to Illegal Copies: The Controversy over Tinkering with Microcomputers in Taiwan, 1980–1984," *IEEE Annals of the History of Computing* 33, no. 2 (April–June 2011): 75–88.

128. Sayers, "History of IBM's International Operations, 1911–2006," 438; Untitled project report and attachments, July 11, 1975 and later, Legal Records/Taiwan/Finance 1971–1975, Box 91, IBM Archives, Somers, New York.

129. Poon, *Competition and Cooperation in Taiwan's Information Technology Industry*, 53–56.

130. Ibid.

131. Alice H. Amsden and Wan-wen Chu, *Beyond Late Development: Taiwan's Upgrading Policies* (Cambridge, Mass.: MIT Press, 2003): 19–76.

132. UMC was Taiwan's first semiconductor manufacturing company. By 2010 it was usually ranked as the 5th largest IC foundry in the world. It positions itself as a services company, rather than as a manufacturing operation. For details see, http://www.umc.com (last accessed 2/7/2011).

133. Jong-Tsong Chiang, "Management of National Technology Programs in a Newly Industrialized Country—Taiwan," *Technovision* 10, no. 8 (1990): 531–554.

134. Mathews and Cho, *Tiger Technology*, 157–202.

135. Ibid., 258.

136. Ibid., 259; International Data Corporation, *Taiwan Multiuser Market Review and Forecast, 1989–1994* (Framingham, Mass.: IDC, September 1990): 2–6, CBI 55, Box 103, Folder 12, Charles Babbage Institute, University of Minnesota, Minneapolis. On industrial parks, see Wen-Hsiung Lee and Wei-Tzen Yang, "The Cradle of Taiwan High Technology Industry Development—Hsinchu Science Park (HSP)," *Technovision* 20 (2000): 55–59 and for its performance in the early 2000s, Chia Chi Sun, "Evaluating and Benchmarking Productive Performances of Six Industries in Taiwan Hsin Chu Industrial Science Park," *Expert Systems with Applications* 38 (2011): 2195–2205; on Taiwanese business-government relations, Sang-Hyup Shin, *European Integration and Foreign Direct Investment in the EU: The Case of the Korean Consumer Electronics Industry* (London: Routledge, 1998): 76–79.

137. On telecommunications, see Tseng Fan-Tung and Mao Chi-Kuo, "Taiwan," in Noam, Komatsuzaki and Conn, *Telecommunicsations in the Pacific Basin*, 315–334.

138. Poon, *Competition and Cooperation in Taiwan's Information Technology Industry*, 55.

139. Ibid., 62.

140. Ibid., 62–63.

141. For example about Acer, Chen, *Made In Taiwan*, 299–306.

142. Yungkai Yang, "The Taiwanese Notebook Computer Production Network in China: Implication for Upgrading of the Chinese Electronics Industry," Personal Computing Industry Center, University of Californaia, Irvine, February 2006.

143. "IT in Taiwan and China," *Economist*, May 29, 2010, pp. 66–67.

144. Poon, *Competition and Cooperation in Taiwan's Information Technology Industry*, 107–142.

145. How firms became globally competitive starting as late comers is just now being examined. For an example of this kind of investigation, see Chia-Wen Lee, Roger Hayter, and David W. Edgington, "Large and Latecomer Firms: The Taiwan Semiconductor Manufacturing Company and Taiwan's Electronics Industry," *Tijdschrift voor Economische en Sociale Geografie* 101, no. 2 (2009): 177–198.

146. Ibid., 171–175.

147. Ibid., 179.

148. Ibid.

149. Jason Dedrick, Kenneth L. Kraemer, and Paul Seever, "Global Market Potential for Information Technology Products and Services," March 2007, Personal Computing Industry Center, University of California-Irvine, p. 11; Tain-Jy Chen, "The Diffusion and Impacts of the Internet and E-Commerce in Taiwan," [undated, circa 2003–04], p. 28, Center for Research on Information Technology and Organization, University of California-Irvine, http://www.crito.uci.edu (last accessed 2/11/2011).

150. Mathews and Cho, *Tiger Technology*, 51–56, 203–244, 260–262.

151. Singapore government statistics, various online sources, but see also International Monetary Fund estimates, 1980–2010 which were published in numerous reports.

152. Hans C. Blomqvist, *Swimming with Sharks: Global and Regional Dimensions of the Singapore Economy* (Singapore: Marshall Csvendish Academic, 2005). For a more critical analysis of problems faced by this nation in the early 2000s, see Manu Bhaskaran, *Re-inventing the Asian Model: The Case of Singapore* (Singapore: Eastern University Press, 2003).

153. Lee Kuan Yew, *From Third World to First: The Singapore Story: 1965–2000* (New York: HarperCollins, 2000): 527.

154. Christopher M. Dent, *The Foreign Economic Policies of Singapore, South Korea and Taiwan* (Cheltenham, U.K.: Edward Elgar, 2002): 61–131.

155. Ibid., 79; Blomqvist, *Swimming with Sharks*, 72–75.

156. Dent, *The Foreign Economic Policies of Singapore, South Korea and Taiwan*, 64.

157. Ibid., 64–66.

158. Hall Hill and Pang Eng Fong, *Technology Exports From A Small, Very Open NIC: The Case of Singapore*, No. 89/6 (Canberra, Australia: Australian National University Research School of Pacific Studies, August 1989): 5.

159. Dent, *The Foreign Economic Policies of Singapore, South Korea and Taiwan*, 82–83; Brian M. Murphy, *The International Politics of New Information Technology* (New York: St. Martin's Press, 1986): 244–245.

160. Francis C.C. Koh, Winston T.H. Koh, and Feichin Ted Tschang, "An Analytical Framework for Science Parks and Technology Districts with An Application to Singapore," *Journal of Business Venturing* 20 (2005): 217–239.

161. Murphy, *The International Politics of New Information Technology*, 246–247.

162. Sayers, "History of IBM's International Operations, 1911–2006," 384–386.

163. Lawrence Loh, "Technological Policy and National Competitiveness," in Toh Mun Heng and Tan Kong Yam (eds.), *Competitiveness of the Singapore Economy: A Strategic Perspective* (Singapore: Singapore University Press, 1998): 46–49.

164. Ho Kong Chong, "Globalization and the Social Fabric of Competitiveness," Heng and Yam, *Competitiveness of the Singapore Economy*, 304.

165. Vijay Gurbaxani, Kenneth L. Kraemer, John Leslie King, Sheryl Jarman, Jason Dedrick, K.S. Ramon, and K.S. Yap, "Government as the Driving Force Toward the Information Society: National Computer Policy in Singapore," *The Information Society* 7, no. 2 (990): 155–185.

166. Thompson S.H. Teo and Margaret Tan, "An Empirical Study of Adopters and Non-Adopters of the Internet in Singapore," *Information and Management* 34 (1998): 339–345.

167. Ibid.

168. National Computer Board, *Singapore IT Usage Survey 94* (Singapore: National Computer Board, 1994).

169. Infocomm Development Authority of Singapore, *Annual Survey on Business Info-comm Usage for 2006* (Singapore: IDA Singapore, 2007): 6.
170. Infocomm Development Authority of Singapore, *Annual Survey on Infocomm Man-power for 2008* (Singapore: IDA Singapore, 2009): 5.
171. Ibid., 342–343.
172. K.E. Haynes, Y.M. Hsing, and R.R. Stough, "Regional Port Dynamics in a Global Economy: The Case of Kaohsiung, Taiwan," *Maritime Policy and Management* 24 (1997): 93–113; Mario Polèse, *The Wealth and Poverty of Regions: Why Cities Matter* (Chicago, Ill.: University of Chicago Press, 2009): 69–75.
173. B.T. Mah, "Towards Making Singapore the Logistics Hub of the Region," *Speeches* (1994): 28.
174. C.A. Airriess, "Regional Production, Information-Communication Technology, and the Developmental State: The Rise of Singapore as a Global Container Hub," *Geoforum* 32 (2001): 235–254.
175. Ibid., 243.
176. J.E. Lee-Patridge, T.S.H. Teo, and V.K.G. Lim, "Information Technology Manage-ment: The Case of the Port of Singapore Authority," *Journal of Strategic Informa-tion Systems* 9 (2000): 97 for the quote, but see whole article, 85–99.
177. Geok Theng Lau and Jessica Voon, "Factors Affecting the Adoption of Electronic Commerce Among Small and Medium Enterprises in Singapore," *Journal of Asian Business* 20, no. 1 (2004): 5–45.
178. Lim Cher Ping, Khine Myint Swe, Timothy Hew, Philip Wong, and Divaharan Shanti, "Exploring Critical Aspects of Information Technologies Integration in Singapore Schools," *Australian Journal of Educational Technology* 19, no. 1 (2003): 1–24.
179. Thompson T.S.H. Teo and C. Ranganthan, "Adopters and Non-Adopters of Business-to-Business Electronic Commerce in Singapore," *Information and Management* 42 (2004): 89–102; Poh Kam Wong, "Global and National Factors Affecting E-Commerce Diffusion in Singapore," *The Information Society* 19 (2003): 19–32.
180. Mark Goh, "Congestion Management and Electronic Road Pricing in Singapore," *Journal of Transport Geography* 10 (2002): 29–38.
181. Eddie C.Y. Kuo, "Singapore," in Noam, Komatsuzaki, and Conn, *Telecommunica-tions in the Pacific Basin*, 265–285.
182. Kenneth C.C. Yang, "Exploring Factors Affecting the Adoption of Mobile Com-merce in Singapore," *Telematics and Informatics* 22 (2005): 257–277, statistics, p. 259.
183. Ibid., 260.
184. Eddie C.Y. Kuo and Linda Low, "Information Economy and Changing Occupa-tional Structure in Singapore," *The Information Society* 17 (2001): 281–293.
185. For an excellent sampling of this industry from around the world and across various industries, see Alfred D. Chandler, Jr., Franco Amatori, and Takashi Hikino (eds.), *Big Business and the Wealth of Nations* (Cambridge: Cambridge Uni-versity Press, 1997). This collection is limited, however to North American and Western European experiences, the focus areas of most business historians in the 1970s–2000.
186. Exemplified by the essays in Noam, Komatsuzaki, and Conn, *Telecommunications in the Pacific Region*, and those in Georgette Wang (ed.), *Treading Different Paths: Informatization in Asian Nations* (Norwood, N.J.: Ablex Publishing Co., 1994).
187. With this observation in mind, one should consult his account of how Asian elec-tronics firms succeeded, Alfred D. Chandler, Jr., *Inventing the Electronic Century:*

The Epic Story of the Consumer Electronics and the Computer Science Industries (New York: Free Press, 2001).

188. Manuel Castells has been defining telematic societies for several decades, encapsulating his broad views on the subject recently in *Communication Power* (Oxford: Oxford University Press, 2009).

189. The turning point came with Paul N. Edwards, *The Closed World: Computers and the Politics of Discourse in Cold War America* (Cambridge, Mass.: MIT Press, 1996).

190. Dedrick and Kraemer, *Asia's Computer Challenge*, 301.

191. Ibid., 302.

192. "Country Review: Asia," August 25, 1998, Gerstner/Trip Files/Asia 1998, Box 248, folder 2, IBM Corporate Archives, Somers, N.Y.

CHAPTER 9

1. John H. Maier, "Information Technology in China," *Asian Survey* 20, no. 8 (August 1980): 864.

2. See for example, James M. Popkin and Partha Iyengar, *IT and the East: How China and India Are Altering the Future of technology and Innovation* (Boston, Mass.: Harvard Business School Press, 2007; Ernest H. Preeg, *India and China: An Advanced Technology Race and How the United States Should Respond* (Arlington, Va.: Manufacturers Alliance, 2008); Marcus Franda, *China and India Online: Information Technology Politics and Diplomacy in the World's Two Largest Nations* (Lanham, Md.: Rowman & Littlefield, 2002).

3. Tarun Khanna, *Billions of Entrepreneurs: How China and India Are Reshaping Their Futures—And Yours* (Boston, Mass.: Harvard Business School Press, 2007): 7.

4. Quoted in Ernest H. Preeg, *India and China: An Advanced Technology Race and How the United States Should Respond* (Arlington, Va.: Manufacturers Alliance/MAPI, 2008): 25.

5. Ibid., 59–68.

6. Franda, *China and India Online*, 20.

7. V.P. Chitale, *Foreign Technology in India* (New Delhi: Economic and Scientific Research Foundation, 1973): 70–71.

8. Ibid., 72.

9. The topic has a large and growing literature not yet in agreement on the numbers of casualties of the famines and political purges, see for example, Frank Dikötter, *Mao's Great Famine: The History of China's Most Devastating Catastrophe, 1958–1962* (New York: Walker & Company, 2010): 324–334.

10. Robert F. Dernberger, "Economic Development and Modernization in Contemporary China: The Attempt to Limit Dependence on the Transfer of Modern Industrial Technology From Abroad and To Control Its Corruption of the Maoist Social Revolution," in Joint Economic Committee, Congress of the United States, *Issues in East-West Commercial Relations* (Washington, D.C.: U.S. Government Printing Office, 1979): 98–104.

11. Genevieve D. Dean, "A Note on Recent Policy Changes," in Richard Baum (ed.), *China's Four Modernizations: The New Technological Revolution* (Boulder, Col.: Westview Press, 1980): 105.

12. Ibid.

13. Dean, "A Note on Recent Policy Changes," 105–108; Shannon R. Brown, "China's Program of Technology Acquisition," Ibid., 155.

14. I relied for much of my account in the next several paragraphs on the primary research and reporting done by Zhang Jiuchun and Zhang Baichun, "Founding of

the Chinese Academy of Sciences' Institute of Computing Technology," *IEEE Annals of the History of Computing* 29, no. 1(January–March 2007): 16–33.

15. Ken W. Sayers, "A Summary History of IBM's International Operations, 1911–2006," October 20, 2006, p. 100, IBM Corporate Archives, Somers, N.Y.

16. Brown, "China's Program of Technology Acquisition," 158.

17. Ibid.

18. Zhang and Zhang, "Founding of the Chinese Academy of Sciences' Institute of Computing Technology," 18–21.

19. Ibid.

20. Ibid., 27.

21. Ibid., 27–28.

22. Ibid., 29.

23. OECD, *Electronic Computers: Gaps in Technology* (Paris: OECD, 1969): 33.

24. P. Russell Nyberg, "Computer Technology in Communist China," *Datamation* (February 1968): 39.

25. Zhang and Zhang, "Founding of the Chinese Academy of Sciences' Institute of Computing Technology," 30.

26. Nyberg, "Computer Technology in Communist China," 43.

27. "World Trade Story," Box 4 "World Trade Corporation Annual Reports/Slide Presentations" Folder 4–9 R-6, IBM Corporate Archives, Somers, N.Y.

28. Bohdan O. Szuprowicz, "China's Computer Industry," *Datamation* (June 1975): 83.

29. Ibid., 84.

30. Ibid.

31. Ibid., 85.

32. Ibid., 85–86.

33. All material for this paragraph drawn from Denis Fred Simon and Detlef Rehn, *Technological Innovation in China: The Case of Shanghai's Electronics Industry* (Cambridge, Mass.: Ballinger, 1988): 49–51.

34. Thomas Finger, "Recent Policy Trends in Industrial Science and Technology," in Baum, *China's Four Modernizations*, 79.

35. Ibid., 80.

36. Ibid.

37. Qiwen Lu, *China's Leap into the Information Age: Innovation and Organization in the Computer Industry* (Oxford: Oxford University Press, 2000): 7–10.

38. Barry Naughton, *The Chinese Economy: Transitions and Growth* (Cambridge, Mass.: MIT Press, 2007): 77–79, 86–90.

39. Harvey L. Garner, "Computing in China, 1978," *Computer* (March 1979): 84.

40. Ibid., 95.

41. While highly reliable numbers are unavailable, these extant data make sense as acquisition of systems occurred all through the 1970s. An earlier report from the mid-1970s had put the number of installed computers in 1973 at "a few hundred machines," leading one observer to reach the obvious conclusion that China "remained at a very low level of computerization," Arthur D. Little, "The World Computer Industry, 1973–1978" February 1974, p. 26, CBI 55, Box 9, folder 33, Charles Babbage Institute, University of Minnesota, Minneapolis.

42. Ibid., 81–96.

43. John H. Maier, "Information Technology in China," *Asian Survey* 20, no. 8 (August 1980): 863.

44. Ibid., 864.

45. Ibid., 865.
46. U.S. Congress, Office of Technology Assessment, *Technology Transfer to China*, OTA-ISC-340 (Washington, D.C.: U.S. Government Printing Office, July 1987): 28–29.
47. A point made effectively by Robert P. Kreps, *Buying Hens Not Eggs: The Acquisition of Communications and Information Technology by the Peoples' Republic of China* (Cambridge, Mass.: Program on Information Resources Policy, Harvard University, 1988)
48. U.S. Congress, Office of Technology Assessment, *International Competitiveness in Electronics*, OTA-ISC-200 (Washington, D.C.: U.S. Government Printing Office, November 1983): 387.
49. William J. Long, "Economic Incentives and International Cooperation: Technology Transfer to the People's Republic of China, 1978–86," *Journal of Peace Research* 28, no. 2 (1991): 180, but see the entire article and its rich bibliography for U.S. Cold War technology transfer policies, 175–189.
50. Jeff X. Zhang and Yan Wang, *The Emerging Market of China's Computer Industry* (Westport, Conn.: Quorum Books, 1995): 19–21.
51. Sayers, "History of IBM's International Operations," 100–101. Sayers reported that IBM had as many as 50 systems installed or on order in 1984 in China, the year IBM established IBM China, Inc. as a wholly-owned local subsidiary, relieving IBM Japan of the responsibility of selling in China, Ibid., 101.
52. Office of Technology Assessment, *International Competitiveness in Electronics*, 96.
53. Zhang and Wang, *China's Computer Industry*, 21.
54. Ibid., 23–24.
55. Ibid., 25–26, based on Chinese sources.
56. Ibid., 26–27.
57. Ibid., 32.
58. Ibid., 33–34.
59. Ibid., 35–42; Qiwen Lu, *China's Leap into the Information Age*, treats each of the major providers with a chapter-long history.
60. See, for example, Ted Tschang and Lan Xue, "The Chinese Software Industry," in Ashish Arora and Alfonso Gambardella (eds.), *From Underdogs to Tigers: The Rise and Growth of the Software Industry in Brazil, China, India, Ireland, and Israel* (Oxford: Oxford University Press, 2005): 131–167.
61. The subject of intellectual property rights and China has been the subject of much discussion, particularly in Europe and North America, where practices in China are seen as a major inhibitor to normal trade relations. For brief introductions to the issues in China, see James M. Pokin and Partha Iyengar, *IT and the East: How China and India are Altering the Future of Technology and Innovation* (Boston, Mass.: Harvard Business School Press, 2007): 44, 58–59, 63–65, 175–176; Zhang and Wang, *China's Computer Industry*, 150–159; Kreps, *Buying Hens Not Eggs*, 65–70.
62. Zhang and Wang, *China's Computer Industry*, 46.
63. S.T. Nandasara and Yoshiki Mikami, "Asian Language Processing: History and Perspectives," *IEEE Annals of the History of Computing* 31, no. 1 (January–March 2009): 4–7.
64. Zhang and Wang, *China's Computer Industry*, 48; Robert Buderi and Gregory T. Huang, *Guanxi (The Art of Relationships): Microsoft, China, and Bill Gates's Plan to Win the Road Ahead* (New York: Simon & Schuster, 2006).
65. Zhang and Wang, *China's Computer Industry*, 50–51; Lu, *China's Leap into the Information Age*, which contains chapter-length studies of the Stone Group Company,

Legend Computer Group Company, Founder Group Company, and China Great Wall Computer Company.

66. Zhang and Wang, *China's Computer Industry*, 52–53.
67. Ibid., 54.
68. Kreps, *Buying Hens Not Eggs*, 65.
69. Zhang and Wang, *China's Computer Industry*, 42.
70. International Data Corporation, "China Computer Industry Review and Forecast, 1986–1991" (Framingham, Mass.: IDC, 1988): 1, CBI 55, Box 60, Folder 24, Charles Babbage Institute, University of Minnesota, Minneapolis.
71. OTA, *Technology Transfer to China*, 96.
72. Ibid., 2.
73. Simon and Rehn discuss the experience of Shanghai, *Technological Innovation in China*, 87–158.
74. For insight into this problem, Simon and Rehn, *Technological Innovation in China*, footnote 33, p. 83, footnote 58, p. 85, but see also Denis Fred Simon and Cong Cao, *China's Emerging Technological Edge: Assessing the Role of High-End Talent* (Cambridge: Cambridge University Press, 2009): 23–39
75. OTA, *Technology Transfer to China*, 100.
76. IDC, "China Computer Industry Review and Forecast, 1986–1991," 2–3; OTA, *Technology Transfer to China*, 93–101.
77. On how these worked, see Lu, *China's Leap into the Information Age*, 177–190.
78. OTA, *Technology Transfer to China*, 157.
79. Dhulin Gu, *China's Industrial Technology: Market Reform and Organizational Change* (London: Routledge, 1999): 17–22.
80. Yongnian Zheng, *Technological Empowerment: The Internet, State, and Society in China* (Stanford, Call.: Stanford University Press, 2008): 50–52.
81. International Data Corporation, "China Computer Industry Review and Forecast, 1986–1991" (Framingham, Mass.: IDC, 1988):7, CBI 55, Box 60, Folder 24, Charles Babbage Institute, University of Minnesota, Minneapolis.
82. Ibid., 8.
83. Kreps, *Buying Hens Not Eggs*, 20.
84. OTA, Technology Transfer to China, 95.
85. Ibid.
86. Joseph Y. Battat, "Transfer of Computer and Data Processing Technologies: First-Hand Experiences of a Foreign Consultant," U.S. Congress, Joint Economic Committee Congress of the United States, *China's Economy Looks Toward the Year 2000*, vol. 2, *Economic Openness in Modernizing China* (Washington, D.C.: U.S. Government Printing Office, 1986): 252.
87. For a series of quotes and specifics from Chinese sources, see Kreps, *Buying Hens Not Eggs*, 28–30.
88. Ibid., 21.
89. Ibid., 66.
90. Ibid., 66.
91. Kenneth L. Kraemer and Jason Dedrick, "National Computer Policy and Development in China," Working Paper PAC-060A (Irvine, Cal.: Center for Research on Information Technology and Organiation, University of California, Irvine, 1994), http://crito.uci.edu/papers (last accessed 8/10/2010).
92. Kreps, *Buying Hens Not Eggs*, 62.
93. Nobel Laureate economist Michael Spence has argued that proper education was one of a small handful of activities essential for the diffusion of modern methods

and technologies in an economy undergoing development, *The Next Convergence: The Future of Economic Growth in a Multispeed World* (New York: Farrar, Straus and Giroux, 2011): 79–82, 113–115.

94. Ibid., 9.
95. Ibid., 13.
96. Ibid., 21.
97. Ibid., 27–28.
98. Ibid., 28–29.
99. Vijay Gurbaxani et al., "Government as the Driving Force Toward the Information Society: National Computer Policy in Singapore," *The Information Society* 7, no. 2 (1990): 155–185; Jason Dedrick and Kenneth L. Kraemer, "Caught In The Middle: Information Technology Policy in Australia," Ibid., 9, no. 4 (1993): 333–363; Kenneth L. Kraemer and Jason Dedrick, "Coordination, Entrepreneurship and Flexibility: Information Technology Policy in Taiwan," (Irving, Cal.: Center for Research on Information Technology and Organization, University of California, Irvine, 1994).
100. Kraemer and Dedrick, "National Computer Policy and Development in China," 32.
101. Gu, *China's Industrial Technology*, 92–110.
102. Chien-Hsun Chen and Hui-Tzu Shih, *High-Tech Industries in China* (Cheltenham, U.K.: Edgard Elgar, 2005): 6.
103. For a description of the import features of the 1990s, see Xu Jingping, "China's International Technology Transfer: The Current Situation, Problems and Future Prospects," in Charles Feinstein and Christopher Howe (eds.), *Chinese Technology Transfer in the 1990s: Current Experience, Historical Problems and International Perspectives* (Cheltenham, U.K.: Edward Elgar, 1997): 88–89.
104. Ibid., 105.
105. Gu, *China's Industrial Technology*, 17–22; Adam Segal, *Digital Dragon: High-Technology Enterprises in China* (Ithaca, N.Y.: Cornell University Press, 2003): 36–38.
106. The major source on the role of these industrial parks for computing and telecommunications is now Haiyang Li (ed.), *Growth of New Technology Ventures in China's Emerging Market* (Cheltenham, U.K.: Edward Elgar, 2006), and in particular, 230–232.
107. Gu, *China's Industrial Technology*, 106–109.
108. Segal, *Digital Dragon*, 3.
109. Neil Gregory, Stanley Nollen, and Stoyan Tenev, *New Industries from New Places* (Washington, D.C.: World Bank, 2009): 141–144.
110. Ernest J. Wilson III, *The Information Revolution and Developing Countries* (Cambridge, Mass.: MIT Press, 2004): 233.
111. Harwit, *China's Telecommunications Revolution*, 79.
112. Kreps, *Buying Hens Not Eggs*, 72.
113. Naughton, *The Chinese Economy*, 343–345.
114. Xiaobai Shen, *The Chinese Road to High Technology: A Study of Telecommunications Switching Technology in the Economic Transition* (London: Macmillan, 1999):41–49, 63–104; Eric Harwit, *China's Telecommunications Revolution* (New York: Oxford University Press, 2008): 18–78.
115. Tony Walker, "Hats Off to the Revolution," *Financial Times*, March 30, 1993, p. 21.
116. Zheng, *Technological Empowerment*, 107–108.
117. In addition to possibly undercounting users in the nation, there is the added communities of Chinese diasporas interacting with fellow nations, Brenda Chan, "The

Internet and New Chinese Migrants," in Andoni Alonso and Pedro J. Oirzabal (eds.), *Diasporas in The New Media Age* (Reno, Nev.: University of Nevada Press, 2010): 225–241; Yu Zhou, "The Migration of Chinese Professionals and the Development of the Chinese ICT Industry," Ibid., 242–264.

118. There are now numerous inventories of how many Internet users are in China and the numbers are not consistent from one to another. However, proportionally they all exhibit the patterns evident in Table 9.9 of late adoption.

119. Harwit, *China's Telecommunications Revolution*, 79.

120. Case studies provide insight into the dynamics involved in China's expanding mobile phone market. For an example, see So and Westland, *Redwired*, 160–177.

121. Ibid., 82–87.

122. For discussion of specific price cuts, Ibid., 92–94.

123. Ibid., 101.

124. Zheng, *Technological Empowerment*, 107–116; Zhou Yongming, *Historicizing Online Politics: Telegraphy, the Internet, and Political Participation in China* (Stanford, Cal.: Stanford University Press, 2006); Guobin Yang, *The Power of the Internet in China: Citizen Activism Online* (New York: Columbia University Press, 2009): 32; Jack Linchuan Qiu, *Working-Class Network Society: Communication Technology and the Information Have-Less in Urban China* (Cambridge, Mass.: MIT Press, 2009): 83–154; Sherman So and J. Christopher Westland, *Redwired: China's Internet Revolution* (London: Marshall Cavendish, 2010).

125. Harwit, *China's Telecommunications Revolution*, 171–172.

126. For summary of patterns of diffusion, Marina Yue Zhang and Bruce W. Stening, *China 2.0: The Transformation of a Emerging Superpower . . . and the New Opportunities* (Singapore: John Wiley & Sons, 2010): 60–63.

127. While all of the citations in this chapter regarding the Internet in China discusses these issues, for a collection of commentary of a much broad scope on modern China, see Thomas Buoye, Kirk Denton, Bruce Dickson, Barry Naughton, and Martin K. Whyte (eds.), *China: Adapting the Past Confronting the Future* (Ann Arbor, Mich.: Center for Chinese Studies, University of Michigan, 2002). Recently problems with Google also surfaced, briefly discussed by Siva Vaidhyanathan, *The Googlization of Everything (And Why We Should Worry)* (Berkeley, Cal.: University of California Press, 2011): 9–10, 74, 116–121, 124–134, 153, 117.

128. Zhang and Stening, *China 2.0*, 79.

129. Ibid., 81–84; Yang, *the Power of the Internet in China*, 51–53.

130. Yang, *The Power of the Internet in China*, 105.

131. Ibid., 107.

132. Ibid., 133.

133. Ibid., 134–138.

134. Ibid., 108–109.

135. Qiu, *Working-Class Network Society*, 22.

136. His argument is that in China there was no crisp digital divide when it came to access to the Internet because there were less expensive technologies that could be afforded by low income users and that this stratum of society was large, growing in number, and active on the Internet, Qiu, *Working-Class Network Society*, 83–154. Similar patterns of Internet access were appearing as well in other parts of Asia, Africa, and Latin America.

137. Ibid., 24–26.

138. Yuval Atsman and Max Magni, "China's Internet Obsession," *McKinsey Quarterly* (March 2010), Online gaming is very popular in China, as across much of Asia, So

and Westland, *Redwired*, 137–159. http://www.mckinseyquarterly.com/Marketing/Digital_Marketing/Chinas_Internet-obsessio . . . (last accessed 3/10/2010).

139. Rongxing Guo, *An Introduction to the Chinese Economy: The Driving Forces Behind Modern Day China* (Singapore: John Wiley & Sons, 2010): 135–151.

140. Naughton, *The Chinese Economy*, 388–392.

141. Ling Zhijun, *The Lenovo Affair: The Growth of China's Computer Giant and Its Take-over of IBM-PC* (Singapore: John Wiley & Sons, 2005); Zhang and Stening, *China 2.0*, 237–240.

142. Naughton, *The Chinese Economy*, 396.

143. Ming Zeng and Peter J. Williamson, Dragons At Your Door: How Chinese Cost Innovation Is Disrupting Global Competition (Boston, Mass.: Harvard Business School Press, 2007): 19, 28–39, 46–49.

144. Shahid Yusuf and Kaoru Nabeshima, "Strengthening China's Technological Capability," Policy Research Working Paper 4309 (Washington, D.C.: The World Bank, August 2007): 5; Martin Schaaper, "An Emerging Knowledge-Based Economy in China? Indicators From OECD Databases," DSTI/DOC (2004)4 (Paris: OECD, March 22, 2004): 13–16.

145. Schaaper, "An Emerging Knowledge-Based Economy in China? Indicators From OECD Databases," 21.

146. Carl J. Dahlman and Jean-Eric Aubert, *China and the Knowledge Economy: Seizing the 21st Century* (Washington, D.C.: World Bank, 2001); Kevin Zhu, Sean Xu, Kenneth L. Kraemer, and Jason Dedrick, "Global Convergence and Local Divergence in e-Commerce: Cross-Country Analysis," in Kenneth L. Kraemer, Jason Dedrick, Nigel P. Melville, and Kevin Zhu (eds.), *Global E-Commerce: Impacts of National Environment and Policy* (Cambridge: Cambridge University Press, 2006): 345–384; Cyrill Eltschinger, *Source Code China: The New Global Hub of IT Outsourcing* (Singapore: John Wiley & Sons, 2007): 2–6; David Sheff, *China Dawn: The Story of a Technology and Business Revolution* (New York: HarperBusiness, 2002); Mathews and Cho, *Tiger Technology*, 29–70; Popkin and Lyengar, *IT and the East*, 27–69; Rebecca A. Fannin, *Silicon Dragon: How China Is Winning the Tech Race* (New York: McGraw-Hill, 2008): 3–18.

147. OECD, "Is China the New Centre for Offshoring of IT and ICT-Enabled Services?" DSTI/ICCP/IE(2006)10/FINAL (Paris: OECD, April 5, 2007): 5.

148. "China Internet Population Hits 384 Million," ICT Statistics Newslog, January 15, 2010, http://www.itu.int/ITU-D/ict/newslog/China+Internet+Population+Hits+384+Million.aspx (last accessed 3/01/2011).

149. "China Has 420M Internet Users and a Need for Speed," ICT Statistics Newslog, August 31, 2010, http://www.itu.int?ITU-D/ict/newslog/China+Has+420M+Internet+Users+And+A+Need+For+Speed.aspx (last accessed 3/01/2011).

150. Dan Schiller, "Poles of Market Growth?: Open Questions About China, Information and the World," *Global Media and Communication* 1, no. 1 (2005): 79–103.

151. Geoffrey Z. Liu, *Evergreen: Bringing Information Resources to Rural China* Washington, D.C.: Council on Library and Information Resources, July 2005); Christine Zhen-Wei Qiang et al., "Rural Informatization in China," World Bank Working Paper No. 172 (Washington, D.C.: World Bank, June 2009).

152. Zhang and Stening, *China 2.0*, 284.

153. Barry Naughton and Adam Segal, "China in Search of a Workable Model: Technology Development in the New Millennium," in William W. Keller and Richard J. Samuels (eds.), *Crisis and Innovation in Asian Technology* (Cambridge: Cambridge University Press, 2003): 186.

CHAPTER 10

1. Patrick French, "Another Country, Another Era," *India Today*, December 26, 2005.
2. These included, for example, the Atomic Energy Commission (AEC), Department of Electronics, and Electronics Commission.
3. Examples include Tarun Khanna, *Billions of Entrepreneurs: How China and India Are Reshaping Their Futures and Yours* (Boston, Mass.: Harvard Business School Press, 2007); Rafiq Dossani, *India Arriving: How the Economic Powerhouse Is Redefining Global Business* (New York: AMACOM, 2008); Patrick French, *India A Portrait* (New York: Alfred A. Knopf, 2011).
4. IBM's ramp up in India was remarkably fast. In 2002, it had several thousand employees; in 2004 it was 9,000, and in 2006, over 43,000. In subsequent years additional expansion occurred as the company established sites in 40 cities and worked with over 2,500 business partners, Sayer, "History of IBM's International Operations," 242–243.
5. For example, Pete Engardio (ed.), *Chindia: China and India Are Revolutionizing Global Business* (New York: McGraw-Hill, 2007).
6. Arvind Panagariya, *India: The Emerging Giant* (New York: Oxford University Press, 2008): 13.
7. Gary Fields, *Poverty, Inequality and Development* (Cambridge: Cambridge University Press, 1980): 204.
8. Panagariya, *India*, 136–138.
9. "India Census: Population Goes Up to 1.21bn," BBC News, South Asia, March 31, 2011, http://www.bbc.co.uk/news/world-south-asia-12916888 (last accessed 3/31/2011).
10. Amiya Kumar Bagchi, "Enemies of an Information Society Through The Ages," in Amiya Kumar Bagchi, Dipankar Sinha, and Barnita Baghi (eds.), *Webs of History: Information, Communication and Technology From Early to Post-Colonial India* (New Delhi: Indian History Conference, 2005): 288–289.
11. R. Srivastave and S.K. Sasikumar, "An Overview of Migration in India, Its Impacts and Key Issues," Paper presented in the Regional Conference on Migration, Development and Pro-Poor Policy Choices in Asia (Dhaka, June 2003).
12. World Bank, *2011: The Little Data Book on Information and Communication Technology* (Washington, D.C.: World Bank, 2011): 104.
13. Sean M. Dougherty, Richard Herd, Thomas Chalaux and Abdul Azeez Erumban, "India's Growth Pattern and Obstacles to Higher Growth," Economic Department Working Papers, No. 623 (Paris: OECD, August 11, 2008): 11.
14. Ibid., entire report.
15. "World Coal Institute-India," http://www.worldcoal.org/pages/content/index.asp?PageID=402 (last accessed 8/15/2011); http://www.iea.org/weo/database_electricity/electricity_access-database.htm (last accessed 8/15/2011); A.S. Pabla, "Power Quality Problems in India," htt/Sessions/1999/Documents/Papers/2_8.pdf (last accessed 10/5/2009); "India Electric Power," http://www.photius.com/countries/india/economy/india_economyelectric_power.html (last accessed 8/18/2011); Panagariya, *India*,382–394.
16. Bagchi, Sinha, and Bagchi, *Webs of History*, 286–288.
17. Geoffrey S. Kirkan, Peter K. Cornelius, Jeffrey D. Sachs, and Klaus Schwab, *The Global Information Technology Report, 2001–2002: Readiness for the Networked World* (New York: Oxford University Press for the World Economic Forum, 2002): quote on p. 220, statistics, p. 221.

18. World Bank, *2006 Information and Communications for Development: Global Trends and Policies* (Washington, D.C.: World Bank, 2006): 2007.
19. Economist Intelligence Unit, *Digital Economy Rankings 2010: Beyond e-Readiness* (London: Economist Intelligence Unit, 2010): 24.
20. Robert Schware, "Software Industry Entry Strategies for Developing Countries: A 'Walking on Two Legs Proposition',", *World Development* 20, no. 2 (1992): 143–164.
21. Panagariya, *India*, 282–210.
22. Dedrick and Kraemer, "India's quest for Self-Reliance in Information Technology," 491–492.
23. For a technical description of this system, see Mohi Mukherjee, "The First Computer in India," in Utpal K. Banerjee (ed.), *Computer Education in India: Past, Present and Future* (New Delhi: Concept Publishing Company, 1996): 13–16.
24. D. Dutta Majumder, "Thoughts on Emergence of IT Activities in India," in Banerjee, *Computer Education in India*, quote p. 4, much of this paragraph was drawn from pp. 1–7.
25. Ken W. Sayers, "A Summary History of IBM's International Operations, 1911–2006," October 20, 2006, p. 240, IBM Corporate Archives, Somers, New York.
26. Majumder, "Thoughts on Emergence of IT Activities in India," 4–7.
27. B. Nag, "Computer Design and Development in India," Banerjee, *Computer Education in India*, 21–25.
28. Banerjee, *Computer Education in India*, contains memoirs and short histories of these and other early training programs.
29. For a rare look into the Indian military's experience learning about computers, see Banerjee, *Computer Education India*, 317–323.
30. J. Roy, "Early Computers," Ibid., 12.
31. B. Nag, "Computer Design and Development in India," Ibid., 21–25.
32. These early inventories can be viewed as quite reliable since there were so few systems to count. The best source for this period continues to be Om Vikas and L. Ravichandran, "Computerization in India: A Statistical Review," *Electronics: Information and Planning* 6 (December 1978): 318–351.
33. Banerjee, *Computer Education in India*, 33–34.
34. The earliest history of computing in India also begins largely with the narrative of events in the 1970s, despite some activities in the 1950s and 1960s, Joseph M. Grieco, *Between Dependency and Autonomy: India's Experience with the International Computer Industry* (Berkeley, Cal.: University of California Press, 1984) and explains the early role of the electronics industry, pp. 20–22.
35. Deepak Lai, "Driving Forces Behind Acceleration of Indian Growth and the Outlook to 2030," Paper for seminar in honor of Angus Maddison, November, 2006, in possession of the author; Panagariya, *India*, 22–77.
36. Panagariya, *India*, 75, second quote, p. 77.
37. Ibid., 77.
38. India, Electronics Committee of India, *Electronics in India* (New Delhi: Electronics Committee of India, 1966): 235.
39. Grieco, *Between Dependency and Autonomy*, 24–26.
40. Ibid., 26–28; Vikas and Ravichandran, "Computerization in India," 318–351.
41. Grieco, *Between Dependency and Autonomy*, 26–28, quote p.28.
42. Grieco's account of the IBM exit from India remains the most complete and best documented, Ibid., 24–25, 46–50, 89–93, 122–123, 139–141. See also, Panagariya, *India*, 60–62.

43. For an introduction to the issue, Chris Freeman and Luc Soefe, *The Economics of Industrial Innovation*, 3rd ed. (Cambridge, Mass.: MIT Press, 1997): 355–358. I have been profoundly influenced about the subject by, among others, William J. Baumol's work and that of his circle of like-minded economists. Two important sources of his influence on my view of national technology evolution are Edward N. Wolff and Sue Ann Batey Blackman, *Productivity and American Leadership: The Long View* (Cambridge, Mass.: MIT Press, 1991) and his broader study with Robert E. Litan and Carl J. Schramm, *Good Capitalism, Bad Capitalism and the Economics of Growth and Prosperity* (New Haven, Conn.: Yale University Press, 2007).

44. Presentations on India by H. Figueroa, "India-Satus Report," undated, but June 1977 and that fall, RG 5 Business Planning/CMC/Meeting Material/1977-6/21, Box 37, folder 6, IBM Archives, Somers, New York.

45. Ibid.

46. Ibid.

47. Ibid.

48. Joseph M. Grieco, "Between Dependency and Autonomy: India's Experience with the International Computer Industry," *International Organization* 36, no. 3 (Summer 1982): 615.

49. The problems are described in considerable detail by Richard Heeks, *India's Software Industry: State Policy, Liberalisation and Industrial Development* (New Delhi: SAGE Publications, 1996): 67–157, 271–311; and for a government description of the rules for acquiring systems that clearly demonstrate the complexity and constraining features of government intervention in the diffusion process, see Government of India, Department of Electronics, *Annual Report 1976–77* (New Delhi: Department of Electronics, 1977): 94–95, copy in CBI 32, Box 679, folder 5, Charles Babbage Institute, University of Minnesota, Minneapolis.

50. Grieco, "Between Dependency and Autonomy: India's Experience with the International Computer Industry," 617.

51. Suma S. Athreye, "The Indian Software Industry," in Ashish Arora and Alfonso Gambardella (eds.), *From Underdogs to Tigers: The Rise and Growth of the Software Industry in Brazil, China, India, Ireland, and Israel* (New York: Oxford University Press, 2005): 22–26.

52. Ibid., 22–23; Gurcharan Das, *India Unbound: From Independence to the Global Information Age* (London: Penguin, 2000): 11, 162–163; Deborah A. Walsh, "An Evaluation of the Maturing Indian Software Industry" (Unpublished Ph.D. dissertation, 2006).

53. For effects on local vendors, such as Wipro, see Steve Hamm, *Bangalore Tiger: How Indian Tech Upstart Wipro Is Rewriting the Rules of Global Competition* (New York: McGraw-Hill, 2007): 36–37, quote p. 36.

54. Panagariya, *India*, 65.

55. For further development of this line of reasoning, see Suma S. Athreye, "The Indian Software Industry," in Ashish Arora and Alonso Gambardella (eds.), *From Underdogs to Tigers: The Rise and Growth of the Software Industry in Brazil, China, India, Ireland, and Israel* (Oxford: Oxford University Press, 2006): 22–23.

56. Grieco, *Between Dependency and Autonomy*, 49.

57. Ibid., 50.

58. Ibid., 104–106; Utpal K. Banerjee, *Information Management in Government* (New Delhi: Concept Publishing, 1984): 67–70. Not clear is what the military did, since in other countries where similar "buy local" campaigns were implemented, they

were normally exempt, or proved able to bypass such regulations, as happened in France, for example.

59. Banerjee, *Information Management in Government*, 67–68.

60. As both an historian and as an experienced IBMer with sales, IT management, and product pricing experience in the firm going back to the mid-1970s, I personally gave much thought to the issue of 1401s and the implication that IBM was "dumping" old hardware into India. Given how products were priced, their margins, and how they were supported, the company would have been much better off selling its newer Systems 360 in India and its executives knew that, and I believe they would have done so if possible. At the time, they had pretty well cleaned out the world of its prior appetite for the popular 1401 and had moved it along into the S/360. So I conclude that the IBMers who worried about how much IT India could absorb genuinely thought the Indians had to work through a phase of IT evolution to become sophisticated enough technologically to use the newer system, which was more expensive, hence, required more jobs to run through them to justify their expense, and required different technical and, just as important, managerial skills in very short supply in the country. Failure along any of those dimensions would have tarnished IBM's reputation in the market, and its business, since its products were leased to customers, therefore, subject to cancellation. In short, this was less about whether Indians could operate more complex machinery, and more about whether they had enough processing of a sophisticated-enough level and in sufficient quantities to warrant use of these newer systems. The fact that they adopted minicomputers and PCs seemed proof to me that the IBMers were right in their assessment of the Indian situation at the time.

61. Grieco, *Between Dependency and Autonomy*, 106–107.

62. Ibid., 107–110.

63. Jason Dedrick and Kenneth L. Kraemer, "India's Quest for Self-Reliance in Information Technology: Costs and Benefits of Government Intervention," *Asian Survey* 33, no. 5 (1993): 463–492; Hans-Peter Brunner, "Building Technological Capacity: A Case Study of the Computer Industry in India, 1975–87," *World Development* 19, no. 12 (1991): 1737–1751.

64. Grieco, *Between Dependency aind Autonomy*, 99–102.

65. Banerjee, *Information Management in Government*, 68–73.

66. Panagariya, *India*, 78, and for the political context, 79–80.

67. Ibid., 92.

68. Eddie J. Girdner, "Economic Liberalization in India, The New Electronics Policy," *Asian Survey* 27, no. 11 (November 1987): 1188–1204.

69. Dedrick and Kraemer, "India's Quest for Self-Reliance in Information Technology: Costs and Benefits of Government Intervention," 463–492.

70. Ibid.; Heeks, *India's Software Industry*, 40–46.

71. For a description of early projects promoted by the public sector, see Utpal K. Banerjee, *Computer Applications for Techno-Economic Development* (New Delhi: Concept Publishing, 1985).

72. Dedrick and Kraemer, "India's Quest for Self-Reliance in Information Technology: Costs and Benefits of Government Intervention," 472.

73. Ibid.

74. Utpal K. Banerjee, *Computer Education in India*, 396–400.

75. Dedrick and Kraemer, "India's Quest for Self-Reliance in Information Technology: Costs and Benefits of Government Intervention," 469, but see the entire article for further discussion of this issue.

76. Arvind Singhal and Everett M. Rogers, *India's Information Revolution* (New Delhi: SAGE Publications, 1989): 190–196.

77. Dedrick and Kraemer, "India's Quest for Self-Reliance in Information Technology: Costs and Benefits of Government Intervention," 463–492.

78. Indian Department of Electronics and NASSCOM in Dedrick and Kraemer, "India's Quest for Self-Reliance in Information Technology: Costs and Benefits of Government Intervention," passim.

79. Dedrick and Kraemer, "India's Quest for Self-Reliance in Information Technology: Costs and Benefits of Government Intervention," 469–470.

80. NASSCOM (National Association of Software and Service Companies), "Indian Software Industry 1990–95," Report to National Software Conference '89 (New Delhi, July 1989).

81. Singh and Rogers, *India's Information Revolution*, 169.

82. Dedrick and Kraemer, "India's Quest for Self-Reliance in Information Technology: Costs and Benefits of Government Intervention," 463–492.

83. Ibid. It is normal for adoption rates to demonstrate high rates in the beginning of the process's cycle. If you had more, adopting one computer would represent a 100 percent increase over the previous period.

84. Ibid.

85. Ashok V. Desai, *India's Telecommunications Industry: History, Analysis, Diagnosis* (New Delhi: SAGE Publications, 2006): 39–41.

86. On his activities and biography, Das, *India Unbound*, 207–210.

87. Desai, *India's Telecommunications Industry*, 49.

88. Ibid., 63.

89. Ibid., 19, 65.

90. Ibid., 142.

91. Das, *India Unbound*, 210.

92. Panagariya, *India*, first quote p. 371, second quote, p. 374.

93. Ibid., 395.

94. Desai, *India's Telecommunications Industry*, 142–144.

95. Arvind Singhal and Everett M. Rogers, *India's Communication Revolution: From Bullock Carts to Cyber Marts* (New Delhi: SAGE Publications, 2001): 58.

96. Nagy Hanna, *Exploring Information Technology for Development: A Case Study for India* (Washington, D.C.: World Bank, 1994): xi.

97. Ibid., xv.

98. Ibid., xvi.

99. Hanna, *Exploring Information Technology for Development*, 9–15.

100. Subhash Bhatnagar, "India's Software Industry," in Vandana Chandra (ed.), *Technology, Adaptation, and Exports: How Some Developing Countries Got It Right* (Washington, D.C.: World Bank 2006): 49–52.

101. Bhatnagar, "India's Software Industry," 51.

102. Grieco, *Between Dependency and Autonomy*, 84–87; M.S. Krishnan, Narayan Ramasubbu, and Ramanath Subremanian, "Evolution of the Indian Software Industry: The Emerging Model of Mobilizing Global Talent," in Swaminathan, *Indian Economic Superpower*, 59–74; Franda, *China and India Online*, 27–30.

103. Bhatnagar, "India's Software Industry," 54–56. Viewed as an important development for the future of Indian IT, Singh and Rogers, *India's Communication Revolution*, 149–166; Capella, Singh, Singh, and Useem, *The India Way*, 35–36. Das devoted considerable attention to the importance of knowledge transfer to India from the United States and elsewhere, *India Unbound*, 325–344; see also Payal

Banerjee, "Indian IT Workers in the U.S.: Race, Gender, and State in the Making of Immigrant Labor" (Unpublished Ph.D. dissertation, 209).

104. U.S. visa data show that by 1992 some 10,000 Indians were working in the United States; a number that kept growing through the 1990s and early twenty-first century. In fact, in 2008 the number exceeded 217,000, and in 2010 (latest year of available data) climbed to nearly 207,000. The majority of all these various workers were employed in consulting, largely in IT and overwhelmingly related to software, U.S. Department of Homeland Security, *Yearbook of Immigration Statistics*, various years, available at http://www.dhs.gov/files/statistics/publications/yearbook.shtm (last accessed 9/3/2011).

105. This organization maintains a short history of its past, mission, and description of its activities at http://www.nasscom.in (last accessed 10/1/2011). It is a major source of information on contemporary IT activities in India.

106. Balaji Parthasarathy, "The Computer Software Industry as a Vehicle of Late Industrialization: Lessons from the Indian Case," *Journal of the Asia Pacific Economy* 15, no. 3 (August 2010): 257.

107. Ibid., 254.

108. Sayers, "History of IBM's International Operations," 242.

109. Parthasarathy, "The Computer Software Industry as a Vehicle of Late Industrialization: Lessons from the Indian Case," 255; Dossani, *India Arriving*, 116–117,148–149; Steve Hamm, *Bangalore Tiger: How Indian Tech Upstart Wipro Is Rewriting the Rules of Global Competition* (New York: McGraw-Hill, 2007): 12.

110. Parthasarathy, "The Computer Software Industry as a Vehicle of Late Industrialization: Lessons from the Indian Case," 256; Singhal and Rogers, *India's Communication Revolution*, 167–176.

111. Anthony P. D'Costa, "Exports, University-Industry Linkages, and Innovation Challenges in Bangalore, India," World Bank Policy Research Working Paper 388, April 2006, http://www-wds.worldbank.org / . . . 451/Rendered/PDF/wps3887.pdf (last accessed 8/21/2008); "India's Silicon Valley," Bussinessweek, undated [circa 2008], http://www.businesweek.com/adsections/indian/infotech/2001/silicon.html (last accessed 1/8/2008.

112. James M. Popkin and Partha Iyengar, *IT and the East: How China and India Are Altering the Future of Technology and Innovation* (Boston, Mass.: Harvard Business School Press, 2007): 125.

113. K.C. Krishnadas "India's Design Centers Buck Economy's Trend," http://www.eetimes.com/article/show/Article.jhtm?articled=18308505 (last accessed 9/5/2011). For a detailed analysis of the managerial skills these entrepreneurs had, and their style of running businesses, see Peter Cappelli, Harbir Singh, Jitendra Singh, and Michael Useem, *The India Way: How India's Top Business Leaders Are Revolutionizing Management* (Boston, Mass.: Harvard Business School Press, 2010).

114. Bhatnagar, "India's Software Industry," 64.

115. See, for example, Panagariya, *India*, 370–382; Desai, *India's Telecommunications Industry*, 139–162, but with suggestions for improvement.

116. Parthasarathy, "The Computer Software Industry as a Vehicle of Late Industrialization: Lessons from the Indian Case," 247–270; see an earlier study by him, "Globalizing Information Technology: The Domestic Policy Context for India's Software Production and Exports," *Iterations*, May 3, 2004, pp. 1–38.

117. With pioneers who worked at these institutes commenting on the start and early history of these efforts in Banerjee, *Computer Education in India*.

118. Cited in Bhatnagar, "India's Software Industry," 61.

119. Ibid., 60–61.
120. Parthasarathy, "The Computer Software Industry as a Vehicle of Late Industriali-zation: Lessons from the Indian Case," 264.
121. Jyoti Vig, "Information Technology and the Indian Economy" (Unpublished Ph.D. dissertation, University of Minnesota, 2011).
122. Mohsin U. Khan, *India and Korea: A Comparison of Electronic Technology Policy* (New Delhi: Har-Anand, 1997): 211–214; Joël Ruet, "Asset Specificity, Partnerships and Global Strategies of Information Technology and Biotechnology Firms in India," in Jean-François Huchet, Xavier Richet, and Joël Ruet (eds.), *Globalization in China, India and Russia: Emergence of National Groups and Global Strategies of Firms* (New Delhi: Academic Foundation, 2007: 299–324.
123. Khan, *India and Korea*, 213–216.
124. Hanna, *Exploiting Information Technology for Development*, xii.
125. Ibid., xiii.
126. Ibid.
127. Das, *India Unbound*, 262–264.
128. Hanna, *Exploiting Information Technology for Development*, 15.
129. Ibid., 17. Little work has been done on IT in Indian banks, however, see Debapros-anna Nandy, "Banking Sector Reforms in India and Performance Evaluation of Commercial Banks" Ph.D. dissertation published by Boca Raton, Fla.: Dissertation.com, 2010): 94–113.
130. Nina Hachigian and Lily Wu, *The Information Revolution in Asia* (Santa Monica, Cal.: RAND Corporation, 2003): 86; for examples of state uses see Banerjee, *Computer Applications for Techno-Economic Development*; M. Hilaria Soundari (ed.), *Indian Agriculture and Information Communications Technology (ICT)* (New Delhi: New Century Publications, 2011); Subhash Bhatnagar, "Information Technology and Development: Foundation and Key Issues," in Subhash Bhatnagar and Robert Schware (eds.), *Information and Communication Technology in Development: Cases From India* (New Delhi: Sage Publications, 2000): 21–23; Naresh Kumar Reddy and Mike Graves, "Electronic Support for Rural Healthcare Workers," Ibid., 35–49; Banerjee, *Information Management in Government*.
131. Department of Information Technology, Ministry of Communications and Infor-mation Technology, Government of India, *India: e-Readiness Assessment Report 2006* (New Delhi: Department of Information Technology, 2006): first quote, p. 72, second quote, p. 73.
132. Hanna, *Exploiting Information Technology for Development*, 18.
133. Singhal and Rogers, *India's Communication Revolution*, 57–61, 232–235; Mascar-enhas, *India's Silicon Plateau*, 119–133; Franda, *China and India Online*, 105–167.
134. Das, *India Unbound*, 213, and for his account of the reforms, 213–243.
135. Peter Wolcott and Seymour Goodman, "Global Diffusion of the Internet I: India: Is the Elephant Learning to Dance?" *Communications of the Association for Informa-tion Systems* 11 (2003): 560–646, especially 560–574. The bibliography in this article is superb as it includes websites on the topic.
136. Ibid., 592; Ben A. Petrazzini and Girija Krishnaswamy, "Socioeconomic Implica-tions of Telecommunications Liberalization: India in the International Context," *The Information Society* 14, no. 1 (1998): 3–18.
137. Wolcott and Goodman, "Global Diffusion of the Internet I," 594.
138. Ibid., 612.
139. Robert R. Miller, "Leapfrogging? India's Information Technology Industry and the Internet," Discussion Paper IFD42 (Washington, D.C.: World Bank, May

2001): 3–4. For a blistering analysis of the failure of the Indian's government's attempt to encourage/discourage deployment of the Internet by a political scientist, see Marcus Franda, *China and India Online: Information Technology Politics and Diplomacy in the World's Two Largest Nations* (Lanham, Md.: Rowman & Littlefield, 2002): 112–138.

140. Ibid., 6.

141. Miller, "Leapfrogging? India's Information Technology Industry and the Internet," 10.

142. Various surveys summarized and annotated in Wolcott and Goodman, "Global Diffusion of the Internet I," 613.

143. Ibid., 618.

144. Ibid.

145. Desai, *India's Telecommunications Industry*, 29–30.

146. French, *India*, 185–186.

147. World Bank, *2011: The Little Data Book on Information and Communication Technology* (Washington, D.C.: World Bank, 2011): 104.

148. For example with the poor in rural India, a report as recently as from 2008, reconfirmed this point: "With regard to using the Internet to undertake business, we found that services such as information on grain prices or using the Internet to distribute agricultural goods were largely unavailable. Even when available, these services were most unused. The villagers were well connected through middlemen to all the villages in the surrounding area with which they did business, and there was little added value to posting prices on the Internet," Dossani, *India Arriving*, 234.

149. Because of IBM's massive presence in the Indian IT community, see Engardio, *Chindia*, 193–197.

150. Jayashankar M. Swaminathan, "Outsourcing," in Jayashankar M. Swaminathan (ed.), *Indian Economic Superpower: Fiction or Future?* (Singapore: World Scientific Publishing, 109): 17–19; Vivek Kulkarni, "Offshore to Win and Not Shrink!," Ibid., 19–57.

151. R.C. Mascarenhas, *India's Silicon Plateau: Development of Information and Communication Technology in Bangalore* (New Delhi: Orient BlackSwan, 2010): 137.

152. Ibid., 138–139; Phillip Cooke and Andrea Piccaluga (eds.), *Regional Development in the Knowledge Economy* (London: Routledge, 2006).

153. Mascarenhas, *India's Silicon Plateau: Development of Information and Communication Technology in Bangalore*,139. Another report provided the following statistics on some employers as of September 2006: TCS had 78,000 employees, Infosys 66,000, Wipro 61,000, IBM 140,000, and Accenture a similar number. While the veracity of these estimates cannot be ensured, even if off by a few thousand per firm, these did represent large enterprises by any standard in any country, and also in India, Dossani, *India Arriving*, 119. For how multinational firms were operating in India in the early 2000s, see Ibid., 123–152.

154. Mascarenhas, *India's Silicon Plateau*, 112–113.

155. The issue of workforce mobility has received some attention, in part because of WTO effects and the recession as well; see, for example, Thomas L. Brewer and Stanley D. Nollen, "Knowledge Transfer to Developing Countries After WTO: Theory and Practice in Information Technology in India," Working Paper 98–14, Carnegie Bosch Institute for Applied Studies in International Management, March 1998; Binod Khadria, *Migration of Highly Skilled Indians: Case Studies of IT and Health Professionals* (Paris: Organization for Economic Co-Operation and Development, April 21, 2004) and a second paper by Khadria, *Human Resources in Science*

and Technology in India and the International Mobility of Highly Skilled Indians (Paris: Organization for Economic Co-Operation and Development, May 27, 2004).

156. T.N. Srinivasan, "Information-Technology-Enabled Services and India's Growth Prospects," in Susan M. Collins and Lael Brainard (eds.), *Brookings Trade Forum 2005: Offshoring White-Collar Work* (Washington, D.C.: Brooks Institution Press, 2006): 225.

157. Quoted in Rafiq Dossani and Martin Kenney, "Tata Consultancy Services and Its 'Global Top Ten' Ambition," Unpublished paper, November 24, 2003, p. 15.

158. Kulkarni, "Offshore to Win and Not Shrink!," 19.

159. Organization for Economic Co-operation and Development, *Is China the New Centre for Offshoring of IT and ICT-Enabled Services?* (Paris: OECD, April 5, 2007): 7.

160. A possibility similarly explored by Amandeep Singh Sandhu, "Globalization of Services and the Making of a New Global Labor Force in India's Silicon Valley" (Unpublished Ph.D. dissertation, University of California, Santa Barbara, 2008). IT and BPO services experienced double-digit growth all through the decade, NASSCOM, *Annual Report 2010–11* (New Delhi: NASSCOM, 2011): 1; Capelli, Singh, Singh, and Useem, *The India Way*, 146–151; Vivek Kulkarni, "Offshore to Win and Not Shrink!," in Swaminathan, *Indian Economic Superpower*, 19–54; In Rolee Aranya, "Globalization and Urban Restructuring of Bangalore, India: Growth of the IT Industry, Its Spacial Dynamics and Local Planning Responses" (Unpublished Ph.D. dissertation, 2004).

161. OECD, *Is China the New Centre for Offshoring of IT and ICT-Enabled Services?*, 10; Rafiq Dossani and Martin Kenney, "The Next Wave of Globalization: Relocating Service Provision to India," *World Development* 35, no. 5 (2007): 772–791.

162. Dossani and Kenney, "The Next Wave of Globalization: Relocating Service Provision to India," 772–791.

163. OECD, *Is China the New Centre for Offshoring of IT and ICT-Enabled Services?*, 14.

164. Ibid., 17.

165. http://www.mit.in/ for both this act and its update of 2008 (last accessed 10/5/2011).

166. B.V. Naidu, *India: Emerging Knowledge Base of the 21st Century* (Delhi: STPI, 2006).

167. Hachigian and Wu, *The Information Revolution in Asia*, 37.

168. For data on the continued growth of this business through 2010, see NASSCOM, *The IT-BPO Sector in India, Strategic Review 2011* (New Delhi: NASSCOM, 2011): 5–7.

169. Dossani and Kenney, "The Next Wave of Globalization," 772–791; Franda, *China and India Online*, 139–167; all the essays in Swaminathan, *Indian Economic Superpower*.

170. Dossani and Kenney, "The Next Wave of Globalization," 782; R. Subramanyan, "IBM Pips HP as Largest MNC IT Employer," *The Economic Times*, May 22, 2005, http://www.economictimes.indiatimes.com/articles/1117786 (last accessed 9/20/2006); Nirmalya Kumar and Phanish Puranam, *India Inside* (Boston, Mass.: Harvard Business School Press, 2012.

171. Dossani and Kenney, "The Next Wave of Globalization," 782.

172. Swaminathan, "Outsourcing," OECD, *Is China the New Centre for Offshoring of IT and ICT-Enabled Services?*, 26.

173. Swaminathan, "Outsourcing," 20–54.

174. Analyzed for all the electronics industry of India and that explains how India proved less successful, see K.J. Joseph, *Industry Under Economic Liberalization: The Case of Indian Electronics* (New Delhi: SAGE publications, 1997): 216–225.

175. Singhal and Rogers, *India's Information Revolution*, 217.

176. A point that continues to be made by observers of the Indian environment. See, for example, Mascarenhas, *India's Silicon Plateau*, 140–150.
177. The term *calculators* was used in the years before the 1950s to refer to people who calculated data, usually mathematics or numbers, as in accounting. In the United States, for example, young women with colleges degrees in mathematics were used in the 1930s and early 1940s as "calculators" to perform the work that eventually was routinely done by computers. The term was also used in other countries then and later, but was pretty much no longer after the 1960s.
178. Two examples of this form of economic activity in 2012 are Israel and Costa Rica, each with a vibrant software industry operating successfully in the global market, and Brazil in business process outsourcing services.
179. For example, Cappelli, Singh, Singh, and Usem, *The India Way* and Dossani, *India Arriving*; others are cited in the bibliographic essay accompanying our book.
180. For examples, see Dipankar Sinha, "On Forgetting History: Information and Communication Technology and the Colonisation of Politics in Post-Colonial India," in Bagchi, Sinha, and Bagchi, *Webs of History*, 252–254; Cappelli, Singh, Singh, and Usem, *The India Way*, 19–47; Singhal and Rogers, *India's Communication Revolution*, 44–51; Heeks, *India's Software Industry*, 33–66; Panagariya, *India*, 3–110; Franda, *China and India Online*, 18–21; Das, *India Unbound*, 28–29, 55–57, 74–80; French, *India*, 126–128; Grieco, *Between Dependency and Autonomy*, 103–149; OECD, "India's Growth Patterns and Obstacles to Higher Growth."

CHAPTER 11

1. Nagy Hanna, Sandor Boyson, and Shakuntala Gunaratne, *The East Asian Miracle and Information Technology: Strategic Management of Technological Learning* (Washington, D.C.: World Bank, 1996): ix.
2. Ian Morris, *Why the West Rules—For Now: The Patterns of History, and What They Reveal About the Future* (New York: Farrar, Straus and Giroux, 2010): 557–622.
3. Discussed in the broadest terms by Jared Diamond, *Guns, Germs, and Steel: The Fates of Human Societies* (New York: W.W. Norton, 2005).
4. Striving also to use the latest technologies, Emily Noelle Ignacio, *Building Diaspora: Filipino Cultural Community Formation on the Internet* (New Brunswick, N.J.: Rutgers University Press, 2005); Merlyna Lim, *Islamic Radicalism and Anti-Americanism in Indonesia: The Role of the Internet* (Washington, D.C.: East-West Center, 2005); David T. Hill and Sen Krishna, *The Internet in Indonesia's New Democracy* (London: Routledge, 2005).
5. J.M. Bennett, Rosemary Broomham, P.M. Murton, T. Pearcey, and R.W. Rutledge (eds.), *Computers in Australia: The Development of a Profession* (Sydney: Hale & Iremonger, 1994); Gerard Goggin (ed.), *Virtual Nation: The Internet in Australia* (New South Wales: University of New South Wales Press, 2004); John Deane, "Connections in the History of Australian Computing," in Arthur Tatnall (ed.), *History of Computing: Learning from the Past* (Berlin: Springer, 2010): 1–12; Ric Allen, "The Australian Computer Market," *Datamation* 11, no. 10 (October 1965): 122–129; "Australian Report," Ibid., 17, no. 3 (February 1, 1973): 37–39; John M. Bennett, "Computers in Australian Universities," Ibid., 11, no. 3 (March 1965): 34–36; Frederick Bland, "Australia: A Sales Performance Review," Ibid., 15, no. 4 (April 1969): 95–109; E.M.U., "The Australian Market," Ibid., 11, no.3 (March 1965): 39–40; Trevor Pearcey, *A History of Australian Computing* (Melbourne: Chisholm Institute of Technology, 1988), but see also his memoirs of early computing, "CSIRAC Down Under: There Was a Machine," *Datamation* 11, no. 3 (March 1965): 37–38.

6. Kristin Thompson, *The Frodo Franchise: The Lord of the Rings and Modern Hollywood* (Berkeley, Cal.: University of California Press, 2008) which includes a discussion of the evolution of the film and gaming industries in New Zealand and the role of IT. Little has been written about computing in this country; but see Hone Heke, "Computing in New Zealand," *Datamation* 11, no. 3 (March 1965): 41.

7. Nina Hachigian and Lily Wu, *The Information Revolution In Asia* (Santa Monica, Cal.: RAND Corporation, 2003): xi–xvi.

8. Even the most recent global ranking reports on business conditions and technologies, demonstrate that through their country profiles. For example, see Klaus Schwab, *The Global Competitiveness Report, 2010–2011* (Geneva: World Economic Forum, 2010); World Bank, *Global Economic Prospects: Technology Diffusion in the Developing World 2008* (Washington, D.C.: World Bank, 2008) and an earlier study by the World Bank, *2006 Information and Communications for Development: Global Trends and Policies* (Washington, D.C.: World Bank, 2006).

9. World Bank, *Global Economic Prospects*, 71–73; Hachigian and Wu, *The Information Revolution in Asia*, 2–4.

10. Hachigian and Wu, *The Information Revolution in Asia*, xvi.

11. For example, see Morris H. Crawford, "Information Technology and Industrialization Policy in the Third World: A Case Study of Singapore, Malaysia, and Indonesia," Program on Information Resources Policy, Harvard University, 1984.

12. A great deal of it became entangled with discussions of technological innovations and diffusion, such as the World Bank studies cited above, but see also Frederic C. Deyo (ed.), *The Political Economy of the New Asian Industrialism* (Ithaca, N.Y.: Cornell University Press, 1987); Peter Evans, *Embedded Autonomy: States and Industrial Transformation* (Princeton, N.J.: Princeton University Press, 1995); Jeremy Grace, Charles Kenny, and Christine Zhen-Wei Qiang, *Information and Communication Technologies and Broad-Based Development: A Partial Review of the Evidence* (Washington, D.C.: World Bank, 2004); Shahid Yusuf, M. Anjum Altaf, and Kaoru Nabeshima (eds.), *Global Production Networking and Technological Change in East Asia* (Washington, D.C.: World Bank, 2004); Shahid Yusuf, *Innovative East Asia: The Future of Growth* (Washington, D.C.: World Bank, 2003); and now two classic studies, Linsu Kim and Richard R. Nelson (eds), *Technology, Learning, and Innovation: Experiences of Newly Industrializing Economies* (Cambridge: Cambridge University Press, 2000) and Chris Freeman and Luc Soefe, *The Economics of Industrial Innovation* (Cambridge, Mass.: MIT Press, 1997); Henry S. Rowen, Marguerite Gong Hancock, and William F. Miller (eds.), *Making IT: The Rise of Asia in High Tech* (Stanford, Cal.: Stanford University Press, 2007). I personally have been influenced by the dozens of articles and various books by Richard R. Nelson.

13. Hachigian and Wu, *The Information Revolution in Asia*, xiv.

14. Ibid., xv.

15. Crawford, "Information Technology and Industrialization Policy in the Third World: A Case Study of Singapore, Malaysia, and Indonesia," 59–60.

16. Ibid., 61.

17. Ibid., 85.

18. Jason Dedrick and Kenneth L. Kraemer, *Asia's Computer Challenge: Threat or Opportunity for the United States and the World?* (New York: Oxford University Press, 1998): 252–253.

19. Ibid., 321. The initial big push in early diffusion of IT had occurred 15 to 20 years earlier.

20. Thomas G. Mahnken, "Conclusion: The Diffusion of the Emerging Revolution in Military Affairs in Asia: A Preliminary Assessment," in Emily O. Goldman and Thomas G. Mahnken (eds.), *The Information Revolution in Military Affairs in Asia* (New York: Palgrave, 2004): 209.
21. For a detailed study of several dozen American industries and how they dealt with these issues, and that involved Asian firms as well, see James W. Cortada, *The Digital Hand*, 3 vols (New York: Oxford University Press, 2004–2008).
22. Grace, Kenny and Qiang, *Information and Communication Technologies and Broad-Based Development*, 10.
23. Dedrick and Kraemer, *Asia's Computer Challenge*, 323.
24. Kenneth L. Kraemer and Jason Dedrick, "Payoffs From Investment in Information Technology: Lessons From the Asia-Pacific Region," October, 1993, http://crito.uci.edu/papers (last accessed 8/10/2010), also available in *World Development* 22, no. 12 (1994): 1921–1931.
25. Economist Intelligence Unit, *Digital Economy Rankings 2010: Beyond e-readiness* (London: Economist Intelligence Unit, 2010): 23–24.
26. Most recently reaffirmed by Economics Nobel Laureate Michael Spence, *The Next Convergence: The Future of Economic Growth in a Multispeed World* (New York: Farrar, Straus and Giroux, 2011).
27. For an example written by two former officials and an academic illustrating the issues involved, see Peter F. Cowhey, Jonathan D. Aronson, and Donald Abelson, *Transforming Global Information and Communication Markets: The Political Economy of Innovation* (Cambridge, Mass.: MIT Press, 2009); but see also the perspective of a distinguished economist too, Dani Rodrik, who approaches the subject the way I essentially do, by looking at institutions and sectors within a global context, *One Economics Many Recipes: Globalization, Institutions, and Economic Growth* (Princeton, N.J.: Princeton University Press, 2007).
28. Fredic C. Deyo, "Coalitions, Institutions, and Linkage Sequencing—Toward a Strategic Capacity Model of East Asian Development," in Dayo, *The Political Economy of the New Asian Industrialism*, 227.
29. Ibid., 228.
30. Hanna, Boyson, and Gunaratne, *The East Asian Miracle and Information Technology*, xiv.
31. Evans, *Embedded Autonomy*, 181–206; Thomas L. Friedman, *The World Is Flat: A Brief History of the Twenty-First Century* (New York: Farrar, Straus and Giroux, 2005, 2006).
32. Jeff Saperstein and Daniel Rouach, *Creating Regional Wealth in the Innovation Economy: Models, Perspectives, and Best Practices* (Upper Saddle River, N.J.: Financial Times, 2002); Shahid Yusuf and Kaoru Nabeshima, *Postindustrial East Asian Cities: Innovation for Growth* (Washington, D.C.: World Bank, 2006).
33. When your author was an IBM salesman in the 1970s, he had potentially 3,000 combinations of products to sell, with hundreds made up of thousands of components. Then came whole new generations of mainframes, minicomputers, personal computers, and myriad consumer electronics, all digital and filled with ever-changing components.
34. Georgette Wang, "Many Paths to Many Destinations," in Georgette Wang (ed.), *Treading Different Paths: Informatization in Asian Nations* (Norwood, N.J.: Ablex Publishing Corp., 1994): 247–258.
35. Hanna, Boyson and Gunaratne, *The East Asian Miracle and Information Technology*, xv–xvi.

36. Dedrick and Kraemer, *Asia's Computer Challenge*, 230–234.
37. Richard R. Nelson, *Technology Institutions and Economic Growth* (Cambridge, Mass.: Harvard University Press, 2005): 111.
38. For a superb and extensive discussion of the issue, I have relied on the findings of Richard G. Lipsey, Kenneth I. Carlaw, and Clifford T. Bekar, *Economic Transformations: General Purpose Technologies and Long Term Economic Growth* (Oxford: Oxford University Press, 2005), and in particular, pp. 499–543.
39. Ideas conveniently summarized in Joseph A. Schumpeter, *Can Capitalism Survive?* (New York: Harper Colophon Books, 1978): 21–46.

CHAPTER 12

1. Carl Shapiro and Hal R. Varian, *Information Rules: A Strategic Guide to the Network Economy* (Boston, Mass.: Harvard Business School Press, 1999): 1–2.
2. Alfred D, Chandler, Jr., *The Visible Hand: The Managerial Revolution in American Business* (Cambridge, Mass.: Harvard University Press, 1977).
3. James Beniger, *The Control Revolution: Technological and Economic Origins of the Information Society* (Cambridge, Mass.: Harvard University Press, 1989).
4. Lars Heide, *Punched-Card Systems and the Early Information Explosion, 1880–1945* (Baltimore, Md.: Johns Hopkins University Press, 2009) and for a study just on the U.S. experience, James W. Cortada, *Before the Computer: IBM, NCR, Burroughs, and Remington Rand and the Industry They Created, 1865–1956* (Princeton, N.J.: Princeton University Press, 1993).
5. Paul Ceruzzi, *A History of Modern Computing* (Cambridge, Mass.: MIT Press, 2003).
6. Danny M. Leipziger, "Forward," in Vandana Chandra (ed.), *Technology Adaptation, and Exports: How Some Developing Countries Got It Right* (Washington, D.C.: World Bank, 2006): xiii.
7. Arnuff Grüber, *The Rise and Fall of Infrastructures: Dynamics of Evolution and Technological Change in Transport* (Heidelberg: Physica-Verlag, 1990): 30.
8. Explored by Kevin Kelly, *What Technology Wants* (New York: Penguin, 2011) and in an earlier book by the same author, *Out of Control: The New Biology of Machines, Social Systems, and the Economic World* (New York: Perseus, 1994).
9. Steve Lohr, "More Jobs Predicted for Machines, Not People," *The New York Times*, October 23, 2011, http://www.nytimes.com/2011/10/24/technology/economists-see-more-job-for-machines-not-people.html?_r=1. . . (last accessed 3/9/2012).
10. Frank Levy and Richard J. Murname, *The New Division of Labor: How Computers Are Creating the Next Job Market* (Princeton, N.J.; Princeton University Press, 2004) and for the perspective that computers are taking jobs away, see Eric Brynjolfsson and Andrew P. McAfee, *Race Against the Machine*, e-book, Amazon.com, 2011 (available only on Kindle as of late 2011).
11. As demonstrated by Ray Kurzwell, *The Singularity Is Near: When Humans Transcend Biology* (New York: Viking, 2005).
12. Everett M. Rogers, *Diffusion of Innovations, Fifth Edition* (New York: Free Press, 2003): 346–348, 456–469.
13. Vandana Chandra and Shashi Kolavalli, "Technology, Adaptation, and Exports—How Some Developing Countries Got It Right," in Vandana Chandra (ed.), *Technology, Adaptation, and Exports: How Some Developing Countries Got It Right* (Washington, D.C.: World Bank, 2006): 1–48.
14. Quite often the decision to continue using human labor instead of computers when technically possible is a political one, not an economic one. Governments, for example, do not want newspaper headlines announcing that they are laying off

thousands of employees; some companies do not either. Governments often want to find ways to employ more people so as to reduce the threat of protests, riots, and civil wars. In such circumstances using computers to improve productivity is of little interest to officials.

15. Technologists often refer to this characteristic of making computers into big and small products, more or less capacity, with standard or interchangeable components as "form factors." However, malleability as I use the term is broader than the device-centric more engineering-oriented phrase "form factor."

16. I have discussed this future elsewhere, James W. Cortada, *Information and the Modern Corporation* (Cambridge, Mass.: MIT Press, 2011): 127–149.

17. K.J. Arrow, "The Economic Implications of Learning by Doing," *Review of Economic Studies* 29 (1962): 155–177.

18. A major theme of the essays in Linsu Kim and Richard R. Nelson (eds.), *Technology, Learning, and Innovation: Experiences of Newly Industrializing Economies* (Cambridge: Cambridge University Press, 2000).

19. For example, V.W. Ruttan, *Technology, Growth, and Development—An Induced Innovation Perspective* (Oxford: Oxford University Press, 2001).

20. Daniel R. Headrick summarizes much of this thinking across thousands of years, *Technology: A World History* (Oxford: Oxford University Press, 2009).

21. I have discussed this issue elsewhere, "Power and Use of Context in Business Management," *Journal of Knowledge Management* 13, no. 3 (2009): 13–27; "The Historian in the Businessplace," in Richard Bond and Pillarisetti Sudhir (eds.), *Perspectives on Life After a History P.h.D.* (Washington, D.C.: American Historical Association, 2005): 47–49; "Learning From History: Leveraging Experience and Context to Improve Organizational Excellence," *Journal of Organizational Excellence* 21, no. 2 (Spring 2002): 23–29; "The Case for Applied History in the World of Business: A Call for Action to Historians," *The Historian* 62, no. 4 (Summer 2000): 835–847.

22. Amar Bhidé, *The Venturesome Economy* (Princeton, N.J.: Princeton University Press, 2008): 324.

23. Ibid., 326, 327.

24. Robert Brenner, *The Economics of Global Turbulence* (London: Verso, 2006): 24, 145, 161, 211.

25. M. Abramovitz, "Catching Up, Forging Ahead, and Falling Behind," *Journal of Economic History* 46, no. 2 (1986): 386–406; William J. Baumol, "Productivity Growth, Convergence, and Welfare: What the Long-run Data Show," *American Economic Review* 76, no. 5 (1986): 1072, 1085; see also his larger study with S.A. Blackman and E.N. Wolff, *Productivity and American Leadership* (Cambridge, Mass.: MIT Press, 1989); Richard R. Nelson, "The U.S. Technology Lead: Where Did It Come From and Where Did It Go?," *Research Policy* 19 (1990): 117–132; see also his article with G. Wright, "The Rise and Fall of American Technological Leadership: The Postwar Era in Historical Perspective," *Journal of Economic Literature* (December 1992): 1931–1964.

26. I discussed the issue in more detail elsewhere, hence did not repeat it here, *How Societies Embrace Information Technology: Lessons for Management and the Rest of Us* (New York: John Wiley & Sons, 2009): 71–127.

27. For examples, Alfred Kleinknecht, *Innovation Patterns in Crisis and Prosperity: Schumpeter's Long Cycle Reconsidered* (London: Macmillan, 1987); Joseph A. Schumpeter, *Business Cycles: A Theoretical, Historical and Statistical Analysis of the Capitalist Process*, 2 vols (New York: McGraw-Hill, 1939); Solomos Solomou,

Phases of Economic Growth, 1850–1973: Kondratieff Waves and Kuznets Swings (Cambridge: Cambridge University Press, 1990); Robert U. Ayres, *Technological Transformations and Long Waves* (Laxenburg: International Institute for Applied Systems Analysis, 1989); J.J. Van Duijn, *The Long Wave in Economic Life* (London: George Allen & Unwin, 1983).

28. First introduced by J. Van Gelderen, "Springvloed: beschouwingen over industriële ontwikkeling en prijsbeweging," *De Nieuwe Tijd* 18 (1913): 4–6.

29. Joseph A. Schumpeter's work of the 1930s is accessible in Thomas A. McGraw's outstanding biography of him, *Prophet of Innovation: Joseph Schumpeter and Creative Destruction* (Cambridge, Mass.: Harvard University Press, 2010).

30. Ayres, *Technological Transformations and Long Waves*, 2.

31. Richard Nelson and Sydney G. Winter, "In Search of a Useful Theory of Innovation," *Research Policy* 6, no. 1 (1977); Davendra Sahal, *Patterns of Technological Innovation* (Reading, Mass.: Addison-Wesley, 1981); Giovanni Dosi, "Technological Paradigms and Technological Trajectories," *Research Policy* 11 (1982): 147ff; and Carlotta Perez, *Technological Revolutions and Financial Capital: The Dynamics of Bubbles and Golden Ages* (Cheltenham: Edward Elgar, 2003).

32. Van Duijn, *The Long Wave in Economic Life*; E. Mandel, *Long Waves of Capitalist Development: The Marxist Explanation* (Cambridge: Cambridge University Press, 1980).

33. Nelson and Winter, "In Search of a Useful Theory of Innovation."

34. James M. Utterback and William J. Abernathy, "A Dynamic Model of Process and Product Innovation," *Omega* (1979): 3; James M. Utterback, *Mastering the Dynamics of Innovation*, 2nd ed. (Boston, Mass.: Harvard Business School Press, 1996).

35. So well demonstrated by Mario Polèse, *The Wealth and Poverty of Regions* (Chicago, Ill.: University of Chicago Press, 2009).

36. Mark D.J. Williams, Rebecca Mayer, and Michael Minges, *Africa's ICT Infrastructure: Building on the Mobile Revolution* (Washington, D.C.: World Bank, 2011): 25–70.

37. Denis G. Campbell, *Egypt Unsh@ckled: Using Social Media to @#:) The System* (Carmanthenshire, Wales: Llyfrau Cambria, 2011).

38. For example, see Ken Auletta, *Googled: The End of the World as We Know It* (New York: Penguin, 2009); Siva Vaidhyanathan, *The Googlization of Everything (And Why We Should Worry)* (Berkeley, Cal.: University of California Press, 2011).

39. Dani Rodrik, *One Economics Many Recipes: Globalization, Institutions, and Economic Growth* (Princeton, N.J.: Princeton University Press, 2007): 151. For the entire paragraph Ibid., pp. 102–152.

40. These lines of investigation have been underway at the IBM Institute for Business Value for over a half decade, resulting in over 50 empirically-based studies, such as James W. Cortada, Ashisha M. Gupta, and Marc Le Noir, *How Nations Thrive in the Information Age* (Somers, N.Y.: IBM Corporation, 2007); James W. Cortada, Sietze Dijkstra, Gerry M. Mooney, and Todd Ramsey, *Government 2020 and the Perpetual Collaboration Mandate* (Somers, N.Y.: IBM Corporation, 2008); Susanne Dirks, Constantin Gurdgiev, and Mary Keeling, *Smarter Cities for Smarter Growth* (Somers, N.Y.: IBM Corporation, 2010); James W. Cortada, Vivian A. Nix, and Lynn Reyes, *Open Government Open Data* (Somers, N.Y.: IBM Corporation, 2011). These and other studies from the center are available at http://www.ibm.com/iibv.

41. Allan Collins and Richard Halverson, *Rethinking Education in the Age of Technology: The Digital Revolution and Schooling in America* (New York: Teachers College Columbia University, 2009): 122–146; Mizuko Ito, *Engineering Play: A Cultural History of Children's Software* (Cambridge, Mass.: MIT Press, 2009); Jane McGonigal, *Reality*

Is Broken: Why Games Make Us Better and How They Can Change the World (New York: Penguin, 2011).

42. For a summary of current work on the topic with an extensive bibliography, see David Williamson Shaffer, *How Computer Games Help Children Learn* (New York: Palgrave, 2006).

43. Ibid. Discussion of the issue is also spreading to the general public. See, for example, Emily Listfield, "Generation Wired," *Parade*, October 9, 2011, pp. 9–10, 12, 14, 19, which is delivered every Sunday morning to millions of subscribers to local newspapers all over the United States.

44. Shaffer, *How Computer Games Help Children Learn*, contains an extensive bibliography that includes citations to the controversies, 215–233.

45. Ian Morris, *Why the West Rules—For Now* (New York: Farrar, Straus and Giroux, 2010): 593.

46. Ibid., 594.

47. Ray Kurzwell, *The Singularity Is Near: When Humans Transcend Biology* (New York: Vintage, 2005); see also Kelly, *What Technology Wants*; W. Brian Arthur, *The Nature of Technology: What It Is and How It Evolves* (New York: Free Press, 2009).

48. Morris, *Why the West Rules—For Now*, 595.

49. "Daily Media Use Among Children and Teens Up Dramatically From Five Years Ago," Press Release, Kaiser Family Foundation, January 20, 2010, http://www. kf.org/entmedia/entmedia012010nr. cfm (last accessed 10/20/2011).

APPENDIX A

1. I was trained as a diplomatic historian, and wrote a dissertation on Spain and the American Civil War, which relied extensively on Spanish and French language sources, and which resulted in many fresh perspectives about the diplomacy of the subject. Over the years I have continued to read American diplomatic history, which is what led to my comment about American historians of diplomacy being linguistically challenged. There are exceptions, of course, but they are few and far between.

2. Paul Ceruzzi, *A History of Modern Computing*, 2nd Edition (Cambridge, Mass.: MIT Press, 2003).

3. Michael S. Mahoney, "The History of Computing in the History of Technology," *Annals of the History of Computing* 10 (1988): 113–125; Martin Campbell-Kelly, *From Airline Reservations to Sonic the Hedgehog: A History of the Software Industry* (Cambridge, Mass.: MIT Press, 2003).

4. Dirk de Wit, *The Shaping of Automation: A Historical Analysis of the Interaction Between Technology and Organization, 1950–1985* (Rotterdam: Erasmus Universiteit, 1994).

5. Nathan L. Ensmenger, "The 'Question of Professionalism' in the Computer Fields," *IEEE Annals of the History of Computing* 23, no. 4 (October-December 2001): 56–74; Thomas Haigh, "The Chromium-Plated Tabulator: Institutionalizing an Electronic Revolution, 1954–1958, Ibid., 75–104.

6. For an excellent demonstration of the global approach and its literature, see Richard N. Langlois and Edward W. Steinmueller, "The Evolution of Competitive Advantage in the Worldwide Semiconductor Industry, 1947–1996," in David C. Mowery and Richard R. Nelson (eds.), *Sources of Industrial Leadership: Studies of Seven Industries* (Cambridge: Cambridge University Press, 1999): 19–78.

7. Joanne Yates, *Structuring the Information Age: Life Insurance and Technology in the Twentieth Century* (Baltimore, Md.: Johns Hopkins University Press, 2005).

8. James W. Cortada, *The Digital Hand*, 3 vols (New York: Oxford University Press, 2004–2008).
9. Paul N. Edwards, *The Closed World: Computers and the Politics of Discourse in Cold War America* (Cambridge, Mass.: MIT Press, 1996). He repeated his willingness to broaden the apatture of a discussion about IT and science in *A Vast Machine: Computer Models, Climate Data, and The Politics of Global Warming* (Cambridge, Mass.: MIT Press, 2010).
10. Martin Campbell-Kelly, *ICL: A Business and Technical History* (Oxford: Clarendon Press, 1989); Pierre E. Mounier-Kuhn, *L'informatique de la seconde guerre mondiale au Plan Calcul en France: L'émergence d'une science* (Paris: Presses de l'université Paris-Sorbonne, 2010).

APPENDIX B

1. He has recently summarized much of his thinking in Manuel Castells, *Communication Power* (Oxford: Oxford University Press, 2009), but see also his earlier extended essay, *The Internet Galaxy: Reflections on the Internet, Business, and Society* (Oxford: Oxford University Press, 2001).
2. For a more detailed description of the rankings and methodology for calculating results, see any of the annual ranking reports, but also, James W. Cortada, Ashish M. Gupta, and Marc Le Noir, *How Nations Thrive in the Information Age* (Somers, N.Y.: IBM Corporation, 2007). For an introductory discussion on how governments use this kind of information both Wave One and Wave Two nations to foster both general economic development and IT diffusion, see James W. Cortada, *How Societies Embrace Information Technology: Lessons for Management and the Rest of Us* (New York: John Wiley & Sons, 2009): 71–127.

BIBLIOGRAPHIC ESSAY

The literature on computing around the world has been increasing sharply since the late 1990s, following, in effect, the deployment activities in various nations. As the amount of information technology in a country increases, so too initially economic and public policy publications, followed by the addition of political science and historical studies. That bibliographic pattern has existed since the 1950s, beginning largely in the United States, extending next to Western Europe, then to Japan and South Korea, and finally to many other countries. Major sources of statistical information about diffusion include the International Monetary Bank (IMF), and more importantly for the poorer countries, the World Bank and for the richest, the Organization for Economic Cooperation and Development (OECD). The United Nations is a good source, as is the U.S. Bureau of the Census and the U.S. Department of Commerce for the United States. Industry information providers have also published a torrent of material over the past half century, including Frost & Sullivan, Input, IDG, and Forrester, to mention only a handful out of hundreds of such enterprises.

There is the massive body of publications that have appeared over the past seven decades on computing from associations (such as the IEEE), academic presses (MIT Press, Cambridge University Press for instance) and commercial publishers (Prentice-Hall, McGraw-Hill, and John Wiley & Sons are the largest). Those that proved particularly useful are cited in detail in the end notes for each chapter. The purpose of this bibliographic essay is to discuss those studies that were central to this book and that should prove useful to those who wish to explore further the issue of IT diffusion. It is by no means definitive; but every citation proved useful. The majority of the materials are organized roughly along the lines of the chapter topics. I focus on major studies—specifically books—but provide detailed citations in the end notes to hundreds of monographic articles.

DIFFUSION ECONOMICS AND TECHNOLOGICAL INNOVATIONS

While historians approaching the history of how a technology spread around the world tell the stories of inventors, vendors, users, and describe the forces of war, politics, and social conditions, at its root, the conversation is about the interplay of technological innovation and economics. For a useful introduction to the twin themes and how they are intertwined read the first eighty pages of Sally H. Clarke, Naomi R. Lamoreaux, and Steven W. Usselman (eds.), *The Challenge of Remaining Innovative: Insights from Twentieth-Century American Business* (Stanford, Cal.: Stanford Business Books, 2009). The most comprehensive discussion of the twin themes applied to information technology (IT) on a global basis is the monumental book by Vernon W. Ruttan, *Technology, Growth, and Development: An Induced Innovation Perspective* (New York: Oxford University Press, 2001). For a more specific economic discussion of events in the United States, there are two important studies, Dale W. Jorgenson, Mun S. Ho, and Kevin J. Stiroh, *Productivity*, vol. 3, *Information Technology and the American Growth Resurgence* (Cambridge, Mass.: MIT Press, 2005) and the short, far more accessible study by Daniel E. Sichel, *The Computer Revolution: An Economic Perspective* (Washington, D.C.: Brookings Institution Press, 1997). The classic work those interested in diffusion consult first before any others is Everett M. Rogers, *Diffusion of Innovations*, now in a fifth edition (New York: Free Press, 2003), which has influenced a generation of observers of technological innovations and economics since the first edition was published in 1962. Because of its influence on so many IT executives, see also Carlota Perez, *Technological Revolutions and Financial Capital: The Dynamics of Bubbles and Golden Ages* (Cheltenham, U.K.: Edgar Elgar, 2002). Studies of global effects of computing on national economies are beginning to appear. A most useful example of such publications is the collection of papers, Daniel Cohen, Pietro Garibaldi, and Stefano Scarpetta (eds.), *The ICT Revolution: Productivity Differences and the Digital Divide* (Oxford: Oxford University Press, 2004). A useful comparative study with a great deal of data from OECD nations OECD, *ICT and Economic Growth: Evidence From OECD Countries, Industries and Firms* (Paris: OECD, 2003).

There has been considerable work done to understand the role of IT in national economic development, a body of material that underpins discussions in every chapter of the book. Richard R. Nelson has profoundly influenced my view of the subject, and so the discourse of this book, specifically, *The Sources of Economic Growth* (Cambridge, Mass.: Harvard University Press, 1996) and his, *Technology Institutions and Economic Growth* (Cambridge, Mass.: Harvard University Press, 2005). Linking his work back to contemporary thinking about how capitalism works is crucial and for that purpose see, William J. Baumol, *The Free-Market Innovation Machine: Analyzing the Growth Miracle of Capitalism* (Princeton, N.J.: Princeton University Press, 2002). Both are highly readable. For an anthology of studies dealing with these issues largely in developing economies there is Yutaka Kurihara, Sadayoshi Takaya, Hisashi Harui, and Hiroshi Kamae, *Information Technology and Economic Development* (Hershey, N.Y.: Information Science Reference, 2008). A useful source of data and trends that includes computing appears from time-to-time from the World Bank,

2006 Information and Communications for Development: Global Trends and Policies (Washington, D.C.: World Bank, 2006). Finally, with respect to long waves, a useful discussion can be found in Christopher Freeman, John Clark, and Lue Soete, *Unemployment and Technical Innovation: A Study of Long Waves and Economic Development* (Westport, Conn.: Greenwood Press, 1982); but see also, Peter Hall and Paschal Preston, *The Carrier Wave: New Information Technology and the Geography of Innovation, 1846–2003* (London: Unwin Hyman, 1988). Long waves go through periods of fashion and in the early years of the twenty-first century are not subjects attractive to economists but with behavioral economics in ascendancy, this may change, which is one reason why I use the notion.

Histories of the evolution of IT are abundant. Begin with a study aimed at a general audience by Martin Campbell-Kelly and William Aspray, *Computer: A History of the Information Machine* (New York: Basic Books, 1996); for a more technical history the most authoritative is by Paul E. Ceruzzi, *A History of Modern Computing* (Cambridge, Mass.: MIT Press, 2003). To augment the story of software, told as both technological and business history, see Martin Campbell-Kelly, *From Airline Reservations to Sonic the Hedgehog: A History of the Software Industry* (Cambridge, Mass.: MIT Press, 2003). See also for a general history of the IT industry, Jeffrey R. Yost, *The Computer Industry* (Westport, Conn.: Greenwood Press, 2005). Because so many of the early developments in computing took place in the United States, all of these books are heavily weighted toward the American experience. For an excellent history of semiconductors, also a topic with a considerable body of literature, and international in scope, see the still useful Ernest Braun and Stuart Macdonald, *Revolution in Miniature: The History and Impact of Semiconductor Electronics* (Cambridge: Cambridge University Press, 1983). As the numerous country studies demonstrated, however, one also has to take into account developments in telecommunications and how they interacted with computing, particularly after the 1960s. Those interactions have not been fully studied together, largely piecemeal, but the literature on one or the other is growing and is cited below. However, to introduce the topic as it applies to the period 1996–2002 on a global basis, see Martin Fransman, *Telecoms in the Internet Age: From Boom to Bust to . . .?* (Oxford: Oxford University Press, 2002).

DIFFUSION OF COMPUTING IN THE UNITED STATES, 1940S–1990S

The best place to start understanding the American story is with two books written by Kenneth Flamm, *Targeting the Computer: Government Support and International Competition* (Washington, D.C.: Brookings Institution, 1987) and its companion volume, *Creating the Computer: Government, Industry, and High Technology* (Washington, D.C.: Brookings Institution, 1988). On the early history of the computer industry, see Arthur L. Norberg, *Computers and Commerce: A Study of Technology and Management at Eckert-Mauchly Computer Company, Engineering Research Associates, and Remington Rand, 1946–1957* (Cambridge,

Mass.: MIT Press, 2005). On the very early history of the technology, see Paul E. Ceruzzi, *Reckoners: The Prehistory of the Digital Computer from Relays to the Stored Program Concept, 1935–1945* (Westport, Conn.: Greenwood Press, 1983); on the migration of these technological innovations into commercial products, see James W. Cortada, *The Computer in the United States: From Laboratory to Market, 1930 to 1960* (Armonk, N.Y.: M.E. Sharpe, 1993). Because of the entrepreneurial feature of American IT evolution, two studies set the context, neither a history, the first about who innovates, by Eric von Hippel, *The Sources of Innovation* (New York: Oxford University Press, 1988) and the second about the role of venture capitalists, Amar Bhidé, *The Venturesome Economy: How Innovation Sustains Prosperity in a More Connected World* (Princeton, N.J.: Princeton University Press, 2008). The special role of IT experts working with the technology is crucial to the story and on their involvement we have a well-documented history by Nathan Ensmenger, *Computers, Programmers, and the Politics of Technical Expertise* (Cambridge, Mass.: MIT Press, 2010). For a discussion of American IT presence in the new century, see Catherine L. Mann, *Accelerating the Globalization of America: The Role for Information Technology* (Washington, D.C.: Institute for International Economics, 2006).

The *IEEE Annals of the History of Computing* has published dozens of memoirs and histories of early American computing, which it continues to do. Many books also are available. Key volumes include I. Bernard Cohen, *Howard Aiken—Portrait of a Computer Pioneer* (Cambridge, Mass.: MIT Press, 1998); G. Pascal Zachary, *Endless Frontier: Vannevar Bush, Engineer of the American Century* (Cambridge, Mass.: MIT Press, 1998); Nancy B. Stern, *From ENIAC to UNIVAC: An Appraisal of the Eckert-Mauchly Computers* (Bedford, Mass.: Digital Press, 1981); William Aspray, *John von Neumann and the Origins of Modern Computing* (Cambridge, Mass.: MIT Press, 1990); Norman Macre, *John von Neumann* (New York: Pantheon Books, 1992); Paul N. Edwards, *The Closed World: Computers in the Politics of Discourse in Cold War America* (Cambridge, Mass.: MIT Press, 1996). I have looked at the demand side of the story, how people used computing in, James W. Cortada, *The Digital Hand*, 3 vols. (New York: Oxford University Press, 2004–2008), and with a team of scholars, Alfred D. Chandler, Jr. and James W. Cortada (eds.), *A Nation Transformed by Information: How Information Has Shaped the United States from Colonial Times to the Present* (New York: Oxford University Press, 2000).

Because of the important role of IBM, understanding its activities is crucial. The standard corporate history is Emerson W. Pugh, *Building IBM: Shaping an Industry and Its Technology* (Cambridge, Mass.: MIT Press, 1995); for technical histories of the early years, see Emerson W. Pugh, *Memories That Shaped an Industry: Decisions Leading to IBM System/360* (Cambridge, Mass.: MIT Press, 1984) and two monumental studies, Charles J. Bashe, Lyle R. Johnson, John H. Palmer, and Emerson W. Pugh, *IBM's Early Computers* (Cambridge, Mass.: MIT Press, 1986) and Emerson W. Pugh, Lyle R. Johnson, and John H. Palmer,

IBM's 360 and Early 370 Systems (Cambridge, Mass.: MIT Press, 1991); Franklin
M. Fisher, James W. McKie, and Richard B. Mancke, *IBM and the U.S. Data
Processing Industry: An Economic History* (New York: Praeger, 1983). For an
account of the company, and its key competitors, before and during the early
years of the computer's arrival, see James W. Cortada, *Before the Computer: IBM,
NCR, Burroughs, and Remington Rand and The Industry They Created, 1865–1956*
(Princeton, N.J.: Princeton University Press, 1993). Jeffery Yost has published a
collection of memoir articles written by IBM executives and scientists over
the past two decades, *The IBM Century: Creating the IT Revolution* (Hoboken,
N.J.: John Wiley & Sons and IEEE Computer Society Press, 2011). As part of
IBM's 100th Anniversary celebration in 2011, the firm published Kevin Maney,
Steve Hamm, and Jeffrey M. O'Brien, *Making the World Work Better: The Ideas
That Shaped A Century and A Company* (Upper Saddle River, N.J.: IBM Press,
2011), which contains much useful information.

The most complete inventories of early computers for anywhere in the
world cover the American experience. Montgomery Phister, Jr. produced two
very useful, highly detailed inventories, *Data Processing Technology and Economics* (Santa Monica, Cal.: Santa Monica Publishing, 1976) and *Data Processing Technology and Economics, 1975–1979 Supplement* (Santa Monica, Cal.: Santa
Monica Publishing Co., 1979). Two other useful inventories from the period
include Martin H. Weik, *A Survey of Domestic Electronic Digital Computing
Systems* (Aberdeen Proving Ground, Md.: Ballistics Research Laboratory, 1955)
and Frederick G. Withington, *The Computer Industry, 1969–1974* (Cambridge,
Mass.: Arthur D. Little, 1969). The latter is a reminder that scores of business
consultancies and information providers tracked various American IT products throughout the second half of the twentieth century. For a growing repos-
itory of such materials, view the collections of the Charles Babbage Institute
at the University of Minnesota, at their website, http://www.cbi.umn.edu
(last accessed June 6, 2011). There is now a portal that points to some 1,000
Internet sites and to over 15 million pages of material on the global history of
information around the world, developed by the IT History Society, http://
www.ithistory.org/resources/overview.php (last accessed December 15, 2011).

DIFFUSION OF COMPUTING IN GREAT BRITAIN, FRANCE, AND GERMANY, 1940S–1990

Because diffusion of computing is so dependent on macroeconomic and social
trends, one has to start by understanding non-IT issues. For broad political
and social context, I relied on two introductory, but highly detailed texts, Norman Davis, *Europe: A History* (Oxford: Oxford University Press, 1996), while it
starts with ancient times, he provides a great deal of coverage on the period
since the start of the Second Industrial Revolution in the second half of the
nineteenth century. For an excellent, authoritative and yet highly readable
economic history of modern Europe, see Barry Eichengreen, *The European*

Economy Since 1945: Coordinated Capitalism and Beyond (Princeton, N.J.: Princeton University Press, 2007). As additional background focusing on the use of precomputer IT, specifically punched-card tabulators in Europe, with particular emphasis on Great Britain, France, and Germany in addition to the United States all before 1945, consult Lars Heide, *Punched-Card Systems and the Early Information Explosion, 1880–1945* (Baltimore, Md.: Johns Hopkins University Press, 2009). For a variety of contributed papers largely dealing with Europe, there is Richard Coopey (ed.), *Information Technology Policy: An International History* (Oxford: Oxford University Press, 2004). A much earlier publication covered similar issues in a far more integrated manner essential to the subject, Brian M. Murphy, *The International Politics of New Information Technology* (New York: St. Martin's, 1986). A nearly impossible to find early history of computing in Europe exists, James Connolly, *History of Computing in Europe* (Paris: IBM World Trade Corporation, 1967). Both of Kenneth Flamm's books cited above also provide extensive coverage of computing in Europe for the period 1930s–1970s.

For a general history of business in Great Britain, a useful survey is David J. Jeremy, *A Business History of Britain 1900–1990s* (Oxford: Oxford University Press, 1998). The essential work on computing in Great Britain is a large study of government computing, beginning with a review of nineteenth century developments, Jon Agar, *The Government Machine: A Revolutionary History of the Computer* (Cambridge, Mass.: MIT Press, 2003). On computing during World War II, see Jack Copeland's two books, *Colossus: The First Electronic Computer* (Oxford: Oxford University Press, 2003) and *Colossus: The Secret's of Bletchley Park's Code Breaking Computers* (Oxford: Oxford University Press, 2006). On early government policy towards the computer industry, begin with John Hendry, *Innovating for Failure: Government Policy and the Early British Computer Industry* (Cambridge, Mass.: MIT Press, 1989) and then for an account that digs deeper into the industry, there is the excellent study by Martin Campbell-Kelly, *ICL: A Business and Technical History* (Oxford: Oxford University Press, 1989). But see also Peter J. Bird, *LEO: The First Business Computer* (Workingham: Hasler Publishing, 1994). An early memoir by a distinguished computer pioneer is essential reading for understanding the early years of British IT, Maurice Wilkes, *Memoirs of a Computer Pioneer* (Cambridge, Mass.: MIT Press, 1985).

On British policies and practices, a variety of publications are useful: *The Prospects for the United Kingdom Computer Industry in the 1970s* (London: Her Majesty's Stationery Office, 1971); Brian Oakley and Kenneth Owen, *Alvey: Britain's Strategic Computing Initiative* (Cambridge, Mass.: MIT Press, 1990); Bide Committee, *Information technology: A Plan for Concerted Action* (London: Her Majesty's Stationery Office, 1986); Arthur Cotterell (ed.), *Advanced Information Technology in the New Industrial Society: The Kingston Seminars* (Oxford: Oxford University Press, 1988). Also useful for understanding both the IT community at large, and public policy and the software business, see G. Sashe,

P. Jowett, and J. McGee, *The Software Industry in the U.K.* (London: London Business School, 1986) and Paul Jowett and Margaret Rothwell, *The Economics of Information Technology* (New York: St. Martin's Press, 1986), a little gem of a small but insightful book.

Moving more to the demand side of diffusion, there is D.C. Pitt and B.C. Smith, *The Computer Revolution in Public Administration: The Impact of Information Technology on Government* (Brighton, Eng.: Wheatsheaf Books, 1984), which has been largely superseded by Agar's volume, but still useful. Christine Bellamy and John A. Taylor, *Governing in the Information Age* (Buckingham: Open University Press, 1998), and for case studies see Norio Kambayashi, *Cultural Influences on IT Use: A UK-Japanese Comparison* (New York: Palgrave Macmillan, 2001) and then extremely well done of specific uses of computing British government agencies, see Helen Margetts, *Information Technology in Government: Britain and America* (London: Routledge, 1999). On British diffusion, there is the National Computing Centre, *Information Technology trends* (Manchester: National Computer Centre, 1986). For a detailed discussion of sources of information on British diffusion, analyzing sources and strengths and weaknesses of data, consult Ian Miles and contributors, *Mapping and Measuring the Information Economy* (London: British Library Board, 1990).

A new generation of British scholars is being trained in the field of IT history, doing research on the use of IT. Three dissertations, in particular, are critical to the story: Marie Hicks, "Compiling Inequalities: Computerization in the British Civil Service and Nationalized Industries, 1940–1979" (Unpublished Ph.D. dissertation, Duke University, 2009), Thomas Lean, "The Making of the Macro': Producers, Mediators, Users and the Development of Popular Microcomputing in Britain (1980–1989)" (Unpublished Ph.D. dissertation, University of Manchester, 2008), and Ian Martin, "Centering the Computer in the Business of Banking: Barclays Bank and Technological Change, 1954–1974" (Unpublished Ph.D. dissertation, University of Manchester, 2010).

We are fortunate to now have a major history of early computing in France written by Pierre E. Mounier-Kuhn, who has been studying French computer uses for nearly two decades, *L'informatique de la seconde guerre mondiale au Plan Calcul en France: L'émergence d'une science* (Paris: Presses de L'université Paris-Sorbonne, 2010). This volume is particularly useful for understanding the role of academics and institutes in the evolution of French information technology and computer science in the 1950s and 1960s. On French policies regarding computers begin with Jacques Jublin and Jean-Michel, Quatrpoint, *French Ordinateurs, de l'affaire Bull à l'assassinat de Plan Calcul* (Paris: Editions Malain Moreau, 1976), but also consult Jean-Pierre Brulé, *L'Informatique Malade de L'etat Du Plan calcul à BULL Nationalisée: Un fiasco de 40 milliards* (Paris: Les Belles Lettres, 1993), which includes the text of the actual plan; E. Cohen, *Le colbertisme "high tech"* (Paris: Hachette, 1992). For an introduction to the French issues in English there is Patrick A. Messerlin, "France," in Benn Steil,

David G. Victor, and Richard R. Nelson (Eds.), *Technological Innovation and Economic Performance* (Princeton, N.J.: Princeton University Press, 2002): 148–177. French telecomunications is covered well in Jacques P. Chamoux, *Télécoms, la fin des privileges* (Paris: Presses Universitaires de France, 1993). The standard work on the French Minitel system is M. Marchang, *La grande aventure du Minitel* (Paris: Librairie Larousse, 1987). One of the first major international studies of computers was published in France, and has a great deal of information on the French experience, Jean-Michel Treille, *L'Économie Mondiale de L'Ordinateur* (Paris: Éditions Du Seuil, 1973). For a collection of papers on the early history of French computing, covering such topics as computers at Bull, the Minitel telephone system, use of an IBM 1620, communications protocols, and more recent topics surrounding viruses, there is Jacques André and Pierre Mounier-Kuhn (Eds.), *Actes du 7e colloque sur l'histoire de l'Informatique et des Transmissions, Espacé Ferrie-École Supérieure et d'Application des Transmissions* (Paris: National Institute for Research on Information and Automation, 2004).

France in the modern age of computers was captured in the discussion surrounding the study prepared by Simon Nora and Alain Mine, *The Computerization of Society: A Report to the President of France* (Cambridge, Mass.: MIT Press, 1980), but which was published two years earlier in French, *L'Informatisation de la Société* (Paris: La Documentation Française, 1978). Historians and economists have almost universally overlooked the four sequel volumes published during the course of 1978 under the same title, and subtitled *Annexes* one through four, that introduced substantial quantities of commentary about IT and data on France, Great Britain, West Germany and the United States, and includes a bibliography of books and articles available in France during the late 1970s dealing with computers. When combined with the overview volume— the one scholars and contemporaries used—one has an important source of information about computing and its business, managerial and policy ramifications for Western Europe.

For discussions of subsequent debates, see Ann Mayère, *Pour Une Economie de L'information* (Paris: Editions du Centre National de la Recherche Scientifique, 1990) and Jean Lojkine, *La Révolution Informationnelle* (Paris: Presses Universitaires de France, 1992); the latter is the more comprehensive of the two works and is most specific about French circumstances. Government policy makers continued to be active in the 1980s and 1990s, therefore, consult G. Thèry, *Les autoroutes de l'Information* (Paris: Commissariat Général du Plan, 1996); P. Lafitte, *La France et la société de L'Information* (Paris: Sénat, 1996); P. Martin-Lalande, *L'INTERNET: un vrai défi pour la France* (Paris: French Government, 1997). Although largely out of scope with the discussion on first wave diffusion in France, see Gunnar Trumbull, *Silicon and the State: French Innovation Policy in the Internet Age* (Washington, D.C.: Brookings Institution Press, 2004), particularly the first chapter. To understand the extent to which the French had become users of computers there is a highly useful study conducted by the government on this issue, Agence de l'Informatique, *L'Etate d'informatisation de la France* (Paris: Economica, 1986).

On Germany it is essential to understand the extent of the problems faced by all Germans in the years immediately following the war. For purposes of computer diffusion and economic situations, a useful start is John Gimbel, *Science, Technology, and Reparations: Exploitation and Plunder in Postwar Germany* (Stanford, Cal.: Stanford University Press, 1990). The standard work on computer technology and their associated producers is by Harmut Petzold, *Moderne Rechenkünstler: Die Industrialisierung der Rechentechnik in Deutschland* (Munich: Verlag C.H. Beck, 1992), but also look at his dissertation, a variation of the 1992 publication for other details, *Rechnende Maschinen: Eine historische Untersuchung ihrer Herstellung und Anwendung vom Kaiserreich bis zur Bundesrepublik* (Dusseldorf: VDI Verlag, 1985), also very much oriented toward history of devices. Also consult an equally useful book with a similar orientation, and that is beautifully illustrated, by Wilfried de Beaclair, *Rechnen mit Maschinen: Eine Bildgeschichte der Rechentechnik* (Berlin: Springer, 2005). Various histories exist on key firms and people. For a series of interviews with computer pioneers, see D. Siefkes, A. Braun, P. Eulenhöfer, H. Stach, and K. Städtler, *Pioniere der Informatik: Ihr Lebensgeschichte im Interview. Interviews mit F.L. Bauer, C. Floyd, J. Weizenbaum, N. Wirth und H. Zemanek* (Berlin: Springer, 1999).

On IBM see Edwin Black, *IBM and the Holocaust: The Strategic Alliance Between Nazi Germany and America's Most Powerful Corporation* (New York: Crown Publishers, 2004) which argues that IBM helped the Nazis implement the holocaust; for a more balanced account, there is T. Driessen, *Von Hollerith zu IBM, Zur Fruhgeschichte der Datenverarbeitungs-technik von 1880 bis 1970 aud wirtschaftswissenschaftlicher Sicht* (Cologne: Muller Botermann, 1987). A team of retired IBM engineers and scientists wrote a series of histories on research and product development by the firm in Germany that, while difficult to find since they were privately published, nonetheless are wealthy in detail: Karl E. Ganzhorn, *The IBM Laboratories Beoblingen: Foundation and Build-up: A Personal Review* (Sindelfingen: Privately printed, 2000); Albert Endres, *Die IBM Laboratorien Böblingen: System-Software-Entwicklung* (Sindelfingen: Privately printed, 2001); Albrecht Blaser, *The Heidelberg Science Center: User Oriented Informatics and Computers in Science: An Overview* (Sindelfingen: Privately printed, 2001); Helmut Painke, *Die IBM Laboraterien Böblingen: System-Entwicklung: Ein Persönlicher Rückblick* (Sindelfingen: Privately printed, 2003); Horst E. Barsuhn and Karl E. Ganzhorn, *The IBM Laboratories Boeblingen: Semiconductor and Chip Development* (Sindelfingen: Privately printed, 2005); R. Beyer et al., *IBM Informsationstechnik für Banken und Sparkassen im 20. Jahhundert* (Sindelfingen: Privately printed, 2006).

On Konrad Zuse, it is essential to begin with his memoirs, which shed light on German computing from the mid-1930s to the late 1960s, *The Computer—My Life* (Berlin: Springer-Verlag, 1993). The original edition appeared in German, *Der Computer—Mein Lebenswerk* (Berlin: Springer-Verlag, 1984). For an early study of his work, there is Karl-Heinz, Czauderna, *Konrad Zuse, der Weg zu seinem Computer Z 3* (Munich: R. Oldenbourg, 1979). On Heinz Nixdorf and his company there are two useful studies: Volker Werb, *Heinz Nixdorf* (Paderborn: Schoeningh Ferdinand, 2007) and a shorter biography by Klaus Kemper, *Heinz Nixdorf: Eine Deutsche Karriere* (Landsberg: Verlasg Moderne Industrie, 1987). On SAP, I found useful a book although not a formal history of the firm, by Hasso Plattner, *Anticipating Change: Secrets Behind the SAP Empire* (Rocklin, Cal.: Prima Publishing, 2000).

On the IT industry, semiconductors and other technologies, there are a group of useful studies. M. Breitenacher et. al., *Elekstrotechnische Industrie* (Munich: IFO-Institut für Wirtschaftsforschung, 1974), focusing on semiconductor and computer vendors of the 1960s and 1970s; on the products available in the early 1960s, see Rolf Haske, *Einführung in die Informations- und Dokumentationstechnik unter besonderer Berücksichtigung der Lochkarten* (Leipzig: Bibliographisches Institute, 1965); on the structure and size of the computer industry, N. Kloten et al., *Der EDC-Market in der Bundes Republik Deutschland* (Tübingen: Mohr, 1976) but see also A. Rösner, *Die Wettbewerbverhältnisse auf der Markit fur electronische Datenverarbeitungsanlagen in der BRD* (Berlin: Duncker-Humblot, 1978). See also Arndt Sorge, Gert Hartman, Malcolm Warner, and Ian Nicholas, *Microelectronics and Manpower in Manufacturing: Applications of Computer Numerical Control in Great Britain and West Germany* (Aldershot, Eng.: Gower Publishing Co., 1983). Because Heinz Zemanek's work as a computer scientist profoundly influenced others throughout the German-speaking part of Europe, he cannot be ignored. For his views on computers, see his book, which includes personal recollections, *Weltmacht Computer: Weltreich der Information* (Munich: Bechtle Verlag, 1991).

German-speaking Europe's experience with computers would not be complete without understanding the experience in Switzerland. Fortunately, there are two excellent sources. The first, a book by Gregor Henger is both a detailed history of computing in that country (machines, users, manufacturers, organizations) written by a journalist and a beautifully illustrated volume with hard-to-find photographs of machines, and rarer, of key individuals, *Informatik in der Schweiz: Eine Erfolgsgeschichte verpasster Chancen* (Zurich: Verlag Neue Zürcher Zeitung, 2008). A more scholarly tome is an anthology of contributed articles largely focused on the uses and technology of computing, each with summaries in German, French, and English, and articles either in German or French, Peter Haber (ed.), *Computergeschichte Schweiz: Eine Bestandesaufnahme/Histoire de L'ordinateur en Suisse: Un état des lieux* (Zurich: Cronos Verlag, 2009).

We cannot assume that post-unified Germany is an extension of the past and just as one has to appreciate what Germany was like in 1945 before looking at computing that followed, one needs to follow the same process on post-1990 Germany about which there is now an extensive literature. Two books prove useful for setting the context for contemporary computing in Germany, W.R. Smyser, *The Economy of United Germany: Colossus at the Crossroads* (New York: St. Martin's Press, 1992) and Claire Annesley, *Postindustrial Germany: Services, Technological Transformation and Knowledge in Unified Germany* (Manchester: Manchester University Press, 2004). For the evolving culture of post-unified Germany, with a great deal of information on diffusion, see Im Blickpunkt, *Informations-Gesellschaft* (Stuttgart: Metzler-Poeschel, 2002).

DIFFUSION OF COMPUTING IN ITALY, NETHERLANDS AND SWEDEN, 1950–1990S

Very little research has been done on the role of computers in Italy prior to the 1990s. To understand the economic reasons for why computing came late to Italy, the basic text for the early years is K.J. Allen and A.A. Stevenson, *An Introduction to the Italian Economy* (London: Martin Robertson, 1974), followed with Paul Ginsberg, *A History of Contemporary Italy, 1943–1988* (London: Penguin, 1990; neither, however, discusses computing. For examples of national government policy statements on stimulating development of computing and the use of IT, see, Ministerio del Bilancio a della Programmazione Economica, *Programma Economico Nazionale 1971–75*, Allegato Secondo *Programma 1966–71: Obiettivi e Risultatti* (Rome: Ministerio del Bilancio, 1972). See also Sabrina Pastrrelli, *Quaderni dell'Ufficio Ricerche Storiche* (Rome: Banca D'Italia, December 2006), which briefly looks retrospectively on the Italian government's investments in R&D.

On early experimental computing projects of the 1940s and 1950s, there are various articles that document the experience. These include Luigi Dadda, "Il Centro di Calcoli Numerici e l'introduzione della discipline informarische al Politecnico di Milano," in *Atti de Convegno Internazionale sulle Sotria e Preistoria del Calcolo Automatico e dell'informatica* (Milan: AICA, August 1991): 7–23 and his memoirs of computing in Italy, *Ricordi di un informatico in La cultura informatica in Italia: Riflessioni e testimonianze sulle origini—1950–1970* (Fondazione Adriano Olivetti, Bollati Boringhiere, April 1993): 61–106; Giuseppe DeMarco, Giovanni Mainetto, Serena Pisani, and Pasquele Savino, "The Early Computers of Italy," *IEEE Annals of the History of Computing* 21, no. 4 (October–December 1999): 29–30; Giuseppe DeMarco et al, "The Early Computers in Italy," 30–32; C. Bonfanti, "L'affaire FINAC tra Manchester e Roma (1935–1955) ed alcuni documenti inediti ad esso relativi," in *Atati del Congresso annuale dell'AICA* (Palermo: AICA, September 1994): 35–64; P. Ercoli, "From FINAC to CINAC," in *Atti del Convegno Internazionale sulla Storia e Preistoria del Calcolo Automatico e dell'informatica* (Milan: AICA, August 1991): 57–68; E.L. Aparo, "Mauro Picone e l'Istituto Nazional per le Applicazioni del Calcolo," in *Attiti del Convegno Internazionale sulla Storia e Preistoria del Calcolo Automatico e dell'informatica* (Milan: AICA, August 1991): 47–55. But see also, J.I.F. D Ker, "A Survey of New West-European Digital Computers," *Computer and Automation*, 12 (1963): 27–28.

As with early computing, we do not have a comprehensive history of the role of Olivetti Corporation or about its involvement with computers. However, there are some useful references, including Camillo Bussolati, Franco Malerba, and Salvatore Torrisi, *L'Evoluzione Delle Industrie Ad Alta Technologia In Italia: Entrata tempestiva, decline e opportunità di recupero* (Bologna: Società Editrice Il Mulino, 1996): 20–28; Salvatore Torrisi, "Discontinuità e Credibilità delle Stragtegie di Ingresso nel Settore Informatico," in Ibid., 102–120. On its new small computer, P.G. Perotto, "Olivetti della P101 in Avanti," in *Atti del Convergno Internazionale sulla Storia e Preistoria del Calcolo Automatico e dell'informatica* (Milan: AICA, August 1991): 218–221 and also his, *Programma 101. L'invenzione del personal computer. Una storia appassionante mai raccontata* (Milan: Sperling & Kupfer). The one major work in English on the company in general is by S. Kicherer, *Olivetti: A Study of the Corporate Management of Design* (London: Trefoil Publications, 1990), but clearly not a full corporate biography. There is also a doctoral dissertation in English on the early history of the company's work with pre-computer data processing, rich in insight and bibliography by Ronald Michael Frazzini, "The Development of Italian Informatics: Technology Development of Olivetti Calculating Machines in the Global Business Machine Market" (Unpublished Ph.D. dissertation, University of Minnesota, 2006).

There is a paucity of publications on the role of IT in Italian society and economy. The few important works include L. Soria, *Informatica: un'occassione perduta* (Turin: Einaudi, 1979); Cristiano Antonelli, *New Information Technology and Industrial Change: The Italian Case* (Dordrecht: Kluwer Academic Publishers, 1988); Di A. Bassanetti, M. Iommi, C. Jona-Lasinio, and F. Zollino, *La crescita dell'economia italiana negli anni novanta tra ritardo tecnologico e rallentamento della produttività*, Temi di discussione 539 (Rome: Banca D'Italia, December 2004); Daniele Archibugi, Sergio Cesaratto and Giorgio Sirilli, "Sources of Innovative Activities and Industrial Organization in Italy," *Research Policy* 20 (1991): 299–313; and for a more theoretical discussion, Giulio Sapelli, *Economia, tecnologia e direzione d'impresa in Italia* (Torino: Piccola Biblioteca Einaudi, 1994); Cinzia Colapinto, *L'Innovazione nel settore informatico in Italia: L'attività di corporate venture capital del Gruppo Olivetti negli anni ottanta*, Working Paper No. 2006-36 (Milan: Università degli Studi de Milano, Dipartimento di Scienze Economiche, Aziendali e Statistiche, November 2006); Camillo Bussolatti, Franco Malerba, and Salvatore Torrisi, *L'Evolucione delle industrie ad alta technologia in Italia: Entrata tempestiva, declino e opportunità di recupero* (Bologna: Società Editrice Il Mulino); Emilio Gerelli, *La Politica per L'Innovazione industriale: Problemi e Proposte* (Milan: Franco Angeli Editore, 1982); Camillo Bussolati, Franco Malerba and Salvsatore Torrisi, *L'evoluzione delle industrie ad alta tecnologia in Italia: Entrata tempestiva, declino e opportunità di recupero* (Bologna: il Mulino, 1996); Franco D'Egidio, *L'Economia digitale el il culture change: Come prosperare nella Nuova Economia* (Milan: FrancoAngeli, 2001). For a good, early exception that did focus on Italy see Assinform, *Rapporto Assin-form sulla situazione dell'informatica in Italia* (Milan: Assinform, n.d.). Important sources on statistics and comparisons with other countries begin with data from the 1980s forward, and can be found in a variety of papers published by the OECD, various economics departments of Italian universities, and the Bank of Italy, with the key publications cited in the endnotes for chapter 4.

For the Netherlands, understanding the shape of its economy is crucial, and for that, there is the useful survey by Jan L. van Zanden, *The Economic History of the Netherlands, 1914–1995: A Small Open Economy in the "Long" Twentieth Century* (London: Routledge, 1998). For a local introduction to Dutch computing, see Jan van den Ende, *Knopen, kaarten en chips* (Voorburg/Heeerlen: Utigave Centraal Bureau voor de Statistiek, 1991). For those who cannot read Dutch the essential work is Jan van Den Ende, *The Turn of the Tide: Computerization in Dutch Society, 1900–1965* (Delft: Delft University Press, 1994), which unfortunately only goes to the 1960s. This book can be supplemented with a series of case studies of how computers were used by Dutch institutions, which include a great deal of additional material on Dutch computing past 1965, Dirk de Wit, *The Shaping of Automation: A Historical Analysis of the Interaction Between Technology and Organization, 1950–1985* (Rotterdam: Erasmus Universiteit, 1994). For results of the use of IT, there is Bart van Ark, *The Productivity Problem of the Dutch Economy: Implications for Economic and Social Policies and Business Strategy*, Research Memorandum GD-66 (Groningen: Groningen Growth and Development Centre, September 2003), http://www.e-biblioteka.It/resursai/ES/memorandumi/gd66.pdf (last accessed November 25, 2011). His bibliography cites much of the economic literature on Dutch productivity and ICT. For a study of computing in a Dutch bank across many decades and thus provides a window into Dutch adoption of IT, see n S. de Boer and J. Frankhuizen, *Een eigenzinnige reus: Veertig jaar automatatisering bij de Rabobank* (Utrecht: Rabobank, 2004). Dutch computing is discussed in small pockets of articles and books on Europe in general and in the monographic journal literature, cited in endnotes to chapter four. There is no comprehensive history of Dutch computing covering the entire period in any language.

The Swedish case requires understanding the local economy and culture, and that of the Nordics as a whole, before one can appreciate digital events as the region has some fundamental distinctions that give it some characteristics different than those of the rest of Western Europe to a sufficient degree that generalizing about local experiences with IT need to be done carefully. Begin with Lars Magnusson, *An Economic History of Sweden* (New York: Routledge, 2000), which is very well written and informed, followed by Subhash Thakur, Michael Keen, Balázs Horváth Cerra, *Sweden's Welfare State: Can the Bumblebee Keep Flying?* (Washington, D.C.: International Monetary Fund, 2003), which surveys all manner of political, social, and economic issues,

Almost all the book length studies of Sweden's computing are in Swedish, which makes them accessible only to a small audience, so for most readers almost irrelevant. For example, there is Hans de Geer, *På väg till datasamhället. Datatekniken I politiken,1946–1963* (Stockholm: Royal School of Technology, 1992), which helps define the early role of public policy and IT. Others on this theme include and the rise of IBM in Sweden, Jan Annerstedt, Lars Forssberg, Sten Henriksson and Kenneth Nilsson, *Datorer och politik: Studier I en ny teckniks politiska effekter på det svenska samhället.* (Lund: University of Lund Press, 1970); Hans De Geer, *På väg till datasamhället* (Stockholm: University of Stockholm, 1992); Kent Lindkvist, *Dataeknik och politik: Datapoliten I Sverige 1945–1982* (Lund: Lund University Press, 1984); Per Lundin, *Designing Democracy: The UTOPIA Project and the Role of Labour Movement in Technological Change, 1981–1986* (Stockholm: University of Stockholm, 2005). A more broad-based survey providing background to the IT story is Tord Jöran Hallberg, *IT-Gryning: Svenk datahistoria från 1840-till 1960-talet* (Pozcal, Poland: Studentlitteratur, 2007). In English there is one very useful publication that provides details on both the supply and demand side of the story, while concentrating on describing how Swedes viewed concepts of IT, Magnus Johnsson, *Smart, Fast and Beautiful: On Rhetoric of Technology and Computing Discourse in Sweden, 1955–1995* (Linköping: Linköping University, 1997).

Swedish academics working with pioneers in local computing have launched a major initiative to interview people about their uses of computers (all in Swedish), but have also published proceedings of conferences that discuss computing across all of the Nordics, including Sweden, essential to understanding the digital role of the country: Janis Bubenko, Jr., John Impagliazzo, and Arne Sølvberg (eds.), *History of Nordic Computing: IFIP WG9.7 Working Conference on the History of Nordic Computing (HiNC1), June 16–18, 2003, Trondheim, Norway* (New York: Springer, 2005); John Impagliazzo, Timo Järva, and Petri Paju (eds.), *History of Nordic Computing 2: Second IFIP WG 9.7 Conference, HiNC2 Turku, Finland, August 21–23, 2007 Revised Selected Papers* (Berlin: Springer, 2009); and John Impagliazzo, Per Lundin and Benkt Wangler (eds.), *History of Nordic Computing 3: Third IFIP WG 9.7 Conference, HiNC3, Stockholm, Sweden, October 18–20, 2010, Revised Selected Papers* (Berlin: Springer, 2011). The role played by higher education has been documented in considerable detail in Janis Bubenko, Jr., Carl Gustaf Jansson, Anita Kollerbaur, Tomas Ohlin, and Louise Yngström, *ICT For People: 40 Years of Academic Development in Stockholm* (Stockholm: DSV, 2006). For statistics on government computing, there are the annual reports (in Swedish) Statskontoret, *Statliga Dataorer 1981: State-Managed Computers* (Stockholm: Statskontoret, published in the 1980s and 1990s), but difficult to obtain outside Sweden, but worth examining. As with so many other European countries there is no comprehensive history of IT for Sweden in a non-Swedish language, although snippets of the story are available in articles and chapters in other pan-European themes, many cited in the end notes. But on the industry, consult Dan Johansson, *The Dynamics of Firm and Industry Growth: The Swedish Computing and Communications Industry* (Stockholm: Department of the Organization and Management, KTH, 2001), and for context, Gunnar Eliasson, *The Macroeconomic Effects of Computer and Communications Technology: Entering a New and Immediate Economy* (Stockholm: The Royal Institute of Technology, 2002). There are no comprehensive corporate histories of Facit or Saab in English that tell the story of their role in computing. However, on Saab there is in Swedish a highly detailed history of its IT in three volumes: Saab, *Bits & Bytes ur Datasaabs historia, Tema D21* (Linköping: Datasaabs vänner, 1994), *Bits & Bytes ur Datasaabs historia, Tema Flyg* (Linköping: Datasaabs vänner, 1995), and *Bits & Bytes ur Datasaabs historia, Tema Bank* (Linköping: Datasaabs vänner, 1996). On Swedish microcomputers, especially Luxor Datorer and its ABC product, there is a detailed two volume history by Roland Sjöström, *Positionering under strategisk osäkerhet. En studie av positionering I en ny bransch* (Linköpings tekniska högskola: Ekonomiska Institutionen, 1996) and *Luxor Datorer och persondatorbranschen* (Linköpings tekniska högskola: Ekonomiska Institutionen, 1996). There is also a major collection of studies on Swedish uses of computing, and the development of local IT, Per Lundin, Niklas Stenlås, and Johan Gribbe (eds.), *Science for Welfare and Warfare: Technology and State Initiative in Cold War Sweden* (Sagamore Beach, Mass.: Watson Publishing International, 2010).

Historians and others have examined developments in Norway and Denmark, and increasingly about Finland that parallels the kinds of materials available about Sweden. For example, on Norway there is the excellent volume edited by Jan Fagerberg, David C. Mowery, and Bart Verspagen (eds.), *Innovation, Path Dependency, and Policy: The Norwegian Case* (Oxford: Oxford University Press, 2009) that discusses many technologies, not just computers, rich in detail and bibliographic references. On the other extreme, we have the celebratory volume published by IBM on the occasion of its seventy-fifth anniversary in Norway that offers little information and is difficult to obtain, Gunnar Nerheim and Helge W. Nordvik, *Ikke bare maskiner: Historien om IBM I Norge, 1935–1985* (Oslo: Universitetsforlaget, 1986). Finally, we have the interesting case of an important historian in Denmark, Lars Heide, who has written on the history of computing initially in his native language then in English, suggesting a new trend in European IT historiography. His important study on Danish history is *Hulkort og edth I Danmark, 1911–1970* (Arhus, Den.: Systime, 1996) and in English on a different topic, *Punched-Card Systems and the Early Information Explosion, 1880–1945* (Baltimore, Md.: Johns Hopkins University Press, 2009).

PAN-EUROPEAN IT INITIATIVES, 1970S–1990S

There is a large body of publications that have appeared over the years in the form of government plans for promoting national and international IT industries and deployment, along with a polemical literature on the role of public policies and the shaping of "information societies" and the "new economy." European historians are just beginning to publish pan-European studies, although still as anthologies of chapters. For a useful one dealing with banking, see Bernardo Bátiz-Lazo, J. Carles Maixés, and Paul Thomes (eds.), *Technological Innovation in Retail Finance: International Historical Perspectives* (London: Routledge, 2011). Both of these collections of materials have not been studied well by historians; nonetheless, they are useful sources to consult. To situate the role of IT, ICT, and other technologies into the bigger picture of Europe's historic effort to integrate its economy during the past six decades, a good place to start is with John Gillingham, *European Integration, 1950–2003: Superstate or New Market Economy?* (Cambridge: Cambridge University Press, 2003), which also includes a superb bibliography on the European Union's recent history. One of the few studies that reviews the IT industry across Europe in its earliest period is by Jean-Michel Treille, *L'Économie mondiale de L'Ordinateur* (Paris: Éditions de Seuil, 1973). The best one volume study of pan-European IT initiatives is by Paul Gannon, *Trojan Horses and National Champions: The Crisis in Europe's Computing and Telecommunications Industry* (London: Apot-Amatic, 1997), which is well informed and detailed. For one of the best, and earliest, OECD analyses of Europe's IT situation as compared to that of other nations, see its report, *Gaps in Technology: Electronic Computers* (Paris: OECD, 1969) followed by a later publication, *Information Activities, Electronics and Telecommunications Technologies: Impact on Employment, Growth and Trade*, vol. 1 (Paris: OECD, 1981), augmented by a third analysis from the OECD, *Trends in the Information Economy* (Paris: OECD, 1986). For points of view on pan-European initiatives at the height of Wave One, there are a number of contemporary accounts, but begin with Commission of the European Communities, *European Society Faced with the Challenge of New Information Technologies: A Community Response* (Brussels: Commission of the European Communities, 1979) followed by Gareth Locksley (ed.), *The Single European Market and the Information and Communication Technologies* (London: Belhaven Press, 1990).

For a sampling of the debates on what to do there are many useful publications, such as E. Arnold and K. Guy, *Parallel Convergence: National Strategies in Information Technologies* (London: Frances Printer, 1986); M. Hobday, *Trends in the Diffusion of Application Specific Integrated Circuits* (Sussex: University of Sussex, 1988) which has considerable material on a Philips-Siemens joint project of the 1980s; European Economic Community, *L'espace scientifique et technologique European dans le contexte international. Resources et conditions de la competitivité de la communauté* (Brussels: EEC, 1988) which includes in its policy debates a great deal of information on the structure, status, and characteristics of the West European information processing industry of the 1980s when it was under competitive duress; Commission of the European Communities, *Esprit '88: Putting Technology to Use*, Part 2 (Amsterdam: North-Holland, 1988) describes the most extensive of the pan-European projects just launching at the time; European Economic Community, *European Community Policies for Semiconductors* (Brussels: EEC, 1989) which offers an analysis of the multiple R&D initiatives around Europe and the results achieved so far; OECD, *Government Policies and the Diffusion of Microelectronics* (Paris: OECD, 1989) describes the changing nature of European government support for R&D in the late 1980s; OECD, *The Changing Role of Government Research Laboratories* (Paris: OECD, 1989) describes the rethinking these institutions were undertaking in the late 1980s, with special attention to semiconductors and robotics; M. Sharp and P. Holmes (eds.), *Strategies for New Technologies* (New York: Phillip Allan, 1989) is an example of the debate underway; but more importantly D. Foray, M. Gibbons, and G. Ferné, *Major R&D Programmes for Information Technologies* (Paris: OECD, 1989) which reviews government policies and makes recommendations regarding European R&D strategies, reflecting the extensive churn governments experienced with their IT strategies in the 1980s; see also the collection of essays on the period in Michael S. Steinberg, *The Technical Challenges and Opportunities of a United Europe* (Savage, Md.: Barnes & Noble Books, 1990); while European Economic Community, *L'industrie Européene de l'electronique et de l'informatique* (Brussels: EEC, 1991) provides a data-rich snapshot of the IT industry within the EEC and the context in which it functioned. For historical perspectives, Franco Malebra's analysis of how product innovation was done in Western Europe for IT during the 1980s, at the height of Wave One, "The Organization of the Innovative Process," in Nathan Rosenberg, Ralph Landou, and David C. Mowery (eds.), *Technology and the Wealth of Nations* (Stanford, Cal.: Stanford University Press, 1992): 247–278; see also the extremely useful and well researched chapter by Dimitris Assimakopoulos, Rebecca Marschan-Piekkari, and Stuart MacDonald, "ESPRIT: Europe's Response to U.S. and Japanese Domination of Information Technology," in Richard Coopey (ed.), *Information Technology Policy: An International Perspective* (New York: Oxford University Press, 2004): 247–263. For a continuation of the debate into the 1990s consult the highly detailed analysis by Robin Mansell and W. Edwards Steinmueller, *Mobilizing the Information Society: Strategies for Growth and Opportunity* (Oxford: Oxford University Press, 2000).

EASTERN EUROPE, 1940S–2000

The first step in understanding computing in Communist Europe is to appreciate how governments, politics, and economics unfolded in the region. Since the fall of communism, archives and other sources of data have become available to scholars, although they are only just starting to examine the materials and publish their results. Several important studies exist, however, that are useful, indeed crucial. For a good political history consult Vladislav M. Zubok, *A Failed Empire: The Soviet Union In The Cold War From Stalin to Gorbachev* (Chapel Hill, N.C.: University of North Carolina Press, 2007, 2009). I suggest consulting the 2009 paperback edition as the author includes a useful summary of events that occurred after Gorbachev. Two highly informed books by Anders Åslund explain the interactions between political and economic events which takes the story down into the Putin era, *Russia's Capitalist Revolution: Why Market Reform Succeeded and Democracy Failed* (Washington, D.C.: Peterson Institute for International Economics, 2007) and *Russia After the Global Economic Crisis* (Washington, D.C.: Peterson Institute for International Economics, 2010). There is a policy-oriented discussion focuses on Soviet economics, politics, and briefly the role of technology and scientific R&D in a short collection of essays, Raj M. Desai and Itzhak Goldberg (eds.), *Can Russia Compete?* (Washington, D.C.: Brookings Institution Press, 2008). Three useful studies of Soviet IT that are starting points for the topic, based on Russian publications. Martin Cave, *Computers and Economic Planning: The Soviet Experience* (Cambridge: Cambridge University Press, 1980) is a respected source that discusses the issues raised in our book in a thoughtful and confident manner for the 1950s–70s. On scientific issues, there is the useful study by Bruce Parrott, *Politics and Technology in the Soviet Union* (Cambridge, Mass.: MIT Press, 1983). Slave Gerovitch, *From Newspeak to Cyberspeak: A History of Soviet Cybernetics* (Cambridge, Mass.: MIT Press, 2002) makes effective use of Soviet newspapers and journals to analyze how policy makers and the technical and scientific community responded to the rise and use of digital computing. An excellent, well-informed Ph.D. dissertation that should have been published is essential to the story: William Keith McHenry, "The Absorption of Computerized Management Information Systems in Soviet Enterprises" (Unpublished Ph.D. dissertation, University of Arizona, 1985). Another very well informed study of all Eastern Europe and the Soviets essential to any analysis of IT in the region is J. Wilczynski, *Technology in Comecon: Acceleration of Technological Progress through Economic Planning and the Market* (London: Macmillan, 1974). The work of McHenry compliments this study in a series of articles he published and his U.S. Congressional testimony, which are cited in the endnotes to chapter 6 of our book. While reading his commentaries, also consult the numerous articles written by Seymour E. Goodman, a long-time "Soviet watcher" who is deeply knowledgeable about computing in Central Europe. While all these publications have excellent bibliographies, they are limited. The only formal bibliography available is very dated, but nonetheless useful for the early years, Eugene Gros, *Russian Books on Automation and Computers* (London: Scientific Information Consultants, 1967). For clearly written, highly informed descriptions of the Ryad computer systems the essential sources are two well-informed articles by Richard W. Judy and Robert W. Clough, "Soviet Computers in the 1980s: A Review of the Hardware," *Advances in Computers* 29 (1989): 261–323 and "Soviet Computing in the 1980s: A Survey of the Software and Its Applications," Ibid., 30 (1990): 223–306. Finally, the only first-person memoir available (so far) in English on early computing written by a distinguished Soviet computer scientist is Boris N. Malinovsky, *Pioneers of Soviet Computing* (published electronically,

2010 http://www.sigcis.org/files/malinovsky2010.pdf and is available for downloading without permission, last accessed 8/5/2010).

I found U.S. Congressional hearings on the Soviet economy during the Cold War to be the best contemporary sources on computing across the U.S.S.R. There were over a half dozen of these, often a combination of testimony and formally prepared studies, I cited in the endnotes. The most useful of all these were U.S. Congress, Joint Economic Committee, *Soviet Economy in a Time of Change* 2 vols (Washington, D.C.: U.S. Government Printing Office, 1979); *Gorbachev's Economic Plans* 2 vols (Washington, D.C.: U.S. Government Printing Office, 1987); another government-sponsored collection, John P. Hardt and Richard F. Kaufman (eds.), *East European Economies: Slow Growth in the 1980s* 2 vols (Washington, D.C.: U.S. Government Printing Office, 1986). Finally, another U.S. Government sponsored study, although not for the Congress, was prepared by D.J. Peterson, *Russia and the Information Revolution* (Santa Monica, Cal.: RAND Security Research Division, 2005). David A. Wellman published an analysis of Soviet computing at the height of the Cold War, reflecting very much American governmental thinking on the subject, *A Chip in the Curtain: Computer Technology in the Soviet Union* (Washington, D.C.: National Defense University Press, 1989).

On the question of computing in Soviet education, although a topic of importance in the post-1985 period after the arrival of the PC, there is a very early publication that served as a Soviet mouthpiece worth consulting, Bryce F. Zender, *Computers and Education in the Soviet Union* (Englewood Cliffs, N.J.: Educational Technology Publications, 1975). An early study of Soviet policies and programs relating to information technology, while difficult to find, is worth examining and is well informed, Günther Zell, *Information und Wirtschaftslenkung in der UdSSR: Eine Analyse des volkswitschaftlichen Informationssystems* (Berlin: Dunker & Humbl0ot, 1980). Although not a history, on the 1980s there is the small, useful monograph by Wilson P. Dizard and S. Blake Swensrud, *Gorbachev's Information Revolution: Controlling Glasnost in a New Electronic Era, Significant Issues Series* 9, no. 8 (Washington, D.C.: The Center for Strategic and International Studies, 1987).

Additionally, there were many books published on Soviet science which are largely out of scope with our book, but also many useful articles on Soviet computing written during the Cold War that appeared in all the expected journals, such as *Communications of the ACM* and later *IEEE Annals of the History of Computing*; most are cited in the endnotes to chapter 6. Also examine the articles in the contributed volume, John Impagliazzo and Eduard Proydakov, *Perspectives on Soviet and Russian Computing. First IFIP Wg 9.7 Conference, Sorucom 2006, Petrozavodsk, Russia, July 3–7, 2006* (Berlin: Springer, 2011).

Understanding East Germany's political and economic environment is essential to the story of IT and the best place to begin is with the first study based on archival research by Jeffrey Kopstein, *The Politics of Economic Decline in East Germany, 1945–1989* (Chapel Hill, N.C.: University of North Carolina Press, 1997). Science and technology in the German Democratic Republic has been the subject of other important studies. The earliest included Vladimir Slamecka, *Science in East Germany* (New York: Columbia University Press, 1963) and Raymond Bentley, *Research and Technology in the Former Democratic Republic* (Boulder, Col.: Westview Press, 1992), which are useful for context but badly out of date. The most useful study is by Raymond G. Stokes, *Constructing Socialism: Technology and Change in East Germany, 1945–1990* (Baltimore, Md.: Johns Hopkins University Press, 2000), now the essential work. Several studies examine more closely the role of computing, beginning with the well-informed one by Gary Lee Geipel, "Politics and Technology in the German Democratic Republic, 1977–1990" (Unpublished Ph.D. dissertation, Columbia University, 1993) and that includes a superb bibliography. Difficult to find, but the only study devoted entirely to the early stages of IT deployment in the GDR is Erich Sobeslavsky and Nikolaus Joachim Lehmann, *Zur Geschichte von Rechentechnik und Datenverarbeitung in der DDR 1946–1969* (Dresden: Hannah-Arendt-Institute für Totalitarismusforschung, 1996) and which should be read with Steffen Werner, *Kybernetik Statt Marx?: Politische Ökonomie und marxistische Philosophie in der DDR unter dem Einfluß der elektronischen Datenverarbeitung* (Stuttgart: Verlag Bonn Aktuell GMBH, 1977). Publications on East Europe and the Soviet Union also contain material on East Germany as well.

In recent years there has been a substantial increase in the amount of high quality research being done on East Germany, resulting in a new wave of publications providing more details on both the development of IT in East Germany and its use. For a recent overview of events in the DDR, see Christine Pieper, *Hochschulinformatik in der Bundesrepublik und der DDR bis 1989/1990* (Stuttgart: Franz Steiner Verlag, 2009); do not ignore its bibliography, it is very current and useful. A number of conferences were held in the early 2000s on the history of computing with published proceedings. On the evolution of computer science and technology in East Germany there is the rich volume edited by Friedrich Naumann and Gabriele Schade (eds.), *Informatik in der DDR—eine Bilanz* (Bonn: Gesellschaft für Informatik, 2006) and a sequel that focused more on applications, Birgit Demuth (ed.), *Informatik in der DDR—Grundlagen und Anwendungen* (Bonn: Gesellschaft für Informatik, 2008). Additionally, there is a volume of contributed essays that looks at both computer science and uses in East Germany, Wolfgang Coy and Peter Schirmbacher (eds.), *Informatik in der DDR: Tagung Berlin 2010* (Berlin: Tagungsband zum 4. Symposium "Informatik in der DDR," 2010).

Computing in Eastern Europe has been studied in bits and fits, with results published in articles and chapters in books, many cited in the endnotes of chapter six. This applies to the entire period 1940s to the early 2000s. On the earliest projects in the region, an entire issue of the *IEEE Annals of the History of Computing* was devoted to individual country studies, volume 21, number 3 (July–September, 1999) and which cites earlier studies that appeared in both this publication and elsewhere. On the latter years of the story with historical background see János Kovács (ed.), *Technological Lag and Intellectual Background: Problems of Transition in East Central Europe* (Aldershot, G.B.: Dartmouth, 1995). Essential to understanding Polish circumstances is Andrzej Tomasz Jarmoszko, "Transformation of the Telecommunication Environment in Poland, 1989–1991" (Unpublished Ph.D. dissertation, University of Arizona, 1992) and for context, with some material on IT post-2000, see Jane Hardy, *Poland's New Capitalism* (New York: Pluto Press, 2009). For a short, well informed analysis of the role of Western technologies across the region, consult Steven W. Popper, *East European Reliance on Technology Imports from the West* (Santa Monica, Cal.: RAND Corporation, 1988). He also wrote a dissertation on Hungary, essentially the only full report in English on the subject: "The Diffusion of Process Innovation in Hungary" (Unpublished Ph.D. dissertation, University of California, Berkeley, 1985). Essential to the subject is also David Turnock, *The East European Economy in Context: Communism and Transition* (London: Routledge, 1997) because it puts IT activities in the post-Communist era into the broader context of economic and political events. Turnock's book, however, should be read along with another, Krzysztof Piech and Slavo Radosevic (eds.), *The Knowledge-Based Economy in Central and Eastern Europe: Countries and Industries in a Process of Change* (New York: Palgrave, 2006), which focuses on the role of information, IT, and knowledge-management themes late in the century. On the important experience of Czechoslovakia, we do not yet have a formal comprehensive history, but there is a useful collection of historical studies on aspects of the story, Jaroslav Folta (ed.), *Computing Technology Past & Future* (Prague: National Technical Museum in Prague and Society for the History of Science and Technology, 2001).

COMPUTING COMES TO JAPAN

The literature on Japan almost equals in volume that on the role of IT in the United States or in Western Europe. However, there are several clear paths one can take into the literature. For relevant introductions to the economic context, consult Masataka Kosaka, *A History of Postwar Japan* (Tokyo International, 1982) which provides considerable detail on the early period; however, that needs to be augmented with the excellent study by Thomas F. Gargill and Takayuki Sakamoto, *Japan Since 1980* (Cambridge: Cambridge University Press, 2008). Three books serve as relevant introductions to business operations in Japan: Rodney Clark, *The Japanese Company* (New Haven, Conn.: Yale University Press, 1979) which remains useful despite its age, Naoto Sasaki, *Management and Industrial Structure in Japan* (Oxford: Pergamon Press, 1981) for a Japanese perspective. For a very broad study, see the well-informed discussion by William K. Tabb, *The Postwar Japanese System: Cultural Economy and*

Economic Transformation (New York, N.Y.: Oxford University Press, 1995). An older volume that had a strong practitioner, "here is how it is done" quality to it that I found useful is by T.F.M. Adams and N. Kobayashi, *The World of Japanese Business: An Authoritative Analysis* (Tokyo: Kodansha International, 1969). A minor classic on these themes not to be overlooked was written by sociologist Ezra F. Vogel, *Japan as Number One: Lessons for America* (Cambridge, Mass.: Harvard University Press, 1979). For a Japanese account of its business practices, see Makoto Ohtsu, *Inside Japanese Business: A Narrative History, 1960–2000* (Armonk, N.Y.: M.E. Sharpe, 2003). To complete one's understanding of the Japanese environment there are two excellent histories of MITI, essential to the study of IT in Japan: Chalmers Johnson, *MITI and the Japanese Miracle: The Growth of Industrial Policy, 1925–1975* (Stanford, Cal.: Stanford University Press, 1982) and Scott Callon, *Divided Sun: MITI and the Breakdown of Japanese High-Tech Industrial Policy, 1975–1993* (Stanford, Cal.: Stanford University Press, 1995). For a broad introduction to Japanese technological practices, see Jeffrey K. Liker, John E. Ettlie, and John C. Campbell (eds.), *Engineered in Japan: Japanese Technology-Management Practices* (New York: Oxford University Press, 1995).

There are several paths into the broad highly interconnected discussion of semiconductors, telecommunications, and computing in Japan. Begin with the detailed studies by Martin Fransman and in the order in which they were published, *The Market and Beyond: Information Technology in Japan* (Cambridge: Cambridge University Press, 1990), his most important work, *Japan's Computer and Communications Industry: The Evolution of Industrial Giants and Global Competitiveness* (New York: Oxford University Press, 1995), *Visions of Innovation: The Firm and Japan* (Oxford: Oxford University Press, 1999) and *Telecoms in the Internet Age: From boom to Bust to ...?* (Oxford: Oxford University Press, 2002). His most recent book, *The New ICT Ecosystem* (Cambridge: Cambridge University Press, 2010) is not about Japan, rather, more globally about Wave Two computing and communications. In addition to their rich offerings of details, each has extensive bibliographic references to the vast literature on Japan. The second author to consult is Marie Anchordoguy's two books, which should also be read in chronological order because her excellent conceptualization of Japanese events evolved over time. Short, to the point, and extremely well informed, her studies also embrace the totality of Japan's complex and diverse IT ecosystem, *Computers Inc.: Japan's Challenge to IBM* (Cambridge, Mass.: Council on East Asian Studies, Harvard University, 1989) and *Reprogramming Japan: The High Tech Crisis under Communitarian Capitalism* (Ithaca, N.Y.: Cornell University Press, 2005), the latter the single best one-volume study on the subject. Paralleling her critique of Japan's competitiveness is a small, overlooked study that deserves serious attention, even a decade after its publication, by Michael E. Porter, Hirotaka Takeuchi, and Mariko Sakakibara, *Can Japan Compete?* (Cambridge, Mass.: Perseus Publishing, 2000). For a detailed sociological perspective that is both well researched and written, see Martin Kenney and Richard Florida, *Beyond Mass Production: The Japanese System and Its Transfer to the U.S.* (New York: Oxford University Press, 1993).

Useful narratives more historical in nature written by historians and others are not as plentiful, but begin with Alfred D. Chandler, Jr., *Inventing the Electronic Century: The Epic Story of the Consumer Electronics and Computer Industries* (Cambridge, Mass.: Harvard University Press, 2005) and Kenneth Flamm, *Creating the Computer: Government, Industry, and High Technology* (Washington, D.C.: Brookings Institution, 1988). Gene Gregory has written a comprehensive survey of the various electronics industries in Japan that is excellent for the early post-World War II decades, *Japanese Electronics Technology: Enterprise and Innovation* (Tokyo: The Japan Times, Ltd., 1985). Also useful, indeed an important history, is Bob Johnstone, *We Were Burning: Japanese Entrepreneurs and the Forging of the Electronic Age* (New York: Basic Books, 1979). While Gregory, Fransman, and Anchordoguy among others discuss the semiconductor industry, another useful treatment is by Yui Kimura, *The Japanese Semiconductor Industry: Structure, Competitive Strategies and Performance* (Greenwich, Conn.: JAI Press, 1988). For PCs and other electronics in Japan and elsewhere in Asia, an essential source is by Jason Dedrick and Kenneth L. Kramer, *Asia's Computer Challenge: Threat or Opportunity for the United States and the World?* (New York: Oxford University Press, 1998). On robotics begin with Frederik L. Schodt, *Inside the Robot Kingdom: Japan, Mechatronics, and the Coming Robotopia* (Tokyo: Kodansha International, 1988), which is a combination historical and contemporary account; while for a serious economic discussion produced by Japanese researchers, see Kuni Sadamoto (ed.), *Robots in the Japanese Economy: Facts about Robots and Their Significance* (Tokyo: Survey Japan, 1981). For a massively detailed collection of data and narrative on the robotics industry, market, and uses, there is the indispensible volume by The Japan Industrial Robot Association, *The Robotics Industry of Japan: Today and Tomorrow* (Tokyo: Fuji Corporation, 1982). On cell phones, see Mizuko Ito, Daisuke Okabe, and Misa Matsuda (eds.), *Personal, Portable, Pedestrian: Mobile Phones in Japanese Life* (Cambridge, Mass.: MIT Press, 2005).

Japan's exceptionalism is often framed in terms of its approach to technology and nowhere is this more so than with such issues as government sponsored IT initiatives, Fifth Generation computing, and software. Jon Sigurdson, *Industry and State Partnership in Japan: The Very Large Scale Integrated Circuits (VLSI) Project* (Lund: Research Policy Institute, 1986) provides a useful overview of the project, while the book that panicked people the most in the West about the Fifth Generation was written by Edward A. Feigenbaum and Pamela McCorduck, *The Fifth Generation: Artificial Intelligence and Japan's Computer Challenge to the World* (Reading, Mass.: Addison-Wesley, 1983). However, the standard work on Japanese software remains M.A. Cusumano, *Japan's Software Factories* (New York: Oxford University Press,

1991). But his book can be used profitably in conjunction with another by two Japanese computer scientists, Yoshiho Matsumoto and Yutaka Ohno, *Japanese Perspectives in Software Engineering* (Singapore: Addison-Wesley, 1989). Other useful resources include Raul Mendez (ed.), *High Performance Computing: Research and Practice in Japan* (New York: John Wiley & Sons, 1992) and Robert S. Cutler (ed.), *Technology Management in Japan: R&D Policy, Industrial Strategies and Current Practice* (Boulder, Col.: Westview Press, 1993). Finally, on the knowledge management dimension of Japan's approach to information and IT, the standard work is Ikujiro Nonaka and Hirotaka Takeuchi, *The Knowledge-Creating Company: How Japanese Companies Create the Dynamics of Innovation* (New York: Oxford University Press, 1995). For one of the only bibliographies on Japanese technology, see Dawn E. Talbot, *Japan's High Technology: An Annotated Guide to English-Language Information Sources* (Phoenix, Ar.: Oryx Press, 1991).

Company histories and memoirs are particularly useful and accessible in English. An historian's account of the Japanese industry at large, and not just about IBM, is by Robert Sobel, *IBM vs. Japan: The Struggle for the Future* (New York: Stein and Day, 1986). CEO of Sony, Akio Morita, wrote *Made in Japan: Akio Morita and Sony* (New York: E.P. Dutton, 1986), essential for the early history of consumer electronics. Gary Katzenstein, *Funny Business: An Outsider's Year in Japan* (New York: Soho Press, 1989) provides a computer engineers view of Sony in the 1980s and for an early history of the firm, John Nathan, *SONY: The Private Life* (Boston, Mass.: Houghton Mifflin, 1919). On NEC, see NEC Corporation, *NEC Corporation, 1899–1999: A Century of Better Products, Better Services* (Tokyo: NEC Corporation, 2002) and *NEC Corporation: The First 80 Years* (Tokyo: NEC Corporation, 1984). A NEC chairman Koji Kobayashi has been a veritable publishing engine of his own about his company's operations, memoirs of his work there, and his views about IT in Japan in general. His key works include, *A Vision of C&C: Computers and Communications* (Cambridge, Mass.: MIT Press, 1986), *Rising to the Challenge: The Autobiography of Koji Kobayashi* (Tokyo: Harcourt Brace Jovanovich Japan, 1989), and *The Rise of NEC: How the World's Greatest C&C Company Is Managed* (Cambridge, Mass.: Blackwell, 1991); none are a substitute for another as each has a great deal of different content. Fijitsu's chairman also wrote a book about his times and company, Taiyu Kobayashi, *Fortune Favors The Brave: Fujitsu: Thirty Years in Computers* (Tokyo: Toyo Keizai Shinposham, 1986). Accounts of Japanese start-up firms are rare so one on an early PC company is welcome by Takeo Miyauchi, *The Flame from Japan: A Story of Success in the Microcomputer Industry* (Tokyo: Cord Computer Corporation, 1982). There is an account of one of Japan's most important mobile telecom firms, John Beck and Mitchell Wade, *DoCoMo: Japan's Wireless Tsunami* (New York: AMACOM, 2003). On Nintendo, see Daniel Sloan, *Playing to Win: Nintendo and the Video Games Industry's Greatest Comeback* (Singapore: John Wiley & Sons, 2011).

The debate about Japan's growing dominance in digital exports in the 1970s and 1980s spawned a large numbers of books critical of the country and Japanese defenders coming to its rescue, and in the process of this global debate, generated considerable amounts of information about IT in Japan and elsewhere almost to the present as the debate continues. Many of the previously cited publications from the 1960s–1990s fit into this debate; however, others proved useful in understanding Japanese IT. A quasi-scholarly discourse on issues with contributions by Japanese observers can be found in Alex S. Edelstein, John E. Bowes, and Sheldon M. Harsel (eds.), *Information Societies: Comparing the Japanese and American Experiences* (Seattle, Wash.: International Communication Center, School of Communications, University of Washington, 1978). A young American business school professor weighed in with a very good analysis of the global rivalry, William H. Davidson, *The Amazing Race: Winning the Technorivalry with Japan* (New York: John Wiley & Sons, 1984). For a well informed explanation of Japanese motives, see Sheridan Tsatsuno, *The Technopolis Strategy: Japan, High Technology, and the Control of the 21st Century* (New York: Prentice Hall, 1986). Particularly useful for understanding the consequences of IT on work in Japan in the 1970s and 1980s, see Tessa Morris-Suzuki, *Beyond Computopia: Information, Automation and Democracy in Japan* (London: Kegan Paul, 1988). For an example of the arguments that Japan is taking over the manufacture and sale of digital products, see Tom Forester, *Silicon Samurai: How Japan Conquered the World's I.T. Industry* (Cambridge, Mass.: Blackwell, 1993), ironically written by a respected expert on computing who seemed to have gotten caught up in the hubris of the debate. Christopher Wood, *The End of Japan Inc. And How the New Japan Will Look* (New York: Simon and Schuster, 1994) explains that Japan's economic miracle was collapsing, describing, in part, IT consequences. For yet another turn in the debate, this time another account of rising Japan, see Mark B. Fuller and John C. Beck, *Japan's Business Renaissance: How the World's Greatest Economy Revived, Renewed, and Reinvented Itself* (New York: McGraw-Hill, 2006), another example of committing oneself to a point of view too close to the events. But back to the earlier debate, for a European perspective, an excellent source is William R. Nester, *European Power and the Japanese Challenge* (New York: New York University Press, 1993). For two earlier defenses of Japan, there is H.J. Welke, *Data Processing in Japan* (Amsterdam: North-Holland Publishing Company, 1982) and Michio Morishima, a Japanese economist, *Why Has Japan "Succeeded"?: Western Technology and the Japanese Ethos* (Cambridge: Cambridge University Press, 1982).

SOUTH KOREA, TAIWAN, AND SINGAPORE

These three nations are often called Asian Tigers because of their aggressive and successful economic transformations from largely rural agricultural societies into industrial powerhouses, and as demonstrated in this book, major producers and users of various information and communications technologies. To understand the role of these societies one needs to understand the Pan-Asian circumstances in which they fit, much as one has to do the same with Japan and the United States on the one hand and on the other various Europeans nations in a pan-Atlantic context. As with Japan, gaining insight into Asian IT activities needs to be viewed initially through the lens of the local telecommunications environment, where indigenous IT skills often originated. Most studies are amalgamations of individually contributed papers and anthologies of chapters. For example, W. Mark Fruin (ed.), *Networks, Markets, and the Pacific Rim: Studies in Strategy* (New York: Oxford University Press, 1998), provides a combination of country cases and pan-Asian discussions. Another useful introduction to telecommunications in the region is found in Eli Noam, Seisuke Komatsuzaki, and Douglas A. Conn (eds.), *Telecommunications in the Pacific Basin: An Evolutionary Approach* (New York: Oxford University Press, 1994). Because of the far more extensive proactive roles played by Asian governments, this topic too needs to be understand in pan-Asian terms and for that a useful collection of papers can be found in Masahiko Aoki, Hyung-Ki Kim, and Masahiro Okuno-Fujiwara (eds.), *The Role of Government In East Asian Economic Development: Comparative Institutional Analysis* (Oxford: Clarendon Press, 1996).

It is essential in understanding South Korea's extensive involvement in ICT to appreciate how its economy evolved since 1950. Begin with Young-Iob Chung, *South Korea in the Fast Lane: Economic Development and Capital Formation* (New York: Oxford University Press, 2007), which contains an enormous amount of economic data and analysis by one of the most regarded experts on that economy. Less academic, but also well informed as a useful companion work, see Myung Oak Kim and Sam Jaffe, *The New Korea: An Inside Look at South Korea's Economic Rise* (New York: AMACOM, 2010). For a Korean perspective, circa 1970s-early 1980s, see Hyung-Koo Lee, *The Korean Economy: Perspectives for the Twenty-First Century* (Albany, N.Y. State University of New York Press, 1996). All three discuss IT and communications industries and deployment. Less relevant but useful for displaying attitudes toward the Korea economy just as it took off in IT matters, see two studies: Edward S. Mason et al. (eds.), *The Economic and Social Modernization of the Republic of Korea* (Cambridge, Mass.: Harvard University Press, 1980) and Parvez Hasan, *Korea: Problems and Issues in a Rapidly Growing Economy* (Baltimore, Md.: Johns Hopkins University Press, 1976). One book was less useful for our purposes because all its issues were picked up by other publications, nonetheless evidently being consulted and well done, Jongryn Mo and Chung-in Moon (eds.), *Democracy and the Korean Economy* (Stanford, Cal.: Hoover Institution Press, 1999). Another book on the same theme has been widely used by I and others to understand political-economic events, John Lie, *Han Unbound: The Political Economy of South Korea* (Stanford, Cal.: Stanford University Press, 1998).

A widely consulted study on Korean technology is Linsu Kim, *Imitation to Innovation: The Dynamics of Korea's Technological Learning* (Boston, Mass.: Harvard Business School Press, 1997), but it should quickly be followed by reading the many articles and other publications by two experts on the region, especially Jason Dedrick and Kenneth L. Kraemer, *Asia's Computer Challenge: Threat or Opportunity for the United States and the World?* (New York: Oxford University Press, 1998). Both volumes cover events in the 1980s and 1990s. Following the strategy of looking at telecommunications as a window into local IT activities, there is the excellent study by James F. Larson, *The Telecommunications Revolution in Korea* (Hong Kong: Oxford University Press, 1995), covering the 1960s-early 1990s. For the period following, a superb, well crafted study was prepared by the World Bank Institute covering the whole era of broadband, cell phones, and the Internet, Joonghae Suh and Derek H.C. Chen (eds.), *Korea as a Knowledge Economy: Evolutionary Process and Lessons Learned* (Washington, D.C.: World Bank, 2007). Reviewing the most current period, there is Tomi T. Ahonen and Jim O'Reilly, *Digital Korea: Convergence of Broadband Internet, 3G Cell Phones, Multiplayer Gaming, Digital TV, Virtual Reality, Electronic Cash, Telematics, Robotics, E-Government and the Intelligent Home* (London: Futuretext, 2007). For Korea's export practices, and even about European affairs, there is Sang-Hyup Shin, *European Integration and Foreign Direct Investment in the EU: The Case of the Korean Consumer Electronics Industry* (London: Routledge, 1998), covering the post 1986 period. The best single volume on South Korea's online gaming activities is Dal Yong Jin's authoritative study, *Korea's Online Gaming Empire* (Cambridge, Mass: MIT Press, 2010). An unusally well-informed unpublished study worth examining was written by Jonathan Clemens, "The Making of a Cyborg Society: South Korea's Information Revolution, 1997–2007" (Unpublished M.A. thesis, University of Hawaii, 2008). Each of these various studies includes extensive bibliographies of relevant articles in both English and in some instances, Korean.

There are a number of corporate histories. Two are of particular value since they concerned one of the largest electronics firms in the world, and provide considerable insight into the semiconductor story: Sea-Jin Chang, *Sony vs. Samsung: The Inside Story of the Electronics Giants' Battle for Global Supremacy* (Singapore: John Wiley & Sons, 2008) and Tony Michell, *Samsung Electronics and the Struggle for Leadership of the Electronics Industry* (Singapore: John Wiley & Sons, 2010). Because of the important role played by digital consumer and industrial products produced all over Asia one should also familiarize themselves with this sub-set of the digital world. Consult Shahid Yusuf, M. Anjum Altaf, and Kaoru Nabeshima (eds.), *Global Production Networking and Technological Change in East Asia* (Washington, D.C.: World Bank, 2004). Regarding North Korea, the single most informative study currently available is Nicholas Eberstadt, *The North Korean Economy: Between Crisis and Catastrophe* (New Brunswick, N.J.: Transaction Publishers, 2007), which includes discussion of ICT. For other sources on both South and North Korea consult the relevant endnotes on the chapter dealing with the Asian tigers.

On Taiwan there is a core set of publications that compliment similar studies done on South Korea and Japan. On the broad issue of Taiwanese economics and role of government, and which includes extensive discussion of IT, begin with Li-min Hsueh, Chen-kuo Hsu, and Dwight H. Perkins, *Industrialization and the State: The Changing Role of the Taiwan Government in the Economy, 1945–1998* (Cambridge, Mass.: Harvard Institute for International Development and Chung-hua Institution for Economic Research, 2001). It also includes an excellent bibliography. For a monograph that focuses more narrowly on IT and its institutional players, there is the quite well informed, clearly articulated volume by the Taiwanese business professor, Teresa Shuk-Ching Poon, *Competition and Cooperation in Taiwan's Information Technology Industry and Industrial Upgrading* (Westport, Conn.: Quorum Books, 2002). A number of studies have been published focusing on Taiwanese semiconductors, the most comprehensive of which is by John A. Mathews and Dong-Sung Cho, *Tiger Technology: The Creation of a Semiconductor Industry in East Asia* (Cambridge: Cambridge University Press, 2000). For a more intimate look at the semiconductor business in managerial and operational terms, there is a collection of contributed chapters written by Taiwanese, Chun-Yen Chang and Po-Lung Yu (eds.), *Made by Taiwan: Booming in the Information Technology Era* (Singapore: World Scientific, 2001). Two articles that set Taiwan and the other Tigers in context with each other are essential to consult. The first is Dilip K. Das, "The Dynamic Growth of the Electronics Industry in Asia," *Journal of Asian Business* 14, no. 4 (1998): 67–99. It is rich in detail, bibliography, and comparative perspectives. The second article is by Henry Wai-Chung Yeung, "From Followers to Market Leaders: Asian Electronics Firms in the Global Economy," *Asia Pacific Viewpoint* 48, no. 1 (April 2007): 1–25, which focuses on Singapore, South Korea, and Taiwan. There are virtually no company histories in English. For Acer Computers there is at least a memoir/history by a company executive, Robert H. Chen, *Made In Taiwan: The Story of Acer Computers* (Taipei: McGraw-Hill, 1997). As with so many other Asian countries, Dedrick and Kraemer include a chapter on Taiwan in their study of the region, *Asia's Computer Challenge*; it is an essential source on Taiwan.

To study the experience of Singapore begin with Christopher M. Dent (ed.), *The Foreign Economic Policies of Singapore, South Korea and Taiwan* (Cheltenham, UK: Edward Elgar, 2002) which is also useful for the other two regions for an explanation of the core issue of export economics. On this nation specifically a useful, short, well informed overview is by Hans Blomqvist, *Swimming with Sharks: Global and Regional Dimensions of the Singapore Economy* (Singapore: Marshall Cavendish International, 2005), followed by Toh Mun Heng and Tan Kong Yam (eds.), *Competitiveness of the Singapore Economy: A Strategic Perspective* (Singapore: Singapore University Press, 1998). All three publications discuss the role of IT. As with the other Asian nations, one should also consult John A. Mathews and Dong-Sung Cho, *Tiger Technology: The Creation of a Semiconductor Industry in East Asia* (Cambridge: Cambridge University Press, 2000), which is rich in details on the entire region. A difficult to find, 52-page monograph, but very useful on both supply and demand sides of the IT diffusion story is Hall Hill and Pang Eng Fong, *Technology Exports From A Small Very Open NIC: The Case of Singapore*, No. 89/6 (Canberra, Aust.: Australian National University Research School of Pacific Studies, 1989). Finally, and the most useful source on extent of diffusion of IT in households and enterprises (including public sector) from 1982 to the present, see National Computer Board, *Singapore IT Usage Survey*, published every two years. The endnotes to my chapter dealing with Singapore also leads one to appropriate websites for survey results since the late 1990s, prior to that date these reports were published as forty or more page pamphlets, and copies are unfortunately difficult to find, but absolutely essential to the history of IT in this city. Perhaps because Singapore is small, thus easy for one or two scholars to study, it has been the subject of many scholarly articles, largely by economists; the major ones became an important source for my study and are cited in the appropriate chapter.

CHINA

Understanding this nation's profound transformation since the 1930s and that of its economy since the 1940s is crucial to any appreciation of the role of IT. Begin with an anthology of materials dealing with culture, history, politics, and economics, among other Chinese topics, in Thomas Buoye, Kirk Denton, Bruce Dickson, Barry Naughton and Martin Whyte (eds.), *China: Adapting the Past Confronting the Future* (Ann Arbor, Mich.: Center for Chinese Studies, University of Michigan, 2002). The standard text on China's economy, and which weaves into the broader economic story the role of various technologies is by Barry Naughton, *The Chinese Economy: Transitions and Growth* (Cambridge, Mass.: MIT Press, 2007). For a survey covering just the past twenty years with extensive commentary on the roles of telecommunications and information technologies, there is the data-rich study by Rongxing Guo, *An Introduction to the Chinese Economy: The Driving Forces Behind Modern Day China* (Singapore: John Wiley & Sons, 2010). A technology-oriented scholarly study of technical talent that offers much in the way of economic perspectives on Chinese ICT is by Denis Fred Simon and Cong Cao, *China's Emerging Technological Edge: Assessing the Role of High-End Talent* (Cambridge: Cambridge University Press, 2009).

There is a swelling body of studies on the role of technologies in modern China, crucial to examine. An early study focusing on the activities of one city is by Denis Fred Simon and Detlef Rehn, *Technological Innovation in China: The Case of Shanghai's Electronics Industry* (Cambridge, Mass.: Ballinger, 1988) and for a collection of case studies of industrial parks and companies, there is very useful monograph by the late Qiwen Lu, *China's Leap into the Information Age: Innovation and Organization in the Computer Industry* (Oxford: Oxford University Press, 2000). The history of a few post-2000 Chinese ICT companies is told by Favid Sheff, *China Dawn: The Story of a Technology and Business Revolution* (New York: HarperBusiness, 2002). Three books provide a more comparative analysis that includes commentary about other technologies besides ICTs: Haiyang Li (ed.), *Growth of New Technology Ventures in China's Emerging Market* (Cheltenham, U.K.: Edward Elgar, 2006); and more focused on IT and with the added benefit of comparing Chinese developments to those in India, Neil Gregory, Stanley Nollen and Stoyan Tenev, *New Industries from New Places: The Emergence of the Software and Hardware Industries in China and India* (Washington, D.C.: World Bank, 2009); and another volume with comparative information on other countries too, William W. Keller and Richard J. Samuels (eds.), *Crisis and Innovation in Asian Technology* (Cambridge: Cambridge University Press, 2003). Two books deal usefully on the current IT environment and are well informed. The first, by Rebecca A. Fannin, *Silicon Dragon: How China Is Winning the Tech Race* (New York: McGraw-Hill, 2008) looks at events at specific companies; Marina Yue Zhang and Bruce W. Stening, provide thoughtful analysis along with a bit of hubris in *China 2.0: The Transformation of an Emerging Superpower . . . and the New Opportunities* (Singapore: John Wiley & Sons, 2010).

There is a growing body of comparative studies looking at multiple countries, mostly in Asia, but also in other parts of the world. Veterans of such studies, Kenneth L. Kraemer, Jason Dedrick, Nigel P. Melville and Kevin Zhu (eds.), analyzed in considerable detail global diffusion of IT through an economic lens, *Global E-Commerce: Impacts of National Environment and Policy* (Cambridge: Cambridge University Press, 2006). An integrated, tight study of IT in China compared to patterns of adoption in Brazil and Ghana as well, can be found in the scholarly study by Ernest J. Wilson III, *The Information Revolution and Developing Countries* (Cambridge, Mass.: MIT Press, 2004). For a more detailed, but equally useful comparative analysis see Ashish Arora and Alonso Gambardella, *From Underdogs to Tigers: The Rise and Growth of the Software Industry in Brazil, China, India, Ireland, and Israel* (Oxford: Oxford University Press, 2005) and James M. Popkin and Partha Lyengar, *IT and the East: How China and India Are Altering the Future of Technology and Innovation* (Boston, Mass.: Harvard Business School Press, 2007).

Specific aspects of China's IT world also exist. The most useful study of the semiconductor business in China is by John A. Mathews and Dong-Sung Cho, *Tiger Technology: The Creation of a Semiconductor Industry in East Asia* (Cambridge: Cambridge University Press, 2000). An early study of telecommunications prior to the effects of the Internet on this industry can be studied in Xiaobai Shen, *The Chinese Road to High Technology: A Study of Telecommunications Switching Technology in the Economic Transition* (London: Macmillan 1999). The first book-length overview of China's IT industry, and that is still highly service-able, is Jeaff X. Xhang and Yan Wang, *The Emerging Market of China's Computer Industry* (Westport, Conn.: Quorum, 1995). On the contemporary scene there are three useful studies. The first, by Robert Buderi and Gregory T. Huang, is useful for gaining insight on the inner-workings of foreign companies oper-ating in China, *Guanxi: Microsoft, China, and Bill Gates's Plan to Win the Road Ahead* (New York: Simon & Schuster, 2006). On the PC and laptop business, including the sale of IBM's PC business to a Chinese firm, we now have the well-informed study by Ling Zhijun, *The Lenovo Affair: The Growth of China's Computer Giant and Its Takeover of IBM-PC* (Singapore: John Wiley & Sons, 2006). Third, on outsourcing, there is only Cyrill Eltschinger, *Source Code China: The New Global Hub of IT Outsourcing* (Singapore: John Wiley & Sons, 2007). To put all these kinds of studies into a rich context turn to Ming Zeng and Peter J. Williamson, *Dragons At Your Door: How Chinese Cost Innovation Is Disrupting Global Competition* (Boston, Mass.: Harvard Business School Press, 2007).

China's experience with the Internet has received considerable attention, and some of the scholarly research on it has been quite good. On the modern telecommunications world in China, begin with what has to be considered an essential study by Eric Harwit, *China's Telecommunications Revolution* (Oxford: Oxford University Press, 2008) and which contains a bibliographic discussion of the topic as well. Then move to a discussion of public policies and programs in Adam Segal, *Digital Dragon: High-Technology Enterprises in China* (Ithaca, N.Y.: Cornell University Press, 2003). For company case studies there is Sher-man So and J. Christopher Westland, *Redwired: China's Internet Revolution* (London: Marshall Cavendish, 2010). The best study done so far on who uses the Internet was done by Jack Linchuan Qiu, using the methods of the sociol-ogist, in *Working-Class Network Society: Communication Technology and the In-formation Have-Less in Urban China* (Cambridge, Mass.: MIT Press, 2009). The political dynamics and use of the Internet is increasingly being studied. Some useful examples include Zhou Yongming, *Historicizing Online Politics: Telegra-phy, the Internet, and Political Participation in China* (Stanford, Cal.: Stanford University Press, 2006); Yongnion Zheng, *Technological Empowerment: The Internet, State, and Society in China* (Stanford, Cal.: Stanford University Press, 2008); and then Guobin Yang, *The Power of the Internet in China: Citizen Ac-tivism Online* (New York: Columbia University Press, 2009). For a useful study of political behavior and use of the Internet for that purpose, consult Michael Chase and James Mulvenon, *You've Got Dissent! Chinese Dissident Use of the Internet and Beijing's Counter-Strategies* (Santa Monica, Cal.: RAND, 2002).

INDIA

To understand India's IT role, a broader appreciation of both IT and the social and economic status of the nation is crucial. For a lively, well-informed introduction to modern India, consult Patrick French, *India: A Portrait* (New

York: Alfred A. Knopf, 2011) then a study by an Indian businessman and writer, Gurcharan Das, *India Unbound: From Independence to the Global Information Age* (New Delhi: Penguin, 2000), which is broader in scope, beginning with 1947. For a deeply informed analysis of the Indian economy from independence to the present, the standard work is by Arvind Panagariya, *India: The Emerging Giant* (New York: Oxford University Press, 2008). For a less rigorous study of recent economic trends, yet full of useful information, there is Jayashankar M. Swaminthan, *Indian Economic Superpower: Fiction or Future?* (Singapore: World Scientific Publishing, 2009).

Various pieces of the IT story and its history have been described largely by economists and political scientists, which tend to be well informed. K.J. Joseph has written a rigorous analysis of economic and political analysis for the first several decades, *Industry Under Economic Liberalization: The Case of Indian Electronics* (New Delhi: SAGE Publications, 1997). One of the most useful sources on the early history of software is by Richard Heeks, *India's Software Industry: State Policy, Liberalisation and Industrial Development* (New Delhi: SAGE Publications, 1996). The closest one gets to a history of IT is by Arvind Singhal and the late Everett M. Rogers, best known for his studies on the diffusion of technologies, with their now outdated study, *India's Information Revolution* (New Delhi: SAGE Publications, 1989). It should be consulted with Heek's book. The best study done on computing and public policy from the 1940s to the mid-1980s is by Joseph M. Grieco, *Between Dependency and Autonomy: India's Experience with the International Computer Industry* (Berkeley, Cal.: University of California Press, 1984). Because telecommunications became part of the IT story in the 1980s, its role in India is central to that of the diffusion of computers. One of the first, and well executed study, is by Arvind Singhal and Everett M. Rogers, *India's Communication Revolution: From Bullock Carts to Cyber Marts* (New Delhi: SAGE Publications, 2001). However, that analysis should be supplemented by that of an expert on Indian communications, Ashok V. Desai, *India's Telecommunications Industry: History, Analysis, Diagnosis* (New Delhi: SAGE Publications, 2006). Also useful for examples and economic insights is Subhash Bhatnagar and Robert Schware (eds.), *Information and Communication Technology in Development: Cases From India* (New Delhi: SAGE Publications, 2000). Because of the title, one other book must be mentioned that does not deal with ICTs, simply because a researcher might go to it, like I did, only to be disappointed, M.V. Desai, *Communication Policies in India* (Paris: UNESCO, 1977); it deals with advertising, television, radio, movies, and other non-computer issues. Another study that only devotes a small portion to communications, dealing with the whole sweep of Indian history is Amiya Kumar Bagchi, Dipankar Sinha, and Barnita Bagchi (eds.), *Webs of History: Information, Communication and Technology From Early to Post-Colonial India* (New Delhi: Manohar, 2005).

The one major source on the diffusion of knowledge about computers in India is a collection of memoirs and brief accounts of various academic and technical institutes, Utpal K. Banerjee (ed.), *Computer Education In India: Past, Present and Future* (New Delhi: Concept Publishing, 1996); it is an essential text for understanding the early history of computing in this country. There are a few studies of uses of IT, often with commentary on public policy, but the topic awaits its historian. In the meanwhile, Utpal K. Banarjee wrote, *Information Management in Government* (New Delhi: Concept Publishing, 1984) and edited a collection of contributed essays, *Computer Applications for Techno-Economic Development* (New Delhi: Concept Publishing, 1985). Recently a collection of essays appeared on various agricultural initiatives undertaken in the early 2000s, M. Hilaria Soundari (ed.), *Indian Agriculture and Information and Communications Technology (ICT)* (New Delhi: New Century Publications, 2011).

The modern IT expansion has now been the subject of numerous publications. Profiles of contemporary IT managerial practices and histories of leading firms include Peter Cappelli, Harbir Singh, Jitendra Singh, and Michael Useem, *The India Way: How India's Top Business Leaders Are Revolutionizing Management* (Boston, Mass.: Harvard Business School Press, 2010); Steve Hamm, *Bangalore Tiger: How Indian Tech Upstart Wipro Is Rewriting the Rules of Global Competition* (New York: McGraw-Hill, 2007); and Rafiq Dossani, *India Arriving: How This Economic Powerhouse Is Redefining Global Business* (New York: AMACOM, 2008). A number of studies looking at both India and China as part of some new mega-global trend include the very well done study by Tarun Khanna, *Billions of Entrepreneurs: How China and India Are Reshaping Their Futures and Yours* (Boston, Mass.: Harvard Business School Press, 2007), which covers many industries, not just IT, but see also James M. Popkin and Partha Iyengar, *IT and the East: How China and India Are Altering the Future of Technology and Innovation* (Boston, Mass.: Harvard Business School Press, 2007); a collection of BusinessWeek articles in Pete Engardio (ed.), *Chindia: How China and India Are Revolutionizing Global Business* (New York: McGraw-Hill, 2007); and an essential scholarly study that focuses on political and international issues, Marcus Franda, *China and India Online: Information Technology Politics and Diplomacy in the World's Two Largest Nations* (Lanham, Md.: Rowman & Littlefield, 2002). On the use of Indian programmers in other countries and as a service export, see Xiang Biao, *Global "Body Shopping": An Indian Labor System in the Information Technology Industry* (Princeton, N.J.: Princeton University Press, 2007). Because Bangalore is currently the epicenter for India's modern IT industry, its role can be explored in two solid studies, James Heitzman, *Network City: Planning the Information Society in Bangalore* (New Delhi: Oxford University Press, 2004) and R.C. Mascarenhas, *India's Silicon Plateau: Development of Information and Communication Technology in Bangalore* (New Delhi: Orient Blackswan, 2010). Since IT is only one part of the city's history, a general history of this urban center, along with an account of its recent IT developments set in the broader context of the city's past can be found in Janaki Nair, *The Promise of the Metropolis: Bangalore's Twentieth Century* (New Delhi: Oxford University Press, 2005). For a discussion of potential future options for the Indian government to pursue, complete with recent case studies of adoption, see R.K. Bagga, Kenneth Keniston, and Rohit Raj Mathur (eds.), *The State, IT and Development* (New Delhi: SAGE Publications, 2005). Finally, about Indian R&D and advanced product development, see Nirmalya Kumar and Phanish Puranam, *India Inside: The Emerging Innovation Challenge to the West* (Boston, Mass.: Harvard Business Review Press, 2012).

A series of important comparative studies with other countries, including India, essential in understanding Indian affairs, are appearing rather quickly. A sampling of these consulted for our project include, Susan M. Collins and Lael Brainard (eds.), *Brookings Trade Forum 2005: Offshoring White-Collar Work* (Washington, D.C.: Brookings Institution Press, 2005); Ashish Arora and Alfonso Gambardella (eds.), *From Underdogs to Tigers: The Rise and Growth of the Software Industry in Brazil, China, India, Ireland, and Israel* (Oxford: Oxford University Press, 2005); Vandana Chandra (ed.), *Technology, Adaptation, and Exports: How Some Developing Countries Got It Right* (Washington, D.C.: World Bank, 2006); and Neil Gregory, Stanley Nollen, and Stoyan Tenev, *New Industries From New Places: The Emergence of the Software and Hardware Industries in China and India* (Washington, D.C.: World Bank, 2009).

PAN-ASIAN IT INITIATIVES, 1960S–2010

Much has been written about Asian economics, and increasingly concerning the role of IT and other technologies, more often than not as comparative studies, rather than as in-depth country accounts. One of the first, and most useful of these, was commissioned by the World Bank, Nagy Hanna, Sandor Boyson, and Shakuntala Gunaratne, *The East Asian Miracle and Information Technology: Strategic Management of Technological Learning* (Washington, D.C.: World Bank, 1996). A study done a few years later for the U.S. National Defense Research Institute focused on similar economic development issues, Nina Hachigian and Lily Wu, *The Information Revolution In Asia* (Santa Monica, Cal.: RAND Corporation, 2003). However, the single best and most comprehensive account of these issues is unquestionably Jason Dedrick and Kenneth L. Kraemer, *Asia's Computer Challenge: Threat or Opportunity for the United States and the World?* (New York: Oxford University Press, 1998).

A great deal of it became entangled with discussions of technological innovations and diffusion, such as the World Bank studies cited in the end notes to chapter 11, but see also Frederic C. Deyo (ed.), *The Political Economy of the New Asian Industrialism* (Ithaca, N.Y.: Cornell University Press, 1987); Peter Evans, *Embedded Autonomy: States and Industrial Transformation* (Princeton, N.J.: Princeton University Press, 1995); Jeremy Grace, Charles Kenny, and Christine Zhen-Wei Qiang, *Information and Communication Technologies and Broad-Based*

Development: A Partial Review of the Evidence (Washington, D.C.: World Bank, 2004); Shahid Yusuf, M. Anjum Altaf, and Kaoru Nabeshima (eds.), *Global Production Networking and Technological Change in East Asia* (Washington, D.C.: World Bank, 2004); Shahid Yusuf, *Innovative East Asia: The Future of Growth* (Washington, D.C.: World Bank, 2003); and now two classic studies, Linsu Kim and Richard R. Nelson (eds), *Technology, Learning, and Innovation: Experiences of Newly Industrializing Economies* (Cambridge: Cambridge University Press, 2000) and Chris Freeman and Luc Soefe, *The Economics of Industrial Innovation* (Cambridge, Mass.: MIT Press, 1997); Henry S. Rowen, Marguerite Gong Hancock, and William F. Miller (eds.), *Making IT: The Rise of Asia in High Tech* (Stanford, Cal.: Stanford University Press, 2007). On telecommunications, the first book to reach out for is by Eli Noam, Seisuke Komatsuzaki, and Douglas A. Conn (eds.), *Telecommunications in the Pacific Basin: An Evolutionary Approach* (New York: Oxford University Press, 1994) which is useful for all the pre-Internet discussions. Finally, an earlier study adds to the discussion from the perspective of political scientists, economists and media experts, Georgette Wang (ed.), *Treading Different Paths: Informatization in Asian Nations* (Norwood, N.J. Ablex Publishing, 1994).

Because the Asian IT experience is now a central topic of discussion about economic development, consulting that literature is essential. For this book I used the following: Richard R. Nelson, *Technology Institutions and Economic Growth* (Cambridge, Mass.: Harvard University Press, 2005); Richard G. Lipsey, Kenneth I. Carlaw, and Clifford T. Bekar, *Economic Transformations: General Purpose Technologies and Long Term Economic Growth* (Oxford: Oxford University Press, 2005); Jeff Saperstein and Daniel Rouach, *Creating Regional Wealth in the Innovation Economy: Models, Perspectives, and Best Practices* (Upper Saddle River, N.J.: Financial Times, 2002); Shahid Yusuf and Kaoru Nabeshima, *Postindustrial East Asian Cities: Innovation for Growth* (Washington, D.C.: World Bank, 2006); Dani Rodrik, *One Economics Many Recipes: Globalization, Institutions, and Economic Growth* (Princeton, N.J.: Princeton University Press, 2007); Peter F. Cowhey, Jonathan D. Aronson, and Donald Abelson, *Transforming Global Information and Communication Markets: The Political Economy of Innovation* (Cambridge, Mass.: MIT Press, 2009); Michael Spence, *The Next Convergence: The Future of Economic Growth in a Multispeed World* (New York: Farrar, Straus and Giroux, 2011).

For counting how many types of IT there are in Asia, there are a variety of publications that continuously appear in newer editions, including those of the World Economic Forum, such as, *The Global Information Technology Report: Readiness for the Networked World* (New York: Oxford University Press, 2002) that comes out under various titles, and authors, and always rich in detail. Publications by the World Bank are essential, for example, *World Bank, 2006 Information and Communications for Development: Global Trends and Policies* (Washington, D.C.: World Bank, 2006) and its *Global Economic*

Prospects: Technology Diffusion in the Developing World, 2008 (Washington, D.C.: World Bank, 2008).

Comparative studies of Asians one to one another and to other areas of the world are plentiful and useful for identifying patterns of adoption, and commonly shared issues. Ernest J. Wilson III, *The Information Revolution and Developing Countries* (Cambridge, Mass.: MIT Press, 2004) is a good example of this kind of analysis. An earlier work that approached similar issues is Jeffrey Henderson, *The Globalization of High Technology Production* (London: Routledge, 1989). Other useful comparative studies include Ashish Arora and Alfonso Gambardella (eds.), *From Underdogs to Tigers: The Rise and Growth of the Software Industry in Brazil, China, India, Ireland, and Israel* (Oxford: Oxford University Press, 2005); Susan M. Collins and Lael Brainard (eds.), *Brookings Trade Forum 2005: Offshoring White-Collar Work* (Washington, D.C.: Brookings Institution Press, 2006), John Zysman and Abraham Newman (eds.), *How Revolutionary Was the Digital Revolution? National Responses, Market Transitions, and Global Technology* (Stanford, Cal.: Stanford University Press, 2006); and Martin Fransman (ed.), *Global Broadband Battles: Why the U.S. and Europe Lag While Asia Leads* (Stanford, Cal.: Stanford University Press, 2006).

On Australia's experience with computing see J.M. Bennett, Rosemary Broomham, P.M. Murton, T. Pearcey, and R.W. Rutledge (eds.), *Computers in Australia: The Development of a Profession* (Sydney: Hale & Iremonger, 1994); Gerard Goggin (ed.), *Virtual Nation: The Internet in Australia* (New South Wales: University of New South Wales Press, 2004); John Deane, "Connections in the History of Australian Computing," in Arthur Tatnall (ed.), *History of Computing: Learning from the Past* (Berlin: Springer, 2010): 1–12. On New Zeeland, see the last chapter of Kristin Thompson, *The Frodo Franchise: The Lord of the Rings and Modern Hollywood* (Berkeley, Cal.: University of California Press, 2008). On the role of IT in Asian military affairs, the standard work is Emily O. Goldman and Thomas G. Mahnken (eds.), *The Information Revolution in Military Affairs in Asia* (New York: Palgrave, 2004). On the most recent Asian entrants into the world of IT, there is almost nothing available, although interest is picking up on the role of the Internet in these nations, for example, Emily Noelle Ignacio, *Building Diaspora: Filipino Cultural Community Formation on the Internet* (New Brunswick, N.J.: Rutgers University Press, 2005); Merlyna Lim, *Islamic Radicalism and Anti-Americanism in Indonesia: The Role of the Internet* (Washington, D.C.: East-West Center, 2005); and David T. Hill and Sen Krishna, *The Internet in Indonesia's New Democracy* (London: Routledge, 2005).

INDEX